GARCH Models

GARCH Models

Structure, Statistical Inference and Financial Applications

Second Edition

Christian Francq

CREST and University of Lille, France

Jean-Michel Zakoian

CREST and University of Lille, France

WILEY

This edition first published 2019

Edition History
John Wiley & Sons (1e, 2010)

Registered Offices
John Wiley & Sons, Inc., 111 River Street, Hoboken, NJ 07030, USA
John Wiley & Sons Ltd, The Atrium, Southern Gate, Chichester, West Sussex, PO19 8SQ, UK

Editorial Office
9600 Garsington Road, Oxford, OX4 2DQ, UK

For details of our global editorial offices, customer services, and more information about Wiley products visit us at www.wiley.com.

Wiley also publishes its books in a variety of electronic formats and by print-on-demand. Some content that appears in standard print versions of this book may not be available in other formats.

Library of Congress Cataloging-in-Publication Data

Names: Francq, Christian, author. | Zakoian, Jean-Michel, author.
Title: GARCH models : structure, statistical inference and financial applications / Christian Francq, Jean-Michel Zakoian.
Other titles: Modèles GARCH. English
Description: 2 edition. | Hoboken, NJ : John Wiley & Sons, 2019. | Includes bibliographical references and index. |
Identifiers: LCCN 2018038962 (print) | LCCN 2019003658 (ebook) | ISBN 9781119313564 (Adobe PDF) | ISBN 9781119313489 (ePub) | ISBN 9781119313571 (hardcover)
Subjects: LCSH: Finance–Mathematical models. | Investments–Mathematical models.
Classification: LCC HG106 (ebook) | LCC HG106 .F7213 2019 (print) | DDC 332.01/5195–dc23
LC record available at https://lccn.loc.gov/2018038962

Cover Design: Wiley
Cover Images: © teekid/E+/Getty Images

Set in 9/11pt TimesLTStd by SPi Global, Chennai, India

10 9 8 7 6 5 4 3 2 1

Contents

Part III Extensions and Applications

Preface to the Second Edition

This edition contains a large number of additions and corrections. The analysis of GARCH models – and more generally volatility models – has undergone various new developments in recent years. There was a need to make the material more complete.

A brief summary of the added material in the second edition is:

1. A new chapter entitled "Parameter-driven volatility models". This chapter is divided in two sections entitled "Stochastic Volatility Models" and "Markov Switching Volatility Models". Two new appendices on "Markov Chains on Countable State Spaces" and "The Kalman Filter" are provided.

2. A new chapter entitled "Alternative Models for the Conditional Variance", replacing and completing the chapter "Asymmetries" of the first version. This chapter contains a new section on "Stochastic Recurrence Equations" and additional material on EGARCH, Log-GARCH, GAS, MIDAS and intraday volatility models among others.

3. A more complete discussion of multivariate GARCH models in Chapter 10. In particular a new section on "Cholesky GARCH" has been added. More emphasis has been given to the inference of multivariate GARCH models, through two new sections entitled "QML Estimation of General MGARCH" and "Looking for Numerically Feasible Estimation Methods".

4. The previous Appendix D entitled "Problems" has been removed, but a new set of corrected problems is available on the webpages of the authors.

5. An up-to-date list of references.

On the other hand, there was not enough space to keep the previous 4th chapter of the 1st edition entitled "Temporal aggregation and weak GARCH models".

The webpage http://christian.francq140.free.fr/Christian-Francq/book-GARCH.html features additional material (codes, data sets, and problems with corrections)

We are indebted to many readers who have used the book and made suggestions for improvements. In particular, we thank Francisco Blasques, Lajos Horvàth, Hamdi Raïssi, Roch Roy, Genaro Sucarrat. We are also indebted to Wiley for their support and assistance in preparing this edition.

Palaiseau, France
September, 2018

Christian Francq
Jean-Michel Zakoian

Preface to the Second Edition

Preface to the First Edition

Autoregressive conditionally heteroscedastic (ARCH) models were introduced by Engle in an article published in *Econometrica* in the early 1980s (Engle, 1982). The proposed application in that article focused on macroeconomic data and one could not imagine, at that time, that the main field of application for these models would be finance. Since the mid 1980s and the introduction of generalized ARCH (or GARCH) models, these models have become extremely popular among both academics and practitioners. GARCH models led to a fundamental changes to the approaches used in finance, through an efficient modeling of volatility (or variability) of the prices of financial assets. In 2003, the Nobel Prize for Economics was jointly awarded to Robert F. Engle and sharing the award with Clive W.J. Granger 'for methods of analyzing economic time series with time-varying volatility (ARCH)'.

Since the late 1980s, numerous extensions of the initial ARCH models have been published (see Bollerslev, 2008, for a (tentatively) exhaustive list). The aim of the present volume is not to review all these models, but rather to provide a panorama, as wide as possible, of current research into the concepts and methods of this field. Along with their development in econometrics and finance journals, GARCH models and their extensions have given rise to new directions for research in probability and statistics. Numerous classes of nonlinear time series models have been suggested, but none of them has generated interest comparable to that in GARCH models. The interest of the academic world in these models is explained by the fact that they are simple enough to be usable in practice, but also rich in theoretical problems, many of them unsolved.

This book is intended primarily for master's students and junior researchers, in the hope of attracting them to research in applied mathematics, statistics or econometrics. For experienced researchers, this book offers a set of results and references allowing them to move towards one of the many topics discussed. Finally, this book is aimed at practitioners and users who may be looking for new methods, or may want to learn the mathematical foundations of known methods.

Some parts of the text have been written for readers who are familiar with probability theory and with time series techniques. To make this book as self-contained as possible, we provide demonstrations of most theoretical results. On first reading, however, many demonstrations can be omitted. Those sections or chapters that are the most mathematically sophisticated and can be skipped without loss of continuity are marked with an asterisk. We have illustrated the main techniques with numerical examples, using real or simulated data. Program codes allowing the experiments to be reproduced are provided in the text and on the authors© web pages. In general, we have tried to maintain a balance between theory and applications.

Readers wishing to delve more deeply into the concepts introduced in this book will find a large collection of exercises along with their solutions. Some of these complement the proofs given in the text.

The book is organized as follows. Chapter 1 introduces the basics of stationary processes and ARMA modeling. The rest of the book is divided into three parts. Part I deals with the standard univariate GARCH model. The main probabilistic properties (existence of stationary solutions, representations, properties of autocorrelations) are presented in Chapter 2. Chapter 3 deals with complementary properties related to mixing, allowing us to characterize the decay of the time dependence. Chapter 4 is devoted to temporal aggregation: it studies the impact of the observation frequency on the properties of GARCH processes.

Part II is concerned with statistical inference. We begin in Chapter 5 by studying the problem of identifying an appropriate model *a priori*. Then we present different estimation methods, starting with the method of least squares in Chapter 6 which, limited to ARCH, offers the advantage of simplicity. The central part of the statistical study is Chapter 7, devoted to the quasi-maximum likelihood method. For these models, testing the nullity of coefficients is not standard and is the subject of Chapter 8. Optimality issues are discussed in Chapter 9, as well as alternative estimators allowing some of the drawbacks of standard methods to be overcome.

Part III is devoted to extensions and applications of the standard model. In Chapter 10, models allowing us to incorporate asymmetric effects in the volatility are discussed. There is no natural extension of GARCH models for vector series, and many multivariate formulations are presented in Chapter 11. Without carrying out an exhaustive statistical study, we consider the estimation of a particular class of models which appears to be of interest for applications. Chapter 12 presents applications to finance. We first study the link between GARCH and diffusion processes, when the time step between two observations converges to zero. Two applications to finance are then presented: risk measurement and the pricing of derivatives.

Appendix A includes the probabilistic properties which are of most importance for the study of GARCH models. Appendix B contains results on autocorrelations and partial autocorrelations. Appendix C provides solutions to the end-of-chapter exercises. Finally, a set of problems and (in most cases) their solutions are provided in Appendix D.

Notation

General notation

$:=$	'is defined as'
x^+, x^-	$\max\{x, 0\}, \max\{-x, 0\}$ (or $\min\{x, 0\}$ in Chapter 10)

Sets and spaces

$\mathbb{N}, \mathbb{Z}, \mathbb{Q}, \mathbb{R}$	positive integers, integers, rational numbers, real numbers
$\mathbb{R}^+$	positive real line
$\mathbb{R}^d$	d-dimensional Euclidean space
D^c	complement of the set $D \subset \mathbb{R}^d$
$[a, b)$	half-closed interval

Matrices

I_d	d-dimensional identity matrix
$\mathcal{M}_{p,q}(\mathbb{R})$	the set of $p \times q$ real matrices

Processes

iid	independent and identically distributed
iid (0,1)	iid centered with unit variance
(X_t) or $(X_t)_{t\in\mathbb{Z}}$	discrete-time process
(ϵ_t)	GARCH process
σ_t^2	conditional variance or volatility
(η_t)	strong white noise with unit variance
κ_η	kurtosis coefficient of η_t
L or B	lag operator
$\sigma\{X_s; s < t\}$ or $X_{\underline{t-1}}$	sigma-field generated by the past of X_t

Functions

$\mathbb{1}_A(x)$	1 if $x \in A$, 0 otherwise
$[x]$	integer part of x
γ_X, ρ_X	autocovariance and autocorrelation functions of (X_t)
$\hat{\gamma}_X, \hat{\rho}_X$	sample autocovariance and autocorrelation

Probability

$\mathcal{N}(m, \Sigma)$	Gaussian law with mean m and covariance matrix Σ
χ_d^2	chi-square distribution with d degrees of freedom
$\chi_d^2(\alpha)$	quantile of order α of the χ_d^2 distribution

$\stackrel{\mathcal{L}}{\to}$	convergence in distribution
a.s.	almost surely
$v_n = o_P(u_n)$	$v_n/u_n \to 0$ in probability
$a \stackrel{o_P(1)}{=} b$	a equals b up to the stochastic order $o_P(1)$

Estimation

$\mathfrak{I}$	Fisher information matrix
$(\kappa_\eta - 1)J^{-1}$	asymptotic variance of the QML
θ_0	true parameter value
Θ	parameter set
θ	element of the parameter set
$\hat{\theta}_n, \hat{\theta}_n^c, \hat{\theta}_{n,f}, \ldots$	estimators of θ_0
$\sigma_t^2 = \sigma_t^2(\theta)$	volatility built with the value θ
$\tilde{\sigma}_t^2 = \tilde{\sigma}_t^2(\theta)$	as σ_t^2 but with initial values
$\ell_t = \ell_t(\theta)$	$-2\log$(conditional variance of ϵ_t)
$\tilde{\ell}_t = \tilde{\ell}_t(\theta)$	approximation of ℓ_t, built with initial values
Var_{as}, Cov_{as}	asymptotic variance and covariance

Some abbreviations

ES	expected shortfall
FGLS	feasible generalized least squares
OLS	ordinary least squares
QML	quasi-maximum likelihood
RMSE	root mean square error
SACR	sample autocorrelation
SACV	sample autocovariance
SPAC	sample partial autocorrelation
VaR	value at risk

1

Classical Time Series Models and Financial Series

The standard time series analysis rests on important concepts such as stationarity, autocorrelation, white noise, innovation, and on a central family of models, the autoregressive moving average (ARMA) models. We start by recalling their main properties and how they can be used. As we shall see, these concepts are insufficient for the analysis of financial time series. In particular, we shall introduce the concept of volatility, which is of crucial importance in finance.

In this chapter, we also present the main stylized facts (unpredictability of returns, volatility clustering and hence predictability of squared returns, leptokurticity of the marginal distributions, asymmetries, etc.) concerning financial series.

1.1 Stationary Processes

Stationarity plays a central part in time series analysis, because it replaces in a natural way the hypothesis of independent and identically distributed (iid) observations in standard statistics.

Consider a sequence of real random variables $(X_t)_{t\in\mathbb{Z}}$, defined on the same probability space. Such a sequence is called a time series, and is an example of a discrete-time stochastic process.

We begin by introducing two standard notions of stationarity.

Definition 1.1 (Strict stationarity) *The process (X_t) is said to be strictly stationary if the vectors $(X_1, \dots, X_k)'$ and $(X_{1+h}, \dots, X_{k+h})'$ have the same joint distribution, for any $k \in \mathbb{N}$ and any $h \in \mathbb{Z}$.*

The following notion may seem less demanding, because it only constrains the first two moments of the variables X_t, but contrary to strict stationarity, it requires the existence of such moments.

Definition 1.2 (Second-order stationarity) *The process (X_t) is said to be second-order stationary if*

(i) $EX_t^2 < \infty, \quad \forall t \in \mathbb{Z}$;

(ii) $EX_t = m, \quad \forall t \in \mathbb{Z}$;

(iii) $\mathrm{Cov}(X_t, X_{t+h}) = \gamma_X(h), \quad \forall t, h \in \mathbb{Z}$.

GARCH Models: Structure, Statistical Inference and Financial Applications, Second Edition. Christian Francq and Jean-Michel Zakoian.

The function $\gamma_X(\cdot)$ $(\rho_X(\cdot) := \gamma_X(\cdot)/\gamma_X(0))$ *is called the autocovariance function (autocorrelation function) of* (X_t).

The simplest example of a second-order stationary process is white noise. This process is particularly important because it allows more complex stationary processes to be constructed.

Definition 1.3 (Weak white noise) *The process* (ϵ_t) *is called weak white noise if, for some positive constant* σ^2:

(i) $E\epsilon_t = 0, \quad \forall t \in \mathbb{Z}$;

(ii) $E\epsilon_t^2 = \sigma^2, \quad \forall t \in \mathbb{Z}$;

(iii) $\mathrm{Cov}(\epsilon_t, \epsilon_{t+h}) = 0, \quad \forall t, h \in \mathbb{Z}, h \neq 0.$

Remark 1.1 (Strong white noise) It should be noted that no independence assumption is made in the definition of weak white noise. The variables at different dates are only uncorrelated, and the distinction is particularly crucial for financial time series. It is sometimes necessary to replace hypothesis (iii) by the stronger hypothesis

(iii′) the variables ϵ_t and ϵ_{t+h} are independent and identically distributed.

The process (ϵ_t) is then said to be strong white noise.

Estimating Autocovariances

The classical time series analysis is centred on the second-order structure of the processes. Gaussian stationary processes are completely characterized by their mean and their autocovariance function. For non-Gaussian processes, the mean and autocovariance give a first idea of the temporal dependence structure. In practice, these moments are unknown and are estimated from a realisation of size n of the series, denoted $X_1, \ldots, X_n$. This step is preliminary to any construction of an appropriate model. To estimate $\gamma(h)$, we generally use the sample autocovariance defined, for $0 \leq h < n$, by

$$\hat{\gamma}(h) = \frac{1}{n} \sum_{j=1}^{n-h} (X_j - \overline{X})(X_{j+h} - \overline{X}) := \hat{\gamma}(-h),$$

where $\overline{X} = (1/n) \sum_{j=1}^{n} X_j$ denotes the sample mean. We similarly define the sample autocorrelation function by $\hat{\rho}(h) = \hat{\gamma}(h)/\hat{\gamma}(0)$ for $|h| < n$.

The previous estimators have finite-sample bias but are asymptotically unbiased. There are other similar estimators of the autocovariance function with the same asymptotic properties (for instance, obtained by replacing $1/n$ by $1/(n-h)$). However, the proposed estimator is to be preferred over others because the matrix $(\hat{\gamma}(i-j))$ is positive semi-definite (see Brockwell and Davis 1991, p. 221).

It is, of course, not recommended to use the sample autocovariances when h is close to n, because too few pairs (X_j, X_{j+h}) are available. Box, Jenkins, and Reinsel (1994, p. 32) suggest that useful estimates of the autocorrelations can only be made if, approximately, $n > 50$ and $h \leq n/4$.

It is often of interest to know – for instance, in order to select an appropriate model – if some or all the sample autocovariances are significantly different from 0. It is then necessary to estimate the covariance structure of those sample autocovariances. We have the following result (see Brockwell and Davis 1991, pp. 222, 226).

Theorem 1.1 (Bartlett's formulas for a strong linear process) *Let* (X_t) *be a linear process satisfying*

$$X_t = \sum_{j=-\infty}^{\infty} \phi_j \epsilon_{t-j}, \qquad \sum_{j=-\infty}^{\infty} |\phi_j| < \infty,$$

where (ϵ_t) is a sequence of iid variables such that

$$E(\epsilon_t) = 0, \quad E(\epsilon_t^2) = \sigma^2, \quad E(\epsilon_t^4) = \kappa_\epsilon \sigma^4 < \infty.$$

Appropriately normalized, the sample autocovariances and autocorrelations are asymptotically normal, with asymptotic variances given by the Bartlett formulas:

$$\lim_{n\to\infty} n\,\mathrm{Cov}\{\hat{\gamma}(h), \hat{\gamma}(k)\} = \sum_{i=-\infty}^{\infty} \gamma(i)\gamma(i+k-h) + \gamma(i+k)\gamma(i-h) + (\kappa_\epsilon - 3)\gamma(h)\gamma(k) \tag{1.1}$$

and

$$\lim_{n\to\infty} n\,\mathrm{Cov}\{\hat{\rho}(h), \hat{\rho}(k)\} = \sum_{i=-\infty}^{\infty} \rho(i)[2\rho(h)\rho(k)\rho(i) - 2\rho(h)\rho(i+k) - 2\rho(k)\rho(i+h) + \rho(i+k-h) + \rho(i-k-h)]. \tag{1.2}$$

Formula (1.2) still holds under the assumptions

$$E\epsilon_t^2 < \infty, \quad \sum_{j=-\infty}^{\infty} |j|\,\phi_j^2 < \infty.$$

In particular, if $X_t = \epsilon_t$ and $E\epsilon_t^2 < \infty$, we have

$$\sqrt{n}\begin{pmatrix}\hat{\rho}(1)\\ \vdots \\ \hat{\rho}(h)\end{pmatrix} \xrightarrow{\mathcal{L}} \mathcal{N}(0, I_h).$$

The assumptions of this theorem are demanding, because they require a strong white noise (ϵ_t). An extension allowing the strong linearity assumption to be relaxed is proposed in Appendix B.2. For many non-linear processes, in particular the ARCH process studies in this book, the asymptotic covariance of the sample autocovariances can be very different from Eq. (1.1) (Exercises 1.6 and 1.8). Using the standard Bartlett formula can lead to specification errors (see Chapter 5).

1.2 ARMA and ARIMA Models

The aim of time series analysis is to construct a model for the underlying stochastic process. This model is then used for analysing the causal structure of the process or to obtain optimal predictions.

The class of ARMA models is the most widely used for the prediction of second-order stationary processes. These models can be viewed as a natural consequence of a fundamental result due to Wold (1938), which can be stated as follows: *any centred, second-order stationary, and 'purely non-deterministic'*[1] *process admits an infinite moving-average representation of the form*

$$X_t = \epsilon_t + \sum_{i=1}^{\infty} c_i \epsilon_{t-i}, \tag{1.3}$$

[1] A stationary process (X_t) is said to be purely nondeterministic if and only if $\bigcap_{n=-\infty}^{\infty} \mathcal{H}_X(n) = \{0\}$, where $\mathcal{H}_X(n)$ denotes, in the Hilbert space of the real, centered, and square integrable variables, the subspace constituted by the limits of the linear combinations of the variables X_{n-i}, $i \geq 0$. Thus, for a purely nondeterministic (or regular) process, the linear past, sufficiently far away in the past, is of no use in predicting future values. See Brockwell and Davis (1991, pp. 187–189) or Azencott and Dacunha-Castelle (1984) for more details.

where (ϵ_t) *is the linear innovation process of* (X_t)*, that is*

$$\epsilon_t = X_t - E(X_t \mid \mathcal{H}_X(t-1)), \tag{1.4}$$

where $\mathcal{H}_X(t-1)$ *denotes the Hilbert space generated by the random variables* $X_{t-1}, X_{t-2}, \ldots$ *. and* $E(X_t|\mathcal{H}_X(t-1))$ *denotes the orthogonal projection of* X_t *onto* $\mathcal{H}_X(t-1)$.[2] The sequence of coefficients (c_i) is such that $\sum_i c_i^2 < \infty$. Note that (ϵ_t) is a weak white noise.

Truncating the infinite sum in Eq. (1.3), we obtain the process

$$X_t(q) = \epsilon_t + \sum_{i=1}^{q} c_i \epsilon_{t-i},$$

called a moving average process of order q, or MA (q). We have

$$\|X_t(q) - X_t\|_2^2 = E\epsilon_t^2 \sum_{i>q} c_i^2 \to 0, \quad \text{as } q \to \infty.$$

It follows that the set of all finite-order moving averages is dense in the set of second-order stationary and purely non-deterministic processes. The class of ARMA models is often preferred to the MA models for parsimony reasons, because they generally require fewer parameters.

Definition 1.4 (ARMA (*p*, *q*) process) *A second-order stationary process* (X_t) *is called ARMA (p, q), where p and q are integers, if there exist real coefficients* $c, a_1, \ldots, a_p, b_1, \ldots, b_q$ *such that*

$$\forall t \in \mathbb{Z}, \quad X_t + \sum_{i=1}^{p} a_i X_{t-i} = c + \epsilon_t + \sum_{j=1}^{q} b_j \epsilon_{t-j}, \tag{1.5}$$

where (ϵ_t) *is the linear innovation process of* (X_t)*.*

This definition entails constraints on the zeros of the autoregressive and moving average polynomials, $a(z) = 1 + \sum_{i=0}^{p} a_i z^i$ and $b(z) = 1 + \sum_{i=0}^{q} b_i z^i$ (Exercise 1.9). The main attraction of this model, and the representations obtained by successively inverting the polynomials $a(\cdot)$ and $b(\cdot)$, is that it provides a framework for deriving the *optimal linear predictions* of the process, in much simpler way than by only assuming the second-order stationarity.

Many economic series display trends, making the stationarity assumption unrealistic. Such trends often vanish when the series is differentiated, once or several times. Let $\Delta X_t = X_t - X_{t-1}$ denote the first-difference series, and let $\Delta^d X_t = \Delta(\Delta^{d-1} X_t)$ (with $\Delta^0 X_t = X_t$) denote the differences of order d.

Definition 1.5 (ARIMA(*p*, *d*, *q*) process) *Let d be a positive integer. The process* (X_t) *is said to be an ARIMA (p, d, q) process if, for* $k = 0, \ldots, d-1$*, the processes* $(\Delta^k X_t)$ *are not second-order stationary, and* $(\Delta^d X_t)$ *is an ARMA (p, q) process.*

The simplest ARIMA process is the ARIMA (0, 1, 0), also called the random walk, satisfying

$$X_t = \epsilon_t + \epsilon_{t-1} + \cdots + \epsilon_1 + X_0, \quad t \geq 1,$$

where ϵ_t is a weak white noise.

For statistical convenience, ARMA (and ARIMA) models are generally used under stronger assumptions on the noise than that of weak white noise. *Strong* ARMA refers to the ARMA model of Definition 1.4 when ϵ_t is assumed to be a strong white noise. This additional assumption allows us to use convenient statistical tools developed in this framework, but considerably reduces the generality of the ARMA class. Indeed, assuming a strong ARMA is tantamount to assuming that (i) the *optimal predictions* of the process are linear ((ϵ_t) being the strong innovation of (X_t)) and (ii) the amplitudes of the

---

[2] In this representation, the equivalence class $E(X_t \mid \mathcal{H}_X(t-1))$ is identified with a random variable.

prediction intervals depend on the horizon but not on the observations. We shall see in the next section how restrictive this assumption can be, in particular for financial time series modelling.

The orders (p, q) of an ARMA process are fully characterized through its autocorrelation function (see Brockwell and Davis 1991, pp. 89–90, for a proof).

Theorem 1.2 (Characterisation of an ARMA process) *Let (X_t) denote a second-order stationary process. We have*

$$\rho(h) + \sum_{i=1}^{p} a_i \rho(h-i) = 0, \quad \text{for all } \mid h \mid > q,$$

if and only if (X_t) is an ARMA (p, q) process.

To close this section, we summarise the method for time series analysis proposed in the famous book by Box and Jenkins (1970). To simplify presentation, we do not consider seasonal series, for which SARIMA models can be considered.

Box–Jenkins Methodology

The aim of this methodology is to find the most appropriate ARIMA (p, d, q) model and to use it for forecasting. It uses an iterative six-stage scheme:

(i) A priori identification of the differentiation order d (or choice of another transformation);

(ii) A priori identification of the orders p and q;

(iii) Estimation of the parameters ($a_1, \ldots, a_p, b_1, \ldots, b_q$ and $\sigma^2 = \text{Var}\, \epsilon_t$);

(iv) Validation;

(v) Choice of a model;

(vi) Prediction.

Although many unit root tests have been introduced in the last 30 years, step (i) is still essentially based on examining the graph of the series. If the data exhibit apparent deviations from stationarity, it will not be appropriate to choose $d = 0$. For instance, if the amplitude of the variations tends to increase, the assumption of constant variance can be questioned. This may be an indication that the underlying process is heteroscedastic.[3] If a regular linear trend is observed, positive or negative, it can be assumed that the underlying process is such that $EX_t = at + b$ with $a \neq 0$. If this assumption is correct, the first-difference series $\Delta X_t = X_t - X_{t-1}$ should not show any trend ($E\Delta X_t = a$) and could be stationary. If no other sign of non-stationarity can be detected (such as heteroscedasticity), the choice $d = 1$ seems suitable. The random walk (whose sample paths may resemble the graph of Figure 1.1), is another example where $d = 1$ is required, although this process does not have any deterministic trend.

Step (ii) is more problematic. The primary tool is the sample autocorrelation function. If, for instance, we observe that $\hat{\rho}(1)$ is far away from 0 but that for any $h > 1$, $\hat{\rho}(h)$ is close to 0,[4] then, from Theorem 1.1, it is plausible that $\rho(1) \neq 0$ and $\rho(h) = 0$ for all $h > 1$. In this case, Theorem 1.2 entails that X_t is an MA(1) process. To identify AR processes, the partial autocorrelation function (see Appendix B.1) plays an analogous role. For mixed models (that is, ARMA (p, q) with $pq \neq 0$), more sophisticated statistics can be used, as will be seen in Chapter 5. Step (ii) often results in the selection of several candidates $(p_1, q_1), \ldots, (p_k, q_k)$ for the ARMA orders. These k models are estimated in step (iii), using, for instance, the least-squares method. The aim of step (iv) is to gauge if the estimated models are reasonably compatible with the data. An important part of the procedure is to examine the residuals

[3] In contrast, a process such that $\text{Var}X_t$ is constant is called (marginally) homoscedastic.

[4] More precisely, for $h > 1$, $\sqrt{n} \mid \hat{\rho}(h) \mid / \sqrt{1 + 2\hat{\rho}^2(1)}$ is a plausible realisation of the $\mid \mathcal{N}(0, 1) \mid$ distribution.

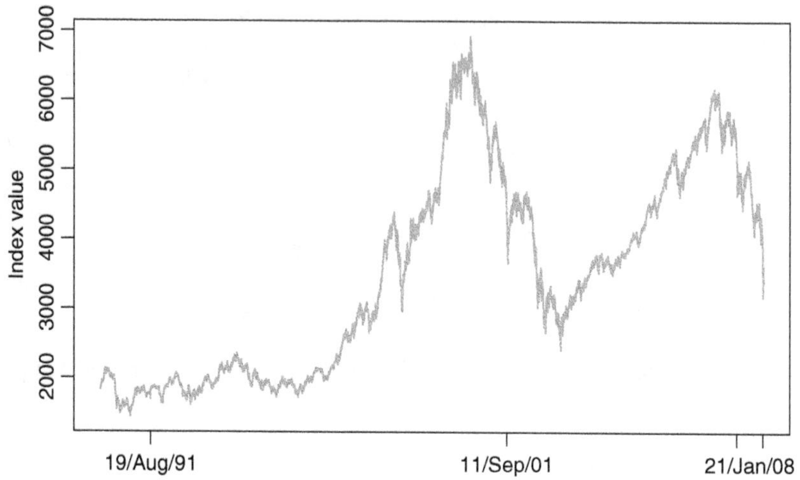

Figure 1.1 CAC 40 index for the period from 1 March 1990 to 15 October 2008 (4702 observations).

which, if the model is satisfactory, should have the appearance of white noise. The correlograms are examined and portmanteau tests are used to decide if the residuals are sufficiently close to white noise. These tools will be described in detail in Chapter 5. When the tests on the residuals fail to reject the model, the significance of the estimated coefficients is studied. Testing the nullity of coefficients sometimes allows the model to be simplified. This step may lead to rejection of all the estimated models, or to consideration of other models, in which case we are brought back to step (i) or (ii). If several models pass the validation step (iv), selection criteria can be used, the most popular being the Akaike (AIC) and Bayesian (BIC) information criteria. Complementing these criteria, the predictive properties of the models can be considered: different models can lead to almost equivalent predictive formulas. The parsimony principle would thus lead us to choose the simplest model, the one with the fewest parameters. Other considerations can also come into play, for instance, models frequently involve a lagged variable at the order 12 for monthly data, but this would seem less natural for weekly data. If the model is appropriate, step (vi) allows us to easily compute the best linear predictions $\hat{X}_t(h)$ at horizon $h = 1, 2, \ldots$. Recall that these linear predictions do not necessarily lead to minimal quadratic errors. Non-linear models, or non-parametric methods, sometimes produce more accurate predictions. Finally, the interval predictions obtained in step (vi) of the Box–Jenkins methodology are based on Gaussian assumptions. Their magnitude does not depend on the data, which for financial series is not appropriate, as we shall see.

1.3 Financial Series

Modelling financial time series is a complex problem. This complexity is not only due to the variety of the series in use (stocks, exchange rates, interest rates, etc.), to the importance of the frequency of observation (second, minute, hour, day, etc.), or to the availability of very large data sets. It is mainly due to the existence of statistical regularities (*stylised facts*) which are common to a large number of financial series and are difficult to reproduce artificially using stochastic models.

Most of these stylised facts were put forward in a paper by Mandelbrot (1963). Since then, they have been documented, and completed, by many empirical studies. They can be observed more or less clearly depending on the nature of the series and its frequency. The properties that we now present are mainly concerned with daily stock prices.

Let p_t denote the price of an asset at time t and let $\epsilon_t = \log(p_t/p_{t-1})$ be the continuously compounded or log return (also simply called the return). The series (ϵ_t) is often close to the series of relative price variations $r_t = (p_t - p_{t-1})/p_{t-1}$, since $\epsilon_t = \log(1 + r_t)$. In contrast to the prices, the returns or relative prices do not depend on monetary units which facilitates comparisons between assets. The following properties have been amply commented upon in the financial literature.

(i) *Non-stationarity of price series.* Samples paths of prices are generally close to a random walk without intercept (see the CAC index series[5] displayed in Figure 1.1). On the other hand, sample paths of returns are generally compatible with the second-order stationarity assumption. For instance, Figures 1.2 and 1.3 show that the returns of the CAC index oscillate around zero.

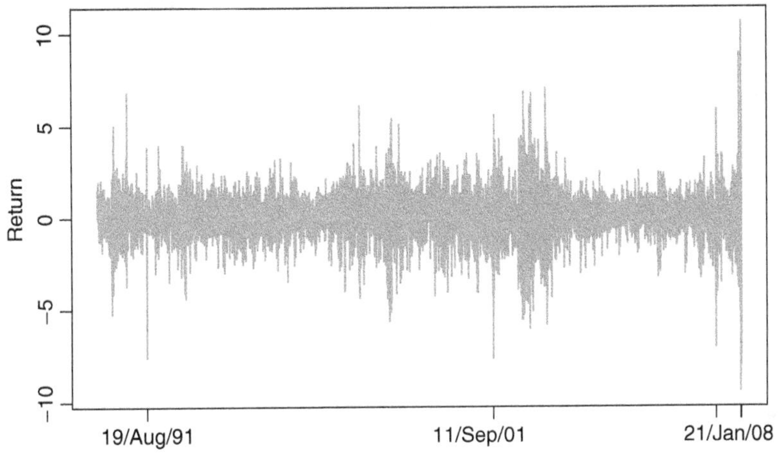

Figure 1.2 CAC 40 returns (2 March 1990 to 15 October 2008). 19 August 1991, Soviet Putsch attempt; 11 September 2001, fall of the Twin Towers; 21 January 2008, effect of the subprime mortgage crisis; 6 October 2008, effect of the financial crisis.

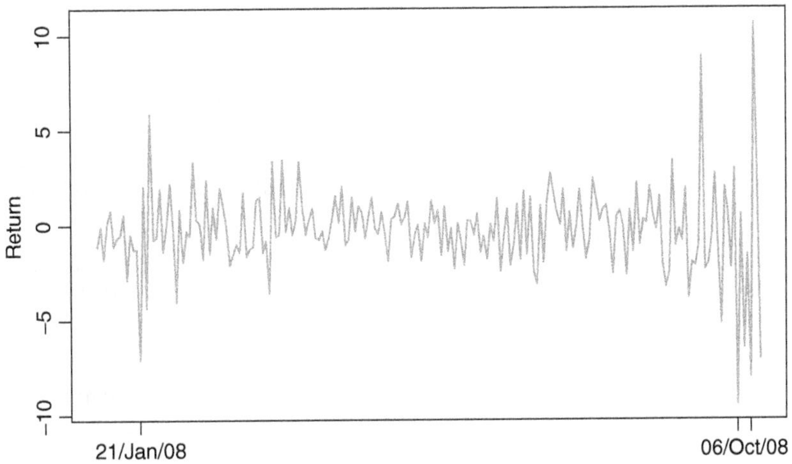

Figure 1.3 Returns of the CAC 40 (2 January 2008 to 15 October 2008).

[5] The CAC 40 index is a linear combination of a selection of 40 shares on the Paris Stock Exchange (CAC stands for 'Cotations Assistées en Continu').

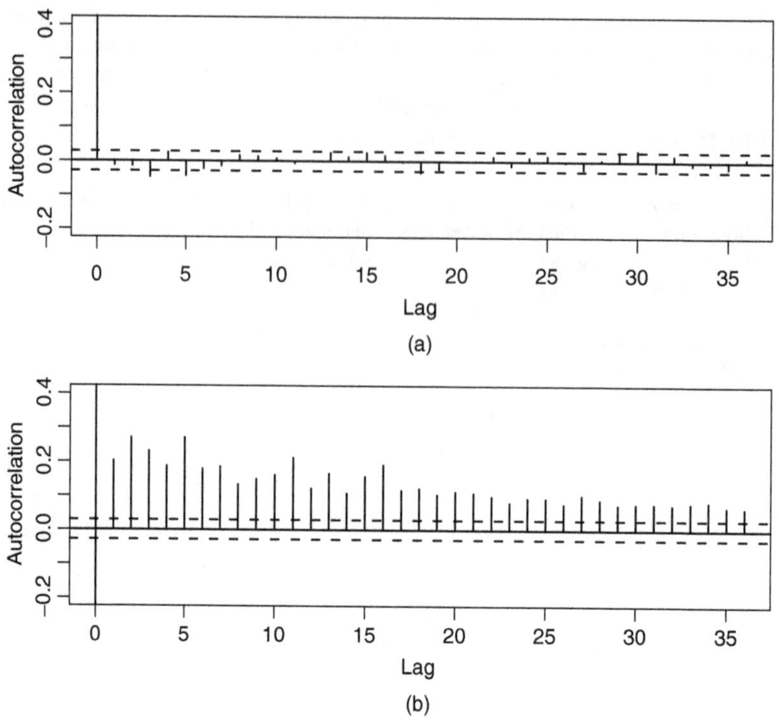

Figure 1.4 Sample autocorrelations of (a) returns and (b) squared returns of the CAC 40 (2 January 2008 to 15 October 2008).

The oscillations vary a great deal in magnitude but are almost constant in average over long sub-periods. The extreme volatility of prices in the last period, induced by the financial crisis of 2008, is worth noting.

(ii) *Absence of autocorrelation for the price variations.* The series of price variations generally displays small autocorrelations, making it close to a white noise. This is illustrated for the CAC in Figure 1.4a. The classical significance bands are used here, as an approximation, but we shall see in Chapter 5 that they must be corrected when the noise is not independent. Note that for intraday series, with very small time intervals between observations (measured in minutes or seconds) significant autocorrelations can be observed due to the so-called *microstructure* effects.

(iii) *Autocorrelations of the squared price returns.* Squared returns (ϵ_t^2) or absolute returns ($|\epsilon_t|$) are generally strongly autocorrelated (see Figure 1.4b). This property is not incompatible with the white noise assumption for the returns, but shows that the white noise is not strong.

(iv) *Volatility clustering.* Large absolute returns $|\epsilon_t|$ tend to appear in clusters. This property is generally visible on the sample paths (as in Figure 1.3). Turbulent (high-volatility) sub-periods are followed by quiet (low-volatility) periods. These sub-periods are recurrent but do not appear in a periodic way (which might contradict the stationarity assumption). In other words, volatility clustering is not incompatible with a homoscedastic (i.e. with a constant variance) marginal distribution for the returns.

(v) *Fat-tailed distributions.* When the empirical distribution of daily returns is drawn, one can generally observe that it does not resemble a Gaussian distribution. Classical tests typically lead to rejection of the normality assumption at any reasonable level. More precisely, the densities have fat tails (decreasing to zero more slowly than $\exp(-x^2/2)$) and are sharply peaked at zero: they are called leptokurtic. A measure of the leptokurticity is the kurtosis coefficient, defined as the ratio of the sample fourth-order moment to the squared sample variance. Asymptotically equal to 3 for Gaussian iid observations, this coefficient is much greater than 3 for returns series. When the time interval over which the returns are computed increases, leptokurticity tends to vanish and the empirical distributions get closer to a Gaussian. Monthly returns, for instance, defined as the sum of daily returns over the month, have a distribution that is much closer to the normal than daily returns. Figure 1.5 compares a kernel estimator of the density of the CAC returns with a Gaussian density. The peak around zero appears clearly, but the thickness of the tails is more difficult to visualise.

(vi) *Leverage effects*. The so-called leverage effect was noted by Black (1976), and involves an asymmetry of the impact of past positive and negative values on the current volatility. Negative returns (corresponding to price decreases) tend to increase volatility by a larger amount than positive returns (price increases) of the same magnitude. Empirically, a positive correlation is often detected between $\epsilon_t^+ = \max(\epsilon_t, 0)$ and $|\epsilon_{t+h}|$ (a price increase should entail future volatility increases), but, as shown in Table 1.1, this correlation is generally less than between $-\epsilon_t^- = \max(-\epsilon_t, 0)$ and $|\epsilon_{t+h}|$.

(vii) *Seasonality*. Calendar effects are also worth mentioning. The day of the week, the proximity of holidays, among other seasonalities, may have significant effects on returns. Following a period of market closure, volatility tends to increase, reflecting the information cumulated during this break. However, it can be observed that the increase is less than if the information had cumulated at constant speed. Let us also mention that the seasonal effect is also very present for intraday series.

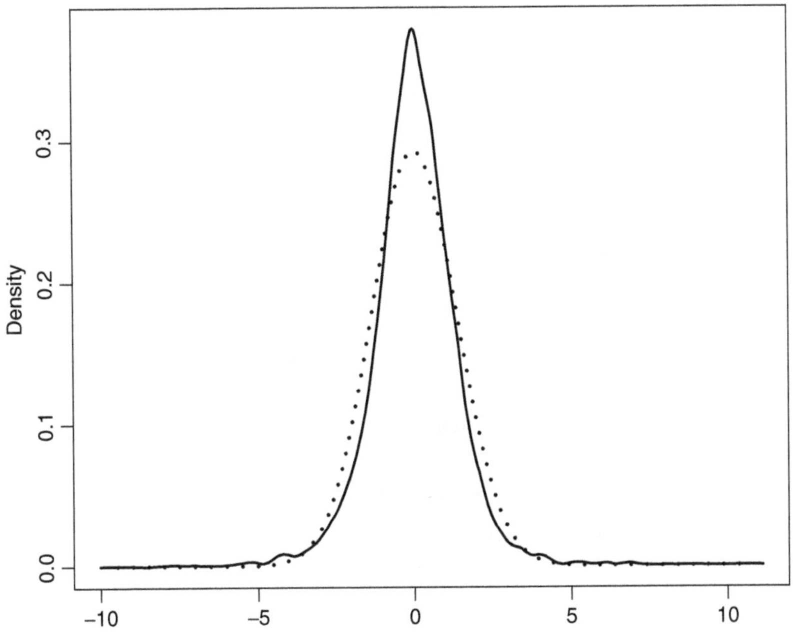

Figure 1.5 Kernel estimator of the CAC 40 returns density (solid line) and density of a Gaussian with mean and variance equal to the sample mean and variance of the returns (dotted line).

Table 1.1 Sample autocorrelations of returns ϵ_t (CAC 40 index, 2 January 2008 to 15 October 2008), of absolute returns $|\epsilon_t|$, sample correlations between ϵ^+_{t-h} and $|\epsilon_t|$, and between $-\epsilon^-_{t-h}$ and $|\epsilon_t|$.

h	1	2	3	4	5	6	7
$\hat{\rho}_\epsilon(h)$	−0.012	−0.014	−0.047	0.025	−0.043	−0.023	−0.014
$\hat{\rho}_{\lvert\epsilon\rvert}(h)$	0.175	0.229	0.235	0.200	0.218	0.212	0.203
$\hat{\rho}(\epsilon^+_{t-h}, \lvert \epsilon_t \rvert)$	0.038	0.059	0.051	0.055	0.059	0.109	0.061
$\hat{\rho}(-\epsilon^-_{t-h}, \lvert \epsilon_t \rvert)$	0.160	0.200	0.215	0.173	0.190	0.136	0.173

We use here the notation $\epsilon_t^+ = \max(\epsilon_t, 0)$ and $\epsilon_t^- = \min(\epsilon_t, 0)$.

1.4 Random Variance Models

The previous properties illustrate the difficulty of financial series modelling. Any satisfactory statistical model for daily returns must be able to capture the main stylised facts described in the previous section. Of particular importance are the leptokurticity, the unpredictability of returns, and the existence of positive autocorrelations in the squared and absolute returns. Classical formulations (such as ARMA models) centred on the second-order structure are inappropriate. Indeed, the second-order structure of most financial time series is close to that of white noise.

The fact that large absolute returns tend to be followed by large absolute returns (whatever the sign of the price variations) is hardly compatible with the assumption of constant *conditional* variance. This phenomenon is called *conditional heteroscedasticity*:

$$\mathrm{Var}(\epsilon_t \mid \epsilon_{t-1}, \epsilon_{t-2}, \dots) \not\equiv \text{const.}$$

Conditional heteroscedasticity is perfectly compatible with stationarity (in the strict and second-order senses), just as the existence of a non-constant conditional mean is compatible with stationarity. The GARCH processes studied in this book will amply illustrate this point.

The models introduced in the econometric literature to account for the very specific nature of financial series (price variations or log-returns, interest rates, etc.) are generally written in the multiplicative form

$$\epsilon_t = \sigma_t \eta_t \tag{1.6}$$

where (η_t) and (σ_t) are real processes such that:

(i) σ_t is measurable with respect to a σ-field, denoted $\mathcal{F}_{t-1}$;

(ii) (η_t) is an iid centred process with unit variance, η_t being independent of $\mathcal{F}_{t-1}$ and $\sigma(\epsilon_u; u < t)$;

(iii) $\sigma_t > 0$.

This formulation implies that the sign of the current price variation (that is, the sign of ϵ_t) is that of η_t, and is independent of past price variations. Moreover, if the first two conditional moments of ϵ_t exist, they are given by

$$E(\epsilon_t \mid \mathcal{F}_{t-1}) = 0, \quad E(\epsilon_t^2 \mid \mathcal{F}_{t-1}) = \sigma_t^2.$$

The random variable σ_t is called the *volatility*[6] of ϵ_t.

It may also be noted that (under existence assumptions)

$$E(\epsilon_t) = E(\sigma_t)E(\eta_t) = 0$$

[6] There is no general agreement concerning the definition of this concept in the literature. Volatility sometimes refers to a conditional standard deviation, and sometimes to a conditional variance.

and

$$\mathrm{Cov}(\epsilon_t, \epsilon_{t-h}) = E(\eta_t)E(\sigma_t\epsilon_{t-h}) = 0, \quad \forall h > 0,$$

which makes (ϵ_t) a weak white noise. The series of squares, on the other hand, generally have non-zero autocovariances: (ϵ_t) is thus not a strong white noise.

The kurtosis coefficient of ϵ_t, if it exists, is related to that of η_t, denoted κ_η, by

$$\frac{E(\epsilon_t^4)}{\{E(\epsilon_t^2)\}^2} = \kappa_\eta \left[1 + \frac{\mathrm{Var}(\sigma_t^2)}{\{E(\sigma_t^2)\}^2}\right]. \tag{1.7}$$

This formula shows that the leptokurticity of financial time series can be taken into account in two different ways: either by using a leptokurtic distribution for the iid sequence (η_t), or by specifying a process (σ_t^2) with a great variability.

Different classes of models can be distinguished depending on the specification adopted for σ_t:

(i) *Conditionally heteroscedastic* (or GARCH-type) processes for which $\mathcal{F}_{t-1} = \sigma(\epsilon_s; s < t)$ is the σ-field generated by the past of ϵ_t. The volatility is here a deterministic function of the past of ϵ_t. Processes of this class differ by the choice of a specification for this function. The standard GARCH models are characterised by a volatility specified as a linear function of the past values of ϵ_t^2. They will be studied in detail in Chapter 2.

(ii) *Stochastic volatility* processes[7] for which $\mathcal{F}_{t-1}$ is the σ-field generated by $\{v_t, v_{t-1}, \dots\}$, where (v_t) is a strong white noise and is independent of (η_t). In these models, volatility is a latent process. The most popular model in this class assumes that the process $\log \sigma_t$ follows an AR(1) of the form

$$\log \sigma_t = \omega + \phi \log \sigma_{t-1} + v_t,$$

where the noises (v_t) and (η_t) are independent.

(iii) *Switching-regime models* for which $\sigma_t = \sigma(\Delta_t, \mathcal{F}_{t-1})$, where (Δ_t) is a latent (unobservable) integer-valued process, independent of (η_t). The state of the variable Δ_t is here interpreted as a regime and, conditionally on this state, the volatility of ϵ_t has a GARCH specification. The process (Δ_t) is generally supposed to be a finite-state Markov chain. The models are thus called Markov-switching models.

1.5 Bibliographical Notes

The time series concepts presented in this chapter are the subject of numerous books. Two classical references are Brockwell and Davis (1991) and Gouriéroux and Monfort (1995, 1996).

The assumption of iid Gaussian price variations has long been predominant in the finance literature and goes back to the dissertation by Bachelier (1900), where a precursor of Brownian motion can be found. This thesis, ignored for a long time until its rediscovery by Kolmogorov in 1931 (see Kahane 1998), constitutes the historical source of the link between Brownian motion and mathematical finance. Nonetheless, it relies on only a rough description of the behaviour of financial series. The stylised facts concerning these series can be attributed to Mandelbrot (1963) and Fama (1965). Based on the analysis of many stock returns series, their studies showed the leptokurticity, hence the non-Gaussianity, of marginal distributions, some temporal dependencies, and non-constant volatilities. Since then, many empirical studies have confirmed these findings. See, for instance, Taylor (2007) for a detailed presentation of the stylised facts of financial times series. In particular, the calendar effects are discussed in detail.

As noted by Shephard (2005), a precursor article on ARCH models is that of Rosenberg (1972). This article shows that the decomposition (1.6) allows the leptokurticity of financial series to be reproduced.

[7] Note, however, that the volatility is also a random variable in GARCH-type processes.

It also proposes some volatility specifications which anticipate both the GARCH and stochastic volatility models. However, the GARCH models to be studied in the next chapters are not discussed in this article. The decomposition of the kurtosis coefficient in (1.7) can be found in Clark (1973).

A number of surveys have been devoted to GARCH models. See, among others, Bollerslev, Chou, and Kroner (1992), Bollerslev, Engle, and Nelson (1994), Pagan (1996), Palm (1996), Shephard (1996), Kim, Shephard, and Chib (1998), Engle (2001, 2002b, 2004), Engle and Patton (2001), Diebold (2004), Bauwens, Laurent, and Rombouts (2006), and Giraitis, Leipus, and Surgailis (2006). Moreover, the books by Gouriéroux (1997) and Xekalaki and Degiannakis (2009) are devoted to GARCH and several books devote a chapter to GARCH: Mills (1993), Hamilton (1994), Franses and van Dijk (2000), Gouriéroux and Jasiak (2001), Franke, Härdle, and Hafner chronological order (2004), McNeil, Frey, and Embrechts (2005), Taylor (2007), Andersen et al. (2009), and Tsay (2010). See also Mikosch (2001).

Although the focus of this book is on financial applications, it is worth mentioning that GARCH models have been used in other areas. Time series exhibiting GARCH-type behaviour have also appeared, for example, in speech signals (Cohen 2004, 2006; Abramson and Cohen 2008), daily and monthly temperature measurements (Tol 1996; Campbell and Diebold 2005; Romilly 2006; Huang, Shiu, and Lin 2008), wind speeds (Ewing, Kruse, and Schroeder 2006), electricity prices (Dupuis 2017) and atmospheric CO_2 concentrations (Hoti, McAleer, and Chan 2005; McAleer and Chan 2006).

Most econometric software (for instance, GAUSS, R, RATS, SAS and SPSS) incorporates routines that permit the estimation of GARCH models. Readers interested in the implementation with Ox may refer to Laurent (2009).

1.6 Exercises

1.1 *(Stationarity, ARMA models, white noises)*
Let (η_t) denote an iid centred sequence with unit variance (and if necessary with a finite fourth-order moment).

1. Do the following models admit a stationary solution? If yes, derive the expectation and the autocorrelation function of this solution.
   (a) $X_t = 1 + 0.5X_{t-1} + \eta_t$;
   (b) $X_t = 1 + 2X_{t-1} + \eta_t$;
   (c) $X_t = 1 + 0.5X_{t-1} + \eta_t - 0.4\eta_{t-1}$.
2. Identify the ARMA models compatible with the following recursive relations, where $\rho(\cdot)$ denotes the autocorrelation function of some stationary process:
   (a) $\rho(h) = 0.4\rho(h-1)$, for all $h > 2$;
   (b) $\rho(h) = 0$, for all $h > 3$;
   (c) $\rho(h) = 0.2\rho(h-2)$, for all $h > 1$.
3. Verify that the following processes are white noises and decide if they are weak or strong.
   (a) $\epsilon_t = \eta_t^2 - 1$;
   (b) $\epsilon_t = \eta_t\eta_{t-1}$;

1.2 *(A property of the sum of the sample autocorrelations)*
Let

$$\hat{\gamma}(h) = \hat{\gamma}(-h) = \frac{1}{n}\sum_{t=1}^{n-h}(X_t - \overline{X}_n)(X_{t+h} - \overline{X}_n), \quad h = 0, \ldots, n-1,$$

denote the sample autocovariances of real observations $X_1, \ldots, X_n$. Set $\hat{\rho}(h) = \hat{\rho}(-h) = \hat{\gamma}(h)/\hat{\gamma}(0)$ for $h = 0, \ldots, n-1$. Show that

$$\sum_{h=1}^{n-1}\hat{\rho}(h) = -\frac{1}{2}.$$

1.3 *(It is impossible to decide whether a process is stationary from a path)*
Show that the sequence $\{(-1)^t\}_{t=0,1,\ldots}$ can be a realisation of a non-stationary process. Show that it can also be a realisation of a stationary process. Comment on the consequences of this result.

1.4 *(Stationarity and ergodicity from a path)*
Can the sequence 0, 1, 0, 1, … be a realisation of a stationary process or of a stationary and ergodic process? The definition of ergodicity can be found in Appendix A.1.

1.5 *(A weak white noise which is not strong)*
Let (η_t) denote an iid $\mathcal{N}(0,1)$ sequence and let k be a positive integer. Set $\epsilon_t = \eta_t\eta_{t-1}\cdots\eta_{t-k}$. Show that (ϵ_t) is a weak white noise, but is not a strong white noise.

1.6 *(Asymptotic variance of sample autocorrelations of a weak white noise)*
Consider the white noise ϵ_t of Exercise 1.5. Compute $\lim_{n\to\infty} n\text{Var}\hat{\rho}(h)$ where $h \neq 0$ and $\hat{\rho}(\cdot)$ denotes the sample autocorrelation function of $\epsilon_1, \ldots, \epsilon_n$. Compare this asymptotic variance with that obtained from the usual Bartlett formula.

1.7 *(ARMA representation of the square of a weak white noise)*
Consider the white noise ϵ_t of Exercise 1.5. Show that ϵ_t^2 follows an ARMA process. Make the ARMA representation explicit when $k = 1$.

1.8 *(Asymptotic variance of sample autocorrelations of a weak white noise)*
Repeat Exercise 1.6 for the weak white noise $\epsilon_t = \eta_t/\eta_{t-k}$, where (η_t) is an iid sequence such that $E\eta_t^4 < \infty$ and $E\eta_t^{-2} < \infty$, and k is a positive integer.

1.9 *(Stationary solutions of an AR(1))*
Let $(\eta_t)_{t\in\mathbb{Z}}$ be an iid centred sequence with variance $\sigma^2 > 0$, and let $a \neq 0$. Consider the AR(1) equation

$$X_t - aX_{t-1} = \eta_t, \quad t \in \mathbb{Z}. \tag{1.8}$$

1. Show that for $|a| < 1$, the infinite sum

$$X_t = \sum_{k=0}^{\infty} a^k \eta_{t-k}$$

converges in quadratic mean and almost surely, and that it is the unique stationary solution of Eq. (1.8).
2. For $|a| = 1$, show that no stationary solution exists.
3. For $|a| > 1$, show that

$$X_t = -\sum_{k=1}^{\infty} \frac{1}{a^k}\eta_{t+k}$$

is the unique stationary solution of Eq. (1.8).
4. For $|a| > 1$, show that the causal representation

$$X_t - \frac{1}{a}X_{t-1} = \epsilon_t, \quad t \in \mathbb{Z}, \tag{1.9}$$

holds, where $(\epsilon_t)_{t\in\mathbb{Z}}$ is a white noise.

1.10 *(Is the S&P 500 a white noise?)*
Figure 1.6 displays the correlogram of the S&P 500 returns from 3 January 1979 to 30 December 2001, as well as the correlogram of the squared returns. Is it reasonable to think that this index is a strong white noise or a weak white noise?

1.11 *(Asymptotic covariance of sample autocovariances)*
Justify the equivalence between (B.18) and (B.14) in the proof of the generalised Bartlett formula of Appendix B.2.

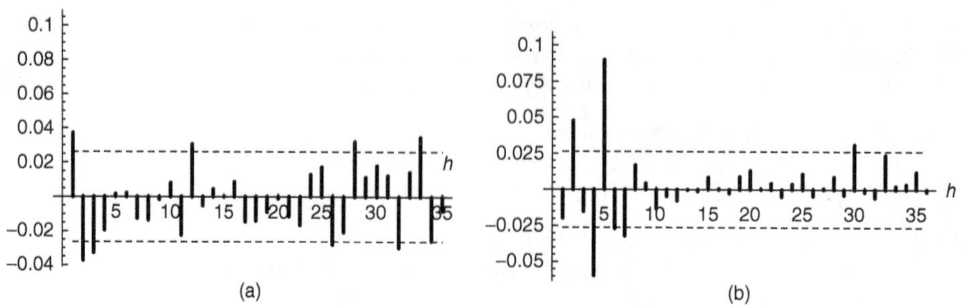

Figure 1.6 Sample autocorrelations $\hat{\rho}(h)$ ($h = 1, \ldots, 36$) of (a) the S&P 500 index from 3 January 1979 to 30 December 2001, and (b) the squared index. The interval between the dashed lines ($\pm 1.96/\sqrt{n}$, where $n = 5804$ is the sample length) should contain approximately 95% of the autocorrelations of a strong white noise.

1.12 *(Asymptotic independence between the $\hat{\rho}(h)$ for a noise)*
Simplify the generalised Bartlett formulas (B.14) and (B.15) when $X = \epsilon$ is a pure white noise.
In an autocorrelogram, consider the random number M of sample autocorrelations falling outside the significance region (at the level 95%, say), among the first m autocorrelations. How can the previous result be used to evaluate the variance of this number when the observed process is a white noise (satisfying the assumptions allowing (B.15) to be used)?

1.13 *(An incorrect interpretation of autocorrelograms)*
Some practitioners tend to be satisfied with an estimated model only if all sample autocorrelations fall within the 95% significance bands. Show, using Exercise 1.12, that based on 20 autocorrelations, say, this approach leads to wrongly rejecting a white noise with a very high probability.

1.14 *(Computation of partial autocorrelations)*
Use the algorithm in (B.7)–(B.9) to compute $r_X(1)$, $r_X(2)$ and $r_X(3)$ as a function of $\rho_X(1)$, $\rho_X(2)$ and $\rho_X(3)$.

1.15 *(Empirical application)*
Download from http://fr.biz.yahoo.com//bourse/accueil.html for instance, a stock index such as the CAC 40. Draw the series of closing prices, the series of returns, the autocorrelation function of the returns, and that of the squared returns. Comment on these graphs.

Part I

Univariate GARCH Models

2

GARCH(p, q) Processes

Autoregressive conditionally heteroscedastic (ARCH) models were introduced by Engle (1982), and their GARCH (generalised ARCH) extension is due to Bollerslev (1986). In these models, the key concept is the *conditional variance*, that is, the variance conditional on the past. In the classical GARCH models, the conditional variance is expressed as a linear function of the squared past values of the series. This particular specification is able to capture the main stylised facts characterising financial series, as described in Chapter 1. At the same time, it is simple enough to allow for a complete study of the solutions. The 'linear' structure of these models can be displayed through several representations that will be studied in this chapter.

We first present definitions and representations of GARCH models. Then we establish the strict and second-order stationarity conditions. Starting with the first-order GARCH model, for which the proofs are easier and the results are more explicit, we extend the study to the general case. We also study the so-called ARCH(∞) models, which allow for a slower decay of squared-return autocorrelations. Then, we consider the existence of moments and the properties of the autocorrelation structure. We conclude this chapter by examining forecasting issues.

2.1 Definitions and Representations

We start with a definition of GARCH processes based on the first two conditional moments.

Definition 2.1 (GARCH(p, q) process) *A process (ϵ_t) is called a GARCH(p, q) process if its first two conditional moments exist and satisfy:*

(i) $E(\epsilon_t \mid \epsilon_u, u < t) = 0, \quad t \in \mathbb{Z}$.

(ii) *There exist constants* ω, α_i, $i = 1, \ldots, q$ *and* β_j, $j = 1, \ldots, p$ *such that*

$$\sigma_t^2 = \mathrm{Var}(\epsilon_t \mid \epsilon_u, u < t) = \omega + \sum_{i=1}^{q} \alpha_i \epsilon_{t-i}^2 + \sum_{j=1}^{p} \beta_j \sigma_{t-j}^2, \quad t \in \mathbb{Z}. \tag{2.1}$$

GARCH Models: Structure, Statistical Inference and Financial Applications, Second Edition. Christian Francq and Jean-Michel Zakoian.

Equation (2.1) can be written in a more compact way as

$$\sigma_t^2 = \omega + \alpha(B)\epsilon_t^2 + \beta(B)\sigma_t^2, \quad t \in \mathbb{Z}, \tag{2.2}$$

where B is the standard backshift operator ($B^i\epsilon_t^2 = \epsilon_{t-i}^2$ and $B^i\sigma_t^2 = \sigma_{t-i}^2$ for any integer i), and α and β are polynomials of degrees q and p, respectively:

$$\alpha(B) = \sum_{i=1}^{q} \alpha_i B^i, \quad \beta(B) = \sum_{j=1}^{p} \beta_j B^j.$$

If $\beta(z) = 0$ we have

$$\sigma_t^2 = \omega + \sum_{i=1}^{q} \alpha_i \epsilon_{t-i}^2 \tag{2.3}$$

and the process is called an ARCH(q) process.[1] By definition, the innovation of the process ϵ_t^2 is the variable $v_t = \epsilon_t^2 - \sigma_t^2$. Substituting in Eq. (2.1) the variables σ_{t-j}^2 by $\epsilon_{t-j}^2 - v_{t-j}$, we get the representation

$$\epsilon_t^2 = \omega + \sum_{i=1}^{r} (\alpha_i + \beta_i)\epsilon_{t-i}^2 + v_t - \sum_{j=1}^{p} \beta_j v_{t-j}, \quad t \in \mathbb{Z}, \tag{2.4}$$

where $r = \max(p, q)$, with the convention $\alpha_i = 0$ ($\beta_j = 0$) if $i > q$ ($j > p$). This equation has the linear structure of an ARMA model, allowing for simple computation of the linear predictions. Under additional assumptions (implying the second-order stationarity of ϵ_t^2), we can state that if (ϵ_t) is GARCH(p, q), then (ϵ_t^2) is an ARMA(r, p) process. In particular, the square of an ARCH(q) process admits, if it is stationary, an AR(q) representation. The ARMA representation will be useful for the estimation and identification of GARCH processes.[2]

Remark 2.1 (Correlation of the squares of a GARCH) We observed in Chapter 1 that a characteristic feature of financial series is that squared returns are autocorrelated, while returns are not. The representation (2.4) shows that GARCH processes are able to capture this empirical fact. If the fourth-order moment of (ϵ_t) is finite, the sequence of the h-order autocorrelations of ϵ_t^2 is the solution of a recursive equation which is characteristic of ARMA models. For the sake of simplicity, consider the GARCH(1, 1) case. The squared process (ϵ_t^2) is ARMA(1, 1), and thus its autocorrelation decreases to zero proportionally to $(\alpha_1 + \beta_1)^h$: for $h > 1$,

$$\mathrm{Corr}(\epsilon_t^2, \epsilon_{t-h}^2) = K(\alpha_1 + \beta_1)^h,$$

where K is a constant independent of h. Moreover, the ϵ_t's are uncorrelated in view of (i) in Definition 2.1.

Definition 2.1 does not directly provide a solution process satisfying those conditions. The next definition is more restrictive but allows explicit solutions to be obtained. The link between the two definitions will be given in Remark 2.5. Let η denote a probability distribution with null expectation and unit variance.

---

[1] This specification quickly turned out to be too restrictive when applied to financial series. Indeed, a large number of past variables have to be included in the conditional variance to obtain a good model fit. Choosing a large value for q is not satisfactory from a statistical point of view because it requires a large number of coefficients to be estimated.

[2] It cannot be used to study the existence of stationary solutions, however, because the process (v_t) is not an iid process.

Definition 2.2 (Strong GARCH(p, q) process) *Let (η_t) be an iid sequence with distribution η. The process (ϵ_t) is called a strong GARCH(p, q) (with respect to the sequence (η_t)) if*

$$\begin{cases} \epsilon_t = \sigma_t \eta_t \\ \sigma_t^2 = \omega + \sum_{i=1}^{q} \alpha_i \epsilon_{t-i}^2 + \sum_{j=1}^{p} \beta_j \sigma_{t-j}^2 \end{cases} \tag{2.5}$$

where the α_i and β_j are non-negative constants, ω is a (strictly) positive constant and $\sigma_t > 0$.

GARCH processes in the sense of Definition 2.1 are sometimes called *semi-strong* following the paper by Drost and Nijman (1993) on temporal aggregation. Substituting ϵ_{t-i} by $\sigma_{t-i}\eta_{t-i}$ in (2.1), we get

$$\sigma_t^2 = \omega + \sum_{i=1}^{q} \alpha_i \sigma_{t-i}^2 \eta_{t-i}^2 + \sum_{j=1}^{p} \beta_j \sigma_{t-j}^2,$$

which can be written as follows:

$$\sigma_t^2 = \omega + \sum_{i=1}^{r} a_i(\eta_{t-i}) \sigma_{t-i}^2, \tag{2.6}$$

where $a_i(z) = \alpha_i z^2 + \beta_i$, $i = 1, \ldots, r$. This representation shows that the volatility process of a strong GARCH is the solution of an autoregressive equation with random coefficients.

Properties of Simulated Paths

Contrary to standard time series models (ARMA), the GARCH structure allows the magnitude of the noise ϵ_t to be a function of its past values. Thus, periods with high-volatility level (corresponding to large values of ϵ_{t-i}^2) will be followed by periods where the fluctuations have a smaller amplitude. Figures 2.1–2.7 illustrate the *volatility clustering* for simulated GARCH models. Large absolute values are not uniformly distributed on the whole period, but tend to cluster. We will see that all these trajectories correspond to strictly stationary processes which, except for the ARCH(1) models of Figures 2.3–2.5, are also second-order stationary. Even if the absolute values can be extremely large, these processes are not explosive, as can be seen from these figures. Higher values of α (theoretically, $\alpha > 3.56$ for the $\mathcal{N}(0, 1)$ distribution, as will be established below) lead to explosive paths. Figures 2.6 and 2.7, corresponding to GARCH(1, 1) models, have been obtained with the same simulated sequence (η_t). As we will see, permuting α and β does not modify the variance of the process but has an effect on the higher-order moments. For instance the simulated process of Figure 2.7, with $\alpha = 0.7$ and $\beta = 0.2$, does not admit a fourth-order moment, in contrast to the process of Figure 2.6. This is reflected by the presence of larger

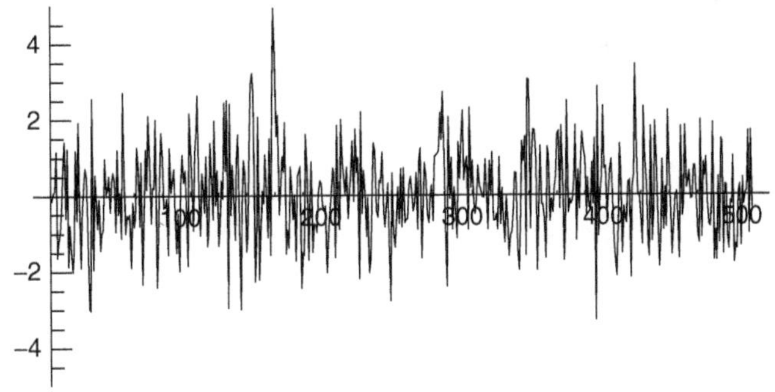

Figure 2.1 Simulation of size 500 of the ARCH(1) process with $\omega = 1$, $\alpha = 0.5$ and $\eta_t \sim \mathcal{N}(0, 1)$.

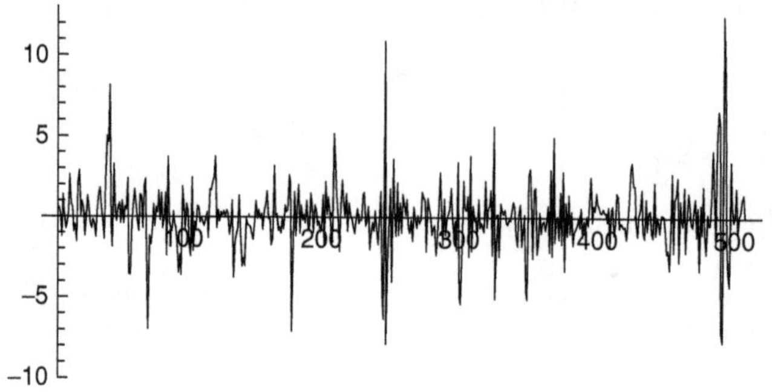

Figure 2.2 Simulation of size 500 of the ARCH(1) process with $\omega = 1$, $\alpha = 0.95$ and $\eta_t \sim \mathcal{N}(0, 1)$.

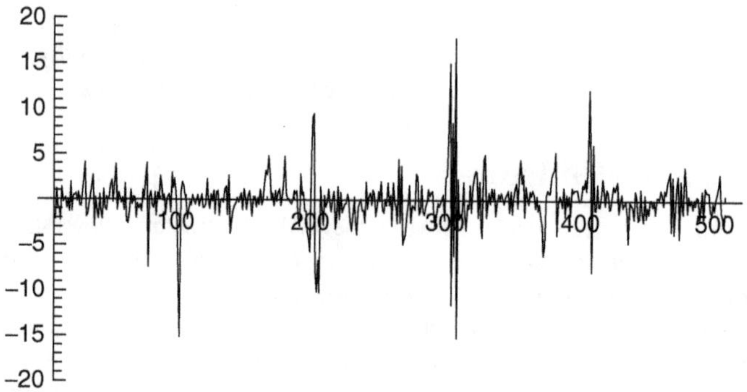

Figure 2.3 Simulation of size 500 of the ARCH(1) process with $\omega = 1$, $\alpha = 1.1$ and $\eta_t \sim \mathcal{N}(0, 1)$.

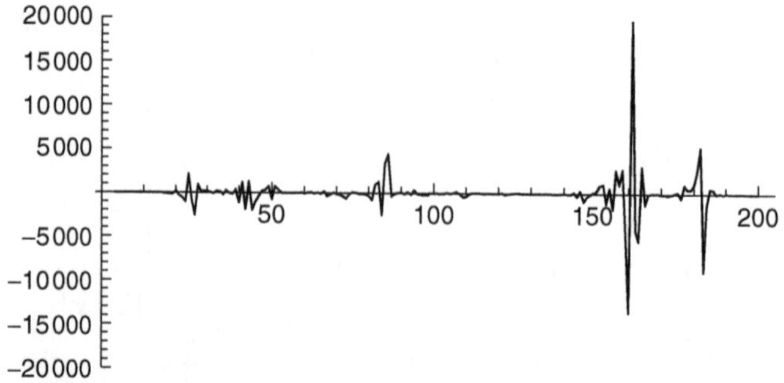

Figure 2.4 Simulation of size 200 of the ARCH(1) process with $\omega = 1$, $\alpha = 3$ and $\eta_t \sim \mathcal{N}(0, 1)$.

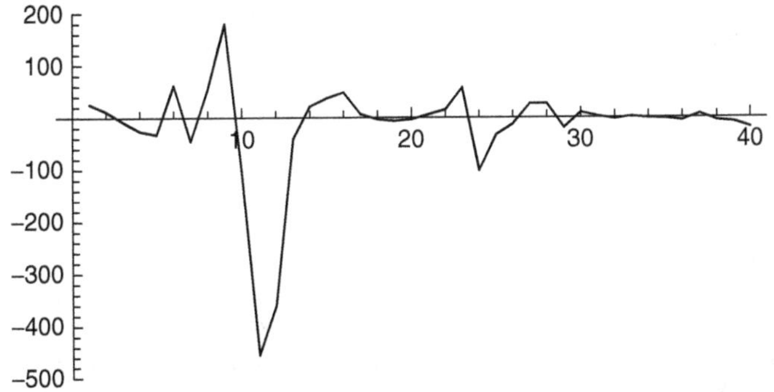

Figure 2.5 Observations 100–140 of Figure 2.4.

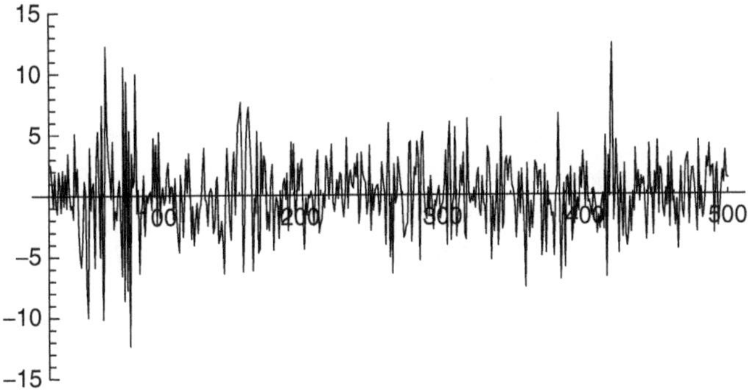

Figure 2.6 Simulation of size 500 of the GARCH(1, 1) process with $\omega = 1$, $\alpha = 0.2$, $\beta = 0.7$ and $\eta_t \sim \mathcal{N}(0, 1)$.

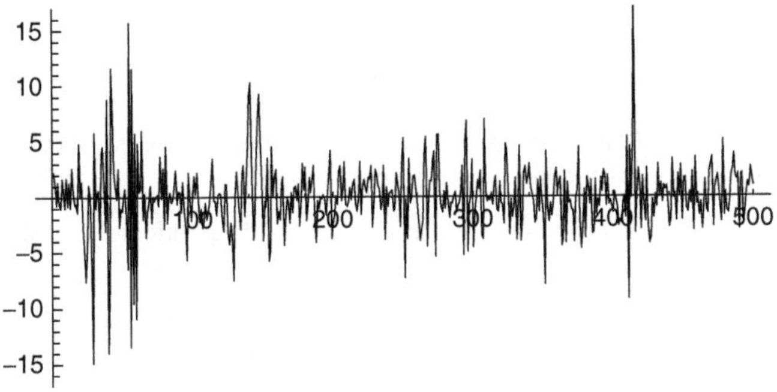

Figure 2.7 Simulation of size 500 of the GARCH(1, 1) process with $\omega = 1$, $\alpha = 0.7$, $\beta = 0.2$ and $\eta_t \sim \mathcal{N}(0, 1)$.

absolute values in Figure 2.7. The two processes are also different in terms of *persistence of shocks*: when β approaches 1, a shock on the volatility has a persistent effect. On the other hand, when α is large, sudden volatility variations can be observed in response to shocks.

2.2 Stationarity Study

This section is concerned with the existence of stationary solutions (in the strict and second-order senses) to model (2.5). We are mainly interested in *non-anticipative* solutions, that is, processes (ϵ_t) such that ϵ_t is a measurable function of the variables η_{t-s}, $s \geq 0$. For such processes, σ_t is independent of the σ-field generated by $\{\eta_{t+h},\ h \geq 0\}$ and ϵ_t is independent of the σ-field generated by $\{\eta_{t+h},\ h > 0\}$. It will be seen that such solutions are also *ergodic*. The concept of ergodicity is discussed in Appendix A.1. We first consider the GARCH(1, 1) model, which can be studied in a more explicit way than the general case. For $x > 0$, let $\log^+ x = \max(\log x, 0)$.

2.2.1 The GARCH(1,1) Case

When $p = q = 1$, model (2.5) has the form

$$\begin{cases} \epsilon_t = \sigma_t \eta_t, & (\eta_t) \text{ iid } (0, 1), \\ \sigma_t^2 = \omega + \alpha \epsilon_{t-1}^2 + \beta \sigma_{t-1}^2, \end{cases} \tag{2.7}$$

with $\omega > 0$, $\alpha \geq 0$, $\beta \geq 0$. Let $a(z) = \alpha z^2 + \beta$.

Theorem 2.1 (Strict stationarity of the strong GARCH(1,1) process) *If*

$$-\infty \leq \gamma := E \log\{\alpha \eta_t^2 + \beta\} < 0, \tag{2.8}$$

then the infinite sum

$$h_t = \left\{ 1 + \sum_{i=1}^{\infty} a(\eta_{t-1}) \dots a(\eta_{t-i}) \right\} \omega \tag{2.9}$$

converges almost surely (a.s.), and the process (ϵ_t) defined by $\epsilon_t = \sqrt{h_t}\eta_t$ is the unique strictly stationary solution of model (2.7). This solution is non-anticipative and ergodic. If $\gamma \geq 0$ and $\omega > 0$, there exists no strictly stationary solution.

Remark 2.2 (On the strict stationarity condition (2.8))

1. When $\omega = 0$ and $\gamma < 0$, it is clear that, in view of condition (2.9), the unique strictly stationary solution is $\epsilon_t = 0$. It is, therefore, natural to impose $\omega > 0$.
2. It may be noted that the condition (2.8) depends on the distribution of η_t and that it is not symmetric in α and β.
3. Examination of the proof below shows that the assumptions $E\eta_t = 0$ and $E\eta_t^2 = 1$, which facilitate the interpretation of the model, are not necessary. It is sufficient to have $E \log^+ \eta_t^2 < \infty$.
4. Condition (2.8) implies $\beta < 1$. Now, if

   $$\alpha + \beta < 1,$$

   then condition (2.8) is satisfied since, by application of the Jensen inequality,

   $$E \log\{a(\eta_t)\} \leq \log E\{a(\eta_t)\} = \log(\alpha + \beta) < 0.$$

5. If (2.8) is satisfied, it is also satisfied for any pair (α_1, β_1) such that $\alpha_1 \leq \alpha$ and $\beta_1 \leq \beta$. In particular, the strict stationarity of a given GARCH(1, 1) model implies that the ARCH(1) model obtained by cancelling β is also stationary.
6. In the ARCH(1) case ($\beta = 0$), the strict stationarity constraint is written as

$$0 \leq \alpha < \exp\{-E(\log \eta_t^2)\}. \tag{2.10}$$

For instance when $\eta_t \sim \mathcal{N}(0,1)$, the condition becomes $\alpha < 3.56$. For a distribution such that $E(\log \eta_t^2) = -\infty$, for instance with a mass at 0, condition (2.10) is always satisfied. For such distributions, a strictly stationary ARCH(1) solution exists whatever the value of α.

Proof of Theorem 2.1 Note that the coefficient $\gamma = E \log \{a(\eta_t)\}$ always exists in $[-\infty, +\infty)$ because $E \log^+ \{a(\eta_t)\} \leq E a(\eta_t) = \alpha + \beta$. Using iteratively the second equation in model (2.7), we get, for $N \geq 1$,

$$\begin{aligned}
\sigma_t^2 &= \omega + a(\eta_{t-1})\sigma_{t-1}^2 \\
&= \omega \left[1 + \sum_{n=1}^{N} a(\eta_{t-1}) \dots a(\eta_{t-n}) \right] + a(\eta_{t-1}) \dots a(\eta_{t-N-1})\sigma_{t-N-1}^2 \\
&:= h_t(N) + a(\eta_{t-1}) \dots a(\eta_{t-N-1})\sigma_{t-N-1}^2.
\end{aligned} \tag{2.11}$$

The limit process $h_t = \lim_{N\to\infty} h_t(N)$ exists in $\overline{\mathbb{R}}^+ = [0, +\infty]$ since the summands are non-negative. Moreover, letting N go to infinity in $h_t(N) = \omega + a(\eta_{t-1})h_{t-1}(N-1)$, we get

$$h_t = \omega + a(\eta_{t-1})h_{t-1}.$$

We now show that h_t is a.s. finite if and only if $\gamma < 0$.

Suppose that $\gamma < 0$. We will use the Cauchy rule for series with non-negative terms.[3] We have

$$[a(\eta_{t-1}) \dots a(\eta_{t-n})]^{1/n} = \exp \left[\frac{1}{n} \sum_{i=1}^{n} \log\{a(\eta_{t-i})\} \right] \to e^{\gamma} \quad \text{a.s.} \tag{2.12}$$

as $n \to \infty$, by application of the strong law of large numbers to the iid sequence $(\log\{a(\eta_t)\})$.[4] The series defined in equality (2.9) thus converges a.s. in $\mathbb{R}$, by application of the Cauchy rule, and the limit process (h_t) takes positive real values. It follows that the process (ϵ_t) defined by

$$\epsilon_t = \sqrt{h_t}\eta_t = \left\{ \omega + \sum_{i=1}^{\infty} a(\eta_{t-1}) \dots a(\eta_{t-i})\omega \right\}^{1/2} \eta_t \tag{2.13}$$

is strictly stationary and ergodic (see Appendix A.1, Theorem A.1). Moreover, (ϵ_t) is a non-anticipative solution of model (2.7).

---

[3] Let $\sum a_n$ be a series with non-negative terms and let $\lambda = \overline{\lim}\, a_n^{1/n}$. Then (i) if $\lambda < 1$ the series $\sum a_n$ converges, (ii) if $\lambda > 1$ the series $\sum a_n$ diverges.

[4] If (X_i) is an iid sequence of random variables admitting an expectation, which can be infinite, then $\frac{1}{n}\sum_{i=1}^n X_i \to EX_1$ a.s. This result, which can be found in Billingsley (1995), follows from the strong law of large numbers for integrable variables: suppose, for instance, that $E(X_i^+) = +\infty$ and let, for any integer $m > 0$, $\tilde{X}_i = X_i^+$ if $0 \leq X_i^+ \leq m$, and $\tilde{X}_i = 0$ otherwise. Then $\frac{1}{n}\sum_{i=1}^n X_i^+ \geq \frac{1}{n}\sum_{i=1}^n \tilde{X}_i \to E\tilde{X}_1$, a.s., by application of the strong law of large numbers to the sequence of integrable variables $\tilde{X}_i$. When m goes to infinity, the increasing sequence $E\tilde{X}_1$ converges to $+\infty$, which allows us to conclude that $\frac{1}{n}\sum_{i=1}^n X_i^+ \to \infty$ a.s.

We now prove the uniqueness. Let $\tilde{\epsilon}_t = \sigma_t\eta_t$ denote another strictly stationary solution. By (2.11) we have

$$\sigma_t^2 = h_t(N) + a(\eta_{t-1})\dots a(\eta_{t-N-1})\sigma_{t-N-1}^2.$$

It follows that

$$\sigma_t^2 - h_t = \{h_t(N) - h_t\} + a(\eta_{t-1})\dots a(\eta_{t-N-1})\sigma_{t-N-1}^2.$$

The term in brackets on the right-hand side tends to 0 a.s. as $N \to \infty$. Moreover, since the series defining h_t converges a.s., we have $a(\eta_{t-1})\dots a(\eta_{t-n}) \to 0$ with probability 1 as $n \to \infty$. In addition, the distribution of σ_{t-N-1}^2 is independent of N by stationarity. Therefore, $a(\eta_{t-1})\dots a(\eta_{t-N-1})\sigma_{t-N-1}^2 \to 0$ in probability as $N \to \infty$. We have proved that $\sigma_t^2 - h_t \to 0$ in probability as $N \to \infty$. This term being independent of N, we necessarily have $h_t = \sigma_t^2$ for any t, a.s.

If $\gamma > 0$, from the convergence in (2.12) and the Cauchy rule, $\sum_{n=1}^{N} a(\eta_{t-1})\dots a(\eta_{t-n}) \to +\infty$, a.s., as $N \to \infty$. Hence, if $\omega > 0$, $h_t = +\infty$ a.s. By (2.11), it is clear that $\sigma_t^2 = +\infty$, a.s. It follows that there exists no a.s. finite solution to model (2.7).

For $\gamma = 0$, we give a proof by contradiction. Suppose there exists a strictly stationary solution (ϵ_t, σ_t^2) of model (2.7). We have, for $n > 0$,

$$\sigma_0^2 \geq \omega \left\{ 1 + \sum_{i=1}^{n} a(\eta_{-1})\dots a(\eta_{-i}) \right\}$$

from which we deduce that $a(\eta_{-1})\dots a(\eta_{-n})\omega$ converges to zero, a.s., as $n \to \infty$, or, equivalently, that

$$\sum_{i=1}^{n} \log a(\eta_i) + \log \omega \to -\infty \quad \text{a.s.} \quad \text{as } n \to \infty. \tag{2.14}$$

By the Chung–Fuchs theorem[5], we have $\limsup \sum_{i=1}^{n} \log a(\eta_i) = +\infty$ with probability 1, which contradicts (2.14).

The next result shows that non-stationary GARCH processes are explosive.

Corollary 2.1 (Conditions of explosion) *For the GARCH(1, 1) model defined by Eq. (2.7) for $t \geq 1$, with initial conditions for ϵ_0 and σ_0,*

$$\gamma > 0 \implies \sigma_t^2 \to +\infty, \quad a.s.\ (t \to \infty).$$

If, in addition, $E \mid \log(\eta_t^2) \mid < \infty$, then

$$\gamma > 0 \implies \epsilon_t^2 \to +\infty, \quad a.s.\ (t \to \infty).$$

Proof. We have

$$\sigma_t^2 \geq \omega \left\{ 1 + \sum_{i=1}^{t-1} a(\eta_{t-1})\dots a(\eta_{t-i}) \right\} \geq \omega a(\eta_{t-1})\dots a(\eta_1). \tag{2.15}$$

Hence,

$$\liminf_{t\to\infty} \frac{1}{t} \log \sigma_t^2 \geq \liminf_{t\to\infty} \frac{1}{t} \sum_{i=1}^{t-1} \log a(\eta_{t-i}) = \gamma.$$

[5] If $X_1, \dots, X_n$ is an iid sequence such that $EX_1 = 0$ and $E|X_1| > 0$ then $\limsup_{n\to\infty} \sum_{i=1}^{n} X_i = +\infty$ and $\liminf_{n\to\infty} \sum_{i=1}^{n} X_i = -\infty$ (see, for instance, Chow and Teicher 1997).

Thus $\log \sigma_t^2 \to \infty$ and $\sigma_t^2 \to \infty$ a.s. as $\gamma > 0$. By the same arguments,

$$\liminf_{t\to\infty} \frac{1}{t} \log \epsilon_t^2 = \liminf_{t\to\infty} \frac{1}{t}(\log \sigma_t^2 + \log \eta_t^2) \geq \gamma + \liminf_{t\to\infty} \frac{1}{t} \log \eta_t^2 = \gamma$$

using Exercise 2.13. The conclusion follows. □

Remark 2.3 (On Corollary 2.1)

1. When $\gamma = 0$, Klüppelberg, Lindner, and Maller (2004) showed that $\sigma_t^2 \to \infty$ in probability.
2. Since, by Jensen's inequality, we have $E\log(\eta_t^2) < \infty$, the restriction $E \mid \log(\eta_t^2) \mid < \infty$ means $E\log(\eta_t^2) > -\infty$. In the ARCH(1) case, this restriction vanishes because the condition $\gamma = E\log \alpha\eta_t^2 > 0$ implies $E\log(\eta_t^2) > -\infty$.

Theorem 2.2 (Second-order stationarity of the GARCH(1,1) process) *Let $\omega > 0$. If $\alpha + \beta \geq 1$, a non-anticipative and second-order stationary solution to the GARCH(1, 1) model does not exist. If $\alpha + \beta < 1$, the process (ϵ_t) defined by (2.13) is second-order stationary. More precisely, (ϵ_t) is a weak white noise. Moreover, there exists no other second-order stationary and non-anticipative solution.*

Proof. If (ϵ_t) is a GARCH(1, 1) process, in the sense of Definition 2.1, which is second-order stationary and non-anticipative, we have

$$E(\epsilon_t^2) = E\{E(\epsilon_t^2 \mid \epsilon_u, u < t)\} = E(\sigma_t^2) = \omega + (\alpha + \beta)E(\epsilon_{t-1}^2),$$

that is,

$$(1 - \alpha - \beta)E(\epsilon_t^2) = \omega.$$

Hence, we must have $\alpha + \beta < 1$. In addition, we get $E(\epsilon_t^2) > 0$. Conversely, suppose $\alpha + \beta < 1$. By Remark 2.2(4), the strict stationarity condition is satisfied. It is thus sufficient to show that the strictly stationary solution defined in (2.13) admits a finite variance. The variable h_t being an increasing limit of positive random variables, the infinite sum and the expectation can be permuted to give,

$$\begin{aligned} E(\epsilon_t^2) = E(h_t) &= \left[1 + \sum_{n=1}^{+\infty} E\{a(\eta_{t-1}) \dots a(\eta_{t-n})\}\right] \omega \\ &= \left[1 + \sum_{n=1}^{+\infty} \{Ea(\eta_t)\}^n\right] \omega \\ &= \left[1 + \sum_{n=1}^{+\infty} (\alpha + \beta)^n\right] \omega = \frac{\omega}{1 - (\alpha + \beta)}. \end{aligned}$$

This proves the second-order stationarity of the solution. Moreover, this solution is a white noise because $E(\epsilon_t) = E\{E(\epsilon_t \mid \epsilon_u, u < t)\} = 0$ and for all $h > 0$,

$$\text{Cov}(\epsilon_t, \epsilon_{t-h}) = E\{\epsilon_{t-h}E(\epsilon_t \mid \epsilon_u, u < t)\} = 0.$$

Let $\tilde{\epsilon}_t = \sqrt{\tilde{h}_t}\eta_t$ denote another second-order and non-anticipative stationary solution. We have

$$\mid h_t - \tilde{h}_t \mid = a(\eta_{t-1}) \dots a(\eta_{t-n}) \mid h_{t-n-1} - \tilde{h}_{t-n-1} \mid,$$

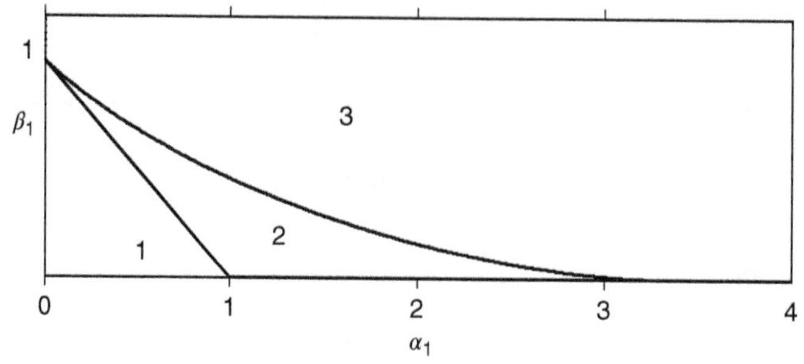

Figure 2.8 Stationarity regions for the GARCH(1, 1) model when $\eta_t \sim \mathcal{N}(0, 1)$: 1, second-order stationarity; 1 and 2, strict stationarity; 3, non-stationarity.

and then

$$\begin{aligned} E \mid h_t - \tilde{h}_t \mid &= E\{a(\eta_{t-1}) \cdots a(\eta_{t-n})\} E \mid h_{t-n-1} - \tilde{h}_{t-n-1} \mid \\ &= (\alpha + \beta)^n E \mid h_{t-n-1} - \tilde{h}_{t-n-1} \mid . \end{aligned}$$

Notice that the second equality uses the fact that the solutions are non-anticipative. This assumption was not necessary to establish the uniqueness of the strictly stationary solution. The expectation of $\mid h_{t-n-1} - \tilde{h}_{t-n-1} \mid$ being bounded by $E \mid h_{t-n-1} \mid + E \mid \tilde{h}_{t-n-1} \mid$, which is finite and independent of n by stationarity, and since $(\alpha + \beta)^n$ tends to 0 when $n \to \infty$, we obtain $E \mid h_t - \tilde{h}_t \mid = 0$ and thus $h_t = \tilde{h}_t$ for all t, a.s. □

Figure 2.8 shows the zones of strict and second-order stationarity for the strong GARCH(1, 1) model when $\eta_t \sim \mathcal{N}(0, 1)$. Note that the distribution of η_t only matters for the strict stationarity. As noted above, the frontier of the strict stationarity zone corresponds to a random walk (for the process $\log(h_t - \omega)$). A similar interpretation holds for the second-order stationarity zone. If $\alpha + \beta = 1$ we have

$$h_t = \omega + h_{t-1} + \alpha h_{t-1}(\eta_{t-1}^2 - 1).$$

Thus, since the last term in this equality is centred and uncorrelated with any variable belonging to the past of h_{t-1}, the process (h_t) is a random walk. The corresponding GARCH process is called *integrated GARCH* (or IGARCH(1, 1)) and will be studied later: it is strictly stationary, has an infinite variance, and a conditional variance which is a random walk (with a positive drift).

2.2.2 The General Case

In the general case of a strong GARCH(p, q) process, the following vector representation will be useful. We have

$$\underline{z}_t = \underline{b}_t + A_t \underline{z}_{t-1}, \tag{2.16}$$

where

$$\underline{b}_t = \underline{b}(\eta_t) = \begin{pmatrix} \omega\eta_t^2 \\ 0 \\ \vdots \\ \omega \\ 0 \\ \vdots \\ 0 \end{pmatrix} \in \mathbb{R}^{p+q}, \quad \underline{z}_t = \begin{pmatrix} \epsilon_t^2 \\ \vdots \\ \epsilon_{t-q+1}^2 \\ \sigma_t^2 \\ \vdots \\ \sigma_{t-p+1}^2 \end{pmatrix} \in \mathbb{R}^{p+q},$$

and

$$A_t = \begin{pmatrix} \alpha_1\eta_t^2 & & \cdots & \alpha_q\eta_t^2 & \beta_1\eta_t^2 & & \cdots & & \beta_p\eta_t^2 \\ 1 & 0 & \cdots & 0 & 0 & & \cdots & & 0 \\ 0 & 1 & \cdots & 0 & 0 & & \cdots & & 0 \\ \vdots & \ddots & \ddots & \vdots & \vdots & \ddots & \ddots & & \vdots \\ 0 & & \cdots \; 1 & 0 & 0 & & \cdots & 0 & 0 \\ \alpha_1 & & \cdots & \alpha_q & \beta_1 & & \cdots & & \beta_p \\ 0 & & \cdots & 0 & 1 & 0 & \cdots & & 0 \\ 0 & & \cdots & 0 & 0 & 1 & \cdots & & 0 \\ \vdots & \ddots & \ddots & \vdots & \vdots & \ddots & \ddots & & \vdots \\ 0 & & \cdots \; 0 & 0 & 0 & & \cdots & 1 & 0 \end{pmatrix} \tag{2.17}$$

is a $(p+q)\times(p+q)$ matrix. In the ARCH(q) case, $\underline{z}_t$ reduces to ϵ_t^2 and its $q-1$ first past values, and A_t to the upper-left block of the above matrix. Equation (2.16) defines a first-order vector autoregressive model, with positive and iid matrix coefficients. The distribution of $\underline{z}_t$ conditional on its infinite past coincides with its distribution conditional on z_{t-1} only, which means that $(\underline{z}_t)$ is a Markov process. Model (2.16) is thus called the *Markov representation* of the GARCH(p, q) model. Iterating Eq. (2.16) yields

$$\underline{z}_t = \underline{b}_t + \sum_{k=1}^{\infty} A_t A_{t-1} \ldots A_{t-k+1} \underline{b}_{t-k}, \tag{2.18}$$

provided that the series exists a.s.. Finding conditions ensuring the existence of this series is the object of what follows. Notice that the existence of the right-hand vector in (2.18) does not ensure that its components are positive. One sufficient condition for

$$\underline{b}_t + \sum_{k=1}^{\infty} A_t A_{t-1} \ldots A_{t-k+1} \underline{b}_{t-k} > 0, \quad \text{a.s.,} \tag{2.19}$$

in the sense that all the components of this vector are strictly positive (but possibly infinite), is that

$$\omega > 0, \quad \alpha_i \geq 0 \quad (i = 1, \ldots, q), \quad \beta_j \geq 0 \ (j = 1, \ldots, p). \tag{2.20}$$

This condition is very simple to use but may not be necessary, as we will see in Section 2.3.2.

Strict Stationarity

The main tool for studying strict stationarity is the concept of the top Lyapunov exponent. Let A be a $(p+q)\times(p+q)$ matrix. The spectral radius of A, denoted by $\rho(A)$, is defined as the greatest modulus of its eigenvalues. Let $\|\cdot\|$ denote any norm on the space of the $(p+q)\times(p+q)$ matrices. We have the following algebra result:

$$\lim_{t\to\infty} \frac{1}{t} \log \|A^t\| = \log \rho(A) \tag{2.21}$$

(Exercise 2.3). This property has the following extension to random matrices.

Theorem 2.3 *Let $\{A_t, t\in\mathbb{Z}\}$ be a strictly stationary and ergodic sequence of random matrices, such that $E\log^+\|A_t\|$ is finite. We have*

$$\lim_{t\to\infty} \frac{1}{t} E(\log \|A_t A_{t-1} \ldots A_1\|) = \gamma = \inf_{t\in\mathbb{N}^*} \frac{1}{t} E(\log \|A_t A_{t-1} \ldots A_1\|), \tag{2.22}$$

γ is called the top Lyapunov exponent and $\exp(\gamma)$ is called the spectral radius of the sequence of matrices $\{A_t, t\in\mathbb{Z}\}$. Moreover,

$$\gamma = \lim_{t\to\infty} a.s. \ \frac{1}{t} \log \|A_t A_{t-1} \ldots A_1\|. \tag{2.23}$$

Remark 2.4 (On the top Lyapunov exponent)

1. It is always true that $\gamma \leq E(\log \|A_1\|)$, with equality in dimension 1.
2. If $A_t = A$ for all $t \in \mathbb{Z}$, we have $\gamma = \log \rho(A)$ in view of Eq. (2.21).
3. All norms on a finite-dimensional space being equivalent, it readily follows that γ is independent of the norm chosen.
4. The equivalence between the two definitions of γ, in equality (2.22), follows from Fekete's lemma (see Exercise 2.11) by noting that $a_t = E \log \|A_t \dots A_1\|$ defines a sub-additive sequence.
5. The equivalence between the definitions of γ can be shown using Kingman's sub-additive ergodic theorem (see Kingman 1973, Theorem 6). The characterisation in condition (2.23) is particularly interesting because its allows us to evaluate this coefficient by simulation. Asymptotic confidence intervals can also be obtained (see Goldsheid 1991).

The following general lemma, which we shall state without proof (see Bougerol and Picard 1992a, Lemma 3.4), is very useful for studying products of random matrices.

Lemma 2.1 *Let $\{A_t, t \in \mathbb{Z}\}$ be an ergodic and strictly stationary sequence of random matrices such that $E \log^+ \|A_t\|$ is finite, endowed with a top Lyapunov exponent γ. Then*

$$\lim_{t\to\infty} a.s. \ \|A_0 \dots A_{-t}\| = 0 \implies \gamma < 0. \tag{2.24}$$

As for ARMA models, we are mostly interested in the non-anticipative solutions (ϵ_t) to model (2.5), that is, those for which ϵ_t belongs to the σ-field generated by $\{\eta_t, \eta_{t-1}, \dots\}$.

Theorem 2.4 (Strict stationarity of the GARCH(*p*, *q*) model) *A necessary and sufficient condition for the existence of a strictly stationary solution to the GARCH(p, q) model (2.5) is that*

$$\gamma < 0,$$

where γ is the top Lyapunov exponent of the sequence $\{A_t, t \in \mathbb{Z}\}$ defined by (2.17). When the strictly stationary solution exists, it is unique, non-anticipative and ergodic.

Proof. We shall use the norm defined by $\|A\| = \sum |a_{ij}|$. For convenience, the norm will be denoted identically whatever the dimension of A. With this convention, this norm is clearly *multiplicative*: $\|AB\| \leq \|A\|\|B\|$ for all matrices A and B such that AB exists.[6] Observe that since the variable η_t has a finite variance, the components of the matrix A_t are integrable. Hence, we have

$$E \log^+ \|A_t\| \leq E\|A_t\| < \infty.$$

First suppose $\gamma < 0$. Then it follows from condition (2.23) that

$$\underline{z}_t(N) = \underline{b}_t + \sum_{n=0}^{N} A_t A_{t-1} \dots A_{t-n} \underline{b}_{t-n-1}$$

converges a.s., when N goes to infinity, to some limit $\underline{z}_t$. Indeed, using the fact that the norm is multiplicative,

$$\|\underline{z}_t(N)\| \leq \|\underline{b}_t\| + \sum_{n=0}^{\infty} \|A_t A_{t-1} \dots A_{t-n}\| \|\underline{b}_{t-n-1}\| \tag{2.25}$$

[6] Other examples of multiplicative norms are the Euclidean norm, $\|A\| = \{\sum a_{ij}^2\}^{1/2} = \{\mathrm{Tr}(A'A)\}^{1/2}$, and the sup norm defined, for any matrix A of size $d \times d$, by $N(A) = \sup \{\|Ax\|;\ x \in \mathbb{R}^d,\ \|x\| \leq 1\}$ where $\|x\| = \sum |x_i|$. A non-multiplicative norm is N_1 defined by $N_1(A) = \max |a_{ij}|$.

and

$$\|A_t \dots A_{t-n}\|^{1/n} \|\underline{b}_{t-n-1}\|^{1/n} = \exp\left[\frac{1}{n}\log \|A_t \dots A_{t-n}\| + \frac{1}{n}\log \|\underline{b}_{t-n-1}\|\right] \xrightarrow{a.s.} e^{\gamma} < 1.$$

To show that $n^{-1}\log \|\underline{b}_{t-n-1}\| \to 0$ a.s. we have used the result proven in Exercise 2.11, which can be applied because

$$E \mid \log \|\underline{b}_{t-n-1}\| \mid \leq \mid \log \omega \mid + E\log^+ \|\underline{b}_{t-n-1}\| \leq \mid \log \omega \mid + E\|\underline{b}_{t-n-1}\| < \infty.$$

It follows that, by the Cauchy rule, $\underline{\tilde{z}}_t$ is well defined in $(\mathbb{R}^{*+})^{p+q}$. Let $\underline{\tilde{z}}_{q+1,t}$ denote the $(q+1)$th component of $\underline{\tilde{z}}_t$. Setting $\epsilon_t = \sqrt{\underline{\tilde{z}}_{q+1,t}}\eta_t$, we define a solution of model (2.5). This solution is non-anticipative since, by (2.18), ϵ_t can be expressed as a measurable function of $\eta_t, \eta_{t-1}, \dots$. By Theorem A.1, together with the ergodicity (η_t), this solution is also strictly stationary and ergodic.

Next we establish the necessary part. From Lemma 2.1, it suffices to prove, by Eq. (2.24), that $\lim_{t\to\infty}$ a.s. $\|A_0 \dots A_{-t}\| = 0$. We shall show that, for $1 \leq i \leq p+q$,

$$\lim_{t\to\infty} A_0 \dots A_{-t} e_i = 0, \quad \text{a.s.,} \tag{2.26}$$

where e_i is the ith element of the canonical base of $\mathbb{R}^{p+q}$. Let (ϵ_t) be a strictly stationary solution of condition (2.5) and let $(\underline{z}_t)$ be defined by condition (2.16). We have, for $t > 0$,

$$\begin{aligned}
\underline{z}_0 &= \underline{b}_0 + A_0 \underline{z}_{-1} \\
&= \underline{b}_0 + \sum_{k=0}^{t-1} A_0 \dots A_{-k}\underline{b}_{-k-1} + A_0 \dots A_{-t}\underline{z}_{-t-1} \\
&\geq \sum_{k=0}^{t-1} A_0 \dots A_{-k}\underline{b}_{-k-1}.
\end{aligned} \tag{2.27}$$

because the coefficients of the matrices A_t, $\underline{b}_0$ and $\underline{z}_t$ are non-negative.[7] It follows that the series $\sum_{k=0}^{t-1} A_0 \dots A_{-k}\underline{b}_{-k-1}$ converges and thus $A_0 \dots A_{-k}\underline{b}_{-k-1}$ tends a.s. to 0 as $k \to \infty$. But since $\underline{b}_{-k-1} = \omega\eta_{-k-1}^2 e_1 + \omega e_{q+1}$, it follows that $A_0 \dots A_{-k}\underline{b}_{-k-1}$ can be decomposed into two positive terms, and we have

$$\lim_{k\to\infty} A_0 \dots A_{-k}\omega\eta_{-k-1}^2 e_1 = 0, \quad \lim_{k\to\infty} A_0 \dots A_{-k}\omega e_{q+1} = 0, \quad \text{a.s.} \tag{2.28}$$

Since $\omega \neq 0$, (2.26) holds for $i = q+1$. Now we use the equality

$$A_{-k}e_{q+i} = \beta_i\eta_{-k}^2 e_1 + \beta_i e_{q+1} + e_{q+i+1}, \quad i = 1, \dots, p, \tag{2.29}$$

with $e_{p+q+1} = 0$ by convention. For $i = 1$, this equality gives

$$0 = \lim_{t\to\infty} A_0 \dots A_{-k}e_{q+1} \geq \lim_{k\to\infty} A_0 \dots A_{-k+1}e_{q+2} \geq 0,$$

hence Eq. (2.26) is true for $i = q+2$, and, by induction, it is also true for $i = q+j, j = 1, \dots, p$ using Eq. (2.29). Moreover, we note that $A_{-k}e_q = \alpha_q\eta_{-k}^2 e_1 + \alpha_q e_{q+1}$, which allows us to see that, from condition (2.28), Eq. (2.26) holds for $i = q$. For the other values of i the conclusion follows from

$$A_{-k}e_i = \alpha_i\eta_{-k}^2 e_1 + \alpha_i e_{q+1} + e_{i+1}, \quad i = 1, \dots, q-1,$$

[7] Here, and in the sequel, we use the notation $x \geq y$, meaning that all the components of the vector x are greater than, or equal to, those of the vector y.

The proof of the uniqueness parallels the arguments given in the case $p = q = 1$. Let (ϵ_t) denote a strictly stationary solution of model (2.5), or equivalently, let $(\underline{z}_t)$ denote a positive and strictly stationary solution of (2.16). For all $N \geq 0$,

$$\underline{z}_t = \tilde{\underline{z}}_t(N) + A_t \dots A_{t-N}\underline{z}_{t-N-1}.$$

Then

$$\|\underline{z}_t - \tilde{\underline{z}}_t\| \leq \|\tilde{\underline{z}}_t(N) - \tilde{\underline{z}}_t\| + \|A_t \dots A_{t-N}\|\|\underline{z}_{t-N-1}\|.$$

The first term on the right-hand side tends to 0 a.s. as $N \to \infty$. In addition, because the series defining $\tilde{\underline{z}}_t$ converges a.s., we have $\|A_t \dots A_{t-N}\| \to 0$ with probability 1 when $n \to \infty$. Moreover, the distribution of $\|\underline{z}_{t-N-1}\|$ is independent of N by stationarity. It follows that $\|A_t \dots A_{t-N}\|\|\underline{z}_{t-N-1}\| \to 0$ in probability as $N \to \infty$. We have shown that $\underline{z}_t - \tilde{\underline{z}}_t \to 0$ in probability when $N \to \infty$. This quantity being independent of N, we necessarily have $\tilde{\underline{z}}_t = \underline{z}_t$ for any t, a.s.. The proof of Theorem 2.4 is now complete. □

Remark 2.5 (On Theorem 2.4 and its proof)

1. This theorem shows, in particular, that the stationary solution of the strong GARCH model is a semi-strong GARCH process, in the sense of Definition 2.1. The converse is not true, however (see the example of Section 4.1.1).

2. Bougerol and Picard (1992b) use a more parsimonious vector representation of the GARCH(p, q) model, based on the vector $\underline{z}_t^* = (\sigma_t^2, \dots, \sigma_{t-p+1}^2, \epsilon_{t-1}^2, \dots, \epsilon_{t-q+1}^2)' \in \mathbb{R}^{p+q+1}$ (Exercise 2.7). However, a drawback of this representation is that it is only defined for $p \geq 1$ and $q \geq 2$.

3. An analogous proof uses the following Markov vector representation based on representation (2.6):

$$\underline{h}_t = \underline{\omega} + B_t \underline{h}_{t-1}, \tag{2.30}$$

with $\underline{\omega} = (\omega, 0, \dots, 0)' \in \mathbb{R}^r$, $\underline{h}_t = (\sigma_t^2, \dots, \sigma_{t-r+1}^2)' \in \mathbb{R}^r$ and

$$B_t = \begin{pmatrix} a_1(\eta_{t-1}) & \dots & a_r(\eta_{t-r}) \\ I_{r-1} & & 0 \end{pmatrix}, \quad a_i(\eta) = \alpha_i \eta^2 + \beta_i,$$

where I_{r-1} is the identity matrix of size $r - 1$. In Theorem 2.4 and its proof, the sequence (A_t) could be replaced by the more parsimonious sequence (B_t). However, the matrices B_t are not independent, and we will require independence to establish subsequent results, like the existence of moments (Theorem 2.9). On the other hand, the representation (2.30) may be more convenient than Eq. (2.16) for deriving other properties such as the mixing properties (see Chapter 3). It is also worth noting (Exercise 2.14) that

$$E\prod_{t=0}^{n} B_t = \prod_{t=0}^{n} EB_t. \tag{2.31}$$

4. To verify the condition $\gamma < 0$, it is sufficient to check that

$$E(\log \|A_t A_{t-1} \dots A_1\|) < 0$$

for some $t > 0$.

5. If a GARCH model admits a strictly stationary solution, any other GARCH model obtained by replacing the α_i and β_j by smaller coefficients will also admit a strictly stationary solution. Indeed, the coefficient γ of the latter model will be smaller than that of the initial model because, with the norm chosen, $0 \leq A \leq B$ implies $\|A\| \leq \|B\|$. In particular, the strict stationarity of a given GARCH process entails that the ARCH process obtained by cancelling the coefficients β_j is also strictly stationary.

6. We emphasise the fact that any strictly stationary solution of a GARCH model is non-anticipative. This is an important difference with ARMA models, for which strictly stationary solutions depending on both past and future values of the noise exist.

The following result provides a simple necessary condition for strict stationarity, in three different forms.

Corollary 2.2 (Consequences of strict stationarity) *Let γ be the top Lyapunov exponent of the sequence $\{A_t, t \in \mathbb{Z}\}$ defined in (2.17). If $\gamma < 0$, we have the following equivalent properties:*

(a) $\sum_{j=1}^{p} \beta_j < 1$;
(b) $1 - \beta_1 z - \cdots - \beta_p z^p = 0 \Longrightarrow |z| > 1$;
(c) $\rho(B) < 1$, *where B is the submatrix of A_t defined by*

$$B = \begin{pmatrix} \beta_1 & \beta_2 & \cdots & & \beta_p \\ 1 & 0 & \cdots & & 0 \\ 0 & 1 & \cdots & & 0 \\ \vdots & \ddots & \ddots & & \vdots \\ 0 & & \cdots & 1 & 0 \end{pmatrix}.$$

Proof. Since all the coefficients of the matrices A_t are non-negative, it is clear that γ is larger than the top Lyapunov exponent of the constant sequence obtained by replacing by 0 the coefficients of the first q rows and of the first q columns of the matrices A_t. But the matrix obtained in this way has the same non-zero eigenvalues as B, and thus has the same spectral radius as B. In view of Remark 2.4(2), it can be seen that

$$\gamma \geq \log\, \rho(B).$$

It follows that $\gamma < 0 \Longrightarrow$ (c). It is easy to show (by induction on p and by computing the determinant with respect to the last column) that, for $\lambda \neq 0$,

$$\det(\lambda I_p - B) = \lambda^p - \lambda^{p-1}\beta_1 - \cdots - \lambda\beta_{p-1} - \beta_p = \lambda^p \mathcal{B}\left(\frac{1}{\lambda}\right), \tag{2.32}$$

where $\mathcal{B}(z) = 1 - \beta_1 z - \cdots - \beta_p z^p$. The equivalence between (b) and (c) is then straightforward. Next, we prove that (a) $\Longleftrightarrow$ (b). We have $\mathcal{B}(0) = 1$ and $\mathcal{B}(1) = 1 - \sum_{j=1}^{p} \beta_j$. Hence, if $\sum_{j=1}^{p} \beta_j \geq 1$ then $\mathcal{B}(1) \leq 0$ and, by a continuity argument, there exists a root of $\mathcal{B}$ in $(0, 1]$. Thus (b) $\Longrightarrow$ (a). Conversely, if $\sum_{j=1}^{p} \beta_j < 1$ and $\mathcal{B}(z_0) = 0$ for a z_0 of modulus less than or equal to 1, then $1 = \sum_{j=1}^{p} \beta_j z_0^j = \left|\sum_{j=1}^{p} \beta_j z_0^j\right| \leq \sum_{j=1}^{p} \beta_j |z_0|^j \leq \sum_{j=1}^{p} \beta_j$, which is impossible. It follows that (a) $\Longrightarrow$ (b) and the proof of the corollary is complete. □

We now give two illustrations allowing us to obtain more explicit stationarity conditions than in the theorem.

Example 2.1 (GARCH(1, 1)) In the GARCH(1, 1) case, we retrieve the strict stationarity condition already obtained. The matrix A_t is written in this case as

$$A_t = (\eta_t^2, 1)'(\alpha_1, \beta_1).$$

We thus have

$$A_t A_{t-1} \dots A_1 = \prod_{k=1}^{t-1} (\alpha_1 \eta_{t-k}^2 + \beta_1) A_t.$$

It follows that

$$\log \|A_t A_{t-1} \dots A_1\| = \sum_{k=1}^{t-1} \log(\alpha_1 \eta_{t-k}^2 + \beta_1) + \log \|A_t\|$$

and, in view of Eq. (2.23) and by the strong law of large numbers, $\gamma = E\log(\alpha_1\eta_t^2 + \beta_1)$. The necessary and sufficient condition for the strict stationarity is then $E\log(\alpha_1\eta_t^2 + \beta_1) < 0$, as obtained above.

Example 2.2 (ARCH(2)) For an ARCH(2) model, the matrix A_t takes the form

$$A_t = \begin{pmatrix} \alpha_1\eta_t^2 & \alpha_2\eta_t^2 \\ 1 & 0 \end{pmatrix}$$

and the stationarity region can be evaluated by simulation. Table 2.1 shows, for different values of the coefficients α_1 and α_2, the empirical means and the standard deviations (in parentheses) obtained for 1000 simulations of size 1000 of $\hat{\gamma} = \frac{1}{1000}\log\|A_{1000}A_{999}\dots A_1\|$. The η_t's have been simulated from a $\mathcal{N}(0,1)$ distribution. Note that in the ARCH(1) case, simulation provides a good approximation of the condition $\alpha_1 < 3.56$, which was obtained analytically. Apart from this case, there exists no explicit strict stationarity condition in terms of the coefficients α_1 and α_2.

Figure 2.9, constructed from these simulations, gives a more precise idea of the strict stationarity region for an ARCH(2) process. We shall establish in Corollary 2.3 a result showing that any strictly stationary GARCH process admits small-order moments. We begin with two lemmas which are of independent interest.

Lemma 2.2 *Let X denote an a.s. positive real random variable. If $EX^r < \infty$ for some $r > 0$ and if $E\log X < 0$, then $EX^s < 1$ for any small enough $s > 0$.*

Proof. The moment-generating function of $Y = \log X$ is defined by $M(u) = Ee^{uY} = EX^u$. The function M is defined and continuous on $[0, r]$ and we have, for $u > 0$,

$$\frac{M(u) - M(0)}{u} = \int g(u, y)\, dP_Y(y)$$

with

$$g(u, y) = \frac{e^{uy} - 1}{u} \uparrow y \quad \text{when } u \downarrow 0.$$

By Beppo Levi's theorem, the right derivative of M at 0 is

$$\int y d\, P_Y(y) = E(\log X) < 0.$$

Since $M(0) = 1$, then $M(s) = EX^s < 1$ for any small enough $s > 0$. □

Table 2.1 Estimations of γ obtained from 1000 simulations of size 1000 in the ARCH(2) case.

	α_1								
α_2	0.25	0.3	1	1.2	1.7	1.8	3.4	3.5	3.6
0	—	—	—	—	—	—	−0.049 (0.071)	−0.018 (0.071)	0.010 (0.071)
0.5	—	—	—	−0.175 (0.040)	−0.021 (0.042)	0.006 (0.044)	—	—	—
1	—	—	−0.011 (0.038)	0.046 (0.038)	—	—	—	—	—
1.75	−0.015 (0.035)	0.001 (0.032)	—	—	—	—	—	—	—

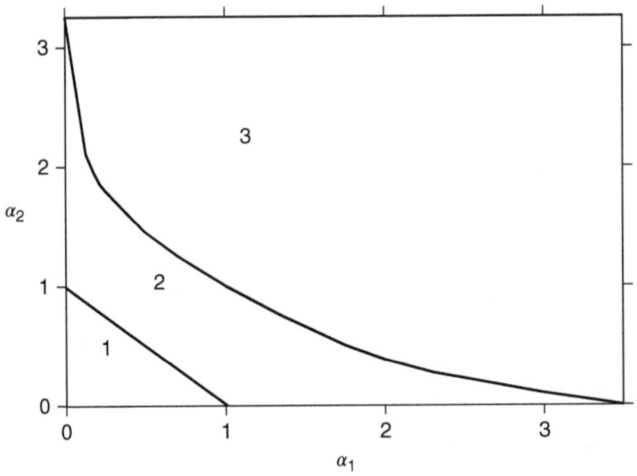

Figure 2.9 Stationarity regions for the ARCH(2) model: 1, second-order stationarity; 1 and 2, strict stationarity; 3, non-stationarity.

The following result, which is stated for any sequence of positive iid matrices, provides another characterisation of the strict stationarity of GARCH models.

Lemma 2.3 *Let $\{A_t\}$ be an iid sequence of positive integrable matrices, with top Lyapunov exponent γ. Then*

$$\gamma < 0 \iff \exists s > 0, \quad \exists k_0 \geq 1, \quad \delta := E(\|A_{k_0} A_{k_0-1} \dots A_1\|^s) < 1.$$

Proof. Suppose $\gamma < 0$. Since $\gamma = \inf_k \frac{1}{k} E(\log \|A_k A_{k-1} \dots A_1\|)$, there exists $k_0 \geq 1$ such that $E(\log \|A_{k_0} A_{k_0-1} \dots A_1\|) < 0$. Moreover,

$$\begin{aligned} E(\|A_{k_0} A_{k_0-1} \dots A_1\|) &= \|E(A_{k_0} A_{k_0-1} \dots A_1)\| \\ &= \|(EA_1)^{k_0}\| \\ &\leq (E\|A_1\|)^{k_0} < \infty \end{aligned} \tag{2.33}$$

using the multiplicative norm $\|A\| = \sum_{i,j} |A(i,j)|$, the positivity of the elements of the A_i, and the independence and equidistribution of the A_i. We conclude, concerning the direct implication, by using Lemma 2.2. The converse does not use the fact that the sequence is iid. If there exist $s > 0$ and $k_0 \geq 1$ such that $\delta < 1$, we have, by Jensen's inequality,

$$\gamma \leq \frac{1}{k_0} E(\log \|A_{k_0} A_{k_0-1} \dots A_1\|) \leq \frac{1}{sk_0} \log \delta < 0.$$

□

Corollary 2.3 *Let γ denote the top Lyapunov exponent of the sequence (A_t) defined in (2.17). Then*

$$\gamma < 0 \implies \exists s > 0, \quad E\sigma_t^{2s} < \infty, \quad E\epsilon_t^{2s} < \infty$$

where $\epsilon_t = \sigma_t \eta_t$ is the strictly stationary solution of the GARCH(p, q) model (2.5).

Proof. The proof of Lemma 2.3 shows that the real s involved in the two previous lemmas can be taken to be less than 1. For $s \in (0, 1]$, $a, b > 0$ we have $\left(\frac{a}{a+b}\right)^s + \left(\frac{b}{a+b}\right)^s \geq 1$ and, consequently,

$\left(\sum_i u_i\right)^s \leq \sum_i u_i^s$ for any sequence of positive numbers u_i. Using this inequality, together with arguments already used (in particular, the fact that the norm is multiplicative), we deduce that the stationary solution defined in condition (2.18) satisfies

$$E\|\underline{z}_t\|^s \leq \|E\underline{b}_1\|^s \left\{ 1 + \sum_{k=0}^{\infty} \delta^k \sum_{i=1}^{k_0} \{E\|A_1\|^s\}^i \right\} < \infty$$

where δ is defined in Lemma 2.3. We conclude by noting that $\sigma_t^{2s} \leq \|\underline{z}_t\|^s$ and $\epsilon_t^{2s} \leq \|\underline{z}_t\|^s$. □

Using Lemma 2.3 and Corollary 2.3 together, it can be seen that for $s \in (0, 1]$,

$$\exists k_0 \geq 1, \quad E(\|A_{k_0} A_{k_0-1} \dots A_1\|^s) < 1 \implies E\epsilon_t^{2s} < \infty. \tag{2.34}$$

The converse is generally true. For instance, we have for $s \in (0, 1]$,

$$\alpha_1 + \beta_1 > 0, \quad E\epsilon_t^{2s} < \infty \implies \lim_{k\to\infty} E(\|A_k A_{k-1} \dots A_1\|^s) = 0 \tag{2.35}$$

(Exercise 2.15).

Second-Order Stationarity

The following theorem gives necessary and sufficient second-order stationarity conditions.

Theorem 2.5 (Second-order stationarity) *If there exists a GARCH(p, q) process, in the sense of Definition 2.1, which is second-order stationary and non-anticipative, and if $\omega > 0$, then*

$$\sum_{i=1}^{q} \alpha_i + \sum_{j=1}^{p} \beta_i < 1. \tag{2.36}$$

Conversely, if condition (2.36) holds, the unique strictly stationary solution of model (2.5) is a weak white noise (and thus is second-order stationary). In addition, there exists no other second-order stationary solution.

Proof. We first show that condition (2.36) is necessary. Let (ϵ_t) be a second-order stationary and non-anticipative GARCH(p, q) process. Then

$$E(\epsilon_t^2) = E\{E(\epsilon_t^2 \mid \epsilon_u, u < t)\} = E(\sigma_t^2)$$

is a positive real number which does not depend on t. Taking the expectation of both sides of equality (2.1), we thus have

$$E(\epsilon_t^2) = \omega + \sum_{i=1}^{q} \alpha_i E(\epsilon_t^2) + \sum_{j=1}^{p} \beta_j E(\epsilon_t^2),$$

that is,

$$\left(1 - \sum_{i=1}^{q} \alpha_i - \sum_{j=1}^{p} \beta_j\right) E(\epsilon_t^2) = \omega. \tag{2.37}$$

Since ω is strictly positive, we must have condition (2.36).

Now suppose that condition (2.36) holds true and let us construct a stationary GARCH solution (in the sense of Definition 2.2). For $t, k \in \mathbb{Z}$, we define $\mathbb{R}^d$-valued vectors as follows:

$$Z_k(t) = \begin{cases} \underline{0} & \text{if } k < 0 \\ \underline{b}_t + A_t Z_{k-1}(t-1) & \text{if } k \geq 0. \end{cases}$$

We have

$$Z_k(t) - Z_{k-1}(t) = \begin{cases} \underline{0} & \text{if } k < 0 \\ \underline{b}_t & \text{if } k = 0 \\ A_t\{Z_{k-1}(t-1) - Z_{k-2}(t-1)\} & \text{if } k > 0. \end{cases}$$

By iterating these relations we get, for $k > 0$,

$$Z_k(t) - Z_{k-1}(t) = A_t A_{t-1} \dots A_{t-k+1} \underline{b}_{t-k}.$$

On the other hand, for the norm $\|C\| = \sum_{i,j} |c_{ij}|$, we have, for any random matrix C with positive coefficients, $E\|C\| = E\sum_{i,j} |c_{ij}| = E\sum_{i,j} c_{ij} = \|E(C)\|$. Hence, for $k > 0$,

$$E\|Z_k(t) - Z_{k-1}(t)\| = \|E(A_t A_{t-1} \dots A_{t-k+1} \underline{b}_{t-k})\|,$$

because the components of the vector $A_t A_{t-1} \dots A_{t-k+1} \underline{b}_{t-k}$ are positive. All terms of the product $A_t A_{t-1} \dots A_{t-k+1} \underline{b}_{t-k}$ are independent (because the process (η_t) is iid, and every term of the product is function of a variable η_{t-i}, the dates $t - i$ being distinct). Moreover, $A := E(A_t)$ and $\underline{b} = E(\underline{b}_t)$ do not depend on t. Finally, for $k > 0$,

$$E\|Z_k(t) - Z_{k-1}(t)\| = \|A^k \underline{b}\| = (1, \dots, 1) A^k \underline{b}$$

because all entries of the vector $A^k \underline{b}$ are positive.

Condition (2.36) implies that the modulus of the eigenvalues of A are strictly less than 1. Indeed, one can verify (Exercise 2.12) that

$$\det(\lambda I_{p+q} - A) = \lambda^{p+q} \left(1 - \sum_{i=1}^{q} \alpha_i \lambda^{-i} - \sum_{j=1}^{p} \beta_j \lambda^{-j}\right). \tag{2.38}$$

Thus if $|\lambda| \geq 1$, using the inequality $|a - b| \geq |a| - |b|$, we get

$$|\det(\lambda I_{p+q} - A)| \geq \left|1 - \sum_{i=1}^{q} \alpha_i \lambda^{-i} - \sum_{j=1}^{p} \beta_j \lambda^{-j}\right| \geq 1 - \sum_{i=1}^{q} \alpha_i - \sum_{j=1}^{p} \beta_j > 0,$$

and then $\rho(A) < 1$. It follows that, in view of the Jordan decomposition, or using (2.21), the convergence $A^k \to 0$ holds at exponential rate when $k \to \infty$. Hence, for any fixed t, $Z_k(t)$ converges both in the L^1 sense, using Cauchy's criterion, and almost surely as $k \to \infty$. Let $\underline{z}_t$ denote the limit of $(Z_k(t))_k$. At fixed k, the process $(Z_k(t))_{t \in \mathbb{Z}}$ is strictly stationary. The limit process $(\underline{z}_t)$ is thus strictly stationary. Finally, it is clear that $\underline{z}_t$ is a solution of Eq. (2.16).

The uniqueness can be shown as in the case $p = q = 1$, using the representation (2.30). □

Remark 2.6 (On the second-order stationarity of GARCH)

1. Under the conditions of Theorem 2.5, the unique stationary solution of model (2.5) is, using (2.37), a white noise of variance

$$\text{Var}(\epsilon_t) = \frac{\omega}{1 - \sum_{i=1}^{q} \alpha_i - \sum_{j=1}^{p} \beta_j}.$$

2. Because the conditions in Theorems 2.4 and 2.5 are necessary and sufficient, we necessarily have

$$\sum_{i=1}^{q} \alpha_i + \sum_{j=1}^{p} \beta_i < 1 \Rightarrow \gamma < 0,$$

since the second-order stationary solution of Theorem 2.5 is also strictly stationary. One can directly check this implication by noting that if condition (2.36) is true, the previous proof shows that the spectral radius $\rho(EA_t)$ is strictly less than 1. Moreover, using a result by Kesten and Spitzer (1984, (1.4)), we always have

$$\gamma \le \log \rho(EA_t). \tag{2.39}$$

IGARCH(p, q) Processes

When

$$\sum_{i=1}^{q} \alpha_i + \sum_{j=1}^{p} \beta_j = 1$$

the model is called an *integrated* GARCH(p, q) or IGARCH(p, q) model (see Engle and Bollerslev 1986). This name is justified by the existence of a unit root in the autoregressive part of representation (2.4) and is introduced by analogy with the integrated ARMA models, or ARIMA. However, this analogy can be misleading: there exists no (strict or second-order) stationary solution of an ARIMA model, whereas an IGARCH model admits a strictly stationary solution under very general conditions. In the univariate case ($p = q = 1$), the latter property is easily shown.

Corollary 2.4 (Strict stationarity of IGARCH(1.1)) *If $P[\eta_t^2 = 1] < 1$ and if $\alpha + \beta = 1$, model (2.7) admits a unique strictly stationary solution.*

Proof. Recall that in the case $p = q = 1$, the matrices A_t can be replaced by $a(\eta_t) = \alpha\eta_t^2 + \beta$. Hence, $\gamma = E \log a(\eta_t) \le \log E\{a(\eta_t)\} = 0$. The inequality is strict unless if $a(\eta_t)$ is a.s. a constant. Since $E\{a(\eta_t)\} = 1$, this constant can only be equal to 1. Thus $\eta_t^2 = 1$ a.s., which is excluded. □

This property extends to the general case under slightly more restrictive conditions on the law of η_t.

Corollary 2.5 *Suppose that the distribution of η_t has an unbounded support and has no mass at 0. Then, if $\sum_{i=1}^{q} \alpha_i + \sum_{j=1}^{p} \beta_j = 1$, model (2.5) admits a unique strictly stationary solution.*

Proof. It is not difficult to show from (2.38) that the spectral radius $\rho(A)$ of the matrix $A = EA_t$ is equal to 1 (Exercise 2.12). It can be shown that the assumptions on the distribution of η_t imply that inequality (2.39) is strict, that is, $\gamma < \log \rho(A)$ (see Kesten and Spitzer 1984, Theorem 2; Bougerol and Picard 1992b, Corollary 2.2). This allows us to conclude by Theorem 2.8. □

Note that this strictly stationary solution has an infinite variance in view of Theorem 2.5.

2.3 ARCH(∞) Representation*

A process (ϵ_t) is called an ARCH(∞) process if there exists a sequence of iid variables (η_t) such that $E(\eta_t) = 0$ and $E(\eta_t^2) = 1$, and a sequence of constants $\phi_i \ge 0$, $i = 1, \dots,$ and $\phi_0 > 0$ such that

$$\epsilon_t = \sigma_t \eta_t, \quad \sigma_t^2 = \phi_0 + \sum_{i=1}^{\infty} \phi_i \epsilon_{t-i}^2. \tag{2.40}$$

This class obviously contains the ARCH(q) process, and we shall see that it more generally contains the GARCH(p, q) process.

2.3.1 Existence Conditions

The existence of a stationary ARCH(∞) process requires assumptions on the sequences (ϕ_i) and (η_t). The following result gives an existence condition.

Theorem 2.6 (Existence of a stationary ARCH(∞) solution) *For any $s \in (0, 1]$, let*

$$A_s = \sum_{i=1}^{\infty} \phi_i^s \quad \text{and} \quad \mu_{2s} = E|\eta_t|^{2s}.$$

Then, if there exists $s \in (0, 1]$ such that

$$A_s \mu_{2s} < 1, \tag{2.41}$$

then there exists a strictly stationary and non-anticipative solution to model (2.40), given by

$$\epsilon_t = \sigma_t \eta_t, \quad \sigma_t^2 = \phi_0 + \phi_0 \sum_{k=1}^{\infty} \sum_{i_1,\ldots i_k \geq 1} \phi_{i_1} \ldots \phi_{i_k} \eta_{t-i_1}^2 \eta_{t-i_1-i_2}^2 \cdots \eta_{t-i_1-\cdots-i_k}^2. \tag{2.42}$$

The process (ϵ_t) defined by (2.42) is the unique strictly stationary and non-anticipative solution of model (2.40) such that $E|\epsilon_t|^{2s} < \infty$.

Proof. Consider the random variable

$$S_t = \phi_0 + \phi_0 \sum_{k=1}^{\infty} \sum_{i_1,\ldots i_k \geq 1} \phi_{i_1} \ldots \phi_{i_k} \eta_{t-i_1}^2 \ldots \eta_{t-i_1-\cdots-i_k}^2, \tag{2.43}$$

taking values in $[0, +\infty]$. Since $s \in (0, 1]$, applying the inequality $(a + b)^s \leq a^s + b^s$ for $a, b \geq 0$ gives

$$S_t^s \leq \phi_0^s + \phi_0^s \sum_{k=1}^{\infty} \sum_{i_1,\ldots i_k \geq 1} \phi_{i_1}^s \ldots \phi_{i_k}^s \eta_{t-i_1}^{2s} \ldots \eta_{t-i_1-\cdots-i_k}^{2s}.$$

Using the independence of the η_t, it follows that

$$\begin{aligned} ES_t^s &\leq \phi_0^s + \phi_0^s \sum_{k=1}^{\infty} \sum_{i_1,\ldots i_k \geq 1} \phi_{i_1}^s \ldots \phi_{i_k}^s E(\eta_{t-i_1}^{2s} \ldots \eta_{t-i_1-\cdots-i_k}^{2s}) \\ &= \phi_0^s \left(1 + \sum_{k=1}^{\infty} (A_s \mu_{2s})^k\right) = \frac{\phi_0^s}{1 - A_s \mu_{2s}}. \end{aligned} \tag{2.44}$$

This shows that S_t is a.s. finite. All the summands being positive, we have

$$\begin{aligned} \sum_{i=1}^{\infty} \phi_i S_{t-i} \eta_{t-i}^2 &= \phi_0 \sum_{i_0=1}^{\infty} \phi_{i_0} \eta_{t-i_0}^2 \\ &\quad + \phi_0 \sum_{i_0=1}^{\infty} \phi_{i_0} \eta_{t-i_0}^2 \sum_{k=1}^{\infty} \sum_{i_1,\ldots i_k \geq 1} \phi_{i_1} \ldots \phi_{i_k} \eta_{t-i_0-i_1}^2 \ldots \eta_{t-i_0-i_1-\cdots-i_k}^2 \\ &= \phi_0 \sum_{k=0}^{\infty} \sum_{i_0,\ldots i_k \geq 1} \phi_{i_0} \ldots \phi_{i_k} \eta_{t-i_0}^2 \ldots \eta_{t-i_0-\cdots-i_k}^2. \end{aligned}$$

Therefore, the following recursive equation holds:

$$S_t = \phi_0 + \sum_{i=1}^{\infty} \phi_i S_{t-i} \eta_{t-i}^2.$$

A strictly stationary and non-anticipative solution of model (2.40) is then obtained by setting $\epsilon_t = S_t^{1/2} \eta_t$. Moreover, $E|\epsilon_t|^{2s} \leq \mu_{2s} \phi_0^s / (1 - A_s \mu_{2s})$ in view of (2.44). Now denote by (ϵ_t) any strictly stationary and non-anticipative solution of model (2.40), such that $E|\epsilon_t|^{2s} < \infty$. For all $q \geq 1$, by q successive substitutions of the ϵ_{t-i}^2 we get

$$\begin{aligned}\sigma_t^2 &= \phi_0 + \phi_0 \sum_{k=1}^{q} \sum_{i_1,\dots i_k \geq 1} \phi_{i_1} \dots \phi_{i_k} \eta_{t-i_1}^2 \dots \eta_{t-i_1-\dots-i_k}^2 \\ &+ \sum_{i_1,\dots i_{q+1} \geq 1} \phi_{i_1} \dots \phi_{i_{q+1}} \eta_{t-i_1}^2 \dots \eta_{t-i_1-\dots-i_q}^2 \epsilon_{t-i_1-\dots-i_{q+1}}^2 \\ &:= S_{t,q} + R_{t,q}.\end{aligned}$$

Note that $S_{t,q} \to S_t$ a.s. when $q \to \infty$, where S_t is defined in (2.43). Moreover, because the solution is non-anticipative, ϵ_t is independent of $\eta_{t'}$ for all $t' > t$. Hence,

$$\begin{aligned}ER_{t,q}^s &\leq \sum_{i_1,\dots i_{q+1} \geq 1} \phi_{i_1}^s \dots \phi_{i_{q+1}}^s E(|\eta_{t-i_1}|^{2s} \dots |\eta_{t-i_1-\dots-i_q}|^{2s} |\epsilon_{t-i_1-\dots-i_{q+1}}|^{2s}) \\ &= (A_s \mu_{2s})^q A_s E|\epsilon_t|^{2s}.\end{aligned}$$

Thus $\sum_{q \geq 1} ER_{t,q}^s < \infty$ since $A_s \mu_{2s} < 1$. Finally, $R_{t,q} \to 0$ a.s. when $q \to \infty$, which implies $\sigma_t^2 = S_t$ a.s.. □

Equality (2.42) is called a Volterra expansion (see Priestley 1988). It shows in particular that, under the conditions of the theorem, if $\phi_0 = 0$, the unique strictly stationary and non-anticipative solution of the ARCH(∞) model is the identically null sequence. An application of Theorem 2.6, obtained for $s = 1$, is that the condition

$$A_1 = \sum_{i=1}^{\infty} \phi_i < 1$$

ensures the existence of a second-order stationary ARCH(∞) process. If

$$(E\eta_t^4)^{1/2} \sum_{i=1}^{\infty} \phi_i < 1,$$

it can be shown (see Giraitis, Kokoszka, and Leipus 2000) that $E\epsilon_t^4 < \infty$, $\text{Cov}(\epsilon_t^2, \epsilon_{t-h}^2) > 0$ for all h and

$$\sum_{h=-\infty}^{+\infty} \text{Cov}(\epsilon_t^2, \epsilon_{t-h}^2) < \infty. \tag{2.45}$$

The fact that the squares have positive autocovariances will be verified later in the GARCH case (see Section 2.5). In contrast to GARCH processes, for which they decrease at exponential rate, the autocovariances of the squares of ARCH(∞) can decrease at the rate $h^{-\gamma}$ with $\gamma > 1$ arbitrarily close to 1. A strictly stationary process of the form (2.40) such that

$$A_1 = \sum_{i=1}^{\infty} \phi_i = 1$$

is called *integrated* ARCH(∞), or IARCH(∞). Notice that an IARCH(∞) process has infinite variance. Indeed, if $E\epsilon_t^2 = \sigma^2 < \infty$, then, by (2.40), $\sigma^2 = \phi_0 + \sigma^2$, which is impossible. From Theorem 2.8, the strictly stationary solutions of IGARCH models (see Corollary 2.5) admit IARCH(∞) representations. The next result provides a condition for the existence of IARCH(∞) processes.

Theorem 2.7 (Existence of IARCH(∞) processes) *If $A_1 = 1$, if η_t^2 has a non-degenerate distribution, if $E \mid \log \eta_t^2 \mid < \infty$ and if, for some $r > 1$, $\sum_{i=1}^{\infty} \phi_i r^i < \infty$, then there exists a strictly stationary and non-anticipative solution to model (2.40) given by (2.42).*

Other integrated processes are the long-memory ARCH, for which the rate of decrease of the ϕ_i is not geometric.

2.3.2 ARCH(∞) Representation of a GARCH

It is sometimes useful to consider the ARCH(∞) representation of a GARCH(p, q) process. For instance, this representation allows the conditional variance σ_t^2 of ϵ_t to be written explicitly as a function of its infinite past. It also allows the positivity condition (2.20) on the coefficients to be weakened. Let us first consider the GARCH(1, 1) model. If $\beta < 1$, we have

$$\sigma_t^2 = \frac{\omega}{1-\beta} + \alpha \sum_{i=1}^{\infty} \beta^{i-1} \epsilon_{t-i-1}^2. \tag{2.46}$$

In this case, we have

$$A_s = \alpha^s \sum_{i=1}^{\infty} \beta^{(i-1)s} = \frac{\alpha^s}{1-\beta^s}.$$

The condition $A_s \mu_{2s} < 1$ thus takes the form

$$\alpha^s \mu_{2s} + \beta^s < 1, \quad \text{for some } s \in (0, 1].$$

For example, if $\alpha + \beta < 1$ this condition is satisfied for $s = 1$. However, second-order stationarity is not necessary for the validity of (2.46). Indeed, if (ϵ_t) denotes the strictly stationary and non-anticipative solution of the GARCH(1, 1) model, then, for any $q \geq 1$,

$$\sigma_t^2 = \omega \sum_{i=1}^{q} \beta^{i-1} + \alpha \sum_{i=1}^{q} \beta^{i-1} \epsilon_{t-i}^2 + \beta^q \sigma_{t-q}^2. \tag{2.47}$$

By Corollary 2.3 there exists $s \in (0, 1[$ such that $E(\sigma_t^{2s}) = c < \infty$. It follows that $\sum_{q\geq 1} E(\beta^q \sigma_{t-q}^2)^s = \beta c/(1-\beta) < \infty$. So $\beta^q \sigma_{t-q}^2$ converges a.s. to 0 and, by letting q go to infinity in (2.47), we get (2.46). More generally, we have the following property.

Theorem 2.8 (ARCH(∞) representation of a GARCH(p, q)) *If (ϵ_t) is the strictly stationary and non-anticipative solution of model (2.5), it admits an ARCH(∞) representation of the form (2.40). The constants ϕ_i are given by*

$$\phi_0 = \frac{\omega}{\mathcal{B}(1)}, \quad \sum_{i=1}^{\infty} \phi_i z^i = \frac{\mathcal{A}(z)}{\mathcal{B}(z)}, \quad z \in \mathbb{C}, \ |z| \leq 1, \tag{2.48}$$

where $\mathcal{A}(z) = \alpha_1 z + \cdots + \alpha_q z^q$ and $\mathcal{B}(z) = 1 - \beta_1 z - \cdots - \beta_p z^p$.

Proof. Rewrite the model in vector form as

$$\underline{\sigma}_t^2 = B \underline{\sigma}_{t-1}^2 + \underline{c}_t,$$

where $\underline{\sigma}_t^2 = (\sigma_t^2, \ldots, \sigma_{t-p+1}^2)'$, $\underline{c}_t = \left(\omega + \sum_{i=1}^q \alpha_i \epsilon_{t-i}^2, 0, \ldots, 0\right)'$ and B is the matrix defined in Corollary 2.2. This corollary shows that, under the strict stationarity condition, we have $\rho(B) < 1$. Moreover, $E\|\underline{c}_t\|^s < \infty$ by Corollary 2.3. Consequently, the components of the vector $\sum_{i=0}^{\infty} B^i \underline{c}_{t-i}$ are a.s. real-valued. We thus have

$$\sigma_t^2 = e' \sum_{i=0}^{\infty} B^i \underline{c}_{t-i}, \quad e = (1, 0, \ldots, 0)'.$$

All that remains is to note that the coefficients obtained in this ARCH(∞) representation coincide with those (2.48). □

The ARCH(∞) representation can be used to weaken the condition (2.20) imposed to ensure the positivity of σ_t^2. Consider the GARCH(p, q) model with $\omega > 0$, without any *a priori* positivity constraint on the coefficients α_i and β_j, and assuming that the roots of the polynomial $\mathcal{B}(z) = 1 - \beta_1 z - \cdots - \beta_p z^p$ have moduli strictly greater than 1. The coefficients ϕ_i introduced in (2.48) are then well defined, and under the assumption

$$\phi_i \geq 0, \quad i = 1, \dots, \tag{2.49}$$

we have

$$\sigma_t^2 = \phi_0 + \sum_{i=1}^{\infty} \phi_i \epsilon_{t-i}^2 \in (0, +\infty].$$

Indeed, $\phi_0 > 0$ because otherwise $\mathcal{B}(1) \leq 0$, which would imply the existence of a root inside the unit circle since $\mathcal{B}(0) = 1$. Moreover, the proofs of the sufficient parts of Theorem 2.4, Lemma 2.3, and Corollary 2.3 do not use the positivity of the coefficients of the matrices A_t. It follows that if the top Lyapunov exponent of the sequence (A_t) is such that $\gamma < 0$, the variable σ_t^2 is a.s. finite-valued. To summarise, the conditions

$$\omega > 0, \quad \gamma < 0, \quad \phi_i \geq 0, \quad i \geq 1 \quad \text{and} \quad (\mathcal{B}(z) = 0 \implies |z| > 1)$$

imply that there exists a strictly stationary and non-anticipative solution to model (2.5). The condition (2.49) is not generally simple to use, however, because it implies an infinity of constraints on the coefficients α_i and β_j. In the ARCH(q) case, it reduces to the condition (2.20), that is, $\alpha_i \geq 0$, $i = 1, \dots, q$. Similarly, in the GARCH(1, 1) case, it is necessary to have $\alpha_1 \geq 0$ and $\beta_1 \geq 0$. However, for $p \geq 1$ and $q > 1$, the condition (2.20) can be weakened (Exercise 2.16).

2.3.3 Long-Memory ARCH

The introduction of long memory into the volatility can be motivated by the observation that the empirical autocorrelations of the squares, or of the absolute values, of financial series decay very slowly in general (see for example Table 2.1). We shall see that it is possible to reproduce this property by introducing ARCH(∞) processes, with a sufficiently slow decay of the modulus of the coefficients ϕ_i.[8] A process (X_t) is said to have *long memory* if it is second-order stationary and satisfies, for $h \to \infty$,

$$|\text{Cov}(X_t, X_{t-h})| \sim K h^{2d-1}, \quad \text{where } d < \frac{1}{2} \tag{2.50}$$

and K is a non-zero constant. An alternative definition relies on distinguishing 'intermediate-memory' processes for which $d < 0$ and thus $\sum_{h=-\infty}^{+\infty} |\text{Cov}(X_t, X_{t-h})| < \infty$, and 'long-memory' processes for which $d \in (0, 1/2[$ and thus $\sum_{h=-\infty}^{+\infty} |\text{Cov}(X_t, X_{t-h})| = \infty$ (see Brockwell and Davis 1991, p. 520). The autocorrelations of an ARMA process decrease at exponential rate when the lag increases. The need for processes with a slower autocovariance decay leads to the introduction of the fractionary ARIMA models. These models are defined through the fractionary difference operator

$$(1 - B)^d = 1 + \sum_{j=1}^{\infty} \frac{(-d)(1-d)\dots(j-1-d)}{j!} B^j, \quad d > -1.$$

Denoting by π_j the coefficient of B^j in this sum, it can be shown that $\pi_j \sim K j^{-d-1}$ when $j \to \infty$, where K is a constant depending on d. An ARIMA(p, d, q) process with $d \in (-0.5, 0.5)$ is defined as a stationary solution of

$$\psi(B)(1 - B)^d X_t = \theta(B)\epsilon_t$$

where ϵ_t is a white noise, and ψ and θ are polynomials of degrees p and q, respectively. If the roots of these polynomials are all outside the unit disk, the unique stationary and purely deterministic solution is causal, invertible and its covariances satisfy condition (2.50) (see Brockwell and Davis 1991,

[8] The slow decrease of the empirical autocorrelations can also be explained by non-stationarity phenomena, such as structural changes or breaks (see, for instance, Mikosch and Stărică 2004).

Theorem 13.2.2.). By analogy with the ARIMA models, the class of FIGARCH(p, d, q) processes is defined by the equations

$$\epsilon_t = \sigma_t \eta_t, \quad \sigma_t^2 = \phi_0 + \left\{ 1 - (1-B)^d \frac{\theta(B)}{\psi(B)} \right\} \epsilon_{t-i}^2, \quad d \in (0, 1[, \quad \phi_0 > 0, \tag{2.51}$$

where ψ and θ are polynomials of degrees p and q, respectively, such that $\psi(0) = \theta(0) = 1$, the roots of ψ have moduli strictly greater than 1 and $\phi_i \geq 0$, where the ϕ_i are defined by

$$\sum_{i=1}^{\infty} \phi_i z^i = 1 - (1-z)^d \frac{\theta(z)}{\psi(z)}.$$

We have $\phi_i \sim K i^{-d-1}$, where K is a positive constant, when $i \to \infty$ and $\sum_{i=1}^{\infty} \phi_i = 1$. The process introduced in (2.51) is thus IARCH(∞), provided it exists. Note that existence of this process cannot be obtained by Theorem 2.7 because, for the FIGARCH model, the ϕ_i decrease more slowly than the geometric rate. The following result, which is a consequence of Theorem 2.6, and whose proof is the subject of Exercise 2.22, provides another sufficient condition for the existence of IARCH(∞) processes.

Corollary 2.6 (Existence of some FIGARCH processes) *If $A_1 = 1$, then condition (2.41) is satisfied if and only if there exists $p^* \in (0, 1]$ such that $A_{p^*} < \infty$ and*

$$\sum_{i=1}^{\infty} \phi_i \log \phi_i + E(\eta_0^2 \log \eta_0^2) \in (0, +\infty]. \tag{2.52}$$

The strictly stationary and non-anticipative solution of model (2.40) is thus given by (2.42) and is such that $E|\epsilon_t|^q < \infty$, for any $q \in [0, 2)$, and $E\epsilon_t^2 = \infty$.

This result can be used to prove the existence of FIGARCH(p,d,q) processes for $d \in (0,1)$ sufficiently close to 1, if the distribution of η_0^2 is assumed to be non-degenerate (hence, $E(\eta_0^2 \log \eta_0^2) > 0$); see Douc, Roueff, and Soulier (2008). The FIGARCH process of Corollary 2.6 does not admit a finite second-order moment. Its square is thus not a long-memory process in the sense of definition (2.50). More generally, it can be shown that the squares of a ARCH(∞) processes do not have the long-memory property. This motivated the introduction of an alternative class, called *linear ARCH* (LARCH) and defined by

$$\epsilon_t = \sigma_t \eta_t, \quad \sigma_t = b_0 + \sum_{i=0}^{\infty} b_i \epsilon_{t-i}, \quad \eta_t \text{ iid}(0, 1). \tag{2.53}$$

Under appropriate conditions, this model is compatible with the long-memory property for ϵ_t^2.

2.4 Properties of the Marginal Distribution

We have seen that, under quite simple conditions, a GARCH(p, q) model admits a strictly stationary solution (ϵ_t). However, the marginal distribution of the process (ϵ_t) is never known explicitly. The aim of this section is to highlight some properties of this distribution through the marginal moments.

2.4.1 Even-Order Moments

We are interested in finding existence conditions for the moments of order $2m$, where m is any positive integer.[9] Let $\otimes$ denote the tensor product, or Kronecker product, and recall that it is defined as follows: for any matrices $A = (a_{ij})$ and B, we have $A \otimes B = (a_{ij}B)$. For any matrix A, let $A^{\otimes m} = A \otimes \cdots \otimes A$. We have the following result.

Theorem 2.9 (2*m* th-order stationarity) *Let $A^{(m)} = E(A_t^{\otimes m})$ where A_t is defined by (2.17). Suppose that $E(\eta_t^{2m}) < \infty$ and that the spectral radius*

$$\rho(A^{(m)}) < 1.$$

Then, for any $t \in \mathbb{Z}$, the series $(\underline{z}_t)$ defined in (2.18) converges in L^m and the process (ϵ_t^2), defined as the first component of $\underline{z}_t$, is strictly stationary and admits moments up to order m. Conversely, if $\rho(A^{(m)}) \geq 1$, there exists no strictly stationary solution (ϵ_t) to model (2.5) such that $E(\epsilon_t^{2m}) < \infty$.

Example 2.3 (Moments of a GARCH(1, 1) process) When $p = q = 1$, the matrix A_t is written as

$$A_t = (\eta_t^2, 1)'(\alpha_1, \beta_1).$$

Hence, all the eigenvalues of the matrix $A^{(m)} = E\{(\eta_t^2, 1)'^{\otimes m}\}(\alpha_1, \beta_1)^{\otimes m}$ are null except one. The non-zero eigenvalue is thus the trace of $A^{(m)}$. It readily follows that the necessary and sufficient condition for the existence of $E(\epsilon_t^{2m})$ is

$$\sum_{i=0}^{m} \binom{m}{i} \alpha_1^i \beta_1^{m-i} \mu_{2i} < 1 \tag{2.54}$$

where $\mu_{2i} = E(\eta_t^{2i})$, $i = 0, \ldots, m$. The moments can be computed recursively, by expanding $E(\underline{z}_t^{\otimes m}) = E(\underline{b}_t + A_t \underline{z}_{t-1})^{\otimes m}$. For the fourth-order moment, a direct computation gives

$$\begin{aligned} E(\epsilon_t^4) &= E(\sigma_t^4)E(\eta_t^4) \\ &= \mu_4\{\omega^2 + 2\omega(\alpha_1 + \beta_1)E(\epsilon_{t-1}^2) + (\beta_1^2 + 2\alpha_1\beta_1)E(\sigma_{t-1}^4) + \alpha_1^2 E(\epsilon_{t-1}^4)\} \end{aligned}$$

and thus

$$E(\epsilon_t^4) = \frac{\omega^2(1 + \alpha_1 + \beta_1)}{(1 - \mu_4\alpha_1^2 - \beta_1^2 - 2\alpha_1\beta_1)(1 - \alpha_1 - \beta_1)} \mu_4,$$

provided that the denominator is positive. Figure 2.10 shows the zones of second-order and fourth-order stationarity for the strong GARCH(1, 1) model when $\eta_t \sim \mathcal{N}(0, 1)$.

This example shows that for a non-trivial GARCH process, that is, when the α_i and β_j are not all equal to zero, the moments cannot exist for any order.

Proof of Theorem 2.9 For $k > 0$, let

$$A_{t,k} = A_t A_{t-1} \cdots A_{t-k+1} \quad \text{and} \quad \underline{z}_{t,k} = A_{t,k}\underline{b}_{t-k},$$

[9] Only even-order moments are considered, because if a symmetry assumption is made on the distribution of η_t, the odd-order moments are null when they exist. If this symmetry assumption is not made, computing these moments seems extremely difficult.

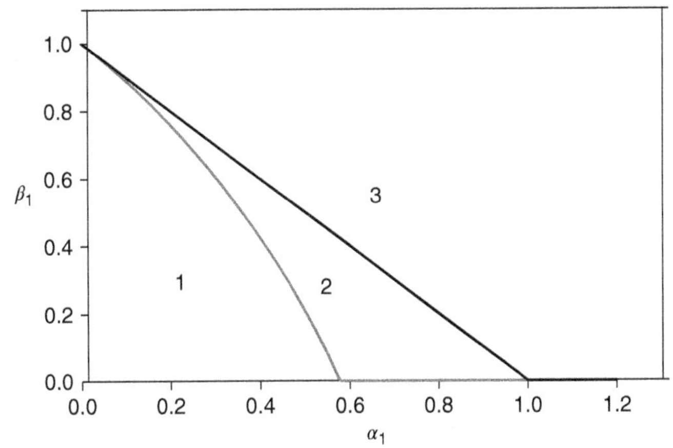

Figure 2.10 Regions of moments existence for the GARCH(1, 1) model: 1, moment of order 4; 1 and 2, moment of order 2; 3, infinite variance.

with the convention that $A_{t,0} = I_{p+q}$ and $\underline{z}_{t,0} = \underline{b}_t$. Notice that the components of $\underline{z}_t := \sum_{k=0}^{\infty} \underline{z}_{t,k}$ are a.s. defined in $[0, +\infty) \cup \{\infty\}$, without any restriction on the model coefficients. Let $\|\cdot\|$ denote the matrix norm such that $\|A\| = \sum_{i,j} |a_{ij}|$. Using the elementary equalities

$$\|A\|\|B\| = \|A \otimes B\| = \|B \otimes A\|$$

and the associativity of the Kronecker product, we obtain, for $k > 0$,

$$E\|\underline{z}_{t,k}\|^m = E\|A_{t,k}\underline{b}_{t-k} \otimes \cdots \otimes A_{t,k}\underline{b}_{t-k}\| = \|E(A_{t,k}\underline{b}_{t-k} \otimes \cdots \otimes A_{t,k}\underline{b}_{t-k})\|,$$

since the elements of the matrix $A_{t,k}\underline{b}_{t-k}$ are positive. For any vector X conformable to the matrix A, we have

$$(AX)^{\otimes m} = A^{\otimes m}X^{\otimes m}$$

by the property of the Kronecker product, $AB \otimes CD = (A \otimes C)(B \otimes D)$, for any matrices such that the products AB and CD are well defined. It follows that

$$E\|\underline{z}_{t,k}\|^m = \|E(A_{t,k}{}^{\otimes m}\underline{b}_{t-k}{}^{\otimes m})\| = \|E(A_t^{\otimes m} \dots A_{t-k+1}^{\otimes m}\underline{b}_{t-k}^{\otimes m})\|. \tag{2.55}$$

Let $\underline{b}^{(m)} = E(\underline{b}_t^{\otimes m})$ and recall that $A^{(m)} = E(A_t^{\otimes m})$. In view of (2.55), we get

$$E\|\underline{z}_{t,k}\|^m = \|(A^{(m)})^k\underline{b}^{(m)}\|,$$

using the independence between the matrices in the product $A_t \dots A_{t-k+1}\underline{b}_{t-k}$ (since A_{t-i} is a function of η_{t-i}). The matrix norm being multiplicative, it follows, using (2.18), that

$$\begin{aligned}
\|\underline{z}_t\|m &= \{E\|\underline{z}_t\|^m\}^{1/m} \\
&\le \sum_{k=0}^{\infty} \|\underline{z}_{t,k}\|_m \\
&\le \left\{\sum_{k=0}^{\infty} \|(A^{(m)})^k\|^{1/m}\right\} \|\underline{b}^{(m)}\|^{1/m}.
\end{aligned} \tag{2.56}$$

If the spectral radius of the matrix $A^{(m)}$ is strictly less than 1, then $\|(A^{(m)})^k\|$ converges to zero at exponential rate when k tends to infinity. In this case, $\underline{z}_t$ is a.s. finite. It is the strictly stationary solution of Eq. (2.16), and this solution belongs to L^m. It is clear, on the other hand, that

$$\|\epsilon_t^2\|_m \leq \|\underline{z}_t\|_m$$

because the norm of $\underline{z}_t$ is greater than that of any of its components. A sufficient condition for the existence of $E(\epsilon_t^{2m})$ is then $\rho(A^{(m)}) < 1$.

Conversely, suppose that (ϵ_t^2) belongs to L^m. For any vectors x and y of the same dimension, let $x \leq y$ mean that the components of $y - x$ are all positive. Then, for any $n \geq 0$,

$$\begin{aligned} E(\underline{z}_t^{\otimes m}) &= E(\underline{z}_{t,0} + \cdots + \underline{z}_{t,n} + A_t \ldots A_{t-n}\underline{z}_{t-n-1})^{\otimes m} \\ &\geq E\left(\sum_{k=0}^{n} \underline{z}_{t,k}\right)^{\otimes m} \\ &\geq \sum_{k=0}^{n} E(\underline{z}_{t,k}^{\otimes m}) \\ &= \sum_{k=0}^{n} (A^{(m)})^k \underline{b}^{(m)} \end{aligned}$$

because all the terms involved in theses expressions are positive. Since the components of $E(\underline{z}_t^{\otimes m})$ are finite, we have

$$\lim_{n\to\infty} (A^{(m)})^n \underline{b}^{(m)} = 0. \tag{2.57}$$

To conclude, it suffices to show that

$$\lim_{n\to\infty} (A^{(m)})^n = 0 \tag{2.58}$$

because this is equivalent to $\rho(A^{(m)}) < 1$. To deduce (2.58) from (2.57), we need to show that for any fixed integer k,

$$\text{the components of } (A^{(m)})^k \underline{b}^{(m)} \text{ are all strictly positive.} \tag{2.59}$$

The previous computations showed that

$$(A^{(m)})^k \underline{b}^{(m)} = E(\underline{z}_{t,k}^{\otimes m}).$$

Given the form of the matrices A_t, the qth component of $\underline{z}_{t,q}$ is the first component of $\underline{b}_{t-q}$, which is not a.s. equal to zero. First suppose that α_q and β_p are both not equal to zero. In this case, the first component of $\underline{z}_{t,k}$ cannot be equal to zero a.s., for any $k \geq q + 1$. Also, still in view of the form of the matrices A_t, the ith component of $\underline{z}_{t,k}$ is the $(i - 1)$th component of $\underline{z}_{t-1,k-1}$ for $i = 2, \ldots, q$. Hence, none of the first q components of $\underline{z}_{t,2q}$ can be equal to zero a.s., and the same property holds for $\underline{z}_{t,k}$ whatever $k \geq 2q$. The same argument shows that none of the last q components of $\underline{z}_{t,k}$ is equal to zero a.s. when $k \geq 2p$. Taking into account the positivity of the variables $\underline{z}_{t,k}$, this shows that in the case $\alpha_q\beta_p \neq 0$, (2.59) holds true for $k \geq \max\{2p, 2q\}$. If $\alpha_q\beta_p = 0$, one can replace $\underline{z}_t$ by a vector of smaller size, obtained by cancelling the component ϵ_{t-q+1}^2 if $\alpha_q = 0$ and the component σ_{t-p+1}^2 if $\beta_p = 0$. The matrix A_t is then replaced by a matrix of smaller size, but with the same non-zero eigenvalues as A_t. Similarly, the matrix $A^{(m)}$ will be replaced by a matrix of smaller size but with the same non-zero eigenvalues. If $\alpha_{q-1}\beta_{p-1} \neq 0$, we are led to the preceding case, otherwise we pursue the dimension reduction.

2.4.2 Kurtosis

An easy way to measure the size of distribution tails is to use the kurtosis coefficient. This coefficient is defined, for a centred (zero-mean) distribution, as the ratio of the fourth-order moment, which is assumed to exist, to the squared second-order moment. This coefficient is equal to 3 for a normal distribution, this value serving as a gauge for the other distributions. In the case of GARCH processes, it is interesting to note the difference between the tails of the marginal and conditional distributions. For a strictly stationary solution, (ϵ_t) of the GARCH(p, q) model defined by (2.5), the conditional moments of order k are proportional to σ_t^{2k}:

$$E(\epsilon_t^{2k} \mid \epsilon_u, u < t) = \sigma_t^{2k} E(\eta_t^{2k}).$$

The kurtosis coefficient of this conditional distribution is thus constant and equal to the kurtosis coefficient of η_t. For a general process of the form

$$\epsilon_t = \sigma_t \eta_t,$$

where σ_t is a measurable function of the past of ϵ_t, η_t is independent of this past and (η_t) is iid centred, the kurtosis coefficient of the stationary marginal distribution is equal, provided that it exists, to

$$\kappa_\epsilon := \frac{E(\epsilon_t^4)}{\{E(\epsilon_t^2)\}^2} = \frac{E\{E(\epsilon_t^4 \mid \epsilon_u, u < t)\}}{[E\{E(\epsilon_t^2 \mid \epsilon_u, u < t)\}]^2} = \frac{E(\sigma_t^4)}{\{E(\sigma_t^2)\}^2} \kappa_\eta$$

where $\kappa_\eta = E\eta_t^4$ denotes the kurtosis coefficient of (η_t). It can thus be seen that the tails of the marginal distribution of (ϵ_t) are fatter when the variance of σ_t^2 is large relative to the squared expectation. The minimum (corresponding to the absence of ARCH effects) is given by the kurtosis coefficient of (η_t),

$$\kappa_\epsilon \geq \kappa_\eta,$$

with equality if and only if σ_t^2 is a.s. constant. In the GARCH(1, 1) case we thus have, from the previous calculations,

$$\kappa_\epsilon = \frac{1 - (\alpha_1 + \beta_1)^2}{1 - (\alpha_1 + \beta_1)^2 - \alpha_1^2(\kappa_\eta - 1)} \kappa_\eta. \tag{2.60}$$

The *excess kurtosis* coefficients of ϵ_t and η_t, relative to the normal distribution, are related by

$$\kappa_\epsilon^* = \kappa_\epsilon - 3 = \frac{6\alpha_1^2 + \kappa_\eta^*\{1 - (\alpha_1 + \beta_1)^2 + 3\alpha_1^2\}}{1 - (\alpha_1 + \beta_1)^2 - 2\alpha_1^2 - \kappa_\eta^* \alpha_1^2}, \quad \kappa_\eta^* = \kappa_\eta - 3.$$

The excess kurtosis of ϵ_t increases with that of η_t and when the GARCH coefficients approach the zone of non-existence of the fourth-order moment. Notice the asymmetry between the GARCH coefficients in the excess kurtosis formula. For the general GARCH(p, q), we have the following result.

Proposition 2.1 (Excess kurtosis of a GARCH(p,q) process) *Let (ϵ_t) denote a GARCH(p, q) process admitting moments up to order 4 and let κ_ϵ^* be its excess kurtosis coefficient. Let $a = \sum_{i=1}^{\infty} \psi_i^2$ where the coefficients ψ_i are defined by*

$$1 + \sum_{i=1}^{\infty} \psi_i z^i = \left\{1 - \sum_{i=1}^{\max(p,q)} (\alpha_i + \beta_i) z^i\right\}^{-1} \left(1 - \sum_{i=1}^{p} \beta_i z^i\right).$$

Then the excess kurtosis of the distribution of ϵ_t relative to the Gaussian is

$$\kappa_\epsilon^* = \frac{6a + \kappa_\eta^*(1 + 3a)}{1 - a(\kappa_\eta^* + 2)}.$$

Proof. The ARMA(max(p, q), p) representation (2.4) implies

$$\epsilon_t^2 = E\epsilon_t^2 + \nu_t + \sum_{i=1}^{\infty} \psi_i \nu_{t-i},$$

where $\sum_{i=1}^{\infty} \mid \psi_i \mid < \infty$ follows from the condition $\sum_{i=1}^{\max(p,q)}(\alpha_i + \beta_i) < 1$, which is a consequence of the existence of $E\epsilon_t^4$. The process $(\nu_t) = (\sigma_t^2(\eta_t^2 - 1))$ is a weak white noise of variance $\text{Var}(\nu_t) = E\sigma_t^4 E(\eta_t^2 - 1)^2 = (\kappa_\eta - 1)E\sigma_t^4$. It follows that

$$\begin{aligned}\text{Var}(\epsilon_t^2) &= \text{Var}(\nu_t) + \sum_{i=1}^{\infty} \psi_i^2 \text{Var}(\nu_{t-i}) = (a+1)(\kappa_\eta - 1)E\sigma_t^4 \\ &= E\epsilon_t^4 - (E\epsilon_t^2)^2 = \kappa_\eta E\sigma_t^4 - (E\epsilon_t^2)^2,\end{aligned}$$

hence

$$E\sigma_t^4 = \frac{(E\epsilon_t^2)^2}{\kappa_\eta - (a+1)(\kappa_\eta - 1)}$$

and

$$\kappa_\epsilon = \frac{\kappa_\eta}{\kappa_\eta - (a+1)(\kappa_\eta - 1)} = \frac{\kappa_\eta}{1 - a(\kappa_\eta - 1)},$$

and the proposition follows. □

It will be seen in Chapter 7 that the Gaussian quasi-maximum likelihood estimator of the coefficients of a GARCH model is consistent and asymptotically normal even if the distribution of the variables η_t is not Gaussian. Since the autocorrelation function of the squares of the GARCH process does not depend on the law of η_t, the autocorrelations obtained by replacing the unknown coefficients by their estimates are generally very close to the empirical autocorrelations. In contrast, the kurtosis coefficients obtained from the theoretical formula, by replacing the coefficients by their estimates and the kurtosis of η_t by 3, can be very far from the coefficients obtained empirically. This is not very surprising since the preceding result shows that the difference between the kurtosis coefficients of ϵ_t computed with a Gaussian and a non-Gaussian distribution for η_t,

$$\kappa_\epsilon^*(\kappa_\eta^*) - \kappa_\epsilon^*(0) = \frac{\kappa_\eta^*(1-a)}{(1-2a)\{1 - a(\kappa_\eta^* + 2)\}},$$

is not bounded as a approaches $(\kappa_\eta^* + 2)^{-1}$.

2.5 Autocovariances of the Squares of a GARCH

We have seen that if (ϵ_t) is a GARCH process which is fourth-order stationary, then (ϵ_t^2) is an ARMA process. It must be noted that this ARMA is very constrained, as can be seen from representation (2.4): the order of the AR part is larger than that of the MA part, and the AR coefficients are greater than those of the MA part, which are positive. We shall start by examining some consequences of these constraints on the autocovariances of (ϵ_t^2). Then we shall show how to compute these autocovariances explicitly.

2.5.1 Positivity of the Autocovariances

For a GARCH(1, 1) model such that $E\epsilon_t^4 < \infty$, the autocorrelations of the squares take the form

$$\rho_{\epsilon^2}(h):=\text{Corr}(\epsilon_t^2, \epsilon_{t-h}^2) = \rho_{\epsilon^2}(1)(\alpha_1 + \beta_1)^{h-1}, \quad h \geq 1, \tag{2.61}$$

where

$$\rho_{\epsilon^2}(1) = \frac{\alpha_1\{1 - \beta_1(\alpha_1 + \beta_1)\}}{1 - (\alpha_1 + \beta_1)^2 + \alpha_1^2} \tag{2.62}$$

(Exercise 2.8). It follows immediately that these autocorrelations are non-negative. The next property generalises this result.

Proposition 2.2 (Positivity of the autocovariances) *If the GARCH(p, q) process (ϵ_t) admits moments of order 4, then*

$$\gamma_{\epsilon^2}(h) = \text{Cov}(\epsilon_t^2, \epsilon_{t-h}^2) \geq 0, \quad \forall h.$$

If, moreover, $\alpha_1 > 0$, then

$$\gamma_{\epsilon^2}(h) > 0, \quad \forall h.$$

Proof. It suffices to show that in the MA(∞) expansion of ϵ_t^2, all the coefficients are non-negative (and strictly positive when $\alpha_1 > 0$). In the notation introduced in condition (2.2), this expansion takes the form

$$\epsilon_t^2 = \{1 - (\alpha + \beta)(1)\}^{-1}\omega + \{1 - (\alpha + \beta)(B)\}^{-1}(1 - \beta(B))u_t := \omega^* + \psi(B)u_t.$$

Noting that $1 - \beta(B) = 1 - (\alpha + \beta)(B) + \alpha(B)$, it suffices to show the non-negativity of the coefficients c_i in the series expansion

$$\frac{\alpha(B)}{1 - (\alpha + \beta)(B)} = \sum_{i=1}^{\infty} c_i B^i.$$

We show by induction that

$$c_i \geq \alpha_1(\alpha_1 + \beta_1)^{i-1}, \quad i \geq 1.$$

Obviously, $c_1 = \alpha_1$. Moreover, with the convention $\alpha_i = 0$ if $i > q$ and $\beta_j = 0$ if $j > p$,

$$c_{i+1} = c_1(\alpha_i + \beta_i) + \ldots + c_i(\alpha_1 + \beta_1) + \alpha_{i+1}.$$

If the inductive assumption holds true at order i, using the positivity of the GARCH coefficients, we deduce that $c_{i+1} \geq \alpha_1(\alpha_1 + \beta_1)^i$. Hence the desired result. □

Remark 2.7 The property that the autocovariances are non-negative is satisfied, more generally, for an ARCH(∞) process of the form (2.40), provided it is fourth-order stationary. It can also be shown that the square of an ARCH(∞) process is *associated*. On these properties see Giraitis, Leipus, and Surgailis (2009) and references therein.

Note that the property of positive autocorrelations for the squares, or for the absolute values, is typically observed on real financial series (see, for instance, the second row of Table 2.1).

2.5.2 The Autocovariances Do Not Always Decrease

Formulas (2.61) and (2.62) show that for a GARCH(1,1) process, the autocorrelations of the squares decrease. An illustration is provided in Figure 2.11. A natural question is whether this property remains true for more general GARCH(p, q) processes. The following computation shows that this is not the case. Consider an ARCH(2) process admitting moments of order 4 (the existence condition and the computation of the fourth-order moment are the subject of Exercise 2.8):

$$\epsilon_t = \sqrt{\omega + \alpha_1 \epsilon_{t-1}^2 + \alpha_2 \epsilon_{t-2}^2}\eta_t.$$

We know that ϵ_t^2 is an AR(2) process, whose autocorrelation function satisfies

$$\rho_{\epsilon^2}(h) = \alpha_1 \rho_{\epsilon^2}(h-1) + \alpha_2 \rho_{\epsilon^2}(h-2), \quad h > 0.$$

It readily follows that

$$\frac{\rho_{\epsilon^2}(2)}{\rho_{\epsilon^2}(1)} = \frac{\alpha_1^2 + \alpha_2(1-\alpha_2)}{\alpha_1}$$

and hence that

$$\rho_{\epsilon^2}(2) < \rho_{\epsilon^2}(1) \Longleftrightarrow \alpha_2(1-\alpha_2) < \alpha_1(1-\alpha_1).$$

The latter equality is, of course, true for the ARCH(1) process ($\alpha_2 = 0$) but is not true for any (α_1, α_2). Figure 2.12 gives an illustration of this non-decreasing feature of the first autocorrelations (and partial autocorrelations). The sequence of autocorrelations is, however, decreasing after a certain lag (Exercise 2.18).

2.5.3 Explicit Computation of the Autocovariances of the Squares

The autocorrelation function of (ϵ_t^2) will play an important role in identifying the orders of the model. This function is easily obtained from the ARMA(max(p, q), p) representation

$$\epsilon_t^2 - \sum_{i=1}^{\max(p,q)} (\alpha_i + \beta_i)\epsilon_{t-i}^2 = \omega + v_t - \sum_{i=1}^{p} \beta_i v_{t-i}.$$

The autocovariance function is more difficult to obtain (Exercise 2.9) because one has to compute

$$Ev_t^2 = E(\eta_t^2 - 1)^2 E\sigma_t^4.$$

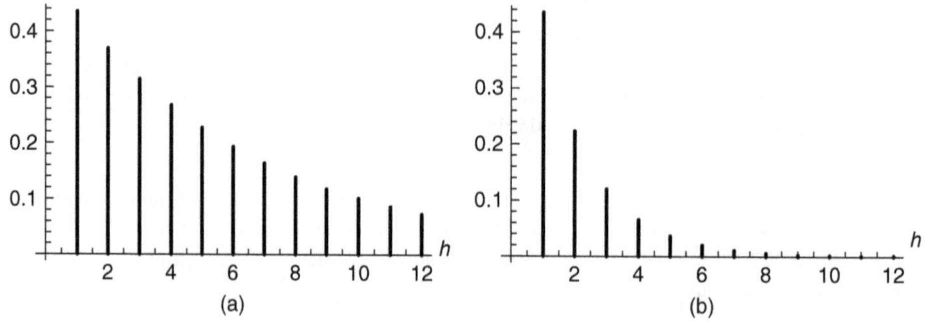

Figure 2.11 Autocorrelation function (a) and partial autocorrelation function (b) of the squares of the GARCH(1, 1) model $\epsilon_t = \sigma_t \eta_t$, $\sigma_t^2 = 1 + 0.3\epsilon_{t-1}^2 + 0.55\sigma_{t-1}^2$, ($\eta_t$) iid $\mathcal{N}(0, 1)$.

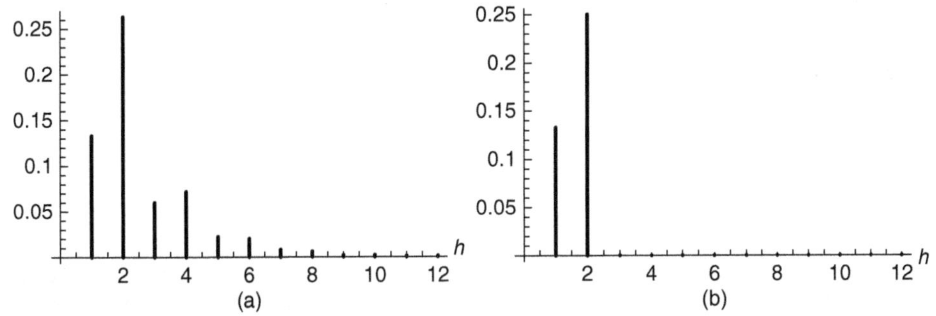

Figure 2.12 Autocorrelation function (a) and partial autocorrelation function (b) of the squares of the ARCH(2) process $\epsilon_t = \sigma_t \eta_t$, $\sigma_t^2 = 1 + 0.1\epsilon_{t-1}^2 + 0.25\epsilon_{t-2}^2$, (η_t) iid $\mathcal{N}(0, 1)$.

One can use the method of Section 2.4.1. Consider the vector representation

$$\underline{z}_t = \underline{b}_t + A_t \underline{z}_{t-1}$$

defined in (2.16) and (2.17). Using the independence between $\underline{z}_t$ and $(\underline{b}_t, A_t)$, together with elementary properties of the Kronecker product $\otimes$, we get

$$\begin{aligned} E\underline{z}_t^{\otimes 2} &= E(\underline{b}_t + A_t\underline{z}_{t-1}) \otimes (\underline{b}_t + A_t\underline{z}_{t-1}) \\ &= E(\underline{b}_t \otimes \underline{b}_t) + E(A_t\underline{z}_{t-1} \otimes \underline{b}_t) + E(\underline{b}_t \otimes A_t\underline{z}_{t-1}) + E(A_t\underline{z}_{t-1} \otimes A_t\underline{z}_{t-1}) \\ &= E\underline{b}_t^{\otimes 2} + E(A_t \otimes \underline{b}_t)E\underline{z}_{t-1} + E(\underline{b}_t \otimes A_t)E\underline{z}_{t-1} + EA_t^{\otimes 2}E\underline{z}_{t-1}^{\otimes 2}. \end{aligned}$$

Thus

$$E\underline{z}_t^{\otimes 2} = (I_{(p+q)^2} - A^{(2)})^{-1}\{\underline{b}^{(2)} + (EA_t \otimes \underline{b}_t + E\underline{b}_t \otimes A_t)\underline{z}^{(1)}\}, \tag{2.63}$$

where

$$A^{(m)} = E(A_t^{\otimes m}), \quad \underline{z}^{(m)} = E\underline{z}_t^{\otimes m}, \quad \underline{b}^{(m)} = E(\underline{b}_t^{\otimes m}).$$

To compute $A^{(m)}$, we can use the decomposition $A_t = \eta_t^2 B + C$, where B and C are deterministic matrices. We then have, letting $\mu_m = E\eta_t^m$,

$$A^{(2)} = E(\eta_t^2 B + C) \otimes (\eta_t^2 B + C) = \mu_4 B^{\otimes 2} + B \otimes C + C \otimes B + C^{\otimes 2}.$$

We obtain $EA_t \otimes \underline{b}_t$ and $E\underline{b}_t \otimes A_t$ similarly. All the components of $\underline{z}^{(1)}$ are equal to $\omega/(1 - \sum \alpha_i - \sum \beta_i)$. Note that for $h > 0$, we have

$$\begin{aligned} E\underline{z}_t \otimes \underline{z}_{t-h} &= E(\underline{b}_t + A_t\underline{z}_{t-1}) \otimes \underline{z}_{t-h} \\ &= \underline{b}^{(1)} \otimes \underline{z}^{(1)} + (A^{(1)} \otimes \mathrm{I}_{p+q})E\underline{z}_t \otimes \underline{z}_{t-h+1}. \end{aligned} \tag{2.64}$$

Let $\underline{e}_1 = (1, 0, \dots, 0) \in \mathbb{R}^{(p+q)^2}$. The following algorithm can be used:

- Define the vectors $\underline{z}^{(1)}$, $\underline{b}^{(1)}$, $\underline{b}^{(2)}$, and the matrices $EA_t \otimes \underline{b}_t$, $E\underline{b}_t \otimes A_t$, $A^{(1)}$, $A^{(2)}$ as a function of α_i, β_i, and ω, μ_4.
- Compute $E\underline{z}_t^{\otimes 2}$ from Eq. (2.63).
- For $h = 1, 2, \dots$, compute $E\underline{z}_t \otimes \underline{z}_{t-h}$ from Eq. (2.64).
- For $h = 0, 1, \dots$, compute $\gamma_{\epsilon^2}(h) = \underline{e}_1 E\underline{z}_t \otimes \underline{z}_{t-h} - (\underline{e}_1\underline{z}^{(1)})^2$.

This algorithm is not very efficient in terms of computation time and memory space, but it is easy to implement.

2.6 Theoretical Predictions

The definition of GARCH processes in terms of conditional expectations allows us to compute the optimal predictions of the process and its square given its infinite past. Let (ϵ_t) be a stationary GARCH (p, q) process, in the sense of Definition 2.1. The optimal prediction (in the L^2 sense) of ϵ_t given its infinite past is 0 by Definition 2.1(i). More generally, for $h+1>0$,

$$E(\epsilon_{t+h} \mid \epsilon_u, u < t) = E\{E(\epsilon_{t+h} \mid \epsilon_{\underline{t+h-1}}) \mid \epsilon_u, u < t\} = 0, \quad t \in \mathbb{Z},$$

which shows that the optimal prediction of any future variable given the infinite past is zero. The main attraction of GARCH models obviously lies not in the prediction of the GARCH process itself but in the prediction of its square. The optimal prediction of ϵ_t^2 given the infinite past of ϵ_t is σ_t^2. More generally, the predictions at horizon $h \geq 0$ are obtained recursively by

$$\begin{aligned} E(\epsilon_{t+h}^2 \mid \epsilon_u, u < t) &= E(\sigma_{t+h}^2 \mid \epsilon_u, u < t) \\ &= \omega + \sum_{i=1}^{q} \alpha_i E(\epsilon_{t+h-i}^2 \mid \epsilon_u, u < t) + \sum_{j=1}^{p} \beta_j E(\sigma_{t+h-j}^2 \mid \epsilon_u, u < t), \end{aligned}$$

with, for $i \leq h$,

$$E(\epsilon_{t+h-i}^2 \mid \epsilon_u, u < t) = E(\sigma_{t+h-i}^2 \mid \epsilon_u, u < t),$$

for $i > h$,

$$E(\epsilon_{t+h-i}^2 \mid \epsilon_u, u < t) = \epsilon_{t+h-i}^2,$$

and for $i \geq h$,

$$E(\sigma_{t+h-i}^2 \mid \epsilon_u, u < t) = \sigma_{t+h-i}^2.$$

These predictions coincide with the optimal linear predictions of the future values of ϵ_t^2 given its infinite past. Note that there exists a more general class of GARCH models (weak GARCH) for which the two types of predictions, optimal and linear optimal, do not necessarily coincide. Weak GARCH representations appear, in particular, when GARCH are temporally aggregated (see Drost and Werker, 1996; Drost, Nijman and Werker 1998), when they are observed with a measurement error as in Gouriéroux, Monfort and Renault (1993) and King, Sentana and Wadhwani (1994), or for the beta-ARCH of Diebolt and Guégan (1991). See Francq and Zakoïan (2000) for other examples.

It is important to note that $E(\epsilon_{t+h}^2 \mid \epsilon_u, u < t) = \text{Var}(\epsilon_{t+h} \mid \epsilon_u, u < t)$ is the conditional variance of the prediction error of ϵ_{t+h}. Hence, the accuracy of the predictions depends on the past: it is particularly low after a turbulent period, that is, when the past values are large in absolute value (assuming that the coefficients α_i and β_j are non-negative). This property constitutes a crucial difference with standard ARMA models, for which the magnitude of the prediction intervals is constant, for a given horizon.

Figures 2.13–2.16, based on simulations, allow us to visualise this difference. In Figure 2.13, obtained from a Gaussian white noise, the predictions at horizon 1 have a constant variance: the confidence interval [−1.96, 1.96] contains roughly 95% of the realisations. Using a constant interval for the next three series, displayed in Figures 2.14–2.16, would imply very bad results. In contrast, the intervals constructed here (for conditionally Gaussian distributions, with zero mean and variance σ_t^2) do contain about 95% of the observations: in the quiet periods a small interval is enough, whereas in turbulent periods, the variability increases and larger intervals are needed.

For a strong GARCH process it is possible to go further, by computing optimal predictions of the powers of ϵ_t^2, provided that the corresponding moments exist for the process (η_t). For instance, computing the predictions of ϵ_t^4 allows us to evaluate the variance of the prediction errors of ϵ_t^2. However, these computations are tedious, the linearity property being lost for such powers.

When the GARCH process is not directly observed but is the innovation of an ARMA process, the accuracy of the prediction at some date t directly depends of the magnitude of the conditional

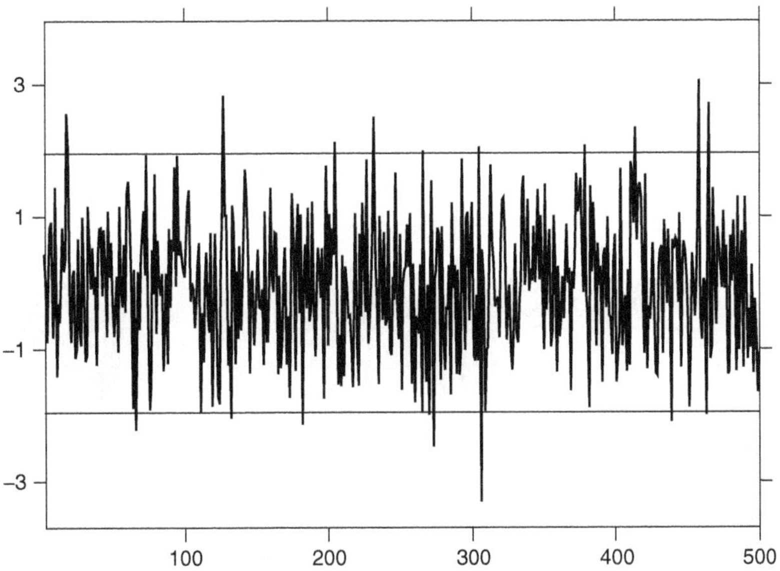

Figure 2.13 Prediction intervals at horizon 1, at 95%, for the strong $\mathcal{N}(0, 1)$ white noise.

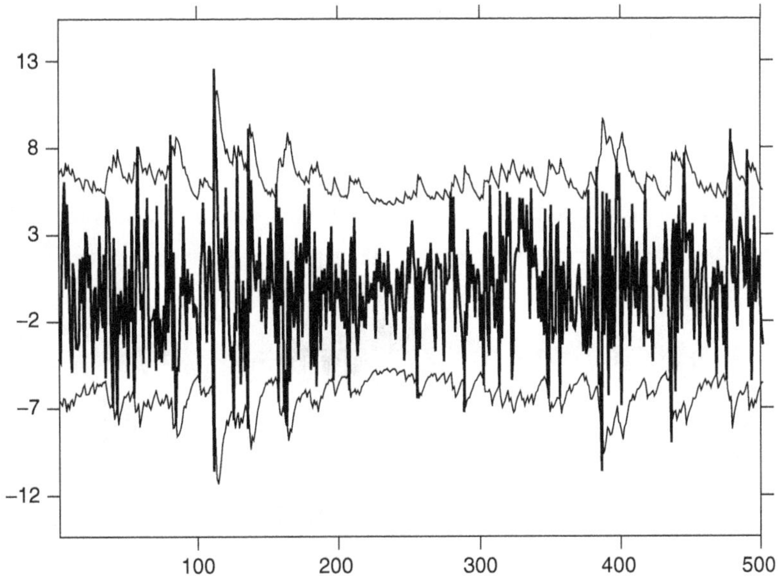

Figure 2.14 Prediction intervals at horizon 1, at 95%, for the GARCH(1, 1) process simulated with $\omega = 1$, $\alpha = 0.1$, $\beta = 0.8$ and $\mathcal{N}(0, 1)$ distribution for (η_t).

heteroscedasticity at this date. Consider, for instance, a stationary AR(1) process, whose innovation is a GARCH(1, 1) process:

$$\begin{cases} X_t = \phi X_{t-1} + \epsilon_t \\ \epsilon_t = \sigma_t \eta_t \\ \sigma_t^2 = \omega + \alpha\epsilon_{t-1}^2 + \beta\sigma_{t-1}^2, \end{cases} \tag{2.65}$$

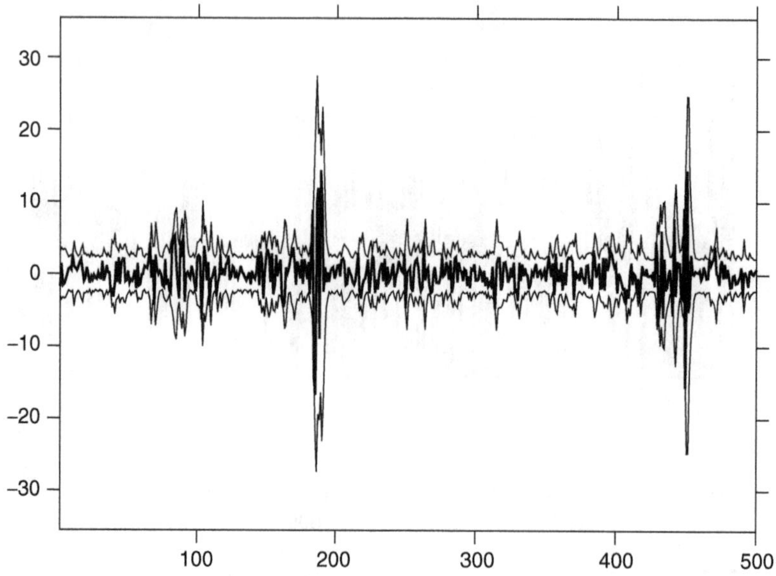

Figure 2.15 Prediction intervals at horizon 1, at 95%, for the GARCH(1, 1) process simulated with $\omega = 1$, $\alpha = 0.6$, $\beta = 0.2$ and $\mathcal{N}(0, 1)$ distribution for (η_t).

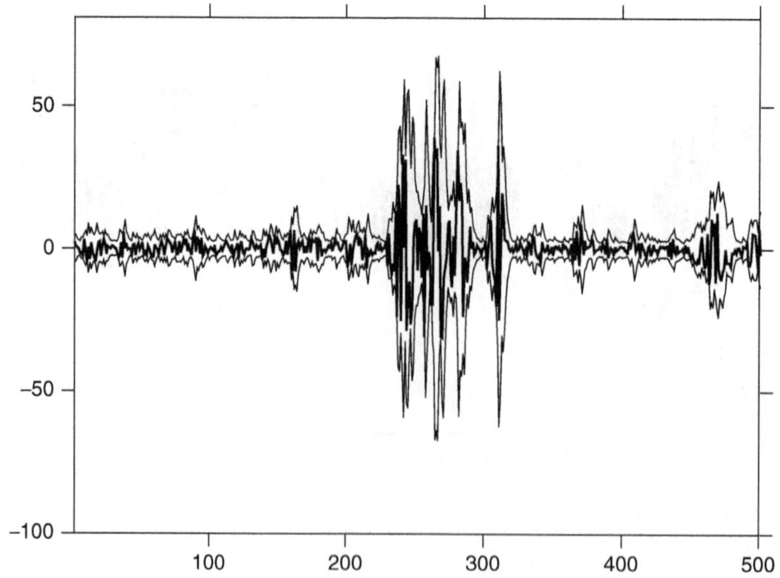

Figure 2.16 Prediction intervals at horizon 1, at 95%, for the GARCH(1, 1) process simulated with $\omega = 1$, $\alpha = 0.7$, $\beta = 0.3$ and $\mathcal{N}(0, 1)$ distribution for (η_t).

where $\omega > 0$, $\alpha \geq 0$, $\beta \geq 0$, $\alpha + \beta \leq 1$ and $|\phi| < 1$. We have, for $h \geq 0$,

$$X_{t+h} = \epsilon_{t+h} + \phi\epsilon_{t+h-1} + \cdots + \phi^h\epsilon_t + \phi^{h+1}X_{t-1}.$$

Hence,

$$E(X_{t+h} \mid X_u, u < t) = \phi^{h+1}X_{t-1}$$

since the past of X_t coincides with that of its innovation ϵ_t. Moreover,

$$\begin{aligned}
\operatorname{Var}(X_{t+h} \mid X_u, u < t) &= \operatorname{Var}\left(\sum_{i=0}^{h} \phi^{h-i}\epsilon_{t+i} \mid \epsilon_u, u < t\right) \\
&= \sum_{i=0}^{h} \phi^{2(h-i)}\operatorname{Var}(\epsilon_{t+i} \mid \epsilon_u, u < t).
\end{aligned}$$

Since $\operatorname{Var}(\epsilon_t \mid \epsilon_u, u < t) = \sigma_t^2$ and, for $i \geq 1$,

$$\begin{aligned}
\operatorname{Var}(\epsilon_{t+i} \mid \epsilon_u, u < t) &= E(\sigma_{t+i}^2 \mid \epsilon_u, u < t) \\
&= \omega + (\alpha + \beta)E(\sigma_{t+i-1}^2 \mid \epsilon_u, u < t) \\
&= \omega\{1 + \cdots + (\alpha + \beta)^{i-1}\} + (\alpha + \beta)^i\sigma_t^2,
\end{aligned}$$

we have

$$\operatorname{Var}(\epsilon_{t+i} \mid \epsilon_u, u < t) = \omega\frac{1 - (\alpha + \beta)^i}{1 - (\alpha + \beta)} + (\alpha + \beta)^i\sigma_t^2, \quad \text{for all } i \geq 0.$$

Consequently,

$$\begin{aligned}
&\operatorname{Var}(X_{t+h} \mid X_u, u < t) \\
&= \left(\sum_{i=0}^{h} \phi^{2(h-i)}\right)\frac{\omega}{1 - (\alpha + \beta)} + \sum_{i=0}^{h} (\alpha + \beta)^i\phi^{2(h-i)}\left\{\sigma_t^2 - \frac{\omega}{1 - (\alpha + \beta)}\right\} \\
&= \frac{\omega(1 - \phi^{2(h+1)})}{\{1 - (\alpha + \beta)\}(1 - \phi^2)} + \left\{\sigma_t^2 - \frac{\omega}{1 - (\alpha + \beta)}\right\}\frac{\phi^{2(h+1)} - (\alpha + \beta)^{(h+1)}}{\phi^2 - (\alpha + \beta)}
\end{aligned}$$

if $\phi^2 \neq \alpha + \beta$ and

$$\operatorname{Var}(X_{t+h} \mid X_u, u < t) = \frac{\omega(1 - \phi^{2(h+1)})}{(1 - \phi^2)^2} + \left\{\sigma_t^2 - \frac{\omega}{1 - (\alpha + \beta)}\right\}(h + 1)\phi^{2h}$$

if $\phi^2 = \alpha + \beta$. The coefficient of $\sigma_t^2 - \frac{\omega}{1-(\alpha+\beta)}$ always being positive, it can be seen that the variance of the prediction at horizon h increases linearly with the difference between the conditional variance at time t and the unconditional variance of ϵ_t. A large negative difference (corresponding to a low-volatility period) thus results in highly accurate predictions. Conversely, the accuracy deteriorates when σ_t^2 is large. When the horizon h increases, the importance of this factor decreases. If h tends to infinity, we retrieve the unconditional variance of X_t:

$$\lim_{h\to\infty} \operatorname{Var}(X_{t+h} \mid X_u, u < t) = \operatorname{Var}(X_t) = \frac{\operatorname{Var}(\epsilon_t)}{1 - \phi^2}.$$

Now we consider two non-stationary situations. If $|\phi| = 1$, and initialising, for instance at 0, all the variables at negative dates (because here the infinite pasts of X_t and ϵ_t do not coincide), the previous formula becomes

$$\operatorname{Var}(X_{t+h} \mid X_u, u < t) = \frac{\omega h}{\{1 - (\alpha + \beta)\}} + \left\{\sigma_t^2 - \frac{\omega}{1 - (\alpha + \beta)}\right\}\frac{1 - (\alpha + \beta)^{(h+1)}}{1 - (\alpha + \beta)}.$$

Thus, the impact of the observations before time t does not vanish as h increases. It becomes negligible, however, compared to the deterministic part which is proportional to h. If $\alpha + \beta = 1$ (IGARCH(1, 1) errors), we have

$$\operatorname{Var}(\epsilon_{t+i} \mid \epsilon_u, u < t) = \omega i + \sigma_t^2, \quad \text{for all } i \geq 0,$$

and it can be seen that the impact of the past variables on the variance of the predictions remains constant as the horizon increases. This phenomenon is called *persistence of shocks* on the volatility. Note, however, that, as in the preceding case, the non-random part of the decomposition of $\operatorname{Var}(\epsilon_{t+i} \mid \epsilon_u, u < t)$ becomes dominant when the horizon tends to infinity. The asymptotic precision of the predictions of ϵ_t is null, and this is also the case for X_t since

$$\operatorname{Var}(X_{t+h} \mid X_u, u < t) \geq \operatorname{Var}(\epsilon_{t+h} \mid \epsilon_u, u < t).$$

2.7 Bibliographical Notes

The strict stationarity of the GARCH(1, 1) model was first studied by Nelson (1990a) under the assumption $E \log^+ \eta_t^2 < \infty$. His results were extended by Klüppelberg, Lindner, and Maller (2004) to the case of $E \log^+ \eta_t^2 = +\infty$. For GARCH($p$, q) models, the strict stationarity conditions were established by Bougerol and Picard (1992b). For model (2.16), where $(A_t, \underline{b}_t)$ is a strictly stationary and ergodic sequence with a logarithmic moment, Brandt (1986) showed that $\gamma < 0$ ensures the existence of a unique strictly stationary solution. In the case where $(A_t, \underline{b}_t)$ is iid, Bougerol and Picard (1992a) established the converse property showing that, under an irreducibility condition, a necessary condition for the existence of a strictly stationary and non-anticipative solution is that $\gamma < 0$. Liu (2006) used representation (2.30) to obtain stationarity conditions for more general GARCH models. The second-order stationarity condition for the GARCH(p, q) model was obtained by Bollerslev (1986), as well as the existence of an ARMA representation for the square of a GARCH (see also Bollerslev 1988). Nelson and Cao (1992) obtained necessary and sufficient positivity conditions for the GARCH(p, q) model. These results were extended by Tsai and Chan (2008). The 'zero-drift' GARCH(1,1) model in which $\omega = 0$ was studied by Hafner and Preminger (2015) and Li et al. (2017). In the latter article it was shown that, though non-stationary, the zero-drift GARCH(1, 1) model is stable when $\gamma = 0$, with its sample paths oscillating randomly between zero and infinity over time.

GARCH are not the only models that generate uncorrelated times series with correlated squares. Another class of interest which shares these characteristics is that of the all-pass time series models (ARMA models in which the roots of the AR polynomials are reciprocals of roots of the MA polynomial and vice versa) studied by Breidt, Davis, and Trindade (2001).

ARCH(∞) models were introduced by Robinson (1991); see Giraitis, Leipus, and Surgailis (2009) for the study of these models. The condition for the existence of a strictly stationary ARCH(∞) process was established by Robinson and Zaffaroni (2006) and Douc, Roueff, and Soulier (2008). The condition for the existence of a second-order stationary solution, as well as the positivity of the autocovariances of the squares, were obtained by Giraitis, Kokoszka, and Leipus (2000). Theorems 2.8 and 2.7 were proven by Kazakevičius and Leipus (2002, 2003). The uniqueness of an ARCH(∞) solution is discussed in Kazakevičius and Leipus (2007). The asymptotic properties of quasi-maximum likelihood estimators (see Chapter 7) were established by Robinson and Zaffaroni (2006). See Doukhan, Teyssière, and Winant (2006) for the study of multivariate extensions of ARCH(∞) models. The introduction of FIGARCH models is due to Baillie, Bollerslev, and Mikkelsen (1996), but the existence of solutions was recently established by Douc, Roueff, and Soulier (2008), where Corollary 2.6 is proven. LARCH(∞) models were introduced by Robinson (1991) and their probability properties studied by Giraitis, Robinson, and Surgailis (2000), Giraitis and Surgailis (2002), Berkes and Horváth (2003a), and Giraitis et al. (2004). The estimation of such models has been studied by Beran and Schützner (2009), Truquet (2014), and Francq and Zakoïan (2010).

The fourth-order moment structure and the autocovariances of the squares of GARCH processes were analysed by Milhøj (1984), Karanasos (1999), and He and Teräsvirta (1999). The necessary and sufficient condition for the existence of even-order moments was established by Ling and McAleer

(2002a), the sufficient part having been obtained by Chen and An 1998. Ling and McAleer (2002b) derived an existence condition for the moment of order s, with $s > 0$, for a family of GARCH processes including the standard model and the extensions presented in Chapter 4. The computation of the kurtosis coefficient for a general GARCH(p, q) model is due to Bai, Russell, and Tiao (2004).

Several authors have studied the tail properties of the stationary distribution. See Mikosch and Stărică (2000), Borkovec and Klüppelberg (2001), Basrak, Davis, and Mikosch (2002), and Davis and Mikosch (2009a,b). See Embrechts, Klüppelberg and Mikosch (1997) for a classical reference on extremal events.

Andersen and Bollerslev (1998) discussed the predictive qualities of GARCH, making a clear distinction between the prediction of volatility and that of the squared returns (Exercise 2.23).

2.8 Exercises

2.1 *(Non-correlation of ϵ_t with any function of its past?)*
For a GARCH process does $\text{Cov}(\epsilon_t, f(\epsilon_{t-h})) = 0$ hold for any function f and any $h > 0$?

2.2 *(Strict stationarity of GARCH(1, 1) for two laws of η_t)*
For the GARCH(1, 1) give an explicit strict stationarity condition in the two following cases: (i) the only possible values of η_t are -1 and 1; (ii) η_t follows a uniform distribution.

2.3 *(Lyapunov coefficient of a constant sequence of matrices)*
Prove equality (2.21) for a diagonalisable matrix. Use the Jordan representation to extend this result to any square matrix.

2.4 *(Lyapunov coefficient of a sequence of matrices)*
Consider the sequence (A_t) defined by $A_t = z_t A$, where (z_t) is an ergodic sequence of real random variables such that $E \log^+ |z_t| < \infty$, and A is a square non-random matrix. Find the Lyapunov coefficient γ of the sequence (A_t) and give an explicit expression for the condition $\gamma < 0$.

2.5 *(Alternative definition of the top Lyapunov exponent)*

1. Show that, everywhere in Theorem 2.3, the matrix product $A_t A_{t-1} \dots A_1$ can be replaced by $A_{-1} A_{-2} \dots A_{-t}$.
2. Justify the first convergence after (2.25).

2.6 *(Multiplicative norms)*
Show the results of footnote 6 on page 28.

2.7 *(Another vector representation of the GARCH(p, q) model)*
Verify that the vector $\underline{z}_t^* = (\sigma_t^2, \dots, \sigma_{t-p+1}^2, \epsilon_{t-1}^2, \dots, \epsilon_{t-q+1}^2)' \in \mathbb{R}^{p+q-1}$ allows us to define, for $p \geq 1$ and $q \geq 2$, a vector representation which is equivalent to those used in this chapter, of the form $\underline{z}_t^* = \underline{b}_t^* + A_t^* \underline{z}_{t-1}^*$.

2.8 *(Fourth-order moment of an ARCH(2) process)*
Show that for an ARCH(2) model, the condition for the existence of the moment of order 4, with $\mu_4 = E\eta_t^4$, is written as

$$\alpha_2 < 1 \quad \text{and} \quad \mu_4 \alpha_1^2 < \frac{1 - \alpha_2}{1 + \alpha_2}(1 - \mu_4 \alpha_2^2).$$

Compute this moment.

2.9 *(Direct computation of the autocorrelations and autocovariances of the square of a GARCH(1, 1) process)*
Find the autocorrelation and autocovariance functions of (ϵ_t^2) when (ϵ_t) is solution of the GARCH(1, 1) model

$$\begin{cases} \epsilon_t = \sigma_t \eta_t \\ \sigma_t^2 = \omega + \alpha \epsilon_{t-1}^2 + \beta \sigma_{t-1}^2, \end{cases}$$

where $(\eta_t) \sim \mathcal{N}(0, 1)$ and $1 - 3\alpha^2 - \beta^2 - 2\alpha\beta > 0$.

2.10 *(Computation of the autocovariance of the square of a GARCH(1,1) process by the general method)*
Use the method of Section 2.5.3 to find the autocovariance function of (ϵ_t^2) when (ϵ_t) is solution of a GARCH(1,1) model. Compare with the method used in Exercise 2.9.

2.11 *(Fekete's lemma for sub-additive sequences)*
A sequence $(a_n)_{n\geq 1}$ is called sub-additive if $a_{n+m} \leq a_n + a_m$ for all n and m. Show that we then have $\lim_{n\to\infty} a_n/n = \inf_{n\geq 1} a_n/n$.

2.12 *(Characteristic polynomial of* EA_t*)*
Let $A = EA_t$, where $\{A_t, t\in \mathbb{Z}\}$ is the sequence defined in Eq. (2.17).

1. Prove equality (2.38).
2. If $\sum_{i=1}^q \alpha_i + \sum_{j=1}^p \beta_j = 1$, show that $\rho(A) = 1$.

2.13 *(A condition for a sequence* X_n *to be* o(n)*)*
Let (X_n) be a sequence of identically distributed random variables, admitting a finite expectation. Show that

$$\frac{X_n}{n} \to 0 \quad \text{when } n\to\infty$$

with probability 1. Prove that the convergence may fail if the expectation of X_n does not exist (an iid sequence with density $f(x) = x^{-2}\mathbb{1}_{x\geq 1}$ may be considered).

2.14 *(Expectation of a product of dependent random variables equal to the product of their expectations)*
Prove equality (2.31).

2.15 *(Necessary condition for the existence of the moment of order 2s)*
Suppose that (ϵ_t) is the strictly stationary solution of model (2.5) with $E\epsilon_t^{2s} < \infty$, for $s\in(0, 1]$. Let

$$\underline{z}_t^{(K)} = \underline{b}_t + \sum_{k=1}^{K} A_t A_{t-1} \dots A_{t-k+1} \underline{b}_{t-k}. \tag{2.66}$$

1. Show that when $K\to\infty$,

$$\|\underline{z}_t^{(K)} - \underline{z}_t^{(K-1)}\|^s \to 0 \text{ a.s.}, \quad E\|\underline{z}_t^{(K)} - \underline{z}_t^{(K-1)}\|^s \to 0.$$

2. Use this result to prove that $E(\|A_k A_{k-1}\dots A_1 \underline{b}_0\|^s) \to 0$ as $k\to\infty$.
3. Let (X_n) a sequence of $\ell\times m$ matrices and $Y = (Y_1, \dots, Y_m)'$ a vector which is independent of (X_n) and such that for all i, $0 < E|Y_i|^s < \infty$. Show that, when $n\to\infty$,

$$E\|X_n Y\|^s \to 0 \implies E\|X_n\|^s \to 0.$$

4. Let $A = EA_t$, $\underline{b} = E\underline{b}_t$ and suppose there exists an integer N such that $A^N \underline{b} > 0$ (in the sense that all elements of this vector are strictly positive). Show that there exists $k_0 \geq 1$ such that

$$E(\|A_{k_0} A_{k_0-1} \dots A_1\|^s) < 1.$$

5. Deduce condition (2.35) from the preceding question.
6. Is the condition $\alpha_1 + \beta_1 > 0$ necessary?

2.16 *(Positivity conditions)*
In the GARCH$(1, q)$ case give a more explicit form for the conditions in (2.49). Show, by taking $q = 2$, that these conditions are less restrictive than (2.20).

2.17 *(A minoration for the first autocorrelations of the square of an ARCH)*
Let (ϵ_t) be an ARCH(q) process admitting moments of order 4. Show that, for $i = 1, \ldots, q$,

$$\rho_{\epsilon^2}(i) \geq \alpha_i.$$

2.18 *(Asymptotic decrease of the autocorrelations of the square of an ARCH(2) process)*
Figure 2.12 shows that the first autocorrelations of the square of an ARCH(2) process, admitting moments of order 4, can be non-decreasing. Show that this sequence decreases after a certain lag.

2.19 *(Convergence in probability to $-\infty$)*
If (X_n) and (Y_n) are two independent sequences of random variables such that $X_n + Y_n \to -\infty$ and $X_n \nrightarrow -\infty$ in probability, then $Y_n \to -\infty$ in probability.

2.20 *(GARCH model with a random coefficient)*
Proceeding as in the proof of Theorem 2.1, study the stationarity of the GARCH(1, 1) model with random coefficient $\omega = \omega(\eta_{t-1})$,

$$\begin{cases} \epsilon_t = \sigma_t \eta_t \\ \sigma_t^2 = \omega(\eta_{t-1}) + \alpha \epsilon_{t-1}^2 + \beta \sigma_{t-1}^2, \end{cases} \tag{2.67}$$

under the usual assumptions and $\omega(\eta_{t-1}) > 0$ a.s. Use the result of Exercise 2.19 to deal with the case $\gamma := E \log a(\eta_t) = 0$.

2.21 *(RiskMetrics model)*
The RiskMetrics model used to compute the value at risk (see Chapter 11) relies on the following equations:

$$\begin{cases} \epsilon_t = \sigma_t \eta_t, & (\eta_t) \text{ iid } \mathcal{N}(0, 1) \\ \sigma_t^2 = \lambda \sigma_{t-1}^2 + (1 - \lambda) \epsilon_{t-1}^2 \end{cases}$$

where $0 < \lambda < 1$. Show that this model has no stationary and non trivial solution.

2.22 *(IARCH(∞) models: proof of Corollary 2.6)*

1. In model (2.40), under the assumption $A_1 = 1$, show that

$$\sum_{i=1}^{\infty} \phi_i \log \phi_i \leq 0 \quad \text{and} \quad E(\eta_0^2 \log \eta_0^2) \geq 0.$$

2. Suppose that condition (2.41) holds. Show that the function $f: [p,1] \mapsto \mathbb{R}$ defined by $f(q) = \log(A_q \mu_{2q})$ is convex. Compute its derivative at 1 and deduce that condition (2.52) holds.
3. Establish the reciprocal and show that $E|\epsilon_t|^q < \infty$ for any $q \in [0, 2)$.

2.23 *(On the predictive power of GARCH)*
In order to evaluate the quality of the prediction of ϵ_t^2 obtained by using the volatility of a GARCH(1, 1) model, econometricians have considered the linear regression

$$\epsilon_t^2 = a + b\sigma_t^2 + u_t,$$

where σ_t^2 is replaced by the volatility estimated from the model. They generally obtained a very small determination coefficient, meaning that the quality of the regression was bad. It this surprising? In order to answer that question compute, under the assumption that $E\epsilon_t^4$ exists, the theoretical R^2 defined by

$$R^2 = \frac{\text{Var}(\sigma_t^2)}{\text{Var}(\epsilon_t^2)}.$$

Show, in particular, that $R^2 < 1/\kappa_\eta$.

3

Mixing*

It will be shown that, under mild conditions, GARCH processes are geometrically ergodic and β-mixing. These properties entail the existence of laws of large numbers and of central limit theorems (see Appendix A), and thus play an important role in the statistical analysis of GARCH processes. This chapter relies on the Markov chain techniques set out, for example, by Meyn and Tweedie (1996).

3.1 Markov Chains with Continuous State Space

Recall that for a Markov chain only the most recent past is of use in obtaining the conditional distribution. More precisely, (X_t) is said to be a *homogeneous Markov chain*, evolving on a space E (called the state space) equipped with a σ-field $\mathcal{E}$, if for all $x \in E$, and for all $B \in \mathcal{E}$,

$$\forall s, t \in \mathbb{N}, \quad \mathbb{P}(X_{s+t} \in B \mid X_r, r < s; X_s = x) := P^t(x, B). \tag{3.1}$$

In this equation, $P^t(x, B)$ corresponds to the *transition probability* of moving from the state x to the set B in t steps. The Markov property refers to the fact that $P^t(x, B)$ does not depend on X_r, $r < s$. The fact that this probability does not depend on s is referred to as *time homogeneity*. For simplicity, we write $P(x, B) = P^1(x, B)$. The function $P : E \times \mathcal{E} \to [0, 1]$ is called *a transition kernel* and satisfies:

(i) $\forall B \in \mathcal{E}$, the function $P(\cdot, B)$ is measurable;

(ii) $\forall x \in E$, the function $P(x, \cdot)$ is a probability measure on $(E, \mathcal{E})$.

The law of the process (X_t) is characterised by an initial probability measure μ and a transition kernel P. For all integers t and all $(t+1)$-tuples $(B_0, \ldots, B_t)$ of elements of $\mathcal{E}$, we set

$$\begin{aligned} &\mathbb{P}_\mu(X_0 \in B_0, \ldots, X_t \in B_t) \\ &= \int_{x_0 \in B_0} \cdots \int_{x_{t-1} \in B_{t-1}} \mu(dx_0) P(x_0, dx_1) \ldots P(x_{t-1}, B_t). \end{aligned} \tag{3.2}$$

In what follows, (X_t) denotes a Markov chain on $E = \mathbb{R}^d$ and $\mathcal{E}$ is the Borel σ-field.

GARCH Models: Structure, Statistical Inference and Financial Applications, Second Edition. Christian Francq and Jean-Michel Zakoian.

Irreducibility and Recurrence

The Markov chain (X_t) is said to be *ϕ-irreducible* for a non-trivial (that is, not identically equal to zero) measure ϕ on $(E, \mathcal{E})$, if

$$\forall B \in \mathcal{E}, \quad \phi(B) > 0 \quad \Rightarrow \quad \forall x \in E, \quad \exists t > 0, \quad P^t(x, B) > 0.$$

If (X_t) is ϕ-irreducible, it can be shown that there exists a *maximal irreducibility measure*, that is, an irreducibility measure M such that all the other irreducibility measures are absolutely continuous with respect to M. If $M(B) = 0$, then the set of points from which B is accessible is also of zero measure (see Meyn and Tweedie 1996, Proposition 4.2.2). Such a measure M is not unique, but the set

$$\mathcal{E}^+ = \{B \in \mathcal{E} \mid M(B) > 0\}$$

does not depend on the maximal irreducibility measure M. For a particular model, finding a measure that makes the chain irreducible may be a non-trivial problem (but see Exercise 3.1 for an example of a time series model for which the determination of such a measure is very simple).

A ϕ-irreducible chain is called *recurrent* if

$$U(x, B) := \sum_{t=1}^{\infty} P^t(x, B) = +\infty, \quad \forall x \in E, \ \forall B \in \mathcal{E}^+,$$

and is called *transient* if

$$\exists (B_j)_j, \quad \mathcal{E} = \bigcup_j B_j, \quad \mathrm{U}(x, B_j) \leq M_j < \infty, \quad \forall x \in E.$$

Note that $U(x, B) = E \sum_{t=1}^{\infty} \mathbb{1}_B(X_t)$ can be interpreted as the average time that the chain spends in B when it starts at x. It can be shown that a ϕ-irreducible chain (X_t) is either recurrent or transient (see Meyn and Tweedie 1996, Theorem 8.3.4). It is said that (X_t) is *positive recurrent* if

$$\limsup_{t \to \infty} P^t(x, B) > 0, \quad \forall x \in E, \forall B \in \mathcal{E}^+.$$

If a ϕ-irreducible chain is not positive recurrent, it is called *null recurrent*. For a ϕ-irreducible chain, positive recurrence is equivalent to the existence of a (unique) *invariant probability measure* (see Meyn and Tweedie 1996, Theorem 18.2.2), that is, a probability π such that

$$\forall B \in \mathcal{E}, \quad \pi(B) = \int P(x, B)\pi(dx).$$

An important consequence of this equivalence is that, for Markov time series, the issue of finding strict stationarity conditions reduces to that of finding conditions for positive recurrence. Indeed, it can be shown (see Exercise 3.2) that for any chain (X_t) with initial measure μ,

$$(X_t) \text{ is stationary} \Longleftrightarrow \mu \text{ is invariant.} \tag{3.3}$$

For this reason, the invariant probability is also called the *stationary probability*.

Small Sets and Aperiodicity

For a ϕ-irreducible chain, there exists a class of sets enjoying properties that are similar to those of the elementary states of a finite state space Markov chain. A set $C \in \mathcal{E}$ is called a *small set*[1] if there exists an integer $m \geq 1$ and a nontrivial measure ν on $\mathcal{E}$ such that

$$\forall x \in C, \quad \forall B \in \mathcal{E}, \qquad P^m(x, B) \geq \nu(B).$$

---

[1] Meyn and Tweedie (1996) introduce a more general notion, called a 'petite set', obtained by replacing, in the definition, the transition probability in m steps by an average of the transition probabilities, $\sum_{m=0}^{\infty} a_m P^m(x, B)$, where (a_m) is a probability distribution.

In the AR(1) case, for instance it is easy to find small sets (see Exercise 3.4). For more sophisticated models, the definition is not sufficient and more explicit criteria are needed. For the so-called Feller chains, we will see below that it is very easy to find small sets. For a general chain, we have the following criterion (see Nummelin 1984, Proposition 2.11): $C \in \mathcal{E}^+$ is a small set if there exists $A \in \mathcal{E}^+$ such that, for all $B \subset A$, $B \in \mathcal{E}^+$, there exists $T > 0$ such that

$$\inf_{x \in C} \sum_{t=0}^{T} P^t(x, B) > 0.$$

If the chain is ϕ-irreducible, it can be shown that there exists a countable cover of E by small sets. Moreover, each set $B \in \mathcal{E}^+$ contains a small set $C \in \mathcal{E}^+$. The existence of small sets allows us to define cycles for ϕ-irreducible Markov chains with general state space, as in the case of countable space chains. More precisely, the *period* is the greatest common divisor (gcd) of the set

$$\{n \geq 1 \mid \forall x \in C, \ \ \forall B \in \mathcal{E}, P^n(x, B) \geq \delta_n \nu(B), \text{ for some } \ \delta_n > 0\},$$

where $C \in \mathcal{E}^+$ is any small set (the gcd is independent of the choice of C). When $d = 1$, the chain is said to be *aperiodic*. Moreover, it can be shown (see Meyn and Tweedie 1996, Theorem 5.4.4) that there exist disjoint sets $D_1, \ldots, D_d \in \mathcal{E}$ such that (with the convention $D_{d+1} = D_1$):

(i) $\forall i = 1, \ldots, d, \forall x \in D_i, P(x, D_{i+1}) = 1$;

(ii) $\phi(E - \bigcup D_i) = 0$.

A necessary and sufficient condition for the aperiodicity of (X_t) is that there exists $A \in \mathcal{E}^+$ such that for all $B \subset A$, $B \in \mathcal{E}^+$, there exists $t > 0$ such that

$$P^t(x, B) > 0 \quad \text{and} \quad P^{t+1}(x, B) > 0 \quad \forall x \in B \tag{3.4}$$

(see Chan 1990, Proposition A1.2).

Geometric Ergodicity and Mixing

In this section, we study the convergence of the probability $\mathbb{P}_\mu(X_t \in \cdot)$ to a probability $\pi(\cdot)$ independent of the initial probability μ, as $t \to \infty$.

It is easy to see that if there exists a probability measure π such that, for an initial measure μ,

$$\forall B \in \mathcal{E}, \quad \mathbb{P}_\mu(X_t \in B) \to \pi(B), \quad \text{when } t \to +\infty, \tag{3.5}$$

where $\mathbb{P}_\mu(X_t \in B)$ is defined in (3.2) (for $(B_0, \ldots, B_t) = (E, \ldots, E, B)$), then the probability π is invariant (see Exercise 3.3). Note also that (3.5) holds for any measure μ if and only if

$$\forall B \in \mathcal{E}, \quad \forall x \in E, \quad P^t(x, B) \to \pi(B), \quad \text{when } \ t \to +\infty.$$

On the other hand, if the chain is irreducible, aperiodic, and admits an invariant probability π, for π-almost all $x \in E$,

$$\| P^t(x, .) - \pi \| \to 0 \quad \text{when } t \to +\infty, \tag{3.6}$$

where $\| \cdot \|$ denotes the total variation norm[2] (see Meyn and Tweedie 1996, Theorem 14.0.1). A chain (X_t) such that the convergence (3.6) holds for all x is said to be *ergodic*. However, this convergence is not sufficient for mixing. We will define a stronger notion of ergodicity.

[2] The total variation norm of a (signed) measure m is defined by $\|m\| = \sup \int f dm$, where the supremum is taken over $\{f : E \to \mathbb{R}, f \text{ measurable and } |f| \leq 1\}$.

The chain (X_t) is called *geometrically ergodic* if there exists $\rho \in (0, 1)$ such that

$$\forall x \in E, \quad \rho^{-t} \parallel P^t(x,.) - \pi \parallel \to 0 \quad \text{when } t \to +\infty. \tag{3.7}$$

Geometric ergodicity entails the so-called *α- and β-mixing*. The general definition of the α- and β-mixing coefficients is given in Appendix A.3.1. For a stationary Markov process, the definition of the α-mixing coefficient reduces to

$$\alpha_X(k) = \sup_{f,g} | \operatorname{Cov}(f(X_0), g(X_k)) |,$$

where the first supremum is taken over the set of the measurable functions f and g such that $|f| \leq 1$, $|g| \leq 1$ (see Bradley 1986, 2005). A general process $X = (X_t)$ is said to be α-mixing (β-mixing) if $\alpha_X(k)$ ($\beta_X(k)$) converges to 0 as $k \to \infty$. Intuitively, these mixing properties characterise the decrease in dependence when past and future become sufficiently far apart. The α-mixing is sometimes called strong mixing, but β-mixing entails strong mixing because $\alpha_X(k) \leq \beta_X(k)$ (see Appendix A.3.1).

Davydov (1973) showed that for an ergodic Markov chain (X_t), of invariant probability measure π,

$$\beta_X(k) = \int \parallel P^k(x,.) - \pi \parallel \pi(dx).$$

It follows that $\beta_X(k) = O(\rho^k)$ if the convergence (3.7) holds. Thus

$$\begin{aligned} &(X_t) \text{ is stationary and geometrically ergodic} \\ \Rightarrow\ &(X_t) \text{ is geometrically } \beta\text{-mixing}. \end{aligned} \tag{3.8}$$

Two Ergodicity Criteria

For particular models, it is generally not easy to directly verify the properties of recurrence, existence of an invariant probability law, and geometric ergodicity. Fortunately, there exist simple criteria on the transition kernel.

We begin by defining the notion of Feller chain. The Markov chain (X_t) is said to be *a Feller chain* if, for all bounded continuous functions g defined on E, the function of x defined by $E(g(X_t) \mid X_{t-1} = x)$ is continuous. For instance, for an AR(1) we have, with obvious notation,

$$E\{g(X_t) \mid X_{t-1} = x\} = E\{g(\theta x + \epsilon_t)\}.$$

The continuity of the function $x \to g(\theta x + y)$ for all y, and its boundedness, ensure, by the Lebesgue dominated convergence theorem, that (X_t) is a Feller chain. For a Feller chain, the compact sets $C \in \mathcal{E}^+$ are small sets (see Feigin and Tweedie 1985).

The following theorem provides an effective way to show the geometric ergodicity (and thus the β-mixing) of numerous Markov processes.

Theorem 3.1 (Feigin and Tweedie (1985, Theorem 1)) *Assume that*

(i) (X_t) is a Feller chain;

(ii) (X_t) is ϕ-irreducible;

(iii) there exists a compact set $A \subset E$ such that $\phi(A) > 0$ and a continuous function $V : E \to \mathbb{R}^+$ satisfying

$$V(x) \geq 1, \quad \forall x \in A, \tag{3.9}$$

and for $\delta > 0$,

$$E\{V(X_t) \mid X_{t-1} = x\} \leq (1 - \delta)V(x), \quad \forall x \notin A. \tag{3.10}$$

Then (X_t) is geometrically ergodic.

This theorem will be applied to GARCH processes in the next section (see also Exercise 3.5 for a bilinear example). In Eq. (3.10), V can be interpreted as an energy function. When the chain is outside the centre A of the state space, the energy dissipates, on average. When the chain lies inside A, the energy is bounded, by the compactness of A and the continuity of V. Sometimes V is called a test function and (iii) is said to be a drift criterion.

Let us explain why these assumptions imply the existence of an invariant probability measure. For simplicity, assume that the test function V takes its values in $[1, +\infty)$, which will be the case for the applications to GARCH models we will present in the next section. Denote by $\mathbf{P}$ the operator which, to a measurable function f in E, associates the function $\mathbf{P}f$ defined by

$$\forall x \in E, \quad \mathbf{P}f(x) = \int_E f(y)P(x, dy) = E\{f(X_t) \mid X_{t-1} = x\}.$$

Let $\mathbf{P}^t$ be the tth iteration of $\mathbf{P}$, obtained by replacing $P(x, dy)$ by $P^t(x, dy)$ in the previous integral. By convention $\mathbf{P}^0 f = f$ and $P^0(x, A) = \mathbb{1}_A$. Equations (3.9) and (3.10) and the boundedness of V by some $M > 0$ on A yield an inequality of the form

$$\mathbf{P}V \leq (1 - \delta)V + b\mathbb{1}_A$$

where $b = M - (1 - \delta)$. Iterating this relation t times, we obtain, for $x_0 \in A$

$$\forall t \geq 0, \quad \mathbf{P}^{t+1}V(x_0) \leq (1 - \delta)\mathbf{P}^t V(x_0) + bP^t(x_0, A). \tag{3.11}$$

It follows (see Exercise 3.6) that there exists a constant $\kappa > 0$ such that for n large enough,

$$Q_n(x_0, A) \geq \kappa \quad \text{where } Q_n(x_0, A) = \frac{1}{n}\sum_{t=1}^{n} P^t(x_0, A). \tag{3.12}$$

The sequence $Q_n(x_0, \cdot)$ being a sequence of probabilities on $(E, \mathcal{E})$, it admits an accumulation point for vague convergence: there exists a measure π of mass less than 1 and a subsequence (n_k) such that for all continuous functions f with compact support,

$$\lim_{k\to\infty} \int_E f(y)Q_{n_k}(x_0, dy) = \int_E f(y)\pi(dy). \tag{3.13}$$

In particular, if we take $f = \mathbb{1}_A$ in this equality, we obtain $\pi(A) \geq \kappa$, thus π is not equal to zero. Finally, it can be shown that π is a probability and that (3.13) entails that π is an invariant probability for the chain (X_t) (see Exercise 3.7).

For some models, the drift criterion (iii) is too restrictive because it relies on transitions in only one step. The following criterion, adapted from Meyn and Tweedie (1996, Theorems 19.1.3, 6.2.9, and 6.2.5), is an interesting alternative relying on the transitions in n steps.

Theorem 3.2 (Geometric Ergodicity Criterion) *Assume that*

(i) (X_t) is an aperiodic Feller chain;

(ii) (X_t) is ϕ-irreducible where the support of ϕ has non-empty interior;

(iii) there exists a compact $C \subset E$, an integer $n \geq 1$, and a continuous function $V : E \to \mathbb{R}^+$ satisfying

$$1 \leq V(x), \quad \forall x \in C, \tag{3.14}$$

and for $\delta > 0$ and $b > 0$,

$$\begin{aligned} E\{V(X_{t+n}) \mid X_{t-1} = x\} &\leq (1 - \delta)V(x), & \forall x \notin C, \\ E\{V(X_{t+n}) \mid X_{t-1} = x\} &\leq b, & \forall x \in C. \end{aligned} \tag{3.15}$$

Then (X_t) is geometrically ergodic.

The compact C of condition (iii) can be replaced by a small set, but the function V must be bounded on C. When (X_t) is not a Feller chain, a similar criterion exists, for which it is necessary to consider such small sets (see Meyn and Tweedie 1996, Theorem 19.1.3).

3.2 Mixing Properties of GARCH Processes

We begin with the ARCH(1) process because this is the only case where the process (ϵ_t) is Markovian.

The ARCH(1) Case

Consider the model

$$\begin{cases} \epsilon_t = \sigma_t \eta_t \\ \sigma_t^2 = \omega + \alpha \epsilon_{t-1}^2, \end{cases} \tag{3.16}$$

where $\omega > 0$, $\alpha \geq 0$ and (η_t) is a sequence of iid (0, 1) variables. The following theorem establishes the mixing property of the ARCH(1) process under the necessary and sufficient strict stationarity condition (see Theorem 2.1 and (2.10)). An extra assumption on the distribution of η_t is required, but this assumption is mild:

Assumption A *The law P_η of the process (η_t) is absolutely continuous, of density f with respect to the Lebesgue measure λ on $(\mathbb{R}, \mathcal{B}(\mathbb{R}))$. We assume that*

$$\inf\{\eta \mid \eta > 0, f(\eta) > 0\} = \inf\{-\eta \mid \eta < 0, f(\eta) > 0\} := \eta^0, \tag{3.17}$$

and that there exists $\tau > 0$ such that

$$(-\eta^0 - \tau, -\eta^0) \cup (\eta^0, \ \eta^0 + \tau) \subset \{f > 0\}.$$

Note that this assumption includes, in particular, the standard case where f is positive over a neighbourhood of 0, possibly over all $\mathbb{R}$. We then have $\eta^0 = 0$. Equality (3.17) implies some (local) symmetry of the law of (η_t). This symmetry facilitates the proof of the following theorem, but it can be omitted (see Exercise 3.8).

Theorem 3.3 (Mixing of the ARCH(1) Model) *Under Assumption A and for*

$$\alpha < e^{-E \log \eta_t^2}, \tag{3.18}$$

the non-anticipative strictly stationary solution of the ARCH(1) model (3.16) is geometrically ergodic, and thus geometrically β-mixing.

Proof. Let $\psi(x) = (\omega + \alpha x^2)^{1/2}$. A process (ϵ_t) satisfying

$$\epsilon_t = \psi(\epsilon_{t-1})\eta_t, \quad t \geq 1,$$

where η_t is independent of ϵ_{t-i}, $i > 0$, is clearly a homogenous Markov chain on $(\mathbb{R}, \mathcal{B}(\mathbb{R}))$, with transition probabilities

$$P(x, B) = \mathbb{P}(\epsilon_1 \in B \mid \epsilon_0 = x) = \int_{\frac{1}{\psi(x)}B} dP_\eta(y).$$

We will show that the conditions of Theorem 3.1 are satisfied. □

Step (i) We have

$$E\{g(\epsilon_t) \mid \epsilon_{t-1} = x\} = E[g\{\psi(x)\eta_t\}].$$

If g is continuous and bounded, the same is true for the function $x \to g\{\psi(x)y\}$, for all y. By the Lebesgue theorem, it follows that (ϵ_t) is a Feller chain.

Step (ii) To show the ϕ-irreducibility of the chain, for some measure ϕ, assume for the moment that $\eta^0 = 0$ in Assumption A. Suppose, for instance, that f is positive on $[0, \tau)$. Let ϕ be the restriction of the Lebesgue measure to the interval $[0, \sqrt{\omega}\tau)$. Since $\psi(x) \geq \sqrt{\omega}$, it can be seen that

$$\phi(B) > 0 \Rightarrow \lambda\left\{\frac{1}{\psi(x)}B \cap [0, \tau)\right\} > 0 \Rightarrow P(x, B) > 0.$$

It follows that the chain (ϵ_t) is ϕ-irreducible. In particular, $\phi = \lambda$ if η_t has a positive density over $\mathbb{R}$.

The proof of the irreducibility in the case $\eta^0 > 0$ is more difficult. First note that

$$E \log \alpha\eta_t^2 = \int_{(-\infty,-\eta^0] \cup [\eta^0,+\infty)} \log(\alpha x^2)f(x)d\lambda(x) \geq \log \alpha(\eta^0)^2.$$

Now $E \log \alpha\, \eta_t^2 < 0$ by (3.18). Thus we have

$$\rho := \alpha(\eta^0)^2 < 1.$$

Let $\tau' \in (0, \tau)$ be small enough such that

$$\rho_1 := \alpha(\eta^0 + \tau')^2 < 1.$$

Iterating the model, we obtain that, for $\epsilon_0 = x$ fixed,

$$\epsilon_t^2 = \omega(\eta_t^2 + \alpha\eta_t^2\eta_{t-1}^2 + \cdots + \alpha^{t-1}\eta_t^2 \ldots \eta_1^2) + \alpha^t\eta_t^2 \ldots \eta_1^2 x^2.$$

It follows that the function

$$Y_t = (\eta_1^2, \ldots, \eta_t^2) \to Z_t = (\eta_1^2, \ldots, \eta_{t-1}^2, \epsilon_t^2)$$

is a diffeomorphism between open subsets of $\mathbb{R}^t$. Moreover, in view of Assumption A, the vector Y_t has a density on $\mathbb{R}^t$. The same is thus true for Z_t, and it follows that, given $\epsilon_0 = x$,

$$\text{the variable } \epsilon_t^2 \text{ has a density with respect to } \lambda. \tag{3.19}$$

We now introduce the event

$$\Xi_t = \bigcap_{s=1}^{t}\{\eta_s \in (-\eta^0 - \tau', -\eta^0) \cup [\eta^0, \eta^0 + \tau']\}. \tag{3.20}$$

Assumption A implies that $\mathbb{P}(\Xi_t) > 0$. Conditional on Ξ_t, we have

$$\epsilon_t^2 \in I_t := \left[\omega(\eta^0)^2\frac{1-\rho^t}{1-\rho} + \rho^t x^2, \omega(\eta^0)^2\frac{1-\rho_1^t}{1-\rho_1} + \rho_1^t x^2\right].$$

Since the bounds of the interval I_t are reached, the intermediate value theorem and (3.19) entail that, given $\epsilon_0 = x$, ϵ_t^2 has, conditionally on Ξ_t, a positive density on I_t. It follows that

$$\epsilon_t \text{ has, conditionally on } \Xi_t, \text{ a positive density on } J_t \tag{3.21}$$

where $J_t = \{x \in \mathbb{R} \mid x^2 \in I_t\}$. Let

$$I = \left[\frac{\omega(\eta^0)^2}{1-\rho}, \frac{\omega(\eta^0 + \tau')^2}{1-\rho_1}\right], \quad J = \{x \in \mathbb{R} \mid x^2 \in I\}$$

and let λ_J be the restriction of the Lebesgue measure to J. We have

$$\begin{aligned}\lambda_J(B) > 0 &\Rightarrow \exists t, \quad \lambda(B \cap J_t) > 0 \\ &\Rightarrow \exists t, \quad \mathbb{P}(\epsilon_t \in B \mid \epsilon_0 = x) \geq \mathbb{P}(\epsilon_t \in B \mid (\epsilon_0 = x) \cap \Xi_t)\mathbb{P}(\Xi_t) > 0.\end{aligned}$$

The chain (ϵ_t) is thus ϕ-irreducible with $\phi = \lambda_J$.

Step (iii) We shall use Lemma 2.2. The variable $\alpha\eta_t^2$ is almost surely positive and satisfies $E(\alpha\eta_t^2) = \alpha < \infty$ and $E \log \alpha\eta_t^2 < 0$, in view of assumption (3.18). Thus, there exists $s > 0$ such that

$$c := \alpha^s \mu_{2s} < 1,$$

where $\mu_{2s} = E\eta_t^{2s}$. The proof of Lemma 2.2 shows that we can assume $s \leq 1$. Let $V(x) = 1 + x^{2s}$. Condition (3.9) is obviously satisfied for all x. Let $0 < \delta < 1 - c$ and let the compact set

$$A = \{x \in \mathbb{R}; \omega^s \mu_{2s} + \delta + (c - 1 + \delta)x^{2s} \geq 0\}.$$

Since A is a nonempty closed interval with centre 0, we have $\phi(A) > 0$. Moreover, by the inequality $(a + b)^s \leq a^s + b^s$ for $a, b \geq 0$ and $s \in [0, 1]$ (see the proof of Corollary 2.3), we have, for $x \notin A$,

$$\begin{aligned}E[V(\epsilon_t) \mid \epsilon_{t-1} = x] &\leq 1 + (\omega^s + \alpha^s x^{2s})\mu_{2s} \\ &= 1 + \omega^s \mu_{2s} + cx^{2s} \\ &< (1 - \delta)V(x),\end{aligned}$$

which proves condition (3.10). It follows that the chain (ϵ_t) is geometrically ergodic. Therefore, in view of property (3.8), the chain obtained with the invariant law as initial measure is geometrically β-mixing. The proof of the theorem is complete.

Remark 3.1 (Case Where the Law of η_t does not have a Density) The condition on the density of η_t is not necessary for the mixing property. Suppose, for example, that $\eta_t^2 = 1$, a.s. (that is, η_t takes the values -1 and 1, with probability 1/2). The strict stationarity condition reduces to $\alpha < 1$, and the strictly stationary solution is $\epsilon_t = \sqrt{\omega/(1-\alpha)}\eta_t$, a.s.. This solution is mixing since it is an independent white noise.

Another pathological example is obtained when η_t has a mass at 0: $\mathbb{P}(\eta_t = 0) = \theta > 0$. Regardless of the value of α, the process is strictly stationary because the right-hand side of inequality condition (3.18) is equal to $+\infty$. A noticeable feature of this chain is the existence of *regeneration* times at which the past is forgotten. Indeed, if $\eta_t = 0$ then $\epsilon_t = 0, \epsilon_{t+1} = \sqrt{\omega}\eta_{t+1}, \ldots$. It is easy to see that the process is then mixing, regardless of α.

The GARCH(1, 1) Case

Let us consider the GARCH(1, 1) model

$$\begin{cases}\epsilon_t = \sigma_t\eta_t \\ \sigma_t^2 = \omega + \alpha\epsilon_{t-1}^2 + \beta\sigma_{t-1}^2,\end{cases} \tag{3.22}$$

where $\omega > 0$, $\alpha \geq 0$, $\beta \geq 0$ and the sequence (η_t) is as in the previous section. In this case (σ_t) is Markovian, but (ϵ_t) is not Markovian when $\beta > 0$. The following result extends Theorem 3.3.

Theorem 3.4 (Mixing of the GARCH(1, 1) Model) *Under Assumption A and if*

$$E \log(\alpha\eta_t^2 + \beta) < 0, \tag{3.23}$$

then the non-anticipative strictly stationary solution of the GARCH(1, 1) model (3.22) is such that the Markov chain (σ_t) is geometrically ergodic and the process (ϵ_t) is geometrically β-mixing.

Proof. If $\alpha = 0$ the strictly stationary solution is iid, and the conclusion of the theorem follows in this case. We now assume that $\alpha > 0$. We first show the conclusions of the theorem that concern the process (σ_t). A homogenous Markov chain (σ_t) is defined on $(\mathbb{R}^+, \mathcal{B}(\mathbb{R}^+))$ by setting, for $t \geq 1$,

$$\sigma_t^2 = \omega + a(\eta_{t-1})\sigma_{t-1}^2, \tag{3.24}$$

where $a(x) = \alpha x^2 + \beta$. Its transition probabilities are given by

$$\forall x > 0, \quad \forall B \in \mathcal{B}(\mathbb{R}^+), \quad P(x, B) = \mathbb{P}(\sigma_1 \in B \mid \sigma_0 = x) = \int_{B_x} dP_\eta(y),$$

where $B_x = \{\eta;\ \{\omega + a(\eta)x^2\}^{1/2} \in B\}$. We show the stated results by checking the conditions of Theorem 3.1.

Step (i) The arguments given in the ARCH(1) case, with

$$E\{g(\sigma_t) \mid \sigma_{t-1} = x\} = E[g\{(\omega + a(\eta_t)x^2)^{1/2}\}],$$

are sufficient to show that (σ_t) is a Feller chain.

Step (ii) To show the irreducibility, note that condition (3.23) implies

$$\rho := a(\eta^0) < 1,$$

since $|\eta_t| \geq \eta^0$ almost surely and $a(\cdot)$ is an increasing function. Let $\tau' \in (0, \tau)$ be small enough such that

$$\rho_1 := a(\eta^0 + \tau') < 1.$$

If $\sigma_0 = x \in \mathbb{R}^+$, we have, for $t > 0$,

$$\sigma_t^2 = \omega[1 + a(\eta_{t-1}) + a(\eta_{t-1})a(\eta_{t-2}) + \cdots + a(\eta_{t-1}) \ldots a(\eta_1)] + a(\eta_{t-1}) \ldots a(\eta_0)x^2.$$

Conditionally on Ξ_t, defined in (3.20), we have

$$\sigma_t^2 \in I_t = \left[\frac{\omega}{1-\rho} + \rho^t\left(x^2 - \frac{\omega}{1-\rho}\right),\ \frac{\omega}{1-\rho_1} + \rho_1^t\left(x^2 - \frac{\omega}{1-\rho_1}\right)\right].$$

Let

$$I = \overline{\lim_{t\to\infty}} I_t = \left[\frac{\omega}{1-\rho},\ \frac{\omega}{1-\rho_1}\right].$$

Then, given $\sigma_0 = x$,

$$\sigma_t \text{ has, conditionally on } \Xi_t, \text{ a positive density on } J_t$$

where $J_t = \{x \in \mathbb{R}^+ \mid x^2 \in I_t\}$. Let λ_J be the restriction of the Lebesgue measure to $J = \left[\sqrt{\frac{\omega}{1-\rho}}, \sqrt{\frac{\omega}{1-\rho_1}}\right]$. We have

$$\begin{aligned}\lambda_J(B) > 0 &\Rightarrow \exists t, \quad \lambda(B \cap J_t) > 0\\ &\Rightarrow \exists t, \quad \mathbb{P}(\sigma_t \in B \mid \sigma_0 = x) \geq \mathbb{P}(\sigma_t \in B \mid (\sigma_0 = x) \cap \Xi_t)\mathbb{P}(\Xi_t) > 0.\end{aligned}$$

The chain (σ_t) is thus ϕ-irreducible with $\phi = \lambda_J$.

Step (iii) We again use Lemma 2.2. By the arguments used in the ARCH(1) case, there exists $s \in [0, 1]$ such that

$$c_1 := E\{a(\eta_{t-1})^s\} < 1.$$

Define the test function by $V(x) = 1 + x^{2s}$, let $0 < \delta < 1 - c_1$ and let the compact set

$$A = \{x \in \mathbb{R}^+;\ \omega^s + \delta + (c_1 - 1 + \delta)x^{2s} \geq 0\}.$$

We have, for $x \notin A$,

$$\begin{aligned} E[V(\sigma_t) \mid \sigma_{t-1} = x] &\leq 1 + \omega^s + c_1 x^{2s} \\ &< (1 - \delta)V(x), \end{aligned}$$

which proves condition (3.10). Moreover, (3.9) is satisfied.

To be able to apply Theorem 3.1, it remains to show that $\phi(A) > 0$ where ϕ is the above irreducibility measure. In view of the form of the intervals I and A, it is clear that, denoting by $\mathring{A}$ the interior of A,

$$\begin{aligned} \phi(A) > 0 &\Longleftrightarrow \sqrt{\frac{\omega}{1-\rho}} \in \mathring{A} \\ &\Longleftrightarrow \omega^s + \delta + (c_1 - 1 + \delta)\left(\frac{\omega}{1-\rho}\right)^s > 0. \end{aligned}$$

Therefore, it suffices to choose δ sufficiently close to $1 - c_1$ so that the last inequality is satisfied. For such a choice of δ, the compact set A satisfies the assumptions of Theorem 3.1. Consequently, the chain (σ_t) is geometrically ergodic. Therefore, the non-anticipative strictly stationary solution (σ_t), satisfying model (3.24) for $t \in \mathbb{Z}$, is geometrically β-mixing.

Step (iv) Finally, we show that the process (ϵ_t) inherits the mixing properties of (σ_t). Since $\epsilon_t = \sigma_t\eta_t$, it is sufficient to show that the process $Y_t = (\sigma_t, \eta_t)'$ enjoys the mixing property. It is clear that (Y_t) is a Markov chain on $\mathbb{R}^+ \times \mathbb{R}$ equipped with the Borel σ-field. Moreover, (Y_t) is strictly stationary because, under condition (3.23), the strictly stationary solution (σ_t) is non-anticipative, thus Y_t is a function of $\eta_t, \eta_{t-1}, \ldots$. Moreover, σ_t is independent of η_t. Thus the stationary law of (Y_t) can be denoted by $\mathbb{P}_Y = \mathbb{P}_\sigma \otimes \mathbb{P}_\eta$ where $\mathbb{P}_\sigma$ denotes the law of σ_t and $\mathbb{P}_\eta$ that of η_t. Let $\tilde{P}^t(y, \cdot)$ the transition probabilities of the chain (Y_t). We have, for $y = (y_1, y_2) \in \mathbb{R}^+ \times \mathbb{R}$, $B_1 \in \mathcal{B}(\mathbb{R}^+)$, $B_2 \in \mathcal{B}(\mathbb{R})$ and $t > 0$,

$$\begin{aligned} \tilde{P}^t(y,\ B_1 \times B_2) &= \mathbb{P}(\sigma_t \in B_1,\ \eta_t \in B_2 \mid \sigma_0 = y_1,\ \eta_0 = y_2) \\ &= \mathbb{P}_\eta(B_2)\mathbb{P}(\sigma_t \in B_1 \mid \sigma_0 = y_1,\ \eta_0 = y_2) \\ &= \mathbb{P}_\eta(B_2)\mathbb{P}(\sigma_t \in B_1 \mid \sigma_1 = \omega + a(y_2)y_1) \\ &= \mathbb{P}_\eta(B_2)P^{t-1}(\omega + a(y_2)y_1,\ B_1). \end{aligned}$$

It follows, since $\mathbb{P}_\eta$ is a probability, that

$$\|\tilde{P}^t(y,\ \cdot) - \mathbb{P}_Y(\cdot)\| = \|P^{t-1}(\omega + a(y_2)y_1,\ \cdot) - \mathbb{P}_\sigma(\cdot)\|.$$

The right-hand side converges to 0 at exponential rate, in view of the geometric ergodicity of (σ_t). It follows that (Y_t) is geometrically ergodic and thus β-mixing. The process (ϵ_t) is also β-mixing, since ϵ_t is a measurable function of Y_t. □

Theorem 3.4 is of interest because it provides a proof of strict stationarity which is completely different from that of Theorem 2.8. A slightly more restrictive assumption on the law of η_t has been required, but the result obtained in Theorem 3.4 is stronger.

The ARCH(q) Case

The approach developed in the case $q = 1$ does not extend trivially to the general case because (ϵ_t) and (σ_t) lose their Markov property when $p > 1$ or $q > 1$. Consider the model

$$\begin{cases} \epsilon_t = \sigma_t \eta_t \\ \sigma_t^2 = \omega + \sum_{i=1}^q \alpha_i \epsilon_{t-i}^2, \end{cases} \tag{3.25}$$

where $\omega > 0$, $\alpha_i \geq 0$, $i = 1, \ldots, q$, and (η_t) is defined as in the previous section. We will once again use the Markov representation

$$\underline{z}_t = \underline{b}_t + A_t \underline{z}_{t-1}, \tag{3.26}$$

where

$$A_t = \begin{pmatrix} \boldsymbol{\alpha}_{1:\,q-1}\eta_t^2 & \alpha_q \eta_t^2 \\ I_{q-1} & 0 \end{pmatrix}, \quad \underline{b}_t = (\omega\eta_t^2,\ 0, \ldots, 0)', \quad \underline{z}_t = (\epsilon_t^2, \ldots, \epsilon_{t-q+1}^2)'.$$

Recall that γ denotes the top Lyapunov exponent of the sequence $\{A_t, t \in \mathbb{Z}\}$.

Theorem 3.5 (Mixing of the ARCH(q) Model) *If η_t has a positive density on a neighbourhood of 0 and $\gamma < 0$, then the non-anticipative strictly stationary solution of the ARCH(q) model (3.25) is geometrically β-mixing.*

Proof. We begin by showing that the non-anticipative and strictly stationary solution $(\underline{z}_t)$ of the model (3.26) is mixing. We will use Theorem 3.2 because a one-step drift criterion is not sufficient.

Using representation (3.26) and the independence between η_t and the past of $\underline{z}_t$, it can be seen that the process $(\underline{z}_t)$ is a Markov chain on $(\mathbb{R}^+)^q$ equipped with the Borel σ-field, with transition probabilities

$$\forall \underline{x} \in (\mathbb{R}^+)^q, \quad \forall B \in \mathcal{B}((\mathbb{R}^+)^q), \quad P(\underline{x}, B) = \mathbb{P}(\underline{b}_1 + A_1 \underline{x} \in B).$$

The Feller property of the chain $(\underline{z}_t)$ is obtained by the arguments employed in the ARCH(1) and GARCH(1, 1) cases, relying on the independence between η_t and the past of $\underline{z}_t$, as well as on the continuity of the function $\underline{x} \to \underline{b}_t + A_t \underline{x}$.

In order to establish the irreducibility, let us consider the transitions in q steps. Starting from $\underline{z}_0 = \underline{x}$, after q transitions, the chain reaches a state $\underline{z}_q$ of the form

$$\underline{z}_q = \begin{pmatrix} \eta_q^2 \psi_q(\eta_{q-1}^2, \ldots, \eta_1^2,\ \underline{x}) \\ \vdots \\ \eta_1^2 \psi_1(\underline{x}) \end{pmatrix},$$

where the functions ψ_i are such that $\psi_i(\cdot) \geq \omega > 0$. Let $\tau > 0$ be such that the density f of η_t be positive on $(-\tau, \tau)$, and let ϕ be the restriction to $[0, \omega\tau^2($^{q} of the Lebesgue measure λ on $\mathbb{R}^q$.

It follows that, for all $B = B_1 \times \cdots \times B_q \in \mathcal{B}((\mathbb{R}^+)^q)$, $\phi(B) > 0$ implies that, for all $\underline{x}, y_1, \ldots, y_q \in (\mathbb{R}^+)^q$, and for all $i = 1, \ldots, q$,

$$\lambda\left\{\frac{1}{\psi_i(y_1, \ldots, y_{i-1},\ \underline{x})} B_i \cap [0, \tau^2)\right\} > 0,$$

which implies in turn that, for all $\underline{x} \in (\mathbb{R}^+)^q$, $P^q(\underline{x}, B) > 0$. We conclude that the chain $(\underline{z}_t)$ is ϕ-irreducible.

The same argument shows that

$$\phi(B) > 0 \Rightarrow \forall k > 0, \quad \forall \underline{x} \in (\mathbb{R}^+)^q, \quad P^{q+k}(\underline{x}, B) > 0.$$

The criterion given in (3.4) can then be checked, which implies that the chain is aperiodic.

We now show that condition (iii) of Theorem 3.2 is satisfied with the test function

$$V(\underline{x}) = 1 + \|\underline{x}\|^s,$$

where $\|\cdot\|$ denotes the norm $\|A\| = \sum |A_{ij}|$ of a matrix $A = (A_{ij})$ and $s \in (0, 1)$ is such that

$$\rho := E(\|A_{k_0}A_{k_0-1}\dots A_1\|^s) < 1$$

for some integer $k_0 \geq 1$. The existence of s and k_0 is guaranteed by Lemma 2.3. Iterating (3.26), we have

$$\underline{z}_{k_0} = \underline{b}_{k_0} + \sum_{k=0}^{k_0-2} A_{k_0}\dots A_{k_0-k}\underline{b}_{k_0-k-1} + A_{k_0}\dots A_1\underline{z}_0.$$

The norm being multiplicative, it follows that

$$\|\underline{z}_{k_0}\|^s \leq \|\underline{b}_{k_0}\|^s + \sum_{k=0}^{k_0-2} \|A_{k_0}\dots A_{k_0-k}\|^s\|\underline{b}_{k_0-k-1}\|^s + \|A_{k_0}\dots A_1\|^s\|\underline{z}_0\|^s.$$

Thus, for all $\underline{x} \in (\mathbb{R}^+)^{p+q}$,

$$\begin{aligned} E(V(\underline{z}_{k_0}) \mid \underline{z}_0 = \underline{x}) &\leq 1 + E\|\underline{b}_{k_0}\|^s + \sum_{k=0}^{k_0-2} E\|A_{k_0}\dots A_{k_0-k}\|^s E\|\underline{b}_{k_0-k-1}\|^s + \rho\|\underline{x}\|^s \\ &:= K + \rho\|\underline{x}\|^s. \end{aligned}$$

The inequality comes from the independence between A_t and $\underline{b}_{t-i}$ for $i > 0$. The existence of the expectations on the right-hand side of the inequality comes from arguments used to show property (2.33). Let $\delta > 0$ such that $1 - \delta > \rho$ and let C be the subset of $(\mathbb{R}^+)^{p+q}$ defined by

$$C = \{\underline{x} \mid (1-\delta-\rho)\|\underline{x}\|^s \leq K - (1-\delta)\}.$$

We have $C \neq \emptyset$ because $K > 1 - \delta$. Moreover, C is compact because $1 - \delta - \rho > 0$. Condition (3.14) is clearly satisfied, V being greater than 1. Moreover, (3.15) also holds true for $n = k_0 - 1$. We conclude that, in view of Theorem 3.2, the chain $\underline{z}_t$ is geometrically ergodic and, when it is initialised with the stationary measure, the chain is stationary and β-mixing.

Consequently, the process (ϵ_t^2), where (ϵ_t) is the non-anticipative strictly stationary solution of model (3.25), is β-mixing, as a measurable function of $\underline{z}_t$. This argument is not sufficient to conclude concerning (ϵ_t). For $k > 0$, let

$$Y_0 = f(\dots, \epsilon_{-1}, \epsilon_0), \quad Z_k = g(\epsilon_k, \epsilon_{k+1}, \dots),$$

where f and g are measurable functions. Note that

$$E(Y_0 \mid \epsilon_t^2, t \in \mathbb{Z}) = E(Y_0 \mid \epsilon_t^2, t \leq 0, \eta_u^2, u \geq 1) = E(Y_0 \mid \epsilon_t^2, t \leq 0).$$

Similarly, we have $E(Z_k \mid \epsilon_t^2, t \in \mathbb{Z}) = E(Z_k \mid \epsilon_t^2, t \geq k)$, and we have independence between Y_0 and Z_k conditionally on (ϵ_t^2). Thus, we obtain

$$\begin{aligned} \mathrm{Cov}(Y_0, Z_k) &= E(Y_0Z_k) - E(Y_0)E(Z_k) \\ &= E\{E(Y_0Z_k \mid \epsilon_t^2, t \in \mathbb{Z})\} \\ &\quad - E\{E(Y_0 \mid \epsilon_t^2, t \in \mathbb{Z})\} \; E\{E(Z_k \mid \epsilon_t^2, t \in \mathbb{Z})\} \\ &= E\{E(Y_0 \mid \epsilon_t^2, t \leq 0)E(Z_k \mid \epsilon_t^2, t \geq k)\} \\ &\quad - E\{E(Y_0 \mid \epsilon_t^2, t \leq 0)\} \; E\{E(Z_k \mid \epsilon_t^2, t \geq k)\} \\ &= \mathrm{Cov}\{E(Y_0 \mid \epsilon_t^2, t \leq 0), \; E(Z_k \mid \epsilon_t^2, t \geq k)\} \\ &:= \mathrm{Cov}\{f_1(\dots, \epsilon_{-1}^2, \epsilon_0^2), \; g_1(\epsilon_k^2, \epsilon_{k+1}^2, \dots)\}. \end{aligned}$$

It follows, in view of the definition (A.5) of the strong mixing coefficients, that

$$\alpha_{\epsilon}(k) \leq \alpha_{\epsilon^2}(k).$$

In view of (A.6), we also have $\beta_{\epsilon}(k) \leq \beta_{\epsilon^2}(k)$. Actually, (A.7) entails that the converse inequalities are always true, so we have $\alpha_{\epsilon^2}(k) = \alpha_{\epsilon}(k)$ and $\beta_{\epsilon^2}(k) = \beta_{\epsilon}(k)$. The theorem is thus shown. □

3.3 Bibliographical Notes

A major reference on ergodicity and mixing of general Markov chains is Meyn and Tweedie (1996). For a more succinct presentation, see Chan (1990), Tjøstheim (1990), and Tweedie (2001). For survey papers on mixing conditions, see Bradley (1986, 2005). We also mention the book by Doukhan (1994) which proposes definitions and examples of other types of mixing, as well as numerous limit theorems.

For vectorial representations of the form (3.26), the Feller, aperiodicity and irreducibility properties were established by Cline and Pu (1998, Theorem 2.2), under assumptions on the error distribution and on the regularity of the transitions.

The geometric ergodicity and mixing properties of the GARCH(p, q) processes were established in the Ph.D. thesis of Boussama (1998), using results of Mokkadem (1990) on polynomial processes. The proofs use concepts of algebraic geometry to determine a subspace of the states on which the chain is irreducible. For the GARCH(1, 1) and ARCH(q) models we did not need such sophisticated notions. The proofs given here are close to those given in Francq and Zakoïan (2006b), which considers more general GARCH(1, 1) models. Mixing properties were obtained by Carrasco and Chen (2002) for various GARCH-type models under stronger conditions than the strict stationarity (for example, $\alpha + \beta < 1$ for a standard GARCH(1, 1); see their Table 1). Meitz and Saikkonen (2008a,b) showed mixing properties under mild moment assumptions for a general class of first-order Markov models, and applied their results to the GARCH(1, 1).

The mixing properties of ARCH(∞) models are studied by Fryzlewicz and Rao (2011). They develop a method for establishing geometric ergodicity which, contrary to the approach of this chapter, does not rely on the Markov chain theory. Other approaches, for instance developed by Ango Nze and Doukhan (2004) and Hörmann (2008), aim to establish probability properties (different from mixing) of GARCH-type sequences, which can be used to establish central limit theorems.

3.4 Exercises

3.1 *(Irreducibility Condition for an AR(1) Process)*
Given a sequence $(\mathcal{E}_t)_{t\in\mathbb{N}}$ of iid centred variables of law $P_{\mathcal{E}}$ which is absolutely continuous with respect to the Lebesgue measure λ on $\mathbb{R}$, let $(X_t)_{t\in\mathbb{N}}$ be the AR(1) process defined by

$$X_t = \theta X_{t-1} + \mathcal{E}_t, \quad t \geq 0$$

where $\theta \in \mathbb{R}$.

(a) Show that if $P_{\mathcal{E}}$ has a positive density over $\mathbb{R}$, then (X_t) constitutes a λ-irreducible chain.
(b) Show that if the density of $\mathcal{E}_t$ is not positive over all $\mathbb{R}$, the existence of an irreducibility measure is not guaranteed.

3.2 *Equivalence between Stationarity and Invariance of the Initial Measure*
Show the equivalence (3.3).

3.3 *(Invariance of the Limit Law)*
Show that if π is a probability such that for all B, $\mathbb{P}_{\mu}(X_t \in B) \to \pi(B)$ when $t \to \infty$, then π is invariant.

3.4 *(Small Sets for AR(1))*
For the AR(1) model of Exercise 3.1, show directly that if the density f of the error term is positive everywhere, then the compacts of the form $[-c, c]$, $c > 0$, are small sets.

3.5 *(From Feigin and Tweedie 1985)*
For the bilinear model

$$X_t = \theta X_{t-1} + b\mathcal{E}_t X_{t-1} + \mathcal{E}_t, \quad t \geq 0,$$

where $(\mathcal{E}_t)$ is as in Exercise 3.1(a), show that if

$$E \mid \theta + b\mathcal{E}_t \mid < 1,$$

then there exists a unique strictly stationary solution and this solution is geometrically ergodic.

3.6 *(Lower Bound for the Empirical Mean of the* $P^t(x_0, A)$*)*
Show inequality (3.12).

3.7 *(Invariant Probability)*
Show the invariance of the probability π satisfying condition (3.13).
Hints: (i) For a function g which is continuous and positive (but not necessarily with compact support), this equality becomes

$$\liminf_{k\to\infty} \int_E g(y) Q_{n_k}(x_0, dy) \geq \int_E g(y)\pi(dy)$$

(see Meyn and Tweedie 1996, Lemma D.5.5).
(ii) For all σ-finite measures μ on $(\mathbb{R}, \mathcal{B}(\mathbb{R}))$ we have

$$\forall B \in \mathcal{B}(\mathbb{R}), \quad \mu(B) = \sup\{\mu(C); \ C \subset B, \ C \text{ compact}\}$$

(see Meyn and Tweedie 1996, Theorem D.3.2).

3.8 *(Mixing of the ARCH(1) Model for an Asymmetric Density)*
Show that Theorem 3.3 remains true when Assumption A is replaced by the following:
The law P_η is absolutely continuous, with density f, with respect to λ. There exists $\tau > 0$ such that

$$(\eta_-^0 - \tau, \eta_-^0) \cup (\eta_+^0, \ \eta_+^0 + \tau) \subset \{f > 0\},$$

where $\eta_-^0 = \sup\{\eta \mid \eta < 0, f(\eta) > 0\}$ and $\eta_+^0 = \inf\{\eta \mid \eta > 0, f(\eta) > 0\}$.

3.9 *(A Result on Decreasing Sequences)*
Show that if u_n is a decreasing sequence of positive real numbers such that $\sum_n u_n < \infty$, we have $\sup_n n u_n < \infty$. Show that this result applies to the proof of Corollary A.3 in Appendix A.

3.10 *(Complements to the Proof of Corollary A.3)*
Complete the proof of Corollary A.3 by showing that the term d_4 is uniformly bounded in t, h and k.

3.11 *(Non-mixing Chain)*
Consider the non-mixing Markov chain defined in Example A.3. Which of the assumptions (i)–(iii) in Theorem 3.1 does the chain satisfy and which does it not satisfy?

4

Alternative Models for the Conditional Variance

Classical GARCH models rely on modelling the conditional variance as a linear function of the squared past innovations. The merits of this specification are its ability to reproduce several important characteristics of financial time series – succession of quiet and turbulent periods, autocorrelation of the squares but absence of autocorrelation of the returns, leptokurticity of the marginal distributions – and the fact that it is sufficiently simple to allow for an extended study of the probability and statistical properties.

The particular functional form of the standard GARCH volatility entails, however, important restrictions. For example, it entails positive autocorrelations of the squares at any lags (see Proposition 2.2). As will see in Part II of this book, the positivity constraints on the GARCH coefficients entail also technical difficulties for the inference. The standard GARCH formulation also does not permit to incorporate exogenous information coming from other time series, for instance macro-economic variables or intraday realised volatilities, possibly observed at different frequencies.

From an empirical point of view, the symmetric form of the classical GARCH model is one of its most obvious drawbacks. Indeed, by construction, the conditional variance only depends on the modulus of the past variables: past positive and negative innovations have the same effect on the current volatility. This property is in contradiction to many empirical studies on series of stocks, showing a negative correlation between the squared current innovation and the past innovations: if the conditional distribution was symmetric in the past variables, such a correlation would be equal to zero. However, conditional asymmetry is a stylised fact: the volatility increase due to a price decrease is generally stronger than that resulting from a price increase of the same magnitude.

The symmetry property of standard GARCH models has the following interpretation in terms of autocorrelations. If the law of η_t is symmetric, and under the assumption that the GARCH process is second-order stationary, we have

$$\operatorname{Cov}(\sigma_t, \epsilon_{t-h}) = 0, \quad h > 0, \tag{4.1}$$

because σ_t is an even function of the ϵ_{t-i}, $i > 0$ (see Exercise 4.1). Introducing the positive and negative components of ϵ_t,

$$\epsilon_t^+ = \max(\epsilon_t, 0), \quad \epsilon_t^- = \min(\epsilon_t, 0),$$

GARCH Models: Structure, Statistical Inference and Financial Applications, Second Edition. Christian Francq and Jean-Michel Zakoian.

Table 4.1 Empirical autocorrelations (CAC 40 series, period 1988–1998).

h	1	2	3	4	5	10	20	40
$\rho(\epsilon_t, \epsilon_{t-h})$	0.030	0.005	−0.032	0.028	−0.046*	0.016	0.003	−0.019
$\rho(\|\epsilon_t\|, \|\epsilon_{t-h}\|)$	0.090*	0.100*	0.118*	0.099*	0.086*	0.118*	0.055*	0.032
$\rho(\epsilon_t^+, \epsilon_{t-h})$	0.011	−0.094*	−0.148*	−0.018	−0.127*	−0.039*	−0.026	−0.064*

*Autocorrelations which are statistically significant at the 5% level, using $1/n$ as an approximation of the autocorrelations variances, for $n = 2385$.

it is easily seen that Eq. (4.1) holds if and only if

$$\mathrm{Cov}(\epsilon_t^+, \epsilon_{t-h}) = \mathrm{Cov}(\epsilon_t^-, \epsilon_{t-h}) = 0, \quad h > 0. \tag{4.2}$$

This characterisation of the symmetry property in terms of autocovariances can be easily tested empirically and is often rejected on financial series. As an example, for the log-returns series ($\epsilon_t = \ln(p_t/p_{t-1})$) of the CAC 40 index presented in Chapter 1, we get the results shown in Table 4.1.

The absence of significant autocorrelations of the returns and the correlation of their modulus or squares, which constitute the basic properties motivating the introduction of GARCH models, is clearly shown for these data. But just as evident is the existence of an asymmetry in the impact of past innovations on the current volatility. More precisely, admitting that the process (ϵ_t) is second-order stationary and can be decomposed as $\epsilon_t = \sigma_t\eta_t$, where (η_t) is an iid sequence and σ_t is a measurable, positive function of the past of ϵ_t, we have

$$\rho(\epsilon_t^+, \epsilon_{t-h}) = K\,\mathrm{Cov}(\sigma_t, \epsilon_{t-h}) = K[\mathrm{Cov}(\sigma_t, \epsilon_{t-h}^+) + \mathrm{Cov}(\sigma_t, \epsilon_{t-h}^-)]$$

where $K > 0$. For the CAC data, except when $h = 1$ for which the autocorrelation is not significant, the estimates of $\rho(\epsilon_t^+, \epsilon_{t-h})$ seem to be significantly negative.[1] Thus

$$\mathrm{Cov}(\sigma_t, \epsilon_{t-h}^+) < \mathrm{Cov}(\sigma_t, -\epsilon_{t-h}^-),$$

which can be interpreted as a higher impact of the past price decreases on the current volatility, compared to the past price increases of the same magnitude. This phenomenon, $\mathrm{Cov}(\sigma_t, \epsilon_{t-h}) < 0$, is known in the finance literature as the *leverage effect*[2]: volatility tends to increase dramatically following bad news (that is, a fall in prices), and to increase moderately (or even to diminish) following good news.

The models we will consider in this chapter aim at circumventing some of the above-mentioned limitations of the standard GARCH models. We start by studying a general class of Stochastic Recurrence Equations (SRE) satisfied by the volatility of most first-order GARCH formulations.

4.1 Stochastic Recurrence Equation (SRE)

We have seen that the volatility σ_t of the standard GARCH(1,1) model $\epsilon_t = \sigma_t\eta_t$ satisfies the SRE

$$\sigma_t^2 = \omega + (\alpha\eta_{t-1}^2 + \beta)\sigma_{t-1}^2.$$

This particular SRE can be iterated quite explicitly, and this was used to derive the strictly stationary solution in closed form (see the proof of Theorem 2.1). In this chapter, we consider volatility models

[1] Recall, however, that for a noise which is conditionally heteroscedastic, the valid asymptotic bounds at the 95% significance level are not $\pm 1.96/\sqrt{n}$ (see Chapter 5).

[2] When the price of a stock falls, the debt–equity ratio of the company increases. This entails an increase of the risk and hence of the volatility of the stock. When the price rises, the volatility also increases, but by a smaller amount.

satisfying other SREs of the form $\sigma_t^2 = f(\eta_{t-1}, \sigma_{t-1}^2)$. In this section, we begin by stating a general result that will be useful to obtain conditions for the existence of a stationary solution of the form $\sigma_t^2 = \varphi(\eta_{t-1}, \eta_{t-2}, \ldots)$ and for the invertibility of the volatility filter, i.e. the possibility to write the volatility $\sigma_t^2 = \psi(\epsilon_{t-1}, \epsilon_{t-2}, \ldots)$ as a function of the past observations.

Let E and F be two closed intervals of $\mathbb{R}$, let $(X_t)_{t\in\mathbb{Z}}$ be a stationary and ergodic process valued in E, and let $g : E\times F \to F$ a function such that $y \mapsto g(x, y)$ is Lipschitz continuous for all $x \in E$. Set

$$\Lambda_t = \sup_{y_1, y_2 \in F, y_1 \neq y_2} \left| \frac{g(X_t, y_1) - g(X_t, y_2)}{y_1 - y_2} \right|.$$

The following result can be seen as a particular case of a much more general theory developed in Bougerol (1993) and Straumann and Mikosch (2006).

Lemma 4.1 *Assume the following two conditions hold: (i) there exists a constant* $c \in F$ *such that* $E \ln^+ |g(X_1, c) - c| < \infty$*, and (ii)* $E \ln^+ \Lambda_1 < \infty$ *and* $E \ln \Lambda_1 < 0$*. Then there exists a unique stationary and ergodic solution* $(h_t)_{t\in\mathbb{Z}}$ *to the equation*

$$h_t = g(X_t, h_{t-1}), \quad \forall t \in \mathbb{Z} \tag{4.3}$$

Moreover, this solution has the form $h_t = \psi(X_t, X_{t-1}, \ldots)$*, where* ψ *is a measurable function. Consider the case where (iii) the* X_t*'s are independent,* $E|g(X_1, c) - c|^r < \infty$ *and* $E\Lambda_1^r < \infty$ *for some* $r > 0$ *and some* $c \in F$*. Under (iii), the stationary solution satisfies*

$$E|h_t|^s < \infty \quad \textit{for some } s > 0. \tag{4.4}$$

For all $t \geq 0$ *and* $h \in F$*, let us define* $\hat{h}_t(h)$ *recursively by*

$$\hat{h}_t(h) = g\{X_t, \hat{h}_{t-1}(h)\}, \quad \forall t \geq 1 \tag{4.5}$$

and $\hat{h}_0(h) = h$*. Under (i) and (ii), there exists* $\rho \in (0, 1)$ *such that* $\rho^{-t} \mid \hat{h}_t(h) - h_t \mid \to 0$ *a.s. as* $t \to \infty$*.*

Proof. For all $t \in \mathbb{Z}$ and $n \in \mathbb{N}$, let

$$h_{t,n} = g(X_t, h_{t-1,n-1}) \tag{4.6}$$

with $h_{t,0} = c$.

Note that

$$h_{t,n} = \psi_n(X_t, X_{t-1}, \ldots, X_{t-n+1}),$$

for some measurable function $\psi_n : (E^n, \mathcal{B}_{E^n}) \to (F, \mathcal{B}_F)$, with the usual notation. For all n, the sequence $(h_{t,n})_{t\in\mathbb{Z}}$ is thus stationary and ergodic (see Theorem A.1). If for all t, the limit $h_t = \lim_{n\to\infty} h_{t,n}$ exists a.s., then by taking the limit of both sides of Eq. (4.6), it can be seen that the process (h_t) is solution of Eq. (4.3). When it exists, the limit is a measurable function of the form $h_t = \psi(X_t, X_{t-1}, \ldots)$,[3] and is therefore stationary and ergodic. To prove the existence of $\lim_{n\to\infty} h_{t,n}$, we will show that, a.s., $(h_{t,n})_{n\in\mathbb{N}}$ is a Cauchy sequence in the complete space F.

We have

$$\left| \frac{h_{t,n} - h_{t,n-1}}{h_{t-1,n-1} - h_{t-1,n-2}} \right| = \left| \frac{g(X_t, h_{t-1,n-1}) - g(X_t, h_{t-1,n-2})}{h_{t-1,n-1} - h_{t-1,n-2}} \right| \leq \Lambda_t.$$

If follows that

$$|h_{t,n} - h_{t,n-1}| \leq \Lambda_t |h_{t-1,n-1} - h_{t-1,n-2}| \leq \Lambda_t \Lambda_{t-1} \cdots \Lambda_{t-n+2} |g(X_{t-n+1}, c) - c|.$$

[3] For the measurability of h_t, one can consider $h_{t,n}$ as functions of $(X_t, X_{t-1}, \ldots)$ and argue that a limit of measurable functions is measurable.

For $n < m$, we thus have

$$|h_{t,m} - h_{t,n}| \leq \sum_{k=0}^{m-n-1} |h_{t,m-k} - h_{t,m-k-1}|$$
$$\leq \sum_{k=0}^{m-n-1} \Lambda_t\Lambda_{t-1} \cdots \Lambda_{t-m+k+2}|g(X_{t-m+k+1}, c) - c|$$
$$\leq \sum_{j=n}^{\infty} \Lambda_t\Lambda_{t-1} \cdots \Lambda_{t-j+1}|g(X_{t-j}, c) - c|. \tag{4.7}$$

Note that

$$\limsup_{j\to\infty} \ln (\Lambda_t\Lambda_{t-1} \cdots \Lambda_{t-j+1}|g(X_{t-j}, c) - c|)^{1/j}$$
$$= \limsup_{j\to\infty} \frac{1}{j} \sum_{k=1}^{j} \ln \Lambda_{t-k+1} + \frac{\ln|g(X_{t-j}, c) - c|}{j} \leq E \ln \Lambda_1$$

under (i) and the first condition of (ii), by using the ergodic theorem and Exercise 4.12. We conclude, from the Cauchy criterion for the convergence of series with positive terms, that

$$\sum_{j=1}^{\infty} \Lambda_t\Lambda_{t-1} \cdots \Lambda_{t-j+1}|g(X_{t-j}, c) - c|$$

is a.s. finite, under (i) and (ii). It follows that $(h_{t,n})_{n\in\mathbb{N}}$ is a.s. a Cauchy sequence in F. The existence of a stationary and ergodic solution to Eq. (4.3) follows.

Assume that there exists another stationary process (h_t^*) such that $h_t^* = g(X_t, h_{t-1}^*)$. For all $N \geq 0$, we have

$$| h_t - h_t^* | \leq \Lambda_t\Lambda_{t-1} \cdots \Lambda_{t-N} | h_{t-N} - h_{t-N}^* | . \tag{4.8}$$

Since $\Lambda_t\Lambda_{t-1} \cdots \Lambda_{t-N} \to 0$ a.s. as $N \to \infty$, and $| h_{t-N} - h_{t-N}^* | = O_P(1)$ by stationarity, the right-hand side of (4.8) tends to zero in probability. Since the left-hand side does not depend on N, we have $P(| h_t - h_t^* | > \epsilon) = 0$ for all $\epsilon > 0$, and thus $P(h_t = h_t^*) = 1$, which establishes the uniqueness.

Note that (iii) is a stronger condition than (i)–(ii). In view of (4.7), we have

$$|h_t - c| \leq |g(X_t, c) - c| + \sum_{j=1}^{\infty} \Lambda_t\Lambda_{t-1} \cdots \Lambda_{t-j+1}|g(X_{t-j}, c) - c|.$$

By the arguments used to prove Corollary 2.3, we thus show that (iii) entails $E|h_t - c|^s$ for some $s > 0$, and (4.4) follows.

Now, note that

$$| h_t - \hat{h}_t(h) | \leq \Lambda_t\Lambda_{t-1} \cdots \Lambda_1 | h_0 - h | .$$

To conclude, it suffices to take ρ such that

$$1 > e^{E \ln \Lambda_1} > \rho > 0,$$

and to remark that

$$\lim_{t\to\infty} \frac{1}{t} \ln \rho^{-t}\Lambda_t\Lambda_{t-1} \cdots \Lambda_1 = -\ln \rho + E \ln \Lambda_1 < 0$$

under (ii). □

Remark 4.1 (Application to the GARCH(1,1)) Let (η_t) be a stationary and ergodic sequence (not necessarily a strong white noise, as in the standard GARCH(1,1)). Let us apply Lemma 4.1 to find a stationarity condition for the equation $\sigma_t^2 = \omega + (\alpha\eta_t^2 + \beta)\sigma_{t-1}^2$, with the usual positivity constraints on the coefficients. Considering the function $g(x, y) = \omega + (\alpha x + \beta)y$, we have $\Lambda_t = \alpha\eta_t^2 + \beta$ and we retrieve the stationarity conditions (i) $E\ln^+ |\eta_t| < \infty$ and (ii) $\gamma := E\ln(\alpha\eta_t^2 + \beta) < 0$ (the first condition ensures that γ is well defined in $[-\infty, \infty)$ and is always satisfied under the usual assumption $E\eta_t^2 = 1$).

Note that the volatility of the stationary solution (ϵ_t) of a GARCH(1,1) also satisfies the SRE $\sigma_t^2 = g(\epsilon_{t-1}^2, \sigma_{t-1}^2)$ with $g(x, y) = \omega + \alpha x + \beta y$. We then have $\Lambda_t = \beta$, and the condition $\beta < 1$ ensures the invertibility of the model.

Remark 4.2 (Application to other GARCH-type models) For the standard GARCH model considered in the previous remark, Lemma 4.1 is not required since the stationarity and invertibility conditions can be found by elementary arguments. By contrast, Lemma 4.1 will be extremely useful to obtain stationary conditions for the QGARCH model of Section 4.6 and invertibility conditions for the exponential GARCH (EGARCH) model of Section 4.2.

4.2 Exponential GARCH Model

The following definition for the EGARCH model mimics that given for the strong GARCH.

Definition 4.1 (EGARCH(*p*, *q*) process) *Let (η_t) be an iid sequence such that $E(\eta_t) = 0$ and $Var(\eta_t) = 1$. Then (ϵ_t) is said to be an EGARCH(p, q) process if it satisfies an equation of the form*

$$\begin{cases} \epsilon_t & = \sigma_t\eta_t \\ \ln\sigma_t^2 & = \omega + \sum_{i=1}^q \alpha_i g(\eta_{t-i}) + \sum_{j=1}^p \beta_j \ln\sigma_{t-j}^2, \end{cases} \tag{4.9}$$

where $\sigma_t > 0$,

$$g(\eta_{t-i}) = \theta\eta_{t-i} + \varsigma(|\eta_{t-i}| - E|\eta_{t-i}|), \tag{4.10}$$

and ω, α_i, β_j, θ and ς are real numbers.

Remark 4.3 (On the EGARCH model)

1. The relation

$$\sigma_t^2 = e^{\omega} \prod_{i=1}^{q} \exp\{\alpha_i g(\eta_{t-i})\} \prod_{j=1}^{p} (\sigma_{t-j}^2)^{\beta_j}$$

shows that, in contrast to the classical GARCH, the volatility has a multiplicative dynamics. The positivity constraints on the coefficients can be avoided, because the logarithm can be of any sign.

2. According to the usual interpretation, however, innovations of large modulus should increase volatility. Assume for instance that $\alpha_1 = 1$. When $\eta_{t-1} < 0$ (that is, when $\epsilon_{t-1} < 0$), a decrease of 1 unit of η_{t-1}, the other variables being unchanged, entails a change of $\varsigma - \theta$ for the variable $\ln\sigma_t^2$. When $\epsilon_{t-1} > 0$, an increase of 1 unit of η_{t-1} entails a change of $\varsigma + \theta$ on the variable $\ln\sigma_t^2$. Therefore, σ_t^2 increases with $|\eta_{t-1}|$, the sign of η_{t-1} being fixed, if and only if $\varsigma - \theta > 0$ and $\varsigma + \theta > 0$, that is if $-\varsigma < \theta < \varsigma$. In the general case, it suffices to impose

$$-\varsigma < \theta < \varsigma, \quad \alpha_i \geq 0, \quad \beta_j \geq 0.$$

3. The asymmetry property is taken into account through the coefficient θ. We have just seen that, when $\alpha_1 = 1$, the effect of a large negative shock (i.e. $\varsigma - \theta$) is larger than the effect of a large positive shock (i.e. $\varsigma + \theta$) iff $\theta < 0$. Thus, we obtain the typical asymmetry property of financial time series.

4. Another difference from the classical GARCH is that the conditional variance is written as a function of the past standardized innovations (that is, divided by their conditional standard deviation), instead of the past innovations. In particular, $\ln \sigma_t^2$ is a strong ARMA$(p, q - q')$ process, where q' is the first integer i such that $\alpha_i \neq 0$, because $(g(\eta_t))$ is a strong white noise, with variance

$$\text{Var}[g(\eta_t)] = \theta^2 + \varsigma^2 \text{Var}(|\eta_t|) + 2\theta\varsigma \text{Cov}(\eta_t, |\eta_t|).$$

5. The specification (4.4) allows for sign effects through $\theta\eta_{t-i}$ and for modulus effects through $\varsigma(|\eta_{t-i}| - E|\eta_{t-i}|)$. This obviously induces an identifiability problem, which can be solved by setting $\alpha_1 = 1$, or alternatively by setting $\varsigma = 1$ in the case $q = 1$. Note also that, to allow different sign effects for the different lags, one could make θ depend on the lag index i, through the formulation

$$\begin{cases} \epsilon_t & = \sigma_t \eta_t \\ \ln \sigma_t^2 & = \omega + \sum_{i=1}^q \alpha_i \{\theta_i \eta_{t-i} + (|\eta_{t-i}| - E|\eta_{t-i}|)\} + \sum_{j=1}^p \beta_j \ln \sigma_{t-j}^2. \end{cases} \tag{4.11}$$

Stationarity and Existence of Moments

As we have seen, specifications of the function $g(\cdot)$ that are different from (4.10) are possible, depending on the kind of empirical properties we are trying to mimic. The following result does not depend on the specification chosen for $g(\cdot)$. It is, however, assumed that $Eg(\eta_t)$ exists and is equal to 0.

Theorem 4.1 (Stationarity of the EGARCH(p, q) process) *Assume that $g(\eta_t)$ is not a.s. equal to zero and that the polynomials $\alpha(z) = \sum_{i=1}^q \alpha_i z^i$ and $\beta(z) = 1 - \sum_{i=1}^p \beta_i z^i$ have no common root, with $\alpha(z)$ not identically null. Then, the EGARCH(p, q) model defined in (4.9) admits a strictly stationary and non-anticipative solution if and only if the roots of $\beta(z)$ are outside the unit circle. This solution is such that $E(\ln \epsilon_t^2)^2 < \infty$ whenever $E(\ln \eta_t^2)^2 < \infty$ and $Eg^2(\eta_t) < \infty$.*

If, in addition,

$$\prod_{i=1}^{\infty} E \exp\{|\lambda_i g(\eta_t)|\} < \infty, \tag{4.12}$$

where λ_i are defined by $\alpha(L)/\beta(L) = \sum_{i=1}^{\infty} \lambda_i L^i$, then (ϵ_t) is a white noise with variance

$$E(\epsilon_t^2) = E(\sigma_t^2) = e^{\omega^*} \prod_{i=1}^{\infty} g_\eta(\lambda_i),$$

where $\omega^* = \omega/\beta(1)$ and $g_\eta(x) = E[\exp\{xg(\eta_t)\}]$.

Proof. We have $\ln \epsilon_t^2 = \ln \sigma_t^2 + \ln \eta_t^2$. Because $\ln \sigma_t^2$ is the solution of an ARMA $(p, q - 1)$ model, with AR polynomial β, the assumptions made on the lag polynomials are necessary and sufficient to express, in a unique way, $\ln \sigma_t^2$ as an infinite-order moving average:

$$\ln \sigma_t^2 = \omega^* + \sum_{i=1}^{\infty} \lambda_i g(\eta_{t-i}), \quad \text{a.s.}$$

It follows that the processes $(\ln \sigma_t^2)$ and $(\ln \epsilon_t^2)$ are strictly stationary. The process $(\ln \sigma_t^2)$ is second-order stationary and, under the assumption $E(\ln \eta_t^2)^2 < \infty$, so is $(\ln \epsilon_t^2)$. Moreover, using the previous expansion,

$$\epsilon_t^2 = \sigma_t^2 \eta_t^2 = e^{\omega^*} \prod_{i=1}^{\infty} \exp\{\lambda_i g(\eta_{t-i})\} \eta_t^2, \quad \text{a.s.} \tag{4.13}$$

Using the fact that the process $g(\eta_t)$ is iid, we get the desired result on the expectation of (ϵ_t^2) (Exercise 4.4). □

Remark 4.4

1. When $\beta_j = 0$ for $j = 1, \ldots, p$ (EARCH(q) model), the coefficients λ_i cancel for $i > q$. Hence, condition (4.12) is always satisfied, provided that $E\exp\{|\alpha_i g(\eta_t)|\} < \infty$, for $i = 1, \ldots, q$. If the tails of the distribution of η_t are not too heavy (the condition fails for the Student t distributions and specification (4.10)), an EARCH(q) process is then stationary, in both the strict and second-order senses, whatever the values of the coefficients α_i.

2. When η_t is $\mathcal{N}(0,1)$ distributed, and if $g(\cdot)$ is such that (4.10) holds, one can verify (Exercise 4.5) that
$$\ln E\ \exp\{|\lambda_i g(\eta_t)| = O(\lambda_i). \tag{4.14}$$
Since λ_i is obtained from the inversion of the polynomial $\beta(\cdot)$, they decrease exponentially fast to zero. It is then easy to check that (4.12) holds true in this case, without any supplementary assumption on the model coefficients. The strict and second-order stationarity conditions thus coincide, contrary to what happened in the standard GARCH case. To compute the second-order moments, classical integration calculus shows that
$$g_\eta(\lambda_i) = \exp\left\{-\lambda_i\varsigma\sqrt{\frac{2}{\pi}}\right\}\left[\exp\left\{\frac{\lambda_i^2(\theta+\varsigma)^2}{2}\right\}\Phi\{\lambda_i(\theta+\varsigma)\}\right.$$
$$\left.+\exp\left\{\frac{\lambda_i^2(\theta-\varsigma)^2}{2}\right\}\Phi\{\lambda_i(\varsigma-\theta)\}\right],$$
where Φ denotes the cumulative distribution function of the $\mathcal{N}(0,1)$.

Theorem 4.2 (Moments of the EGARCH(p, q) process) *Let m be a positive integer. Under the conditions of Theorem 4.1 and if*
$$\mu_{2m} = E(\eta_t^{2m}) < \infty, \quad \prod_{i=1}^{\infty} E\ \exp\{|m\lambda_i g(\eta_t)|\} < \infty,$$
(ϵ_t^2) *admits a moment of order m given by*
$$E(\epsilon_t^{2m}) = \mu_{2m} e^{m\omega^*} \prod_{i=1}^{\infty} g_\eta(m\lambda_i).$$

Proof. The result straightforwardly follows from (4.13) and Exercise 4.4. □

The previous computation shows that in the Gaussian case, moments exist at any order. This shows that the leptokurticity property may be more difficult to capture with EGARCH than with standard GARCH models.

Assuming that $E(\ln\eta_t^2)^2 < \infty$, the autocorrelation structure of the process $(\ln\epsilon_t^2)$ can be derived by taking advantage of the ARMA form of the dynamics of $\ln\sigma_t^2$. Indeed, replacing the terms in $\ln\sigma_{t-j}^2$ by $\ln\epsilon_{t-j}^2 - \ln\eta_{t-j}^2$, we get
$$\ln\epsilon_t^2 = \omega + \ln\eta_t^2 + \sum_{i=1}^{q}\alpha_i g(\eta_{t-i}) + \sum_{j=1}^{p}\beta_j\ln\epsilon_{t-j}^2 - \sum_{j=1}^{p}\beta_j\ln\eta_{t-j}^2.$$

Let

$$v_t = \ln \epsilon_t^2 - \sum_{j=1}^{p} \beta_j \ln \epsilon_{t-j}^2 = \omega + \ln \eta_t^2 + \sum_{i=1}^{q} \alpha_i g(\eta_{t-i}) - \sum_{j=1}^{p} \beta_j \ln \eta_{t-j}^2.$$

One can easily verify that (v_t) has finite variance. Since v_t only depends on a finite number r ($r = \max(p, q)$) of past values of η_t, it is clear that $\text{Cov}(v_t, v_{t-k}) = 0$ for $k > r$. It follows that (v_t) is an MA(r) process (with intercept) and thus that $(\ln \epsilon_t^2)$ is an ARMA(p, r) process. This result is analogous to that obtained for the classical GARCH models, for which an ARMA(r, p) representation was exhibited for ϵ_t^2. Apart from the inversion of the integers r and p, it is important to note that the noise of the ARMA equation of a GARCH is the strong innovation of the square, whereas the noise involved in the ARMA equation of an EGARCH is generally not the strong innovation of $\ln \epsilon_t^2$. Under this limitation, the ARMA representation can be used to identify the orders p and q and to estimate the parameters β_j and α_i (although the latter do not explicitly appear in the representation).

The autocorrelations of (ϵ_t^2) can be obtained from formula (4.13). Provided the moments exist we have, for $h > 0$,

$$E(\epsilon_t^2 \epsilon_{t-h}^2) = E\left\{ e^{2\omega^*} \prod_{i=1}^{h-1} \exp\{\lambda_i g(\eta_{t-i})\} \eta_t^2 \eta_{t-h}^2 \exp\{\lambda_h g(\eta_{t-h})\} \right.$$
$$\left. \times \prod_{i=h+1}^{\infty} \exp\{(\lambda_i + \lambda_{i-h}) g(\eta_{t-i})\} \right\}$$
$$= e^{2\omega^*} \left\{ \prod_{i=1}^{h-1} g_\eta(\lambda_i) \right\} E(\eta_{t-h}^2 \exp\{\lambda_h g(\eta_{t-h})\}) \prod_{i=h+1}^{\infty} g_\eta(\lambda_i + \lambda_{i-h}),$$

the first product being replaced by 1 if $h = 1$. For $h > 0$, this leads to

$$\text{Cov}(\epsilon_t^2, \epsilon_{t-h}^2) = e^{2\omega^*} \left[\prod_{i=1}^{h-1} g_\eta(\lambda_i) E(\eta_{t-h}^2 \exp\{\lambda_h g(\eta_{t-h})\}) \prod_{i=h+1}^{\infty} g_\eta(\lambda_i + \lambda_{i-h}) \right.$$
$$\left. - \prod_{i=1}^{\infty} \{g_\eta(\lambda_i)\}^2 \right].$$

Invertibility

The existence of a stationary solution (ϵ_t) to a time series model is often obtained by showing that, for some white noise (η_t) and some measurable function ψ,

$$\epsilon_t = \psi(\eta_u, u \leq t).$$

The model is said to be invertible if the reverse holds true, i.e. if

$$\eta_t = \phi(\epsilon_u, u \leq t)$$

for some measurable function ϕ.

Under the stationarity condition given in Theorem 4.1, the model (4.9) expresses $\ln \sigma_t^2$ as a function of the unobserved innovations $\{\eta_u, u < t\}$. To use the EGARCH model in practice (for instance for predicting ϵ_t^2 by a function of $\{\epsilon_u, u < t\}$), it is necessary to be able to express $\ln \sigma_t^2$ as a function of the past observations. Take the example of the EGARCH(1,1) model, which can be rewritten as

$$\begin{cases} \epsilon_t &= \sigma_t \eta_t \\ \ln \sigma_t^2 &= \omega + \theta \eta_{t-1} + \varsigma \mid \eta_{t-1} \mid + \beta \ln \sigma_{t-1}^2. \end{cases} \tag{4.15}$$

Replacing η_t by $\epsilon_t/\sigma_t = \epsilon_t e^{-(\ln \sigma_t^2)/2}$, one can see that the EGARCH(1,1) satisfies the SRE

$$\ln \sigma_t^2 = g(\epsilon_{t-1}, \ln \sigma_{t-1}^2), \quad g(x,y) = \omega + (\theta x + \varsigma \mid x \mid)e^{-\frac{y}{2}} + \beta y.$$

We will thus apply Lemma 4.1. Assuming $\beta \in [0, 1)$ and $-\varsigma < \theta < \varsigma$, which appear as reasonable restrictions for the applications (see 2. of Remark 4.3), $\ln \sigma_t^2$ belongs to the interval $F = [\omega/(1-\beta), \infty)$. We also have $\theta x + \varsigma \mid x \mid \geq 0$ and

$$\sup_{y \in F} \left| \frac{\partial}{\partial y} g(x,y) \right| = \sup_{y \in F} \left| -\frac{1}{2}(\theta x + \varsigma \mid x \mid)e^{-y/2} + \beta \right|$$
$$= \max \left\{ \beta, \frac{1}{2}(\theta x + \varsigma \mid x \mid)e^{-\frac{\omega}{2(1-\beta)}} - \beta \right\}.$$

Lemma 4.1 then entails the following result, due to Straumann and Mikosch (2006).

Theorem 4.3 (Invertibility of the EGARCH(1, 1) model) *Let (ϵ_t) be the stationary and ergodic solution to the EGARCH(1,1) model (4.15) with $\beta \in [0, 1)$ and $-\varsigma < \theta < \varsigma$. For $h \in F = [\omega/(1-\beta), \infty)$, let $\hat{h}_1(h) = h$ and $\hat{h}_t(h) = \ln \hat{\sigma}_t^2(h) = \omega + (\theta \epsilon_{t-1} + \varsigma \mid \epsilon_{t-1} \mid)e^{-\frac{\hat{h}_{t-1}(h)}{2}} + \beta \hat{h}_{t-1}(h)$, for $t = 2, 3, \ldots$*

Under the condition

$$E \ln \max \left\{ \beta, \frac{1}{2}(\theta \epsilon_1 + \varsigma \mid \epsilon_1 \mid)e^{-\frac{\omega}{2(1-\beta)}} - \beta \right\} < 0, \tag{4.16}$$

there exists $\rho \in (0, 1)$ such that $\rho^{-t}|\sigma_t^2 - \hat{\sigma}_t^2(h)| \to 0$ a.s. as $t \to \infty$.

To illustrate the importance of the invertibility issue, consider the EGARCH(1,1) model (4.15), with η_t iid $\mathcal{N}(0, 1)$ and two parameter sets:

$$\text{Design A} : \omega = -0.3, \ \theta = -0.02, \ \varsigma = 0.20, \ \beta = 0.97$$
$$\text{Design B} : \omega = -4, \ \theta = -0.02, \ \varsigma = 5.8, \ \beta = 0.2.$$

The two models are stationary because $|\beta| < 1$. Figure 4.1 shows that, for Design A (a), the initial value required for computing recursively the approximation of the volatility has no effect asymptotically.

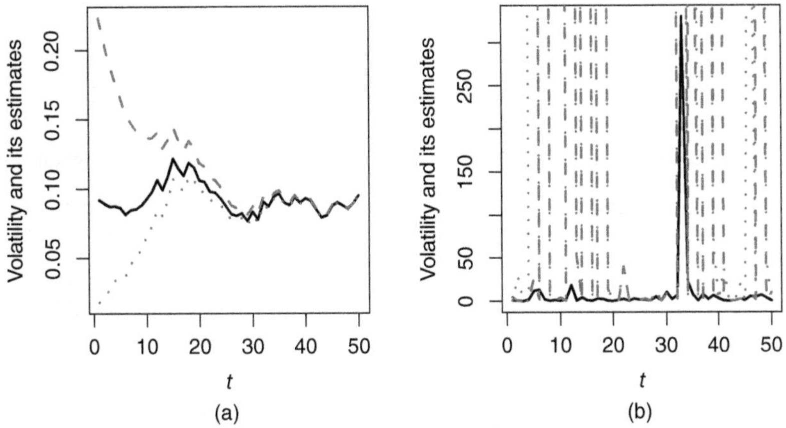

Figure 4.1 Volatility σ_t^2 (in full line) and volatility estimates $\hat{\sigma}_t^2(h)$ (in dashed and dotted lines) for two different initial values h, when the simulated EGARCH model is invertible (Design A, a) or non-invertible (Design B, b).

In that case, the volatility at time t, when t is large, can be accurately approximated by a function of the available observations $\epsilon_1, \dots, \epsilon_{t-1}$. The situation is dramatically different in Design B (b). The estimate $\hat{\sigma}_t^2(h)$ varies considerably with h and may be completely different from $\sigma_t^2(h)$ (for visibility reasons, we have not been able to represent the more extreme values of $\hat{\sigma}_t^2(h)$ in the graph), even for large t.

It must be emphasised that, even if an EGARCH model with parameters as in Design B is a stationary and ergodic process, it cannot be used to recover the volatility from the past observations. In this case, such a model is not very useful. In practice, when an EGARCH is fitted on a real series, it is thus important to assess if the estimated model corresponds to an invertible model (for instance by evaluating the left-hand side of Eq. (4.16) using Monte Carlo simulations, and/or by studying how $\hat{\sigma}_t^2(h)$ varies with t and h). The invertibility is also crucial to be able to consistently estimate the parameter of an EGARCH (see Wintenberger 2013).

4.3 Log-GARCH Model

A formulation which, at first sight, seems very close to the EGARCH is the Log-GARCH, initially defined as follows.

Definition 4.2 (Standard Log-GARCH(p, q) process) *Let (η_t) be an iid sequence of random variables such that $E(\eta_t) = 0$ and $Var(\eta_t) = 1$. Then (ϵ_t) is called a Log-GARCH(p, q) process if it satisfies an equation of the form*

$$\begin{cases} \epsilon_t &= \sigma_t \eta_t \\ \ln \sigma_t^2 &= \omega + \sum_{i=1}^q \alpha_i \ln \epsilon_{t-i}^2 + \sum_{j=1}^p \beta_j \ln \sigma_{t-j}^2, \end{cases} \tag{4.17}$$

where ω, α_i and β_i are real numbers.

Remark 4.5 (On the Log-GARCH model)

1. As for the EGARCH, the dynamics being specified on the log-volatility, which may take positive and negative values, it is not necessary to impose positivity constraints on the coefficients.

2. If the polynomial $1 - \sum_{j=1}^p \beta_j z^j$ has its roots outside the unit circle, one can write the volatility as an explicit function of the past observations:

$$\ln \sigma_t^2 = \frac{\omega}{1 - \sum_{j=1}^p \beta_j} + \sum_{i=1}^{\infty} c_i \ln \epsilon_{t-i}^2, \quad \text{where} \quad \sum_{k=1}^{\infty} c_k z^k = \frac{\sum_{i=1}^q \alpha_i z^i}{1 - \sum_{j=1}^p \beta_j z^j}, \mid z \mid \leq 1.$$

   Therefore, contrary to the EGARCH, the invertibility of the Log-GARCH is not an issue.

3. Definition (4.17) requires $\epsilon_{t-i}^2 \neq 0$. Therefore, the Log-GARCH model of Definition 4.2 is incompatible with a series of returns that contains, or may contain, zeroes. Consequently, it is necessary to assume that $P(\eta_t = 0) = 0$.

4. Model (4.17) is not able to take into account the typical asymmetry properties of the stock returns since the volatility does not depend on the sign of the past returns.

To cope with the last two points, one can consider the following asymmetric extension of the Log-GARCH, which nests the standard Log-GARCH as a special case.

Definition 4.3 (Extended Log-GARCH(p, q) process) *Let (η_t) be an iid sequence of random variables such that $E(\eta_t) = 0$ and $Var(\eta_t) = 1$. Then (ϵ_t) is called Extended Log-GARCH(p, q) process if it* *satisfies an equation of the form*

$$\begin{cases} \epsilon_t &= \sigma_t \eta_t \\ \ln \sigma_t^2 &= \omega_t + \sum_{i=1}^q \alpha_{i,t} \ln \epsilon_{t-i}^2 + \sum_{j=1}^p \beta_j \ln \sigma_{t-j}^2 \end{cases} \tag{4.18}$$

with $\omega_t = \omega + \sum_{i=1}^q \omega_{i-} \mathbb{1}_{\{\epsilon_{t-i}<0\}}$ *and* $\alpha_{i,t} = \alpha_{i+} \mathbb{1}_{\{\epsilon_{t-i}>0\}} + \alpha_{i-} \mathbb{1}_{\{\epsilon_{t-i}<0\}}$, *where* ω, ω_{i-}, α_{i+}, α_{i-}, *and* β_i *are real numbers, and* $\sigma_t > 0$.

Remark 4.6 (On the Extended Log-GARCH model)

1. In (4.18), by making the convention

$$0 \times \ln \epsilon_{t-i}^2 = 0$$

whatever the value of ϵ_{t-i} (i.e. even if $\epsilon_{t-i} = 0$), the model a priori accommodates series (ϵ_t) that contain zeroes, contrary to the standard Log-GARCH (4.17). Therefore, model (4.18) does not require the assumption $P(\eta_t = 0) = 0$.

2. The model (4.18) is stable by scaling, in the sense that if (ϵ_t) is an extended Log-GARCH process, then for any $c \neq 0$, the process $(c\epsilon_t)$ is also an extended Log-GARCH process. This stability-by-scaling property also holds true for the sub-model $\ln \sigma_t^2 = \omega_t + \sum_{i=1}^q \alpha_i \ln \epsilon_{t-i}^2 + \sum_{j=1}^p \beta_j \ln \sigma_{t-j}^2$, but it does not hold for the model $\ln \sigma_t^2 = \omega + \sum_{i=1}^q \alpha_{i,t} \ln \epsilon_{t-i}^2 + \sum_{j=1}^p \beta_j \ln \sigma_{t-j}^2$. Indeed, in the latter case, we have

$$\begin{cases} c\epsilon_t = c\sigma_t \eta_t \\ \ln c^2 \sigma_t^2 = \omega_t + \sum_{i=1}^q \alpha_{i,t} \ln c^2 \epsilon_{t-i}^2 + \sum_{j=1}^p \beta_j \ln c^2 \sigma_{t-j}^2 \end{cases}$$

where

$$\omega_t = \ln c^2 + \omega - \sum_{i=1}^q \alpha_{i,t} \ln c^2 - \sum_{j=1}^p \beta_j \ln c^2$$

is not constant, in general.

4.3.1 Stationarity of the Extended Log-GARCH Model

First consider the case $p = q = 1$, which can be handled more explicitly than the general case. Write ω_-, α_+, α_-, and β instead of ω_{1-}, α_{1+}, α_{1-}, and β_1. The volatility of the model can be rewritten as

$$\begin{aligned} \ln \sigma_t^2 &= \omega + \alpha_+ \mathbb{1}_{\{\epsilon_{t-1} \neq 0\}} \ln \epsilon_{t-1}^2 + \beta \ln \sigma_{t-1}^2 \\ &\quad + (\alpha_- - \alpha_+) \mathbb{1}_{\{\epsilon_{t-1}<0\}} \ln \epsilon_{t-1}^2 + \omega_- \mathbb{1}_{\{\epsilon_{t-1}<0\}}. \end{aligned}$$

Assume that

$$E \ln^+ \mid \mathbb{1}_{\{\eta_t \neq 0\}} \ln \eta_t^2 \mid < \infty, \tag{4.19}$$

and

$$|\alpha_+ + \beta|^{\pi_+} |\alpha_- + \beta|^{\pi_-} |\beta|^{\pi_0} < 1 \tag{4.20}$$

where $\pi_+ = P(\eta_0 > 0)$, $\pi_- = P(\eta_0 < 0)$, and $\pi_0 = P(\eta_0 = 0)$. Recall that, in our notations, we make the convention that $0 \times \ln x = 0$ even if $x = 0$. Therefore, the case $P(\eta_t = 0) > 0$ is not precluded by Eq. (4.19). Let

$$\omega_t = \omega + \omega_- \mathbb{1}_{\{\eta_{t-1}<0\}}, \quad \alpha_t = \alpha_+ \mathbb{1}_{\{\eta_{t-1}>0\}} + \alpha_- \mathbb{1}_{\{\eta_{t-1}<0\}}$$

and σ_t^2 defined by

$$\ln \sigma_t^2 = \omega_t + \alpha_t \ln \eta_{t-1}^2 + \sum_{j=0}^{\infty} (\omega_{t-j-1} + \alpha_{t-j-1} \ln \eta_{t-j-2}^2) \prod_{i=0}^{j} (\alpha_{t-i} + \beta). \tag{4.21}$$

By the Cauchy rule, the series (4.21) converges absolutely with probability one because

$$E \ln \mid \alpha_t + \beta \mid = \ln |\beta + \alpha_+|^{\pi_+} |\beta + \alpha_-|^{\pi_-} |\beta|^{\pi_0} < 0$$

under (4.20), and

$$\limsup_{j\to\infty} \frac{\ln \mid \omega_{t-j} + \alpha_{t-j} \ln \eta_{t-j-1}^2 \mid}{j} \leq 0.$$

The last inequality comes from the fact that $E \ln^+ \mid \omega_t + \alpha_t \ln \eta_{t-1}^2 \mid < \infty$ (see Exercise 4.12), which is implied by (4.19). Now note that the process $\epsilon_t = \sigma_t \eta_t$, where σ_t^2 is defined by Eq. (4.21) is an extended Log-GARCH(1,1) process of the form (4.18), because $\{\epsilon_t = 0\}$ iff $\{\eta_t = 0\}$ and $\{\epsilon_t > 0\}$ iff $\{\eta_t > 0\}$.

Now consider the general extended Log-GARCH(p, q) process. Because coefficients equal to zero can always be added, it is not restrictive to assume $p > 1$ and $q > 1$. Let the vectors

$$\begin{aligned}
\boldsymbol{\omega}_- &= (\omega_{1-}, \ldots, \omega_{q-})' \in \mathbb{R}^q, \quad \boldsymbol{\alpha}_+ = (\alpha_{1+}, \ldots, \alpha_{q+})' \in \mathbb{R}^q, \\
\boldsymbol{\alpha}_- &= (\alpha_{1-}, \ldots, \alpha_{q-})' \in \mathbb{R}^q, \quad \boldsymbol{\beta} = (\beta_1, \ldots, \beta_p)' \in \mathbb{R}^p, \\
\boldsymbol{\epsilon}_{t,q}^+ &= (\mathbb{1}_{\{\epsilon_t>0\}} \ln \epsilon_t^2, \ldots, \mathbb{1}_{\{\epsilon_{t-q+1}>0\}} \ln \epsilon_{t-q+1}^2)' \in \mathbb{R}^q, \\
\boldsymbol{\epsilon}_{t,q}^- &= (\mathbb{1}_{\{\epsilon_t<0\}} \ln \epsilon_t^2, \ldots, \mathbb{1}_{\{\epsilon_{t-q+1}<0\}} \ln \epsilon_{t-q+1}^2)' \in \mathbb{R}^q, \\
\boldsymbol{z}_t &= (\boldsymbol{\epsilon}_{t,q}^+, \boldsymbol{\epsilon}_{t,q}^-, \ln \sigma_t^2, \ldots, \ln \sigma_{t-p+1}^2)' \in \mathbb{R}^{2q+p}, \\
\boldsymbol{b}_t &= ((\omega_t + \ln \eta_t^2)\mathbb{1}_{\{\eta_t>0\}}, \mathbf{0}_{q-1}', (\omega_t + \ln \eta_t^2)\mathbb{1}_{\{\eta_t<0\}}, \mathbf{0}_{q-1}', \omega_t, \mathbf{0}_{p-1}')' \in \mathbb{R}^{2q+p},
\end{aligned}$$

and the matrix

$$\boldsymbol{C}_t = \begin{pmatrix}
\mathbb{1}_{\{\eta_t>0\}}\boldsymbol{\alpha}_+' & \mathbb{1}_{\{\eta_t>0\}}\boldsymbol{\alpha}_-' & \mathbb{1}_{\{\eta_t>0\}}\boldsymbol{\beta}' \\
\boldsymbol{I}_{q-1} \quad \mathbf{0}_{q-1} & \mathbf{0}_{(q-1)\times q} & \mathbf{0}_{(q-1)\times p} \\
\mathbb{1}_{\{\eta_t<0\}}\boldsymbol{\alpha}_+' & \mathbb{1}_{\{\eta_t<0\}}\boldsymbol{\alpha}_-' & \mathbb{1}_{\{\eta_t<0\}}\boldsymbol{\beta}' \\
\mathbf{0}_{(q-1)\times q} & \boldsymbol{I}_{q-1} \quad \mathbf{0}_{q-1} & \mathbf{0}_{(q-1)\times p} \\
\boldsymbol{\alpha}_+' & \boldsymbol{\alpha}_-' & \boldsymbol{\beta}' \\
\mathbf{0}_{(p-1)\times q} & \mathbf{0}_{(p-1)\times q} & \boldsymbol{I}_{p-1} \quad \mathbf{0}_{p-1}
\end{pmatrix},$$

where $\boldsymbol{I}_k$ denote the $k \times k$ identity matrix. Model (4.18) is rewritten in matrix form as

$$\boldsymbol{z}_t = \boldsymbol{C}_t \boldsymbol{z}_{t-1} + \boldsymbol{b}_t. \tag{4.22}$$

Let γ be the top Lyapunov exponent of the sequence $\{\boldsymbol{C}_t, t \in \mathbb{Z}\}$,

$$\gamma = \lim_{t\to\infty} a.s. \frac{1}{t} \ln \| \boldsymbol{C}_t \boldsymbol{C}_{t-1} \ldots \boldsymbol{C}_1 \| = \inf_{t\geq 1} \frac{1}{t} E(\ln \| \boldsymbol{C}_t \boldsymbol{C}_{t-1} \ldots \boldsymbol{C}_1 \|). \tag{4.23}$$

Theorem 4.4 (Strict stationarity of the Extended Log-GARCH(p, q)) *Assume (4.19). A sufficient condition for the existence of a strictly stationary and non-anticipative solution of the extended Log-GARCH(p, q) model (4.18) is $\gamma < 0$, where γ is defined by (4.23).*

This stationary and non-anticipative solution, when $\gamma < 0$, is unique and ergodic.

Proof. Since the random variable $\|C_0\|$ is bounded, we have $E\ln^+\|C_0\| < \infty$. The moment condition (4.19) entails that we also have $E\ln^+\|b_0\| < \infty$. When $\gamma < 0$, Cauchy's root test shows that, a.s., the series

$$z_t = b_t + \sum_{n=0}^{\infty} C_t C_{t-1} \dots C_{t-n} b_{t-n-1} \tag{4.24}$$

converges absolutely for all t and satisfies Eq. (4.22). A strictly stationary solution to model (4.18) is then obtained as $\epsilon_t = \exp\left\{\frac{1}{2}z_{2q+1,t}\right\}\eta_t$, where $z_{i,t}$ denotes the ith element of z_t. This solution is non-anticipative and ergodic, as a measurable function of $\{\eta_u, u \le t\}$.

We now prove that (4.24) is the unique non-anticipative solution of Eq. (4.22) when $\gamma < 0$. Let (z_t^*) be a strictly stationary process satisfying $z_t^* = C_t z_{t-1}^* + b_t$. For all $N \ge 0$,

$$z_t^* = z_t(N) + C_t \dots C_{t-N} z_{t-N-1}^*, \quad z_t(N) = b_t + \sum_{n=0}^{N} C_t C_{t-1} \dots C_{t-n} b_{t-n-1}.$$

We then have

$$\| z_t - z_t^* \| \le \left| \sum_{n=N+1}^{\infty} C_t C_{t-1} \dots C_{t-n} b_{t-n-1} \right| + \| C_t \dots C_{t-N} \| \| z_{t-N-1}^* \| .$$

The first term in the right-hand side tends to 0 a.s. when $N \to \infty$. The second term tends to 0 in probability because, in view of the first equality in (4.23), $\gamma < 0$ entails that $\|C_t \dots C_{t-N}\| \to 0$ a.s. and the distribution of $\| z_{t-N-1}^* \|$ is independent of N by stationarity. We have shown that $z_t - z_t^* \to 0$ in probability when $N \to \infty$. This quantity being independent of N, we have $z_t = z_t^*$ a.s. for any t. □

Example 4.1 (The Log-GARCH(1,1) case) In the case $p = q = 1$, we still have Eq. (4.22) with obvious modifications of the definitions of z_t, C_t and b_t. We have

$$C_t C_{t-1} \dots C_1 = \begin{pmatrix} \mathbb{1}_{\{\eta_t > 0\}} \\ \mathbb{1}_{\{\eta_t < 0\}} \\ 1 \end{pmatrix} \begin{pmatrix} \alpha_+ & \alpha_- & \beta \end{pmatrix} \prod_{i=1}^{t-1} (\alpha_+ \mathbb{1}_{\{\eta_i > 0\}} + \alpha_- \mathbb{1}_{\{\eta_i < 0\}} + \beta)$$

and

$$\gamma = E\ln|\alpha_+ \mathbb{1}_{\{\eta_0 > 0\}} + \alpha_- \mathbb{1}_{\{\eta_0 < 0\}} + \beta| = \ln(|\alpha_+ + \beta|^{\pi_+} |\alpha_- + \beta|^{\pi_-} |\beta|^{\pi_0}).$$

The condition $\gamma < 0$ is thus equivalent to (4.20).

Remark 4.7 (The condition $\gamma < 0$ is not necessary) Assume, for instance $p = q = 1$ with $\omega_{1-} = 0$, $\alpha_+ = \alpha_- = \alpha$, and $P(\eta_1 = 0) = 0$. In that case, $\gamma < 0$ is equivalent to $|\alpha + \beta| < 1$. In addition, assume that $\eta_0^2 = 1$ a.s. Then when $\alpha + \beta \ne 1$, there exists a stationary solution to model (4.18) defined by $\epsilon_t = \exp(c/2)\eta_t$, with $c = \omega/(1 - \alpha - \beta)$.

Example 4.2 (Another particular case) In the case $\boldsymbol{\alpha}_+ = \boldsymbol{\alpha}_- = \boldsymbol{\alpha}$, we have

$$\ln \sigma_t^2 = \omega_t + \sum_{i=1}^{q} \alpha_i \mathbb{1}_{\eta_{t-i} \ne 0} \ln \eta_{t-i}^2 + \sum_{j=1}^{r} (\beta_j + \alpha_j \mathbb{1}_{\eta_{t-j} \ne 0}) \ln \sigma_{t-j}^2,$$

with $r = \max\{p, q\}$ and the convention $\alpha_i = 0$ for $i > q$ and $\beta_i = 0$ for $i > p$. If $P(\eta_t = 0) = 0$, then $\ln \sigma_t^2$ satisfies an ARMA-type equation with time-varying intercept, of the form

$$\left\{ 1 - \sum_{i=1}^{r} (\alpha_i + \beta_i) B^i \right\} \ln \sigma_t^2 = \omega_t + \sum_{i=1}^{q} \alpha_i B^i v_t,$$

where $v_t = \ln \eta_t^2$. This equation is a standard ARMA(r, q) equation in the symmetric case $\omega_{1-} = \cdots = \omega_{q-} = 0$ and under the moment condition $E(\ln \eta_t^2)^2 < \infty$, but these two assumptions are not needed. It is well known that this ARMA-type equation admits a non-degenerate and non-anticipative stationary solution if and only if the roots of the polynomial $1 - \sum_{j=1}^{r}(\beta_j + \alpha_j)z^j$ lie outside the unit circle.

4.3.2 Existence of Moments and Log-Moments

In Corollary 2.3, we have seen that for the GARCH models, the strict stationarity condition entails the existence of a moment of order $s > 0$ for $|\epsilon_t|$. The following Lemma shows that this is also the case for $| \mathbb{1}_{\epsilon_t \neq 0} \ln \epsilon_t^2 |$ in the Log-GARCH model, when the condition (4.19) is slightly reinforced.

Proposition 4.1 (Existence of a fractional log-moment) *Assume that $\gamma < 0$ and that $E|\mathbb{1}_{\eta_{t-i} \neq 0} \ln \eta_0^2|^{s_0} < \infty$ for some $s_0 > 0$. Let ϵ_t be the strictly stationary solution of Eq. (4.18). There exists $s > 0$ such that $E|\mathbb{1}_{\epsilon_t \neq 0} \ln \epsilon_t^2|^s < \infty$ and $E|\ln \sigma_t^2|^s < \infty$.*

Proof. Let X be a random variable such that $X > 0$ a.s. and $EX^r < \infty$ for some $r > 0$. Lemma 2.2 shows that if $E \ln X < 0$, then for $s > 0$ sufficiently small we have $EX^s < 1$. Noting that the random variables $\|\boldsymbol{C}_t\|$ are independent and bounded, we have $E \| \boldsymbol{C}_t \dots \boldsymbol{C}_1 \| \leq (E \| \boldsymbol{C}_1 \|)^t < \infty$ for all t. Note also that if $\gamma < 0$, then there exists t such that $E(\ln \| \boldsymbol{C}_t \dots \boldsymbol{C}_1 \|) < 0$. Therefore, Lemma 2.2 shows that we have $E \| \boldsymbol{C}_{k_0} \cdots \boldsymbol{C}_1 \|^s < 1$ for some $s \in (0, 1)$ and some $k_0 \geq 1$. In view of (4.24), the c_r-inequality and standard arguments (see Corollary 2.3) entail that $E \| \mathbf{z}_t \|^s < \infty$, provided $E \| \boldsymbol{b}_t \|^s < \infty$, which holds true when $s \leq s_0$. □

We now give conditions for the existence of higher-order log-moments, restricting ourselves to the Log-GARCH(1,1) case. We have the Markovian representation

$$\ln \sigma_t^2 = a_t \ln \sigma_{t-1}^2 + u_t, \tag{4.25}$$

with

$$u_t = \omega_t + (\alpha_+ \mathbb{1}_{\{\eta_{t-1} > 0\}} + \alpha_- \mathbb{1}_{\{\eta_{t-1} < 0\}}) \ln \eta_{t-1}^2,$$

$$a_t = \alpha_+ \mathbb{1}_{\{\eta_{t-1} > 0\}} + \alpha_- \mathbb{1}_{\{\eta_{t-1} < 0\}} + \beta.$$

Under the conditions (4.19) and (4.20), Eq. (4.25) admits the solution

$$\ln \sigma_t^2 = u_t + \sum_{i=0}^{\infty} \left(\prod_{j=0}^{i} a_{t-j} \right) u_{t-i-1}. \tag{4.26}$$

The next proposition gives conditions for the existence of $E|\ln \sigma_t^2|^m$ and $E|\mathbb{1}_{\epsilon_t \neq 0} \ln \epsilon_t^2|^m$ for $m \geq 1$.

Proposition 4.2 (Existence of log-moments of order m) *Let $m \geq 1$, then for the Extended Log-GARCH (1,1) model, the conditions $E|\mathbb{1}_{\eta_0 \neq 0} \ln \eta_0^2|^m < \infty$ and*

$$\pi_+ |\alpha_+ + \beta|^m + \pi_- |\alpha_- + \beta|^m + \pi_0 |\beta|^m < 1 \tag{4.27}$$

entail $E|\mathbb{1}_{\epsilon_t \neq 0} \ln \epsilon_t^2|^m < \infty$.

Proof. By Jensen's inequality, (4.27) (i.e. the condition $E|a_t|^m < 1$) entails (4.20) (i.e. the condition $m^{-1} E \ln |a_t|^m < 0$). With $\rho := \| a_1 \|_m < 1$ and $K = \| u_1 \|_m$, the solution (4.26) satisfies

$$\| \ln \sigma_t^2 \|_m \leq \| u_t \|_m + \sum_{i=1}^{\infty} (E|a_1|^m)^{i/m} \| u_1 \|_m \leq K \sum_{i=0}^{\infty} \rho^i < \infty.$$

The conclusion follows. □

In the extended Log-GARCH(1,1) model with $\alpha_- > \max\{0, \alpha_+\}$ and $\omega_- \geq 0$, negative returns impact future volatilities more importantly than positive returns of the same magnitude when this magnitude is large (at least larger than 1), but the effect of the sign can be reversed for small returns in absolute value. A measure of the average leverage effect can be defined through the covariance between η_{t-1} and the current log-volatility.

Proposition 4.3 (Average leverage effect in the Log-GARCH(1,1)) *Consider the extended Log-GARCH(1,1) model under the conditions*

$$E|\mathbb{1}_{\eta_0 \neq 0} \ln \eta_0^2|^m < \infty \quad \textit{and} \quad (4.27) \textit{ for } m = 1 \textit{ and } m = 2.$$

Then

$$\begin{aligned}\mathrm{Cov}(\eta_{t-1}, \ln \sigma_t^2) &= \omega_- E(\mathbb{1}_{\eta_0<0}\eta_0) + \alpha_+ E(\mathbb{1}_{\eta_0>0}\eta_0 \ln \eta_0^2) + \alpha_- E(\mathbb{1}_{\eta_0<0}\eta_0 \ln \eta_0^2) \\ &\quad + \{\alpha_+ E(\mathbb{1}_{\eta_0>0}\eta_0) + \alpha_- E(\mathbb{1}_{\eta_0<0}\eta_0)\} E \ln \sigma_0^2, \end{aligned} \tag{4.28}$$

where

$$E \ln \sigma_0^2 = \frac{\omega + \omega_- \pi_- + \alpha_+ E(\mathbb{1}_{\eta_0>0} \ln \eta_0^2) + \alpha_- E(\mathbb{1}_{\eta_0<0} \ln \eta_0^2)}{1 - \pi_+(\alpha_+ + \beta) - \pi_-(\alpha_- + \beta) - \pi_0 \beta}.$$

If the covariance (4.28) is negative, the leverage effect is present: past negative innovations tend to increase the log-volatility, and hence the volatility, more than past positive innovations. The sign of the covariance depends not only on all the GARCH coefficients but also on the innovations distribution. Interestingly, the leverage effect may hold with $\alpha_+ > \alpha_-$ and/or $\omega_- < 0$.

To illustrate the moment conditions on ϵ_t, as well as the computation of the autocorrelations of (ϵ_t^2), let us come back to the standard Log-GARCH(p, q) model

$$\begin{cases} \epsilon_t &= \sigma_t \eta_t, \\ \ln \sigma_t^2 &= \omega_0 + \sum_{i=1}^q \alpha_{0i} \ln \epsilon_{t-i}^2 + \sum_{j=1}^p \beta_{0j} \ln \sigma_{t-j}^2, \end{cases} \tag{4.29}$$

where (η_t) is iid $\mathcal{N}(0, 1)$. The process $(\ln \sigma_t^2)$ thus satisfies

$$\mathcal{A}_{\theta_0}(B) \ln \sigma_t^2 = \omega_0 + \mathcal{C}_{\theta_0}(B) \ln \eta_t^2 \tag{4.30}$$

where $\mathcal{A}_\theta(z) = 1 - \sum_{i=1}^r (\alpha_i + \beta_i) z^i$, $\mathcal{C}_\theta(z) = \sum_{i=1}^q \alpha_i z^i$, $r = \max\{p, q\}$, $\alpha_i = 0$ for $i > q$ and $\beta_i = 0$ for $i > p$. Assuming $\mathcal{A}_\theta(z) \neq 0$ for all $|z| \leq 1$, we have

$$\ln \sigma_t^2 = c + \sum_{i=1}^{\infty} \pi_i \ln \eta_{t-i}^2,$$

with

$$c = \frac{\omega_0}{\mathcal{A}_{\theta_0}(1)}, \quad \sum_{i=1}^{\infty} \pi_i z^i = \mathcal{A}_{\theta_0}^{-1}(z) \mathcal{C}_{\theta_0}(z).$$

We thus have

$$\sigma_t^2 = e^c \prod_{i=1}^{\infty} (\eta_{t-i}^2)^{\pi_i}, \quad \epsilon_{t-h}^2 = e^c \eta_{t-h}^2 \prod_{j=1}^{\infty} (\eta_{t-h-j}^2)^{\pi_j}.$$

Since $\eta_t \sim \mathcal{N}(0, 1)$, we have

$$\mu(s) := E|\eta_1|^s = \frac{2^{s/2}}{\sqrt{\pi}} \Gamma\left(\frac{1+s}{2}\right) \quad \text{for } s > -1.$$

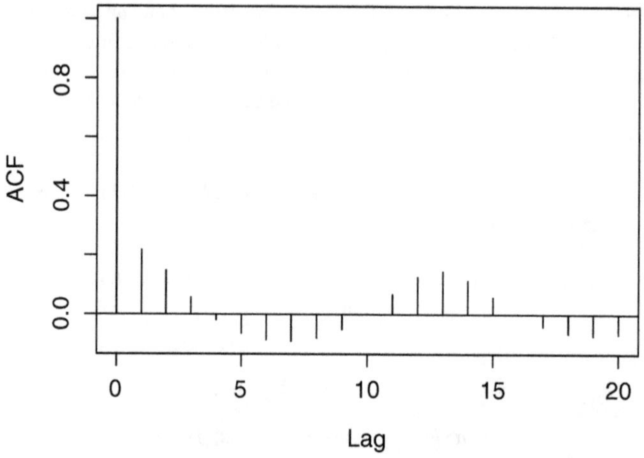

Figure 4.2 Theoretical autocorrelation function of the squares of a Log-GARCH(2,1) model.

For $s \leq -1$, the moment $\mu(s)$ is infinite. This allows to compute

$$E\epsilon_t^2 = e^c \prod_{i=1}^{\infty} \mu(2\pi_i), \quad E\epsilon_t^4 = \mu(4)e^{2c} \prod_{i=1}^{\infty} \mu(4\pi_i),$$

$$E\epsilon_t^2\epsilon_{t-1}^2 = e^{2c}\mu\{2(1+\pi_1)\} \prod_{i=1}^{\infty} \mu\{2(\pi_i + \pi_{i+1})\}$$

and

$$E\epsilon_t^2\epsilon_{t-h}^2 = e^{2c} \left\{ \prod_{j=1}^{h-1} \mu(2\pi_j) \right\} \mu\{2(1+\pi_h)\} \prod_{i=1}^{\infty} \mu\{2(\pi_i + \pi_{i+h})\}$$

for $h \geq 1$. Noting that $\ln \mu(s) = O(s)$ as $s \to 0$, these moments exist if and only if $\pi_i > -1/4$ for all i, which is somewhat a strange moment condition. Another astonishing result is that, if $\underline{\pi} := \inf_i \pi_i < 0$, then the moment $E|\epsilon_t|^d < \infty$ if and only if $d < -1/(2\underline{\pi})$. If $\underline{\pi} \geq 0$, then $E|\epsilon_t|^d < \infty$ for all d.

Figure 4.2 displays the theoretical autocorrelation function (ACF) of the squares of a Log-GARCH(2,1) model (4.29) with $\alpha_{01} = 0.06$, $\beta_{01} = 1.64$, and $\beta_{02} = -0.95$. Note the existence of negative autocorrelations. The Log-GARCH is thus able to take into account lagged countercyclical heteroscedasticity, i.e. the fact that a high volatility period could entail a low volatility period at a certain horizon. According to Proposition 2.2, this is not possible with standard GARCH models.

4.3.3 Relations with the EGARCH Model

Take the example of an EGARCH(1,1) model, which can be written as

$$\begin{cases} \tilde{\epsilon}_t &= \tilde{\sigma}_t \tilde{\eta}_t \\ \ln \tilde{\sigma}_t^2 &= \tilde{\omega} + \tilde{\gamma}_+ \tilde{\eta}_{t-1}^+ - \tilde{\gamma}_- \tilde{\eta}_{t-1}^- + \tilde{\beta} \ln \tilde{\sigma}_{t-1}^2, \end{cases} \tag{4.31}$$

where the $\tilde{\eta}_t$'s are iid with $E\tilde{\eta}_t^2 = 1$, $|\tilde{\beta}| < 1$, and with the notation $x^+ = \max\{x, 0\}$ and $x^- = \min\{x, 0\}$. The volatility of the EGARCH is given by

$$\ln \tilde{\sigma}_t^2 = \sum_{i=0}^{\infty} \tilde{\beta}^i \{\tilde{\omega} + \tilde{\gamma}_+ \tilde{\eta}_{t-i-1}^+ - \tilde{\gamma}_- \tilde{\eta}_{t-i-1}^-\}.$$

Now consider a Log-GARCH(1,1) of the form

$$\begin{cases} \epsilon_t &= \sigma_t \eta_t \\ \ln \sigma_t^2 &= \omega + \omega_- \mathbb{1}_{\eta_{t-1}<0} + \alpha \ln \eta_{t-1}^2 + \tilde{\beta} \ln \sigma_{t-1}^2, \end{cases} \tag{4.32}$$

with the parameters $\alpha \neq 0$, $\omega + \alpha c_+ = \tilde{\omega}$, $\omega_- = -\alpha(c_- - c_+)$, and the iid sequence

$$\eta_t = e^{\frac{c+}{2}} e^{\frac{\tilde{\gamma}+}{2\alpha}|\tilde{\eta}_t|} \mathbb{1}_{\tilde{\eta}_t \geq 0} - e^{\frac{c-}{2}} e^{\frac{\tilde{\gamma}-}{2\alpha}|\tilde{\eta}_t|} \mathbb{1}_{\tilde{\eta}_t<0},$$

with constants c_+ and c_- to be chosen later. The volatility of this Log-GARCH is given by

$$\begin{aligned} \ln \sigma_t^2 &= \sum_{i=0}^{\infty} \tilde{\beta}^i \{\omega + \omega_- \mathbb{1}_{\eta_{t-i-1}<0} + \alpha \ln \eta_{t-i-1}^2\} \\ &= \sum_{i=0}^{\infty} \tilde{\beta}^i \{\omega + (\omega_- + \alpha c_-) \mathbb{1}_{\tilde{\eta}_{t-i-1}<0} + \alpha c_+ \mathbb{1}_{\tilde{\eta}_{t-i-1}\geq 0} + \tilde{\gamma}_+ \tilde{\eta}_{t-i-1}^+ - \tilde{\gamma}_- \tilde{\eta}_{t-i-1}^-\} \\ &= \sum_{i=0}^{\infty} \tilde{\beta}^i \{\tilde{\omega} + \tilde{\gamma}_+ \tilde{\eta}_{t-i-1}^+ - \tilde{\gamma}_- \tilde{\eta}_{t-i-1}^-\}. \end{aligned}$$

Note that, although the processes (ϵ_t) and $(\tilde{\epsilon}_t)$ are not the same, the volatility of the Log-GARCH is equal to the volatility of the EGARCH. In other words, the class of the Extended Log-GARCH volatilities includes that of the EGARCH volatilities, and it can be shown that the inclusion is strict (see Figure 4.3).

More precisely, we have the following result.

Proposition 4.4

(i) For any EGARCH(1,1) process $\tilde{\epsilon}_t = \tilde{\sigma}_t \tilde{\eta}_t$ satisfying condition (4.31), there exists an extended Log-GARCH(1,1) process $\epsilon_t = \sigma_t \eta_t$ satisfying condition (4.32), with the same volatility process $\sigma_t = \tilde{\sigma}_t$, and η_t measurable with respect to $\tilde{\eta}_t$. If $Ee^{s_0|\tilde{\eta}_1|} < \infty$ for some $s_0 > 0$, then (η_t) can be chosen such that $E\eta_1^2 = 1$.

(ii) Conversely, there exist extended Log-GARCH processes $\epsilon_t = \sigma_t \eta_t$ for which there is no non-anticipative EGARCH process $\tilde{\epsilon}_t = \sigma_t \tilde{\eta}_t$ with the same volatility process σ_t and with $\tilde{\eta}_t$ measurable with respect to η_t.

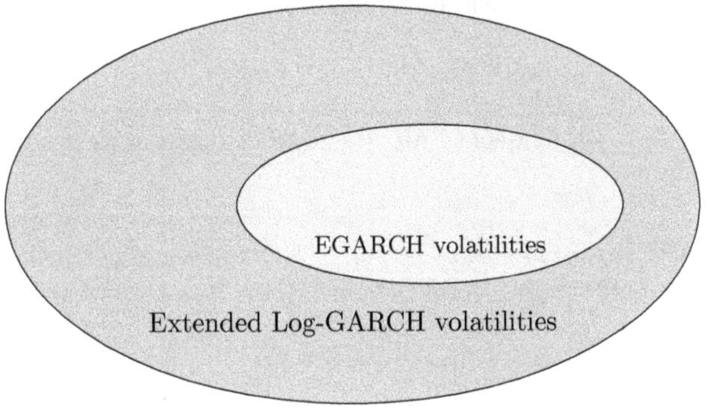

Figure 4.3 The set of the Extended Log-GARCH volatilities contains the set of the EGARCH volatilities.

Proof. Let us complete the proof of (i). Assume that $Ee^{s_0|\tilde{\eta}_1|} < \infty$ for $s_0 > 0$. It suffices to choose α such that $\gamma_+/\alpha < s_0$ and $\gamma_-/\alpha < s_0$, and then c_+ and c_- such that $E\eta_t^2 = 1$.

Now we turn to (ii). Let (ϵ_t) denote an extended Log-GARCH(1,1) process satisfying model (4.18), with $\alpha_{1+} \neq \alpha_{1-}$, and sufficiently general so that the support of the law of $\ln \sigma_1^2$ contains at least three different values. Also assume that $\ln \eta_1^2$ has a finite variance (which entails in particular that $P(\eta_1 = 0) = 0$). Denote by σ_t the volatility of this Log-GARCH. We proceed by contradiction. Suppose there exists a non-anticipative EGARCH(p, q) process satisfying $\tilde{\epsilon}_t = \sigma_t \tilde{\eta}_t$ with $\tilde{\eta}_t = f(\eta_t)$ for some measurable function f. We thus have

$$\begin{aligned}
\ln \sigma_t^2 &= \omega + \omega_- 1_{\eta_{t-1}<0} + (\alpha_{1+} 1_{\{\eta_{t-1}>0\}} + \alpha_{1-} 1_{\{\eta_{t-1}<0\}}) \ln \eta_{t-1}^2 \\
&\quad + (\alpha_{1+} 1_{\{\eta_{t-1}>0\}} + \alpha_{1-} 1_{\{\eta_{t-1}<0\}} + \beta) \ln \sigma_{t-1}^2 \\
&= \tilde{\omega} + (\gamma_+ 1_{\tilde{\eta}_{t-1}>0} + \gamma_- 1_{\tilde{\eta}_{t-1}<0}) \mid \tilde{\eta}_{t-1} \mid + \tilde{\beta} \ln \sigma_{t-1}^2 \\
&\quad + \sum_{j=2}^{p} \tilde{\beta}_j \ln \sigma_{t-j}^2 + \sum_{k=2}^{q} \gamma_{k+} \tilde{\eta}_{t-k}^+ + \gamma_{k-} \tilde{\eta}_{t-k}^-,
\end{aligned}$$

which entails

$$a(\eta_{t-1}) = b_{t-2} + c(\eta_{t-1}) \ln \sigma_{t-1}^2,$$

where $a(\eta_{t-1})$ and $c(\eta_{t-1})$ denotes random variables function of η_{t-1}, and b_{t-2} a variable belonging to the σ-field $\mathcal{F}_{t-2}$ generated by the η_{t-2-j} with $j \geq 0$. We have

$$\begin{aligned}
0 &= \text{var}\{a(\eta_{t-1}) - b_{t-2} - c(\eta_{t-1}) \ln \sigma_{t-1}^2 \mid \mathcal{F}_{t-2}\} \\
&= \text{var}\{a(\eta_{t-1})\} + \ln^2 \sigma_{t-1}^2 \text{var}\{c(\eta_{t-1})\} - 2 \ln \sigma_{t-1}^2 \text{cov}\{a(\eta_{t-1}), c(\eta_{t-1})\},
\end{aligned}$$

from which it follows that $\ln \sigma_{t-1}^2$ takes at most two values, or $a(\eta_{t-1})$ and $c(\eta_{t-1})$ are a.s. constant. This contradicts the above assumptions. □

This proposition shows that the class of the Log-GARCH models generates a richer class of volatilities than the EGARCH.

4.4 Threshold GARCH Model

A natural way to introduce asymmetry is to specify the conditional variance as a function of the positive and negative parts of the past innovations. Recall that

$$\epsilon_t^+ = \max(\epsilon_t, 0), \quad \epsilon_t^- = \min(\epsilon_t, 0)$$

and note that $\epsilon_t = \epsilon_t^+ + \epsilon_t^-$. The threshold GARCH (TGARCH) class of models introduces a threshold effect into the volatility.

Definition 4.4 (TGARCH(p, q) process) *Let (η_t) be an iid sequence of random variables such that $E(\eta_t) = 0$ and $Var(\eta_t) = 1$. Then (ϵ_t) is called a threshold GARCH(p, q) process if it satisfies an equation of the form*

$$\begin{cases} \epsilon_t = \sigma_t \eta_t \\ \sigma_t = \omega + \sum_{i=1}^{q} \alpha_{i,+} \epsilon_{t-i}^+ - \alpha_{i,-} \epsilon_{t-i}^- + \sum_{j=1}^{p} \beta_j \sigma_{t-j}, \end{cases} \tag{4.33}$$

where ω, $\alpha_{i,+}$, $\alpha_{i,-}$, and β_i are real numbers.

Remark 4.8 (On the TGARCH model)

1. Under the constraints
$$\omega > 0, \quad \alpha_{i,+} \geq 0, \quad \alpha_{i,-} \geq 0, \quad \beta_i \geq 0, \tag{4.34}$$
the variable σ_t is always strictly positive and represents the conditional standard deviation of ϵ_t. In general, the conditional standard deviation of ϵ_t is $|\sigma_t|$: imposing the positivity of σ_t is not required (contrary to the classical GARCH models, based on the specification of σ_t^2).

2. The GJR-GARCH model (named for Glosten, Jagannathan, and Runkle 1993) is a variant, defined by
$$\sigma_t^2 = \omega + \sum_{i=1}^{q} \alpha_i \epsilon_{t-i}^2 + \gamma_i \epsilon_{t-i}^2 \mathbb{1}_{\{\epsilon_{t-i}>0\}} + \sum_{j=1}^{p} \beta_j \sigma_{t-j}^2,$$
which corresponds to squaring the variables involved in the second equation of (4.33).

3. Through the coefficients $\alpha_{i,+}$ and $\alpha_{i,-}$, the current volatility depends on both the modulus and the sign of past returns. The model is flexible, allowing the lags i of the past returns to display different asymmetries. Note also that this class contains, as special cases, models displaying no asymmetry, whose properties are very similar to those of the standard GARCH. Such models are obtained for $\alpha_{i,+} = \alpha_{i,-} := \alpha_i$ $(i = 1, \ldots, q)$ and take the form
$$\sigma_t = \omega + \sum_{i=1}^{q} \alpha_i |\epsilon_{t-i}| + \sum_{j=1}^{p} \beta_j \sigma_{t-j}$$
(since $|\epsilon_t| = \epsilon_t^+ - \epsilon_t^-$). This specification is called absolute value GARCH (AVGARCH). Whether it is preferable to model the conditional variance or the conditional standard deviation is an open issue. However, it must be noted that for regression models with non-Gaussian and heteroscedastic errors, one can show that estimators of the noise variance based on the absolute residuals are more efficient than those based on the squared residuals (see Davidian and Carroll 1987).

Figure 4.4 depicts the major difference between GARCH and TGARCH models. The so-called 'news impact curves' display the impact of the innovations at time $t-1$ on the volatility at time t, for first-order models. In this figure, the coefficients have been chosen in such a way that the marginal

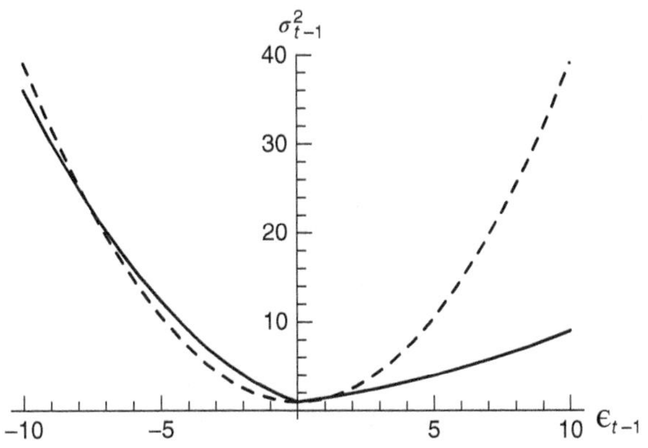

Figure 4.4 News impact curves for the ARCH(1) model, $\epsilon_t = \sqrt{1 + 0.38\epsilon_{t-1}^2}\eta_t$ (dashed line), and the TARCH(1) model, $\epsilon_t = (1 - 0.5\epsilon_{t-1}^- + 0.2\epsilon_{t-1}^+)\eta_t$ (solid line).

variances of ϵ_t in the two models coincide. In this TARCH example, in accordance with the properties of financial time series, negative past values of ϵ_{t-1} have more impact on the volatility than positive values of the same magnitude. The impact is, of course, symmetrical in the ARCH case.

TGARCH models display linearity properties similar to those encountered for the GARCH. Under the positivity constraints (4.34), we have

$$\epsilon_t^+ = \sigma_t \eta_t^+, \quad \epsilon_t^- = \sigma_t \eta_t^-, \tag{4.35}$$

which allows us to write the conditional standard deviation in the form

$$\sigma_t = \omega + \sum_{i=1}^{\max\{p,q\}} a_i(\eta_{t-i})\sigma_{t-i} \tag{4.36}$$

where $a_i(z) = \alpha_{i,+}z^+ - \alpha_{i,-}z^- + \beta_i$, $i = 1, \dots, \max\{p, q\}$. The dynamics of σ_t is thus given by a random coefficient autoregressive model.

Stationarity of the TGARCH(1, 1) Model

The study of the stationarity properties of the TGARCH(1, 1) model is based on Eq. (4.36) and follows from similar arguments to the GARCH(1, 1) case. The strict stationarity condition is written as

$$E[\ln(\alpha_{1,+}\eta_t^+ - \alpha_{1,-}\eta_t^- + \beta_1)] < 0. \tag{4.37}$$

In particular, for the TARCH(1) model ($\beta_1 = 0$), we have

$$\ln(\alpha_{1,+}\eta_t^+ - \alpha_{1,-}\eta_t^-) = \ln(\alpha_{1,+})\mathbb{1}_{\{\eta_t>0\}} + \ln(\alpha_{1,-})\mathbb{1}_{\{\eta_t<0\}} + \ln|\eta_t|.$$

Hence, if the distribution of (η_t) is symmetric, the expectation of the two indicator variables is equal to 1/2 and the strict stationarity condition reduces to

$$\alpha_{1,+}\alpha_{1,-} < e^{-2E\ln|\eta_t|}.$$

Exercise 4.8 shows that the second-order stationarity condition is

$$E[(\alpha_{1,+}\eta_t^+ - \alpha_{1,-}\eta_t^- + \beta_1)^2] < 1. \tag{4.38}$$

This condition can be made explicit in terms of the first two moments of η_t^+ and η_t^-. For instance, if η_t is $\mathcal{N}(0, 1)$ distributed, we get

$$\frac{1}{2}(\alpha_{1,+}^2 + \alpha_{1,-}^2) + \frac{2\beta_1}{\sqrt{2\pi}}(\alpha_{1,+} + \alpha_{1,-}) + \beta_1^2 < 1. \tag{4.39}$$

Of course, the second-order stationarity condition is more restrictive than the strict stationarity condition (see Figure 4.5).

Under the second-order stationarity condition, it is easily seen that the property of symmetry (4.1) is generally violated. For instance if the distribution of η_t is symmetric, we have, for the TARCH(1) model:

$$\text{Cov}(\sigma_t, \epsilon_{t-1}) = \alpha_{1,+}E(\epsilon_{t-1}^+)^2 - \alpha_{1,-}E(\epsilon_{t-i}^-)^2 = (\alpha_{1,+} - \alpha_{1,-})E(\epsilon_{t-1}^+)^2 \neq 0$$

whenever $\alpha_{1,+} \neq \alpha_{1,-}$.

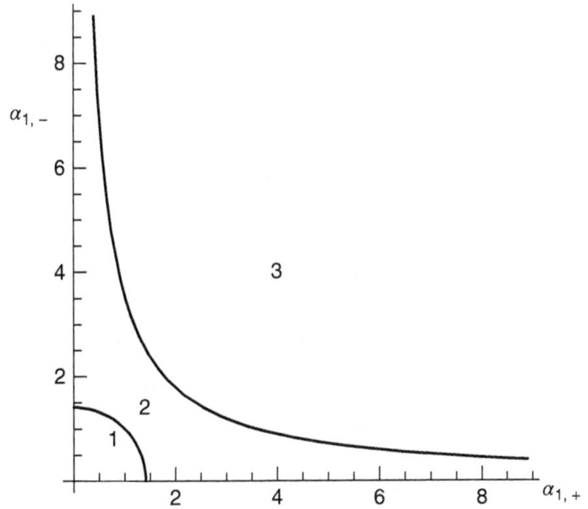

Figure 4.5 Stationarity regions for the TARCH(1) model with $\eta_t \sim \mathcal{N}(0,1)$: 1, second-order stationarity; 1 and 2, strict stationarity; 3, non-stationarity.

Strict Stationarity of the TGARCH(p, q) Model

The study of the general case relies on a representation analogous to Eq. (2.16), obtained by replacing, in the vector $\underline{z}_t$, the variables ϵ_{t-i}^2 by $(\epsilon_{t-i}^+, -\epsilon_{t-i}^-)'$, the σ_{t-i}^2 by σ_{t-i}, and by an adequate modification of $\underline{b}_t$ and A_t. Specifically, using (4.35), we get

$$\underline{z}_t = \underline{b}_t + A_t \underline{z}_{t-1}, \tag{4.40}$$

where

$$\underline{b}_t = \underline{b}(\eta_t) = \begin{pmatrix} \omega\eta_t^+ \\ -\omega\eta_t^- \\ 0 \\ \vdots \\ \omega \\ 0 \\ \vdots \\ 0 \end{pmatrix} \in \mathbb{R}^{p+2q}, \quad \underline{z}_t = \begin{pmatrix} \epsilon_t^+ \\ -\epsilon_t^- \\ \vdots \\ \epsilon_{t-q+1}^+ \\ -\epsilon_{t-q+1}^- \\ \sigma_t \\ \vdots \\ \sigma_{t-p+1} \end{pmatrix} \in \mathbb{R}^{p+2q},$$

and

$$A_t = \begin{pmatrix} \eta_t^+ \boldsymbol{\alpha}_{1:q-1} & \alpha_{q,+}\eta_t^+ & \alpha_{q,-}\eta_t^+ & \eta_t^+ \boldsymbol{\beta}_{1:p-1} & \beta_p \eta_t^+ \\ -\eta_t^- \boldsymbol{\alpha}_{1:q-1} & -\alpha_{q,+}\eta_t^- & -\alpha_{q,-}\eta_t^- & -\eta_t^- \boldsymbol{\beta}_{1:p-1} & -\beta_p \eta_t^- \\ \mathbb{I}_{2q-2} & 0_{2q-2\times 1} & 0_{2q-2\times 1} & 0_{2q-2\times p-1} & 0_{2q-2\times 1} \\ \boldsymbol{\alpha}_{1:q-1} & \alpha_{q,+} & \alpha_{q,-} & \boldsymbol{\beta}_{1:p-1} & \beta_p \\ 0_{p-1\times 2q-2} & 0_{p-1\times 1} & 0_{p-1\times 1} & \mathbb{I}_{p-1} & 0_{p-1\times 1} \end{pmatrix} \tag{4.41}$$

is a matrix of size $(p+2q)\times(p+2q)$,

$$\boldsymbol{\alpha}_{1:q-1} = (\alpha_{1,+}, \alpha_{1,-}, \dots, \alpha_{q-1,+}, \alpha_{q-1,-}) \in \mathbb{R}^{2q-2},$$
$$\boldsymbol{\beta}_{1:p-1} = (\beta_1, \cdots, \beta_{p-1}) \in \mathbb{R}^{p-1}.$$

The following result is analogous to that obtained for the strict stationarity of the GARCH(p, q).

Theorem 4.5 (Strict stationarity of the TGARCH(*p*,*q*) model) *A necessary and sufficient condition for the existence of a strictly stationary and non-anticipative solution of the TGARCH (p, q) model (4.33) and (4.34) is that $\gamma < 0$, where γ is the top Lyapunov exponent of the sequence $\{A_t, t \in \mathbb{Z}\}$ defined by (4.41).*

This stationary and non-anticipative solution, when $\gamma < 0$, is unique and ergodic.

Proof. The sufficient part of the proof of Theorem 2.4 can be straightforwardly adapted. As for the necessary part, note that the coefficients of the matrices A_t, $\underline{b}_t$, and $\underline{z}_t$ are positive. This allows us to show, as was done previously, that $A_0 \dots A_{-k}\underline{b}_{-k-1}$ tends to 0 a.s. when $k \to \infty$. But since $\underline{b}_{-k-1} = \omega\eta^+_{-k-1}e_1 - \omega\eta^-_{-k-1}e_2 + \omega e_{2q+1}$, using the positivity, we have

$$\lim_{k\to\infty} A_0 \dots A_{-k}\omega\eta^+_{-k-1}e_1 = \lim_{k\to\infty} A_0 \dots A_{-k}\omega\eta^-_{-k-1}e_2$$

$$= \lim_{k\to\infty} A_0 \dots A_{-k}\omega e_{2q+1} = 0, \quad \text{a.s.}$$

It follows that $\lim_{k\to\infty} A_0 \dots A_{-k}e_i = 0$ a.s. for $i = 1, \dots 2q+1$ by induction, as in the GARCH case. □

Numerical evaluation, by means of simulation, of the Lyapunov coefficient γ can be time-consuming because of the large size of the matrices A_t. A condition involving matrices of smaller dimensions can sometimes be obtained. Suppose that the asymmetric effects have a factorisation of the form $\alpha_{i-} = \theta\alpha_{i+}$ for all lags $i = 1, \dots, q$. In this constrained model, the asymmetry is summarised by only one parameter $\theta \neq 1$, the case $\theta > 1$ giving more importance to the negative returns.

Theorem 4.6 (Strict stationarity of the constrained TGARCH(*p*, *q*) model) *A necessary and sufficient condition for the existence of a strictly stationary and non-anticipative solution of the TGARCH(p, q) model (4.33), in which the coefficients ω, $\alpha_{i,-}$, and $\alpha_{i,+}$ satisfy the positivity conditions (4.34) and the $q-1$ constraints (for some constant θ)*

$$\alpha_{1,-} = \theta\alpha_{1,+}, \quad \alpha_{2,-} = \theta\alpha_{2,+}, \quad \dots, \quad \alpha_{q,-} = \theta\alpha_{q,+},$$

is that $\gamma^* < 0$, where γ^* is the top Lyapunov exponent of the sequence of $(p+q-1)\times(p+q-1)$ matrices $\{A^*_t, t \in \mathbb{Z}\}$ defined by

$$A^*_t = \begin{pmatrix} 0 & \cdots & 0 & \theta(\eta_{t-1}) & 0 & \cdots & 0 \\ & \mathbb{I}_{q-2} & & 0_{q-2\times p+1} & & & \\ \alpha_{2,+} & \cdots & \alpha_{q,+} & \alpha_{1,+}\theta(\eta_{t-1})+\beta_1 & \beta_2 & \cdots & \beta_p \\ & 0_{p-1\times q-1} & & \mathbb{I}_{p-1} & & 0_{p-1\times 1} & \end{pmatrix}, \tag{4.42}$$

where $\theta(\eta_t) = \eta^+_t - \theta\eta^-_t$. This stationary and non-anticipative solution, when $\gamma^* < 0$, is unique and ergodic.

Proof. If the constrained TGARCH model admits a stationary solution (σ_t, ϵ_t), then a stationary solution exists for the model

$$\underline{z}^*_t = \underline{b}^*_t + A^*_t\underline{z}^*_{t-1}, \tag{4.43}$$

where

$$\underline{b}^*_t = \begin{pmatrix} 0_{q-1} \\ \omega \\ 0_{p-1} \end{pmatrix} \in \mathbb{R}^{p+q-1}, \quad \underline{z}^*_t = \begin{pmatrix} \epsilon^+_{t-1} - \theta\epsilon^-_{t-1} \\ \vdots \\ \epsilon^+_{t-q+1} - \theta\epsilon^-_{t-q+1} \\ \sigma_t \\ \vdots \\ \sigma_{t-p+1} \end{pmatrix} \in \mathbb{R}^{p+q-1}.$$

Conversely, if (4.43) admits a stationary solution, then the constrained TGARCH model admits the stationary solution (σ_t, ϵ_t) defined by $\sigma_t = \underline{z}^*_t(q)$ (the qth component of $\underline{z}^*_t$) and $\epsilon_t = \sigma_t\eta_t$. Thus the constrained TGARCH model admits a strictly stationary solution if and only if model (4.43) has a strictly stationary solution. It can be seen that $\lim_{k\to\infty} A_0^* \cdots A_{-k}^* \underline{b}_{-k-1}^* = 0$ implies $\lim_{k\to\infty} A_0^* \cdots A_{-k}^* e_i = 0$ for $i = 1, \ldots, p+q-1$, using the independence of the matrices in the product $A_0^* \cdots A_{-k}^* \underline{b}_{-k-1}^*$ and noting that, in the case where $\theta(\eta_t)$ is not a.s. equal to zero, the qth component of $\underline{b}_{-k-1}^*$, the first and $(q+1)$th components of $A_{-k}^* \underline{b}_{-k-1}^*$, the second and $(q+2)$th components of $A_{-k+1}^* A_{-k}^* \underline{b}_{-k-1}^*$, etc., are strictly positive with non-zero probability. In the case where $\theta(\eta_t) = 0$, the first $q-1$ rows of $A_0^* \cdots A_{-q+2}^*$ are null, which obviously shows that $\lim_{k\to\infty} A_0^* \cdots A_{-k}^* e_i = 0$ for $i = 1, \ldots, q-1$. For $i = q, \ldots, p+q-1$, the argument used in the case $\theta(\eta_t) \neq 0$ remains valid. The rest of the proof is similar to that of Theorem 2.4. □

*m*th-Order Stationarity of the TGARCH(*p*, *q*) Model

Contrary to the standard GARCH model, the odd-order moments are not more difficult to obtain than the even-order ones for a TGARCH model. The existence condition for such moments is provided by the following theorem.

Theorem 4.7 (*m*th-order stationarity) *Let m be a positive integer. Suppose that $E(|\eta_t|^m) < \infty$. Let $A^{(m)} = E(A_t^{\otimes m})$ where A_t is defined by Eq. (4.41). If the spectral radius*

$$\rho(A^{(m)}) < 1,$$

then, for any $t \in \mathbb{Z}$, the infinite sum $(\underline{z}_t)$ is a strictly stationary solution of (4.41) which converges in L^m and the process (ϵ_t), defined by $\epsilon_t = e_{2q+1}' \underline{z}_t \eta_t$, is a strictly stationary solution of the TGARCH(p, q) model defined by Eq. (4.33), and admits moments up to order m.

Conversely, if $\rho(A^{(m)}) \geq 1$, there exists no strictly stationary solution (ϵ_t) of Eq. (4.33) satisfying the positivity conditions (4.34) and the moment condition $E(|\epsilon_t|^m) < \infty$.

The proof of this theorem is identical to that of Theorem 2.9.

Kurtosis of the TGARCH(1, 1) Model

For the TGARCH(1, 1) model with positive coefficients, the condition for the existence of $E|\epsilon_t|^m$ can be obtained directly. Using the representation

$$\sigma_t = \omega + a(\eta_{t-1})\sigma_{t-1}, \quad a(\eta) = \alpha_{1,+}\eta^+ - \alpha_{1,-}\eta^- + \beta_1,$$

we find that $E\sigma_t^m$ exists and satisfies

$$E\sigma_t^m = \sum_{k=0}^{m} \binom{m}{k} \omega^k E a^{m-k}(\eta_{t-1}) E\sigma_t^{m-k}$$

if and only if

$$Ea^m(\eta_t) < 1. \tag{4.44}$$

If this condition is satisfied for $m = 4$, then the kurtosis coefficient exists. Moreover, if $\eta_t \sim \mathcal{N}(0, 1)$, we get

$$\kappa_\epsilon = 3\frac{E\sigma_t^4}{(E\sigma_t^2)^2},$$

and, using the notation $a_i = Ea^i(\eta_t)$, the moments can be computed successively as

$$\begin{aligned}
a_1 &= \frac{1}{\sqrt{2\pi}}(\alpha_{1,+} + \alpha_{1,-}) + \beta_1, \\
E\sigma_t &= \frac{\omega}{1 - a_1}, \\
a_2 &= \frac{1}{2}(\alpha_{1,+}^2 + \alpha_{1,-}^2) + \frac{2}{\sqrt{2\pi}}\beta_1(\alpha_{1,+} + \alpha_{1,-}) + \beta_1^2, \\
E\sigma_t^2 &= \frac{\omega^2\{1 + a_1\}}{\{1 - a_1\}\{1 - a_2\}}, \\
a_3 &= \sqrt{\frac{2}{\pi}}(\alpha_{1,+}^3 + \alpha_{1,-}^3) + \frac{3}{2}\beta_1(\alpha_{1,+}^2 + \alpha_{1,-}^2) + \frac{3}{\sqrt{2\pi}}\beta_1^2(\alpha_{1,+} + \alpha_{1,-}) + \beta_1^3, \\
E\sigma_t^3 &= \frac{\omega^3\{1 + 2a_1 + 2a_2 + a_1a_2\}}{\{1 - a_1\}\{1 - a_2\}\{1 - a_3\}}, \\
a_4 &= \frac{3}{2}(\alpha_{1,+}^4 + \alpha_{1,-}^4) + 4\sqrt{\frac{2}{\pi}}\beta_1(\alpha_{1,+}^3 + \alpha_{1,-}^3) + \beta_1^4 \\
&\quad + 3\beta_1^2(\alpha_{1,+}^2 + \alpha_{1,-}^2) + \frac{4}{\sqrt{2\pi}}\beta_1^3(\alpha_{1,+} + \alpha_{1,-}), \\
E\sigma_t^4 &= \frac{\omega^4\{1 + 3a_1 + 5a_2 + 3a_1a_2 + 3a_3 + 5a_1a_3 + 3a_2a_3 + a_1a_2a_3\}}{\{1 - a_1\}\{1 - a_2\}\{1 - a_3\}}.
\end{aligned}$$

Many moments of the TGARCH(1, 1) can be obtained similarly, such as the autocorrelations of the absolute values (Exercise 4.9) and squares, but the calculations can be tedious.

4.5 Asymmetric Power GARCH Model

The following class is very general and contains the standard GARCH, the TGARCH, and the Log-GARCH.

Definition 4.5 (APARCH(p, q) process) *Let (η_t) be a sequence of iid variables such that $E(\eta_t) = 0$ and $Var(\eta_t) = 1$. The process (ϵ_t) is called an asymmetric power GARCH(p, q) if it satisfies an equation of the form*

$$\begin{cases} \epsilon_t = \sigma_t\eta_t \\ \sigma_t^\delta = \omega + \sum_{i=1}^q \alpha_i(|\epsilon_{t-i}| - \varsigma_i\epsilon_{t-i})^\delta + \sum_{j=1}^p \beta_j\sigma_{t-j}^\delta, \end{cases} \tag{4.45}$$

where $\omega > 0$, $\delta > 0$, $\alpha_i \geq 0$, $\beta_i \geq 0$, and $|\varsigma_i| \leq 1$.

Remark 4.9 (On the APARCH model)

1. The standard GARCH(p, q) is obtained for $\delta = 2$ and $\varsigma_1 = \cdots = \varsigma_q = 0$.
2. To study the role of the parameter ς_i, let us consider the simplest case, the asymmetric ARCH(1) model. We have

$$\sigma_t^2 = \begin{cases} \omega + \alpha_1(1 - \varsigma_1)^2\epsilon_{t-1}^2, & \text{if } \epsilon_{t-1} \geq 0, \\ \omega + \alpha_1(1 + \varsigma_1)^2\epsilon_{t-1}^2, & \text{if } \epsilon_{t-1} \leq 0. \end{cases} \tag{4.46}$$

Hence, the choice of $\varsigma_i > 0$ ensures that negative innovations have more impact on the current volatility than positive ones of the same modulus. Similarly, for more complex APARCH models, the constraint $\varsigma_i \geq 0$ is a natural way to capture the typical asymmetric property of financial series.

3. Since
$$\alpha_i|1 \pm \varsigma_i|^\delta \epsilon_{t-i}^\delta = \alpha_i|\varsigma_i|^\delta|1 \pm 1/\varsigma_i|^\delta \epsilon_{t-i}^\delta,$$
$|\varsigma_i| \leq 1$ is a non-restrictive identifiability constraint.

4. If $\delta = 1$, the model reduces to the TGARCH model. Using $\ln \sigma_t = \lim_{\delta \to 0}(\sigma_t^\delta - 1)/\delta$, one can interpret the Log-GARCH model as the limit of the APARCH model when $\delta \to 0$. The novelty of the APARCH model is in the introduction of the parameter δ. Note that autocorrelations of the absolute returns are often larger than autocorrelations of the squares. The introduction of the power δ increases the flexibility of GARCH-type models and allows the *a priori* selection of an arbitrary power to be avoided.

Noting that $\{\epsilon_{t-i} > 0\} = \{\eta_{t-i} > 0\}$, one can write

$$\sigma_t^\delta = \omega + \sum_{i=1}^{\max\{p,q\}} a_i(\eta_{t-i})\sigma_{t-i}^\delta \tag{4.47}$$

where

$$\begin{aligned} a_i(z) &= \alpha_i(|\,z\,| - \varsigma z)^\delta + \beta_i \\ &= \alpha_i(1 - \varsigma_i)^\delta|z|^\delta \mathbb{1}_{\{z>0\}} + \alpha_i(1 + \varsigma)^\delta|z|^\delta \mathbb{1}_{\{z<0\}} + \beta_i, \end{aligned}$$

for $i = 1, \ldots, \max\{p, q\}$.

Stationarity of the APARCH(1, 1) Model

Relation (4.47) is an extension of (2.6) which allows us to obtain the stationarity conditions, as in the classical GARCH(1, 1) case. The necessary and sufficient strict stationarity condition is thus

$$E \ln\{\alpha_1(1 - \varsigma_1)^\delta|\eta_t|^\delta \mathbb{1}_{\{\eta_t>0\}} + \alpha_1(1 + \varsigma_1)^\delta|\eta_t|^\delta \mathbb{1}_{\{\eta_t<0\}} + \beta_1\} < 0. \tag{4.48}$$

For the APARCH(1, 0) model, we have

$$\begin{aligned} &\ln\{\alpha_1(1 - \varsigma_1)^\delta|\eta_t|^\delta \mathbb{1}_{\{\eta_t>0\}} + \alpha_1(1 + \varsigma_1)^\delta|\eta_t|^\delta \mathbb{1}_{\{\eta_t<0\}}\} \\ &= \ln(1 - \varsigma_1)^\delta \mathbb{1}_{\{\eta_t>0\}} + \ln(1 + \varsigma_1)^\delta \mathbb{1}_{\{\eta_t<0\}} + \ln \alpha_1|\eta_t|^\delta, \end{aligned}$$

showing that, if the distribution of (η_t) symmetric, the strict stationarity condition reduces to

$$|1 - \varsigma_1|^{\delta/2}|1 + \varsigma_1|^{\delta/2}\alpha_1 < e^{-E \ln|\eta_t|^\delta}.$$

Note that in the limit case where $|\varsigma_1| = 1$, the model is strictly stationary for any value of α_1, as might be expected. Under condition (4.48), the strictly stationary solution is given by

$$\epsilon_t = \sigma_t\eta_t, \quad \sigma_t^\delta = \omega + \sum_{k=1}^{\infty} a_1(\eta_t)\cdots a_1(\eta_{t-k+1})\omega.$$

Assuming $E|\eta_t|^\delta < \infty$, the condition for the existence of $E\epsilon_t^\delta$ (and of $E\sigma_t^\delta$) is

$$Ea_1(\eta_t) = \alpha_1\{(1 - \varsigma_1)^\delta E\eta_t^\delta \mathbb{1}_{\{\eta_t>0\}} + (1 + \varsigma_1)^\delta E|\eta_t|^\delta \mathbb{1}_{\{\eta_t<0\}}\} + \beta_1 < 1, \tag{4.49}$$

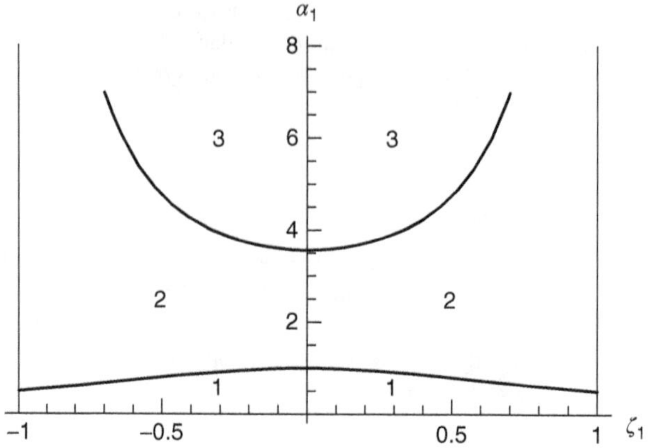

Figure 4.6 Stationarity regions for the APARCH(1,0) model with $\eta_t \sim \mathcal{N}(0, 1)$: 1, second-order stationarity; 1 and 2, strict stationarity; 3, non-stationarity.

which reduces to

$$\frac{1}{2}E|\eta_t|^\delta \alpha_1\{(1+\varsigma_1)^\delta + (1-\varsigma_1)^\delta\} + \beta_1 < 1$$

when the distribution of (η_t) symmetric, with

$$E|\eta_t|^\delta = \sqrt{\frac{2^\delta}{\pi}}\Gamma\left(\frac{1+\delta}{2}\right)$$

when η_t is Gaussian (Γ denoting the Euler gamma function). Figure 4.6 shows the strict and second-order stationarity regions of the APARCH(1, 0) model when η_t is Gaussian.

Obviously, if $\delta \geq 2$, condition (4.49) is sufficient (but not necessary) for the existence of a strictly stationary and second-order stationary solution to the APARCH(1, 1) model. If $\delta \leq 2$, condition (4.49) is necessary (but not sufficient) for the existence of a second-order stationary solution.

4.6 Other Asymmetric GARCH Models

Among other asymmetric GARCH models, which we will not study in detail, let us mention the qualitative threshold ARCH (QTARCH) model, and the quadratic GARCH model (QGARCH or GQARCH). The first-order model of this class, the QGARCH(1, 1), is defined by

$$\epsilon_t = \sigma_t\eta_t, \quad \sigma_t^2 = \omega + \alpha\epsilon_{t-1}^2 + \varsigma\epsilon_{t-1} + \beta\sigma_{t-1}^2, \tag{4.50}$$

where (η_t) is a strong white noise with unit variance, $\alpha \geq 0$ and $\beta \geq 0$.

Remark 4.10 (Positivity condition) The function $x \mapsto \alpha x^2 + \varsigma x$ has its minimum at $x = -\varsigma/2\alpha$, and this minimum is $-\varsigma^2/4\alpha$. A condition ensuring the positivity of σ_t^2 is thus $\omega > -\varsigma^2/4\alpha$. This can also be seen by writing

$$\sigma_t^2 = \omega - \varsigma^2/4\alpha + (\sqrt{\alpha}\epsilon_{t-1} + \varsigma/2\sqrt{\alpha})^2 + \beta\sigma_{t-1}^2.$$

Note that in this case

$$\sigma_t^2 \geq \underline{\sigma}^2 := \frac{\omega - \varsigma^2/4\alpha}{1-\beta}\text{a.s.} \tag{4.51}$$

Remark 4.11 (Leverage effect) The asymmetric effect is taken into account through the coefficient ς. A negative coefficient entails that negative returns have a bigger impact on the volatility of the next period than positive ones. A small price increase, such that the return is less than $-\varsigma/2\alpha$ with $\varsigma > 0$, can even produce less volatility than a zero return. This is a distinctive feature of this model, compared to the EGARCH, TGARCH, or GJR-GARCH for which, by appropriately constraining the parameters, the volatility at time t is minimal in the absence of price movement at time $t-1$.

The condition

$$\alpha + \beta < 1$$

is clearly necessary for the existence of a non-anticipative and second-order stationary solution. To obtain a sufficient condition, we use the approach of Section 4.1. The QGARCH(1, 1) satisfies the SRE

$$\sigma_t^2 = g(\eta_{t-1}, \sigma_{t-1}^2), \quad g(x, y) = \omega + \alpha x^2 y + \tau x \sqrt{y} + \beta y.$$

With the notation of Lemma 4.1 and using (4.51), we have

$$\Lambda_t \leq \alpha \eta_{t-1}^2 + \frac{|\tau \eta_{t-1}|}{2\sqrt{\underline{\sigma}^2}} + \beta.$$

Lemma 4.1 thus entails the following result.

Theorem 4.8 (Stationarity of the QGARCH(1, 1) model) *If $\omega > -\varsigma^2/4\alpha$ and*

$$E \ln \left(\alpha \eta_{t-1}^2 + \frac{|\tau|}{2\sqrt{\underline{\sigma}^2}} |\eta_{t-1}| + \beta \right) < 0,$$

there exists a unique strictly stationary and ergodic solution to the QGARCH(1, 1) model (4.50). Moreover, this solution satisfies $E|\epsilon_t|^s < \infty$ for some $s > 0$.

The second equality in (4.50) cannot be easily expanded because of the presence of $\epsilon_{t-1}^2 = \sigma_{t-1}^2 \eta_{t-1}^2$ and $\epsilon_{t-1} = \sigma_{t-1}\eta_{t-1}$. It is, therefore, not possible to obtain an explicit solution as a function of the η_{t-i}.

Many other asymmetric GARCH models have been introduced. Complex asymmetric responses to past values may be considered. For instance, in the model

$$\sigma_t = \omega + \alpha|\epsilon_{t-1}| \mathbb{1}_{\{|\epsilon_{t-1}| \leq \gamma\}} + \alpha_+ \epsilon_{t-1} \mathbb{1}_{\{\epsilon_{t-1} > \gamma\}} - \alpha_- \epsilon_{t-1} \mathbb{1}_{\{\epsilon_{t-1} < -\gamma\}}, \quad \alpha, \alpha_+, \alpha_-, \gamma \geq 0,$$

asymmetry is only present for large innovations (whose amplitude is larger than the threshold γ).

4.7 A GARCH Model with Contemporaneous Conditional Asymmetry

A common feature of the GARCH models studied up to now is the decomposition

$$\epsilon_t = \sigma_t \eta_t,$$

where σ_t is a positive variable and (η_t) is an iid process. The various models differ by the specification of σ_t as a measurable function of ϵ_{t-i} for $i > 0$. This type of formulation implies several important restrictions:

(i) The process (ϵ_t) is a martingale difference.

(ii) The positive and negative parts of ϵ_t have the same volatility, up to a multiplicative factor.

(iii) The kurtosis and skewness of the conditional distribution of ϵ_t are constant.

Property (ii) is an immediate consequence of the equalities in (4.35). Property (iii) expresses the fact that the conditional law of ϵ_t has the same 'shape' (symmetric or asymmetric, unimodal or polymodal, with or without heavy tails) as the law of η_t.

It can be shown empirically that these properties are generally not satisfied by financial time series. Estimated kurtosis and skewness coefficients of the conditional distribution often present large variations in time. Moreover, property (i) implies that $\text{Cov}(\epsilon_t, z_{t-1}) = 0$, for any variable $z_{t-1} \in L^2$ which is a measurable function of the past of ϵ_t. In particular, one must have

$$\forall h > 0, \quad \text{Cov}(\epsilon_t, \epsilon_{t-h}^+) = \text{Cov}(\epsilon_t, \epsilon_{t-h}^-) = 0 \tag{4.52}$$

or equivalently,

$$\forall h > 0, \quad \text{Cov}(\epsilon_t^+, \epsilon_{t-h}^+) = \text{Cov}(-\epsilon_t^-, \epsilon_{t-h}^+), \quad \text{Cov}(\epsilon_t^+, \epsilon_{t-h}^-) = \text{Cov}(-\epsilon_t^-, \epsilon_{t-h}^-). \tag{4.53}$$

We emphasise the difference between (4.52) and the characterisation (4.2) of the asymmetry studied previously. When Eq. (4.52) does not hold, one can speak of *contemporaneous asymmetry* since the variables ϵ_t^+ and $-\epsilon_t^-$, of the current date, do not have the same conditional distribution.

For the CAC index series, Table 4.2 completes Table 4.1, by providing the cross empirical autocorrelations of the positive and negative parts of the returns.

Without carrying out a formal test, comparison of rows 1 and 3 (or 2 and 4) shows that the leverage effect is present, whereas comparison of rows 3 and 4 shows that property (4.28) does not hold.

A class of GARCH-type models allowing the two kinds of asymmetry is defined as follows. Let

$$\epsilon_t = \sigma_{t,+}\eta_t^+ + \sigma_{t,-}\eta_t^-, \quad t \in \mathbb{Z}, \tag{4.54}$$

where $\{\eta_t\}$ is centred, η_t is independent of $\sigma_{t,+}$ and $\sigma_{t,-}$, and

$$\begin{cases} \sigma_{t,+} = \alpha_{0,+} + \sum_{i=1}^{q} \alpha_{i,+}^+ \epsilon_{t-i}^+ - \alpha_{i,+}^- \epsilon_{t-i}^- + \sum_{j=1}^{p} \beta_{j,+}^+ \sigma_{t-j,+} + \beta_{j,+}^- \sigma_{t-j,-} \\ \sigma_{t,-} = \alpha_{0,-} + \sum_{i=1}^{q} \alpha_{i,-}^+ \epsilon_{t-i}^+ - \alpha_{i,-}^- \epsilon_{t-i}^- + \sum_{j=1}^{p} \beta_{j,-}^+ \sigma_{t-j,+} + \beta_{j,-}^- \sigma_{t-j,-} \end{cases}$$

where $\alpha_{i,+}^+, \alpha_{i,-}^+, \ldots, \beta_{j,-}^- \geq 0$, $\alpha_{0,+}, \alpha_{0,-} > 0$. Without loss of generality, it can be assumed that $E(\eta_t^+) = E(-\eta_t^-) = 1$.

As an immediate consequence of the positivity of $\sigma_{t,+}$ and $\sigma_{t,-}$, we obtain

$$\epsilon_t^+ = \sigma_{t,+}\eta_t^+ \quad \text{and} \quad \epsilon_t^- = \sigma_{t,-}\eta_t^-, \tag{4.55}$$

which is crucial for the study of this model.

Table 4.2 Empirical autocorrelations (CAC 40, for the period 1988–1998).

h	1	2	3	4	5	10	20	40
$\rho(\epsilon_t^+, \epsilon_{t-h}^+)$	0.037	−0.006	−0.013	0.029	−0.039*	0.017	0.023	0.001
$\rho(-\epsilon_t^-, \epsilon_{t-h}^+)$	−0.013	−0.035	−0.019	−0.025	−0.028	−0.007	−0.020	0.017
$\rho(\epsilon_t^+, -\epsilon_{t-h}^-)$	0.026	0.088*	0.135*	0.047*	0.088*	0.056*	0.049*	0.065*
$\rho(\epsilon_t^-, \epsilon_{t-h}^-)$	0.060*	0.074*	0.041*	0.070*	0.027	0.077*	0.015	−0.008*

*Parameters that are statistically significant at the level 5%, using $1/n$ as an approximation for the autocorrelations variance, for $n = 2385$.

Thus, $\sigma_{t,+}$ and $\sigma_{t,-}$ can be interpreted as the volatilities of the positive and negative parts of the noise (up to a multiplicative constant, since we did not specify the variances of η_t^+ and η_t^-). In general, the non-anticipative solution of this model, when it exists, is not a martingale difference because

$$E(\epsilon_t \mid \epsilon_{t-1}, \ldots) = (\sigma_{t,+} - \sigma_{t,-})E(\eta_t^+) \neq 0.$$

An exception is, of course, the situation where the parameters of the dynamics of $\sigma_{t,+}$ and $\sigma_{t,-}$ coincide, in which case we obtain model (4.33).

A simple computation shows that the kurtosis coefficient of the conditional law of ϵ_t is given by

$$\kappa_t = \frac{\sum_{k=0}^{4} \binom{4}{k} \sigma_{t,+}^k \sigma_{t,-}^{4-k} c(k, 4-k)}{\left[\sum_{k=0}^{2} \binom{2}{k} \sigma_{t,+}^k \sigma_{t,-}^{2-k} c(k, 2-k)\right]^2}, \tag{4.56}$$

where $c(k, l) = E[\{\eta_t^+ - E(\eta_t^+)\}^k \{\eta_t^- - E(\eta_t^-)\}^l]$, provided that $E(\eta_t^4) < \infty$. A similar computation can be done for the conditional skewness, showing that the shape of the conditional distribution varies in time, in a more important way than for classical GARCH models.

Methods analogous to those developed for the other GARCH models allow us to obtain existence conditions for the stationary and non-anticipative solutions (references are given at the end of the chapter). In contrast to the GARCH models analysed previously, the stationary solution (ϵ_t) is not always a white noise.

4.8 Empirical Comparisons of Asymmetric GARCH Formulations

We will restrict ourselves to the simplest versions of the GARCH introduced in this chapter and consider their fit to the series of CAC 40 index returns, r_t, over the period 1988–1998 consisting of 2385 values.

Descriptive Statistics

Figure 4.7 displays the first 500 values of the series. The volatility clustering phenomenon is clearly evident. The correlograms in Figure 4.8 indicate absence of autocorrelation. However, squared returns present significant autocorrelations, which is another sign that the returns are not independent. Ljung–Box portmanteau tests, such as those available in SAS (see Table 4.3; Chapter 5 gives more details on these tests), confirm the visual analysis provided by the correlograms. Figure 4.9a, compared

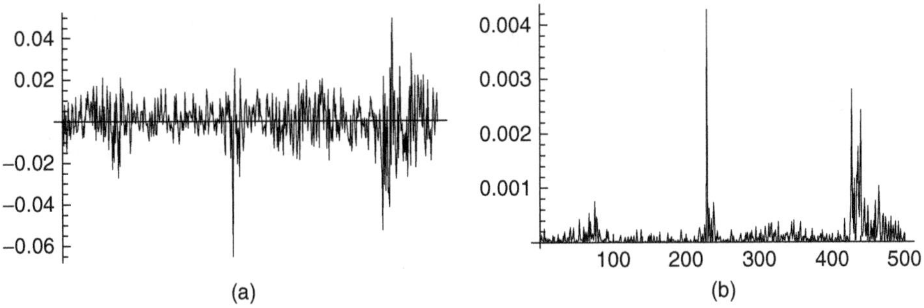

Figure 4.7 The first 500 values of the CAC 40 index (a) and of the squared index (b).

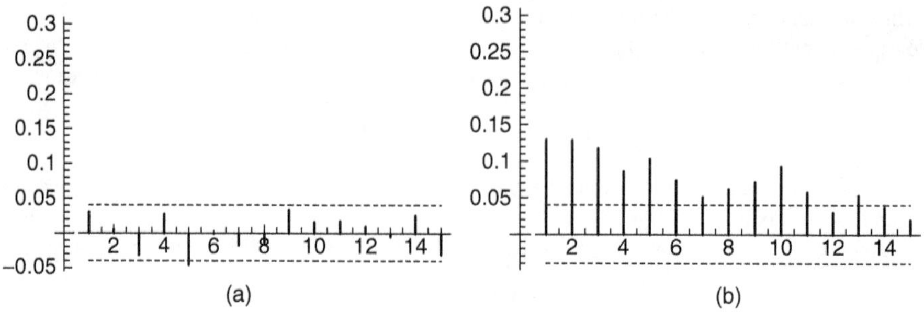

Figure 4.8 Correlograms of the CAC 40 index (a) and the squared index (b). Dashed lines correspond to $\pm 1.96/\sqrt{n}$.

Table 4.3 Portmanteau test of the white noise hypothesis for the CAC 40 series (a) and for the squared index (b).

a: Autocorrelation check for white noise

To Lag	Chi-square	DF	Pr > Khi2	Autocorrelations					
6	11.51	6	0.0737	0.030	0.005	-0.032	0.028	-0.046	-0.001
12	16.99	12	0.1499	-0.018	-0.014	0.034	0.016	0.017	0.010
18	21.22	18	0.2685	-0.005	0.025	-0.031	-0.009	-0.003	0.006
24	27.20	24	0.2954	-0.023	0.003	-0.010	0.030	-0.027	-0.015

b: Autocorrelation check for white noise

To Lag	Chi-square	DF	Pr > Khi2	Autocorrelations					
6	165.90	6	<0.0001	0.129	0.127	0.117	0.084	0.101	0.074
12	222.93	12	<0.0001	0.051	0.060	0.070	0.092	0.058	0.030
18	238.11	18	<0.0001	0.053	0.036	0.020	0.041	0.002	0.013
24	240.04	24	<0.0001	0.006	0.024	0.013	0.003	0.001	-0.002

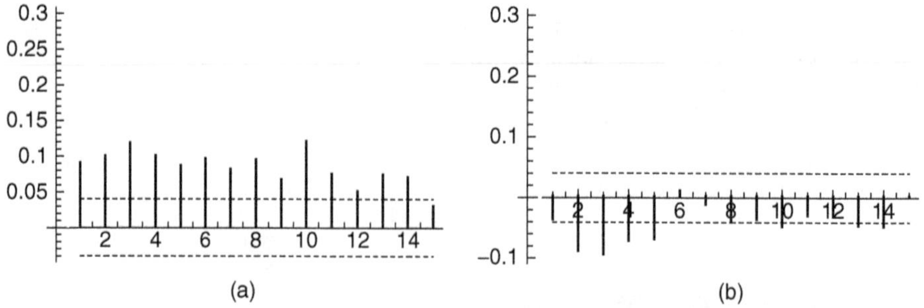

Figure 4.9 Correlogram $h \mapsto \hat{\rho}(|r_t|, |r_{t-h}|)$ of the absolute CAC 40 returns (a) and cross correlograms $h \mapsto \hat{\rho}(|r_t|, r_{t-h})$ measuring the leverage effects (b).

to Figure 4.8b, seems to indicate that the absolute returns are slightly more strongly correlated than the squares. Figure 4.9b displays empirical correlations between the series $|r_t|$ and r_{t-h}. It can be seen that these correlations are negative, which implies the presence of leverage effects (more accentuated, apparently, for lags 2 and 3 than for lag 1).

Fit by Symmetric and Asymmetric GARCH Models

We will consider the classical GARCH(1, 1) model and the simplest asymmetric models (which are the most widely used). Using the AUTOREG and MODEL procedures of SAS, the estimated models are

GARCH(1, 1) model

$$\begin{cases} r_t = \underset{(2\times10^{-4})}{5\times10^{-4}} + \epsilon_t, \quad \epsilon_t = \sigma_t\eta_t, \quad \eta_t \sim \mathcal{N}(0,1) \\ \sigma_t^2 = \underset{(2\times10^{-6})}{8\times10^{-6}} + \underset{(0.02)}{0.09}\ \epsilon_{t-1}^2 + \underset{(0.02)}{0.84}\ \sigma_{t-1}^2 \end{cases} \tag{4.57}$$

EGARCH(1, 1) model

$$\begin{cases} r_t = \underset{(2\times10^{-4})}{4\times10^{-4}} + \epsilon_t, \quad \epsilon_t = \sigma_t\eta_t, \quad \eta_t \sim \mathcal{N}(0,1) \\ \ln\sigma_t^2 = \underset{(0.15)}{-0.64} + \underset{(0.03)}{0.15}\ (\underset{(0.14)}{-0.53}\ \eta_{t-1} + |\eta_{t-1}| - \sqrt{2/\pi}) \\ \qquad + \underset{(0.02)}{0.93}\ \ln\sigma_{t-1}^2 \end{cases} \tag{4.58}$$

QGARCH(1, 1) model

$$\begin{cases} r_t = \underset{(2\times10^{-4})}{3\times10^{-4}} + \epsilon_t, \quad \epsilon_t = \sigma_t\eta_t, \quad \eta_t \sim \mathcal{N}(0,1) \\ \sigma_t^2 = \underset{(2\times10^{-6})}{9\times10^{-6}} + \underset{(0.01)}{0.07}\ \epsilon_{t-1}^2 - \underset{(2\times10^{-4})}{9\times10^{-4}}\ \epsilon_{t-1} + \underset{(0.03)}{0.85}\ \sigma_{t-1}^2 \end{cases} \tag{4.59}$$

GJR-GARCH(1, 1) model

$$\begin{cases} r_t = \underset{(2\times10^{-4})}{4\times10^{-4}} + \epsilon_t, \quad \epsilon_t = \sigma_t\eta_t, \quad \eta_t \sim \mathcal{N}(0,1) \\ \sigma_t^2 = \underset{(2\times10^{-6})}{1\times10^{-5}} + \underset{(0.02)}{0.13}\ \epsilon_{t-1}^2 - \underset{(0.02)}{0.10}\ \epsilon_{t-1}^2 \mathbb{1}_{\{\epsilon_{t-1}>0\}} + \underset{(0.03)}{0.84}\ \sigma_{t-1}^2 \end{cases} \tag{4.60}$$

TGARCH(1, 1) model

$$\begin{cases} r_t = \underset{(2\times10^{-4})}{4\times10^{-4}} + \epsilon_t, \quad \epsilon_t = \sigma_t\eta_t, \quad \eta_t \sim \mathcal{N}(0,1) \\ \sigma_t = \underset{(2\times10^{-4})}{8\times10^{-4}} + \underset{(0.01)}{0.03}\ \epsilon_{t-1}^+ - \underset{(0.02)}{0.12}\ \epsilon_{t-1}^- + \underset{(0.02).}{0.87}\ \sigma_{t-1} \end{cases} \tag{4.61}$$

Interpretation of the Estimated Coefficients

Note that all the estimated models are stationary. The standard GARCH(1, 1) admits a fourth-order moment since, in view of the computation on p. 45, we have $3\alpha^2+\beta^2+2\alpha\beta<1$. It is thus possible to compute the variance and kurtosis in this estimated model (which are, respectively, equal to 1.3×10^{-4} and 3.49 for the standard GARCH(1,1)). Given the ARMA(1, 1) representation for ϵ_t^2, we have $\rho_{\epsilon^2}(h)=(\hat{\alpha}+\hat{\beta})\rho_{\epsilon^2}(h-1)$ for any $h>1$. Since $\hat{\alpha}+\hat{\beta}=0.09+0.84$ is close to 1, the decay of $\rho_{\epsilon^2}(h)$

Table 4.4 Likelihoods of the different models for the CAC 40 series.

	GARCH	EGARCH	QGARCH	GJR-GARCH	TGARCH
$\ln L_n$	7393	7404	7404	7406	7405

to zero will be slow when $h \to \infty$, which can be interpreted as a sign of strong persistence of shocks.[4]

Note that in the EGARCH model the parameter $\theta = -0.53$ is negative, implying the presence of the leverage effect. A similar interpretation can be given to the negative sign of the coefficient of ϵ_{t-1} in the QGARCH model, and to that of $\epsilon_{t-1}^2 \mathbb{1}_{\{\epsilon_{t-1}>0\}}$ in the GJR-GARCH model. In the TGARCH model, the leverage effect is present since $\alpha_{1,-} > \alpha_{1,+} > 0$.

The TGARCH model seems easier to interpret than the other asymmetric models. The volatility (that is, the conditional standard deviation) is the sum of four terms. The first is the intercept $\omega = 8 \times 10^{-4}$. The term $\omega/(1 - \beta_1) = 0.006$ can be interpreted as a 'minimal volatility', obtained by assuming that all the innovations are equal to zero. The next two terms represent the impact of the last observation, distinguishing the sign of this observation, on the current volatility. In the estimated model, the impact of a positive value is 3.5 times less than that of a negative one. The last coefficient measures the importance of the last volatility. Even in absence of news, the decay of the volatility is slow because the coefficient $\beta_1 = 0.87$ is rather close to 1.

Likelihood Comparisons

Table 4.4 gives the log-likelihood, $\ln L_n$, of the observations for the different models. One cannot directly compare the log-likelihood of the standard GARCH(1, 1) model, which has one parameter less, with that of the other models, but the log-likelihoods of the asymmetric models, which all have five parameters, can be compared. The largest likelihood is observed for the GJR threshold model, but, the difference being very slight, it is not clear that this model is really superior to the others.

Resemblances between the Estimated Volatilities

Figure 4.10 shows that the estimated volatilities for the five models are very similar. It follows that the different specifications produce very similar prediction intervals (see Figure 4.11).

Distances between Estimated Models

Differences can, however, be discerned between the various specifications. Table 4.5 gives an insight into the distances between the estimated volatilities for the different models. From this point of view, the TGARCH and EGARCH models are very close and are also the most distant from the standard GARCH. The QGARCH model is the closest to the standard GARCH. Rather surprisingly, the TGARCH and GJR-GARCH models appear quite different. Indeed, the GJR-GARCH is a threshold model for the conditional variance and the TGARCH is a similar model for the conditional standard deviation.

Figure 4.12 confirms the results of Table 4.5. The left-hand scatterplot shows

$$(\sigma^2_{t,\text{TGARCH}} - \sigma^2_{t,\text{GARCH}}, \sigma^2_{t,\text{EGARCH}} - \sigma^2_{t,\text{GARCH}}), \quad t = 1, \dots, n,$$

and the right-hand one

$$(\sigma^2_{t,\text{TGARCH}} - \sigma^2_{t,\text{GARCH}}, \sigma^2_{t,\text{GJR-GARCH}} - \sigma^2_{t,\text{GARCH}}), \quad t = 1, \dots, n.$$

---

[4] In the strict sense, and for any reasonable specification, shocks are non-persistent because $\partial\sigma^2_{t+h}/\partial\epsilon_t \to 0$ a.s., but we wish to express the fact that, in some sense, the decay to 0 is slow.

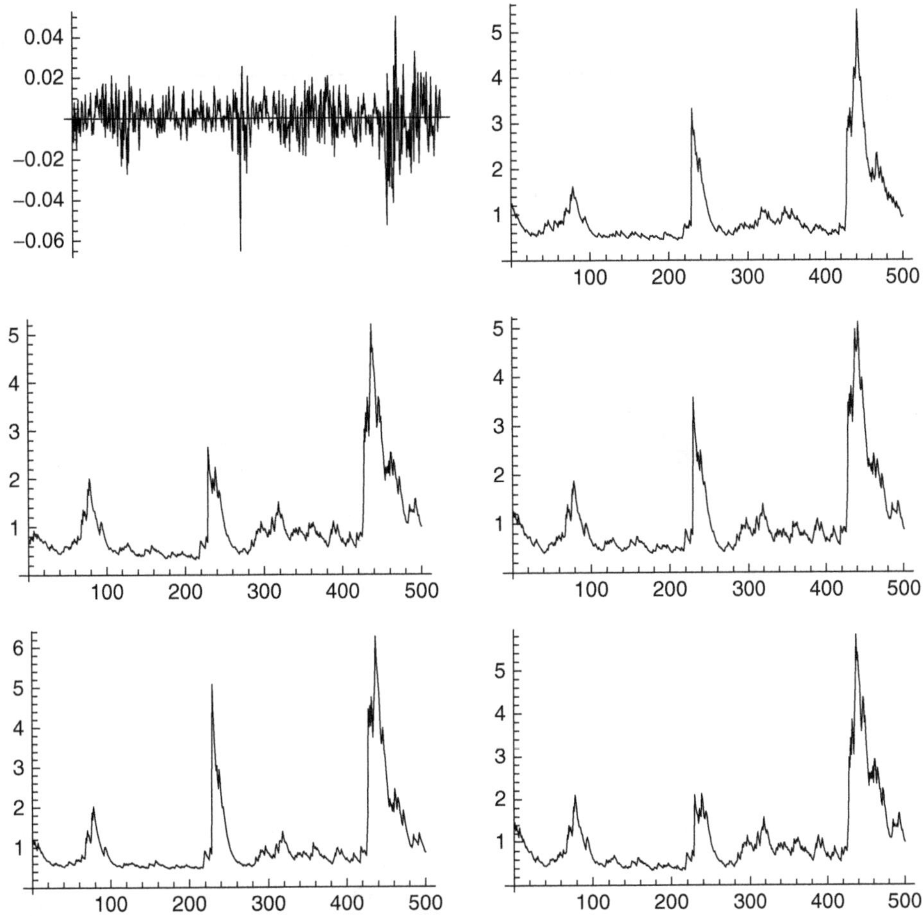

Figure 4.10 From left to right and top to bottom, graph of the first 500 values of the CAC 40 index and estimated volatilities ($\times 10^4$) for the GARCH(1, 1), EGARCH(1, 1), QGARCH(1, 1), GJR-GARCH (1, 1), and TGARCH(1, 1) models.

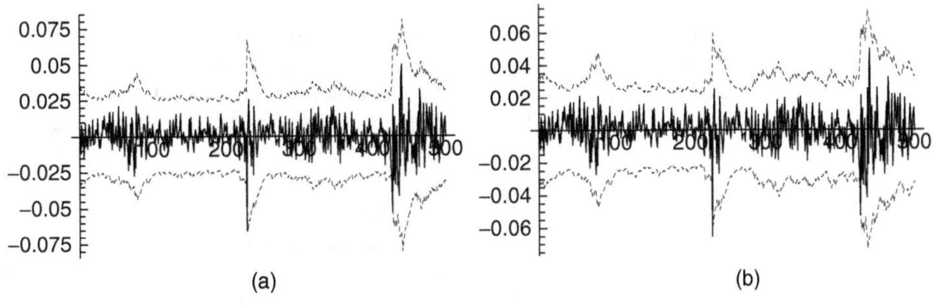

Figure 4.11 Returns r_t of the CAC 40 index (solid lines) and confidence intervals $\overline{r} \pm 3\sigma_t$ (dotted lines), where $\overline{r}$ is the empirical mean of the returns over the whole period 1988–1998 and σ_t is the estimated volatility in the standard GARCH(1, 1) model (a) and in the EGARCH(1, 1) model (b).

Table 4.5 Means of the squared differences between the estimated volatilities ($\times 10^{10}$).

	GARCH	EGARCH	QGARCH	GJR	TGARCH
GARCH	0	10.98	3.58	7.64	12.71
EGARCH	10.98	0	3.64	6.47	1.05
QGARCH	3.58	3.64	0	3.25	4.69
GJR	7.64	6.47	3.25	0	9.03
TGARCH	12.71	1.05	4.69	9.03	0

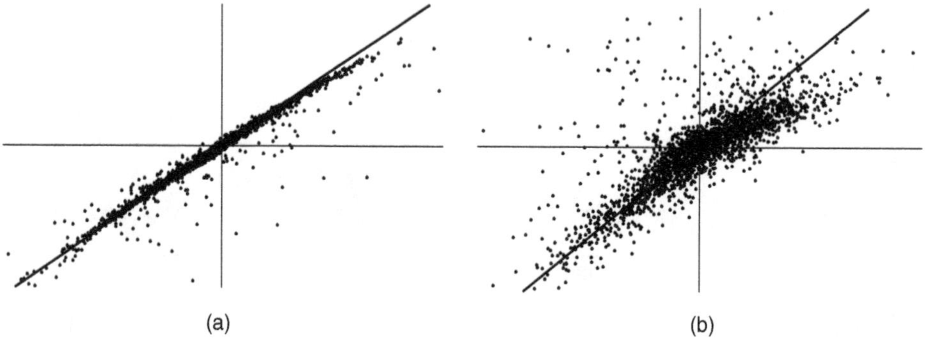

Figure 4.12 Comparison of the estimated volatilities of the EGARCH and TARCH models (a), and of the TGARCH and GJR-GARCH models (b). The estimated volatilities are close when the scatterplot is elongated (see text).

The left-hand graph shows that the difference between the estimated volatilities of the TGARCH and the standard GARCH, denoted by $\sigma^2_{t,\mathrm{TGARCH}} - \sigma^2_{t,\mathrm{GARCH}}$, is always very close to the difference between the estimated volatilities of the EGARCH and the standard GARCH, denoted by $\sigma^2_{t,\mathrm{EGARCH}} - \sigma^2_{t,\mathrm{GARCH}}$ (the difference from the standard GARCH is introduced to make the graphs more readable). The right-hand graph shows much more important differences between the TGARCH and GJR-GARCH specifications.

Comparison between Implied and Sample Values of the Persistence and of the Leverage Effect

We now wish to compare, for the different models, the theoretical autocorrelations $\rho(|r_t|, |r_{t-h}|)$ and $\rho(|r_t|, r_{t-h})$ to the empirical ones. The theoretical autocorrelations being difficult – if not impossible – to obtain analytically, we used simulations of the estimated model to approximate these theoretical autocorrelations by their empirical counterparts. The length of the simulations, 50 000, seemed sufficient to obtain good accuracy (this was confirmed by comparing the empirical and theoretical values when the latter were available).

Figure 4.13 shows satisfactory results for the standard GARCH model, as far as the autocorrelations of absolute values are concerned. Of course, this model is not able to reproduce the correlations induced by the leverage effect. Such autocorrelations are adequately reproduced by the TGARCH model, as can be seen from the top and bottom right panels. The autocorrelations for the other asymmetric models are not reproduced here but are very similar to those of the TGARCH. The negative correlations between r_t and the r_{t-h} appear similar to the empirical ones.

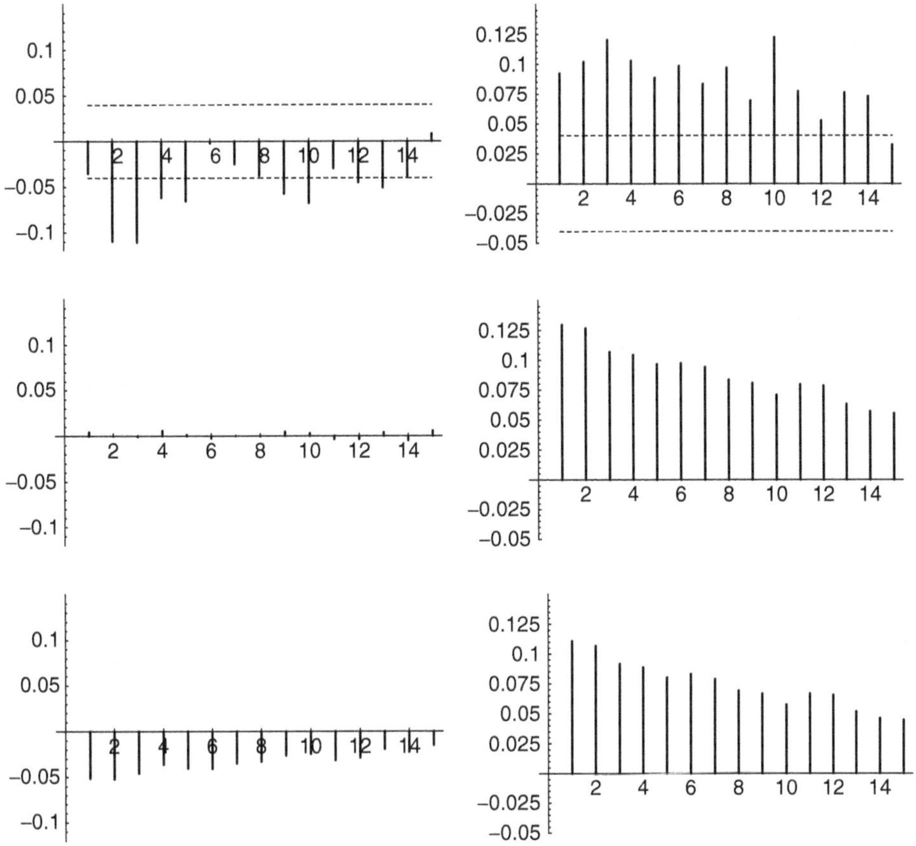

Figure 4.13 Correlogram $h \mapsto \rho(|r_t|, |r_{t-h}|)$ of the absolute values (left) and cross correlogram $h \mapsto \rho(|r_t|, r_{t-h})$ measuring the leverage effect (right), for the CAC 40 series (top), for the standard GARCH (middle), and for the TGARCH (bottom) estimated on the CAC 40 series.

Implied and Empirical Kurtosis

Table 4.6 shows that the theoretical variances obtained from the estimated models are close to the observed variance of the CAC 40 index. In contrast, the estimated kurtosis values are all much below the observed value.

In all these five models, the conditional distribution of the returns is assumed to be $\mathcal{N}(0, 1)$. This choice may be inadequate, which could explain the discrepancy between the estimated theoretical and

Table 4.6 Variance ($\times 10^4$) and kurtosis of the CAC 40 index and of simulations of length 50 000 of the five estimated models.

	CAC 40	GARCH	EGARCH	QGARCH	GJR	TGARCH
Kurtosis	5.9	3.5	3.4	3.3	3.6	3.4
Variance	1.3	1.3	1.3	1.3	1.3	1.3

Table 4.7 Number of CAC returns outside the limits $\overline{r} \pm 3\hat{\sigma}_t$ (THEO being the theoretical number when the conditional distribution is $\mathcal{N}(0, \hat{\sigma}_t^2)$.

THEO	GARCH	EGARCH	QGARCH	GJR	TGARCH
6	17	13	14	15	13

the empirical kurtosis. Moreover, the normality assumption is clearly rejected by statistical tests, such as the Kolmogorov–Smirnov test, applied to the standardised returns. A leptokurtic distribution is observed for those standardised returns.

Table 4.7 reveals a large number of returns outside the interval $[\overline{r} - 3\hat{\sigma}_t, \overline{r} + 3\hat{\sigma}_t]$, whatever the specification used for $\hat{\sigma}_t$. If the conditional law were Gaussian and if the conditional variance were correctly specified, the probability of one return falling outside the interval would be $2\{1 - \Phi(3)\} = 0.0027$, which would correspond to an average of 6 values out of 2385.

Asymmetric GARCH Models with Non-Gaussian Innovations

To take into account the leptokurtic shape of the residuals distribution, we re-estimated the five GARCH models with a Student t distribution – whose parameter is estimated – for η_t.

For instance, the new estimated TGARCH model is

$$\begin{cases} r_t = \underset{(2\times 10^{-4})}{5 \times 10^{-4}} + \epsilon_t, \quad \epsilon_t = \sigma_t \eta_t, \quad \eta_t \sim t(9.7) \\ \sigma_t = \underset{(1\times 10^{-4})}{4 \times 10^{-4}} + \underset{(0.01)}{0.03}\, \epsilon_{t-1}^+ - \underset{(0.02)}{0.10}\, \epsilon_{t-1}^- + \underset{(0.02)}{0.90}\, \sigma_{t-1}. \end{cases} \tag{4.62}$$

It can be seen that the estimated volatility is quite different from that obtained with the normal distribution (see Table 4.8).

Model with Interventions

Analysis of the residuals show that the values observed at times $t = 228$, 682, and 845 are scarcely compatible with the selected model. There are two ways to address this issue: one could either research a new specification that makes those values compatible with the model, or treat these three values as outliers for the selected model.

Table 4.8 Means of the squares of the differences between the estimated volatilities ($\times 10^{10}$) for the models with Student innovations and the TGARCH model with Gaussian innovations (model (4.33) denoted TGARCH^N).

	GARCH	EGARCH	QGARCH	GJR	TGARCH	TGARCH^N
GARCH	0	5.90	2.72	5.89	7.71	15.77
EGARCH	5.90	0	2.27	5.08	0.89	8.92
QGARCH	2.72	2.27	0	2.34	3.35	9.64
GJR	5.89	5.08	2.34	0	7.21	11.46
TGARCH	7.71	0.89	3.35	7.21	0	7.75
TGARCH^N	15.77	8.92	9.64	11.46	7.75	0

In the first case, one could replace the $\mathcal{N}(0,1)$ distribution of the noise η_t with a more appropriate (leptokurtic) one. The first difficulty with this is that no distribution is evident for these data (it is clear that distributions of Student t or generalised error type would not provide good approximations of the distribution of the standardised residuals). The second difficulty is that changing the distribution might considerably enlarge the confidence intervals. Take the example of a 99% confidence interval at horizon 1. The initial interval $[\overline{r} - 2.57\hat{\sigma}_t, \overline{r} + 2.57\hat{\sigma}_t]$ simply becomes the dilated interval $[\overline{r} - t_{0.995}\hat{\sigma}_t, \overline{r} + t_{0.995}\hat{\sigma}_t]$ with $t_{0.995} \gg 2.57$, provided that the estimates $\hat{\sigma}_t$ are not much affected by the change of conditional distribution. Even if the new interval does contain 99% of returns, there is a good chance that it will be excessively large for most of the data.

So for this first case, we should ideally change the prediction formula for σ_t so that the estimated volatility is larger for the three special data (the resulting smaller standardized residuals $(r_t - \overline{r})/\sqrt{n}$ would become consistent with the $\mathcal{N}(0,1)$ distribution), without much changing volatilities estimated for other data. Finding a reasonable model that achieves this change seems quite difficult.

We have, therefore, opted for the second approach, treating these three values as outliers. Conceptually, this amounts to assuming that the model is not appropriate in certain circumstances. One can imagine that exceptional events occurred shortly before the three dates $t = 228$, 682, and 845. Other special events may occur in the future, and our model will be unable to anticipate the changes in volatility induced by these extraordinary events. The ideal would be to know the values that returns would have had if these exceptional event had not occurred, and to work with these corrected values. This is, of course, not possible, and we must also estimate the adjusted values. We will use an intervention model, assuming that only the returns of the three dates would have changed in the absence of the above-mentioned exceptional events. Other types of interventions can, of course, be envisaged. To estimate what would have been the returns of the three dates in the absence of exceptional events, we can add these three values to the parameters of the likelihood. This can easily be done using a SAS program (see Table 4.9).

4.9 Models Incorporating External Information

GARCH-X Models

The usual GARCH-type models predict the squared returns ϵ_t^2 by means of the past returns only $\{\epsilon_u, u < t\}$. It is, however, often the case that some extra information is available, under the form of a vector $\boldsymbol{x}_{t-1}$ of exogenous covariates, such as the daily volume of transactions, or high frequency intraday data, or even series of other returns. To take advantage of the extra information – in order to improve the prediction of the squares – GARCH models that are augmented with additional explanatory variables (GARCH-X models in short), can be considered. A simple version of GARCH-X is of the form

$$\begin{cases} \epsilon_t = \sigma_t \eta_t \\ \sigma_t^2 = \omega + \alpha \epsilon_{t-1}^2 + \beta \sigma_{t-1}^2 + \boldsymbol{\pi}' \boldsymbol{x}_{t-1}, \end{cases} \tag{4.63}$$

where $\boldsymbol{x}_t = (x_{1,t}, \ldots, x_{r,t})'$ is a vector of r exogenous covariates. The term 'exogenous' does not refer to the concepts of weak or strong exogeneity introduced in the econometric literature, but is employed because the dynamics if the vector $\boldsymbol{x}_t$ is not specified by the GARCH-X model (4.63).

To ensure $\sigma_t^2 > 0$ with probability one, assume that the covariates are a.s. positive and that the coefficients satisfy $\alpha \geq 0, \beta \geq 0, \omega > 0$, and $\pi = (\pi_1, \ldots, \pi_r)' \geq 0$ componentwise. In the GARCH-type models, the sequence (η_t) is traditionally assumed to be iid(0, 1), but the assumption is not necessary for the following result.

Theorem 4.9 (Strict stationarity of the GARCH-X(1, 1)) *Assume*

Assumption A: $(\eta_t, \boldsymbol{x}_t')$ *is a strictly stationary and ergodic process, and there exists* $s > 0$ *such that* $E|\eta_1|^s < \infty$ *and* $E \| \boldsymbol{x}_1 \|^s < \infty$.

If $\gamma := E \ln a(\eta_1) < 0$, with $a(z) = \alpha z^2 + \beta$, the GARCH-X model (4.63) admits a unique strictly stationary, non-anticipative and ergodic solution, given by

$$\sigma_t^2 = \omega_t + \sum_{k=1}^{\infty} \left\{ \prod_{i=1}^{k} a(\eta_{t-i}) \right\} \omega_{t-k}, \quad \text{where } \omega_t = \omega + \boldsymbol{\pi}' \boldsymbol{x}_{t-1}.$$

When $\gamma \geq 0$, there exists no stationary solution to (4.63).

Proof. Note that **A** entails that $E \ln^+ \omega_t < \infty$. Therefore, $\limsup_{k\to\infty} \ln \omega_{t-k}/k \leq 0$ (see Exercise 4.12). By the Cauchy rule, the condition $\gamma < 0$ thus implies the existence of the series defining σ_t^2. Now, if

Table 4.9 SAS program for the fitting of a TGARCH(1, 1) model with interventions.

```
/* Data reading */
 data cac;
 infile 'c:\enseignement\PRedess\Garch\cac8898.dat';
 input indice; date=_n_;
 run;
/* Estimation of a TGARCH(1,1) model */
   proc model data = cac ;
      /* Initial values are attributed to the parameters */
       parameters cacmod1 -0.075735 cacmod2 -0.064956
           cacmod3 -0.0349778 omega .000779460
              alpha_plus 0.034732 alpha_moins 0.12200 beta 0.86887
                  intercept .000426280 ;
      /* The index is regressed on a constant and 3 interventions
           are made*/
       if (_obs_  = 682) then indice= cacmod1;
       else if (_obs_  = 228) then indice= cacmod2;
            else if (_obs_  = 845) then indice= cacmod3;
                 else indice = intercept ;
      /* The conditional variance is modeled by a TGARCH */
      if (_obs_ = 1) then
       if((alpha_plus+alpha_moins)/sqrt(2*constant('pi'))+beta=1)
           then
        h.indice = (omega + (alpha_plus/2+alpha_moins/2+beta)
            *sqrt(mse.indice))**2 ;
       2else h.indice = (omega/(1-(alpha_plus+alpha_moins)/
           sqrt(2*constant('pi'))-beta))**2;
      else
       if zlag(-resid.indice) > 0 then h.indice = (omega
                                 + alpha_plus*zlag(-resid.indice)
                                 + beta*zlag(sqrt(h.indice)))**2 ;
       else h.indice = (omega - alpha_moins*zlag(-resid.indice)
               + beta*zlag(sqrt(h.indice)))**2 ;
      /* The model is fitted and the normalized residuals are
          stored in a SAS table*/
      outvars nresid.indice;
      fit indice / method = marquardt fiml out=residtgarch ;
   run ; quit ;
```

$\sigma_t^2 < \infty$, then by arguments of the proof of Theorem 2.1, for all t the term $\prod_{i=1}^k a(\eta_{t-i})$ tends a.s. to 0 as $k \to \infty$. By Lemma 2.1 this implies $\gamma < 0$. The uniqueness of the solution and the other results are proved as in Theorem 2.1. □

It is interesting to note that, under Assumption **A**, the stationarity condition of (ϵ_t) is not modified by the presence of the exogenous variables. Note that if

$$E(\eta_t|\mathcal{F}_{t-1}) = 0 \quad \text{and} \quad E(\eta_t^2|\mathcal{F}_{t-1}) = 1,$$

where $\mathcal{F}_{t-1}$ is the information set generated by $\{\epsilon_u, \boldsymbol{x}_u : u < t\}$, then σ_t^2 is the conditional variance of ϵ_t given $\mathcal{F}_{t-1}$.

HEAVY Models

Intraday high-frequency data, and thus realised volatility measures RM_t (such as the realized variance), are increasingly available. This extra information can be included as exogenous variable $\boldsymbol{x}_t = \text{RM}_t$ in model (4.63), which generally greatly improves the prediction of ϵ_t^2. The drawback of the GARCH-X model is however that the dynamics of the covariate is not specified. This forces the practitioner to put in the volatility equation of σ_{t+h}^2 only variables that will be available (or predictable) at time $t+h$, which is restrictive if the horizon of prediction h is larger than 1. To specify the dynamics of (RM_t), Shephard and Sheppard (2010) proposed the high-frequency-based volatility (HEAVY) model, in which Model (4.63)

$$\begin{cases} \epsilon_t = \sigma_t \eta_t \\ \sigma_t^2 = \omega + \alpha\epsilon_{t-1}^2 + \beta\sigma_{t-1}^2 + \pi \text{RM}_{t-1}, \end{cases} \tag{4.64}$$

is completed by the Multiplicative Error Model (MEM)

$$\begin{cases} \text{RM}_t = \mu_t z_t \\ \mu_t = \omega^* + \alpha^* \text{RM}_{t-1} + \beta^* \mu_{t-1}, \end{cases} \tag{4.65}$$

where (z_t) is a stationary sequence of positive variables such that $E(z_t|\mathcal{F}_{t-1}) = 1$, $\omega^* > 0$, $\alpha^* \geq 0$ and $\beta^* \geq 0$.

Theorem 4.10 (Strict stationarity of the HEAVY model) *Let $\mathcal{F}_t$ be the sigma-field generated by $\{\eta_u, z_u : u \leq t\}$. Assume that $z_t \geq 0$ a.s. and that $\{(\eta_t, z_t - 1)', \mathcal{F}_t\}$ is a stationary and ergodic martingale difference. Let $a(\eta) = \alpha\eta^2 + \beta$ and $a^*(z) = \alpha^* z^2 + \beta^*$. If*

$$\gamma := E \ln a(\eta_1) < 0 \quad \text{and} \quad \gamma^* := E \ln a^*(z_1) < 0,$$

then the HEAVY model (4.64) and (4.65) admits a unique strictly stationary, non-anticipative, and ergodic solution, and $\sigma_t^2 = E(\epsilon_t^2 \mid \mathcal{F}_{t-1})$. When $\gamma^* \geq 0$, there exists no stationary solution to Eq. (4.65) and no stationary solution to Eq. (4.64). When $\gamma \geq 0$, there exists no stationary solution to Eq. (4.64).

Proof. Note that $\mu_t = \omega^* + a^*(z_{t-1})\mu_{t-1}$, which is similar to (2.11). By the arguments of the proof of Theorem 2.1, it can thus be shown that (4.65) admits a strictly stationary solution (which is then unique, non-anticipative, and ergodic) if and only if $\gamma^* < 0$. We conclude by Theorem 4.9. □

Realised-GARCH Models

An alternative model has been proposed by Hansen, Huang, and Shek (2012) (see also Hansen and Huang, 2016). In this model, called realised-GARCH, a GARCH-X equation similar to Eq. (4.64), or alternatively a Log-GARCH-X equation of the form

$$\begin{cases} \epsilon_t &= \sigma_t \eta_t \\ \ln \sigma_t^2 &= \omega + \beta \ln \sigma_{t-1}^2 + \pi \ln \mathrm{RM}_{t-1}, \end{cases} \tag{4.66}$$

is completed by a measurement equation for the realised volatility measure, that can take the form

$$\ln \mathrm{RM}_t = \xi + \varphi \ln \sigma_t^2 + \tau(\eta_t) + u_t, \tag{4.67}$$

where (u_t) is a strong white noise independent of (η_t). Note that the dynamics are written on the logarithm of the volatility and of the realised volatility measure, in order to avoid positivity constraints. The term $\tau(\eta_t)$ is assumed to be centred, for instance of the form $\tau(z) = \tau_1 z + \tau_2(z^2 - 1)$, and can thus take into account the leverage effect. Defining a strong white noise (v_t) by $v_t = \tau(\eta_{t-1}) + u_{t-1}$, one can see that $\ln \sigma_t^2$ satisfies the AR(1) model

$$\ln \sigma_t^2 = \omega + \pi\xi + (\beta + \pi\varphi) \ln \sigma_{t-1}^2 + \pi v_t,$$

from which stationarity conditions follow. Note, however, that the realised-GARCH is not a stochastic volatility model (see Chapter 12) because σ_t^2 is measurable with respect to the sigma-field $\mathcal{F}_{t-1}$ generated by $\{\mathrm{RM}_u : u < t\}$.

Intraday Scale Models

Visser (2011) proposed a model that directly links the daily returns and the intraday price movements. Let $p_t(u)$ be the price of an asset at the intraday time $u \in [0, 1]$ of the day $t \in \mathbb{N}$. Consider the continuous-time log-return process $R_t(u) = \ln p_t(u)/p_t(0)$. Note that the open-to-close daily return is $\epsilon_t = R_t(1)$. If, for example, $R_t(u) = \sigma_t W_t(u)$ where $\{W_t(\cdot)\}_t$ is an iid sequence of standard Brownian motions, then we have the ARCH-type equation $\epsilon_t = \sigma_t \eta_t$ where (η_t) is iid $\mathcal{N}(0, 1)$. In this model, the realised volatility – usually defined as the sum of the squared five-minute returns – should be a good proxy of the volatility since

$$\sum_{k=1}^{[1/h]} [R_t(kh) - R_t(kh - h)]^2 = \sigma_t^2 \sum_{k=1}^{[1/h]} [W_t(kh) - W_t(kh - h)]^2 \to \sigma_t^2 \tag{4.68}$$

in mean square as $h \to 0$ (see Exercise 4.13). More generally, denoting by $D[0, 1]$ the set of the real-valued function on [0,1] which are right continuous with left limits (càdlàg), if one assumes

$$R_t(u) = \sigma_t \Upsilon_t(u) \quad \text{where } (\Upsilon_t) \text{ is an iid sequence of random elements of } D[0, 1],$$

and

$$H : D[0, 1] \to (0, \infty) \text{ is positively homogeneous,}$$

that is, $H(cR) = cH(R)$ for all $c \geq 0$ and $R \in D[0, 1]$, then the realized measure $\mathrm{RM}_t = H(R_t(\cdot))$ satisfies the ARCH, or more precisely the MEM-type equation

$$\mathrm{RM}_t = H\{\sigma_t \Upsilon_t(\cdot)\} = \sigma_t z_t, \quad \text{where the } z_t = H(\Upsilon_t) \text{ are iid.}$$

Noting that $E(\mathrm{RM}_t^2 | \mathcal{F}_{t-1}) = \sigma_t^2 E z_1^2$, Visser (2011) showed that, up to the unknown scaling constant $E z_1^2$, the parameters giving the dynamics of the daily volatility σ_t can also be estimated from a MEM model on the sequence (RM_t). The resulting estimators are generally more accurate than those based on (ϵ_t), intuitively because the variance of RM_t is often much smaller than that of $\epsilon_t^2 - \sigma_t^2$.

MIDAS Models

The previous models aim at explaining a daily variable (the squared returns) by an intraday variable. More generally, MIxed DAta Sampling (MIDAS) regression is a technique that can be used when independent variables appear at higher frequencies than the dependent variable (see Wang and Ghysels (2015) and the references therein). Engle, Ghysels, and Sohn (2013) proposed a GARCH-MIDAS model with a short-run volatility component g_t and a long-run volatility component τ_t. The short-run component g_t is the volatility of a unit-variance GARCH(1,1) process, defined by

$$\begin{cases} \epsilon_t = \sqrt{g_t}\eta_t \\ g_t = (1-\alpha-\beta) + \alpha\epsilon_{t-1}^2 + \beta g_{t-1}, \end{cases} \tag{4.69}$$

with $\alpha > 0$, $\beta \geq 0$, and $\alpha + \beta < 1$. As usual, (η_t) denotes an iid (0,1) sequence. The observed sequence of returns (r_t) is then assumed to follow an ARCH-type model with the error term ϵ_t and a volatility τ_t driven by past realised volatilities $\text{RV}_t = \sum_{j=0}^{N-1} r_{t-j}^2$:

$$\begin{cases} r_t = \sqrt{\tau_t}\epsilon_t \\ \tau_t = m + \theta\sum_{k=1}^{K} \varphi_k \text{RV}_{t-k}. \end{cases} \tag{4.70}$$

The parameters m and θ are positive. The weights φ_k are positive, and their sum is equal to 1. For instance, one can take

$$\varphi_k = \varphi_k(\rho) = \frac{\rho^k}{\sum_{j=1}^{K} \rho^j}, \quad \rho \in (0,1).$$

Note that

$$\tau_t = m + \theta\sum_{k=1}^{K} \varphi_k \sum_{j=0}^{N-1} r_{t-k-j}^2 := \omega + \sum_{i=1}^{q} \alpha_i r_{t-i}^2$$

with $\omega = m$ and $q = K + N - 1$. The first ARCH coefficient is $\alpha_1 = \theta\varphi_1$ and the last one is $\alpha_q = \theta\varphi_K$. Compared to a standard ARCH(q), the error term ϵ_t is not iid(0,1), but is a GARCH(1,1). The stationarity condition remains, however, the same (see e.g. Remark 4.1, or Theorem 4.9 showing that the iid assumption can be replaced by an assumption of the form **A**).

4.10 Models Based on the Score: GAS and Beta-t-(E)GARCH

Assume an iid sample $X_1, \dots, X_n$ of density f_θ depending on an unknown parameter $\theta \in \mathbb{R}^d$. It is rarely possible to find analytically the maximum likelihood estimator $\hat{\theta}$ maximising the log-likelihood

$$\ell_n(\theta) = \sum_{i=1}^{n} \ln f_\theta(X_i).$$

Starting from a current approximation $\theta^{(k)}$ of $\hat{\theta}$, a Taylor expansion shows that under some regularity conditions

$$0 = \frac{\partial \ell_n(\hat{\theta})}{\partial\theta} \simeq \frac{\partial \ell_n(\theta^{(k)})}{\partial\theta} - \hat{I}_n(\hat{\theta} - \theta^{(k)}),$$

where $\hat{I}_n = -\frac{\partial^2 \ell_n(\theta^{(k)})}{\partial\theta\partial\theta'}$ is an estimate of the Fisher information matrix and $\frac{\partial \ell_n(\theta)}{\partial\theta}$ is called the score. Assuming $\hat{I}_n$ invertible, we thus have $\hat{\theta} \simeq \theta^{(k)} + \hat{I}_n^{-1}\frac{\partial \ell_n(\theta^{(k)})}{\partial\theta}$, which suggests to update the approximation of $\hat{\theta}$ by the so-called Newton–Raphson iteration

$$\theta^{(k+1)} = \theta^{(k)} + \hat{I}_n^{-1}\frac{\partial \ell_n(\theta^{(k)})}{\partial\theta}. \tag{4.71}$$

Now consider a time series whose conditional distribution is driven by a time-varying parameter

$$X_t|\mathcal{F}_{t-1} \sim f_{\theta_t}, \tag{4.72}$$

θ_t being a measurable function of the sigma-field $\mathcal{F}_{t-1}$ generated by the past values $\{X_u, u < t\}$. Roughly speaking, one can say that the larger $\ln f_{\theta_t}(X_t)$ is, the more plausible the time series model (i.e. the time-varying parameter θ_t). If $\ln f_{\theta_t}(X_t)$ is not maximal, the score $\partial \ln f_{\theta_t}(X_t)/\partial\theta_t$ is not null, and one may want to improve the 'likelihood' of the time series model by updating θ_t using an equation based on the score, in the spirit of the Newton–Raphson iteration (4.71). Creal, Koopman, and Lucas (2013) proposed an updating equation of the form

$$\theta_{t+1} = \omega + \beta\theta_t + \alpha S(\theta_t)\frac{\partial \ln f_{\theta_t}(X_t)}{\partial \theta_t}, \tag{4.73}$$

where $S(\theta_t)$ is a scaling factor which may be, similar to Eq. (4.71), the inverse of conditional information matrix

$$I_{t-1} = -E\left(\frac{\partial^2 \ln f_{\theta_t}(X_t)}{\partial\theta_t\partial\theta_t'} \mid \mathcal{F}_{t-1}\right).$$

The Eqs. (4.72) and (4.73) define a Generalized Autoregressive Score (GAS) model. The parameters ω, α and β involved in (4.73) do not appear in the Newton–Raphson iteration (4.71). An argument for including those parameters can be found in Blasques, Koopman and Lucas (2015).

Take the example of a volatility model $\epsilon_t = \sigma_t\eta_t$ with Gaussian innovations. In this example the time-varying parameter is $\theta_t = \sigma_t^2$ and the conditional distribution of ϵ_t is

$$f_{\sigma_t}(\epsilon_t) = \frac{1}{\sigma_t}\phi\left(\frac{\epsilon_t}{\sigma_t}\right),$$

with ϕ the $\mathcal{N}(0, 1)$ density. The score and conditional information are thus

$$\frac{\partial \ln f_{\sigma_t}(\epsilon_t)}{\partial\sigma_t^2} = \frac{\epsilon_t^2 - \sigma_t^2}{2\sigma_t^4}, \qquad I_{t-1} = \frac{1}{2\sigma_t^4}.$$

The updating Eq. (4.73) becomes, with $S(\sigma_t^2) = I_{t-1}^{-1}$,

$$\sigma_{t+1}^2 = \omega + \beta\sigma_t^2 + \alpha(\epsilon_t^2 - \sigma_t^2), \tag{4.74}$$

which is a standard GARCH(1,1) equation. The updating equation depends drastically on the assumed conditional distribution. Assume, for instance that, instead of a Gaussian, the conditional distribution is a Student with $\nu > 2$ degrees of freedom, normalised to obtain a unit variance. The updating Eq. (4.73) becomes

$$\sigma_{t+1}^2 = \omega + \beta\sigma_t^2 + \alpha\left\{\frac{\nu + 1}{\nu - 2 + \frac{\epsilon_t^2}{\sigma_t^2}}\epsilon_t^2 - \sigma_t^2\right\}, \tag{4.75}$$

which is the Beta-t-GARCH of Harvey and Chakravarty (2008) (see Exercise 4.14).[5] Note that when ν is small, an extreme value of ϵ_t^2 has a lower impact on the volatility (4.75) than on the volatility (4.74), in agreement with the fact that an extreme value is more likely to occur with a Student than with a Gaussian distribution. In some sense, GAS models are motivated by a kind of generalization of the likelihood principle, that selects a time-varying parameter on the basis of its score.

This approach is thus an elegant way to introduce non-linear volatility models. The choice of a particular specification, for example a standard GARCH(1,1) with Student innovations or a Beta-t-GARCH model, should however be data-dependent. Indeed, there is a priori no reason to think that Nature chooses

[5] Creal, Koopman and Lucas (2013) and Harvey adn Chakravarty (2008) also proposed alternative formulations. In particular, in the latter reference, the Beta-t-GARCH is obtained when the time-varying parameter is $\theta_t = \ln\sigma_t^2$ instead of $\theta_t = \sigma_t^2$.

with higher probability a given model rather than another. Note also that, similar to the EGARCH (which is a particular GAS model), GAS models such as (4.75) are potentially subject to invertibility issues (see Blasques et al. 2018).

4.11 GARCH-type Models for Observations Other Than Returns

GARCH modelling of the volatility of financial returns has inspired a variety of formulations for other types of data (e.g. durations between transactions, volumes, integer, or functional data). We briefly review some of them.

Duration Models

Many financial data – in particular intra-daily high-frequency data – are irregularly spaced in time. Several approaches have been proposed to tackle this problem. Engle and Russell (1998) introduced the so-called Autoregressive Conditional Duration (ACD) model for durations between events (such as trades, quotes, price changes).

Let t_i be the time at which the ith event of interest occurs, and denote by $x_i = t_i - t_{i-1}$ the ith duration between two consecutive events. The ACD model is specified as

$$\begin{cases} x_i = \psi_i z_i \\ \psi_i = \alpha_0 + \sum_{j=1}^{p} \alpha_j x_{i-j} + \sum_{\ell=1}^{q} \beta_\ell \psi_{i-\ell}, \end{cases} \tag{4.76}$$

where p and q are non-negative integers. Since the duration x_i is necessarily non-negative, it is common to assume that (z_i) is a sequence of non-negative iid random variables with $E(z_i) = 1$, and to impose $\alpha_0 > 0$, $\alpha_j \geq 0$, $\beta_\ell \geq 0$.

Let $\mathcal{F}_{i-1}$ be the information set consisting of all information up to and including time t_{i-1}. It is clear that, if the roots of the polynomial $\beta(L) = 1 - \sum_{\ell=1}^{q} \beta_\ell L^\ell$ are outside the unit circle, and if z_i is independent of $\mathcal{F}_{i-1}$, the variable ψ_i can be interpreted as the conditional mean of the i-th duration,

$$\psi_i = E(x_i \mid \mathcal{F}_{i-1}).$$

The existence of a strictly stationary non-anticipative solution to model (4.76) follows straightforwardly from the study of GARCH models. For instance, when $p = q = 1$, the condition $E \ln(\alpha_1 z_1 + \beta_1) < 0$ ensures the existence of a strictly stationary and non-anticipative solution. Higher-order models can be studied by rewriting the conditional duration as $\psi_i = \alpha_0 + \sum_{j=1}^{\max(p,q)} (\alpha_j z_{i-j} + \beta_j)\psi_{i-j}$, and by deducing a Markov vector representation.

Estimation of the parameters, $\theta_0 = (\alpha_0, \ldots, \alpha_p, \beta_1, \ldots, \beta_q)'$, can be performed using the QML approach. Gouriéroux, Monfort, and Trognon (1984) showed that, in order to estimate the parameters of a correctly specified conditional mean model, the QML estimator is consistent regardless of the true innovations distribution if and only if the QML is based on a distribution belonging to the linear exponential family. In particular, the QML based on the standard exponential distribution will provide consistent estimation (under regularity conditions that we will not discuss here) of θ_0. The exponential QML is obtained by minimising on some compact set of θ's the criterion

$$\sum_{i=1}^{n} \ln \psi_i(\theta) + \frac{x_i}{\psi_i(\theta)},$$

where the $\psi_i(\theta)$s are computed recursively using initial values for the x_is. Note that this criterion coincides with the Gaussian QML criterion of a GARCH(p, q) model, provided the x_is are replaced by the squared returns ϵ_i^2. This is not surprising since an ACD process is nothing else than the square of a

GARCH process: $x_i = \epsilon_i^2$ with

$$\begin{cases} \epsilon_i = \sqrt{\psi_i}\eta_i \\ \psi_i = \alpha_0 + \sum_{j=1}^{p} \alpha_j \epsilon_{i-j}^2 + \sum_{\ell=1}^{q} \beta_\ell \psi_{i-\ell}, \end{cases}$$

where $\eta_i = z_i s_i$, (s_i) being an iid sequence, independent of the sequence (z_i), and uniformly distributed on $\{+1, -1\}$. It follows that standard packages for estimating GARCH models can be directly used to estimate ACD models. Other distributions with a positive support, like the Weibull distribution considered by Engle and Russell (1998), or the Burr distribution used by Grammig and Maurer (2000), will not produce consistent QML estimators in general.

In practical applications, ACD models encounter some weaknesses. For instance, zero durations are often ignored. Another difficulty is that intraday transactions of a stock often exhibit diurnal patterns. In other words, the strict stationarity assumption in not realistic. Different approaches have been proposed to remove the diurnal patterns in particular using deterministic functions of the time of the day (see e.g. Engle and Russell 1998; Tsay 2005).

Note that the ACD model can also be used to model non-negative time series. An example is the range of an asset during a trading day, which can be seen as a measure of its volatility (see Parkinson 1980). The Conditional Autoregressive Range (CARR) model introduced by Chou (2005) is an ACD model for such data. These models can also be seen as particular instances of the multiplicative error model (MEM) introduced by Engle (2002b).

INGARCH Models

Time series of counts are often observed in the real world, in particular in finance (see e.g. Rydberg and Shephard 2003). Heinen (2003) considered a model called Autoregressive Conditional Poisson (ACP) and applied it to the daily number of price change durations of \$0.75 on the IBM stock. Ferland, Latour, and Oraichi (2006) further studied the ACP model, under the name INteger-valued GARCH (INGARCH). A process (X_t), valued in $\mathbb{N}$, satisfies an INGARCH(p, q) model if the conditional distribution of X_t given its past values is Poisson with intensity parameter

$$\lambda_t = \omega + \sum_{i=1}^{q} \alpha_i X_{t-i} + \sum_{j=1}^{p} \beta_j \lambda_{t-j}, \tag{4.77}$$

where $\omega > 0$, $\alpha_i \geq 0$, and $\beta_j \geq 0$ for $i = 1, \ldots, q$ and $j = 1, \ldots, p$. The name INGARCH comes from the fact that the conditional variance (4.77) resembles the volatility of a GARCH. Note, however, that, contrary to a GARCH, the conditional mean of X_t is not zero (but is equal to λ_t). Other count time series models can be obtained by assuming alternative discrete conditional distributions, with other time-varying parameters. Note, however, that studying the probabilistic structure of count time series models, in particular giving conditions for ergodicity, may be a difficult issue (see Fokianos, Rahbek, and Tjøstheim 2009; Tjøstheim 2012; Davis et al. 2016).

Theorem 4.11 (Strict stationarity and ergodicity of the INGARCH) *There exists a stationary and ergodic solution to the INGARCH(p, q) model (4.77) such that $E\lambda_t < \infty$ if and only if*

$$\sum_{i=1}^{q} \alpha_i + \sum_{j=1}^{p} \beta_j < 1. \tag{4.78}$$

Proof. Let F_λ be the cdf of the Poisson distribution of parameter λ, and let F_λ^- be its generalised inverse. First note that we have

$$\lambda \leq \lambda^* \Rightarrow F_\lambda(x) \geq F_{\lambda^*}(x), \quad \forall x \in \mathbb{R} \Rightarrow F_\lambda^-(u) \leq F_{\lambda^*}^-(u), \quad \forall u \in (0, 1). \tag{4.79}$$

We now construct a stationary sequence (X_t) such that

$$P(X_t \leq x \mid X_u, u < t) = F_{\lambda_t}(x), \tag{4.80}$$

where λ_t satisfies condition (4.77) with the constraint (4.78).

Let (U_t) be an iid sequence of random variables uniformly distributed in [0, 1]. For $t \in \mathbb{Z}$, let $X_t^{(k)} = \lambda_t^{(k)} = 0$ when $k \leq 0$ and, for $k > 0$, let

$$X_t^{(k)} = F^-_{\lambda_t^{(k)}}(U_t), \quad \lambda_t^{(k)} = \omega + \sum_{i=1}^{q} \alpha_i X_{t-i}^{(k-i)} + \sum_{j=1}^{p} \beta_j \lambda_{t-j}^{(k-j)}. \tag{4.81}$$

Note that, for $k \geq 2$, $\lambda_t^{(k)} = \psi_k(U_{t-1}, \dots, U_{t-k+1})$, where $\psi_k : [0, 1]^{k-1} \to [0, \infty)$ is a measurable function. Therefore, for any k, the sequences $(\lambda_t^{(k)})_t$ and $(X_t^{(k)})_t$ are stationary and ergodic.

To show the existence of a solution to (4.80), we will show that

$$\lambda_t = \lim_{k\to\infty} \lambda_t^{(k)} \text{ exists almost surely in } [0, +\infty). \tag{4.82}$$

Indeed, under this condition, for all t we have

$$X_t^{(k)} = F^-_{\lambda_t^{(k)}}(U_t) \to X_t := F^-_{\lambda_t}(U_t) \quad \text{a.s. as } k \to \infty.$$[6]

The limit λ_t then satisfies (4.77) and is measurable with respect to $\{X_u, u < t\}$ (because $\sum_{j=1}^q \beta_j < 1$ under (4.78), and thus, by the arguments given in Corollary 2.2, the polynomial $1 - \sum_{j=1}^q \beta_j z^j$ has its root outside the unit circle). We have then defined a stationary sequence (X_t) satisfying (4.80).

It thus remains to show (4.82) under (4.78). Using (4.79), it can be easily shown by induction that, almost surely

$$0 \leq \lambda_t^{(k-1)} \leq \lambda_t^{(k)}, \quad \forall k.$$

Moreover, $EX_t^{(k)} = E\lambda_t^{(k)}$ exists for any fixed k. We thus have $\rho \in (0, 1)$ such that

$$0 \leq E(\lambda_t^{(k)} - \lambda_t^{(k-1)}) = \sum_{i=1}^{r} (\alpha_i + \beta_i) E(\lambda_{t-i}^{(k-i)} - \lambda_{t-i}^{(k-i-1)}) \leq K\rho^k, \quad \forall k \geq 1.$$

In the case $p = q = 1$, one can take $\rho = \alpha_1 + \beta_1$ since

$$E(\lambda_t^{(k)} - \lambda_t^{(k-1)}) = (\alpha_1 + \beta_1) E(\lambda_{t-1}^{(k-1)} - \lambda_{t-1}^{(k-2)}) = (\alpha_1 + \beta_1)^{k-1}\omega.$$

In the case $r > 1$, using the matrix representation

$$E(\underline{\lambda}_t^{(k)} - \underline{\lambda}_t^{(k-1)}) = AE(\underline{\lambda}_t^{(k-1)} - \underline{\lambda}_t^{(k-2)}),$$

with

$$\underline{\lambda}_t^{(k)} = \begin{pmatrix} \lambda_t^{(k)} \\ \vdots \\ \lambda_t^{(k-r+1)} \end{pmatrix}, \quad A = \begin{pmatrix} \alpha_1 + \beta_1 & \dots & \alpha_r + \beta_r \\ \boldsymbol{I}_{r-1} & & 0_{r-1} \end{pmatrix}$$

one can take $\rho = \rho(A)$ (using Exercise 2.12). This entails that the sequence $\{\lambda_t^{(k)}\}_k$ converges in L^1 and almost surely under (4.78). Moreover, since $\lambda_t = \psi(U_{t-1}, \dots)$, where $\psi : [0, 1]^\infty \to [0, \infty)$ is a measurable function, the sequence (λ_t) is ergodic. Conversely, it is obvious to see that (4.78) is necessary for the existence of a stationary solution to (4.80) such that $EX_t = E\lambda_t < \infty$. □

[6] Indeed, if $\lambda_k \to \lambda$ as $k \to \infty$, then $\lim_{k\to\infty} = F^-_{\lambda_k}(x) = F^-_{\lambda}(x)$ if x does not belong to the countable set of the discontinuity points of F^-_{λ}.

Functional GARCH Models

Given the increasing amount of financial data at a high, or ultra-high frequency, it makes sense to describe such high-resolution tick data as functions. A functional version of the ARCH model was proposed in Hörmann, Horváth, and Reeder (2013) and recently extended by Aue et al. (2017). Their approach relies on a daily segmentation of the data, which is more realistic than only one continuous time process. More precisely, it is assumed that we have a functional time series $\{\epsilon_k(t), 1 \leq k \leq T, 0 \leq t \leq 1\}$, where $\epsilon_k(t)$ are intraday log-returns on day k at time t. For some lag h (for instance 1 or 5 minutes), we thus have $\epsilon_k(t) = \ln p_k(t) - \ln p_k(t-h)$, where $\{p_k(t)\}$ is the underlying price process. For convenience, the price $p_k(t)$ is supposed to be defined on the set $[-h, 1]$, so that $\epsilon_k(t)$ is defined for t belonging to $[0, 1]$.

To define sequences of real-valued functions with domain $[0, 1]$, we use the following notation. Let $\mathcal{H} = L^2[0, 1]$ denote the Hilbert space of square integrable functions with norm $\| x\|_{\mathcal{H}} = \left(\int_0^1 x^2(s)ds\right)^{1/2}$, which is generated by the inner product $\langle x, y\rangle_{\mathcal{H}} = \int_0^1 x(s)y(s)ds$, for $x, y \in \mathcal{H}$. If $x, y \in \mathcal{H}$, xy stands for point-wise multiplication, $xy(s) = x(s)y(s)$ for $s \in [0, 1]$. By $\mathcal{H}^+$ we denote the set of non-negative functions in $\mathcal{H}$. A non-negative integral operator α, mapping $\mathcal{H}^+$ on $\mathcal{H}^+$ is defined, for some integrable kernel function $K_\alpha \geq 0$ defined on $[0, 1]^2$, by

$$\alpha[x](t) = \int K_\alpha(t, s)x(s)ds, \quad x \in \mathcal{H}, t \in [0, 1].$$

A *functional ARCH(1)* process $\{\epsilon_k(t), k \in \mathbb{Z}, 0 \leq t \leq 1\}$ can be defined as the solution of the model

$$\begin{cases} \epsilon_k = \sigma_k \eta_k \\ \sigma_k^2 = \omega + \alpha[\epsilon_{k-1}^2], \end{cases} \tag{4.83}$$

where $\omega \in \mathcal{H}^+$, α is a non-negative integral operator with kernel K_α, and $\{\eta_k\}$ is an iid sequence of random elements of $\mathcal{H}$.

Different choices are possible for the kernel K_α, leading to different interpretations. For a constant kernel, $K_\alpha(t, s) = a$ for some positive constant a, we have $\alpha(\epsilon_{k-1}^2)(t) = a \int_0^1 \epsilon_{k-1}^2(s)ds$. It follows that the volatility at time t of day k has the form

$$\sigma_k^2(t) = \omega(t) + a \int_0^1 \epsilon_{k-1}^2(s)ds,$$

in which the effects of t and k are separate. In this simple specification, the volatility of day k depends on time t through a deterministic function. Now suppose that $K_\alpha(t, s) = a\phi(s - t)$, where ϕ denotes the Gaussian density, a a positive constant. It follows that

$$\sigma_k^2(t) = \omega(t) + a \int_0^1 \phi(s - t)\epsilon_{k-1}^2(s)ds.$$

Hence, the volatility at time t of day k depends on the returns of day $k - 1$, but with maximal weight around time t.

As in usual GARCH models, it is necessary to constrain the parameters to obtain a strictly stationary solution. Hörmann, Horváth, and Reeder (2013) showed that if (i) $\eta_k \in L^4[0, 1]$, (ii) the operator α is bounded and

$$E\{K(\eta_1^2)\}^s < 1, \quad \text{for some } s < 1,$$

where $K(\eta_1^2) = \left(\int \int \alpha^2(s, t)\eta_1^4(s)dsdt\right)^{1/2}$, then the functional ARCH(1) model admits a unique strictly stationary solution. Moreover, this solution is non-anticipative, with

$$\sigma_k^2 = g(\eta_{k-1}, \eta_{k-2}, \ldots)$$

for some measurable functional $g:\mathcal{H}^{\mathbb{N}} \mapsto \mathcal{H}$. Consequently, the strictly stationary solution is also ergodic.

The intra-day correlations of the stationary solutions can be characterised. Assuming that $E\eta_1(t) = 0, E\eta_1^2(t) < \infty$ and $E\sigma_1^2(t) < \infty$ for all $t \in [0, 1]$, we find that $E\epsilon_k(t) = E(\epsilon_k(t) \mid \mathcal{F}_{k-1}) = 0$ and for $s, t \in [0, 1]$

$$\mathrm{Cov}(\epsilon_k(t), \epsilon_k(s) \mid \mathcal{F}_{k-1}) = \sigma_k(t)\sigma_k(s)C_\eta(s, t),$$

where $\mathcal{F}_{k-1}$ is the σ-field generated by the sequence $\{\eta_i, i < t\}$, and $C_\eta(s, t) = \mathrm{Cov}(\eta_k(t), \eta_k(s))$.

To estimate the model, an identifiability assumption is required. It can be assumed that $E \parallel \eta_1^2 \parallel < \infty$ and $E\eta_1^2(\cdot) = 1$. Similar to standard GARCH models, the functional ARCH(1) admits a 'weak' functional AR(1) representation. Noting that the expectation of $v_k = \epsilon_k^2 - \sigma_k^2$ is the zero function, we get

$$\epsilon_k^2 = \omega + \alpha[\epsilon_{k-1}^2] + v_k,$$

which can serve as a basis for estimating α and ω (see Hörmann, Horváth, and Reeder 2013 for details).

4.12 Complementary Bibliographical Notes

The asymmetric reaction of the volatility to past positive and negative shocks has been well documented since the articles by Black (1976) and Christie (1982). These articles use the leverage effect to explain the fact that the volatility tends to overreact to price decreases, compared to price increases of the same magnitude. Other explanations, related to the existence of time-dependent risk premia, have been proposed; see, for instance Campbell and Hentschel (1992), Bekaert and Wu (2000) and references therein. More recently, Avramov, Chordia, and Goyal (2006) advanced an explanation founded on the volume of the daily exchanges. Empirical evidence of asymmetry has been given in numerous studies: see, for example, Engle and Ng (1993), Glosten, Jagannathan, and Runkle (1993), Nelson (1991), Wu (2001), and Zakoïan (1994).

The introduction of 'news impact curves' providing a visualisation of the different forms of volatility is due to Pagan and Schwert (1990) and Engle and Ng (1993). The AVGARCH model was introduced by Taylor (1986) and Schwert (1989). The EGARCH model was introduced and studied by Nelson (1991). The GJR-GARCH model was introduced by Glosten, Jagannathan, and Runkle (1993). The Log-GARCH model of Definition 4.2 has been introduced independently by Geweke (1986), Pantula (1986), and Milhøj (1987). The idea behind the extension proposed in Definition 4.3, which authorises the presence of zero observations, is due to Sucarrat and Grønneberg (2016) (see also Hautsch, Malec, and Schienle 2014). The stationarity results of Section 4.3 are extensions of results obtained by Francq, Wintenberger, and Zakoïan (2013, 2017) for variants of the Log-GARCH model precluding zero observations. The TGARCH model was introduced and studied by Zakoïan (1994). This model is inspired by the threshold models of Tong (1978) and Tong and Lim (1980), which are used for the conditional mean. See Gonçalves and Mendes Lopes (1994, 1996) for the stationarity study of the TGARCH model. An extension was proposed by Rabemananjara and Zakoïan (1993) in which the volatility coefficients are not constrained to be positive. The TGARCH model was also extended to the case of a non-zero threshold by Hentschel (1995), and to the case of multiple thresholds by Liu, Li, and Li (1997). Model (4.45), with $\delta = 2$ and $\varsigma_1 = \cdots = \varsigma_q$, was studied by Straumann (2005). This model is called 'asymmetric GARCH' (AGARCH(p, q)) by Straumann, but the acronym AGARCH, which has been employed for several other models, is ambiguous. A variant is the double-threshold ARCH (DTARCH) model of Li and Li (1996), in which the thresholds appear both in the conditional mean and the conditional variance. Specifications making the transition variable continuous were proposed by Hagerud (1997), González-Rivera (1998), and Taylor (2004). Various classes of models rely on Box–Cox transformations of the volatility: the APARCH model was proposed by Higgins and Bera (1992) in its symmetric form (NGARCH model) and then generalized by Ding, Granger, and Engle (1993); another generalisation is that of Hwang and Kim (2004). The qualitative threshold ARCH model was proposed by Gouriéroux and Monfort (1992), the quadratic ARCH model by Sentana (1995). The conditional density was modelled by Hansen (1994). The contemporaneous asymmetric GARCH model of Section 4.7 was proposed by El Babsiri and Zakoïan (2001). In this article, the strict and second-order stationarity conditions were established and the statistical inference was studied. Comparisons of asymmetric GARCH models were proposed by Awartani and Corradi (2005), Chen, Gerlach, and So (2006), and Hansen and Lunde (2005). GARCH-X Models have been studied by Han and Kristensen (2014) and Francq and Thieu (2015), among others. Francq and Sucarrat (2017) proposed a numerically effective estimator of a multivariate Log-GARCH model with covariates of any signs. The class of GAS models are developed by Creal, Koopman, and Lucas (2011, 2013) and Harvey (2013). The website http://www.gasmodel.com/ contains papers and computer codes devoted to the GAS models. References addressing both the spacing of the data and the price changes are Bauwens and Giot (2003), Engle (2001), Ghysels and Jasiak (1998), Russell and Engle (2005), Meddahi et al. (2006). Estimation methods for functional GARCH models were developed by Hörman et al. (2013), Aue et al. (2017), and Cerovecki et al. (2019).

4.13 Exercises

4.1 *(Non-correlation between the volatility and past values when the law of η_t is symmetric)*
Prove the symmetry property (4.1).

4.2 *(The expectation of a product of independent variables is not always the product of the expectations)*
Find a sequence X_i of independent real random variables such that $Y = \prod_{i=1}^{\infty} X_i$ exists a.s., EY and EX_i exist for all i, and $\prod_{i=1}^{\infty} EX_i$ exists, but such that $EY \neq \prod_{i=1}^{\infty} EX_i$.

4.3 *(Convergence of an infinite product entails convergence of the infinite sum of logarithms)*
Prove that, under the assumptions of Theorem 4.1, condition (4.12) entails the absolute convergence of the series of general term $\ln g_\eta(\lambda_i)$.

4.4 *(Variance of an EGARCH)*
Complete the proof of Theorem 4.1 by showing in detail that (4.13) entails the desired result on $E\epsilon_t^2$.

4.5 *(A Gaussian EGARCH admits a variance)*
Show that, for an EGARCH with Gaussian innovations, condition (4.14) for the existence of a second-order moment is satisfied.

4.6 *(ARMA representation for the logarithm of the square of an EGARCH)*
Compute the ARMA representation of $\ln \epsilon_t^2$ when ϵ_t is an EGARCH(1, 1) process with η_t Gaussian. Provide an explicit expression by giving numerical values for the EGARCH coefficients.

4.7 *(β-mixing of an EGARCH)*
Using Exercise 3.5, give simple conditions for an EGARCH(1, 1) process to be geometrically β-mixing.

4.8 *(Stationarity of a TGARCH)*
Establish the second-order stationarity condition (4.39) of a TGARCH(1, 1) process.

4.9 *(Autocorrelation of the absolute value of a TGARCH)*
Compute the autocorrelation function of the absolute value of a TGARCH(1, 1) process when the noise η_t is Gaussian. Would this computation be feasible for a standard GARCH process?

4.10 *(A TGARCH is an APARCH)*
Check that the results obtained for the APARCH(1, 1) model can be used to retrieve those obtained for the TGARCH(1, 1) model.

4.11 *(Study of a threshold model)*
Consider the model

$$\epsilon_t = \sigma_t \eta_t, \quad \ln \sigma_t = \omega + \alpha_+ \eta_{t-1} \mathbb{1}_{\{\eta_{t-1}>0\}} - \alpha_- \eta_{t-1} \mathbb{1}_{\{\eta_{t-1}<0\}}.$$

To which class does this model belong? Which constraints is it natural to impose on the coefficients? What are the strict and second-order stationarity conditions? Compute $\mathrm{Cov}(\sigma_t, \epsilon_{t-1})$ in the case where $\eta_t \sim \mathcal{N}(0, 1)$, and verify that the model can capture the leverage effect.

4.12 *(A condition for $\limsup X_n/n \le 0$)*
Show that if (X_n) is an identically distributed sequence of real variables such that $EX_1^+ < \infty$, then $\limsup_{n\to\infty} X_n/n \le 0$ a.s.

4.13 *(Quadratic variations of the Brownian motion)*
Show Equality (4.68).

4.14 *(Beta-t-GARCH)*
The density of a Student distribution with $\nu > 2$ degrees of freedom and variance σ^2 is

$$f_\sigma(x) = \frac{1}{\sigma\sqrt{\frac{\nu-2}{\nu}}} f\left(\frac{x}{\sigma\sqrt{\frac{\nu-2}{\nu}}}\right)$$

where

$$f(x) = \frac{1}{\sqrt{\nu\pi}} \frac{\Gamma\left(\frac{\nu+1}{2}\right)}{\Gamma\left(\frac{\nu}{2}\right)} \left(1 + \frac{x^2}{\nu}\right)^{-\frac{\nu+1}{2}}.$$

Compute $\partial \ln f_\sigma(x)/\partial\sigma^2$ and deduce (4.75). *Hint:* if $X \sim f$, then $\frac{X^2/\nu}{1+X^2/\nu}$ follows a Beta distribution with mean $1/(\nu+1)$ and variance $2\nu/\{(\nu+1)^2(\nu+3)\}$.

Part II

Statistical Inference

5

Identification

In this chapter, we consider the problem of selecting an appropriate GARCH or ARMA-GARCH model for given observations $X_1, \dots, X_n$ of a centred stationary process. A large part of the theory of finance rests on the assumption that prices follow a random walk. The price variation process, $X = (X_t)$, should thus constitute a martingale difference sequence, and should coincide with its innovation process, $\epsilon = (\epsilon_t)$. The first question addressed in this chapter, in Section 5.1, will be the test of this property, at least a consequence of it: absence of correlation. The problem is far from trivial because standard tests for non-correlation are actually valid under an independence assumption. Such an assumption is too strong for GARCH processes which are dependent though uncorrelated.

If significant sample autocorrelations are detected in the price variations – in other words, if the random walk assumption cannot be sustained – the practitioner will try to fit an ARMA(P, Q) model to the data before using a GARCH(p, q) model for the residuals. Identification of the orders (P, Q) will be treated in Section 5.2, identification of the orders (p, q) in Section 5.3. Tests of the ARCH effect (and, more generally, Lagrange multiplier (LM) tests) will be considered in Section 5.4.

5.1 Autocorrelation Check for White Noise

Consider the GARCH(p, q) model

$$\begin{cases} \epsilon_t = \sigma_t \eta_t \\ \sigma_t^2 = \omega + \sum_{i=1}^{q} \alpha_i \epsilon_{t-i}^2 + \sum_{j=1}^{p} \beta_j \sigma_{t-j}^2 \end{cases} \tag{5.1}$$

with (η_t) a sequence of iid centred variables with unit variance, $\omega > 0$, $\alpha_i \geq 0$ ($i = 1, \dots, q$), $\beta_j \geq 0$ ($j = 1, \dots, p$). We saw in Section 2.2 that, whatever the orders p and q, the non-anticipative second-order stationary solution of model (5.1) is a white noise, that is, a centred process whose theoretical autocorrelations $\rho(h) = E\epsilon_t\epsilon_{t+h}/E\epsilon_t^2$ satisfy $\rho(h) = 0$ for all $h \neq 0$.

Given observations $\epsilon_1, \dots, \epsilon_n$, the theoretical autocorrelations of a centred process (ϵ_t), are generally estimated by the sample autocorrelations (SACRs)

$$\hat{\rho}(h) = \frac{\hat{\gamma}(h)}{\hat{\gamma}(0)}, \quad \hat{\gamma}(h) = \hat{\gamma}(-h) = n^{-1} \sum_{t=1}^{n-h} \epsilon_t \epsilon_{t+h}, \tag{5.2}$$

GARCH Models: Structure, Statistical Inference and Financial Applications, Second Edition. Christian Francq and Jean-Michel Zakoian.

for $h = 0, 1, \dots, n-1$. According to Theorem 1.1, if (ϵ_t) is an iid sequence of centred random variables with finite variance, then

$$\sqrt{n}\hat{\rho}(h) \xrightarrow{\mathcal{L}} \mathcal{N}(0, 1),$$

for all $h \neq 0$. For a strong white noise, the SACRs thus lie between the confidence bounds $\pm 1.96/\sqrt{n}$ with a probability of approximately 95% when n is large. In standard software, these bounds at the 5% level are generally displayed with dotted lines, as in Figure 5.2. These significance bands are not valid for a weak white noise, in particular for a GARCH process (Exercises 5.3 and 5.4). Valid asymptotic bands are derived in the next section.

5.1.1 Behaviour of the Sample Autocorrelations of a GARCH Process

Let $\hat{\rho}_m = (\hat{\rho}(1), \dots, \hat{\rho}(m))'$ denote the vector of the first m SACRs, based on n observations of the GARCH(p, q) process defined by model (5.1). Let $\hat{\gamma}_m = (\hat{\gamma}(1), \dots, \hat{\gamma}(m))'$ denote a vector of sample autocovariances (SACVs).

Theorem 5.1 (Asymptotic distributions of the SACVs and SACRs) *If (ϵ_t) is the non-anticipative and stationary solution of the GARCH(p, q) model (5.1) and if $E\epsilon_t^4 < \infty$, then, when $n \to \infty$,*

$$\sqrt{n}\hat{\gamma}_m \xrightarrow{\mathcal{L}} \mathcal{N}(0, \Sigma_{\hat{\gamma}_m}) \quad \textit{and} \quad \sqrt{n}\hat{\rho}_m \xrightarrow{\mathcal{L}} \mathcal{N}(0, \Sigma_{\hat{\rho}_m} := \{E\epsilon_t^2\}^{-2}\Sigma_{\hat{\gamma}_m}),$$

where

$$\Sigma_{\hat{\gamma}_m} = \begin{pmatrix} E\epsilon_t^2\epsilon_{t-1}^2 & E\epsilon_t^2\epsilon_{t-1}\epsilon_{t-2} & \dots & E\epsilon_t^2\epsilon_{t-1}\epsilon_{t-m} \\ E\epsilon_t^2\epsilon_{t-1}\epsilon_{t-2} & E\epsilon_t^2\epsilon_{t-2}^2 & & \vdots \\ \vdots & & \ddots & \\ E\epsilon_t^2\epsilon_{t-1}\epsilon_{t-m} & \dots & & E\epsilon_t^2\epsilon_{t-m}^2 \end{pmatrix}$$

is non-singular. If the law of η_t is symmetric, then $\Sigma_{\hat{\gamma}_m}$ is diagonal.

Note that $\Sigma_{\hat{\rho}_m} = I_m$ when (ϵ_t) is a strong white noise, in accordance with Theorem 1.1.

Proof. Let $\tilde{\gamma}_m = (\tilde{\gamma}(1), \dots, \tilde{\gamma}(m))'$, where $\tilde{\gamma}(h) = n^{-1}\sum_{t=1}^n \epsilon_t\epsilon_{t-h}$. Since, for m fixed,

$$\| \sqrt{n}\hat{\gamma}_m - \sqrt{n}\tilde{\gamma}_m \|_2 = \frac{1}{\sqrt{n}} \left\{ \sum_{h=1}^m E\left(\sum_{t=1}^h \epsilon_t\epsilon_{t-h} \right)^2 \right\}^{1/2} \leq \frac{1}{\sqrt{n}} \sum_{h=1}^m \sum_{t=1}^h \|\epsilon_t\|_4^2 \to 0$$

as $n \to \infty$, the asymptotic distribution of $\sqrt{n}\hat{\gamma}_m$ coincides with that of $\sqrt{n}\tilde{\gamma}_m$. Let h and k belong to $\{1, \dots, m\}$. By stationarity,

$$\begin{aligned} \mathrm{Cov}\{\sqrt{n}\tilde{\gamma}(h), \sqrt{n}\tilde{\gamma}(k)\} &= \frac{1}{n} \sum_{t,s=1}^n \mathrm{Cov}(\epsilon_t\epsilon_{t-h}, \epsilon_s\epsilon_{s-k}) \\ &= \frac{1}{n} \sum_{\ell=-n+1}^{n-1} (n - |\ell|)\, \mathrm{Cov}(\epsilon_t\epsilon_{t-h}, \epsilon_{t+\ell}\epsilon_{t+\ell-k}) \\ &= E\epsilon_t^2\epsilon_{t-h}\epsilon_{t-k} \end{aligned}$$

because

$$\mathrm{Cov}(\epsilon_t\epsilon_{t-h}, \epsilon_{t+\ell}\epsilon_{t+\ell-k}) = \begin{cases} E\epsilon_t^2\epsilon_{t-h}\epsilon_{t-k} & \text{if } \ell = 0, \\ 0 & \text{otherwise.} \end{cases}$$

From this, we deduce the expression for $\Sigma_{\hat{\gamma}_m}$. From the Cramér–Wold theorem,[1] the asymptotic normality of $\sqrt{n}\tilde{\gamma}_m$ will follow from showing that for all non-zero $\lambda = (\lambda_1, \dots, \lambda_m)' \in \mathbb{R}^m$,

$$\sqrt{n}\lambda'\tilde{\gamma}_m \xrightarrow{\mathcal{L}} \mathcal{N}(0, \lambda'\Sigma_{\hat{\gamma}_m}\lambda). \tag{5.3}$$

Let $\mathcal{F}_t$ denote the σ-field generated by $\{\epsilon_u, u \leq t\}$. We obtain the convergence (5.3) by applying a central limit theorem (CLT) to the sequence $(\epsilon_t \sum_{i=1}^m \lambda_i \epsilon_{t-i}, \mathcal{F}_t)_t$, which is a stationary, ergodic, and square integrable martingale difference (see Corollary A.1).

The asymptotic behaviour of $\hat{\rho}_m$ immediately follows from that of $\hat{\gamma}_m$ (as in Exercise 5.3).

Reasoning by contradiction, suppose that $\Sigma_{\hat{\gamma}_m}$ is singular. Then, because this matrix is the covariance of the vector $(\epsilon_t\epsilon_{t-1}, \dots, \epsilon_t\epsilon_{t-m})'$, there exists an exact linear combination of the components of $(\epsilon_t\epsilon_{t-1}, \dots, \epsilon_t\epsilon_{t-m})'$ that is equal to zero. For some $i_0 \geq 1$, we then have $\epsilon_t\epsilon_{t-i_0} = \sum_{i=i_0+1}^m \lambda_i\epsilon_t\epsilon_{t-i}$, that is, $\epsilon_{t-i_0}\mathbb{1}_{\{\eta_t\neq 0\}} = \sum_{i=i_0+1}^m \lambda_i\epsilon_{t-i}\mathbb{1}_{\{\eta_t\neq 0\}}$. Hence,

$$\begin{aligned} E\epsilon_{t-i_0}^2\mathbb{1}_{\{\eta_t\neq 0\}} &= \sum_{i=i_0+1}^m \lambda_i E(\epsilon_{t-i_0}\epsilon_{t-i}\mathbb{1}_{\{\eta_t\neq 0\}}) \\ &= \sum_{i=i_0+1}^m \lambda_i E(\epsilon_{t-i_0}\epsilon_{t-i})\mathbb{P}(\eta_t \neq 0) = 0 \end{aligned}$$

which is absurd. It follows that $\Sigma_{\hat{\gamma}_m}$ is non-singular.

When the law of η_t is symmetric, the diagonal form of $\Sigma_{\hat{\gamma}_m}$ is a consequence of property (7.24) in Chapter 7. See Exercises 5.5 and 5.6 for the GARCH (1, 1) case. □

A consistent estimator $\hat{\Sigma}_{\hat{\gamma}_m}$ of $\Sigma_{\hat{\gamma}_m}$ is obtained by replacing the generic term of $\Sigma_{\hat{\gamma}_m}$ by

$$n^{-1}\sum_{t=1}^n \epsilon_t^2\epsilon_{t-i}\epsilon_{t-j},$$

with, by convention, $\epsilon_s = 0$ for $s < 1$. Clearly, $\hat{\Sigma}_{\hat{\rho}_m} := \hat{\gamma}^{-2}(0)\hat{\Sigma}_{\hat{\gamma}_m}$ is a consistent estimator of $\Sigma_{\hat{\rho}_m}$ and is almost surely invertible for n large enough. This can be used to construct asymptotic significance bands for the SACRs of a GARCH process.

Practical Implementation

The following R code allows us to draw a given number of autocorrelations $\hat{\rho}(i)$ and the significance bands $\pm 1.96\sqrt{\hat{\Sigma}_{\hat{\rho}_m}(i,i)/n}$.

```
 # autocorrelation function
gamma<-function(x,h){ n<-length(x); h<-abs(h);x<-x-mean(x)
gamma<-sum(x[1:(n-h)]*x[(h+1):n])/n }
rho<-function(x,h) rho<-gamma(x,h)/gamma(x,0)
# acf function with significance bands of a weak white noise
nl.acf<-function(x,main=NULL,method="NP"){
n<-length(x); nlag<-as.integer(min(10*log10(n),n-1))
acf.val<-sapply(c(1:nlag),function(h) rho(x,h)); x2<-x^2
var<-1+(sapply(c(1:nlag),function(h) gamma(x2,h)))/gamma(x,0)^2
band<-sqrt(var/n)
minval<-1.2*min(acf.val,-1.96*band,-1.96/sqrt(n))
maxval<-1.2*max(acf.val,1.96*band,1.96/sqrt(n))
acf(x,xlab="Lag",ylab="SACR",ylim=c(minval,maxval),main=main)
lines(c(1:nlag),-1.96*band,lty=1,col="red")
lines(c(1:nlag),1.96*band,lty=1,col="red") }
```

[1] For any sequence (Z_n) of random vectors of size d, $Z_n \xrightarrow{\mathcal{L}} Z$ if and only if, for all $\lambda \in \mathbb{R}^d$, we have $\lambda' Z_n \xrightarrow{\mathcal{L}} \lambda' Z$.

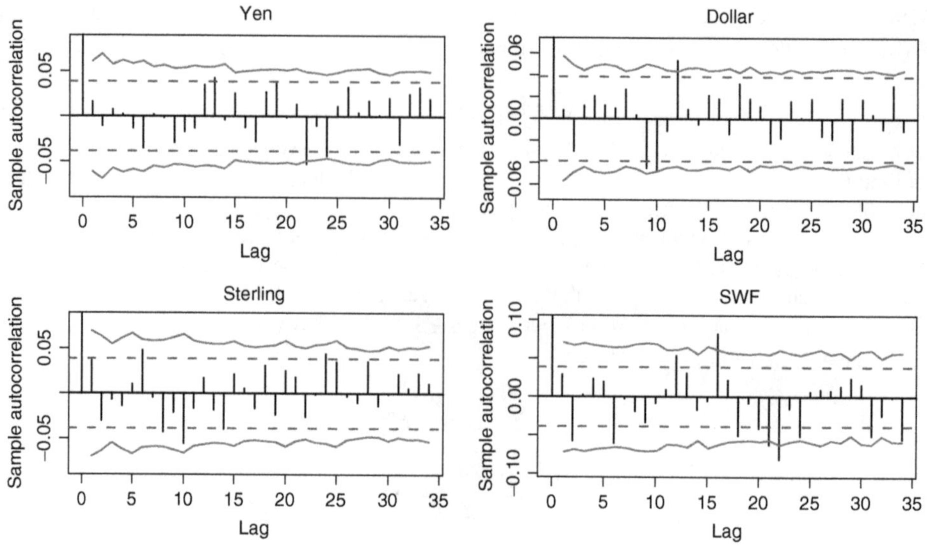

Figure 5.1 SACR of exchange rates against the euro, standard significance bands for the SACRs of a strong white noise (dotted lines) and significance bands for the SACRs of a semi-strong white noise (solid lines).

In Figure 5.1 we have plotted the SACRs and their significance bands for daily series of exchange rates of the dollar, pound, yen and Swiss franc against the euro, for the period from 4 January 1999 to 22 January 2009. It can be seen that the SACRs are often outside the standard significance bands $\pm 1.96/\sqrt{n}$, which leads us to reject the strong white noise assumption for all these series. On the other hand, most of the SACRs are inside the significance bands shown as solid lines, which is in accordance with the hypothesis that the series are realizations of semi-strong white noises.

5.1.2 Portmanteau Tests

The standard portmanteau test for checking that the data is a realization of a strong white noise is that of Ljung and Box (1978). It involves computing the statistic

$$Q_m^{LB} := n(n+2)\sum_{i=1}^{m} \hat{\rho}^2(i)/(n-i)$$

and rejecting the strong white noise hypothesis if Q_m^{LB} is greater than the $(1-\alpha)$-quantile of a χ^2_m.[2]

Portmanteau tests are constructed for checking non-correlation, but the asymptotic distribution of the statistics is no longer χ^2_m when the series departs from the strong white noise assumption. For instance, these tests are not robust to conditional heteroscedasticity. In the GARCH framework, we may wish to simultaneously test the nullity of the first m autocorrelations using more robust portmanteau statistics.

[2] The asymptotic distribution of Q_m^{LB} is χ^2_m. The Box and Pierce (1970) statistic $Q_m^{BP} := n\sum_{i=1}^{m} \hat{\rho}^2(i)$ has the same asymptotic distribution, but the Q_m^{LB} statistic is believed to perform better for finite samples.

Theorem 5.2 (Corrected portmanteau test in the presence of ARCH) *Under the assumptions of Theorem 5.1, the portmanteau statistic*

$$Q_m = n\hat{\rho}'_m \hat{\Sigma}^{-1}_{\hat{\rho}_m} \hat{\rho}_m$$

has an asymptotic χ^2_m *distribution.*

Proof. It suffices to use Theorem 5.1 and the following result: if $X_n \xrightarrow{\mathcal{L}} \mathcal{N}(0, \Sigma)$, with Σ non-singular, and if $\hat{\Sigma}_n \to \Sigma$ in probability, then $X'_n \hat{\Sigma}^{-1}_n X_n \xrightarrow{\mathcal{L}} \chi^2_m$. □

A portmanteau test of asymptotic level α based on the first m SACRs involves rejecting the hypothesis that the data are generated by a GARCH process if Q_m is greater than the $(1 - \alpha)$-quantile of a χ^2_m.

5.1.3 Sample Partial Autocorrelations of a GARCH

Denote by r_m $(\hat{r}_m)$ the vector of the m first partial autocorrelations (sample partial autocorrelations (SPACs)) of the process (ϵ_t). By Theorem B.3, we know that for a weak white noise, the SACRs and SPACs have the same asymptotic distribution. This applies in particular to a GARCH process. Consequently, under the hypothesis of GARCH white noise with a finite fourth-order moment, consistent estimators of $\Sigma_{\hat{r}_m}$ are

$$\hat{\Sigma}^{(1)}_{\hat{r}_m} = \hat{\Sigma}_{\hat{\rho}_m} \quad \text{or} \quad \hat{\Sigma}^{(2)}_{\hat{r}_m} = \hat{J}_m \hat{\Sigma}_{\hat{\rho}_m} \hat{J}'_m,$$

where $\hat{J}_m$ is the matrix obtained by replacing $\rho_X(1), \dots, \rho_X(m)$ by $\hat{\rho}_X(1), \dots, \hat{\rho}_X(m)$ in the Jacobian matrix J_m of the mapping $\rho_m \mapsto r_m$, and $\hat{\Sigma}_{\hat{\rho}_m}$ is the consistent estimator of $\Sigma_{\hat{\rho}_m}$ defined after Theorem 5.1.

Although it is not current practice, one can test the simultaneous nullity of several theoretical partial autocorrelations using portmanteau tests based on the statistics

$$Q^{r,BP}_m = n\hat{r}'_m \hat{r}_m \quad \text{and} \quad Q^r_m = n\hat{r}'_m (\hat{\Sigma}^{(i)}_{\hat{r}_m})^{-1} \hat{r}_m$$

with, for instance, $i = 2$. From Theorem B.3, under the strong white noise assumption, the statistics $Q^{r,BP}_m$, Q^{BP}_m, and Q^{LB}_m have the same χ^2_m asymptotic distribution. Under the hypothesis of a pure GARCH process, the statistics Q^r_m and Q_m also have the same χ^2_m asymptotic distribution.

5.1.4 Numerical Illustrations

Standard Significance Bounds for the SACRs are Not Valid

The right-hand graph of Figure 5.2 displays the sample correlogram of a simulation of size $n = 5000$ of the GARCH(1, 1) white noise

$$\begin{cases} \epsilon_t = \sigma_t \eta_t \\ \sigma^2_t = 1 + 0.3\epsilon^2_{t-1} + 0.55\sigma^2_{t-1}, \end{cases} \tag{5.4}$$

where (η_t) is a sequence of iid $\mathcal{N}(0, 1)$ variables. It is seen that the SACRs of order 2 and 4 are sharply outside the 95% significance bands computed under the strong white noise assumption. An inexperienced practitioner could be tempted to reject the hypothesis of white noise, in favour of a more complicated ARMA model whose residual autocorrelations would lie between the significance bounds $\pm 1.96/\sqrt{n}$. To avoid this type of specification error, one has to be conscious that the bounds $\pm 1.96/\sqrt{n}$ are not valid for the SACRs of a GARCH white noise. In our simulation, it is possible to compute exact asymptotic bounds at the 95% level (Exercise 5.4). In the right-hand graph of Figure 5.2, these bounds are drawn in thick dotted lines. All the SACRs are now inside, or very slightly outside, those bounds. If we had been given the data, with no prior information, this graph would have given us no grounds on which to reject the simple hypothesis that the data is a realization of a GARCH white noise.

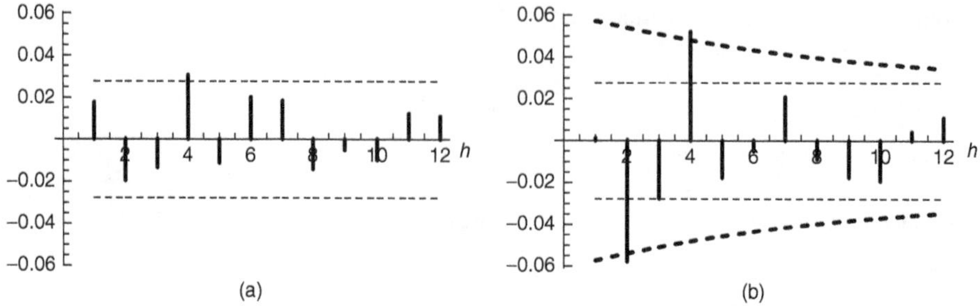

Figure 5.2 SACRs of a simulation of a strong white noise (a) and of the GARCH(1, 1) white noise (5.4) (b). Approximately 95% of the SACRs of a strong white noise should lie inside the thin dotted lines $\pm 1.96/\sqrt{n}$. Approximately 95% of the SACRs of a GARCH(1, 1) white noise should lie inside the thick dotted lines.

Estimating the Significance Bounds of the SACRs of a GARCH

Of course, in real situations, the significance bounds depend on unknown parameters, and thus cannot be easily obtained. It is, however, possible to estimate them in a consistent way, as described in Section 5.1.1. For a simulation of model (5.4) of size $n = 5000$, Figure 5.3 shows as thin dotted lines the estimation thus obtained of the significance bounds at the 5% level. The estimated bounds are fairly close to the exact asymptotic bounds.

The SPACs and Their Significance Bounds

Figure 5.4 shows the SPACs of the simulation (5.4) and the estimated significance bounds of the $\hat{r}(h)$, at the 5% level (based on $\hat{\Sigma}^{(2)}_{\hat{r}_m}$). By comparing Figures 5.3 and 5.4, it can be seen that the SACRs and SPACs of the GARCH simulation look much alike. This is not surprising in view of Theorem B.4.

Portmanteau Tests of Strong White Noise and of Pure GARCH

Table 5.1 displays p-values of white noise tests based on Q_m and the usual Ljung–Box statistics, for the simulation of (5.4). Apart from the test with $m = 4$, the Q_m tests do not reject, at the 5% level, the hypothesis that the data comes from a GARCH process. On the other hand, the Ljung–Box tests clearly reject the strong white noise assumption.

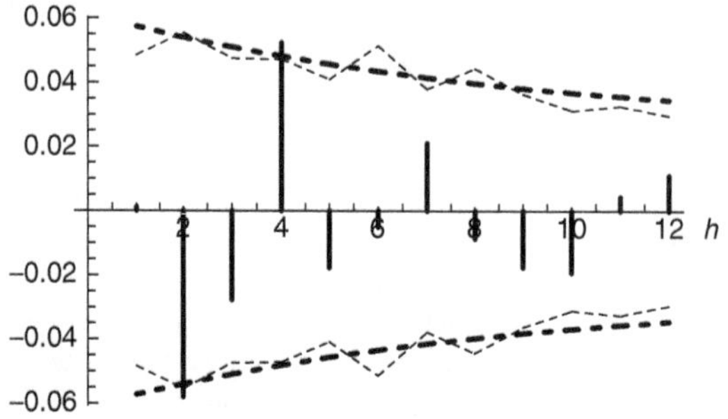

Figure 5.3 Sample autocorrelations of a simulation of size $n = 5000$ of the GARCH(1, 1) white noise (5.4). Approximately 95% of the SACRs of a GARCH(1, 1) white noise should lie inside the thin dotted lines. The exact asymptotic bounds are shown as thick dotted lines.

Table 5.1 Portmanteau tests on a simulation of size $n = 5000$ of the GARCH(1, 1) white noise (5.4).

m	1	2	3	4	5	6	7	8	9	10	11	12
Tests based on Q_m, for the hypothesis of GARCH white noise												
$\hat{\rho}(m)$	0.00	−0.06	−0.03	0.05	−0.02	0.00	0.02	−0.01	−0.02	−0.02	0.00	0.01
$\hat{\sigma}_{\hat{\rho}(m)}$	0.025	0.028	0.024	0.024	0.021	0.026	0.019	0.023	0.019	0.016	0.017	0.015
Q_m	0.00	4.20	5.49	10.19	10.90	10.94	12.12	12.27	13.16	14.61	14.67	15.20
$\mathbb{P}(\chi^2_m > Q_m)$	0.9637	0.1227	0.1391	0.0374	0.0533	0.0902	0.0967	0.1397	0.1555	0.1469	0.1979	0.2306
Usual tests, for the strong white noise hypothesis												
$\hat{\rho}(m)$	0.00	−0.06	−0.03	0.05	−0.02	0.00	0.02	−0.01	−0.02	−0.02	0.00	0.01
$\hat{\sigma}_{\hat{\rho}(m)}$	0.014	0.014	0.014	0.014	0.014	0.014	0.014	0.014	0.014	0.014	0.014	0.014
Q_m^{LB}	0.01	16.78	20.59	34.18	35.74	35.86	38.05	38.44	39.97	41.82	41.91	42.51
$\mathbb{P}(\chi^2_m > Q_m^{LB})$	0.9365	0.0002	0.0001	0.0000	0.0000	0.0000	0.0000	0.0000	0.0000	0.0000	0.0000	0.0000

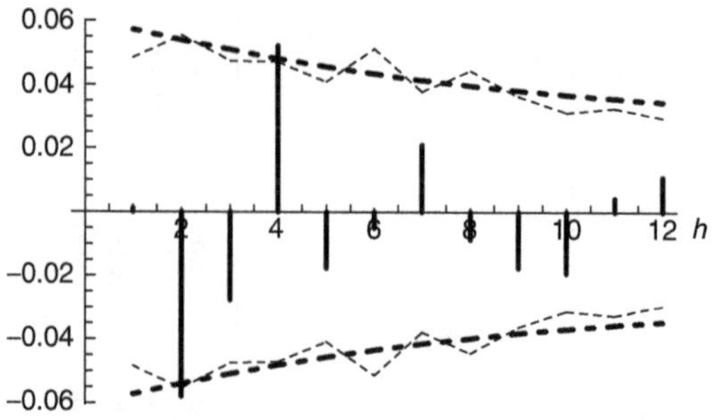

Figure 5.4 Sample partial autocorrelations of a simulation of size $n = 5000$ of the GARCH(1, 1) white noise (5.4). Approximately 95% of the SPACs of a GARCH(1, 1) white noise should lie inside the thin dotted lines. The exact asymptotic bounds are shown as thick dotted lines.

Portmanteau Tests Based on Partial Autocorrelations

Table 5.2 is similar to Table 5.1, but presents portmanteau tests based on the SPACs. As expected, the results are very close to those obtained for the SACRs.

An Example Showing that Portmanteau Tests Based on the SPACs Can Be More Powerful than Those Based on the SACRs

Consider a simulation of size $n = 100$ of the strong MA(2) model

$$X_t = \eta_t + 0.56\eta_{t-1} - 0.44\eta_{t-2}, \quad \eta_t \text{ iid } \mathcal{N}(0, 1). \tag{5.5}$$

By comparing the top two and bottom two parts of Table 5.3, we note that the hypotheses of strong white noise and pure GARCH are better rejected when the SPACs, rather than the SACRs, are used. This follows from the fact that, for this MA(2), only two theoretical autocorrelations are not equal to 0, whereas many theoretical partial autocorrelations are far from 0. For the same reason, the results would have been inverted if, for instance, an AR(1) alternative had been considered.

5.2 Identifying the ARMA Orders of an ARMA-GARCH

Assume that the tools developed in Section 5.1 lead to rejection of the hypothesis that the data is a realisation of a pure GARCH process. It is then sensible to look for an ARMA(P, Q) model with GARCH innovations. The problem is then to choose (or identify) plausible orders for the model

$$X_t - \sum_{i=1}^{P} a_i X_{t-i} = \epsilon_t - \sum_{i=1}^{Q} b_i \epsilon_{t-i} \tag{5.6}$$

under standard assumptions (the AR and MA polynomials having no common root and having roots outside the unit disk, with $a_P b_Q \neq 0$, $E\epsilon_t^4 < \infty$), where (ϵ_t) is a GARCH white noise of the form (5.1).

5.2.1 Sample Autocorrelations of an ARMA-GARCH

Recall that an MA(Q) satisfies $\rho_X(h) = 0$ for all $h > Q$, whereas an AR(P) satisfies $r_X(h) = 0$ for all $h > P$. The SACRs and SPACs thus play an important role in identifying the orders P and Q.

Table 5.2 As Table 5.1, for tests based on partial autocorrelations instead of autocorrelations.

m	1	2	3	4	5	6	7	8	9	10	11	12
GARCH white noise tests based on Q_m^r												
$\hat{r}(m)$	0.00	−0.06	−0.03	0.05	−0.02	0.00	0.02	−0.01	−0.01	−0.02	0.00	0.01
$\hat{\sigma}_{\hat{r}(m)}$	0.025	0.028	0.024	0.024	0.021	0.026	0.019	0.023	0.019	0.016	0.017	0.015
Q_m^r	0.00	4.20	5.49	9.64	10.65	10.650	11.92	12.24	12.77	14.24	14.24	14.67
$\mathbb{P}(\chi_m^2 > Q_m^r)$	0.9637	0.1227	0.1393	0.0470	0.0587	0.0998	0.1032	0.1407	0.1735	0.1623	0.2200	0.2599
Strong white noise tests based on $Q_m^{r,LB}$												
$\hat{r}(m)$	0.02	−0.01	−0.01	−0.02	0.00	0.01	0.02	−0.01	−0.01	−0.02	0.00	0.01
$\hat{\sigma}_{\hat{r}(m)}$	0.014	0.014	0.014	0.014	0.014	0.014	0.014	0.014	0.014	0.014	0.014	0.014
$Q_m^{r,LB}$	0.01	16.77	20.56	32.55	34.76	34.76	37.12	37.94	38.84	40.71	40.71	41.20
$\mathbb{P}(\chi_m^2 > Q_m^{r,LB})$	0.9366	0.0002	0.0001	0.0000	0.0000	0.0000	0.0000	0.0000	0.0000	0.0000	0.0000	0.0000

Table 5.3 White noise portmanteau tests on a simulation of size $n = 100$ of the MA(2) model (5.5).

m	1	2	3	4	5	6
Tests of GARCH white noise based on autocorrelations						
Q_m	1.6090	4.5728	5.5495	6.2271	6.2456	6.4654
$\mathbb{P}(\chi^2_m > Q_m)$	0.2046	0.1016	0.1357	0.1828	0.2830	0.3731
Tests of GARCH white noise based on partial autocorrelations						
Q^r_m	1.6090	5.8059	9.8926	16.7212	21.5870	25.3162
$\mathbb{P}(\chi^2_m > Q^r_m)$	0.2046	0.0549	0.0195	0.0022	0.0006	0.0003
Tests of strong white noise based on autocorrelations						
Q^{LB}_m	3.4039	8.4085	9.8197	10.6023	10.6241	10.8905
$\mathbb{P}(\chi^2_m > Q^{LB}_m)$	0.0650	0.0149	0.0202	0.0314	0.0594	0.0918
Tests of strong white noise based on partial autocorrelations						
$Q^{r,BP}_m$	3.3038	10.1126	15.7276	23.1513	28.4720	32.6397
$\mathbb{P}(\chi^2_m > Q^{r,BP}_m)$	0.0691	0.0064	0.0013	0.0001	0.0000	0.0000

Invalidity of the Standard Bartlett Formula and Modified Formula

The validity of the usual Bartlett formula rests on assumptions including the strong white noise hypothesis (Theorem 1.1) which are obviously incompatible with GARCH errors. We shall see that this formula leads to an underestimation of the variances of the SACRs and SPACs, and thus to erroneous ARMA orders. We shall only consider the SACRs because Theorem B.2 shows that the asymptotic behaviour of the SPACs easily follows from that of the SACRs.

We assume throughout that the law of η_t is symmetric. By Theorem B.5, the asymptotic behaviour of the SACRs is determined by the generalised Bartlett formula (B.15). This formula involves the theoretical autocorrelations of (X_t) and (ϵ_t^2), as well as the ratio $\kappa_\epsilon - 1 = \gamma_{\epsilon^2}(0)/\gamma_\epsilon^2(0)$. More precisely, using Remark 1 of Theorem 7.2.2 in Brockwell and Davis (1991), the generalised Bartlett formula is written as

$$\lim_{n\to\infty} n \operatorname{Cov}\{\hat{\rho}_X(i), \hat{\rho}_X(j)\} = v_{ij} + v^*_{ij},$$

where

$$v_{ij} = \sum_{\ell=1}^{\infty} w_i(\ell) w_j(\ell), \quad v^*_{ij} = (\kappa_\epsilon - 1) \sum_{\ell=1}^{\infty} \rho_{\epsilon^2}(\ell) w_i(\ell) w_j(\ell), \tag{5.7}$$

and

$$w_i(\ell) = \{2\rho_X(i)\rho_X(\ell) - \rho_X(\ell + i) - \rho_X(\ell - i)\}.$$

The following result shows that the standard Bartlett formula always underestimates the asymptotic variances of the sample autocorrelations in presence of GARCH errors.

Proposition 5.1 *Under the assumptions of Theorem B.5, if the linear innovation process (ϵ_t) is a GARCH process with η_t symmetrically distributed, then*

$$v^*_{ii} \geq 0 \quad \textit{for all } i > 0.$$

If, moreover, $\alpha_1 > 0$, $\text{Var}(\eta_t^2) > 0$ *and* $\sum_{h=-\infty}^{+\infty} \rho_X(h) \neq 0$, *then*

$$v_{ii}^* > 0 \quad \textit{for all } i > 0.$$

Proof. From Proposition 2.2, we have $\rho_{\epsilon^2}(\ell) \geq 0$ for all ℓ, with strict inequality when $\alpha_1 > 0$. It thus follows immediately from the modified Bartett formula (5.7) that $v_{ii}^* \geq 0$. When $\alpha_1 > 0$, this inequality is strict unless if $\kappa_\epsilon = 1$ or $w_i(\ell) = 0$ for all $\ell \geq 1$, that is,

$$2\rho_X(i)\rho_X(\ell) = \rho_X(\ell + i) + \rho_X(\ell - i).$$

Suppose this relations holds, and note that it is also satisfied for all $\ell \in \mathbb{Z}$. Moreover, summing over ℓ, we obtain

$$2\rho_X(i) \sum_{-\infty}^{\infty} \rho_X(\ell) = \sum_{-\infty}^{\infty} \rho_X(\ell + i) + \rho_X(\ell - i) = 2 \sum_{-\infty}^{\infty} \rho_X(\ell).$$

Because the sum of all autocorrelations is supposed to be non-zero, we thus have $\rho_X(i) = 1$. Taking $\ell = i$ in the previous relation, we thus find that $\rho_X(2i) = 1$. Iterating this argument yields $\rho_X(ni) = 1$, and letting n go to infinity gives a contradiction. Finally, one cannot have $\kappa_\epsilon = 1$ because

$$\text{Var}(\epsilon_t^2) = (E\epsilon_t^2)^2(\kappa_\epsilon - 1) = Eh_t^2\text{Var}(\eta_t^2) + \text{Var}h_t \geq \omega^2\text{Var}(\eta_t^2) > 0.$$

□

Consider, by way of illustration, the ARMA(2,1)-GARCH(1, 1) process defined by

$$\begin{cases} X_t - 0.8X_{t-1} + 0.8X_{t-2} = \epsilon_t - 0.8\epsilon_{t-1} \\ \epsilon_t = \sigma_t\eta_t, \quad \eta_t \text{ iid } \mathcal{N}(0, 1) \\ \sigma_t^2 = 1 + 0.2\epsilon_{t-1}^2 + 0.6\sigma_{t-1}^2. \end{cases} \tag{5.8}$$

Figure 5.5 shows the theoretical autocorrelations and partial autocorrelations for this model. The bands shown as solid lines should contain approximately 95% of the SACRs and SPACs, for a realisation of size $n = 1000$ of this model. These bands are obtained from formula (B.15), the autocorrelations of (ϵ_t^2) being computed as in Section 2.5.3. The bands shown as dotted lines correspond to the standard Bartlett formula (still at the 95% level). It can be seen that using this formula, which is erroneous in the presence of GARCH, would lead to identification errors because it systematically underestimates the variability of the sample autocorrelations (Proposition 5.1).

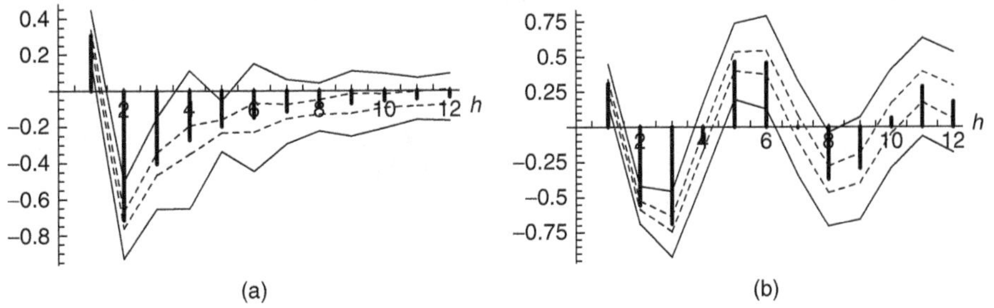

Figure 5.5 Autocorrelations (a) and partial autocorrelations (b) for model (5.8). Approximately 95% of the SACRs (SPACs) of a realisation of size $n = 1000$ should lie between the bands shown as solid lines. The bands shown as dotted lines correspond to the standard Bartlett formula.

Algorithm for Estimating the Generalised Bands

In practice, the autocorrelations of (X_t) and (ϵ_t^2), as well as the other theoretical quantities involved in the generalized Bartlett formula (B.15), are obviously unknown. We propose the following algorithm for estimating such quantities:

1. Fit an AR(p_0) model to the data $X_1, \dots, X_n$ using an information criterion for the selection of the order p_0.
2. Compute the autocorrelations $\rho_1(h)$, $h = 1, 2, \dots$, of this AR(p_0) model.
3. Compute the residuals $e_{p_0+1}, \dots, e_n$ of this estimated AR(p_0).
4. Fit an AR(p_1) model to the squared residuals $e_{p_0+1}^2, \dots, e_n^2$, again using an information criterion for p_1.
5. Compute the autocorrelations $\rho_2(h)$, $h = 1, 2, \dots$, of this AR(p_1) model.
6. Estimate $\lim_{n\to\infty} n \operatorname{Cov}\{\hat{\rho}(i), \hat{\rho}(j)\}$ by $\hat{v}_{ij} + \hat{v}_{ij}^*$, where

$$\hat{v}_{ij} = \sum_{\ell=-\ell_{\max}}^{\ell_{\max}} \rho_1(\ell)[2\rho_1(i)\rho_1(j)\rho_1(\ell) - 2\rho_1(i)\rho_1(\ell+j) - 2\rho_1(j)\rho_1(\ell+i) + \rho_1(\ell+j-i) + \rho_1(\ell-j-i)],$$

$$\hat{v}_{ij}^* = \frac{\hat{\gamma}_{\epsilon^2}(0)}{\hat{\gamma}_\epsilon^2(0)} \sum_{\ell=-\ell_{\max}}^{\ell_{\max}} \rho_2(\ell)[2\rho_1(i)\rho_1(j)\rho_1^2(\ell) - 2\rho_1(j)\rho_1(\ell)\rho_1(\ell+i) - 2\rho_1(i)\rho_1(\ell)\rho_1(\ell+j) + \rho_1(\ell+i)\{\rho_1(\ell+j) + \rho_1(\ell-j)\}],$$

$$\hat{\gamma}_{\epsilon^2}(0) = \frac{1}{n-p_0}\sum_{t=p_0+1}^{n} e_t^4 - \hat{\gamma}_\epsilon^2(0), \quad \hat{\gamma}_\epsilon^2(0) = \frac{1}{n-p_0}\sum_{t=p_0+1}^{n} e_t^2,$$

and $\ell_{\max}$ is a truncation parameter, numerically determined so as to have $|\rho_1(\ell)|$ and $|\rho_2(\ell)|$ less than a certain tolerance (for instance, 10^{-5}) for all $\ell > \ell_{\max}$.

This algorithm is fast when the Durbin–Levinson algorithm is used to fit the AR models. Figure 5.6 shows an application of this algorithm (using the BIC information criterion).

5.2.2 Sample Autocorrelations of an ARMA-GARCH Process When the Noise is Not Symmetrically Distributed

The generalised Bartlett formula (B.15) holds under condition (B.13), which may not be satisfied if the distribution of the noise η_t, in the GARCH equation, is not symmetric. We shall consider the asymptotic

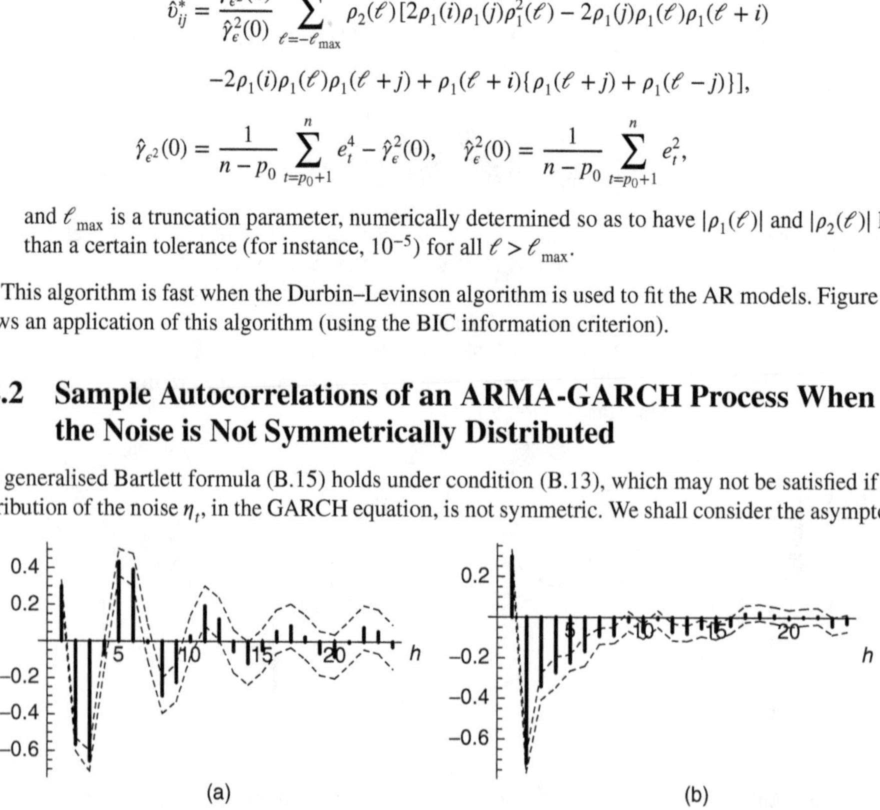

Figure 5.6 SACRs (a) and SPACs (b) of a simulation of size $n = 1000$ of model (5.8). The dotted lines are the estimated 95% confidence bands.

behaviour of the SACVs and SACRs for very general linear processes whose innovation (ϵ_t) is a weak white noise. Retaining the notation of Theorem B.5, the following property allows the asymptotic variance of the SACRs to be interpreted as the spectral density at 0 of a vector process (see, for instance, Brockwell and Davis 1991, for the concept of spectral density). Let $\hat{\gamma}_{0:m} = (\hat{\gamma}(0), \dots, \hat{\gamma}(m))'$.

Theorem 5.3 *Let $(X_t)_{t\in\mathbb{Z}}$ be a real stationary process satisfying*

$$X_t = \sum_{j=-\infty}^{\infty} \psi_j \epsilon_{t-j}, \qquad \sum_{j=-\infty}^{\infty} |\psi_j| < \infty,$$

where $(\epsilon_t)_{t\in\mathbb{Z}}$ is a weak white noise such that $E\epsilon_t^4 < \infty$. Let $\Upsilon_t = X_t(X_t, X_{t+1}, \dots, X_{t+m})'$, $\Gamma_\Upsilon(h) = E\Upsilon_t^*\Upsilon_{t+h}^{*\prime}$ and

$$f_{\Upsilon^*}(\lambda) := \frac{1}{2\pi} \sum_{h=-\infty}^{+\infty} e^{-ih\lambda}\Gamma_\Upsilon(h),$$

the spectral density of the process $\Upsilon^* = (\Upsilon_t^*)$, $\Upsilon_t^* = \Upsilon_t - E\Upsilon_t$. Then we have

$$\lim_{n\to\infty} n\text{Var}\hat{\gamma}_{0:m} := \Sigma_{\hat{\gamma}_{0:m}} = 2\pi f_{\Upsilon^*}(0). \tag{5.9}$$

Proof. By stationarity and application of the Lebesgue-dominated convergence theorem,

$$\begin{aligned} n\text{Var}\hat{\gamma}_{0:m} + o(1) &= n\text{Cov}\left(\frac{1}{n}\sum_{t=1}^{n}\Upsilon_t^*, \frac{1}{n}\sum_{s=1}^{n}\Upsilon_s^*\right) \\ &= \sum_{h=-n+1}^{n-1}\left(1 - \frac{|h|}{n}\right)\text{Cov}(\Upsilon_t^*, \Upsilon_{t+h}^*) \\ &\to \sum_{h=-\infty}^{+\infty}\Gamma_\Upsilon(h) = 2\pi f_{\Upsilon^*}(0) \end{aligned}$$

as $n \to \infty$. □

The matrix $\Sigma_{\hat{\gamma}_{0:m}}$ involved in (5.9) is called the *long-run variance* in the econometric literature, as a reminder that it is the limiting variance of a sample mean. Several methods can be considered for long-run variance estimation.

(i) The naive estimator based on replacing the $\Gamma_\Upsilon(h)$ by the $\hat{\Gamma}_\Upsilon(h)$ in $f_{\Upsilon^*}(0)$ is inconsistent (Exercise 1.2). However, a consistent estimator can be obtained by weighting the $\hat{\Gamma}_\Upsilon(h)$, using a weight close to 1 when h is very small compared to n, and a weight close to 0 when h is large. Such an estimator is called heteroscedastic and autocorrelation consistent (HAC) in the econometric literature.

(ii) A consistent estimator of $f_{\Upsilon^*}(0)$ can also be obtained using the smoothed periodogram (see Brockwell and Davis 1991, Section 10.4).

(iii) For a vector AR(r),

$$\mathcal{A}_r(B)Y_t := Y_t - \sum_{i=1}^{r} A_i Y_{t-i} = Z_t, \quad (Z_t) \text{ white noise with variance } \Sigma_Z,$$

the spectral density at 0 is

$$f_Y(0) = \frac{1}{2\pi}\mathcal{A}_r^{-1}(1)\Sigma_Z\mathcal{A}_r^{-1\prime}(1).$$

A vector AR model is easily fitted, even a high-order AR, using a multivariate version of the Durbin–Levinson algorithm (see Brockwell and Davis 1991, p. 422). The following method can thus be proposed:

1. Fit AR(r) models, with $r = 0, 1 \ldots, R$, to the data $\Upsilon_1 - \overline{\Upsilon}_n, \ldots, \Upsilon_{n-m} - \overline{\Upsilon}_n$ where $\overline{\Upsilon}_n = (n-m)^{-1}\sum_{t=1}^{n-m}\Upsilon_t$.
2. Select a value r_0 by minimising an information criterion, for instance the BIC.
3. Take

$$\hat{\Sigma}_{\hat{\gamma}_{0:m}} = \hat{\mathcal{A}}_{r_0}^{-1}(1)\hat{\Sigma}_{r_0}\hat{\mathcal{A}}_{r_0}^{-1\prime}(1),$$

   with obvious notation.

In our applications, we used method (iii).

5.2.3 Identifying the Orders (P, Q)

Order determination based on the sample autocorrelations and partial autocorrelations in the mixed ARMA(P, Q) model is not an easy task. Other methods, such as the corner method, presented in the next section, and the epsilon algorithm, rely on more convenient statistics.

The Corner Method

Denote by $D(i, j)$ the $j \times j$ Toeplitz matrix

$$D(i,j) = \begin{pmatrix} \rho_X(i) & \rho_X(i-1) & \cdots & \rho_X(i-j+1) \\ \rho_X(i+1) & & & \\ \vdots & & & \\ \rho_X(i+j-1) & \cdots & \rho_X(i+1) & \rho_X(i) \end{pmatrix}$$

and let $\Delta(i, j)$ denote its determinant. Since $\rho_X(h) = \sum_{i=1}^{P} a_i\rho_X(h-i) = 0$, for all $h > Q$, it is clear that $D(i, j)$ is not a full-rank matrix if $i > Q$ and $j > P$. More precisely, P and Q are minimal orders (that is, (X_t) does not admit an ARMA(P', Q') representation with $P' < P$ or $Q' < Q$) if and only if

$$\begin{cases} \Delta(i,j) = 0 \;\; \forall i > Q \;\text{ and }\; \forall j > P, \\ \Delta(i,P) \neq 0 \;\; \forall i \geq Q, \\ \Delta(Q,j) \neq 0 \;\; \forall j \geq P. \end{cases} \tag{5.10}$$

The minimal orders P and Q are thus characterised by the following table:

(T1)

$i \backslash j$	1	2	· · ·	Q	$Q+1$	·	·	·	·
1	ρ_1	ρ_2	· · ·	ρ_q	ρ_{q+1}	·	·	·	·
·									
·									
·									
P				×	×	×	×	×	×
$P+1$				×	0	0	0	0	0
				×	0	0	0	0	0
				×	0	0	0	0	0
				×	0	0	0	0	0

where $\Delta(j, i)$ is at the intersection of row i and column j, and × denotes a non-zero element.

The orders P and Q are thus characterised by a corner of zeros in table $(T1)$, hence the term 'corner method'. The entries in this table are easily obtained using the recursion on j given by

$$\Delta(i,j)^2 = \Delta(i+1,j)\Delta(i-1,j) + \Delta(i,j+1)\Delta(i,j-1), \tag{5.11}$$

and letting $\Delta(i,0) = 1$, $\Delta(i,1) = \rho_X(|i|)$.

Denote by $\hat{D}(i,j)$, $\hat{\Delta}(i,j)$, $(\widehat{T1})$, … the items obtained by replacing $\{\rho_X(h)\}$ by $\{\hat{\rho}_X(h)\}$ in $D(i,j)$, $\Delta(i,j)$, $(T1)$, …. Only a finite number of SACRs $\hat{\rho}_X(1),\dots,\hat{\rho}_X(K)$ are available in practice, which allows $\hat{\Delta}(j,i)$ to be computed for $i \geq 1$, $j \geq 1$ and $i+j \leq K+1$. Table $(\widehat{T1})$ is thus triangular. Because $\hat{\Delta}(j,i)$ consistently estimates $\Delta(j,i)$, the orders P and Q are characterised by a corner of small values in table $(\widehat{T1})$. However, the notion of 'small value' in $(\widehat{T1})$ is not precise enough.[3]

It is preferable to consider the Studentised statistics defined, for $i = -K,\dots,K$ and $j = 0,\dots,K-|i|+1$, by

$$t(i,j) = \sqrt{n}\frac{\hat{\Delta}(i,j)}{\hat{\sigma}_{\hat{\Delta}(i,j)}}, \quad \hat{\sigma}^2_{\hat{\Delta}(i,j)} = \frac{\partial\hat{\Delta}(i,j)}{\partial\rho'_K}\hat{\Sigma}_{\hat{\rho}_K}\frac{\partial\hat{\Delta}(i,j)}{\partial\rho_K}, \tag{5.12}$$

where $\hat{\Sigma}_{\hat{\rho}_K}$ is a consistent estimator of the asymptotic covariance matrix of the first K SACRs, which can be obtained by the algorithm of Section 5.2.1 or by that of Section 5.2.2, and where the Jacobian $\frac{\partial\hat{\Delta}(i,j)}{\partial\rho'_K} = \left(\frac{\partial\hat{\Delta}(i,j)}{\partial\rho_X(1)},\dots,\frac{\partial\hat{\Delta}(i,j)}{\partial\rho_X(K)}\right)$ is obtained from the differentiation of (5.11):

$$\frac{\partial\hat{\Delta}(i,0)}{\partial\rho_X(k)} = 0 \quad \text{for} \quad i = -K-1,\dots,K-1 \quad \text{and} \quad k = 1,\dots,K;$$

$$\frac{\partial\hat{\Delta}(i,1)}{\partial\rho_X(k)} = \mathbb{1}_{\{k\}}(|i|) \quad \text{for } i = -K,\dots,K \quad \text{and} \quad k = 1,\dots,K;$$

$$\frac{\partial\hat{\Delta}(i,j+1)}{\partial\rho_X(k)} = \frac{2\hat{\Delta}(i,j)\frac{\partial\hat{\Delta}(i,j)}{\partial\rho_X(k)} - \hat{\Delta}(i+1,j)\frac{\partial\hat{\Delta}(i-1,j)}{\partial\rho_X(k)} - \hat{\Delta}(i-1,j)\frac{\partial\hat{\Delta}(i+1,j)}{\partial\rho_X(k)}}{\hat{\Delta}(i,j-1)} - \frac{\{\hat{\Delta}(i,j)^2 - \hat{\Delta}(i+1,j)\,\hat{\Delta}(i-1,j)\}\frac{\partial\hat{\Delta}(i,j-1)}{\partial\rho_X(k)}}{\hat{\Delta}^2(i,j-1)}$$

for $k = 1,\dots,K$, $i = -K+j,\dots,K-j$ and $j = 1,\dots,K$.

When $\Delta(i,j) = 0$, the statistic $t(i,j)$ asymptotically follows a $\mathcal{N}(0,1)$ (provided, in particular, that EX_t^4 exists). If, in contrast, $\Delta(i,j) \neq 0$ then $\sqrt{n}\,|\,t(i,j)\,| \to \infty$ *a.s.* when $n \to \infty$. We can reject the hypothesis $\Delta(i,j) = 0$ at level α if $|t(i,j)|$ is beyond the $(1-\alpha/2)$-quantile of a $\mathcal{N}(0,1)$. We can also automatically detect a corner of small values in the table giving the $t(i,j)$, if no entry in this corner is greater than this $(1-\alpha/2)$-quantile in absolute value. This practice does not correspond to any formal test at level α, but allows a small number of plausible values to be selected for the orders P and Q.

Illustration of the Corner Method

For a simulation of size $n = 1000$ of the ARMA(2,1)-GARCH(1, 1) model (5.8), we obtain the following table:

```
.p.|.q..1....2....3....4....5....6....7....8....9...10...11...12...
 1 |  17.6-31.6-22.6 -1.9 11.5  8.7 -0.1 -6.1 -4.2  0.5  3.5  2.1
 2 |  36.1 20.3 12.2  8.7  6.5  4.9  4.0  3.3  2.5  2.1  1.8
 3 |  -7.8 -1.6 -0.2  0.5  0.7 -0.7  0.8 -1.4  1.2 -1.1
 4 |   5.2  0.1  0.4  0.3  0.6 -0.1 -0.3  0.5 -0.2
```

[3] Comparing $\hat{\Delta}(i,j)$ and $\hat{\Delta}(i',j')$ for $j \neq j'$ (that is, entries of different rows in table $(\widehat{T1})$) is all the more difficult as these are determinants of matrices of different sizes.

```
 5 |  -3.7  0.4 -0.1 -0.5  0.4 -0.2  0.2 -0.2
 6 |   2.8  0.6  0.5  0.4  0.2  0.4  0.2
 7 |  -2.0 -0.7  0.2  0.0 -0.4 -0.3
 8 |   1.7  0.8  0.0  0.2  0.2
 9 |  -0.6 -1.2 -0.5 -0.2
10 |   1.4  0.9 -0.2
11 |  -0.2 -1.2
12 |   1.2
```

A corner of values which can be viewed as plausible realisations of the $\mathcal{N}(0, 1)$ can be observed. This corner corresponds to the rows 3, 4, … and the columns 2, 3, …, leading us to select the ARMA(2, 1) model. The automatic detection routine for corners of small values gives:

```
ARMA(P,Q) MODELS FOUND WITH GIVEN SIGNIFICANCE LEVEL
    PROBA        CRIT             MODELS FOUND
  0.200000       1.28     ( 2, 8)    ( 3, 1)    (10, 0)
  0.100000       1.64     ( 2, 1)    ( 8, 0)
  0.050000       1.96     ( 1,10)    ( 2, 1)    ( 7, 0)
  0.020000       2.33     ( 0,11)    ( 1, 9)    ( 2, 1)    ( 6, 0)
  0.010000       2.58     ( 0,11)    ( 1, 8)    ( 2, 1)    ( 6, 0)
  0.005000       2.81     ( 0,11)    ( 1, 8)    ( 2, 1)    ( 5, 0)
  0.002000       3.09     ( 0,11)    ( 1, 8)    ( 2, 1)    ( 5, 0)
  0.001000       3.29     ( 0,11)    ( 1, 8)    ( 2, 1)    ( 5, 0)
  0.000100       3.72     ( 0, 9)    ( 1, 7)    ( 2, 1)    ( 5, 0)
  0.000010       4.26     ( 0, 8)    ( 1, 6)    ( 2, 1)    ( 4, 0)
```

We retrieve not only the orders $(P, Q) = (2, 1)$ of the simulated model but also other plausible orders. This is not surprising since the ARMA(2, 1) model can be well approximated by other ARMA models, such as an AR(6), an MA(11) or an ARMA(1, 8) (but in practice, the ARMA(2, 1) should be preferred for parsimony reasons).

5.3 Identifying the GARCH Orders of an ARMA-GARCH Model

The Box–Jenkins methodology described in Chapter 1 for ARMA models can be adapted to GARCH(p, q) models. In this section, we consider only the identification problem. First, suppose that the observations are drawn from a pure GARCH. The choice of a small number of plausible values for the orders p and q can be achieved in several steps, using various tools:

(i) inspection of the sample autocorrelations and SPACs of $\epsilon_1^2, \dots, \epsilon_n^2$;

(ii) inspection of statistics that are functions of the sample autocovariances of ϵ_t^2 (corner method, epsilon algorithm, etc.);

(iii) use of information criteria (AIC, BIC, etc.);

(iv) tests of the significance of certain coefficients;

(v) analysis of the residuals.

Steps (iii) and (v), and to a large extent step (iv), require the estimation of models and are used to validate or modify them. Estimation of GARCH models will be studied in detail in the forthcoming chapters.

Step (i) relies on the ARMA representation for the square of a GARCH process. In particular, if (ϵ_t) is an ARCH(q) process, then the theoretical partial autocorrelation function $r_{\epsilon^2}(\cdot)$ of (ϵ_t^2) satisfies

$$r_{\epsilon^2}(h) = 0, \qquad \forall h > q.$$

For mixed models, the corner method can be used.

5.3.1 Corner Method in the GARCH Case

To identify the orders of a GARCH(p, q) process, one can use the fact that (ϵ_t^2) follows an ARMA($\tilde{P}, \tilde{Q}$) with $\tilde{P} = \max(p, q)$ and $\tilde{Q} = p$. In the case of a pure GARCH, $(\epsilon_t) = (X_t)$ is observed. The asymptotic variance of the SACRs of $\epsilon_1^2, \dots, \epsilon_n^2$ can be estimated by the method described in Section 5.2.2. The table of Studentised statistics for the corner method follows, as described in the previous section. The problem is then to detect at least one corner of normal values starting from the row $\tilde{P} + 1$ and the column $\tilde{Q} + 1$ of the table, under the constraints $\tilde{P} \geq 1$ (because $\max(p, q) \geq q \geq 1$) and $\tilde{P} \geq \tilde{Q}$. This leads to a selection of GARCH(p, q) models such that $(p, q) = (\tilde{Q}, \tilde{P})$ when $\tilde{Q} < \tilde{P}$ and $(p, q) = (\tilde{Q}, 1)$, $(p, q) = (\tilde{Q}, 2), \dots, (p, q) = (\tilde{Q}, \tilde{P})$ when $\tilde{Q} \geq \tilde{P}$.

In the ARMA-GARCH case, the ϵ_t is unobserved but can be approximated by the ARMA residuals. Alternatively, to avoid the ARMA estimation, residuals from fitted ARs, as described in steps 1 and 3 of the algorithm of Section 5.2.1, can be used.

5.3.2 Applications

A Pure GARCH

Consider a simulation of size $n = 5000$ of the GARCH(2, 1) model

$$\begin{cases} \epsilon_t = \sigma_t \eta_t \\ \sigma_t^2 = \omega + \alpha \epsilon_{t-1}^2 + \beta_1 \sigma_{t-1}^2 + \beta_2 \sigma_{t-2}^2, \end{cases} \tag{5.13}$$

where (η_t) is a sequence of iid $\mathcal{N}(0, 1)$ variables, $\omega = 1$, $\alpha = 0.1$, $\beta_1 = 0.05$ and $\beta_2 = 0.8$.

The table of Studentised statistics for the corner method is as follows:

.max(p,q).	.p..1...	.2...	.3...	.4...	.5...	.6...	.7...	.8...	.9...	.10...	.11...	.12...	.13...	.14...	.15...
1	5.3	2.9	5.1	2.2	5.3	5.9	3.6	3.7	2.9	2.9	3.4	1.4	5.8	2.4	3.0
2	-2.4	-3.5	2.4	-4.4	2.2	-0.7	0.6	-0.7	-0.3	0.4	1.1	-2.5	2.8	-0.2	
3	4.9	2.4	0.7	1.7	0.7	-0.8	0.2	0.4	0.3	0.3	0.7	1.4	1.4		
4	-0.4	-4.3	-1.8	-0.6	1.0	-0.6	0.4	-0.4	0.5	-0.6	0.4	-1.1			
5	4.6	2.4	0.6	0.9	0.8	0.5	0.3	-0.4	-0.5	0.5	-0.8				
6	-3.1	-1.7	1.4	-0.8	-0.3	0.3	0.3	-0.5	0.5	0.4					
7	3.1	1.2	0.3	0.6	0.3	0.2	0.5	0.1	-0.7						
8	-1.0	-1.3	-0.7	-0.5	0.8	-0.5	0.3	-0.6							
9	1.5	0.3	0.2	0.7	-0.5	0.5	-0.7								
10	-1.7	0.1	0.3	-0.7	-0.6	0.5									
11	1.8	1.2	0.6	0.7	-1.0										
12	1.6	-1.3	-1.4	-1.1											
13	4.2	2.3	1.4												
14	-1.2	-0.6													
15	1.4														

A corner of plausible $\mathcal{N}(0, 1)$ values is observed starting from the row $\tilde{P} + 1 = 3$ and the column $\tilde{Q} + 1 = 3$, which corresponds to GARCH(p, q) models such that $(\max(p, q), p) = (2, 2)$, that is, $(p, q) = (2, 1)$ or $(p, q) = (2, 2)$. A small number of other plausible values are detected for (p, q).

```
GARCH(p,q) MODELS FOUND WITH GIVEN SIGNIFICANCE LEVEL
  PROBA       CRIT               MODELS FOUND
 0.200000     1.28     ( 3, 1)    ( 3, 2)    ( 3, 3)    ( 1,13)
 0.100000     1.64     ( 3, 1)    ( 3, 2)    ( 3, 3)    ( 2, 4)     ( 0,13)
 0.050000     1.96     ( 2, 1)    ( 2, 2)    ( 0,13)
 0.020000     2.33     ( 2, 1)    ( 2, 2)    ( 1, 5)    ( 0,13)
 0.010000     2.58     ( 2, 1)    ( 2, 2)    ( 1, 4)    ( 0,13)
 0.005000     2.81     ( 2, 1)    ( 2, 2)    ( 1, 4)    ( 0,13)
 0.002000     3.09     ( 2, 1)    ( 2, 2)    ( 1, 4)    ( 0,13)
 0.001000     3.29     ( 2, 1)    ( 2, 2)    ( 1, 4)    ( 0,13)
 0.000100     3.72     ( 2, 1)    ( 2, 2)    ( 1, 4)    ( 0,13)
 0.000010     4.26     ( 2, 1)    ( 2, 2)    ( 1, 4)    ( 0, 5)
```

An ARMA-GARCH

Let us resume the simulation of size $n = 1000$ of the ARMA(2, 1)-GARCH(1, 1) model (5.8). The table of Studentised statistics for the corner method, applied to the SACRs of the observed process, was presented in Section 5.2.3. A small number of ARMA models, including the ARMA(2, 1), were selected. Let $e_{1+p_0}, \dots, e_n$ denote the residuals when an AR(p_0) is fitted to the observations, the order p_0 being selected using an information criterion.[4] Applying the corner method again, but this time on the SACRs of the squared residuals $e^2_{1+p_0}, \dots, e^2_n$, and estimating the covariances between the SACRs by the multivariate AR spectral approximation, as described in Section 5.2.2, we obtain the following table:

```
.max(p,q).|.p..1....2....3....4....5....6....7....8....9...10...11...12...
        1 |   4.5  4.1  3.5  2.1  1.1  2.1  1.2  1.0  0.7  0.4 -0.2  0.9
        2 |  -2.7  0.3 -0.2  0.1 -0.4  0.5 -0.2  0.2 -0.1  0.4 -0.2
        3 |   1.4 -0.2  0.0 -0.2  0.2  0.3 -0.2  0.1 -0.2  0.1
        4 |  -0.9  0.1  0.2  0.2 -0.2  0.2  0.0 -0.2 -0.1
        5 |   0.3 -0.4  0.2 -0.2  0.1  0.1 -0.1  0.1
        6 |  -0.7  0.4 -0.2  0.2 -0.1  0.1 -0.1
        7 |   0.0 -0.1 -0.2  0.1 -0.1 -0.2
        8 |  -0.1  0.1 -0.1 -0.2 -0.1
        9 |  -0.3  0.1 -0.1 -0.1
       10 |   0.1 -0.2 -0.1
       11 |  -0.4  0.2
       12 |  -1.0
```

A corner of values compatible with the $\mathcal{N}(0, 1)$ is observed starting from row 2 and column 2, which corresponds to a GARCH(1, 1) model. Another corner can be seen below row 2, which corresponds to a GARCH(0, 2) = ARCH(2) model. In practice, in this identification step, at least these two models would be selected. The next step would be the estimation of the selected models, followed by a validation step involving testing the significance of the coefficients, examining the residuals and comparing the models via information criteria. This validation step allows a final model to be retained which can be used for prediction purposes.

```
GARCH(p,q) MODELS FOUND WITH GIVEN SIGNIFICANCE LEVEL
  PROBA       CRIT               MODELS FOUND
 0.200000     1.28     ( 1, 1)    ( 0, 3)
 0.100000     1.64     ( 1, 1)    ( 0, 2)
```

[4] One can also use the innovations algorithm of Brockwell and Davis (1991, p. 172) for rapid fitting of MA models. Alternatively, one of the previously selected ARMA models, for instance the ARMA(2, 1), can be used to approximate the innovations.

```
0.050000     1.96    ( 1, 1)    ( 0, 2)
0.020000     2.33    ( 1, 1)    ( 0, 2)
0.010000     2.58    ( 1, 1)    ( 0, 2)
0.005000     2.81    ( 0, 1)
0.002000     3.09    ( 0, 1)
0.001000     3.29    ( 0, 1)
0.000100     3.72    ( 0, 1)
0.000010     4.26    ( 0, 1)
```

5.4 Lagrange Multiplier Test for Conditional Homoscedasticity

To test linear restrictions on the parameters of a model, the most widely used tests are the Wald test, the LM test, and likelihood ratio (LR) test. The LM test, also referred to as the Rao test or the score test, is attractive because it only requires estimation of the restricted model (unlike the Wald and LR tests which will be studied in Chapter 8), which is often much easier than estimating the unrestricted model. We start by deriving the general form of the LM test. Then we present an LM test for conditional homoscedasticity in Section 5.4.2.

5.4.1 General Form of the LM Test

Consider a parametric model, with true parameter value $\theta_0 \in \mathbb{R}^d$, and a null hypothesis

$$H_0 : R\theta_0 = r,$$

where R is a given $s \times d$ matrix of full rank s, and r is a given $s \times 1$ vector. This formulation allows one to test, for instance, whether the first s components of θ_0 are null (it suffices to set $R = [I_s : 0_{s\times(d-s)}]$ and $r = 0_s$). Let $\ell_n(\theta)$ denote the log-likelihood of observations $X_1, \ldots, X_n$. We assume the existence of unconstrained and constrained (by H_0) maximum likelihood estimators, respectively, satisfying

$$\hat{\theta} = \arg\sup_{\theta} \ell_n(\theta) \quad \text{and} \quad \hat{\theta}^c = \arg\sup_{\theta: R\theta = r} \ell_n(\theta).$$

Under some regularity assumptions (which will be discussed in detail in Chapter 7 for the GARCH(p, q) model), the score vector satisfies a CLT and we have

$$\frac{1}{\sqrt{n}} \frac{\partial}{\partial\theta} \ell_n(\theta_0) \xrightarrow{\mathcal{L}} \mathcal{N}(0, \mathfrak{I}) \quad \text{and} \quad \sqrt{n}(\hat{\theta} - \theta_0) \xrightarrow{\mathcal{L}} \mathcal{N}(0, \mathfrak{I}^{-1}), \tag{5.14}$$

where $\mathfrak{I}$ is the Fisher information matrix. To derive the constrained estimator, we introduce the Lagrangian

$$\mathcal{L}(\theta, \lambda) = \ell_n(\theta) - \lambda'(R\theta - r).$$

We have

$$(\hat{\theta}^c, \hat{\lambda}) = \arg\sup_{(\theta,\lambda)} \mathcal{L}(\theta, \lambda).$$

The first-order conditions give

$$R'\hat{\lambda} = \frac{\partial}{\partial\theta} \ell_n(\hat{\theta}^c) \quad \text{and} \quad R\hat{\theta}^c = r.$$

The second convergence in (5.14) thus shows that under H_0,

$$\sqrt{n}R(\hat{\theta} - \hat{\theta}^c) = \sqrt{n}R(\hat{\theta} - \theta_0) \xrightarrow{\mathcal{L}} \mathcal{N}(0, R\mathfrak{I}^{-1}R'). \tag{5.15}$$

Using the convention $a \stackrel{c}{=} b$ for $a = b + c$, asymptotic expansions entail, under usual regularity conditions (more rigorous statements will be given in Chapter 7),

$$0 = \frac{1}{\sqrt{n}}\frac{\partial}{\partial\theta}\ell_n(\hat{\theta}) \stackrel{o_P(1)}{=} \frac{1}{\sqrt{n}}\frac{\partial}{\partial\theta}\ell_n(\theta_0) - \mathfrak{I}\sqrt{n}(\hat{\theta} - \theta_0),$$

$$\frac{1}{\sqrt{n}}\frac{\partial}{\partial\theta}\ell_n(\hat{\theta}^c) \stackrel{o_P(1)}{=} \frac{1}{\sqrt{n}}\frac{\partial}{\partial\theta}\ell_n(\theta_0) - \mathfrak{I}\sqrt{n}(\hat{\theta}^c - \theta_0),$$

which, by subtraction, gives

$$\sqrt{n}(\hat{\theta} - \hat{\theta}^c) \stackrel{o_P(1)}{=} \mathfrak{I}^{-1}\frac{1}{\sqrt{n}}\frac{\partial}{\partial\theta}\ell_n(\hat{\theta}^c) = \mathfrak{I}^{-1}\frac{1}{\sqrt{n}}R'\hat{\lambda}. \tag{5.16}$$

Finally, results (5.15) and (5.16) imply

$$\frac{1}{\sqrt{n}}\hat{\lambda} \stackrel{o_P(1)}{=} (R\mathfrak{I}^{-1}R')^{-1}\sqrt{n}R(\hat{\theta} - \theta_0) \xrightarrow{\mathcal{L}} \mathcal{N}\{0, (R\mathfrak{I}^{-1}R')^{-1}\}$$

and then

$$\frac{1}{\sqrt{n}}\frac{\partial}{\partial\theta}\ell_n(\hat{\theta}^c) = R'\frac{1}{\sqrt{n}}\hat{\lambda} \xrightarrow{\mathcal{L}} \mathcal{N}\{0, R'(R\mathfrak{I}^{-1}R')^{-1}R\}.$$

Thus, under H_0, the test statistic

$$LM_n := \frac{1}{n}\hat{\lambda}'R\hat{\mathfrak{I}}^{-1}R'\hat{\lambda} = \frac{1}{n}\frac{\partial}{\partial\theta'}\ell_n(\hat{\theta}^c)\hat{\mathfrak{I}}^{-1}\frac{\partial}{\partial\theta}\ell_n(\hat{\theta}^c) \tag{5.17}$$

asymptotically follows a χ_s^2, provided that $\hat{\mathfrak{I}}$ is an estimator converging in probability to $\mathfrak{I}$. In general one can take

$$\hat{\mathfrak{I}} = -\frac{1}{n}\frac{\partial^2\ell_n(\hat{\theta}^c)}{\partial\theta\,\partial\theta'}.$$

The critical region of the LM test at the asymptotic level α is $\{LM_n > \chi_s^2(1 - \alpha)\}$.

The Case Where the LM_n Statistic Takes the Form nR^2

Implementation of an LM test can sometimes be extremely simple. Consider a non-linear conditionally homoscedastic model in which a dependent variable Y_t is related to its past values and to a vector of exogenous variables X_t by $Y_t = F_{\theta_0}(W_t) + \epsilon_t$, where ϵ_t is iid $(0, \sigma_0^2)$ and $W_t = (X_t, Y_{t-1}, \dots)$. Assume, in addition, that W_t and ϵ_t are independent. We wish to test the hypothesis

$$H_0 : \psi_0 = 0$$

where

$$\theta_0 = \begin{pmatrix} \beta_0 \\ \psi_0 \end{pmatrix}, \quad \beta_0 \in \mathbb{R}^{d-s}, \quad \psi_0 \in \mathbb{R}^s.$$

To retrieve the framework of the previous section, let $R = [0_{s \times (d-s)} : I_s]$ and note that

$$\frac{\partial}{\partial\psi}\ell_n(\hat{\theta}^c) = R\frac{\partial}{\partial\theta}\ell_n(\hat{\theta}^c) = RR'\hat{\lambda} = \hat{\lambda} \quad \text{and} \quad \frac{1}{\sqrt{n}}\hat{\lambda} \xrightarrow{\mathcal{L}} \mathcal{N}(0, \Sigma_\lambda),$$

where $\Sigma_\lambda = (\mathfrak{I}^{22})^{-1}$ and $\mathfrak{I}^{22} = R\mathfrak{I}^{-1}R'$ is the bottom right-hand block of $\mathfrak{I}^{-1}$. Suppose that $\sigma_0^2 = \sigma^2(\beta_0)$ does not depend on ψ_0. With a Gaussian likelihood (Exercise 5.9) we have

$$\frac{1}{\sqrt{n}}\hat{\lambda} = \frac{1}{\sqrt{n}\hat{\sigma}^{c2}}\sum_{t=1}^{n}\epsilon_t(\hat{\theta}^c)\frac{\partial}{\partial\psi}F_{\hat{\theta}^c}(W_t) = \frac{1}{\sqrt{n}\hat{\sigma}^{c2}}\mathbf{F}'_\psi\hat{\mathbf{U}}^c,$$

where $\epsilon_t(\theta) = Y_t - F_\theta(W_t)$, $\hat{\sigma}^{c2} = \sigma^2(\hat{\beta}^c)$, $\hat{\epsilon}_t^c = \epsilon_t(\hat{\theta}^c)$, $\hat{\mathbf{U}}^c = (\hat{\epsilon}_1^c, \dots, \hat{\epsilon}_n^c)'$ and

$$\mathbf{F}'_\psi = \left(\frac{\partial F_{\hat{\theta}^c}(W_1)}{\partial\psi} \quad \cdots \quad \frac{\partial F_{\hat{\theta}^c}(W_n)}{\partial\psi}\right).$$

Partition $\mathfrak{I}$ into blocks as

$$\mathfrak{I} = \begin{pmatrix}\mathfrak{I}_{11} & \mathfrak{I}_{12}\\ \mathfrak{I}_{21} & \mathfrak{I}_{22}\end{pmatrix},$$

where $\mathfrak{I}_{11}$ and $\mathfrak{I}_{22}$ are square matrices of respective sizes $d-s$ and s. Under the assumption that the information matrix $\mathfrak{I}$ is block-diagonal (that is, $\mathfrak{I}_{12} = 0$), we have $\mathfrak{I}^{22} = \mathfrak{I}_{22}^{-1}$ where $\mathfrak{I}_{22} = R\mathfrak{I}R'$, which entails $\Sigma_\lambda = \mathfrak{I}_{22}$. We can then choose

$$\hat{\Sigma}_\lambda = \frac{1}{\hat{\sigma}^{c4}}\frac{1}{n}\hat{\mathbf{U}}^{c\prime}\hat{\mathbf{U}}^c\frac{1}{n}\mathbf{F}'_\psi\mathbf{F}_\psi \overset{o_P(1)}{=} \frac{1}{\hat{\mathbf{U}}^{c\prime}\hat{\mathbf{U}}^c}\mathbf{F}'_\psi\mathbf{F}_\psi$$

as a consistent estimator of Σ_λ. We end up with

$$\mathrm{LM}_n = n\frac{\hat{\mathbf{U}}^{c\prime}\mathbf{F}_\psi(\mathbf{F}'_\psi\mathbf{F}_\psi)^{-1}\mathbf{F}'_\psi\hat{\mathbf{U}}^c}{\hat{\mathbf{U}}^{c\prime}\hat{\mathbf{U}}^c}, \tag{5.18}$$

which is nothing other than n times the uncentred determination coefficient in the regression of $\hat{\epsilon}_t^c$ on the variables $\partial F_{\hat{\theta}^c}(W_t)/\partial\psi_i$ for $i = 1, \dots, s$ (Exercise 5.10).

LM Test with Auxiliary Regressions

We extend the previous framework by allowing $\mathfrak{I}_{12}$ to be not equal to zero. Assume that σ^2 does not depend on θ. In view of Exercise 5.9, we can then estimate Σ_λ by[5]

$$\hat{\Sigma}_\lambda^* = \frac{1}{\hat{\mathbf{U}}^{c\prime}\hat{\mathbf{U}}^c}(\mathbf{F}'_\psi\mathbf{F}_\psi - \mathbf{F}'_\psi\mathbf{F}_\beta(\mathbf{F}'_\beta\mathbf{F}_\beta)^{-1}\mathbf{F}'_\beta\mathbf{F}_\psi),$$

where

$$\mathbf{F}'_\beta = \left(\frac{\partial F_{\hat{\theta}^c}(W_1)}{\partial\beta} \cdots \frac{\partial F_{\hat{\theta}^c}(W_n)}{\partial\beta}\right).$$

Suppose the model is linear under the constraint H_0, so that

$$\hat{\mathbf{U}}^c = \mathbf{Y} - \mathbf{F}_\beta\hat{\beta}^c \quad \text{and} \quad \hat{\sigma}^{c2} = \hat{\mathbf{U}}^{c\prime}\hat{\mathbf{U}}^c/n$$

with

$$\mathbf{Y} = (Y_1, \dots, Y_n)' \quad \text{and} \quad \hat{\beta}^c = (\mathbf{F}'_\beta\mathbf{F}_\beta)^{-1}\mathbf{F}'_\beta\mathbf{Y},$$

up to some negligible terms.

Now consider the linear regression

$$\mathbf{Y} = \mathbf{F}_\beta\beta^* + \mathbf{F}_\psi\psi^* + \mathbf{U}. \tag{5.19}$$

[5] For a partitioned invertible matrix $A = \begin{pmatrix}A_{11} & A_{12}\\ A_{21} & A_{22}\end{pmatrix}$, where A_{11} and A_{22} are invertible square blocks, the bottom right-hand block of A^{-1} is written as $A^{22} = (A_{22} - A_{21}A_{11}^{-1}A_{12})^{-1}$ (Exercise 6.7).

Exercise 5.10 shows that, in this auxiliary regression, the LM statistic for testing the hypothesis

$$H_0^* : \psi^* = 0$$

is given by

$$\begin{aligned}\mathrm{LM}_n^* &= n^{-1}(\hat{\sigma}^c)^{-4}\hat{\mathbf{U}}^{c\prime}\mathbf{F}_\psi\hat{\Sigma}_\lambda^{*-1}\mathbf{F}_\psi'\hat{\mathbf{U}}^c \\ &= (\hat{\sigma}^c)^{-2}\hat{\mathbf{U}}^{c\prime}\mathbf{F}_\psi(\mathbf{F}_\psi'\mathbf{F}_\psi - \mathbf{F}_\psi'\mathbf{F}_\beta(\mathbf{F}_\beta'\mathbf{F}_\beta)^{-1}\mathbf{F}_\beta'\mathbf{F}_\psi)^{-1}\mathbf{F}_\psi'\hat{\mathbf{U}}^c.\end{aligned}$$

This statistic is precisely the LM test statistic for the hypothesis $H_0 : \psi = 0$ in the initial model. From Exercise 5.10, the LM test statistic of the hypothesis $H_0^* : \psi^* = 0$ in model (5.19) can also be written as

$$\mathrm{LM}_n^* = n\frac{\hat{\mathbf{U}}^{c\prime}\hat{\mathbf{U}}^c - \hat{\mathbf{U}}'\hat{\mathbf{U}}}{\hat{\mathbf{U}}^{c\prime}\hat{\mathbf{U}}^c}, \tag{5.20}$$

where $\hat{\mathbf{U}} = \mathbf{Y} - \mathbf{F}_\beta\hat{\beta}^* - \mathbf{F}_\psi\hat{\psi}^* =: \mathbf{Y} - \mathbf{F}\hat{\theta}^*$, with $\hat{\theta}^* = (\mathbf{F}'\mathbf{F})^{-1}\mathbf{F}'\mathbf{Y}$. We finally obtain the so-called Breusch–Godfrey form of the LM statistic by interpreting LM_n^* as n times the determination coefficient of the auxiliary regression

$$\hat{\mathbf{U}}^c = \mathbf{F}_\beta\gamma + \mathbf{F}_\psi\psi^{**} + \mathbf{V}, \tag{5.21}$$

where $\hat{\mathbf{U}}^c$ is the vector of residuals in the regression of $\mathbf{Y}$ on the columns of $\mathbf{F}_\beta$.

Indeed, in the two regressions (5.19) and (5.21), the vector of residuals is $\hat{\mathbf{V}} = \hat{\mathbf{U}}$, because $\hat{\beta}^* = \hat{\beta}^c + \hat{\gamma}$ and $\hat{\psi}^* = \hat{\psi}^{**}$. Finally, we note that the determination coefficient is centred (in other words, it is the R^2 which is provided by standard statistical software) when a column of $\mathbf{F}_\beta$ is constant.

Quasi-LM Test

When $\ell_n(\theta)$ is no longer supposed to be the log-likelihood, but only the quasi-log-likelihood (a thorough study of the quasi-likelihood for GARCH models will be made in Chapter 7), the equations can in general be replaced by

$$\frac{1}{\sqrt{n}}\frac{\partial}{\partial\theta}\ell_n(\theta_0) \xrightarrow{\mathcal{L}} \mathcal{N}(0, I) \quad \text{and} \quad \sqrt{n}(\hat{\theta} - \theta_0) \xrightarrow{\mathcal{L}} \mathcal{N}(0, J^{-1}IJ^{-1}), \tag{5.22}$$

where

$$I = \lim_{n\to\infty}\frac{1}{n}\mathrm{Var}\frac{\partial}{\partial\theta}\ell_n(\theta_0), \quad J = \lim_{n\to\infty}\frac{1}{n}\frac{\partial^2}{\partial\theta\,\partial\theta'}\ell_n(\theta_0) \quad \text{a.s.}$$

It is then recommended that the statistic (5.17) be replaced by the more complex, but more robust, expression

$$\begin{aligned}\mathrm{LM}_n &= \frac{1}{n}\hat{\lambda}'R\hat{J}^{-1}R'(R\hat{J}^{-1}\hat{I}\hat{J}^{-1}R')^{-1}R\hat{J}^{-1}R'\hat{\lambda} \\ &= \frac{1}{n}\frac{\partial}{\partial\theta'}\ell_n(\hat{\theta}^c)\hat{J}^{-1}R'(R\hat{J}^{-1}\hat{I}\hat{J}^{-1}R')^{-1}R\hat{J}^{-1}\frac{\partial}{\partial\theta}\ell_n(\hat{\theta}^c),\end{aligned} \tag{5.23}$$

where $\hat{I}$ and $\hat{J}$ are consistent estimators of I and J. A consistent estimator of J is obviously obtained as a sample mean. Estimating the long-run variance I requires more involved methods, such as those described in Section 5.2.2 (HAC or other methods).

5.4.2 LM Test for Conditional Homoscedasticity

Consider testing the conditional homoscedasticity assumption

$$H_0 : \alpha_{01} = \cdots = \alpha_{0q} = 0$$

in the ARCH(q) model

$$\begin{cases} \epsilon_t = \sigma_t \eta_t, & \eta_t \text{ iid}(0,1) \\ \sigma_t^2 = \omega_0 + \sum_{i=1}^{q} \alpha_{0i}\epsilon_{t-i}^2, & \omega_0 > 0, \ \alpha_{0i} \geq 0. \end{cases}$$

At the parameter value $\theta = (\omega, \alpha_1, \ldots, \alpha_q)$ the quasi-log-likelihood is written, neglecting unimportant constants, as

$$\ell_n(\theta) = -\frac{1}{2}\sum_{t=1}^{n}\left\{\frac{\epsilon_t^2}{\sigma_t^2(\theta)} + \log \sigma_t^2(\theta)\right\}, \quad \sigma_t^2(\theta) = \omega + \sum_{i=1}^{q} \alpha_i \epsilon_{t-i}^2,$$

with the convention $\epsilon_{t-i} = 0$ for $t \leq 0$. The constrained quasi-maximum likelihood estimator is[6]

$$\hat{\theta}^c = (\hat{\omega}^c, 0, \ldots, 0), \quad \text{where } \hat{\omega}^c = \sigma_t^2(\theta^c) = \frac{1}{n}\sum_{t=1}^{n} \epsilon_t^2.$$

At $\theta_0 = (\omega_0, \ldots, 0)$, the score vector satisfies

$$\begin{aligned}
\frac{1}{\sqrt{n}}\frac{\partial \ell_n(\theta_0)}{\partial \theta} &= \frac{1}{2\sqrt{n}}\sum_{t=1}^{n}\frac{1}{\sigma_t^4(\theta_0)}\{\epsilon_t^2 - \sigma_t^2(\theta_0)\}\frac{\partial \sigma_t^2(\theta_0)}{\partial \theta} \\
&= \frac{1}{2\sqrt{n}}\sum_{t=1}^{n}\frac{1}{\omega_0}(\eta_t^2 - 1)\begin{pmatrix} 1 \\ \epsilon_{t-1}^2 \\ \vdots \\ \epsilon_{t-q}^2 \end{pmatrix} \\
&\xrightarrow{\mathcal{L}} \mathcal{N}(0, I), \quad I = \frac{\kappa_\eta - 1}{4\omega_0^2}\begin{pmatrix} 1 & \boldsymbol{\omega}_0 \\ \boldsymbol{\omega}_0' & I_{22} \end{pmatrix}
\end{aligned}$$

under H_0, where $\boldsymbol{\omega}_0 = (\omega_0, \ldots, \omega_0)' \in \mathbb{R}^q$ and I_{22} is a matrix whose diagonal elements are $\omega_0^2\kappa_\eta$ with $\kappa_\eta = E\eta_t^4$, and whose other entries are equal to ω_0^2. The bottom right-hand block of I^{-1} is thus

$$I^{22} = \frac{4\omega_0^2}{\kappa_\eta - 1}\{I_{22} - \boldsymbol{\omega}_0\boldsymbol{\omega}_0'\}^{-1} = \frac{4\omega_0^2}{\kappa_\eta - 1}\{(\kappa_\eta - 1)\omega_0^2 I_q\}^{-1} = \frac{4}{(\kappa_\eta - 1)^2}I_q. \tag{5.24}$$

In addition, we have

$$\begin{aligned}
\frac{1}{n}\frac{\partial^2 \ell_n(\theta_0)}{\partial\theta\partial\theta'} &= \frac{1}{2n}\sum_{t=1}^{n}\frac{1}{\sigma_t^6(\theta_0)}\{2\epsilon_t^2 - \sigma_t^2(\theta_0)\}\frac{\partial \sigma_t^2(\theta_0)}{\partial \theta}\frac{\partial \sigma_t^2(\theta_0)}{\partial \theta'} \\
&= \frac{1}{2n}\sum_{t=1}^{n}\frac{1}{\omega_0^2}(2\eta_t^2 - 1)\begin{pmatrix} 1 \\ \epsilon_{t-1}^2 \\ \vdots \\ \epsilon_{t-q}^2 \end{pmatrix}(1, \epsilon_{t-1}^2, \ldots, \epsilon_{t-q}^2) \\
&\to J = \frac{2}{\kappa_\eta - 1}I, \quad \text{a.s.}
\end{aligned}$$

[6] Indeed, the function $\sigma^2 \mapsto x/\sigma^2 + n\log\sigma^2$ reaches its minimum at $\sigma^2 = x/n$.

From the equality (5.23), using estimators of I and J such that $\hat{J} = 2/(\hat{\kappa}_\eta - 1)\hat{I}$, we obtain

$$\text{LM}_n = \frac{2}{\hat{\kappa}_\eta - 1}\frac{1}{n}\hat{\lambda}' R\hat{J}^{-1}R'\hat{\lambda} = \frac{1}{n}\frac{\partial}{\partial\theta'}\ell_n(\hat{\theta}^c)\hat{I}^{-1}\frac{\partial}{\partial\theta}\ell_n(\hat{\theta}^c).$$

Using the relation (5.24) and noting that

$$\frac{\partial}{\partial\theta}\ell_n(\hat{\theta}^c) = \begin{pmatrix} 0 \\ \frac{\partial}{\partial\alpha}\ell_n(\hat{\theta}^c)\end{pmatrix}, \quad \frac{\partial}{\partial\alpha}\ell_n(\hat{\theta}^c) = \frac{1}{2}\sum_{t=1}^{n}\frac{1}{\hat{\omega}^c}\left(\frac{\epsilon_t^2}{\hat{\omega}^c} - 1\right)\begin{pmatrix}\epsilon_{t-1}^2 \\ \vdots \\ \epsilon_{t-q}^2\end{pmatrix},$$

we obtain

$$\text{LM}_n = \frac{1}{n}\frac{\partial}{\partial\alpha'}\ell_n(\hat{\theta}^c)\hat{I}^{22}\frac{\partial}{\partial\alpha}\ell_n(\hat{\theta}^c) = \frac{1}{n}\sum_{h=1}^{q}\left\{\frac{1}{\hat{\kappa}_\eta - 1}\sum_{t=1}^{n}\left(\frac{\epsilon_t^2}{\hat{\omega}^c} - 1\right)\frac{\epsilon_{t-h}^2}{\hat{\omega}^c}\right\}^2. \tag{5.25}$$

Equivalence with a Portmanteau Test

Using

$$\sum_{t=1}^{n}\left(\frac{\epsilon_t^2}{\hat{\omega}^c} - 1\right) = 0 \quad \text{and} \quad \frac{1}{n}\sum_{t=1}^{n}\left(\frac{\epsilon_t^2}{\hat{\omega}^c} - 1\right)^2 = \hat{\kappa}_\eta - 1,$$

it follows from relation (5.25) that

$$\text{LM}_n = n\sum_{h=1}^{q}\hat{\rho}_{\epsilon^2}^2(h), \tag{5.26}$$

which shows that the LM test is equivalent to a portmanteau test on the squares.

Expression in Terms of R^2

To establish a connection with the linear model, write

$$\frac{\partial}{\partial\theta}\ell_n(\hat{\theta}^c) = n^{-1}X'Y,$$

where Y is the $n \times 1$ vector $1 - \epsilon_t^2/\hat{\omega}^c$, and X is the $n \times (q+1)$ matrix with first column $1/2\hat{\omega}^c$ and $(i+1)$th column $\epsilon_{t-i}^2/2\hat{\omega}^c$. Estimating I by $(\hat{\kappa}_\eta - 1)n^{-1}X'X$, where $\hat{\kappa}_\eta - 1 = n^{-1}Y'Y$, we obtain

$$\text{LM}_n = n\frac{Y'X(X'X)^{-1}X'Y}{Y'Y}, \tag{5.27}$$

which can be interpreted as n times the determination coefficient in the linear regression of Y on the columns of X. Because the determination coefficient is invariant by linear transformation of the variables (Exercise 5.11), we simply have $\text{LM}_n = nR^2$ where R^2 is the determination coefficient[7] of the regression of ϵ_t^2 on a constant and q lagged variables $\epsilon_{t-1}^2, \dots, \epsilon_{t-q}^2$. Under the null hypothesis of conditional homoscedasticity, LM_n asymptotically follows a χ_q^2. The version of the LM statistic given in expression (5.27) differs from the one given in expression (5.25) because relation (5.24) is not satisfied when I is replaced by $(\hat{\kappa}_\eta - 1)n^{-1}X'X$.

[7] We mean here the centered determination coefficient (the one usually given by standard software) not the uncentred one as was the case in Section 5.4.1. There is sometimes confusion between these coefficients in the literature.

5.5 Application to Real Series

Consider the returns of the CAC 40 stock index from 2 March 1990 to 29 December 2006 (4245 observations) and of the FTSE 100 index of the London Stock Exchange from 3 April 1984 to 3 April 2007 (5812 observations). The correlograms for the returns and squared returns are displayed in Figure 5.7. The bottom correlograms of Figure 5.7, as well as the portmanteau tests of Table 5.4, clearly show that, for the two indices, the strong white noise assumption cannot be sustained. These portmanteau tests can be considered as versions of LM tests for conditional homoscedasticity (see Section 5.4.2). Table 5.5 displays the nR^2 version of the LM test of Section 5.4.2. Note that the two versions of the LM statistic are quite different but lead to the same unambiguous conclusions: the hypothesis of no ARCH effect must be rejected, as well as the hypothesis of absence of autocorrelation for the CAC 40 or FTSE 100 returns.

The first correlogram of Figure 5.7 and the first part of Table 5.6 lead us to think that the CAC 40 series is fairly compatible with a weak white noise structure (and hence with a GARCH structure). Recall that the 95% significance bands, shown as dotted lines on the upper correlograms of Figure 5.7, are valid under the strong white noise assumption but may be misleading for weak white noises (such as GARCH). The second part of Table 5.6 displays classical Ljung–Box tests for non-correlation. It may be noted that the CAC 40 returns series does not pass the classical portmanteau tests.[8] This does not mean, however, that the white noise assumption should be rejected. Indeed, we know that such classical portmanteau tests are invalid for conditionally heteroscedastic series.

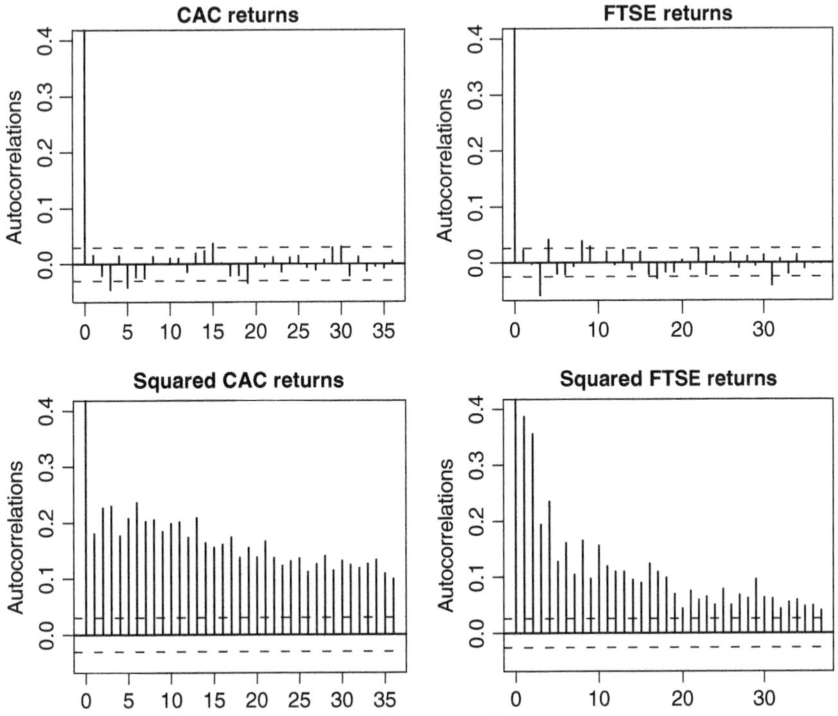

Figure 5.7 Correlograms of returns and squared returns of the CAC 40 index (2 March 1990 to 29 December 2006) and the FTSE 100 index (3 April 1984 to 3 April 2007).

[8] Classical portmanteau tests are those provided by standard commercial software, in particular those of the table entitled 'Autocorrelation Check for White Noise' of the ARIMA procedure in SAS.

Table 5.4 Portmanteau tests on the squared CAC 40 returns (2 March 1990 to 29 December 2006) and FTSE 100 returns (3 April 1984 to 3 April 2007).

m	1	2	3	4	5	6	7	8	9	10	11	12
Tests for non-correlation of the squared CAC 40												
$\hat{\rho}_{\epsilon^2}(m)$	0.181	0.226	0.231	0.177	0.209	0.236	0.202	0.206	0.184	0.198	0.201	0.173
$\hat{\sigma}_{\hat{\rho}_{\epsilon^2}(m)}$	0.030	0.030	0.030	0.030	0.030	0.030	0.030	0.030	0.030	0.030	0.030	0.030
Q_m^{LB}	138.825	356.487	580.995	712.549	896.465	1133.276	1307.290	1486.941	1631.190	1798.789	1970.948	2099.029
p-value	0.000	0.000	0.000	0.000	0.000	0.000	0.000	0.000	0.000	0.000	0.000	0.000
Tests for non-correlation of the squared FTSE 100												
$\hat{\rho}_{\epsilon^2}(m)$	0.386	0.355	0.194	0.235	0.127	0.161	0.160	0.151	0.115	0.148	0.141	0.135
$\hat{\sigma}_{\hat{\rho}_{\epsilon^2}(m)}$	0.026	0.026	0.026	0.026	0.026	0.026	0.030	0.030	0.030	0.030	0.030	0.030
Q_m^{LB}	867.573	1601.808	1820.314	2141.935	2236.064	2387.596	964.803	1061.963	1118.258	1211.899	1296.512	1374.324
p-value	0.000	0.000	0.000	0.000	0.000	0.000	0.000	0.000	0.000	0.000	0.000	0.000

Table 5.5 LM tests for conditional homoscedasticity of the CAC 40 and FTSE 100.

m	1	2	3	4	5	6	7	8	9
				Tests for absence of ARCH for the CAC 40					
LM_n	138.7	303.3	421.7	451.7	500.8	572.4	600.3	621.6	629.7
p-value	0.000	0.000	0.000	0.000	0.000	0.000	0.000	0.000	0.000
				Tests for absence of ARCH for the FTSE 100					
LM_n	867.1	1157.3	1157.4	1220.8	1222.4	1236.6	1237.0	1267.0	1267.3
p-value	0.000	0.000	0.000	0.000	0.000	0.000	0.000	0.000	0.000

Table 5.7 is the analog of Table 5.6 for the FTSE 100 index. Conclusions are more disputable in this case. Although some p-values of the upper part of Table 5.7 are slightly $<5\%$, one cannot exclude the possibility that the FTSE 100 index is a weak (GARCH) white noise. On the other hand, the assumption of strong white noise can be categorically rejected, the p-values (bottom of Table 5.7) being almost equal to zero. Table 5.8 confirms the identification of an ARMA(0, 0) process for the CAC 40. Table 5.9 would lead us to select an ARMA(0, 0), ARMA(1, 1), AR(3) or MA(3) model for the FTSE 100. Recall that this a priori identification step should be completed by an estimation of the selected models, followed by a validation step. For the CAC 40, Table 5.10 indicates that the most reasonable GARCH model is simply the GARCH(1, 1). For the FTSE 100, plausible models are the GARCH(2, 1), GARCH(2, 2), GARCH(2, 3), or ARCH(4), as can be seen from Table 5.11. The choice between these models is the object of the estimation and validation steps.

5.6 Bibliographical Notes

In this chapter, we have adapted tools generally employed to deal with the identification of ARMA models. Correlograms and partial correlograms are studied in depth in the book by Brockwell and Davis (1991). In particular, they provide a detailed proof for the Bartlett formula giving the asymptotic behaviour of the sample autocorrelations of a strong linear process. The generalised Bartlett formula (B.15) was established by Francq and Zakoïan (2009c). The textbook by Li (2004) can serve as a reference for the various portmanteau adequacy tests, as well as Godfrey (1988) for the LM tests. It is now well known that tools generally used for the identification of ARMA models should not be directly used in presence of conditional heteroscedasticity, or other forms of dependence in the linear innovation process (see, for instance, Diebold 1986; Romano and Thombs 1996; Berlinet and Francq 1997; Francq, Roy, and Zakoïan 2005; Boubacar Maïnassara and Saussereau 2018). The corner method was proposed by Béguin, Gouriéroux, and Monfort (1980) for the identification of mixed ARMA models. There are many alternatives to the corner method, in particular the epsilon algorithm (see Berlinet 1984) and the generalised autocorrelations of Glasbey (1982).

Additional references on tests of ARCH effects are Engle (1982, 1984), Bera and Higgins (1997), and Li (2004).

In this chapter, we have assumed the existence of a fourth-order moment for the observed process. When only the second-order moment exists, Basrak, Davis, and Mikosch (2002) showed in particular that the sample autocorrelations converge very slowly. When even the second-order moment does not exist, the sample autocorrelations have a degenerate asymptotic distribution.

Table 5.6 Portmanteau tests on the CAC 40 (2 March 1990 to 29 December 2006).

m	1	2	3	4	5	6	7	8	9	10	11	12	13	14	15	16
Tests of GARCH white noise based on Q_m																
$\hat{\rho}(m)$	0.016	−0.020	−0.045	0.015	−0.041	−0.023	−0.025	0.014	0.000	0.011	0.010	−0.014	0.020	0.024	0.037	0.001
$\hat{\sigma}_{\hat{\rho}(m)}$	0.041	0.044	0.044	0.041	0.043	0.044	0.042	0.043	0.041	0.042	0.042	0.041	0.043	0.040	0.040	0.040
Q_m	0.587	1.431	5.544	6.079	9.669	10.725	12.076	12.475	12.476	12.718	12.954	13.395	14.214	15.563	18.829	18.833
p-value	0.443	0.489	0.136	0.193	0.085	0.097	0.098	0.131	0.188	0.240	0.296	0.341	0.359	0.341	0.222	0.277
Usual tests for strong white noise																
$\hat{\rho}(m)$	0.016	−0.020	−0.045	0.015	−0.041	−0.023	−0.025	0.014	0.000	0.011	0.010	−0.014	0.020	0.024	0.037	0.001
$\hat{\sigma}_{\hat{\rho}(m)}$	0.030	0.030	0.030	0.030	0.030	0.030	0.030	0.030	0.030	0.030	0.030	0.030	0.030	0.030	0.030	0.030
Q_m^{LB}	1.105	2.882	11.614	12.611	19.858	22.134	24.826	25.629	25.629	26.109	26.579	27.397	29.059	31.497	37.271	37.279
p-value	0.293	0.237	0.009	0.013	0.001	0.001	0.001	0.001	0.002	0.004	0.005	0.007	0.006	0.005	0.001	0.002

Table 5.7 Portmanteau tests on the FTSE 100 (3 April 1984 to 3 April 2007).

m	1	2	3	4	5	6	7	8	9	10	11	12	13	14	15	16
Tests of GARCH white noise based on Q_m																
$\hat{\rho}(m)$	0.023	−0.002	−0.059	0.041	−0.021	−0.021	−0.006	0.039	0.029	0.000	0.019	−0.003	0.023	−0.013	0.019	−0.022
$\hat{\sigma}_{\hat{\rho}(m)}$	0.057	0.055	0.044	0.047	0.039	0.042	0.037	0.042	0.036	0.041	0.038	0.037	0.037	0.036	0.035	0.039
Q_m	0.618	0.624	7.398	10.344	11.421	12.427	12.527	15.796	18.250	18.250	19.250	19.279	20.700	21.191	22.281	23.483
p-value	0.432	0.732	0.060	0.035	0.044	0.053	0.085	0.045	0.032	0.051	0.057	0.082	0.079	0.097	0.101	0.101
Usual tests for strong white noise																
$\hat{\rho}(m)$	0.023	−0.002	−0.059	0.041	−0.021	−0.021	−0.006	0.039	0.029	0.000	0.019	−0.003	0.023	−0.013	0.019	−0.022
$\hat{\sigma}_{\hat{\rho}(m)}$	0.026	0.026	0.026	0.026	0.026	0.026	0.026	0.026	0.026	0.026	0.026	0.026	0.026	0.026	0.026	0.026
Q_m^{LB}	3.019	3.047	23.053	32.981	35.442	38.088	38.294	47.019	51.874	51.874	54.077	54.139	57.134	58.098	60.173	62.882
p-value	0.082	0.218	0.000	0.000	0.000	0.000	0.000	0.000	0.000	0.000	0.000	0.000	0.000	0.000	0.000	0.000

Table 5.8 Studentised statistics for the corner method for the CAC 40 series and selected ARMA orders.

```
.p.|.q..1....2....3....4....5....6....7....8....9...10...11...12...13...14...15...
 1 |   0.8 -0.9 -2.0  0.7 -1.9 -1.0 -1.2  0.6  0.0  0.5  0.5 -0.7  0.9  1.2  1.8
 2 |   0.9  0.8  1.1 -1.1  1.1 -0.3  0.8  0.2 -0.4  0.2  0.4  0.0  0.7 -0.1
 3 |  -2.0  1.1 -0.9 -1.0 -0.6  0.8 -0.5  0.4 -0.1  0.3  0.3  0.4  0.5
 4 |  -0.8 -1.1  1.0 -0.4  0.7 -0.5  0.2  0.4  0.4  0.3 -0.2 -0.3
 5 |  -2.0  1.1 -0.6  0.7 -0.6  0.3 -0.3  0.2  0.0  0.3  0.3
 6 |   1.0 -0.3 -0.8 -0.5 -0.3  0.2  0.3  0.1 -0.2  0.3
 7 |  -1.1  0.7 -0.4  0.2 -0.3  0.3 -0.3  0.3 -0.3
 8 |  -0.4  0.0 -0.3  0.3 -0.1 -0.1 -0.3  0.4
 9 |  -0.1 -0.2 -0.1  0.3 -0.1 -0.1 -0.3
10 |  -0.4  0.2 -0.3  0.2 -0.3 0.3
11 |   0.5  0.4  0.2 -0.1  0.2
12 |   0.8  0.1 -0.3 -0.3
13 |   1.0  0.8  0.5
14 |  -1.1 -0.2
15 |   1.8
ARMA(P,Q) MODELS FOUND WITH GIVEN SIGNIFICANCE LEVEL
 PROBA      CRIT              MODELS FOUND
0.200000    1.28    ( 1, 1)
0.100000    1.64    ( 1, 1)
0.050000    1.96    ( 0, 3)    ( 1, 1)    ( 5, 0)
0.020000    2.33    ( 0, 0)
0.010000    2.58    ( 0, 0)
0.005000    2.81    ( 0, 0)
0.002000    3.09    ( 0, 0)
0.001000    3.29    ( 0, 0)
0.000100    3.72    ( 0, 0)
0.000010    4.26    ( 0, 0)
```

Table 5.9 Studentised statistics for the corner method for the FTSE 100 series and selected ARMA orders.

.p. | .q.	1	2	3	4	5	6	7	8	9	10	11	12	13	14	15
1	0.8	-0.1	-2.6	1.7	-1.0	-1.0	-0.3	1.8	1.6	0.0	1.1	-0.2	1.3	-0.7	1.2
2	0.1	0.8	1.2	0.2	1.0	0.3	0.9	1.0	0.6	-0.9	0.5	-0.8	0.6	-0.4	
3	-2.6	1.2	-0.7	-0.6	-0.7	0.8	-0.4	0.5	0.7	0.3	-0.3	-0.1	0.2		
4	-1.8	0.3	0.6	-0.7	0.6	0.0	-0.4	0.4	0.6	-0.3	0.1	-0.1			
5	-1.1	1.1	-0.7	0.6	-0.6	0.5	-0.3	0.5	0.5	0.1	0.2				
6	1.1	0.5	-0.8	0.2	-0.4	0.6	0.5	0.5	0.4	0.2					
7	0.0	0.9	-0.2	-0.3	0.0	0.5	0.5	0.4	0.3						
8	-1.6	0.7	-0.3	0.2	-0.4	0.4	-0.4	0.3							
9	1.4	0.5	0.6	0.5	0.4	0.3	0.2								
10	0.0	-0.9	-0.4	-0.2	-0.1	0.0									
11	1.2	0.6	0.0	0.0	0.1										
12	0.2	-0.8	0.0	0.0											
13	1.3	0.6	0.1												
14	0.5	-0.6													
15	1.1														

ARMA(P,Q) MODELS FOUND WITH GIVEN SIGNIFICANCE LEVEL

PROBA	CRIT	MODELS FOUND		
0.200000	1.28	(0,13)	(1, 1)	(9, 0)
0.100000	1.64	(0, 8)	(1, 1)	(4, 0)
0.050000	1.96	(0, 3)	(1, 1)	(3, 0)
0.020000	2.33	(0, 3)	(1, 1)	(3, 0)
0.010000	2.58	(0, 3)	(1, 1)	(3, 0)
0.005000	2.81	(0, 0)		
0.002000	3.09	(0, 0)		
0.001000	3.29	(0, 0)		
0.000100	3.72	(0, 0)		
0.000010	4.26	(0, 0)		

Table 5.10 Studentised statistics for the corner method for the squared CAC 40 series and selected GARCH orders.

.max(p,q).	.p..1	2	3	4	5	6	7	8	9	10	11	12	13	14	15...
1	5.2	5.4	5.0	5.3	4.6	4.7	5.4	4.6	4.5	4.1	3.2	3.9	3.7	5.2	3.9
2	-4.6	0.6	0.9	-1.4	0.2	1.0	-0.4	0.5	-0.6	0.2	0.4	-1.0	1.3	0.6	
3	3.5	0.9	0.8	0.9	0.7	0.5	-0.2	-0.4	0.4	0.4	0.5	0.9	0.8		
4	-4.0	-1.5	-0.9	-0.4	0.0	0.4	-0.4	0.2	-0.3	-0.2	0.2	0.3			
5	4.2	0.2	0.8	0.1	0.3	0.3	0.3	0.3	0.2	0.2	0.2				
6	-5.1	1.1	-0.6	0.4	-0.3	0.3	0.4	-0.2	-0.1	0.2					
7	2.5	-0.3	-0.4	-0.4	0.3	0.4	0.2	0.1	0.2						
8	-3.5	0.5	0.3	0.3	-0.3	-0.1	-0.1	0.2							
9	1.4	-0.9	0.4	-0.3	0.3	-0.1	0.1								
10	-3.4	0.3	-0.5	-0.2	-0.2	0.2									
11	1.5	0.4	0.5	0.2	0.2										
12	-2.4	-1.0	-0.9	0.3											
13	3.7	1.9	0.9												
14	-0.6	-0.1													
15	0.1														

GARCH(p,q) MODELS FOUND WITH GIVEN SIGNIFICANCE LEVEL

PROBA	CRIT	MODELS FOUND		
0.200000	1.28	(2, 1)	(2, 2)	(0,13)
0.100000	1.64	(2, 1)	(2, 2)	(0,13)
0.050000	1.96	(1, 1)	(0,13)	
0.020000	2.33	(1, 1)	(0,13)	
0.010000	2.58	(1, 1)	(0,13)	
0.005000	2.81	(1, 1)	(0,13)	
0.002000	3.09	(1, 1)	(0,13)	
0.001000	3.29	(1, 1)	(0,13)	
0.000100	3.72	(1, 1)	(0, 6)	
0.000010	4.26	(1, 1)	(0, 6)	

Table 5.11 Studentised statistics for the corner method for the squared FTSE 100 series and selected GARCH orders.

```
.max(p,q).|.p..1....2....3....4....5....6....7....8....9...10...11...12...13...14...15...
        1 |   5.7 11.7  5.8 12.9  2.2  3.8  2.5  2.9  2.8  3.9  2.3  2.9  1.9  3.6  2.3
        2 |  -5.2  3.3 -2.9  2.2 -4.6  4.3 -1.9  1.5 -1.7  2.7 -1.0 -0.2  0.3 -0.2
        3 |  -0.1 -7.7  1.3 -0.2  0.6 -2.3  0.5  0.3  0.6  1.6  0.5  0.3  0.2
        4 |  -8.5  4.2 -0.1 -0.4 -0.1  1.2 -0.3 -0.7 -0.3  1.7  0.1  0.1
        5 |  -0.3 -1.6  0.5 -0.2  0.6  0.9  0.7 -0.2  0.8  1.4 -0.1
        6 |  -1.9  1.6  0.6  1.4  0.9  0.4 -0.7  0.9 -1.4  1.2
        7 |   0.7 -1.0 -1.0 -0.8  0.3 -0.6  0.5  0.6  1.1
        8 |  -1.2  0.7 -0.3  0.5 -0.6  0.7 -0.8 -0.5
        9 |  -0.3 -1.0  0.5  0.1 -1.3 -0.4  1.1
       10 |  -1.6  1.2 -0.8  0.9 -0.9  1.1
       11 |   0.6  0.7  0.7  0.2  1.1
       12 |   1.8 -0.4 -0.9 -1.2
       13 |   1.2  0.9  0.8
       14 |   0.3 -0.9
       15 |   0.8
 GARCH(p,q) MODELS FOUND WITH GIVEN SIGNIFICANCE LEVEL
   PROBA      CRIT             MODELS FOUND
  0.200000     1.28    ( 1, 6)    ( 0,12)
  0.100000     1.64    ( 1, 4)    ( 0,12)
  0.050000     1.96    ( 2, 3)    ( 0, 4)
  0.020000     2.33    ( 2, 1)    ( 2, 2)    ( 0, 4)
  0.010000     2.58    ( 2, 1)    ( 2, 2)    ( 0, 4)
  0.005000     2.81    ( 2, 1)    ( 2, 2)    ( 0, 4)
  0.002000     3.09    ( 2, 1)    ( 2, 2)    ( 0, 4)
  0.001000     3.29    ( 2, 1)    ( 2, 2)    ( 0, 4)
  0.000100     3.72    ( 2, 1)    ( 2, 2)    ( 0, 4)
  0.000010     4.26    ( 2, 1)    ( 2, 2)    ( 1, 3)    ( 0, 4)
```

Concerning the HAC estimators of a long-run variance matrix, see, for instance, Andrews (1991) and Andrews and Monahan (1992). The method based on the spectral density at 0 of an AR model follows from Berk (1974). A comparison with the HAC method is proposed in den Hann and Levin (1997).

5.7 Exercises

5.1 *(Asymptotic behaviour of the SACVs of a martingale difference)*
Let (ϵ_t) denote a martingale difference sequence such that $E\epsilon_t^4 < \infty$ and $\hat{\gamma}(h) = n^{-1}\sum_{t=1}^{n} \epsilon_t \epsilon_{t+h}$. By applying Corollary A.1, derive the asymptotic distribution of $n^{1/2}\hat{\gamma}(h)$ for $h \neq 0$.

5.2 *(Asymptotic behaviour of $n^{1/2}\hat{\gamma}(1)$ for an ARCH(1) process)*
Consider the stationary non-anticipative solution of an ARCH(1) process

$$\begin{cases} \epsilon_t = \sigma_t \eta_t \\ \sigma_t^2 = \omega + \alpha \epsilon_{t-1}^2, \end{cases} \tag{5.28}$$

where (η_t) is a strong white noise with unit variance and $\mu_4 \alpha^2 < 1$ with $\mu_4 = E\eta_t^4$. Derive the asymptotic distribution of $n^{1/2}\hat{\gamma}(1)$.

5.3 *(Asymptotic behaviour of $n^{1/2}\hat{\rho}(1)$ for an ARCH(1) process)*
For the ARCH(1) model of Exercise 5.2, derive the asymptotic distribution of $n^{1/2}\hat{\rho}(1)$. What is the asymptotic variance of this statistic when $\alpha = 0$? Draw this asymptotic variance as a function of α and conclude accordingly.

5.4 *(Asymptotic behaviour of the SACRs of a GARCH(1, 1) process)*
For the GARCH(1, 1) model of Exercise 2.8, derive the asymptotic distribution of $n^{1/2}\hat{\rho}(h)$, for $h \neq 0$ fixed.

5.5 *(Moment of order 4 of a GARCH(1, 1) process)*
For the GARCH(1, 1) model of Exercise 2.8, compute $E\epsilon_t \epsilon_{t+1} \epsilon_s \epsilon_{s+2}$.

5.6 *(Asymptotic covariance between the SACRs of a GARCH(1, 1) process)*
For the GARCH(1, 1) model of Exercise 2.8, compute

$$\text{Cov}\{n^{1/2}\hat{\rho}(1), n^{1/2}\hat{\rho}(2)\}.$$

5.7 *(First five SACRs of a GARCH(1, 1) process)*
Evaluate numerically the asymptotic variance of the vector $\sqrt{n}\hat{\rho}_5$ of the first five SACRs of the GARCH(1, 1) model defined by

$$\begin{cases} \epsilon_t = \sigma_t \eta_t, \quad \eta_t \text{ iid } \mathcal{N}(0, 1) \\ \sigma_t^2 = 1 + 0.3\epsilon_{t-1}^2 + 0.55\sigma_{t-1}^2. \end{cases}$$

5.8 *(Generalised Bartlett formula for an MA(q)-ARCH(1) process)*
Suppose that X_t follows an MA(q) of the form

$$X_t = \epsilon_t - \sum_{i=1}^{q} b_i \epsilon_{t-i},$$

where the error term is an ARCH(1) process

$$\epsilon_t = \sigma_t \eta_t, \quad \sigma_t^2 = \omega + \alpha \epsilon_{t-1}^2, \quad \eta_t \text{ iid } \mathcal{N}(0, 1), \quad \alpha^2 < 1/3.$$

How is the generalised Bartlett formula (B.15) expressed for $i = j > q$?

5.9 *(Fisher information matrix for dynamic regression model)*
In the regression model $Y_t = F_{\theta_0}(W_t) + \epsilon_t$ introduced in Section 5.4.1, suppose that (ϵ_t) is a $\mathcal{N}(0, \sigma_0^2)$ white noise. Suppose also that the regularity conditions entailing the two convergences (5.14) hold. Give an explicit form to the blocks of the matrix I, and consider the case where σ^2 does not depend on θ.

5.10 *(LM tests in a linear regression model)*
Consider the regression model

$$Y = X_1\beta_1 + X_2\beta_2 + U,$$

where $Y = (Y_1, \dots, Y_n)$ is the dependent vector variable, X_i is an $n \times k_i$ matrix of explicative variables with rank k_i $(i = 1, 2)$, and the vector U is a $\mathcal{N}(0, \sigma^2 I_n)$ error term. Derive the LM test of the hypothesis $H_0 : \beta_2 = 0$. Consider the case $X_1'X_2 = 0$ and the general case.

5.11 *(Centred and uncentred* R^2*)*
Consider the regression model

$$Y_t = \beta_1 X_{t1} + \cdots + \beta_k X_{tk} + \epsilon_t, \quad t = 1, \dots, n,$$

where the ϵ_t are iid, centred, and have a variance $\sigma^2 > 0$. Let $Y = (Y_1, \dots, Y_n)'$ be the vector of dependent variables, $X = (X_{ij})$ the $n \times k$ matrix of explanatory variables, $\epsilon = (\epsilon_1, \dots, \epsilon_n)'$ the vector of the error terms and $\beta = (\beta_1, \dots, \beta_k)'$ the parameter vector. Let $P_X = X(X'X)^{-1}X'$ denote the orthogonal projection matrix on the vector subspace generated by the columns of X.

The uncentred determination coefficient is defined by

$$R_{nc}^2 = \frac{\|\hat{Y}\|^2}{\|Y\|^2}, \quad \hat{Y} = P_X Y \tag{5.29}$$

and the (centred) determination coefficient is defined by

$$R^2 = \frac{\|\hat{Y} - \bar{y}e\|^2}{\|Y - \bar{y}e\|^2}, \quad e = (1, \dots, 1)', \quad \bar{y} = \frac{1}{n}\sum_{t=1}^{n} Y_t. \tag{5.30}$$

Let T denote a $k \times k$ invertible matrix, c a number different from 0 and d any number. Let $\tilde{Y} = cY + de$ and $\tilde{X} = XT$. Show that if $\bar{y} = 0$ and if e belongs to the vector subspace generated by the columns of X, then R_{nc}^2 defined by relation (5.29) is equal to the determination coefficient in the regression of $\tilde{Y}$ on the columns of $\tilde{X}$.

6

Estimating ARCH Models by Least Squares

The simplest estimation method for ARCH models is that of ordinary least squares (OLSs). This estimation procedure has the advantage of being numerically simple, but has two drawbacks: (i) the OLS estimator is not efficient and is outperformed by methods based on the likelihood or on the quasi-likelihood that will be presented in the next chapters; (ii) in order to provide asymptotically normal estimators, the method requires moments of order 8 for the observed process. An extension of the OLS method, the feasible generalised least squares (FGLS) method, suppresses the first drawback and attenuates the second by providing estimators that are asymptotically as accurate as the quasi-maximum likelihood under the assumption that moments of order 4 exist. Note that the least-squares methods are of interest in practice because they provide initial estimators for the optimisation procedure that is used in the quasi-maximum likelihood method.

We begin with the unconstrained OLS and FGLS estimators. Then, in Section 6.3, we will see how to take into account positivity constraints on the parameters.

6.1 Estimation of ARCH(q) models by Ordinary Least Squares

In this section, we consider the OLS estimator of the ARCH(q) model:

$$\begin{cases} \epsilon_t = \sigma_t \eta_t, \\ \sigma_t^2 = \omega_0 + \sum_{i=1}^{q} \alpha_{0i} \epsilon_{t-i}^2 \quad \text{with } \omega_0 > 0, \quad \alpha_{0i} \geq 0, i = 1, \dots, q, \end{cases} \tag{6.1}$$

$$(\eta_t) \text{ is an iid sequence, } E(\eta_t) = 0, \ \ \mathrm{Var}(\eta_t) = 1.$$

The OLS method uses the AR representation on the squares of the observed process. No assumption is made on the law of η_t.

The true value of the vector of the parameters is denoted by $\theta_0 = (\omega_0, \alpha_{01}, \dots, \alpha_{0q})'$, and we denote by θ a generic value of the parameter.

From (6.1), we obtain the AR (q) representation

$$\epsilon_t^2 = \omega_0 + \sum_{i=1}^{q} \alpha_{0i}\, \epsilon_{t-i}^2 + u_t, \tag{6.2}$$

where $u_t = \epsilon_t^2 - \sigma_t^2 = (\eta_t^2 - 1)\sigma_t^2$. The sequence $(u_t, \mathcal{F}_t)_t$ constitutes a martingale difference when $E\epsilon_t^2 = \sigma_t^2 < \infty$, denoting by $\mathcal{F}_t$ the σ-field generated by $\{\epsilon_s : s \leq t\}$.

Assume that we observe $\epsilon_1, \dots, \epsilon_n$, a realisation of length n of the process (ϵ_t), and let $\epsilon_0, \dots, \epsilon_{1-q}$ be initial values. For instance, the initial values can be chosen equal to zero. Introducing the vector

$$Z_{t-1}' = (1, \epsilon_{t-1}^2, \dots, \epsilon_{t-q}^2),$$

in view of (6.2), we obtain the system

$$\epsilon_t^2 = Z_{t-1}'\theta_0 + u_t, \qquad t = 1, \dots, n, \tag{6.3}$$

which can be written as

$$Y = X\theta_0 + U,$$

with the $n \times (q+1)$ matrix

$$X = \begin{pmatrix} 1 & \epsilon_0^2 & \dots & \epsilon_{-q+1}^2 \\ \vdots & & & \\ 1 & \epsilon_{n-1}^2 & \dots & \epsilon_{n-q}^2 \end{pmatrix} = \begin{pmatrix} Z_0' \\ \vdots \\ Z_{n-1}' \end{pmatrix}$$

and the $n \times 1$ vectors

$$Y = \begin{pmatrix} \epsilon_1^2 \\ \vdots \\ \epsilon_n^2 \end{pmatrix}, \qquad U = \begin{pmatrix} u_1 \\ \vdots \\ u_n \end{pmatrix}.$$

Assume that the matrix $X'X$ is invertible, or equivalently that X has full column rank (we will see that this is always the case asymptotically, and thus for n large enough). The OLS estimator of θ_0 follows:

$$\hat{\theta}_n := \arg\min_{\theta} \|Y - X\theta\|^2 = (X'X)^{-1}X'Y. \tag{6.4}$$

Under assumptions OLS1 and OLS2 below, the variance of u_t exists and is constant. The OLS estimator of $\sigma_0^2 = \mathrm{Var}(u_t)$ is

$$\hat{\sigma}^2 = \frac{1}{n-q-1}\|Y - X\hat{\theta}_n\|^2 = \frac{1}{n-q-1}\sum_{t=1}^{n}\left\{\epsilon_t^2 - \hat{\omega} - \sum_{i=1}^{q}\hat{\alpha}_i\epsilon_{t-i}^2\right\}^2.$$

Remark 6.1 (OLS estimator of a GARCH model) An OLS estimator can also be defined for a GARCH (p, q) model, but the estimator is not explicit, because ϵ_t^2 does not satisfy an AR model when $p \neq 0$ (see Exercise 7.5).

To establish the consistency of the OLS estimators of θ_0 and σ_0^2, we must consider the following assumptions:

OLS1. (ϵ_t) is the non-anticipative strictly stationary solution of model (6.1), and $\omega_0 > 0$.

OLS2. $E\epsilon_t^4 < +\infty$.

OLS3. $\mathbb{P}(\eta_t^2 = 1) \neq 1$.

Explicit conditions for assumptions OLS1 and OLS2 were given in Chapter 2. Assumption OLS3 that the law of η_t is non-degenerate allows us to identify the parameters. The assumption also guarantees the invertibility of $X'X$ for n large enough.

Theorem 6.1 (Consistency of the OLS estimator of an ARCH model) *Let $(\hat{\theta}_n)$ be a sequence of estimators satisfying (6.4). Under assumptions OLS1–OLS3, almost surely*

$$\hat{\theta}_n \to \theta_0, \quad \hat{\sigma}_n^2 \to \sigma_0^2, \quad as\ n \to \infty.$$

Proof. The proof consists of several steps.

(i) We have seen (Theorem 2.4) that (ϵ_t), the unique nonanticipative stationary solution of the model, is ergodic. The process (Z_t) is also ergodic because Z_t is a measurable function of $\{\epsilon_{t-i}, i \geq 0\}$. The ergodic theorem (see Theorem A.2) then entails that

$$\frac{1}{n}X'X = \frac{1}{n}\sum_{t=1}^{n} Z_{t-1}Z'_{t-1} \to E(Z_{t-1}Z'_{t-1}), \text{a.s.}, \quad \text{as } n \to \infty. \tag{6.5}$$

The existence of the expectation is guaranteed by assumption OLS3. Note that the initial values are involved only in a fixed number of terms of the sum, and thus they do not matter for the asymptotic result. Similarly, we have

$$\frac{1}{n}X'U = \frac{1}{n}\sum_{t=1}^{n} Z_{t-1}u_t \to E(Z_{t-1}u_t), \text{a.s.}, \quad \text{as } n \to \infty.$$

(ii) The invertibility of the matrix $EZ_{t-1}Z'_{t-1} = EZ_tZ'_t$ is shown by contradiction. Assume that there exists a non-zero vector c of $\mathbb{R}^{q+1}$ such that $c'EZ_tZ'_tc = 0$. Thus $E\{c'Z_t\}^2 = 0$, and it follows that $c'Z_t = 0$ a.s. Therefore, there exists a linear combination of the variables $\epsilon_t^2, \dots, \epsilon_{t-q+1}^2$ which is a.s. equal to a constant. Without loss of generality, one can assume that, in this linear combination, the coefficient of $\epsilon_t^2 = \eta_t^2\sigma_t^2$ is 1. Thus η_t^2 is a.s. a measurable function of the variables $\epsilon_{t-1}, \dots, \epsilon_{t-q}$. However, the solution being non-anticipative, η_t^2 is independent of these variables. This implies that η_t^2 is a.s. equal to a constant. This constant is necessarily equal to 1, but this leads to a contradiction with OLS3. Thus $E(Z_{t-1}Z'_{t-1})$ is invertible.

(iii) The innovation of ϵ_t^2 being $u_t = \epsilon_t^2 - \sigma_t^2 = \epsilon_t^2 - E(\epsilon_t^2 \mid \mathcal{F}_{t-1})$, we have the orthogonality relations

$$E(u_t) = E(u_t\epsilon_{t-1}^2) = \dots = E(u_t\epsilon_{t-q}^2) = 0$$

that is

$$E(Z_{t-1}u_t) = 0.$$

(iv) Point (ii) shows that $n^{-1}X'X$ is a.s. invertible, for n large enough and that, almost surely, as $n \to \infty$,

$$\hat{\theta}_n - \theta_0 = \left(\frac{X'X}{n}\right)^{-1}\frac{X'U}{n} \to \{E(Z_{t-1}Z'_{t-1})\}^{-1}E(Z_{t-1}u_t) = 0. \qquad \square$$

For the asymptotic normality of the OLS estimator, we need the following additional assumption.

OLS4: $E(\epsilon_t^8) < +\infty$.

Consider the $(q+1)\times(q+1)$ matrices

$$A = E(Z_{t-1}Z'_{t-1}), \qquad B = E(\sigma_t^4 Z_{t-1}Z'_{t-1}).$$

The invertibility of A was established in the proof of Theorem 6.1, and the invertibility of B is shown by the same argument, noting that $c'\sigma_t^2 Z_{t-1} = 0$ if and only if $c'Z_{t-1} = 0$ because $\sigma_t^2 > 0$ a.s. The following result establishes the asymptotic normality of the OLS estimator.

$$\text{Let } \kappa_\eta = E\eta_t^4.$$

Theorem 6.2 (Asymptotic normality of the OLS estimator) *Under assumptions OLS1–OLS4,*

$$\sqrt{n}(\hat{\theta}_n - \theta_0) \xrightarrow{\mathcal{L}} \mathcal{N}(0, \quad (\kappa_\eta - 1)A^{-1}BA^{-1}).$$

Proof. In view of (6.3), we have

$$\begin{aligned}
\hat{\theta}_n &= \left(\frac{1}{n}\sum_{t=1}^{n} Z_{t-1}Z'_{t-1}\right)^{-1}\left(\frac{1}{n}\sum_{t=1}^{n} Z_{t-1}\epsilon_t^2\right) \\
&= \left(\frac{1}{n}\sum_{t=1}^{n} Z_{t-1}Z'_{t-1}\right)^{-1}\left\{\frac{1}{n}\sum_{t=1}^{n} Z_{t-1}(Z'_{t-1}\theta_0 + u_t)\right\} \\
&= \theta_0 + \left(\frac{1}{n}\sum_{t=1}^{n} Z_{t-1}Z'_{t-1}\right)^{-1}\left\{\frac{1}{n}\sum_{t=1}^{n} Z_{t-1}u_t\right\}.
\end{aligned}$$

Thus

$$\sqrt{n}(\hat{\theta}_n - \theta_0) = \left(\frac{1}{n}\sum_{t=1}^{n} Z_{t-1}Z'_{t-1}\right)^{-1}\left\{\frac{1}{\sqrt{n}}\sum_{t=1}^{n} Z_{t-1}u_t\right\}. \tag{6.6}$$

Let $\lambda \in \mathbb{R}^{q+1}$, $\lambda \neq 0$. The sequence $(\lambda' Z_{t-1}u_t, \mathcal{F}_t)$ is a square-integrable ergodic stationary martingale difference, with variance

$$\begin{aligned}
\text{Var}(\lambda' Z_{t-1}u_t) &= \lambda' E(Z_{t-1}Z'_{t-1}u_t^2)\lambda = \lambda' E\{Z_{t-1}Z'_{t-1}(\eta_t^2 - 1)^2\sigma_t^4\}\lambda \\
&= (\kappa_\eta - 1)\lambda' B\lambda.
\end{aligned}$$

By the CLT (see Corollary A.1), we obtain that for all $\lambda \neq 0$,

$$\frac{1}{\sqrt{n}}\sum_{t=1}^{n} \lambda' Z_{t-1}u_t \xrightarrow{\mathcal{L}} \mathcal{N}(0, \quad (\kappa_\eta - 1)\lambda' B\lambda).$$

Using the Cramér–Wold device, it follows that

$$\frac{1}{\sqrt{n}}\sum_{t=1}^{n} Z_{t-1}u_t \xrightarrow{\mathcal{L}} \mathcal{N}(0, \quad (\kappa_\eta - 1)B). \tag{6.7}$$

The conclusion follows from (6.5)–(6.7). □

Remark 6.2 (Estimation of the information matrices) Consistent estimators $\hat{A}$ and $\hat{B}$ of the matrices A and B are obtained by replacing the theoretical moments by their empirical counterparts,

$$\hat{A} = \frac{1}{n}\sum_{t=1}^{n} Z_{t-1}Z'_{t-1}, \quad \hat{B} = \frac{1}{n}\sum_{t=1}^{n} \hat{\sigma}_t^4 Z_{t-1}Z'_{t-1},$$

where $\hat{\sigma}_t^2 = Z'_{t-1}\hat{\theta}_n$. The fourth-order moment of the process $\eta_t = \epsilon_t/\sigma_t$ is also consistently estimated by $\hat{\mu}_4 = n^{-1}\sum_{t=1}^{n}(\epsilon_t/\hat{\sigma}_t)^4$. Finally, a consistent estimator of the asymptotic variance of the OLS estimator is defined by

$$\widehat{\text{Var}}_{as}\{\sqrt{n}(\hat{\theta}_n - \theta_0)\} = (\hat{\mu}_4 - 1)\hat{A}^{-1}\hat{B}\,\hat{A}^{-1}.$$

Table 6.1 Strict stationarity and moment conditions for the ARCH(1) model when η_t follows the $\mathcal{N}(0,1)$ distribution or the Student t distribution (normalised in such a way that $E\eta_t^2 = 1$).

	Strict stationarity	$E\epsilon_t^2 < \infty$	$E\epsilon_t^4 < \infty$	$E\epsilon_t^8 < \infty$
Normal	$\alpha_{01} < 3.562$	$\alpha_{01} < 1$	$\alpha_{01} < 0.577$	$\alpha_{01} < 0.312$
t_3	$\alpha_{01} < 7.389$	$\alpha_{01} < 1$	Not satisfied	Not satisfied
t_5	$\alpha_{01} < 4.797$	$\alpha_{01} < 1$	$\alpha_{01} < 0.333$	Not satisfied
t_9	$\alpha_{01} < 4.082$	$\alpha_{01} < 1$	$\alpha_{01} < 0.488$	$\alpha_{01} < 0.143$

Table 6.2 Asymptotic variance of the OLS estimator of an ARCH(1) model with $\omega_0 = 1$, when $\eta_t \sim \mathcal{N}(0,1)$.

α_{01}	0.1	0.2	0.3
$\text{Var}_{as}\{\sqrt{n}(\hat{\theta}_n - \theta_0)\}$	$\begin{pmatrix} 3.98 & -1.85 \\ -1.85 & 2.15 \end{pmatrix}$	$\begin{pmatrix} 8.03 & -5.26 \\ -5.26 & 5.46 \end{pmatrix}$	$\begin{pmatrix} 151.0 & -106.5 \\ -106.5 & 77.6 \end{pmatrix}$

Example 6.1 (ARCH(1)) When $q = 1$, the moment conditions OLS2 and OLS4 take the form $\kappa_\eta \alpha_{01}^2 < 1$ and $\mu_8 \alpha_{01}^4 < 1$ (see (2.54)). We have

$$A = \begin{pmatrix} 1 & E\epsilon_{t-1}^2 \\ E\epsilon_{t-1}^2 & E\epsilon_{t-1}^4 \end{pmatrix}, \quad B = \begin{pmatrix} E\sigma_t^4 & E\sigma_t^4\epsilon_{t-1}^2 \\ E\sigma_t^4\epsilon_{t-1}^2 & E\sigma_t^4\epsilon_{t-1}^4 \end{pmatrix},$$

with

$$E\epsilon_t^2 = \frac{\omega_0}{1-\alpha_{01}}, \quad E\epsilon_t^4 = \kappa_\eta E\sigma_t^4 = \frac{\omega_0^2(1+\alpha_{01})}{(1-\kappa_\eta\alpha_{01}^2)(1-\alpha_{01})}\kappa_\eta.$$

The other terms of the matrix B are obtained by expanding $\sigma_t^4 = (\omega_0 + \alpha_{01}\epsilon_{t-1}^2)^2$ and calculating the moments of order 6 and 8 of ϵ_t^2.

Table 6.1 shows, for different laws of the iid process, that the moment conditions OLS2 and OLS4 impose strong constraints on the parameter space.

Table 6.2 displays numerical values of the asymptotic variance, for different values of α_{01} and $\omega_0 = 1$, when η_t follows the normal $\mathcal{N}(0,1)$. The asymptotic accuracy of $\hat{\theta}_n$ becomes very low near the boundary of the domain of existence of $E\epsilon_t^8$. The OLS method can, however, be used for higher values of α_{01}, because the estimator remains consistent when $\alpha_{01} < 3^{-1/2} = 0.577$, and thus can provide initial values for an algorithm maximising the likelihood.

6.2 Estimation of ARCH(q) Models by Feasible Generalised Least Squares

In a linear regression model when, conditionally on the exogenous variables, the errors are heteroscedastic, the FGLS estimator is asymptotically more accurate than the OLS estimator. Note that in (6.3), the errors u_t are, conditionally on Z_{t-1}, heteroscedastic with conditional variance $\text{Var}(u_t \mid Z_{t-1}) = (\kappa_\eta - 1)\sigma_t^4$.

For all $\theta = (\omega, \alpha_1, \dots, \alpha_q)'$, let

$$\sigma_t^2(\theta) = \omega + \sum_{i=1}^{q} \alpha_i \epsilon_{t-i}^2 \quad \text{and} \quad \hat{\Omega} = \text{diag}(\sigma_1^{-4}(\hat{\theta}_n), \dots, \sigma_n^{-4}(\hat{\theta}_n)).$$

The FGLS estimator is defined by

$$\tilde{\theta}_n = (X'\hat{\Omega}X)^{-1}X'\hat{\Omega}Y.$$

Theorem 6.3 (Asymptotic properties of the FGLS estimator) *Under assumptions OLS1–OLS3 and if $\alpha_{0i} > 0$, $i = 1, \dots, q$,*

$$\tilde{\theta}_n \to \theta_0, \quad \text{a.s.}, \qquad \sqrt{n}(\tilde{\theta}_n - \theta_0) \xrightarrow{\mathcal{L}} \mathcal{N}(0, \quad (\kappa_\eta - 1)J^{-1}),$$

where $J = E(\sigma_t^{-4} Z_{t-1} Z'_{t-1})$ is positive definite.

Proof. It can be shown that J is positive definite by the argument used in Theorem 6.1.

We have

$$\begin{aligned}
\tilde{\theta}_n &= \left(\frac{1}{n}\sum_{t=1}^{n} \sigma_t^{-4}(\hat{\theta}_n) Z_{t-1} Z'_{t-1}\right)^{-1} \left(\frac{1}{n}\sum_{t=1}^{n} \sigma_t^{-4}(\hat{\theta}_n) Z_{t-1} \epsilon_t^2\right) \\
&= \left(\frac{1}{n}\sum_{t=1}^{n} \sigma_t^{-4}(\hat{\theta}_n) Z_{t-1} Z'_{t-1}\right)^{-1} \left\{\frac{1}{n}\sum_{t=1}^{n} \sigma_t^{-4}(\hat{\theta}_n) Z_{t-1} (Z'_{t-1}\theta_0 + u_t)\right\} \\
&= \theta_0 + \left(\frac{1}{n}\sum_{t=1}^{n} \sigma_t^{-4}(\hat{\theta}_n) Z_{t-1} Z'_{t-1}\right)^{-1} \left\{\frac{1}{n}\sum_{t=1}^{n} \sigma_t^{-4}(\hat{\theta}_n) Z_{t-1} u_t\right\}. \qquad (6.8)
\end{aligned}$$

A Taylor expansion around θ_0 yields, with $\sigma_t^2 = \sigma_t^2(\theta_0)$,

$$\sigma_t^{-4}(\hat{\theta}_n) = \sigma_t^{-4} - 2\sigma_t^{-6}(\theta^*) \frac{\partial \sigma_t^2}{\partial \theta'}(\theta^*)(\hat{\theta}_n - \theta_0), \qquad (6.9)$$

where θ^* is between $\hat{\theta}_n$ and θ_0. Note that, for all θ, $\frac{\partial \sigma_t^2}{\partial \theta}(\theta) = Z_{t-1}$. It follows that

$$\frac{1}{n}\sum_{t=1}^{n} \sigma_t^{-4}(\hat{\theta}_n) Z_{t-1} Z'_{t-1} = \frac{1}{n}\sum_{t=1}^{n} \sigma_t^{-4} Z_{t-1} Z'_{t-1} - \frac{2}{n}\sum_{t=1}^{n} \sigma_t^{-6}(\theta^*) Z_{t-1} Z'_{t-1} \times Z'_{t-1}(\hat{\theta}_n - \theta_0).$$

The first term on the right-hand side of the equality converges a.s. to J by the ergodic theorem. The second term converges a.s. to 0 because the OLS estimator is consistent and

$$\begin{aligned}
\left\| \frac{1}{n}\sum_{t=1}^{n} \sigma_t^{-6}(\theta^*) Z_{t-1} Z'_{t-1} \times Z'_{t-1}(\hat{\theta}_n - \theta_0) \right\| &\le \left(\frac{1}{n}\sum_{t=1}^{n} \|\sigma_t^{-2}(\theta^*) Z_{t-1}\|^3\right) \|\hat{\theta}_n - \theta_0\| \\
&\le K\|\hat{\theta}_n - \theta_0\|,
\end{aligned}$$

for n large enough. The constant bound K is obtained by arguing that the components of $\hat{\theta}_n$, and thus those of θ^*, are strictly positive for n large enough (because $\hat{\theta}_n \to \theta_0$ almost surely). Thus, we have $\sigma_t^{-2}(\theta^*)\epsilon_{t-i}^2 < 1/\theta_i^*$, for $i = 1, \dots, q$, and finally, $\|\sigma_t^{-2}(\theta^*) Z_{t-1}\|$ is bounded. We have shown that

$$\left(\frac{1}{n}\sum_{t=1}^{n} \sigma_t^{-4}(\hat{\theta}_n) Z_{t-1} Z'_{t-1}\right)^{-1} \to J^{-1}. \qquad (6.10)$$

For the term in braces in (6.8), we have

$$\begin{aligned}
&\frac{1}{n}\sum_{t=1}^{n}\sigma_t^{-4}(\hat{\theta}_n)Z_{t-1}u_t \\
&= \frac{1}{n}\sum_{t=1}^{n}\sigma_t^{-4}Z_{t-1}u_t - \frac{2}{n}\sum_{t=1}^{n}\sigma_t^{-6}(\theta^*)Z_{t-1}u_t \times Z'_{t-1}(\hat{\theta}_n - \theta_0) \\
&\to 0 \text{ almost surely,} \qquad (6.11)
\end{aligned}$$

noting that $E(\sigma_t^{-4}Z_{t-1}u_t) = 0$ and

$$\begin{aligned}
&\left\| \frac{2}{n}\sum_{t=1}^{n}\sigma_t^{-6}(\theta^*)Z_{t-1}u_t \times Z'_{t-1}(\hat{\theta}_n - \theta_0) \right\| \\
&= \left\| \frac{2}{n}\sum_{t=1}^{n}\sigma_t^{-6}(\theta^*)Z_{t-1}\sigma_t^2(\theta_0)(\eta_t^2 - 1) \times Z'_{t-1}(\hat{\theta}_n - \theta_0) \right\| \\
&\leq K\left(\frac{1}{n}\sum_{t=1}^{n} | \eta_t^2 - 1 | \right) \|\hat{\theta}_n - \theta_0\| \to 0, \quad \text{a.s.}
\end{aligned}$$

Thus, we have shown that $\tilde{\theta}_n \to \theta_0$, a.s.

Using (6.8), (6.10), and (6.11), we have

$$\begin{aligned}
\sqrt{n}(\tilde{\theta}_n - \theta_0) &= (J^{-1} + R_n)\left\{ \frac{1}{\sqrt{n}}\sum_{t=1}^{n}\sigma_t^{-4}Z_{t-1}u_t \right\} \\
&- \frac{2}{n}(J^{-1} + R_n)\sum_{t=1}^{n}\sigma_t^{-6}(\theta^*)Z_{t-1}u_t \times Z'_{t-1}\sqrt{n}(\hat{\theta}_n - \theta_0),
\end{aligned}$$

where $R_n \to 0$, a.s. A new expansion around θ_0 gives

$$\sigma_t^{-6}(\theta^*) = \sigma_t^{-6} - 3\sigma_t^{-8}(\theta^{**})Z'_{t-1}(\hat{\theta}_n - \theta_0), \qquad (6.12)$$

where θ^{**} is between θ^* and θ_0. It follows that

$$\begin{aligned}
&\sqrt{n}(\tilde{\theta}_n - \theta_0) \\
&= (J^{-1} + R_n)\left\{ \frac{1}{\sqrt{n}}\sum_{t=1}^{n}\sigma_t^{-2}Z_{t-1}(\eta_t^2 - 1) \right\} \\
&\quad - \frac{2}{n}(J^{-1} + R_n)\sum_{t=1}^{n}\sigma_t^{-4}Z_{t-1}(\eta_t^2 - 1) \times Z'_{t-1}\sqrt{n}(\hat{\theta}_n - \theta_0) \\
&\quad + \frac{6}{n^{3/2}}(J^{-1} + R_n)\sum_{t=1}^{n}\sigma_t^{-8}(\theta^{**})Z_{t-1}u_t \times \{Z'_{t-1}\sqrt{n}(\hat{\theta}_n - \theta_0)\} \times \{Z'_{t-1}\sqrt{n}(\theta^* - \theta_0)\} \\
&:= S_{n1} + S_{n2} + S_{n3}. \qquad (6.13)
\end{aligned}$$

The CLT applied to the ergodic and square-integrable stationary martingale difference $\sigma_t^{-2}Z_{t-1}(\eta_t^2 - 1)$ shows that S_{n1} converges in distribution to a Gaussian vector with zero mean and variance

$$J^{-1}E\{\sigma_t^{-4}(\eta_t^2 - 1)^2 Z_{t-1}Z'_{t-1}\}J^{-1} = (\kappa_\eta - 1)J^{-1}$$

(see Corollary A.1). Moreover,

$$\frac{1}{n}\sum_{t=1}^{n}\sigma_t^{-4}Z_{t-1}(\eta_t^2-1)\times Z_{t-1}'\sqrt{n}(\hat{\theta}_n-\theta_0)$$

$$=\left\{\frac{1}{\sqrt{n}}\sum_{t=1}^{n}\sigma_t^{-4}Z_{t-1}(\eta_t^2-1)\right\}(\hat{\omega}_n-\omega_0)$$

$$+\sum_{j=1}^{q}\left\{\frac{1}{\sqrt{n}}\sum_{t=1}^{n}\sigma_t^{-4}Z_{t-1}(\eta_t^2-1)\epsilon_{t-j}^2\right\}(\hat{\alpha}_{nj}-\alpha_{0j}).$$

The two terms in braces are bounded in probability by the CLT. Moreover, the terms $(\hat{\omega}_n-\omega_0)$ and $(\hat{\alpha}_{nj}-\alpha_{0j})$ converge a.s. to 0. It follows that S_{n2} tends to 0 in probability. Finally, by arguments already used and because θ^* is between $\hat{\theta}_n$ and θ_0,

$$\|S_{n3}\|\leq K\|J^{-1}+R_n\|\|(\hat{\theta}_n-\theta_0)\|^2\frac{1}{\sqrt{n}}\sum_{t=1}^{n}|\eta_t^2-1|\to 0,$$

in probability. Using (6.12), we have shown the convergence in law. □

The moment condition required for the asymptotic normality of the FGLS estimator is $E(\epsilon_t^4)<\infty$. For the OLS estimator, we had the more restrictive condition $E\epsilon_t^8<\infty$. Moreover, when this eighth-order moment exists, the following result shows that the OLS estimator is asymptotically less accurate than the FGLS estimator.

Theorem 6.4 (OLS versus FGLS asymptotic covariance matrices) *Under assumptions OLS1–OLS4, the matrix*

$$A^{-1}BA^{-1}-J^{-1}$$

is positive semi-definite.

Proof. Let $D=\sigma_t^2A^{-1}Z_{t-1}-\sigma_t^{-2}J^{-1}Z_{t-1}$. Then

$$\begin{aligned}E(DD')&=A^{-1}E(\sigma_t^4Z_{t-1}Z_{t-1}')A^{-1}+J^{-1}E(\sigma_t^{-4}Z_{t-1}Z_{t-1}')J^{-1}\\&\quad-A^{-1}E(Z_{t-1}Z_{t-1}')J^{-1}-J^{-1}E(Z_{t-1}Z_{t-1}')A^{-1}\\&=A^{-1}BA^{-1}-J^{-1}\end{aligned}$$

and the result follows. □

We will see in Chapter 7 that the asymptotic variance of the FGLS estimator coincides with that of the quasi-maximum likelihood estimator (but the asymptotic normality of the latter is obtained without moment conditions). This result explains why quasi-maximum likelihood is preferred to OLS (and even to FGLS) for the estimation of ARCH (and GARCH) models. Note, however, that the OLS estimator often provides a good initial value for the optimisation algorithm required for the quasi-maximum likelihood method.

6.3 Estimation by Constrained Ordinary Least Squares

Negative components are not precluded in the OLS estimator $\hat{\theta}_n$ defined by (6.4) (see Exercise 6.3). When the estimate has negative components, predictions of the volatility can be negative. In order to avoid this problem, we consider the constrained OLS estimator defined by

$$\hat{\theta}_n^c = \arg \min_{\theta \in [0,\infty)^{q+1}} Q_n(\theta), \quad Q_n(\theta) = \frac{1}{n} \|Y - X\theta\|^2.$$

The existence of $\hat{\theta}_n^c$ is guaranteed by the continuity of the function Q_n and the fact that

$$\{nQ_n(\theta)\}^{1/2} \geq \|X\theta\| - \|Y\| \to \infty$$

as $\|\theta\| \to \infty$ and $\theta \geq 0$, whenever X has non-zero columns. Note that the latter condition is satisfied at least for n is large enough (see Exercise 6.5).

6.3.1 Properties of the Constrained OLS Estimator

The following theorem gives a condition for equality between the constrained and unconstrained estimators. The theorem is stated in the ARCH case but is true in a much more general framework.

Theorem 6.5 (Equality between constrained and unconstrained OLS) *If X is of rank $q+1$, the constrained and unconstrained estimators coincide, $\hat{\theta}_n^c = \hat{\theta}_n$, if and only if $\hat{\theta}_n \in [0, +\infty)^{q+1}$.*

Proof. Since $\hat{\theta}_n$ and $\hat{\theta}_n^c$ are obtained by minimising the same function $Q_n(\cdot)$, and since $\hat{\theta}_n^c$ minimises this function on a smaller set, we have $Q_n(\hat{\theta}_n) \leq Q_n(\hat{\theta}_n^c)$. Moreover, $\hat{\theta}_n^c \in [0, +\infty)^{q+1}$, and we have $Q_n(\theta) \geq Q_n(\hat{\theta}_n^c)$, for all $\theta \in [0, +\infty)^{q+1}$.

Suppose that the unconstrained estimation $\hat{\theta}_n$ belongs to $[0, +\infty)^{q+1}$. In this case, $Q_n(\hat{\theta}_n) = Q_n(\hat{\theta}_n^c)$. Because the unconstrained solution is unique, $\hat{\theta}_n^c = \hat{\theta}_n$.

The converse is trivial. □

We now give a way to obtain the constrained estimator from the unconstrained estimator.

Theorem 6.6 (Constrained OLS as a projection of OLS) *If X has rank $q+1$, the constrained estimator $\hat{\theta}_n^c$ is the orthogonal projection of $\hat{\theta}_n$ on $[0, +\infty)^{q+1}$ with respect to the metric $X'X$, that is,*

$$\hat{\theta}_n^c = \arg \min_{\theta \in [0,+\infty)^{q+1}} (\hat{\theta}_n - \theta)' X' X (\hat{\theta}_n - \theta). \tag{6.14}$$

Proof. If we denote by P the orthogonal projector on the columns of X, and $M = I_n - P$, we have

$$\begin{aligned} nQ(\theta) &= \|Y - X\theta\|^2 = \|P(Y - X\theta)\|^2 + \|M(Y - X\theta)\|^2 \\ &= \|X(\hat{\theta}_n - \theta)\|^2 + \|MY\|^2, \end{aligned}$$

using properties of projections, Pythagoras's theorem, and $PY = X\hat{\theta}_n$. The constrained estimation $\hat{\theta}_n^c$ thus solves (6.14). Note that, since X has full column rank, a norm is well defined by $\|x\|_{X'X} = \sqrt{x'X'Xx}$. The characterisation (6.14) is equivalent to

$$\theta_n^c \in [0, +\infty)^{q+1}, \quad \|\hat{\theta}_n - \hat{\theta}_n^c\|_{X'X} \leq \|\hat{\theta}_n - \theta\|_{X'X}, \quad \forall \theta \in [0, +\infty)^{q+1}. \tag{6.15}$$

Since $[0, +\infty)^{q+1}$ is convex, $\hat{\theta}_n^c$ exists, is unique, and is the $X'X$-orthogonal projection of $\hat{\theta}_n$ on $[0, +\infty)^{q+1}$. This projection is characterised by

$$\hat{\theta}_n^c \in [0, +\infty)^{q+1} \quad \text{and} \quad \langle \hat{\theta}_n - \hat{\theta}_n^c, \hat{\theta}_n^c - \theta \rangle_{X'X} \geq 0, \quad \forall \theta \in [0, +\infty)^{q+1} \tag{6.16}$$

(see Exercise 6.9). This characterisation shows that, when $\hat{\theta}_n \notin [0, +\infty)^{q+1}$, the constrained estimation $\hat{\theta}_n^c$ must lie at the boundary of $[0, +\infty)^{q+1}$. Otherwise, it suffices to take $\theta \in [0, +\infty)^{q+1}$ between $\hat{\theta}_n^c$ and $\hat{\theta}_n$ to obtain a scalar product equal to -1. □

The characterisation (6.15) allows us to easily obtain the strong consistency of the constrained estimator.

Theorem 6.7 (Consistency of the constrained OLS estimator) *Under the assumptions of Theorem 6.1, almost surely,*

$$\hat{\theta}_n^c \to \theta_0 \quad \text{as } n \to \infty.$$

Proof. Since $\theta_0 \in [0, +\infty)^{q+1}$, in view of (6.15), we have

$$\|\hat{\theta}_n - \hat{\theta}_n^c\|_{X'X/n} \leq \|\hat{\theta}_n - \theta_0\|_{X'X/n}.$$

It follows that, using the triangle inequality,

$$\|\hat{\theta}_n^c - \theta_0\|_{X'X/n} \leq \|\hat{\theta}_n^c - \hat{\theta}_n\|_{X'X/n} + \|\hat{\theta}_n - \theta_0\|_{X'X/n} \leq 2\|\hat{\theta}_n - \theta_0\|_{X'X/n}.$$

Since, in view of Theorem 6.1, $\hat{\theta}_n \to \theta_0$ a.s. and $X'X/n$ converges a.s. to a positive definite matrix, it follows that $\|\hat{\theta}_n - \theta_0\|_{X'X/n} \to 0$ and thus that $\|\hat{\theta}_n - \hat{\theta}_n^c\|_{X'X/n} \to 0$ a.s. Using Exercise 6.12, the conclusion follows. □

6.3.2 Computation of the Constrained OLS Estimator

We now give an explicit way to obtain the constrained estimator. We have already seen that if all the components of the unconstrained estimator $\hat{\theta}_n$ are positive, we have $\hat{\theta}_n^c = \hat{\theta}_n$. Now suppose that one component of $\hat{\theta}_n$ is negative, for instance the last one. Let

$$X = (X^{(1)}, X^{(2)}), \quad X^{(1)} = \begin{pmatrix} 1 & \epsilon_0^2 & \dots & \epsilon_{-q+2}^2 \\ 1 & \epsilon_1^2 & \dots & \epsilon_{-q+3}^2 \\ \vdots & & & \\ 1 & \epsilon_{n-1}^2 & \dots & \epsilon_{n-q+1}^2 \end{pmatrix},$$

and

$$\hat{\theta}_n = (X'X)^{-1}X'Y = \begin{pmatrix} \hat{\theta}_n^{(1)} \\ \hat{\alpha}_q \end{pmatrix}, \quad \tilde{\theta}_n = \begin{pmatrix} \tilde{\theta}_n^{(1)} \\ 0 \end{pmatrix} = \begin{pmatrix} (X^{(1)\prime}X^{(1)})^{-1}X^{(1)\prime}Y \\ 0 \end{pmatrix}.$$

Note that $\hat{\theta}_n^{(1)} \neq \tilde{\theta}_n^{(1)}$ in general (see Exercise 6.11).

Theorem 6.8 (Explicit form of the constrained estimator) *Assume that X has rank $q + 1$ and $\hat{\alpha}_q < 0$. Then*

$$\tilde{\theta}_n^{(1)} \in [0, +\infty)^q \iff \hat{\theta}_n^c = \tilde{\theta}_n.$$

Proof. Let $P^{(1)} = X^{(1)}(X^{(1)\prime}X^{(1)})^{-1}X^{(1)\prime}$ be the projector on the columns of $X^{(1)}$ and let $M^{(1)} = I - P^{(1)}$. We have

$$\tilde{\theta}_n'X' = (Y'X^{(1)}(X^{(1)\prime}X^{(1)})^{-1}, 0)\begin{pmatrix} X^{(1)\prime} \\ X^{(2)\prime} \end{pmatrix} = Y'P^{(1)},$$

$$\tilde{\theta}_n'X'X = (Y'X^{(1)}, Y'P^{(1)}X^{(2)}),$$

$$\hat{\theta}_n'X'X = Y'X = (Y'X^{(1)}, Y'X^{(2)}),$$

$$(\hat{\theta}_n' - \tilde{\theta}_n')X'X = (0, Y'M^{(1)}X^{(2)}).$$

Because $\hat{\theta}_n'e_{q+1} < 0$, with $e_{q+1} = (0, \dots, 0, 1)'$, we have $(\hat{\theta}_n' - \tilde{\theta}_n')e_{q+1} < 0$. This can be written as

$$(\hat{\theta}_n' - \tilde{\theta}_n')X'X(X'X)^{-1}e_{q+1} < 0,$$

or alternatively

$$Y'M^{(1)}X^{(2)}\{(X'X)^{-1}\}_{q+1,q+1} < 0.$$

Thus $Y'M^{(1)}X^{(2)} < 0$. It follows that for all $\theta = (\theta^{(1)'}, \theta^{(2)})'$ such that $\theta^{(2)} \in [0, \infty)$,

$$\begin{aligned}\langle \hat{\theta}_n - \tilde{\theta}_n, \tilde{\theta}_n - \theta \rangle_{X'X} &= (\hat{\theta}_n - \tilde{\theta}_n)'X'X(\tilde{\theta}_n - \theta) \\ &= (0, Y'M^{(1)}X^{(2)}) \begin{pmatrix} \tilde{\theta}_n^{(1)} - \theta^{(1)} \\ -\theta^{(2)} \end{pmatrix} \\ &= -\theta^{(2)}Y'M^{(1)}X^{(2)} \geq 0.\end{aligned}$$

In view of (6.16), we have $\hat{\theta}_n^c = \tilde{\theta}_n$ because $\tilde{\theta}_n \in [0, +\infty)^{q+1}$. □

6.4 Bibliographical Notes

The OLS method was proposed by Engle (1982) for ARCH models. The asymptotic properties of the OLS estimator were established by Weiss (1984, 1986), in the ARMA-GARCH framework, under eighth-order moments assumptions. Pantula (1989) also studied the asymptotic properties of the OLS method in the AR(1)-ARCH(q) case, and he gave an explicit form for the asymptotic variance. The FGLS method was developed, in the ARCH case, by Bose and Mukherjee (2003) (see also Gouriéroux 1997). Aknouche (2012) developed a multistage weighted least squares estimate which has the same asymptotic properties as the FGLS but avoids the moment conditions of the OLS. The convexity results used for the study of the constrained estimator can be found, for instance, in Moulin and Fogelman-Soulié (1979).

6.5 Exercises

6.1 *(Estimating the ARCH(q) for q = 1, 2, ...)*
Describe how to use the Durbin algorithm (B.7)–(B.9) to estimate an ARCH(q) model by OLS.

6.2 *(Explicit expression for the OLS estimator of an ARCH process)*
With the notation of Section 6.1, show that, when X has rank q, the estimator $\hat{\theta} = (X'X)^{-1}X'Y$ is the unique solution of the minimisation problem

$$\hat{\theta} = \arg \min_{\theta \in \mathbb{R}^{q+1}} \sum_{t=1}^{n} (\epsilon_t^2 - Z_{t-1}'\theta)^2.$$

6.3 *(OLS estimator with negative values)*
Give a numerical example (with, for instance, $n = 2$) showing that the unconstrained OLS estimator of the ARCH(q) parameters (with, for instance, $q = 1$) can take negative values.

6.4 *(Unconstrained and constrained OLS estimator of an ARCH(2) process)*
Consider the ARCH(2) model

$$\begin{cases} \epsilon_t = \sigma_t \eta_t \\ \sigma_t^2 = \omega + \alpha_1 \epsilon_{t-1}^2 + \alpha_2 \epsilon_{t-2}^2. \end{cases}$$

Let $\hat{\theta} = (\hat{\omega}, \hat{\alpha}_1, \hat{\alpha}_2)'$ be the unconstrained OLS estimator of $\theta = (\omega, \alpha_1, \alpha_2)'$. Is it possible to have

1. $\hat{\alpha}_1 < 0$?
2. $\hat{\alpha}_1 < 0$ and $\hat{\alpha}_2 < 0$?
3. $\hat{\omega} < 0$, $\hat{\alpha}_1 < 0$ and $\hat{\alpha}_2 < 0$?

Let $\hat{\theta}^c = (\hat{\omega}^c, \hat{\alpha}_1^c, \hat{\alpha}_2^c)'$ be the OLS constrained estimator with $\hat{\alpha}_1^c \geq 0$ and $\hat{\alpha}_2^c \geq 0$. Consider the following numerical example with $n = 3$ observations and two initial values: $\epsilon_{-1}^2 = 0$, $\epsilon_0^2 = 1$, $\epsilon_1^2 = 0$, $\epsilon_2^2 = 1/2$, $\epsilon_3^2 = 1/2$. Compute $\hat{\theta}$ and $\hat{\theta}^c$ for these observations.

6.5 *(The columns of the matrix X are non-zero)*
Show that if $\omega_0 > 0$, the matrix X cannot have a column equal to zero for n large enough.

6.6 *(Estimating an AR(1) with ARCH(q) errors)*
Consider the model

$$X_t = \phi_0 X_{t-1} + \epsilon_t, \qquad |\phi_0| < 1,$$

where (ϵ_t) is the strictly stationary solution of model (6.1) under the condition $E\epsilon_t^4 < \infty$. Show that the OLS estimator of ϕ is consistent and asymptotically normal. Is the assumption $E\epsilon_t^4 < \infty$ necessary in the case of iid errors?

6.7 *(Inversion of a block matrix)*
For a matrix partitioned as $A = \begin{bmatrix} A_{11} & A_{12} \\ A_{21} & A_{22} \end{bmatrix}$, show that the inverse (when it exists) is of the form

$$A^{-1} = \begin{bmatrix} A_{11}^{-1} + A_{11}^{-1} A_{12} F A_{21} A_{11}^{-1} & -A_{11}^{-1} A_{12} F \\ -F A_{21} A_{11}^{-1} & F \end{bmatrix},$$

where $F = (A_{22} - A_{21} A_{11}^{-1} A_{12})^{-1}$.

6.8 *(Does the OLS asymptotic variance depend on ω_0?)*

1. Show that for an ARCH(q) model $E(\epsilon_t^{2m})$ is proportional to ω_0^m (when it exists).
2. Using Exercise 6.7, show that, for an ARCH(q) model, the asymptotic variance of the OLS estimator of the α_{0i} does not depend on ω_0.
3. Show that the asymptotic variance of the OLS estimator of ω_0 is proportional to ω_0^2.

6.9 *(Properties of the projections on closed convex sets)*
Let E be an Hilbert space, with a scalar product $\langle \cdot, \cdot \rangle$ and a norm $\| \cdot \|$. When $C \subset E$ and $x \in E$, it is said that $x^* \in C$ is a best approximation of x on C if $\|x - x^*\| = \min_{y \in C} \|x - y\|$.

1. Show that if C is closed and convex, x^* exists and is unique. This point is then called the projection of x on C.
2. Show that x^* satisfies the so-called variational inequalities:

$$\forall y \in C, \quad \langle x^* - x, x^* - y \rangle \leq 0 \tag{6.17}$$

and prove that x^* is the unique point of C satisfying these inequalities.

6.10 *(Properties of the projections on closed convex cones)*
Recall that a subset K of the vectorial space is a cone if for all $y \in K$ and for all $\lambda \geq 0$, we have $\lambda y \in K$. Let K be a closed convex cone of the Hilbert space E.

1. Show that the projection x^* of $x \in E$ on K (see Exercise 6.9) is characterised by

$$\begin{cases} \langle x - x^*, x^* \rangle & = & 0, \\ \langle x - x^*, y \rangle & \leq & 0, \quad \forall y \in K. \end{cases} \tag{6.18}$$

2. Show that x^* satisfies
   (a) $\forall x \in E, \forall \lambda \geq 0, (\lambda x)^* = \lambda x^*$.
   (b) $\forall x \in E, \|x\|^2 = \|x^*\|^2 + \|x - x^*\|^2$, thus $\|x^*\| \leq \|x\|$.

6.11 *(OLS estimation of a subvector of parameters)*
Consider the linear model $Y = X\theta + U$ with the usual assumptions. Let M_2 be the matrix of the orthogonal projection on the orthogonal subspace of $X^{(2)}$, where $X = (X^{(1)}, X^{(2)})$. Show that the OLS estimator of $\theta^{(1)}$ (where $\theta = (\theta^{(1)'}, \theta^{(2)'})'$, with obvious notation) is $\hat{\theta}_n^{(1)} = (X^{(1)'}M_2X^{(1)})^{-1}X^{(1)'}M_2Y$.

6.12 *(A matrix result used in the proof of Theorem 6.7)*
Let (J_n) be a sequence of symmetric $k \times k$ matrices converging to a positive definite matrix J. Let (X_n) be a sequence of vectors in $\mathbb{R}^k$ such that $X_n' J_n X_n \to 0$. Show that $X_n \to 0$.

6.13 *(Example of constrained estimator calculus)*
Take the example of Exercise 6.3 and compute the constrained estimator.

7

Estimating GARCH Models by Quasi-Maximum Likelihood

The quasi-maximum likelihood (QML) method is particularly relevant for GARCH models because it provides consistent and asymptotically normal estimators for *strictly stationary* GARCH processes under mild regularity conditions, but with no moment assumptions on the observed process. By contrast, the least-squares methods of the previous chapter require moments of order 4 at least.

In this chapter, we study in details the conditional QML method (conditional on initial values). We first consider the case when the observed process is pure GARCH. We present an iterative procedure for computing the Gaussian log-likelihood, conditionally on fixed or random initial values. The likelihood is written as if the law of the variables η_t were Gaussian $\mathcal{N}(0, 1)$ (we refer to pseudo- or quasi-likelihood), but this assumption is not necessary for the strong consistency of the estimator. In the second part of the chapter, we will study the application of the method to the estimation of ARMA–GARCH models. The asymptotic properties of the quasi-maximum likelihood estimator (QMLE) are established at the end of the chapter.

7.1 Conditional Quasi-Likelihood

Assume that the observations $\epsilon_1, \dots, \epsilon_n$ constitute a realisation (of length n) of a GARCH(p, q) process, more precisely a non-anticipative strictly stationary solution of

$$\begin{cases} \epsilon_t = \sqrt{h_t}\eta_t \\ h_t = \omega_0 + \sum_{i=1}^{q} \alpha_{0i}\epsilon_{t-i}^2 + \sum_{j=1}^{p} \beta_{0j}h_{t-j}, \quad \forall t \in \mathbb{Z}, \end{cases} \tag{7.1}$$

where (η_t) is a sequence of iid variables of variance 1, $\omega_0 > 0$, $\alpha_{0i} \geq 0$ $(i = 1, \dots, q)$, and $\beta_{0j} \geq 0$ $(j = 1, \dots, p)$. The orders p and q are assumed known. The vector of the parameters

$$\theta = (\theta_1, \dots, \theta_{p+q+1})' := (\omega, \alpha_1, \dots, \alpha_q, \beta_1, \dots, \beta_p)' \tag{7.2}$$

GARCH Models: Structure, Statistical Inference and Financial Applications, Second Edition. Christian Francq and Jean-Michel Zakoian.

belongs to a parameter space of the form

$$\Theta \subset (0, +\infty) \times [0, \infty)^{p+q}. \tag{7.3}$$

The true value of the parameter is unknown, and is denoted by

$$\theta_0 = (\omega_0, \alpha_{01}, \dots, \alpha_{0q}, \beta_{01}, \dots, \beta_{0p})'.$$

To write the likelihood of the model, a distribution must be specified for the iid variables η_t. Here we do not make any assumption on the distribution of these variables, but we work with a function called the (Gaussian) quasi-likelihood, which, conditionally on some initial values, coincides with the likelihood when the η_t are distributed as standard Gaussian. Given initial values $\epsilon_0, \dots, \epsilon_{1-q}, \tilde{\sigma}_0^2, \dots, \tilde{\sigma}_{1-p}^2$ to be specified below, the conditional Gaussian quasi-likelihood is given by

$$L_n(\theta) = L_n(\theta; \epsilon_1, \dots, \epsilon_n) = \prod_{t=1}^{n} \frac{1}{\sqrt{2\pi\tilde{\sigma}_t^2}} \exp\left(-\frac{\epsilon_t^2}{2\tilde{\sigma}_t^2}\right),$$

where the $\tilde{\sigma}_t^2$ are recursively defined, for $t \geq 1$, by

$$\tilde{\sigma}_t^2 = \tilde{\sigma}_t^2(\theta) = \omega + \sum_{i=1}^{q} \alpha_i \epsilon_{t-i}^2 + \sum_{j=1}^{p} \beta_j \tilde{\sigma}_{t-j}^2. \tag{7.4}$$

For a given value of θ, under the second-order stationarity assumption, the unconditional variance (corresponding to this value of θ) is a reasonable choice for the unknown initial values:

$$\epsilon_0^2 = \cdots = \epsilon_{1-q}^2 = \sigma_0^2 = \cdots = \sigma_{1-p}^2 = \frac{\omega}{1 - \sum_{i=1}^{q} \alpha_i - \sum_{j=1}^{p} \beta_j}. \tag{7.5}$$

Such initial values are, however, not suitable for IGARCH models, in particular, and more generally when the second-order stationarity is not imposed. Indeed, the constant (7.5) would then take negative values for some values of θ. In such a case, suitable initial values are

$$\epsilon_0^2 = \cdots = \epsilon_{1-q}^2 = \tilde{\sigma}_0^2 = \cdots = \tilde{\sigma}_{1-p}^2 = \omega \tag{7.6}$$

or

$$\epsilon_0^2 = \cdots = \epsilon_{1-q}^2 = \tilde{\sigma}_0^2 = \cdots = \tilde{\sigma}_{1-p}^2 = \epsilon_1^2. \tag{7.7}$$

A QMLE of θ is defined as any measurable solution $\hat{\theta}_n$ of

$$\hat{\theta}_n = \arg\max_{\theta \in \Theta} L_n(\theta).$$

Taking the logarithm, it is seen that maximising the likelihood is equivalent to minimising, with respect to θ,

$$\tilde{\mathbf{l}}_n(\theta) - n^{-1} \sum_{t=1}^{n} \tilde{\ell}_t, \quad \text{where} \quad \tilde{\ell}_t = \tilde{\ell}_t(\theta) = \frac{\epsilon_t^2}{\tilde{\sigma}_t^2} + \log \tilde{\sigma}_t^2 \tag{7.8}$$

and $\tilde{\sigma}_t^2$ is defined by (7.4). A QMLE is thus a measurable solution of the equation

$$\hat{\theta}_n = \arg\min_{\theta \in \Theta} \tilde{\mathbf{l}}_n(\theta). \tag{7.9}$$

It will be shown that the choice of the initial values is unimportant for the asymptotic properties of the QMLE. However, in practice this choice may be important. Note that other methods are possible for generating the sequence $\tilde{\sigma}_t^2$; for example, by taking $\tilde{\sigma}_t^2 = c_0(\theta) + \sum_{i=1}^{t-1} c_i(\theta)\epsilon_{t-i}^2$, where the $c_i(\theta)$ are recursively computed (see Berkes, Horváth, and Kokoszka 2003b). Note that for computing $\tilde{\mathbf{l}}_n(\theta)$, this procedure involves a number of operations of order n^2, whereas the one we propose involves a number of order n. It will be convenient to approximate the sequence $(\tilde{\epsilon}_t(\theta))$ by an ergodic stationary sequence. Assuming that the roots of $\mathcal{B}_\theta(z)$ are outside the unit disk, the non-anticipative and ergodic strictly stationary sequence $(\sigma_t^2)_t = \{\sigma_t^2(\theta)\}_t$ is defined as the solution of

$$\sigma_t^2 = \omega + \sum_{i=1}^{q} \alpha_i \epsilon_{t-i}^2 + \sum_{j=1}^{p} \beta_j \sigma_{t-j}^2, \qquad \forall t. \tag{7.10}$$

Note that $\sigma_t^2(\theta_0) = h_t$.

Likelihood Equations

Likelihood equations are obtained by canceling the derivative of the criterion $\tilde{\mathbf{l}}_n(\theta)$ with respect to θ, which gives

$$\frac{1}{n}\sum_{t=1}^{n}\{\epsilon_t^2 - \tilde{\sigma}_t^2\}\frac{1}{\tilde{\sigma}_t^4}\frac{\partial \tilde{\sigma}_t^2}{\partial \theta} = 0. \tag{7.11}$$

These equations can be interpreted as orthogonality relations, for large n. Indeed, as will be seen in the next section, the left-hand side of Eq. (7.11) has the same asymptotic behaviour as

$$\frac{1}{n}\sum_{t=1}^{n}\{\epsilon_t^2 - \sigma_t^2\}\frac{1}{\sigma_t^4}\frac{\partial \sigma_t^2}{\partial \theta},$$

the impact of the initial values vanishing as $n \to \infty$.

The innovation of ϵ_t^2 is $v_t = \epsilon_t^2 - h_t^2$. Thus, under the assumption that the expectation exists, we have

$$E_{\theta_0}\left(v_t \frac{1}{\sigma_t^4(\theta_0)}\frac{\partial \sigma_t^2(\theta_0)}{\partial \theta}\right) = 0,$$

because $\frac{1}{\sigma_t^4(\theta_0)}\frac{\partial \sigma_t^2(\theta_0)}{\partial \theta}$ is a measurable function of the ϵ_{t-i}, $i > 0$. This result can be viewed as the asymptotic version of condition (7.11) at θ_0, using the ergodic theorem.

7.1.1 Asymptotic Properties of the QMLE

In this chapter, we will use the matrix norm defined by $\|A\| = \sum |a_{ij}|$ for all matrices $A = (a_{ij})$. The spectral radius of a square matrix A is denoted by $\rho(A)$.

Strong Consistency

Recall that model (7.1) admits a strictly stationary solution if and only if the sequence of matrices $\mathbf{A}_0 = (A_{0t})$, where

$$A_{0t} = \begin{pmatrix}
\alpha_{01}\eta_t^2 & & \cdots & & \alpha_{0q}\eta_t^2 & \beta_{01}\eta_t^2 & & \cdots & & \beta_{0p}\eta_t^2 \\
1 & 0 & \cdots & & 0 & 0 & & \cdots & & 0 \\
0 & 1 & \cdots & & 0 & 0 & & \cdots & & 0 \\
\vdots & \ddots & \ddots & & \vdots & \vdots & \ddots & \ddots & & \vdots \\
0 & & \cdots & 1 & 0 & 0 & & \cdots & 0 & 0 \\
\alpha_{01} & & \cdots & & \alpha_{0q} & \beta_{01} & & \cdots & & \beta_{0p} \\
0 & & \cdots & & 0 & 1 & 0 & \cdots & & 0 \\
0 & & \cdots & & 0 & 0 & 1 & \cdots & & 0 \\
\vdots & \ddots & \ddots & & \vdots & \vdots & \ddots & \ddots & & \vdots \\
0 & & \cdots & 0 & 0 & 0 & & \cdots & 1 & 0
\end{pmatrix},$$

admits a strictly negative top Lyapunov exponent, $\gamma(\mathbf{A}_0) < 0$, where

$$\gamma(\mathbf{A}_0) := \inf_{t \in \mathbb{N}^*} \frac{1}{t} E(\log \|A_{0t}A_{0t-1} \dots A_{01}\|)$$
$$= \lim_{t\to\infty} \text{a.s.} \frac{1}{t} \log \|A_{0t}A_{0t-1} \dots A_{01}\|. \tag{7.12}$$

Let

$$\mathcal{A}_\theta(z) = \sum_{i=1}^{q} \alpha_i z^i \quad \text{and} \quad \mathcal{B}_\theta(z) = 1 - \sum_{j=1}^{p} \beta_j z^j.$$

By convention, $\mathcal{A}_\theta(z) = 0$ if $q = 0$ and $\mathcal{B}_\theta(z) = 1$ if $p = 0$. To show strong consistency, the following assumptions are used.

A1: $\theta_0 \in \Theta$ and Θ is compact.

A2: $\gamma(\mathbf{A}_0) < 0$ and for all $\theta \in \Theta$, $\sum_{j=1}^{p} \beta_j < 1$.

A3: η_t^2 has a non-degenerate distribution and $E\eta_t^2 = 1$.

A4: If $p > 0$, $\mathcal{A}_{\theta_0}(z)$ and $\mathcal{B}_{\theta_0}(z)$ have no common roots, $\mathcal{A}_{\theta_0}(1) \neq 0$, and $\alpha_{0q} + \beta_{0p} \neq 0$.

Note that, by Corollary 2.2, the second part of assumption A2 implies that the roots of $\mathcal{B}_\theta(z)$ are outside the unit disk. Thus, a non-anticipative and ergodic strictly stationary sequence $(\sigma_t^2)_t$ is defined by (7.10). Similarly, define

$$\mathbf{l}_n(\theta) = \mathbf{l}_n(\theta; \epsilon_n, \epsilon_{n-1} \dots,) = n^{-1} \sum_{t=1}^{n} \ell_t, \quad \ell_t = \ell_t(\theta) = \frac{\epsilon_t^2}{\sigma_t^2} + \log \sigma_t^2.$$

Example 7.1 (Parameter space of a GARCH(1, 1) process) In the case of a GARCH(1, 1) process, assumptions A1 and A2 hold true when, for instance, the parameter space is of the form

$$\Theta = [\delta, 1/\delta] \times [0, 1/\delta] \times [0, 1-\delta],$$

where $\delta \in (0, 1)$ is a constant, small enough so that the true value $\theta_0 = (\omega_0, \alpha_0, \beta_0)'$ belongs to Θ. Figure 7.1 displays, in the plane (α, β), the zones of strict stationarity (when η_t is $\mathcal{N}(0, 1)$ distributed) and of second-order stationarity, as well as an example of a parameter space Θ (the grey zone) compatible with assumptions A1 and A2.

The first result states the strong consistency of $\hat{\theta}_n$. The proof of this theorem, and of the next ones, is given in Section 7.4.

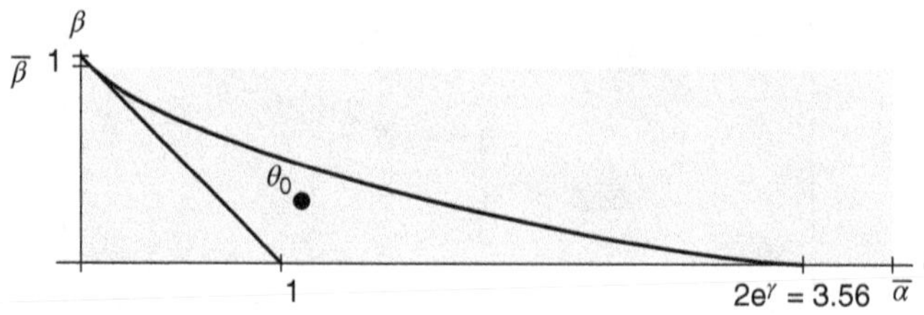

Figure 7.1 GARCH(1, 1): zones of strict and second-order stationarity and parameter space $\Theta = [\underline{\omega}, \overline{\omega}] \times [0, \overline{\alpha}] \times [0, \overline{\beta}]$.

Theorem 7.1 (Strong consistency of the QMLE) *Let $(\hat{\theta}_n)$ be a sequence of QMLEs satisfying condition (7.9), with initial conditions (7.6) or (7.7). Under assumptions A1–A4, almost surely*

$$\hat{\theta}_n \to \theta_0, \quad \text{as } n \to \infty.$$

Remark 7.1

1. It is not assumed that the true value of the parameter θ_0 belongs to the interior of Θ. Thus, the theorem allows to handle cases where some coefficients, α_i or β_j, are null.
2. It is important to note that the strict stationarity condition is only assumed at θ_0, not over all Θ. In view of Corollary 2.2, the condition $\sum_{j=1}^{p} \beta_j < 1$ is weaker than the strict stationarity condition.
3. Assumption A4 disappears in the ARCH case. In the general case, this assumption allows for an overidentification of either of the two orders, p or q, but not of both. We then consistently estimate the parameters of a GARCH($p-1$, q) (or GARCH(p, $q-1$)) process if an overparameterised GARCH(p, q) model is used.
4. When $p \neq 0$, assumption A4 precludes the case where all the α_{0i} are zero. In such a case, the strictly stationary solution of the model is the strong white noise, which can be written in multiple forms. For instance, a strong white noise of variance 1 can be written in the GARCH(1, 1) form with $\sigma_t^2 = \sigma^2(1-\beta) + 0 \times \epsilon_{t-1}^2 + \beta\sigma_{t-1}^2$.
5. The assumption of absence of a common root, in A4, is restrictive only if $p > 1$ and $q > 1$. Indeed, if $q = 1$, the unique root of $\mathcal{A}_{\theta_0}(z)$ is 0, and we have $\mathcal{B}_{\theta_0}(0) \neq 0$. If $p = 1$ and $\beta_{01} \neq 0$, the unique root of $\mathcal{B}_{\theta_0}(z)$ is $1/\beta_{01} > 0$ (if $\beta_{01} = 0$, the polynomial does not admit any root). Because the coefficients α_{0i} are positive this value cannot be a zero of $\mathcal{A}_{\theta_0}(z)$.
6. The assumption $E\eta_t = 0$ is not required for the consistency (and asymptotic normality) of the QMLE of a GARCH. The conditional variance of ϵ_t is thus, in general, only proportional to h_t: $\text{Var}(\epsilon_t \mid \epsilon_u, u < t) = \{1 - (E\eta_t)^2\} h_t$. The assumption $E\eta_t^2 = 1$ is made for identifiability reasons and is not restrictive provided that $E\eta_t^2 < \infty$.

Asymptotic Normality

The following additional assumptions are considered.

A5: $\theta_0 \in \overset{\circ}{\Theta}$, where $\overset{\circ}{\Theta}$ denotes the interior of Θ.

A6: $\kappa_\eta = E\eta_t^4 < \infty$.

The limiting distribution of $\hat{\theta}_n$ is given by the following result.

Theorem 7.2 (Asymptotic normality of the QMLE) *Under assumptions A1–A6,*

$$\sqrt{n}(\hat{\theta}_n - \theta_0) \xrightarrow{\mathcal{L}} \mathcal{N}(0, (\kappa_\eta - 1)J^{-1}),$$

where

$$J := E_{\theta_0}\left(\frac{\partial^2 \ell_t(\theta_0)}{\partial\theta\,\partial\theta'}\right) = E_{\theta_0}\left(\frac{1}{\sigma_t^4(\theta_0)} \frac{\partial\sigma_t^2(\theta_0)}{\partial\theta} \frac{\partial\sigma_t^2(\theta_0)}{\partial\theta'}\right) \tag{7.13}$$

is a positive definite matrix.

Remark 7.2

1. Assumption A5 is standard and entails the first-order condition (at least asymptotically). Indeed, if $\hat{\theta}_n$ is consistent, it also belongs to the interior of Θ, for large n. At this maximum, the derivative of the objective function cancels. However, assumption A5 is restrictive because it precludes, for instance, the case $\beta_{01}=0$.

2. When one or several components of θ_0 are null, assumption A5 is not satisfied, and the theorem cannot be used. It is clear that, in this case, the asymptotic distribution of $\sqrt{n}(\hat{\theta}_n-\theta_0)$ cannot be normal because the estimator is constrained. If, for instance $\beta_{01}=0$, the distribution of $\sqrt{n}(\hat{\beta}_1-\beta_{01})$ is concentrated in $[0,\infty)$, for all n, and thus cannot be asymptotically normal. This kind of 'boundary' problem is the object of a specific study in Chapter 8.

3. Assumption A6 does not concern ϵ_t^2 and does not preclude the IGARCH case. Only a fourth-order moment assumption on η_t is required. This assumption is clearly necessary for the existence of the variance of the score vector $\partial\ell_t(\theta_0)/\partial\theta$. In the proof of this theorem, it is shown that

$$E_{\theta_0}\left\{\frac{\partial\ell_t(\theta_0)}{\partial\theta}\right\}=0,\quad \text{Var}_{\theta_0}\left\{\frac{\partial\ell_t(\theta_0)}{\partial\theta}\right\}=(\kappa_\eta-1)J.$$

4. In the ARCH case ($p=0$), the asymptotic variance of the QMLE reduces to that of the FGLS estimator (see Theorem 6.3). Indeed, in this case, we have $\partial\sigma_t^2(\theta)/\partial\theta=Z_{t-1}$. Theorem 6.3 requires, however, the existence of a fourth-order moment for the observed process, whereas there is no moment assumption for the asymptotic normality of the QMLE. Moreover, Theorem 6.4 shows that the QMLE of an ARCH(q) is asymptotically more accurate than that of the ordinary least square (OLS) estimator.

7.1.2 The ARCH(1) Case: Numerical Evaluation of the Asymptotic Variance

Consider the ARCH(1) model

$$\epsilon_t=\{\omega_0+\alpha_0\epsilon_{t-1}^2\}^{1/2}\eta_t,$$

with $\omega_0>0$ and $\alpha_0>0$, and suppose that the variables η_t satisfy assumption A3. The unknown parameter is $\theta_0=(\omega_0,\alpha_0)'$. In view of condition (2.10), the strict stationarity constraint A2 is written as

$$\alpha_0<\exp\{-E(\log\eta_t^2)\}.$$

Assumption A1 holds true if, for instance, the parameter space is of the form $\Theta=[\delta,1/\delta]\times[0,1/\delta]$, where $\delta>0$ is a constant, chosen sufficiently small so that θ_0 belongs to Θ. By Theorem 7.1, the QMLE of θ_0 is then strongly consistent. Since $\partial\tilde{\sigma}_t^2/\partial\theta=(1,\epsilon_{t-1}^2)'$, the QMLE $\hat{\theta}_n=(\hat{\omega}_n,\hat{\alpha}_n)'$ is characterised by the normal equation

$$\frac{1}{n}\sum_{t=1}^{n}\frac{\epsilon_t^2-\hat{\omega}_n-\hat{\alpha}_n\epsilon_{t-1}^2}{(\hat{\omega}_n+\hat{\alpha}_n\epsilon_{t-1}^2)^2}\begin{pmatrix}1\\ \epsilon_{t-1}^2\end{pmatrix}=0$$

with, for instance $\epsilon_0^2=\epsilon_1^2$. This estimator does not have an explicit form and must be obtained numerically. Theorem 7.2, which provides the asymptotic distribution of the estimator, only requires the extra assumption that θ_0 belongs to $\overset{\circ}{\Theta}=(\delta,1/\delta)\times(0,1/\delta)$. Thus, if $\alpha_0=0$ (that is, if the model is conditionally homoscedastic), the estimator remains consistent but is no longer asymptotically normal. Matrix J takes the form

$$J=E_{\theta_0}\begin{bmatrix}\frac{1}{(\omega_0+\alpha_0\epsilon_{t-1}^2)^2} & \frac{\epsilon_{t-1}^2}{(\omega_0+\alpha_0\epsilon_{t-1}^2)^2}\\ \frac{\epsilon_{t-1}^2}{(\omega_0+\alpha_0\epsilon_{t-1}^2)^2} & \frac{\epsilon_{t-1}^4}{(\omega_0+\alpha_0\epsilon_{t-1}^2)^2}\end{bmatrix},$$

Table 7.1 Asymptotic variance for the QMLE of an ARCH(1) process with $\eta_t \sim \mathcal{N}(0, 1)$.

	$\omega_0 = 1, \alpha_0 = 0.1$	$\omega_0 = 1, \alpha_0 = 0.5$	$\omega_0 = 1, \alpha_0 = 0.95$
$\text{Var}_{as}\{\sqrt{n}(\hat{\theta}_n - \theta_0)\}$	$\begin{pmatrix} -3.46 - 1.34 \\ -1.34 - 1.87 \end{pmatrix}$	$\begin{pmatrix} -4.85 - 2.15 \\ -2.15 - 3.99 \end{pmatrix}$	$\begin{pmatrix} -6.61 - 2.83 \\ -2.83 - 6.67 \end{pmatrix}$

Table 7.2 Comparison of the empirical and theoretical asymptotic variances, for the QMLE of the parameter $\alpha_0 = 0.9$ of an ARCH(1), when $\eta_t \sim \mathcal{N}(0, 1)$.

n	$\overline{\alpha}_n$	RMSE(α)	$\{\text{Var}_{as}[\sqrt{n}(\hat{\alpha}_n - \alpha_0)]\}^{1/2}/\sqrt{n}$	$\hat{P}[\hat{\alpha}_n \geq 1]$
100	0.85221	0.25742	0.25014	0.266
250	0.88336	0.16355	0.15820	0.239
500	0.89266	0.10659	0.11186	0.152
1000	0.89804	0.08143	0.07911	0.100

and the asymptotic variance of $\sqrt{n}(\hat{\theta}_n - \theta_0)$ is

$$\text{Var}_{as}\{\sqrt{n}(\hat{\theta}_n - \theta_0)\} = (\kappa_\eta - 1)J^{-1}.$$

Table 7.1 displays numerical evaluations of this matrix. An estimation of J is obtained by replacing the expectations by empirical means, obtained from simulations of length 10 000, when η_t is $\mathcal{N}(0, 1)$ distributed. This experiment is repeated 1 000 times to obtain the results presented in the table.

In order to assess, in finite samples, the quality of the asymptotic approximation of the variance of the estimator, the following Monte Carlo experiment is conducted. For the value θ_0 of the parameter, and for a given length n, N samples are simulated, leading to N estimations $\hat{\theta}_n^{(i)}$ of θ_0, $i = 1, \dots, N$. We denote by $\overline{\theta}_n = (\overline{\omega}_n, \overline{\alpha}_n)'$ their empirical mean. The root mean squared error (RMSE) of estimation of α is denoted by

$$\text{RMSE}(\alpha) = \left\{ \frac{1}{N} \sum_{i=1}^{N} (\hat{\alpha}_n^{(i)} - \overline{\alpha}_n)^2 \right\}^{1/2}$$

and can be compared to $\{\text{Var}_{as}[\sqrt{n}(\hat{\alpha}_n - \alpha_0)]\}^{1/2}/\sqrt{n}$, the latter quantity being evaluated independently, by simulation. A similar comparison can obviously be made for the parameter ω. For $\theta_0 = (0.2, 0.9)'$ and $N = 1000$, Table 7.2 displays the results, for different sample length n.

The similarity between columns 3 and 4 is quite satisfactory, even for moderate sample sizes. The last column gives the empirical probability (that is, the relative frequency within the N samples) that $\hat{\alpha}_n$ is greater than 1 (which is the limiting value for second-order stationarity). These results show that, even if the mean of the estimations is close to the true value for large n, the variability of the estimator remains high. Finally, note that the length $n = 1000$ remains realistic for financial series.

7.1.3 The Non-stationary ARCH(1)

When the strict stationarity constraint is not satisfied in the ARCH(1) case, that is, when

$$\alpha_0 \geq \exp\{-E \log \eta_t^2\}, \tag{7.14}$$

one can define an ARCH(1) process starting with initial values. For a given value ϵ_0, we define

$$\epsilon_t = h_t^{1/2}\eta_t, \quad h_t = \omega_0 + \alpha_0 \epsilon_{t-1}^2, \quad t = 1, 2, \dots, \tag{7.15}$$

where $\omega_0 > 0$ and $\alpha_0 > 0$, with the usual assumptions on the sequence (η_t). As already noted, σ_t^2 converges to infinity almost surely when

$$\alpha_0 > \exp\{-E \log \eta_t^2\}, \tag{7.16}$$

and only in probability when the inequality (7.14) is an equality (see Corollary 2.1 and Remark 2.3 following it). Is it possible to estimate the coefficients of such a model? The answer is only partly positive: it is possible to consistently estimate the coefficient α_0, but the coefficient ω_0 cannot be consistently estimated. The practical impact of this result thus appears to be limited, but because of its theoretical interest, the problem of estimating coefficients of non-stationary models deserves attention. Consider the QMLE of an ARCH(1), that is to say a measurable solution of

$$(\hat{\omega}_n, \hat{\alpha}_n) = \arg\min_{\theta\in\Theta} \frac{1}{n} \sum_{t=1}^{n} \ell_t(\theta), \quad \ell_t(\theta) = \frac{\epsilon_t^2}{\sigma_t^2(\theta)} + \log \sigma_t^2(\theta), \tag{7.17}$$

where $\theta = (\omega, \alpha)$, Θ is a compact set of $(0, \infty)^2$, and $\sigma_t^2(\theta) = \omega + \alpha\epsilon_{t-1}^2$ for $t = 1, \ldots, n$ (starting with a given initial value for ϵ_0^2). The almost sure convergence of ϵ_t^2 to infinity will be used to show the strong consistency of the QMLE of α_0. The following lemma completes Corollary 2.1 and gives the rate of convergence of ϵ_t^2 to infinity under condition (7.16).

Lemma 7.1 *Define the ARCH(1) model by equation (7.15) with any initial condition $\epsilon_0^2 \geq 0$. The non-stationarity condition (7.16) is assumed. Then, almost surely, as $n \to \infty$,*

$$\frac{1}{h_n} = o(\rho^n) \quad \text{and} \quad \frac{1}{\epsilon_n^2} = o(\rho^n)$$

for any constant ρ such that

$$1 > \rho > \exp\{-E \log \eta_t^2\}/\alpha_0. \tag{7.18}$$

This result entails the strong consistency and asymptotic normality of the QMLE of α_0.

Theorem 7.3 *Consider the assumptions of Lemma 7.1 and the QMLE defined by equation (7.17). Assume $\theta_0 = (\omega_0, \alpha_0) \in \Theta$. Then*

$$\hat{\alpha}_n \to \alpha_0 \quad \text{a.s.}, \tag{7.19}$$

and when θ_0 belongs to the interior of Θ,

$$\sqrt{n}(\hat{\alpha}_n - \alpha_0) \xrightarrow{\mathcal{L}} \mathcal{N}\{0, (\kappa_\eta - 1)\alpha_0^2\} \tag{7.20}$$

as $n \to \infty$.

In the proof of this theorem, it is shown that the score vector satisfies

$$\frac{1}{\sqrt{n}} \sum_{t=1}^{n} \frac{\partial}{\partial\theta} \ell_t(\theta_0) \xrightarrow{\mathcal{L}} \mathcal{N}\left\{0, J = (\kappa_\eta - 1)\begin{pmatrix} 0 & 0 \\ 0 & \alpha_0^{-2} \end{pmatrix}\right\}.$$

In the standard statistical inference framework, the variance J of the score vector is (proportional to) the Fisher information. According to the usual interpretation, the form of the matrix J shows that, asymptotically and for almost all observations, the variations of the log-likelihood $n^{-1/2}\sum_{t=1}^{n} \log \ell_t(\theta)$ are insignificant when θ varies from (ω_0, α_0) to $(\omega_0 + h, \alpha_0)$ for small h. In other words, the limiting log-likelihood is flat at the point (ω_0, α_0) in the direction of variation of ω_0. Thus, minimising this limiting function does not allow θ_0 to be found. This leads us to think that the QML of ω_0 is likely to be inconsistent when the strict stationarity condition is not satisfied. Figure 7.2 displays numerical results illustrating the performance of the QMLE in finite samples. For different values of the parameters, 100 replications

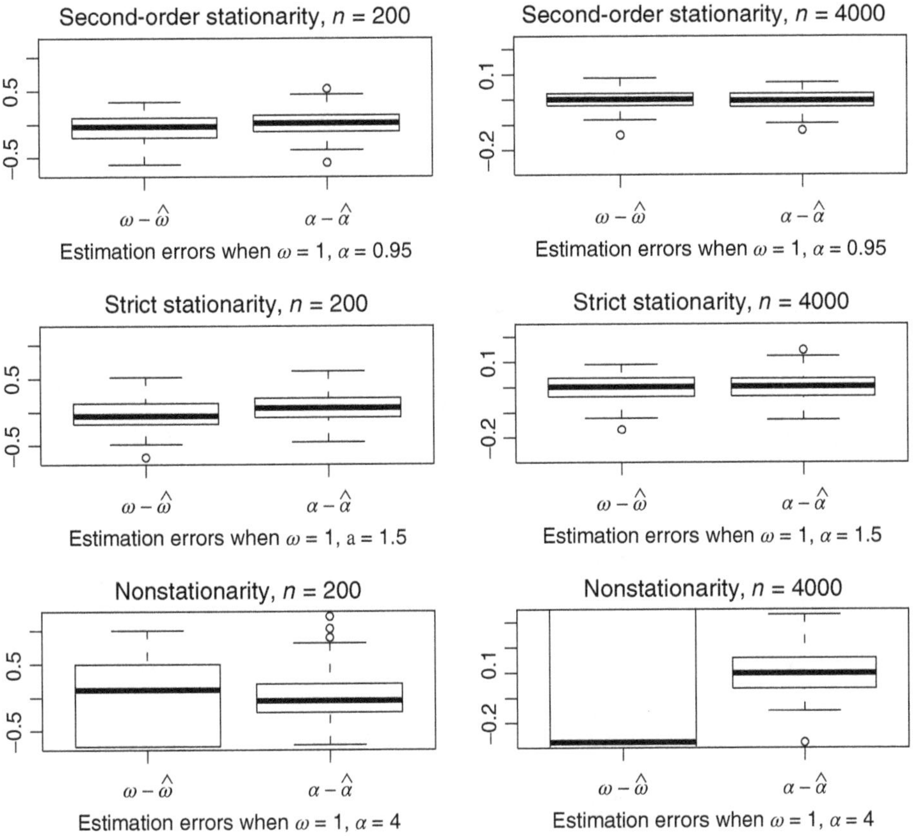

Figure 7.2 Box-plots of the QML estimation errors for the parameters ω_0 and α_0 of an ARCH(1) process, with $\eta_t \sim \mathcal{N}(0, 1)$.

of the ARCH(1) model have been generated, for the sample sizes $n = 200$ and $n = 4000$. The top panels of the figure correspond to a second-order stationary ARCH(1), with parameter $\theta_0 = (1, 0.95)$. The panels in the middle correspond to a strictly stationary ARCH(1) of infinite variance, with $\theta_0 = (1, 1.5)$. The results obtained for these two cases are similar, confirming that second-order stationarity is not necessary for estimating an ARCH. The bottom panels, corresponding to the explosive ARCH(1) with parameter $\theta_0 = (1, 4)$, confirm the asymptotic results concerning the estimation of α_0. They also illustrate the failure of the QML to estimate ω_0 under the nonstationarity condition (7.16). The results even deteriorate when the sample size increases.

7.2 Estimation of ARMA–GARCH Models by Quasi-Maximum Likelihood

In this section, the previous results are extended to cover the situation where the GARCH process is not directly observed, but constitutes the innovation of an observed ARMA process. This framework is relevant because, even for financial series, it is restrictive to assume that the observed series is the realisation of a noise. From a theoretical point of view, it will be seen that the extension to the ARMA–GARCH case is far from trivial. Assume that the observations $X_1, \dots, X_n$ are generated by a strictly stationary non-anticipative solution of the ARMA(P, Q)-GARCH(p, q) model

$$\begin{cases} X_t - c_0 = \sum_{i=1}^{P} a_{0i}(X_{t-i} - c_0) + e_t - \sum_{j=1}^{Q} b_{0j}e_{t-j} \\ e_t = \sqrt{h_t}\eta_t \\ h_t = \omega_0 + \sum_{i=1}^{q} \alpha_{0i}e_{t-i}^2 + \sum_{j=1}^{p} \beta_{0j}h_{t-j}, \end{cases} \tag{7.21}$$

where (η_t) and the coefficients ω_0, α_{0i} and β_{0j} are defined as in model (7.1). The orders P, Q, p, q are assumed known. The vector of the parameters is denoted by

$$\varphi = (\vartheta', \theta')' = (c, a_1, \dots a_P, b_1, \dots, b_Q, \theta')',$$

where θ is defined as in equality (7.2). The parameter space is

$$\Phi \subset \mathbb{R}^{P+Q+1} \times (0, +\infty) \times [0, \infty)^{p+q}.$$

The true value of the parameter is denoted by

$$\varphi_0 = (\vartheta_0', \theta_0')' = (c_0, a_{01}, \dots a_{0P}, b_{01}, \dots, b_{0Q}, \theta_0')'.$$

We still employ a Gaussian quasi-likelihood conditional on initial values. If $q \geq Q$, the initial values are

$$X_0, \dots, X_{1-(q-Q)-P}, \tilde{\epsilon}_{-q+Q}, \dots, \tilde{\epsilon}_{1-q}, \tilde{\sigma}_0^2, \dots, \tilde{\sigma}_{1-p}^2.$$

These values (the last p of which are positive) are assumed to be fixed, but they could depend on the parameter and/or on the observations. For any ϑ, the values of $\tilde{\epsilon}_t(\vartheta)$, for $t = -q + Q + 1, \dots, n$, and then, for any θ, the values of $\tilde{\sigma}_t^2(\varphi)$, for $t = 1, \dots, n$, can thus be computed from

$$\begin{cases} \tilde{\epsilon}_t = \tilde{\epsilon}_t(\vartheta) = X_t - c - \sum_{i=1}^{P} a_i(X_{t-i} - c) + \sum_{j=1}^{Q} b_j\tilde{\epsilon}_{t-j} \\ \tilde{\sigma}_t^2 = \tilde{\sigma}_t^2(\varphi) = \omega + \sum_{i=1}^{q} \alpha_i\tilde{\epsilon}_{t-i}^2 + \sum_{j=1}^{p} \beta_j\tilde{\sigma}_{t-j}^2. \end{cases} \tag{7.22}$$

When $q < Q$, the fixed initial values are

$$X_0, \dots, X_{1-(q-Q)-P}, \epsilon_0, \dots, \epsilon_{1-Q}, \tilde{\sigma}_0^2, \dots, \tilde{\sigma}_{1-p}^2.$$

Conditionally on these initial values, the Gaussian log-likelihood is given by

$$\tilde{\mathbf{l}}_n(\varphi) = n^{-1}\sum_{t=1}^{n} \tilde{\ell}_t, \qquad \tilde{\ell}_t = \tilde{\ell}_t(\phi) = \frac{\tilde{\epsilon}_t^2(\vartheta)}{\tilde{\sigma}_t^2(\varphi)} + \log \tilde{\sigma}_t^2(\varphi).$$

A QMLE is defined as a measurable solution of the equation

$$\hat{\varphi}_n = \arg\min_{\varphi \in \Phi} \tilde{\mathbf{l}}_n(\varphi).$$

Strong Consistency

Let $a_\vartheta(z) = 1 - \sum_{i=1}^{P} a_i z^i$ and $b_\vartheta(z) = 1 - \sum_{j=1}^{Q} b_j z^j$. Standard assumptions are made on these AR and MA polynomials, and assumption A1 is modified as follows:

A7: $\varphi_0 \in \Phi$ and Φ is compact.

A8: For all $\varphi \in \Phi$, $a_\vartheta(z)b_\vartheta(z) = 0$ implies $|z| > 1$.

A9: $a_{\vartheta_0}(z)$ and $b_{\vartheta_0}(z)$ have no common roots, $a_{0P} \neq 0$ or $b_{0Q} \neq 0$.

Under assumptions A2 and A8, (X_t) is supposed to be the unique strictly stationary nonanticipative solution of model (7.21). Let $\epsilon_t = \epsilon_t(\vartheta) = a_\vartheta(B)b_\vartheta^{-1}(B)(X_t - c)$ and $\ell_t = \ell_t(\varphi) = \epsilon_t^2/\sigma_t^2 + \log \sigma_t^2$, where $\sigma_t^2 = \sigma_t^2(\varphi)$ is the non-anticipative and ergodic strictly stationary solution of (7.10). Note that $e_t = \epsilon_t(\vartheta_0)$ and $h_t = \sigma_t^2(\varphi_0)$. The following result is an extension of Theorem 7.1.

Theorem 7.4 (Consistency of the QMLE) *Let $(\hat{\varphi}_n)$ be a sequence of QMLEs. Assume that $E\eta_t = 0$. Then, under assumptions A2–A4 and A7–A9, almost surely*

$$\hat{\varphi}_n \to \varphi_0, \quad as\ n \to \infty.$$

Remark 7.3

1. As in the pure GARCH case, the theorem does not impose a finite variance for e_t (and thus for X_t). In the pure ARMA case, where $e_t = \eta_t$ admits a finite variance, this theorem reduces to a standard result concerning ARMA models with iid errors (see Brockwell and Davis 1991, p. 384).
2. Apart from the condition $E\eta_t = 0$, the conditions required for the strong consistency of the QMLE are not stronger than in the pure GARCH case.

Asymptotic Normality When the Moment of Order 4 Exists

So far, the asymptotic results of the QMLE (consistency and asymptotic normality in the pure GARCH case, consistency in the ARMA–GARCH case) have not required any moment assumption on the observed process (for the asymptotic normality in the pure GARCH case, a moment of order 4 is assumed for the iid process, not for ϵ_t). One might think that this will be the same for establishing the asymptotic normality in the ARMA–GARCH case. The following example shows that this is not the case.

Example 7.2 (Non-existence of J without moment assumption) Consider the AR(1)-ARCH(1) model

$$X_t = a_{01}X_{t-1} + e_t, \quad e_t = \sqrt{h_t}\eta_t, \quad h_t = \omega_0 + \alpha_0 e_{t-1}^2 \tag{7.23}$$

where $|a_{01}| < 1$, $\omega_0 > 0$, $\alpha_0 \geq 0$, and the distribution of the iid sequence (η_t) is defined, for $a > 1$, by

$$\mathbb{P}(\eta_t = a) = \mathbb{P}(\eta_t = -a) = \frac{1}{2a^2}, \quad \mathbb{P}(\eta_t = 0) = 1 - \frac{1}{a^2}.$$

Then the process (X_t) is always stationary, for any value of α_0 (because $\exp\{-E(\log \eta_t^2)\} = +\infty$; see the strict stationarity constraint (2.10)). By contrast, X_t does not admit a moment of order 2 when $\alpha_0 \geq 1$ (see Theorem 2.2). The first component of the (normalised) score vector is

$$\begin{aligned}\frac{\partial \ell_t(\theta_0)}{\partial a_1} &= \left(1 - \frac{e_t^2}{\sigma_t^2}\right)\left(\frac{1}{h_t}\frac{\partial \sigma_t^2(\theta_0)}{\partial a_1}\right) + \frac{2e_t}{h_t}\frac{\partial \epsilon_t(\theta_0)}{\partial a_1} \\ &= -2\alpha_0(1 - \eta_t^2)\left(\frac{e_{t-1}X_{t-2}}{h_t}\right) - 2\frac{\eta_t X_{t-1}}{\sqrt{h_t}}.\end{aligned}$$

We have

$$E\left\{\alpha_0(1 - \eta_t^2)\left(\frac{e_{t-1}X_{t-2}}{h_t}\right) + \frac{\eta_t X_{t-1}}{\sqrt{h_t}}\right\}^2$$

$$\geq E\left[\left\{\alpha_0(1-\eta_t^2)\left(\frac{e_{t-1}X_{t-2}}{h_t}\right)+\frac{\eta_t X_{t-1}}{\sqrt{h_t}}\right\}^2 \middle| \eta_{t-1}=0\right]\mathbb{P}(\eta_{t-1}=0)$$

$$=\frac{a_{01}^2}{\omega_0}\left(1-\frac{1}{a^2}\right)E(X_{t-2}^2)$$

since, first, $\eta_{t-1}=0$ entails $\epsilon_{t-1}=0$ and $X_{t-1}=a_{01}X_{t-2}$, and second, η_{t-1} and X_{t-2} are independent. Consequently, if $EX_t^2=\infty$ and $a_{01}\neq 0$, the score vector does not admit a variance.

This example shows that it is not possible to extend the result of asymptotic normality obtained in the GARCH case to the ARMA–GARCH models without additional moment assumptions. This is not surprising because for ARMA models (which can be viewed as limits of ARMA–GARCH models when the coefficients α_{0i} and β_{0j} tend to 0) the asymptotic normality of the QMLE is shown with second-order moment assumptions. For an ARMA with infinite variance innovations, the consistency of the estimators may be faster than in the standard case and the asymptotic distribution belongs to the class of the α-stable laws, but is non-Gaussian in general. We show the asymptotic normality with a moment assumption of order 4. Recall that, by Theorem 2.9, this assumption is equivalent to $\rho\{E(A_{0t}\otimes A_{0t})\}<1$. We make the following assumptions:

A10 $\rho\{E(A_{0t}\otimes A_{0t})\}<1$ and, for all $\theta\in\Theta$, $\sum_{j=1}^{p}\beta_j<1$.

A11 $\varphi_0\in\overset{\circ}{\Phi}$, where $\overset{\circ}{\Phi}$ denotes the interior of Φ.

A12 There exists no set Λ of cardinality 2 such that $\mathbb{P}(\eta_t\in\Lambda)=1$.

Assumption A10 implies that $\kappa_\eta=E(\eta_t^4)<\infty$ and makes assumption A2 superfluous. The identifiability assumption A12 is slightly stronger than the first part of assumption A3 when the distribution of η_t is not symmetric. We are now in a position to state conditions ensuring the asymptotic normality of the QMLE of an ARMA–GARCH model.

Theorem 7.5 (Asymptotic normality of the QMLE) *Assume that $E\eta_t=0$ and that assumptions A3, A4, and A8–A12 hold true. Then*

$$\sqrt{n}(\hat{\varphi}_n-\varphi_0)\xrightarrow{\mathcal{L}}\mathcal{N}(0,\Sigma),$$

where $\Sigma=\mathcal{J}^{-1}\mathcal{I}\mathcal{J}^{-1}$,

$$\mathcal{I}=E_{\varphi_0}\left(\frac{\partial\ell_t(\varphi_0)}{\partial\varphi}\frac{\partial\ell_t(\varphi_0)}{\partial\varphi'}\right),\qquad \mathcal{J}=E_{\varphi_0}\left(\frac{\partial^2\ell_t(\varphi_0)}{\partial\varphi\partial\varphi'}\right).$$

If, in addition, the distribution of η_t is symmetric, we have

$$\mathcal{I}=\begin{pmatrix}I_1 & 0\\ 0 & I_2\end{pmatrix},\qquad \mathcal{J}=\begin{pmatrix}J_1 & 0\\ 0 & J_2\end{pmatrix},$$

with

$$I_1=(\kappa_\eta-1)E_{\varphi_0}\left(\frac{1}{\sigma_t^4}\frac{\partial\sigma_t^2}{\partial\vartheta}\frac{\partial\sigma_t^2}{\partial\vartheta'}(\varphi_0)\right)+4E_{\varphi_0}\left(\frac{1}{\sigma_t^2}\frac{\partial\epsilon_t}{\partial\vartheta}\frac{\partial\epsilon_t}{\partial\vartheta'}(\varphi_0)\right),$$

$$I_2=(\kappa_\eta-1)E_{\varphi_0}\left(\frac{1}{\sigma_t^4}\frac{\partial\sigma_t^2}{\partial\theta}\frac{\partial\sigma_t^2}{\partial\theta'}(\varphi_0)\right),$$

$$J_1=E_{\varphi_0}\left(\frac{1}{\sigma_t^4}\frac{\partial\sigma_t^2}{\partial\vartheta}\frac{\partial\sigma_t^2}{\partial\vartheta'}(\varphi_0)\right)+E_{\varphi_0}\left(\frac{2}{\sigma_t^2}\frac{\partial\epsilon_t}{\partial\vartheta}\frac{\partial\epsilon_t}{\partial\vartheta'}(\varphi_0)\right),$$

$$J_2=E_{\varphi_0}\left(\frac{1}{\sigma_t^4}\frac{\partial\sigma_t^2}{\partial\theta}\frac{\partial\sigma_t^2}{\partial\theta'}(\varphi_0)\right).$$

Remark 7.4

1. It is interesting to note that if η_t has a symmetric law, then the asymptotic variance Σ is block-diagonal, which is interpreted as an asymptotic independence between the estimators of the ARMA coefficients and those of the GARCH coefficients. The asymptotic distribution of the estimators of the ARMA coefficients depends, however, on the GARCH coefficients (in view of the form of the matrices I_1 and J_1 involving the derivatives of σ_t^2). On the other hand, still when the distribution of I and J is symmetric, the asymptotic accuracy of the estimation of the GARCH parameters is not affected by the ARMA part: the lower left block $J_2^{-1}I_2J_2^{-1}$ of Σ depends only on the GARCH coefficients. The block-diagonal form of Σ may also be of interest for testing problems of joint assumptions on the ARMA and GARCH parameters.

2. Assumption A11 imposes the strict positivity of the GARCH coefficients and it is easy to see that this assumption constrains only the GARCH coefficients. For any value of ϑ_0, the restriction of Φ to its first $P+Q+1$ coordinates can be chosen sufficiently large so that its interior contains ϑ_0 and assumption A8 is satisfied.

3. In the proof of the theorem, the symmetry of the iid process distribution is used to show the following result, which is of independent interest.

   If the distribution of η_t is symmetric then,

   $$\forall j, \quad E\{g(\epsilon_t^2, \epsilon_{t-1}^2, \ldots)\epsilon_{t-j}f\left(\epsilon_{t-j-1}, \epsilon_{t-j-2}, \ldots\right)\} = 0, \tag{7.24}$$

   provided this expectation exists (see Exercise 7.1).

Example 7.3 (Numerical evaluation of the asymptotic variance) Consider the AR(1)-ARCH(1) model defined by equation (7.23). In the case where η_t follows the $\mathcal{N}(0, 1)$ law, condition A10 for the existence of a moment of order 4 is written as $3\alpha_0^2 < 1$, that is, $\alpha_0 < 0.577$ (see (2.54)). In the case where η_t follows the $\chi^2(1)$ distribution, normalised in such a way that $E\eta_t = 0$ and $E\eta_t^2 = 1$, this condition is written as $15\alpha_0^2 < 1$, that is, $\alpha_0 < 0.258$. To simplify the computation, assume that $\omega_0 = 1$ is known. Table 7.3 provides a numerical evaluation of the asymptotic variance Σ, for these two distributions and for different values of the parameters a_0 and α_0. It is clear that the asymptotic variance of the two parameters strongly depends on the distribution of the iid process. These experiments confirm the independence of the asymptotic distributions of the AR and ARCH parameters in the case where the distribution of η_t is symmetric. They reveal that the independence does not hold when this assumption is relaxed. Note the strong impact of the ARCH coefficient on the asymptotic variance of the AR coefficient. On the other hand, the simulations confirm that in the case where the distribution is symmetric, the AR coefficient has no impact on the asymptotic accuracy of the ARCH coefficient. When the distribution is not symmetric, the impact, if there is any, is very weak. For the computation of the expectations involved in the matrix Σ, see Exercise 7.8. In particular, the values corresponding to $\alpha_0 = 0$ (AR(1) without ARCH effect) can be analytically computed. Note also that the results obtained for the asymptotic variance of the estimator of the ARCH coefficient in the case $a_0 = 0$ do not coincide with those of Table 7.2. This is not surprising because in this table ω_0 is not supposed to be known.

7.3 Application to Real Data

In this section, we employ the QML method to estimate GARCH(1, 1) models on daily returns of 11 stock market indices, namely the CAC, DAX, DJA, DJI, DJT, DJU, FTSE, Nasdaq, Nikkei, SMI, and S&P 500 indices. The observations cover the period from 2 January 1990 to 22 January 2009[1] (except for those indices for which the first observation is after 1990). The GARCH(1, 1) model has been chosen

[1] For the Nasdaq an outlier has been eliminated because the base price was reset on the trading day following 31 December 1993.

Table 7.3 Matrices Σ of asymptotic variance of the estimator of (a_0, α_0) for an AR(1)-ARCH(1), when $\omega_0 = 1$ is known and the distribution of η_t is $\mathcal{N}(0, 1)$ or normalised $\chi^2(1)$.

	$\alpha_0 = 0$	$\alpha_0 = 0.1$	$\alpha_0 = 0.25$	$x \in [0, 1]$
$a_0 = 0$				
$\eta_t \sim \mathcal{N}(0, 1)$	$\begin{pmatrix} 1.00 & 0.00 \\ 0.00 & 0.67 \end{pmatrix}$	$\begin{pmatrix} 1.14 & 0.00 \\ 0.00 & 1.15 \end{pmatrix}$	$\begin{pmatrix} 1.20 & 0.00 \\ 0.00 & 1.82 \end{pmatrix}$	$\begin{pmatrix} 1.08 & 0.00 \\ 0.00 & 2.99 \end{pmatrix}$
$\eta_t \sim \chi^2(1)$	$\begin{pmatrix} 1.00 & 0.54 \\ 0.54 & 0.94 \end{pmatrix}$	$\begin{pmatrix} 1.70 & -1.63 \\ -1.63 & 8.01 \end{pmatrix}$	$\begin{pmatrix} 2.78 & -1.51 \\ -1.51 & 18.78 \end{pmatrix}$	–
$a_0 = -0.5$				
$\eta_t \sim \mathcal{N}(0, 1)$	$\begin{pmatrix} 0.75 & 0.00 \\ 0.00 & 0.67 \end{pmatrix}$	$\begin{pmatrix} 0.82 & 0.00 \\ 0.00 & 1.15 \end{pmatrix}$	$\begin{pmatrix} 0.83 & 0.00 \\ 0.00 & 1.82 \end{pmatrix}$	$\begin{pmatrix} 0.72 & 0.00 \\ 0.00 & 2.99 \end{pmatrix}$
$\eta_t \sim \chi^2(1)$	$\begin{pmatrix} 0.75 & -0.40 \\ -0.40 & 0.94 \end{pmatrix}$	$\begin{pmatrix} 1.04 & -0.99 \\ -0.99 & 8.02 \end{pmatrix}$	$\begin{pmatrix} 1.41 & -0.78 \\ -0.78 & 18.85 \end{pmatrix}$	–
$a_0 = -0.9$				
$\eta_t \sim \mathcal{N}(0, 1)$	$\begin{pmatrix} 0.19 & 0.00 \\ 0.00 & 0.67 \end{pmatrix}$	$\begin{pmatrix} 0.19 & 0.00 \\ 0.00 & 1.15 \end{pmatrix}$	$\begin{pmatrix} 0.18 & 0.00 \\ 0.00 & 1.82 \end{pmatrix}$	$\begin{pmatrix} 0.13 & 0.00 \\ 0.00 & 2.98 \end{pmatrix}$
$\eta_t \sim \chi^2(1)$	$\begin{pmatrix} 0.19 & -0.10 \\ -0.10 & 0.94 \end{pmatrix}$	$\begin{pmatrix} 0.20 & -0.19 \\ -0.19 & 8.01 \end{pmatrix}$	$\begin{pmatrix} 0.21 & -0.12 \\ -0.12 & 18.90 \end{pmatrix}$	–

because it constitutes the reference model, by far the most commonly used in empirical studies. However, in Chapter 8, we will see that it can be worth considering models with higher orders p and q.

Table 7.4 displays the estimators of the parameters ω, α, β, together with their estimated standard deviations. The last column gives estimates of $\rho_4 = (\alpha + \beta)^2 + (E\eta_0^4 - 1)\alpha^2$, obtained by replacing the unknown parameters by their estimates and $E\eta_0^4$ by the empirical mean of the fourth-order moment of the standardised residuals. We have $E\epsilon_0^4 < \infty$ if and only if $\rho_4 < 1$. The estimates of the GARCH coefficients are quite homogenous over all the series and are similar to those usually obtained in empirical studies of daily returns. The coefficients α are close to 0.1, and the coefficients β are close to 0.9, which indicates a strong persistence of the shocks on the volatility. The sum $\alpha + \beta$ is greater than 0.98 for 10 of the 11 series, and greater than 0.96 for all the series. Since $\alpha + \beta < 1$, the assumption of second-order stationarity cannot be rejected, for any series (see Section 8.1). A fortiori, by Remark 2.6, the strict stationarity cannot be rejected. The existence of moments of order 4, $E\epsilon_t^4 < \infty$, is questionable for all the series because $(\hat{\alpha} + \hat{\beta})^2 + (\widehat{E\eta_0^4} - 1)\hat{\alpha}^2$ is extremely close to 1. Recall, however, that consistency and asymptotic normality of the QMLE do not require any moment on the observed process but do require strict stationarity.

7.4 Proofs of the Asymptotic Results*

We denote by $K > 0$ and $\rho \in [0, 1)$ generic constants whose values can vary from line to line. As an example, one can write for $0 < \rho_1 < 1$ and $0 < \rho_2 < 1$, $i_1 \geq 0$, $i_2 \geq 0$,

$$0 < K \sum_{i \geq i_1} \rho_1^i + K \sum_{i \geq i_2} i\rho_2^i \leq K\rho^{\min(i_1, i_2)}.$$

Table 7.4 GARCH(1, 1) models estimated by QML for 11 indices.

Index	ω	α	β	ρ_4
CAC	0.033 (0.009)	0.090 (0.014)	0.893 (0.015)	1.0067
DAX	0.037 (0.014)	0.093 (0.023)	0.888 (0.024)	1.0622
DJA	0.019 (0.005)	0.088 (0.014)	0.894 (0.014)	0.9981
DJI	0.017 (0.004)	0.085 (0.013)	0.901 (0.013)	1.0020
DJT	0.040 (0.013)	0.089 (0.016)	0.894 (0.018)	1.0183
DJU	0.021 (0.005)	0.118 (0.016)	0.865 (0.014)	1.0152
FTSE	0.013 (0.004)	0.091 (0.014)	0.899 (0.014)	1.0228
Nasdaq	0.025 (0.006)	0.072 (0.009)	0.922 (0.009)	1.0021
Nikkei	0.053 (0.012)	0.100 (0.013)	0.880 (0.014)	0.9985
SMI	0.049 (0.014)	0.127 (0.028)	0.835 (0.029)	1.0672
S&P 500	0.014 (0.004)	0.084 (0.012)	0.905 (0.012)	1.0072

The estimated standard deviations are given in parentheses. $\rho_4 = (\hat{\alpha} + \hat{\beta})^2 + (\widehat{E\eta_0^4} - 1)\hat{\alpha}^2$.

Proof of Theorem 7.1 The proof is based on a vectorial autoregressive representation of order 1 of the vector $\underline{\sigma}_t^2 = (\sigma_t^2, \sigma_{t-1}^2, \dots, \sigma_{t-p+1}^2)$, analogous to that used for the study of stationarity. Assumption A2 allows us to write $\underline{\sigma}_t^2$ as a series depending on the infinite past of the variable ϵ_t^2. It can be shown that the initial values are not important asymptotically, using the fact that, under the strict stationarity assumption, ϵ_t^2 necessarily admits a moment order s, with $s > 0$. This property also allows us to show that the expectation of $\ell_t(\theta_0)$ is well defined in $\mathbb{R}$ and that $E_{\theta_0}(\ell_t(\theta)) - E_{\theta_0}(\ell_t(\theta_0)) \geq 0$, which guarantees that the limit criterion is minimised at the true value. The difficulty is that $E_{\theta_0}(\ell_t^+(\theta))$ can be equal to $+\infty$. Assumptions A3 and A4 are crucial to establishing the identifiability: the former assumption precludes the existence of a constant linear combination of $\epsilon_{t-j}^2, j \geq 0$. The assumption of absence of common root is also used. The ergodicity of $\ell_t(\theta)$ and a compactness argument conclude the proof.

It will be convenient to rewrite equation (7.10) in matrix form. We have

$$\underline{\sigma}_t^2 = \underline{c}_t + B\underline{\sigma}_{t-1}^2, \tag{7.25}$$

where

$$\underline{\sigma}_t^2 = \begin{pmatrix} \sigma_t^2 \\ \sigma_{t-1}^2 \\ \vdots \\ \sigma_{t-p+1}^2 \end{pmatrix}, \quad \underline{c}_t = \begin{pmatrix} \omega + \sum_{i=1}^{q} \alpha_i \epsilon_{t-i}^2 \\ 0 \\ \vdots \\ 0 \end{pmatrix}, \quad B = \begin{pmatrix} \beta_1 & \beta_2 & \cdots & \beta_p \\ 1 & 0 & \cdots & 0 \\ \vdots & & & \\ 0 & \cdots & 1 & 0 \end{pmatrix}. \tag{7.26}$$

We will establish the following intermediate results.

(a) $\lim_{n\to\infty} \sup_{\theta\in\Theta} |\mathbf{l}_n(\theta) - \tilde{\mathbf{l}}_n(\theta)| = 0$, a.s..

(b) $(\exists t \in \mathbb{Z}$ such that $\sigma_t^2(\theta) = \sigma_t^2(\theta_0) P_{\theta_0}$ a.s.$) \Longrightarrow \theta = \theta_0$.

(c) $E_{\theta_0}|\ell_t(\theta_0)| < \infty$, and if $\theta \neq \theta_0$, $E_{\theta_0}\ell_t(\theta) > E_{\theta_0}\ell_t(\theta_0)$.

(d) For any $\theta \neq \theta_0$, there exists a neighbourhood $V(\theta)$ such that

$$\liminf_{n\to\infty} \inf_{\theta^*\in V(\theta)} \tilde{\mathbf{l}}_n(\theta^*) > E_{\theta_0}\ell_1(\theta_0), \quad \text{a.s.}$$

(a) Asymptotic irrelevance of the initial values
In view of Corollary 2.2, the condition $\sum_{j=1}^{p} \beta_j < 1$ of assumption A2 implies that $\rho(B) < 1$. The compactness of Θ implies that

$$\sup_{\theta\in\Theta} \rho(B) < 1. \tag{7.27}$$

Iterating (7.25), we thus obtain

$$\underline{\sigma}_t^2 = \underline{c}_t + B\underline{c}_{t-1} + B^2\underline{c}_{t-2} + \cdots + B^{t-1}\underline{c}_1 + B^t\underline{\sigma}_0^2 = \sum_{k=0}^{\infty} B^k \underline{c}_{t-k}. \tag{7.28}$$

Let $\underline{\tilde{\sigma}}_t^2$ be the vector obtained by replacing σ_{t-i}^2 by $\tilde{\sigma}_{t-i}^2$ in $\underline{\sigma}_t^2$, and let $\underline{\tilde{c}}_t$ be the vector obtained by replacing $\epsilon_0^2, \dots, \epsilon_{1-q}^2$ by the initial values (7.6) or (7.7). We have

$$\underline{\tilde{\sigma}}_t^2 = \underline{c}_t + B\underline{c}_{t-1} + \cdots + B^{t-q-1}\underline{c}_{q+1} + B^{t-q}\underline{\tilde{c}}_q + \cdots + B^{t-1}\underline{\tilde{c}}_1 + B^t\underline{\tilde{\sigma}}_0^2. \tag{7.29}$$

From inequality (7.27), it follows that almost surely

$$\begin{aligned}\sup_{\theta\in\Theta} \|\underline{\sigma}_t^2 - \underline{\tilde{\sigma}}_t^2\| &= \sup_{\theta\in\Theta} \left\| \left\{ \sum_{k=1}^{q} B^{t-k}(\underline{c}_k - \underline{\tilde{c}}_k) + B^t(\underline{\sigma}_0^2 - \underline{\tilde{\sigma}}_0^2) \right\} \right\| \\ &\le K\rho^t, \quad \forall t.\end{aligned} \tag{7.30}$$

For $x > 0$ we have $\log x \le x - 1$. It follows that, for $x, y > 0$, $\left|\log \frac{x}{y}\right| \le \frac{|x-y|}{\min(x,y)}$. We thus have almost surely, using inequality (7.30),

$$\begin{aligned}\sup_{\theta\in\Theta} |\mathbf{l}_n(\theta) - \tilde{\mathbf{l}}_n(\theta)| &\le n^{-1} \sum_{t=1}^{n} \sup_{\theta\in\Theta} \left\{ \left| \frac{\tilde{\sigma}_t^2 - \sigma_t^2}{\tilde{\sigma}_t^2 \sigma_t^2} \right| \epsilon_t^2 + \left| \log\left(\frac{\sigma_t^2}{\tilde{\sigma}_t^2}\right) \right| \right\} \\ &\le \left\{ \sup_{\theta\in\Theta} \frac{1}{\omega^2} \right\} K n^{-1} \sum_{t=1}^{n} \rho^t \epsilon_t^2 + \left\{ \sup_{\theta\in\Theta} \frac{1}{\omega} \right\} K n^{-1} \sum_{t=1}^{n} \rho^t.\end{aligned} \tag{7.31}$$

The existence of a moment of order $s > 0$ for ϵ_t^2, deduced from assumption A1 and Corollary 2.3, allows us to show that $\rho^t \epsilon_t^2 \to 0$ a.s. (see Exercise 7.2). Using Cesàro's lemma, point (a) follows.

(b) Identifiability of the parameter
Assume that $\sigma_t^2(\theta) = \sigma_t^2(\theta_0)$, P_{θ_0} almost surely. By Corollary 2.2, the polynomial $\mathcal{B}_\theta(B)$ is invertible under assumption A2. Using equality (7.10), we obtain

$$\left\{ \frac{\mathcal{A}_\theta(B)}{\mathcal{B}_\theta(B)} - \frac{\mathcal{A}_{\theta_0}(B)}{\mathcal{B}_{\theta_0}(B)} \right\} \epsilon_t^2 = \frac{\omega_0}{\mathcal{B}_{\theta_0}(1)} - \frac{\omega}{\mathcal{B}_\theta(1)} \quad \text{a.s. } \forall t.$$

If the operator in B between braces were not null, then there would exist a constant linear combination of the ϵ_{t-j}^2, $j \ge 0$. Thus the linear innovation of the process (ϵ_t^2) would be equal to zero. Since the distribution of η_t^2 is non-degenerate, in view of assumption A3,

$$\epsilon_t^2 - E_{\theta_0}(\epsilon_t^2 \mid \epsilon_{t-1}^2, \dots) = \sigma_t^2(\theta_0)(\eta_t^2 - 1) \ne 0, \quad \text{with positive probability.}$$

We thus have

$$\frac{\mathcal{A}_\theta(z)}{\mathcal{B}_\theta(z)} = \frac{\mathcal{A}_{\theta_0}(z)}{\mathcal{B}_{\theta_0}(z)}, \quad \forall |z| \le 1 \text{ and } \frac{\omega}{\mathcal{B}_\theta(1)} = \frac{\omega_0}{\mathcal{B}_{\theta_0}(1)}. \tag{7.32}$$

Under assumption A4 (absence of common root), it follows that $\mathcal{A}_\theta(z) = \mathcal{A}_{\theta_0}(z)$, $\mathcal{B}_\theta(z) = \mathcal{B}_{\theta_0}(z)$ and $\omega = \omega_0$. We have thus shown (b).

(c) The limit criterion is minimised at the true value
The limit criterion may not be integrable at some points, but $E_{\theta_0}\mathbf{l}_n(\theta) = E_{\theta_0}\ell_t(\theta)$ is well defined in $\mathbb{R} \cup \{+\infty\}$ because, with the notation $x^- = \max(-x, 0)$ and $x^+ = \max(x, 0)$,

$$E_{\theta_0}\ell_t^-(\theta) \leq E_{\theta_0}\log^-\sigma_t^2 \leq \max\{0, -\log\omega\} < \infty.$$ [2]

It is, however, possible to have $E_{\theta_0}\ell_t(\theta) = \infty$ for some values of θ. This occurs, for instance when $\theta = (\omega, 0, \ldots, 0)$ and (ϵ_t) is an IGARCH such that $E_{\theta_0}\epsilon_t^2 = \infty$. We will see that this cannot occur at θ_0, meaning that the limit criterion is integrable at θ_0. To establish this result, we have to show that $E_{\theta_0}\ell_t^+(\theta_0) < \infty$. Using Jensen's inequality and, once again, the existence of a moment of order $s > 0$ for ϵ_t^2, we obtain

$$E_{\theta_0}\log^+\sigma_t^2(\theta_0) < \infty$$

because

$$E_{\theta_0}\log\sigma_t^2(\theta_0) = E_{\theta_0}\frac{1}{s}\log\{\sigma_t^2(\theta_0)\}^s \leq \frac{1}{s}\log E_{\theta_0}\{\sigma_t^2(\theta_0)\}^s < \infty.$$

Thus

$$E_{\theta_0}\ell_t(\theta_0) = E_{\theta_0}\left\{\frac{\sigma_t^2(\theta_0)\eta_t^2}{\sigma_t^2(\theta_0)} + \log\sigma_t^2(\theta_0)\right\} = 1 + E_{\theta_0}\log\sigma_t^2(\theta_0) < \infty.$$

Having already established that $E_{\theta_0}\ell_t^-(\theta_0) < \infty$, it follows that $E_{\theta_0}\ell_t(\theta_0)$ is well defined in $\mathbb{R}$. Since for all $x > 0$, $\log x \leq x - 1$ with equality if and only if $x = 1$, we have

$$\begin{aligned} E_{\theta_0}\ell_t(\theta) - E_{\theta_0}\ell_t(\theta_0) &= E_{\theta_0}\log\frac{\sigma_t^2(\theta)}{\sigma_t^2(\theta_0)} + E_{\theta_0}\frac{\sigma_t^2(\theta_0)\eta_t^2}{\sigma_t^2(\theta)} - E_{\theta_0}\eta_t^2 \\ &= E_{\theta_0}\log\frac{\sigma_t^2(\theta)}{\sigma_t^2(\theta_0)} + E_{\theta_0}\frac{\sigma_t^2(\theta_0)}{\sigma_t^2(\theta)} - 1 \\ &\geq E_{\theta_0}\left\{\log\frac{\sigma_t^2(\theta)}{\sigma_t^2(\theta_0)} + \log\frac{\sigma_t^2(\theta_0)}{\sigma_t^2(\theta)}\right\} = 0 \end{aligned} \tag{7.33}$$

with equality if and only if $\sigma_t^2(\theta_0)/\sigma_t^2(\theta) = 1$ P_{θ_0}-a.s., that is, in view of (b), if and only if $\theta = \theta_0$.[3]

(d) Compactness of Θ and ergodicity of $(\ell_t(\theta))$
For all $\theta \in \Theta$ and any positive integer k, let $V_k(\theta)$ be the open ball of centre θ and radius $1/k$. Because of (a), we have

$$\begin{aligned} \liminf_{n\to\infty} \inf_{\theta^*\in V_k(\theta)\cap\Theta} \tilde{\mathbf{l}}_n(\theta^*) &\geq \liminf_{n\to\infty} \inf_{\theta^*\in V_k(\theta)\cap\Theta} \mathbf{l}_n(\theta^*) - \limsup_{n\to\infty}\sup_{\theta\in\Theta}|\mathbf{l}_n(\theta) - \tilde{\mathbf{l}}_n(\theta)| \\ &\geq \liminf_{n\to\infty} n^{-1}\sum_{t=1}^n \inf_{\theta^*\in V_k(\theta)\cap\Theta} \ell_t(\theta^*). \end{aligned}$$

To obtain the convergence of this empirical mean, the standard ergodic theorem cannot be applied (see Theorem A.2) because we have seen that $\ell_t(\theta^*)$ is not necessarily integrable, except at θ_0. We thus use a modified version of this theorem, which allows for an ergodic and strictly stationary sequence of variables admitting an expectation in $\mathbb{R} \cup \{+\infty\}$ (see Exercise 7.3). This version of the ergodic theorem can be applied to $\{\ell_t(\theta^*)\}$, and thus to $\{\inf_{\theta^*\in V_k(\theta)\cap\Theta}\ell_t(\theta^*)\}$ (see Exercise 7.4), which allows us to conclude that

$$\liminf_{n\to\infty} n^{-1}\sum_{t=1}^n \inf_{\theta^*\in V_k(\theta)\cap\Theta} \ell_t(\theta^*) = E_{\theta_0}\inf_{\theta^*\in V_k(\theta)\cap\Theta}\ell_1(\theta^*).$$

[2] We use here the fact that $(f+g)^- \leq g^-$ for $f \geq 0$, and that if $f \leq g$ then $f^- \geq g^-$.

[3] To show (7.33) it can be assumed that $E_{\theta_0}|\log\sigma_t^2(\theta)| < \infty$ and that $E_{\theta_0}|\epsilon_t^2/\sigma_t^2(\theta)| < \infty$ (in order to use the linearity property of the expectation), otherwise $E_{\theta_0}\ell_t(\theta) = +\infty$ and the relation is trivially satisfied.

By Beppo Levi's theorem, $E_{\theta_0} \inf_{\theta^* \in V_k(\theta)\cap\Theta} \ell_1(\theta^*)$ increases to $E_{\theta_0}\ell_1(\theta)$ as $k \to \infty$. Given relation (7.33), we have shown (d).

The conclusion of the proof uses a compactness argument. First note that for any neighbourhood $V(\theta_0)$ of θ_0,

$$\limsup_{n\to\infty} \inf_{\theta^* \in V(\theta_0)} \tilde{\mathbf{l}}_n(\theta^*) \leq \lim_{n\to\infty} \tilde{\mathbf{l}}_n(\theta_0) = \lim_{n\to\infty} \mathbf{l}_n(\theta_0) = E_{\theta_0}\ell_1(\theta_0). \tag{7.34}$$

The compact set Θ is covered by the union of an arbitrary neighbourhood $V(\theta_0)$ of θ_0 and the set of the neighbourhoods $V(\theta)$ satisfying (d), $\theta \in \Theta \backslash V(\theta_0)$. Thus, there exists a finite subcover of Θ of the form $V(\theta_0), V(\theta_1), \ldots, V(\theta_k)$, where, for $i = 1, \ldots, k$, $V(\theta_i)$ satisfies (d). It follows that

$$\inf_{\theta\in\Theta} \tilde{\mathbf{l}}_n(\theta) = \min_{i=0,1,\ldots,k} \inf_{\theta\in\Theta\cap V(\theta_i)} \tilde{\mathbf{l}}_n(\theta).$$

The relations (d) and (7.34) show that, almost surely, $\hat{\theta}_n$ belongs to $V(\theta_0)$ for n large enough. Since this is true for any neighbourhood $V(\theta_0)$, the proof is complete. □

Proof of Theorem 7.2 The proof of this theorem is based on a standard Taylor expansion of criterion (7.8) at θ_0. Since $\hat{\theta}_n$ converges to θ_0, which lies in the interior of the parameter space by assumption A5, the derivative of the criterion is equal to zero at $\hat{\theta}_n$. We thus have

$$\begin{aligned} 0 &= n^{-1/2} \sum_{t=1}^{n} \frac{\partial}{\partial\theta} \tilde{\ell}_t(\hat{\theta}_n) \\ &= n^{-1/2} \sum_{t=1}^{n} \frac{\partial}{\partial\theta} \tilde{\ell}_t(\theta_0) + \left(\frac{1}{n} \sum_{t=1}^{n} \frac{\partial^2}{\partial\theta_i \partial\theta_j} \tilde{\ell}_t(\theta_{ij}^*) \right) \sqrt{n}(\hat{\theta}_n - \theta_0) \end{aligned} \tag{7.35}$$

where the θ_{ij}^* are between $\hat{\theta}_n$ and θ_0. It will be shown that

$$n^{-1/2} \sum_{t=1}^{n} \frac{\partial}{\partial\theta} \tilde{\ell}_t(\theta_0) \xrightarrow{\mathcal{L}} \mathcal{N}(0, (\kappa_\eta - 1)J), \tag{7.36}$$

and that

$$n^{-1} \sum_{t=1}^{n} \frac{\partial^2}{\partial\theta_i \partial\theta_j} \tilde{\ell}_t(\theta_{ij}^*) \to J(i,j) \text{ in probability.} \tag{7.37}$$

The proof of the theorem immediately follows. We will split the proof of convergences (7.36) and (7.37) into several parts:

(a) $E_{\theta_0} \left\| \frac{\partial \ell_t(\theta_0)}{\partial\theta} \frac{\partial \ell_t(\theta_0)}{\partial\theta'} \right\| < \infty$, $E_{\theta_0} \left\| \frac{\partial^2 \ell_t(\theta_0)}{\partial\theta\,\partial\theta'} \right\| < \infty$.

(b) J is invertible and $\mathrm{Var}_{\theta_0} \left\{ \frac{\partial \ell_t(\theta_0)}{\partial\theta} \right\} = \{\kappa_\eta - 1\}J$.

(c) There exists a neighbourhood $\mathcal{V}(\theta_0)$ of θ_0 such that, for all $i, j, k \in \{1, \ldots, p+q+1\}$,

$$E_{\theta_0} \sup_{\theta\in\mathcal{V}(\theta_0)} \left| \frac{\partial^3 \ell_t(\theta)}{\partial\theta_i \partial\theta_j \partial\theta_k} \right| < \infty.$$

(d) $\left\| n^{-1/2} \sum_{t=1}^{n} \left\{ \frac{\partial \ell_t(\theta_0)}{\partial\theta} - \frac{\partial \tilde{\ell}_t(\theta_0)}{\partial\theta} \right\} \right\|$ and $\sup_{\theta\in\mathcal{V}(\theta_0)} \left\| n^{-1} \sum_{t=1}^{n} \left\{ \frac{\partial^2 \ell_t(\theta)}{\partial\theta\,\partial\theta'} - \frac{\partial^2 \tilde{\ell}_t(\theta)}{\partial\theta\,\partial\theta'} \right\} \right\|$ tend in probability to 0 as $n \to \infty$.

(e) $n^{-1/2} \sum_{t=1}^{n} \frac{\partial}{\partial\theta} \ell_t(\theta_0) \xrightarrow{\mathcal{L}} \mathcal{N}(0, (\kappa_\eta - 1)J)$.

(f) $n^{-1} \sum_{t=1}^{n} \frac{\partial^2}{\partial\theta_i \partial\theta_j} \ell_t(\theta_{ij}^*) \to J(i,j)$ a.s.

(a) Integrability of the derivatives of the criterion at θ_0.
Since $\ell_t(\theta) = \epsilon_t^2/\sigma_t^2 + \log \sigma_t^2$, we have

$$\frac{\partial \ell_t(\theta)}{\partial \theta} = \left\{1 - \frac{\epsilon_t^2}{\sigma_t^2}\right\}\left\{\frac{1}{\sigma_t^2}\frac{\partial \sigma_t^2}{\partial \theta}\right\}, \tag{7.38}$$

$$\frac{\partial^2 \ell_t(\theta)}{\partial \theta\, \partial \theta'} = \left\{1 - \frac{\epsilon_t^2}{\sigma_t^2}\right\}\left\{\frac{1}{\sigma_t^2}\frac{\partial^2 \sigma_t^2}{\partial \theta\, \partial \theta'}\right\} + \left\{2\frac{\epsilon_t^2}{\sigma_t^2} - 1\right\}\left\{\frac{1}{\sigma_t^2}\frac{\partial \sigma_t^2}{\partial \theta}\right\}\left\{\frac{1}{\sigma_t^2}\frac{\partial \sigma_t^2}{\partial \theta'}\right\}. \tag{7.39}$$

At $\theta = \theta_0$, the variable $\epsilon_t^2/\sigma_t^2 = \eta_t^2$ is independent of σ_t^2 and its derivatives. To show (a), it thus suffices to show that

$$E_{\theta_0}\left\|\frac{1}{\sigma_t^2}\frac{\partial \sigma_t^2}{\partial \theta}(\theta_0)\right\| < \infty, \quad E_{\theta_0}\left\|\frac{1}{\sigma_t^2}\frac{\partial^2 \sigma_t^2}{\partial \theta\, \partial \theta'}(\theta_0)\right\| < \infty, \quad E_{\theta_0}\left\|\frac{1}{\sigma_t^4}\frac{\partial \sigma_t^2}{\partial \theta}\frac{\partial \sigma_t^2}{\partial \theta'}(\theta_0)\right\| < \infty. \tag{7.40}$$

In view of relation (7.28), we have

$$\frac{\partial \underline{\sigma}_t^2}{\partial \omega} = \sum_{k=0}^{\infty} B^k \underline{1}, \qquad \frac{\partial \underline{\sigma}_t^2}{\partial \alpha_i} = \sum_{k=0}^{\infty} B^k \underline{\epsilon}_{t-k-i}^2, \tag{7.41}$$

$$\frac{\partial \underline{\sigma}_t^2}{\partial \beta_j} = \sum_{k=1}^{\infty}\left\{\sum_{i=1}^{k} B^{i-1} B^{(j)} B^{k-i}\right\} \underline{c}_{t-k}, \tag{7.42}$$

where $\underline{1} = (1, 0, \dots, 0)'$, $\underline{\epsilon}_t^2 = (\epsilon_t^2, 0, \dots, 0)'$, and $B^{(j)}$ is a $p \times p$ matrix with 1 in position $(1, j)$ and zeros elsewhere. Note that, in view of the positivity of the coefficients and relations (7.41) and (7.42), the derivatives of σ_t^2 are positive or null. In view of relations (7.41), it is clear that $\partial \sigma_t^2/\partial \omega$ is bounded. Since $\sigma_t^2 \geq \omega > 0$, the variable $\{\partial \sigma_t^2/\partial \omega\}/\sigma_t^2$ is also bounded. This variable thus possesses moments of all orders. In view of the second equality in relations (7.41) and of the positivity of all the terms involved in the sums, we have

$$\alpha_i \frac{\partial \underline{\sigma}_t^2}{\partial \alpha_i} = \sum_{k=0}^{\infty} B^k \alpha_i \underline{\epsilon}_{t-k-i}^2 \leq \sum_{k=0}^{\infty} B^k \underline{c}_{t-k} = \underline{\sigma}_t^2.$$

It follows that

$$\frac{1}{\sigma_t^2}\frac{\partial \sigma_t^2}{\partial \alpha_i} \leq \frac{1}{\alpha_i}. \tag{7.43}$$

The variable $\sigma_t^{-2}(\partial \sigma_t^2/\partial \alpha_i)$ thus admits moments of all orders at $\theta = \theta_0$. In view of (7.42) and $\beta_j B^{(j)} \leq B$, we have

$$\beta_j \frac{\partial \underline{\sigma}_t^2}{\partial \beta_j} \leq \sum_{k=1}^{\infty}\left\{\sum_{i=1}^{k} B^{i-1} B\, B^{k-i}\right\} \underline{c}_{t-k} = \sum_{k=1}^{\infty} k B^k \underline{c}_{t-k}. \tag{7.44}$$

Using inequality (7.27), we have $\|B^k\| \leq K\rho^k$ for all k. Moreover, ϵ_t^2 having a moment of order $s \in (0, 1)$, the variable $\underline{c}_t(1) = \omega + \sum_{i=1}^q \alpha_i \epsilon_{t-i}^2$ has the same moment.[4] Using in addition (7.44), the inequality $\sigma_t^2 \geq \omega + B^k(1, 1)\underline{c}_{t-k}(1)$ and the relation $x/(1 + x) \leq x^s$ for all $x \geq 0$,[5] we obtain

$$E_{\theta_0}\frac{1}{\sigma_t^2}\frac{\partial \sigma_t^2}{\partial \beta_j} \leq E_{\theta_0}\frac{1}{\beta_j}\sum_{k=1}^{\infty}\frac{kB^k(1, 1)\underline{c}_{t-k}(1)}{\omega + B^k(1, 1)\underline{c}_{t-k}(1)}$$

---

[4] We use the inequality $(a + b)^s \leq a^s + b^s$ for all $a, b \geq 0$ and any $s \in (0, 1]$. Indeed, $x^s \geq x$ for all $x \in [0, 1]$, and if $a + b > 0$, $\left(\frac{a}{a+b}\right)^s + \left(\frac{b}{a+b}\right)^s \geq \frac{a}{a+b} + \frac{b}{a+b} = 1$.

[5] If $x \geq 1$ then $x^s \geq 1 \geq x/(1 + x)$. If $0 \leq x \leq 1$ then $x^s \geq x \geq x/(1 + x)$.

$$\leq \frac{1}{\beta_j} \sum_{k=1}^{\infty} k E_{\theta_0} \left\{ \frac{B^k(1,1)\underline{c}_{t-k}(1)}{\omega} \right\}^s$$

$$\leq \frac{K^s}{\omega^s \beta_j} E_{\theta_0} \{\underline{c}_{t-k}(1)\}^s \sum_{k=1}^{\infty} k\rho^{sk} \leq \frac{K}{\beta_j}. \tag{7.45}$$

Under assumption A5 we have $\beta_{0j} > 0$ for all j, which entails that the first expectation in (7.40) exists.

We now turn to the higher-order derivatives of σ_t^2. In view of the first equality of relations (7.41), we have

$$\frac{\partial^2 \underline{\sigma}_t^2}{\partial \omega^2} = \frac{\partial^2 \underline{\sigma}_t^2}{\partial \omega \partial \alpha_i} = 0 \quad \text{and} \quad \frac{\partial^2 \underline{\sigma}_t^2}{\partial \omega \partial \beta_j} = \sum_{k=1}^{\infty} \left\{ \sum_{i=1}^{k} B^{i-1} B^{(j)} B^{k-i} \right\} \underline{1}. \tag{7.46}$$

We thus have

$$\beta_i \frac{\partial^2 \underline{\sigma}_t^2}{\partial \omega \partial \beta_j} \leq \sum_{k=1}^{\infty} k B^k \underline{1},$$

which is a vector of finite constants (since $\rho(B) < 1$). It follows that $\partial^2 \sigma_t^2(\theta_0)/\partial\omega\partial\theta_i$ is bounded, and thus admits moments of all orders. It is of course the same for $\{\partial^2 \sigma_t^2(\theta_0)/\partial\omega\partial\theta_i\}/\sigma_t^2(\theta_0)$. The second equality of (7.41) gives

$$\frac{\partial^2 \underline{\sigma}_t^2}{\partial \alpha_i \partial \alpha_j} = 0 \quad \text{and} \quad \frac{\partial^2 \underline{\sigma}_t^2}{\partial \alpha_i \partial \beta_j} = \sum_{k=1}^{\infty} \left\{ \sum_{i=1}^{k} B^{i-1} B^{(j)} B^{k-i} \right\} \underline{\epsilon}_{t-k-i}^2. \tag{7.47}$$

The arguments used to show the inequality (7.45) then show that

$$E_{\theta_0} \frac{\partial^2 \sigma_t^2 / \partial \alpha_i \partial \beta_j}{\sigma_t^2} \leq \frac{K}{\beta_j}.$$

This entails that $\{\partial^2 \sigma_t^2(\theta_0)/\partial\alpha_i\partial\theta\}/\sigma_t^2(\theta_0)$ is integrable. Differentiating relation (7.42) with respect to $\beta_{j'}$, we obtain

$$\begin{aligned}
\beta_j \beta_{j'} \frac{\partial^2 \underline{\sigma}_t^2}{\partial \beta_j \partial \beta_{j'}} &= \beta_j \beta_{j'} \sum_{k=2}^{\infty} \left[\sum_{i=2}^{k} \left\{ \left(\sum_{\ell=1}^{i-1} B^{\ell-1} B^{(j')} B^{i-1-\ell} \right) B^{(j)} B^{k-i} \right\} \right. \\
&\quad \left. + \sum_{i=1}^{k-1} \left\{ B^{i-1} B^{(j)} \left(\sum_{\ell=1}^{k-i} B^{\ell-1} B^{(j')} B^{k-i-\ell} \right) \right\} \right] \underline{c}_{t-k} \\
&\leq \sum_{k=2}^{\infty} \left[\sum_{i=2}^{k} (i-1) B^k + \sum_{i=1}^{k-1} (k-i) B^k \right] \underline{c}_{t-k} \\
&= \sum_{k=2}^{\infty} k(k-1) B^k \underline{c}_{t-k}
\end{aligned} \tag{7.48}$$

because $\beta_j B^{(j)} \leq B$. As for (7.45), it follows that

$$E_{\theta_0} \frac{\partial^2 \sigma_t^2 / \partial \beta_j \partial \beta_j'}{\sigma_t^2} \leq \frac{K}{\beta_j \beta_{j'}}$$

and the existence of the second expectation in (7.40) is proven.

Since $\{\partial\sigma_t^2/\partial\omega\}/\sigma_t^2$ is bounded, and since by inequality (7.43) the variables $\{\partial\sigma_t^2/\partial\alpha_i\}/\sigma_t^2$ are bounded at θ_0, it is clear that

$$E_{\theta_0}\left\|\frac{1}{\sigma_t^4(\theta_0)}\frac{\partial\sigma_t^2(\theta_0)}{\partial\theta_i}\frac{\partial\sigma_t^2(\theta_0)}{\partial\theta}\right\| < \infty,$$

for $i = 1, \ldots, q+1$. With the notation and arguments already used to show (7.45), and using the elementary inequality $x/(1+x) \le x^{s/2}$ for all $x \ge 0$, Minkowski's inequality implies that

$$\left\{E_{\theta_0}\left(\frac{1}{\sigma_t^2(\theta_0)}\frac{\partial\sigma_t^2(\theta_0)}{\partial\beta_j}\right)^2\right\}^{1/2} \le \frac{1}{\beta_{0j}}\sum_{k=1}^{\infty} k\left\{E_{\theta_0}\left(\frac{B^k(1,1)\underline{c}_{t-k}(1)}{\omega_0}\right)^s\right\}^{1/2} < \infty.$$

Finally, the Cauchy–Schwarz inequality entails that the third expectation of (7.40) exists.

(b) Invertibility of J and connection with the variance of the criterion derivative. Using (a), and once again, the independence between $\eta_t^2 = \epsilon_t^2/\sigma_t^2(\theta_0)$ and σ_t^2 and its derivatives, we have by relation (7.38),

$$E_{\theta_0}\left\{\frac{\partial\ell_t(\theta_0)}{\partial\theta}\right\} = E_{\theta_0}(1-\eta_t^2)E_{\theta_0}\left\{\frac{1}{\sigma_t^2(\theta_0)}\frac{\partial\sigma_t^2(\theta_0)}{\partial\theta}\right\} = 0.$$

Moreover, in view of the moment conditions (7.40), J exists and satisfies relation (7.13). We also have

$$\begin{aligned}\mathrm{Var}_{\theta_0}\left\{\frac{\partial\ell_t(\theta_0)}{\partial\theta}\right\} &= E_{\theta_0}\left\{\frac{\partial\ell_t(\theta_0)}{\partial\theta}\frac{\partial\ell_t(\theta_0)}{\partial\theta'}\right\}\\ &= E\{(1-\eta_t^2)^2\}E_{\theta_0}\left\{\frac{\partial\sigma_t^2(\theta_0)/\partial\theta}{\sigma_t^2(\theta_0)}\frac{\partial\sigma_t^2(\theta_0)/\partial\theta'}{\sigma_t^2(\theta_0)}\right\}\\ &= \{\kappa_\eta - 1\}J. \end{aligned} \quad (7.49)$$

Assume now that J is singular. Then there exists a non-zero vector λ in $\mathbb{R}^{p+q+1}$ such that $\lambda'\{\partial\sigma_t^2(\theta_0)/\partial\theta\} = 0$ a.s.[6] In view of relation (7.10) and the stationarity of $\{\partial\sigma_t^2(\theta_0)/\partial\theta\}_t$, we have

$$0 = \lambda'\frac{\partial\sigma_t^2(\theta_0)}{\partial\theta} = \lambda'\begin{pmatrix}1\\ \epsilon_{t-1}^2\\ \vdots\\ \epsilon_{t-q}^2\\ \sigma_{t-1}^2(\theta_0)\\ \vdots\\ \sigma_{t-p}^2(\theta_0)\end{pmatrix} + \sum_{j=1}^{p}\beta_j\lambda'\frac{\partial\sigma_{t-j}^2(\theta_0)}{\partial\theta} = \lambda'\begin{pmatrix}1\\ \epsilon_{t-1}^2\\ \vdots\\ \epsilon_{t-q}^2\\ \sigma_{t-1}^2(\theta_0)\\ \vdots\\ \sigma_{t-p}^2(\theta_0)\end{pmatrix}.$$

Let $\lambda = (\lambda_0, \lambda_1, \ldots, \lambda_{q+p})'$. It is clear that $\lambda_1 = 0$, otherwise ϵ_{t-1}^2 would be measurable with respect to the σ-field generated by $\{\eta_u, u < t-1\}$. For the same reason, we have $\lambda_2 = \cdots = \lambda_{2+i} = 0$ if $\lambda_{q+1} = \cdots = \lambda_{q+i} = 0$. Consequently, $\lambda \neq 0$ implies the existence of a GARCH$(p-1, q-1)$ representation. By the arguments used to show relations (7.32), assumption A4 entails that this is impossible. It follows that $\lambda' J\lambda = 0$ implies $\lambda = 0$, which completes the proof of (b).

[6] We have

$$\lambda' J\lambda = E\left[\frac{1}{\sigma_t^4(\theta_0)}\left(\lambda'\frac{\partial\sigma_t^2(\theta_0)}{\partial\theta}\right)^2\right] = 0$$

if and only if $\sigma_t^{-2}(\theta_0)(\lambda'\partial\sigma_t^2(\theta_0)/\partial\theta)^2 = 0$ a.s., that is, if and only if $(\lambda'\partial\sigma_t^2(\theta_0)/\partial\theta)^2 = 0$ a.s.

(c) Uniform integrability of the third-order derivatives of the criterion. Differentiating relation (7.39), we obtain

$$\begin{aligned}
\frac{\partial^3 \ell_t(\theta)}{\partial\theta_i\partial\theta_j\partial\theta_k} &= \left\{1-\frac{\epsilon_t^2}{\sigma_t^2}\right\}\left\{\frac{1}{\sigma_t^2}\frac{\partial^3\sigma_t^2}{\partial\theta_i\partial\theta_j\partial\theta_k}\right\} \\
&+\left\{2\frac{\epsilon_t^2}{\sigma_t^2}-1\right\}\left\{\frac{1}{\sigma_t^2}\frac{\partial\sigma_t^2}{\partial\theta_i}\right\}\left\{\frac{1}{\sigma_t^2}\frac{\partial^2\sigma_t^2}{\partial\theta_j\partial\theta_k}\right\} \\
&+\left\{2\frac{\epsilon_t^2}{\sigma_t^2}-1\right\}\left\{\frac{1}{\sigma_t^2}\frac{\partial\sigma_t^2}{\partial\theta_j}\right\}\left\{\frac{1}{\sigma_t^2}\frac{\partial^2\sigma_t^2}{\partial\theta_i\partial\theta_k}\right\} \\
&+\left\{2\frac{\epsilon_t^2}{\sigma_t^2}-1\right\}\left\{\frac{1}{\sigma_t^2}\frac{\partial\sigma_t^2}{\partial\theta_k}\right\}\left\{\frac{1}{\sigma_t^2}\frac{\partial^2\sigma_t^2}{\partial\theta_i\partial\theta_j}\right\} \\
&+\left\{2-6\frac{\epsilon_t^2}{\sigma_t^2}\right\}\left\{\frac{1}{\sigma_t^2}\frac{\partial\sigma_t^2}{\partial\theta_i}\right\}\left\{\frac{1}{\sigma_t^2}\frac{\partial\sigma_t^2}{\partial\theta_j}\right\}\left\{\frac{1}{\sigma_t^2}\frac{\partial\sigma_t^2}{\partial\theta_k}\right\}. \qquad (7.50)
\end{aligned}$$

We begin by studying the integrability of $\{1-\epsilon_t^2/\sigma_t^2\}$. This is the most difficult term to deal with. Indeed, the variable ϵ_t^2/σ_t^2 is not uniformly integrable on Θ: at $\theta=(\omega, 0')$, the ratio ϵ_t^2/σ_t^2 is integrable only if $E\epsilon_t^2$ exists. We will, however, show the integrability of $\{1-\epsilon_t^2/\sigma_t^2\}$ uniformly in θ in the neighbourhood of θ_0. Let Θ^* be a compact set which contains θ_0 and which is contained in the interior of Θ ($\forall\theta\in\Theta^*$, we have $\theta\geq\theta_*>0$ component by component). Let B_0 be the matrix B (defined in (7.26)) evaluated at the point $\theta=\theta_0$. For all $\delta>0$, there exists a neighbourhood $\mathcal{V}(\theta_0)$ of θ_0, included in Θ^*, such that for all $\theta\in\mathcal{V}(\theta_0)$,

$$B_0 \leq (1+\delta)B \quad (\text{i.e. } B_0(i,j)\leq(1+\delta)B(i,j) \text{ for all } i \text{ and } j).$$

Note that, since $\mathcal{V}(\theta_0)\subset\Theta^*$, we have $\sup_{\theta\in\mathcal{V}(\theta_0)} 1/\alpha_i<\infty$. From relation (7.28), we obtain

$$\sigma_t^2 = \omega\sum_{k=0}^{\infty}B^k(1,1)+\sum_{i=1}^{q}\alpha_i\left\{\sum_{k=0}^{\infty}B^k(1,1)\epsilon_{t-k-i}^2\right\}$$

and, again using $x/(1+x)\leq x^s$ for all $x\geq 0$ and all $s\in(0,1)$,

$$\begin{aligned}
\sup_{\theta\in\mathcal{V}(\theta_0)}\frac{\sigma_t^2(\theta_0)}{\sigma_t^2} &\leq \sup_{\theta\in\mathcal{V}(\theta_0)}\left\{\frac{\omega_0\sum_{k=0}^{\infty}B_0^k(1,1)}{\omega}+\sum_{i=1}^{q}\alpha_{0i}\left(\sum_{k=0}^{\infty}\frac{B_0^k(1,1)\epsilon_{t-k-i}^2}{\omega+\alpha_iB^k(1,1)\epsilon_{t-k-i}^2}\right)\right\} \\
&\leq K+\sum_{i=1}^{q}\sup_{\theta\in\mathcal{V}(\theta_0)}\left\{\frac{\alpha_{0i}}{\alpha_i}\sum_{k=0}^{\infty}\frac{B_0^k(1,1)}{B^k(1,1)}\left(\frac{\alpha_iB^k(1,1)\epsilon_{t-k-i}^2}{\omega}\right)^s\right\} \\
&\leq K+K\sum_{i=1}^{q}\sum_{k=0}^{\infty}(1+\delta)^k\rho^{ks}\epsilon_{t-k-i}^{2s}. \qquad (7.51)
\end{aligned}$$

If s is chosen such that $E\epsilon_t^{2s}<\infty$ and, for instance, $\delta=(1-\rho^s)/(2\rho^s)$, then the expectation of the previous series is finite. It follows that there exists a neighbourhood $\mathcal{V}(\theta_0)$ of θ_0 such that

$$E_{\theta_0}\sup_{\theta\in\mathcal{V}(\theta_0)}\frac{\epsilon_t^2}{\sigma_t^2}=E_{\theta_0}\sup_{\theta\in\mathcal{V}(\theta_0)}\frac{\sigma_t^2(\theta_0)}{\sigma_t^2}<\infty.$$

Using inequality (7.51), keeping the same choice of δ but taking s such that $E\epsilon_t^{4s} < \infty$, the triangle inequality gives

$$\left\| \sup_{\theta \in \mathcal{V}(\theta_0)} \frac{\epsilon_t^2}{\sigma_t^2} \right\|_2 = \kappa_\eta^{1/2} \left\| \sup_{\theta \in \mathcal{V}(\theta_0)} \frac{\sigma_t^2(\theta_0)}{\sigma_t^2} \right\|_2$$

$$\leq \kappa_\eta^{1/2} K + \kappa_\eta^{1/2} K q \sum_{k=0}^{\infty} (1+\delta)^k \rho^{ks} \| \epsilon_t^{2s} \|_2 < \infty. \tag{7.52}$$

Now consider the second term in braces in the right-hand side of equality (7.50). Differentiating relations (7.46), (7.47), and (7.48), with the arguments used to show inequality (7.43), we obtain

$$\sup_{\theta \in \Theta^*} \frac{1}{\sigma_t^2} \frac{\partial^3 \sigma_t^2}{\partial \theta_{i_1} \partial \theta_{i_2} \partial \theta_{i_3}} \leq K,$$

when the indices i_1, i_2 and i_3 are not all in $\{q+1, q+2, \ldots, q+1+p\}$ (that is, when the derivative is taken with respect to at least one parameter different from the β_j). Using again the arguments used to show relations (7.44) and (7.48), and then inequality (7.45), we obtain

$$\beta_i \beta_j \beta_k \frac{\partial^3 \sigma_t^2}{\partial \beta_i \partial \beta_j \partial \beta_k} \leq \sum_{k=3}^{\infty} k(k-1)(k-2) B^k(1,1) \underline{c}_{t-k}(1),$$

$$\sup_{\theta \in \Theta^*} \frac{1}{\sigma_t^2} \frac{\partial^3 \sigma_t^2}{\partial \beta_i \partial \beta_j \partial \beta_k} \leq K \left\{ \sup_{\theta \in \Theta^*} \frac{1}{\omega^s \beta_i \beta_j \beta_k} \right\} \sum_{k=3}^{\infty} k(k-1)(k-2) \rho^{ks} \left\{ \sup_{\theta \in \Theta^*} \underline{c}_{t-k}(1) \right\}^s$$

for any $s \in (0, 1)$. Since $E_{\theta_0} \{\sup_{\theta \in \Theta^*} \underline{c}_{t-k}(1)\}^{2s} < \infty$ for some $s > 0$, it follows that

$$E_{\theta_0} \sup_{\theta \in \Theta^*} \left| \frac{1}{\sigma_t^2} \frac{\partial^3 \sigma_t^2}{\partial \theta_i \partial \theta_j \partial \theta_k} \right|^2 < \infty. \tag{7.53}$$

It is easy to see that in this inequality the power 2 can be replaced by any power d:

$$E_{\theta_0} \sup_{\theta \in \Theta^*} \left| \frac{1}{\sigma_t^2} \frac{\partial^3 \sigma_t^2}{\partial \theta_i \partial \theta_j \partial \theta_k} \right|^d < \infty.$$

Using the Cauchy–Schwarz inequality, and the moment conditions (7.52) and (7.53), we obtain

$$E_{\theta_0} \sup_{\theta \in \mathcal{V}(\theta_0)} \left| \left\{ 1 - \frac{\epsilon_t^2}{\sigma_t^2} \right\} \left\{ \frac{1}{\sigma_t^2} \frac{\partial^3 \sigma_t^2}{\partial \theta_i \partial \theta_j \partial \theta_k} \right\} \right| < \infty.$$

The other terms in braces in the right-hand side of equality (7.50) are handled similarly. We show in particular that

$$E_{\theta_0} \sup_{\theta \in \Theta^*} \left| \frac{1}{\sigma_t^2} \frac{\partial^2 \sigma_t^2}{\partial \theta_i \partial \theta_j} \right|^d < \infty, \quad E_{\theta_0} \sup_{\theta \in \Theta^*} \left| \frac{1}{\sigma_t^2} \frac{\partial \sigma_t^2}{\partial \theta_i} \right|^d < \infty, \tag{7.54}$$

for any integer d. With the aid of Hölder's inequality, this allows us to establish, in particular, that

$$E_{\theta_0} \sup_{\theta \in \mathcal{V}(\theta_0)} \left| \left\{ 2 - 6\frac{\epsilon_t^2}{\sigma_t^2} \right\} \left\{ \frac{1}{\sigma_t^2} \frac{\partial \sigma_t^2}{\partial \theta_i} \right\} \left\{ \frac{1}{\sigma_t^2} \frac{\partial \sigma_t^2}{\partial \theta_j} \right\} \left\{ \frac{1}{\sigma_t^2} \frac{\partial \sigma_t^2}{\partial \theta_k} \right\} \right|$$

$$\leq \left\| \sup_{\theta \in \mathcal{V}(\theta_0)} \left| 2 - 6\frac{\epsilon_t^2}{\sigma_t^2} \right| \right\|_2 \max_i \left\| \sup_{\theta \in \Theta^*} \left| \frac{1}{\sigma_t^2} \frac{\partial \sigma_t^2}{\partial \theta_i} \right| \right\|_6^3 < \infty.$$

Thus we obtain (c).

(d) Asymptotic decrease of the effect of the initial values.
Using relation (7.29), we obtain the analogs of (7.41) and (7.42) for the derivatives of $\underline{\tilde{\sigma}}_t^2$:

$$\frac{\partial \underline{\tilde{\sigma}}_t^2}{\partial \omega} = \sum_{k=0}^{t-1-q} B^k \underline{1} + \sum_{k=1}^{q} B^{t-k} \frac{\partial \underline{\tilde{c}}_k}{\partial \omega} + B^t \frac{\partial \underline{\tilde{\sigma}}_0^2}{\partial \omega}, \tag{7.55}$$

$$\frac{\partial \underline{\tilde{\sigma}}_t^2}{\partial \alpha_i} = \sum_{k=0}^{t-1-q} B^k \underline{\epsilon}_{t-k-i}^2 + \sum_{k=1}^{q} B^{t-k} \frac{\partial \underline{\tilde{c}}_k}{\partial \alpha_i} + B^t \frac{\partial \underline{\tilde{\sigma}}_0^2}{\partial \alpha_i}, \tag{7.56}$$

$$\frac{\partial \underline{\tilde{\sigma}}_t^2}{\partial \beta_j} = \sum_{k=1}^{t-1-q} \left\{ \sum_{i=1}^{k} B^{i-1} B^{(j)} B^{k-i} \right\} \underline{c}_{t-k} + \sum_{k=1}^{q} \left\{ \sum_{i=1}^{t-k} B^{i-1} B^{(j)} B^{t-k-i} \right\} \underline{\tilde{c}}_k, \tag{7.57}$$

where $\partial \underline{\tilde{\sigma}}_0^2 / \partial \omega$ is equal to $(0, \dots, 0)'$ when the initial conditions are given by (7.7), and is equal to $(1, \dots, 1)'$ when the initial conditions are given by (7.6). The second-order derivatives have similar expressions. The compactness of Θ and the fact that $\rho(B) < 1$ allow us to claim that, almost surely,

$$\sup_{\theta \in \Theta} \left\| \frac{\partial \sigma_t^2}{\partial \theta} - \frac{\partial \tilde{\sigma}_t^2}{\partial \theta} \right\| < K\rho^t, \quad \sup_{\theta \in \Theta} \left\| \frac{\partial^2 \sigma_t^2}{\partial \theta \partial \theta'} - \frac{\partial^2 \tilde{\sigma}_t^2}{\partial \theta \partial \theta'} \right\| < K\rho^t, \quad \forall t. \tag{7.58}$$

Using (7.30), we obtain

$$\left| \frac{1}{\sigma_t^2} - \frac{1}{\tilde{\sigma}_t^2} \right| = \left| \frac{\tilde{\sigma}_t^2 - \sigma_t^2}{\sigma_t^2 \tilde{\sigma}_t^2} \right| \leq \frac{K\rho^t}{\sigma_t^2}, \quad \frac{\sigma_t^2}{\tilde{\sigma}_t^2} \leq 1 + K\rho^t. \tag{7.59}$$

Since

$$\frac{\partial \tilde{\ell}_t(\theta)}{\partial \theta} = \left\{ 1 - \frac{\epsilon_t^2}{\tilde{\sigma}_t^2} \right\} \left\{ \frac{1}{\tilde{\sigma}_t^2} \frac{\partial \tilde{\sigma}_t^2}{\partial \theta} \right\} \quad \text{and} \quad \frac{\partial \ell_t(\theta)}{\partial \theta} = \left\{ 1 - \frac{\epsilon_t^2}{\sigma_t^2} \right\} \left\{ \frac{1}{\sigma_t^2} \frac{\partial \sigma_t^2}{\partial \theta} \right\},$$

we have, using inequalities (7.59) and the first inequality in (7.58),

$$\begin{aligned}
\left| \frac{\partial \ell_t(\theta_0)}{\partial \theta_i} - \frac{\partial \tilde{\ell}_t(\theta_0)}{\partial \theta_i} \right| &= \left| \left\{ \frac{\epsilon_t^2}{\tilde{\sigma}_t^2} - \frac{\epsilon_t^2}{\sigma_t^2} \right\} \left\{ \frac{1}{\sigma_t^2} \frac{\partial \sigma_t^2}{\partial \theta_i} \right\} + \left\{ 1 - \frac{\epsilon_t^2}{\tilde{\sigma}_t^2} \right\} \left\{ \frac{1}{\sigma_t^2} - \frac{1}{\tilde{\sigma}_t^2} \right\} \left\{ \frac{\partial \sigma_t^2}{\partial \theta_i} \right\} \right. \\
&\quad \left. + \left\{ 1 - \frac{\epsilon_t^2}{\tilde{\sigma}_t^2} \right\} \left\{ \frac{1}{\tilde{\sigma}_t^2} \right\} \left\{ \frac{\partial \sigma_t^2}{\partial \theta_i} - \frac{\partial \tilde{\sigma}_t^2}{\partial \theta_i} \right\} \right| (\theta_0) \\
&\leq K\rho^t (1 + \eta_t^2) \left| 1 + \left\{ \frac{1}{\sigma_t^2(\theta_0)} \frac{\partial \sigma_t^2(\theta_0)}{\partial \theta_i} \right\} \right|.
\end{aligned}$$

It follows that

$$\left| n^{-1/2} \sum_{t=1}^{n} \left\{ \frac{\partial \ell_t(\theta_0)}{\partial \theta_i} - \frac{\partial \tilde{\ell}_t(\theta_0)}{\partial \theta_i} \right\} \right| \leq K n^{-1/2} \sum_{t=1}^{n} \rho^t (1 + \eta_t^2) \left| 1 + \frac{1}{\sigma_t^2(\theta_0)} \frac{\partial \sigma_t^2(\theta_0)}{\partial \theta_i} \right|. \tag{7.60}$$

Markov's inequality, (7.40), and the independence between η_t and $\sigma_t^2(\theta_0)$ imply that, for all $\varepsilon > 0$,

$$\begin{aligned}
&\mathbb{P}\left(n^{-1/2} \sum_{t=1}^{n} \rho^t (1 + \eta_t^2) \left| 1 + \frac{1}{\sigma_t^2(\theta_0)} \frac{\partial \sigma_t^2(\theta_0)}{\partial \theta} \right| > \varepsilon \right) \\
&\quad \leq \frac{2}{\varepsilon} \left(1 + E_{\theta_0} \left| \frac{1}{\sigma_t^2(\theta_0)} \frac{\partial \sigma_t^2(\theta_0)}{\partial \theta} \right| \right) n^{-1/2} \sum_{t=1}^{n} \rho^t \to 0,
\end{aligned}$$

which, by inequality (7.60), shows the first part of (d).

Now consider the asymptotic impact of the initial values on the second-order derivatives of the criterion in a neighbourhood of θ_0. In view of (7.39) and the previous computations, we have

$$\begin{aligned}
&\sup_{\theta\in\mathcal{V}(\theta_0)}\left| n^{-1}\sum_{t=1}^{n}\left\{\frac{\partial^2\ell_t(\theta)}{\partial\theta_i\partial\theta_j}-\frac{\partial^2\tilde{\ell}_t(\theta)}{\partial\theta_i\partial\theta_j}\right\}\right| \\
&\leq n^{-1}\sum_{t=1}^{n}\sup_{\theta\in\mathcal{V}(\theta_0)}\left|\left\{\frac{\epsilon_t^2}{\tilde{\sigma}_t^2}-\frac{\epsilon_t^2}{\sigma_t^2}\right\}\left\{\frac{1}{\sigma_t^2}\frac{\partial^2\sigma_t^2}{\partial\theta_i\partial\theta_j}\right\}\right. \\
&\quad+\left\{1-\frac{\epsilon_t^2}{\tilde{\sigma}_t^2}\right\}\left\{\left(\frac{1}{\sigma_t^2}-\frac{1}{\tilde{\sigma}_t^2}\right)\frac{\partial^2\sigma_t^2}{\partial\theta_i\partial\theta_j}+\frac{1}{\tilde{\sigma}_t^2}\left(\frac{\partial^2\sigma_t^2}{\partial\theta_i\partial\theta_j}-\frac{\partial^2\tilde{\sigma}_t^2}{\partial\theta_i\partial\theta_j}\right)\right\} \\
&\quad+\left\{2\frac{\epsilon_t^2}{\sigma_t^2}-2\frac{\epsilon_t^2}{\tilde{\sigma}_t^2}\right\}\left\{\frac{1}{\sigma_t^2}\frac{\partial\sigma_t^2}{\partial\theta_i}\right\}\left\{\frac{1}{\sigma_t^2}\frac{\partial\sigma_t^2}{\partial\theta_j}\right\} \\
&\quad+\left\{2\frac{\epsilon_t^2}{\tilde{\sigma}_t^2}-1\right\}\left\{\left(\frac{1}{\sigma_t^2}-\frac{1}{\tilde{\sigma}_t^2}\right)\frac{\partial\sigma_t^2}{\partial\theta_i}+\frac{1}{\tilde{\sigma}_t^2}\left(\frac{\partial\sigma_t^2}{\partial\theta_i}-\frac{\partial\tilde{\sigma}_t^2}{\partial\theta_i}\right)\right\}\left\{\frac{1}{\sigma_t^2}\frac{\partial\sigma_t^2}{\partial\theta_j}\right\} \\
&\quad\left.+\left\{2\frac{\epsilon_t^2}{\tilde{\sigma}_t^2}-1\right\}\left\{\frac{1}{\tilde{\sigma}_t^2}\frac{\partial\tilde{\sigma}_t^2}{\partial\theta_i}\right\}\left\{\left(\frac{1}{\sigma_t^2}-\frac{1}{\tilde{\sigma}_t^2}\right)\frac{\partial\sigma_t^2}{\partial\theta_j}+\frac{1}{\tilde{\sigma}_t^2}\left(\frac{\partial\sigma_t^2}{\partial\theta_j}-\frac{\partial\tilde{\sigma}_t^2}{\partial\theta_j}\right)\right\}\right| \\
&\leq Kn^{-1}\sum_{t=1}^{n}\rho^t\Upsilon_t,
\end{aligned}$$

where

$$\Upsilon_t=\sup_{\theta\in\mathcal{V}(\theta_0)}\left\{1+\frac{\epsilon_t^2}{\sigma_t^2}\right\}\left\{1+\frac{1}{\sigma_t^2}\frac{\partial^2\sigma_t^2}{\partial\theta_i\partial\theta_j}+\frac{1}{\sigma_t^2}\frac{\partial\sigma_t^2}{\partial\theta_i}\frac{1}{\sigma_t^2}\frac{\partial\sigma_t^2}{\partial\theta_j}\right\}.$$

In view of (7.52), (7.54), and Hölder's inequality, it can be seen that, for a certain neighbourhood $\mathcal{V}(\theta_0)$, the expectation of Υ_t is a finite constant. Using Markov's inequality once again, the second convergence of (d) is then shown.

(e) CLT for martingale increments.
The conditional score vector is obviously centred, which can be seen from equality (7.38), using the fact that $\sigma_t^2(\theta_0)$ and its derivatives belong to the σ-field generated by $\{\epsilon_{t-i}, i\geq 0\}$, and the fact that $E_{\theta_0}(\epsilon_t^2\mid\epsilon_u, u<t)=\sigma_t^2(\theta_0)$:

$$E_{\theta_0}\left(\frac{\partial}{\partial\theta}\ell_t(\theta_0)\mid\epsilon_u,u<t\right)=\frac{1}{\sigma_t^4(\theta_0)}\left(\frac{\partial}{\partial\theta}\sigma_t^2(\theta_0)\right)E_{\theta_0}(\sigma_t^2(\theta_0)-\epsilon_t^2\mid\epsilon_u,u<t)=0.$$

Note also that by relation (7.49), $\mathrm{Var}_{\theta_0}(\partial\ell_t(\theta_0)/\partial\theta)$ is finite. In view of the invertibility of J and the assumptions on the distribution of η_t (which entail $0<\kappa_\eta-1<\infty$), this covariance matrix is non-degenerate. It follows that, for all $\lambda\in\mathbb{R}^{p+q+1}$, the sequence $\left\{\lambda'\frac{\partial}{\partial\theta}\ell_t(\theta_0),\epsilon_t\right\}_t$ is a square integrable ergodic stationary martingale difference. Corollary A.1 and the Cramér–Wold theorem (see, for example, Billingsley 1995, pp. 383, 476 and 360) entail (e).

(f) Use of a second Taylor expansion and of the ergodic theorem.
Consider the Taylor expansion (7.35) of the criterion at θ_0. We have, for all i and j,

$$n^{-1}\sum_{t=1}^{n}\frac{\partial^2}{\partial\theta_i\partial\theta_j}\ell_t(\theta_{ij}^*)=n^{-1}\sum_{t=1}^{n}\frac{\partial^2}{\partial\theta_i\partial\theta_j}\ell_t(\theta_0)+n^{-1}\sum_{t=1}^{n}\frac{\partial}{\partial\theta'}\left\{\frac{\partial^2}{\partial\theta_i\partial\theta_j}\ell_t(\tilde{\theta}_{ij})\right\}(\theta_{ij}^*-\theta_0),\quad(7.61)$$

where $\tilde{\theta}_{ij}$ is between θ^*_{ij} and θ_0. The almost sure convergence of $\tilde{\theta}_{ij}$ to θ_0, the ergodic theorem and (c) imply that almost surely

$$\limsup_{n\to\infty}\left\|n^{-1}\sum_{t=1}^{n}\frac{\partial}{\partial\theta'}\left\{\frac{\partial^2}{\partial\theta_i\partial\theta_j}\ell_t(\tilde{\theta}_{ij})\right\}\right\| \le \limsup_{n\to\infty} n^{-1}\sum_{t=1}^{n}\sup_{\theta\in\mathcal{V}(\theta_0)}\left\|\frac{\partial}{\partial\theta'}\left\{\frac{\partial^2}{\partial\theta_i\partial\theta_j}\ell_t(\theta)\right\}\right\|$$
$$= E_{\theta_0}\sup_{\theta\in\mathcal{V}(\theta_0)}\left\|\frac{\partial}{\partial\theta'}\left\{\frac{\partial^2}{\partial\theta_i\partial\theta_j}\ell_t(\theta)\right\}\right\| < \infty.$$

Since $\|\theta^*_{ij}-\theta_0\|\to 0$ almost surely, the second term on the right-hand side of equality (7.61) converges to 0 with probability 1. By the ergodic theorem, the first term on the right-hand side of equality (7.61) converges to $J(i,j)$. To complete the proof of Theorem 7.2, it suffices to apply Slutsky's lemma. In view of (d), (e) and (f) we obtain convergences (7.36) and (7.37).

Proof of the Results of Section 7.1.3

Proof of Lemma 7.1 We have

$$\rho^n h_n = \rho^n\omega_0\left\{1+\sum_{t=1}^{n-1}\alpha_0^t\eta_{n-1}^2\cdots\eta_{n-t}^2\right\}+\rho^n\alpha_0^n\eta_{n-1}^2\cdots\eta_1^2\epsilon_0^2$$
$$\ge \rho^n\omega_0\prod_{t=1}^{n-1}\alpha_0\eta_t^2.$$

Thus

$$\liminf_{n\to\infty}\frac{1}{n}\log\rho^n h_n \ge \lim_{n\to\infty}\frac{1}{n}\left\{\log\rho\omega_0+\sum_{t=1}^{n-1}\log\rho\alpha_0\eta_t^2\right\} = E\log\rho\alpha_0\eta_1^2 > 0,$$

using relation (7.18) for the latter inequality. It follows that $\log\rho^n h_n$, and thus $\rho^n h_n$, tend almost surely to $+\infty$ as $n\to\infty$. The first convergence is shown. The second one is established similarly by noting that $\log\eta_n^2/n\to 0$ a.s. when $n\to\infty$, because $E|\log\eta_1^2|<\infty$ (see Exercice 2.13). □

Proof of (7.19) Note that

$$(\hat{\omega}_n,\hat{\alpha}_n)=\arg\max_{\theta\in\Theta} Q_n(\theta),$$

where

$$Q_n(\theta)=\frac{1}{n}\sum_{t=1}^{n}\{\ell_t(\theta)-\ell_t(\theta_0)\}.$$

We have

$$Q_n(\theta)=\frac{1}{n}\sum_{t=1}^{n}\eta_t^2\left\{\frac{\sigma_t^2(\theta_0)}{\sigma_t^2(\theta)}-1\right\}+\log\frac{\sigma_t^2(\theta)}{\sigma_t^2(\theta_0)}$$
$$=\frac{1}{n}\sum_{t=1}^{n}\eta_t^2\frac{(\omega_0-\omega)+(\alpha_0-\alpha)\epsilon_{t-1}^2}{\omega+\alpha\epsilon_{t-1}^2}+\log\frac{\omega+\alpha\epsilon_{t-1}^2}{\omega_0+\alpha_0\epsilon_{t-1}^2}.$$

For all $\theta\in\Theta$, we have $\alpha\neq 0$. Letting

$$O_n(\alpha)=\frac{1}{n}\sum_{t=1}^{n}\eta_t^2\frac{(\alpha_0-\alpha)}{\alpha}+\log\frac{\alpha}{\alpha_0}$$

and

$$d_t = \frac{\alpha(\omega_0 - \omega) - \omega(\alpha_0 - \alpha)}{\alpha(\omega + \alpha\epsilon_{t-1}^2)},$$

we have

$$Q_n(\theta) - O_n(\alpha) = \frac{1}{n}\sum_{t=1}^{n} \eta_t^2 d_{t-1} + \frac{1}{n}\sum_{t=1}^{n} \log \frac{(\omega + \alpha\epsilon_{t-1}^2)\alpha_0}{(\omega_0 + \alpha_0\epsilon_{t-1}^2)\alpha} \to 0 \quad \text{a.s.}$$

since, by Lemma 7.1, $\epsilon_t^2 \to \infty$ almost surely as $t \to \infty$. It is easy to see that this convergence is uniform on the compact set Θ:

$$\lim_{n\to\infty} \sup_{\theta\in\Theta} |Q_n(\theta) - O_n(\alpha)| = 0 \quad \text{a.s.} \tag{7.62}$$

Let α_0^- and α_0^+ be two constants such that $\alpha_0^- < \alpha_0 < \alpha_0^+$. It can always be assumed that $0 < \alpha_0^-$. With the notation $\hat{\sigma}_\eta^2 = n^{-1}\sum_{t=1}^n \eta_t^2$, the solution of

$$\alpha_n^* = \arg\min_\alpha O_n(\alpha)$$

is $\alpha_n^* = \alpha_0\hat{\sigma}_\eta^2$. This solution belongs to the interval (α_0^-, α_0^+) when n is large enough. In this case

$$\alpha_n^{**} = \arg \min_{\alpha\notin(\alpha_0^-,\alpha_0^+)} O_n(\alpha)$$

is one of the two extremities of the interval (α_0^-, α_0^+), and thus

$$\lim_{n\to\infty} O_n(\alpha_n^{**}) = \min\{\lim_{n\to\infty} O_n(\alpha_0^-), \lim_{n\to\infty} O_n(\alpha_0^+)\} > 0.$$

This result and convergence (7.62) show that almost surely

$$\lim_{n\to\infty} \min_{\theta\in\Theta,\alpha\notin(\alpha_0^-,\alpha_0^+)} Q_n(\theta) > 0.$$

Since $\min_\theta Q_n(\theta) \le Q_n(\theta_0) = 0$, it follows that

$$\lim_{n\to\infty} \arg\min_{\theta\in\Theta} Q_n(\theta) \in (0, \infty) \times (\alpha_0^-, \alpha_0^+).$$

Since (α_0^-, α_0^+) is an interval which contains α_0 and can be arbitrarily small, we obtain the result. □

To prove the asymptotic normality of the QMLE, we need the following intermediate result.

Lemma 7.2 *Under the assumptions of Theorem 7.3, we have*

$$\sum_{t=1}^{\infty} \sup_{\theta\in\Theta} \left| \frac{\partial}{\partial\omega} \ell_t(\theta) \right| < \infty \quad a.s., \tag{7.63}$$

$$\sum_{t=1}^{\infty} \sup_{\theta\in\Theta} \left\| \frac{\partial^2}{\partial\omega\partial\theta} \ell_t(\theta) \right\| < \infty \quad a.s., \tag{7.64}$$

$$\frac{1}{n}\sum_{t=1}^{n} \sup_{\theta\in\Theta} \left| \frac{\partial^2}{\partial\alpha^2} \ell_t(\omega, \alpha_0) - \frac{1}{\alpha_0^2} \right| = o(1) \quad a.s., \tag{7.65}$$

$$\frac{1}{n}\sum_{t=1}^{n} \sup_{\theta\in\Theta} \left| \frac{\partial^3}{\partial\alpha^3} \ell_t(\theta) \right| = O(1) \quad a.s. \tag{7.66}$$

Proof Using Lemma 7.1, there exists a real random variable K and a constant $\rho \in (0, 1)$ independent of θ and of t such that

$$\left| \frac{\partial}{\partial\omega} \ell_t(\theta) \right| = \left| \frac{-(\omega_0 + \alpha_0\epsilon_{t-1}^2)\eta_t^2}{(\omega + \alpha\epsilon_{t-1}^2)^2} + \frac{1}{\omega + \alpha\epsilon_{t-1}^2} \right| \le K\rho^t(\eta_t^2 + 1).$$

Since it has a finite expectation, the series $\sum_{t=1}^{\infty} K\rho^t(\eta_t^2+1)$ is almost surely finite. This shows result (7.63), and result (7.64) follows similarly. We have

$$\begin{aligned}\frac{\partial^2 \ell_t(\omega,\alpha_0)}{\partial\alpha^2} - \frac{1}{\alpha_0^2} &= \left\{2\frac{(\omega_0+\alpha_0\epsilon_{t-1}^2)\eta_t^2}{\omega+\alpha_0\epsilon_{t-1}^2} - 1\right\}\frac{\epsilon_{t-1}^4}{(\omega+\alpha_0\epsilon_{t-1}^2)^2} - \frac{1}{\alpha_0^2}\\ &= (2\eta_t^2-1)\frac{\epsilon_{t-1}^4}{(\omega+\alpha_0\epsilon_{t-1}^2)^2} - \frac{1}{\alpha_0^2} + r_{1,t}\\ &= 2(\eta_t^2-1)\frac{1}{\alpha_0^2} + r_{1,t} + r_{2,t}\end{aligned}$$

where

$$\sup_{\theta\in\Theta}|r_{1,t}| = \sup_{\theta\in\Theta}\left|\frac{2(\omega_0-\omega)\eta_t^2}{(\omega+\alpha_0\epsilon_{t-1}^2)}\frac{\epsilon_{t-1}^4}{(\omega+\alpha_0\epsilon_{t-1}^2)^2}\right| = o(1) \quad \text{a.s.}$$

and

$$\begin{aligned}\sup_{\theta\in\Theta}|r_{2,t}| &= \sup_{\theta\in\Theta}\left|(2\eta_t^2-1)\left\{\frac{\epsilon_{t-1}^4}{(\omega+\alpha_0\epsilon_{t-1}^2)^2} - \frac{1}{\alpha_0^2}\right\}\right|\\ &= \sup_{\theta\in\Theta}\left|(2\eta_t^2-1)\left\{\frac{\omega^2+2\alpha_0\epsilon_{t-1}^2}{\alpha_0^2(\omega+\alpha_0\epsilon_{t-1}^2)^2}\right\}\right|\\ &= o(1) \quad \text{a.s.}\end{aligned}$$

as $t\to\infty$. Thus convergence (7.65) is shown. To show convergence (7.66), it suffices to note that

$$\begin{aligned}\left|\frac{\partial^3}{\partial\alpha^3}\ell_t(\theta)\right| &= \left|\left\{2-6\frac{(\omega_0+\alpha_0\epsilon_{t-1}^2)\eta_t^2}{\omega+\alpha\epsilon_{t-1}^2}\right\}\left(\frac{\epsilon_{t-1}^2}{\omega+\alpha\epsilon_{t-1}^2}\right)^3\right|\\ &\le \left\{2+6\left(\frac{\omega_0}{\omega}+\frac{\alpha_0}{\alpha}\right)\eta_t^2\right\}\frac{1}{\alpha^3}.\end{aligned}$$

□

Proof of (7.20) We remark that we do not know, a priori, if the derivative of the criterion is equal to zero at $\hat\theta_n = (\hat\omega_n, \hat\alpha_n)$, because we only have the convergence of $\hat\alpha_n$ to α_0. Thus the minimum of the criterion could lie at the boundary of Θ, even asymptotically. By contrast, the partial derivative with respect to the second coordinate must asymptotically vanish at the optimum, since $\hat\alpha_n \to \alpha_0$ and $\theta_0 \in \overset{\circ}{\Theta}$. A Taylor expansion of the derivative of the criterion thus gives

$$\begin{pmatrix}\frac{1}{\sqrt{n}}\sum_{t=1}^n \frac{\partial}{\partial\omega}\ell_t(\hat\theta_n)\\ 0\end{pmatrix} = \frac{1}{\sqrt{n}}\sum_{t=1}^n \frac{\partial}{\partial\theta}\ell_t(\theta_0) + J_n\sqrt{n}(\hat\theta_n-\theta_0), \tag{7.67}$$

where J_n is a 2×2 matrix whose elements are of the form

$$J_n(i,j) = \frac{1}{n}\sum_{t=1}^n \frac{\partial^2}{\partial\theta_i\partial\theta_j}\ell_t(\theta_{i,j}^*)$$

with $\theta_{i,j}^*$ between $\hat\theta_n$ and θ_0. By Lemma 7.1, which shows that $\epsilon_t^2\to\infty$ almost surely, and by the central limit theorem of Lindeberg for martingale increment (see Corollary A.1),

$$\frac{1}{\sqrt{n}}\sum_{t=1}^n \frac{\partial}{\partial\alpha}\ell_t(\theta_0) = \frac{1}{\sqrt{n}}\sum_{t=1}^n (1-\eta_t^2)\frac{\epsilon_{t-1}^2}{\omega_0+\alpha_0\epsilon_{t-1}^2}$$

$$= \frac{1}{\sqrt{n}} \sum_{t=1}^{n} (1 - \eta_t^2) \frac{1}{\alpha_0} + o_P(1)$$

$$\xrightarrow{\mathcal{L}} \mathcal{N}\left(0, \frac{\kappa_\eta - 1}{\alpha_0^2}\right). \tag{7.68}$$

Relation (7.64) of Lemma 7.2 and the compactness of Θ show that

$$J_n(2,1)\sqrt{n}(\hat{\omega}_n - \omega_0) \to 0 \quad \text{a.s.} \tag{7.69}$$

By a Taylor expansion of the function

$$\alpha \mapsto \frac{1}{n} \sum_{t=1}^{n} \frac{\partial^2}{\partial \alpha^2} \ell_t(\omega_{2,2}^*, \alpha),$$

we obtain

$$J_n(2,2) = \frac{1}{n} \sum_{t=1}^{n} \frac{\partial^2}{\partial \alpha^2} \ell_t(\omega_{2,2}^*, \alpha_0) + \frac{1}{n} \sum_{t=1}^{n} \frac{\partial^3}{\partial \alpha^3} \ell_t(\omega_{2,2}^*, \alpha^*)(\alpha_{2,2}^* - \alpha_0),$$

where α^* is between $\alpha_{2,2}^*$ and α_0. Using results (7.65), (7.66), and (7.19), we obtain

$$J_n(2,2) \to \frac{1}{\alpha_0^2} \quad \text{a.s.} \tag{7.70}$$

We conclude using the second row of (7.67), and also using (7.68), (7.69), and (7.70). □

Proof of Theorem 7.4 The proof follows the steps of the proof of Theorem 7.1. We will show the following points:

(a) $\lim_{n\to\infty} \sup_{\varphi\in\Phi} |\mathbf{l}_n(\varphi) - \tilde{\mathbf{l}}_n(\varphi)| = 0$, a.s.

(b) ($\exists t \in \mathbb{Z}$ such that $\epsilon_t(\vartheta) = \epsilon_t(\vartheta_0)$ and $\sigma_t^2(\varphi) = \sigma_t^2(\varphi_0) P_{\varphi_0}$ a.s.) $\Longrightarrow \varphi = \varphi_0$.

(c) If $\varphi \neq \varphi_0$, $E_{\varphi_0} \ell_t(\varphi) > E_{\varphi_0} \ell_t(\varphi_0)$.

(d) For any $\varphi \neq \varphi_0$ there exists a neighbourhood $V(\varphi)$ such that

$$\liminf_{n\to\infty} \inf_{\varphi^* \in V(\varphi)} \tilde{\mathbf{l}}_n(\varphi^*) > E_{\varphi_0} \ell_1(\varphi_0), \quad \text{a.s.}$$

(a) Nullity of the asymptotic impact of the initial values.
Equations (7.10)–(7.28) remain valid under the convention that $\epsilon_t = \epsilon_t(\vartheta)$. Equation (7.29) must be replaced by

$$\underline{\tilde{\sigma}}_t^2 = \underline{\tilde{c}}_t + B\underline{\tilde{c}}_{t-1} + \cdots + B^{t-1}\underline{\tilde{c}}_1 + B^t \underline{\tilde{\sigma}}_0^2, \tag{7.71}$$

where $\underline{\tilde{c}}_t = \left(\omega + \sum_{i=1}^{q} \alpha_i \tilde{\epsilon}_{t-i}^2, 0, \ldots, 0\right)'$, the 'tilde' variables being initialised as indicated before. Assumptions A7 and A8 imply that,

$$\text{for any } k \geq 1 \text{ and } 1 \leq i \leq q, \quad \sup_{\varphi\in\Phi} |\epsilon_{k-i} - \tilde{\epsilon}_{k-i}| \leq K\rho^k, \quad \text{a.s.} \tag{7.72}$$

It follows that almost surely

$$\|\underline{c}_k - \underline{\tilde{c}}_k\| \leq \sum_{i=1}^{q} |\alpha_i| |\tilde{\epsilon}_{k-i}^2 - \epsilon_{k-i}^2|$$

$$\leq \sum_{i=1}^{q} |\alpha_i||\tilde{\epsilon}_{k-i} - \epsilon_{k-i}| \left(| \, 2\epsilon_{k-i} \, | + | \, \tilde{\epsilon}_{k-i} - \epsilon_{k-i} \, |\right)$$

$$\leq K\rho^k \left(\sum_{i=1}^{q} |\epsilon_{k-i}| + 1 \right)$$

and thus, by relations (7.28), (7.71) and (7.27),

$$\|\underline{\sigma}_t^2 - \underline{\tilde{\sigma}}_k^2\| = \left\| \sum_{k=0}^{t} B^{t-k}(\underline{c}_k - \underline{\tilde{c}}_k) + B^t(\underline{\sigma}_0^2 - \underline{\tilde{\sigma}}_0^2) \right\|$$

$$\leq K \sum_{k=0}^{t} \rho^{t-k}\rho^k \left(\sum_{i=1}^{q} |\epsilon_{k-i}| + 1 \right) + K\rho^t$$

$$\leq K\rho^t \sum_{k=-q}^{t} (|\epsilon_k| + 1). \tag{7.73}$$

Similarly, we have that almost surely $|\tilde{\epsilon}_t^2 - \epsilon_t^2| \leq K\rho^t(|\epsilon_t| + 1)$. The difference between the theoretical log-likelihoods with and without initial values can thus be bounded as follows:

$$\sup_{\varphi \in \Phi} |\mathbf{l}_n(\varphi) - \tilde{\mathbf{l}}_n(\varphi)| \leq n^{-1} \sum_{t=1}^{n} \sup_{\varphi \in \Phi} \left\{ \left| \frac{\tilde{\sigma}_t^2 - \sigma_t^2}{\tilde{\sigma}_t^2 \sigma_t^2} \right| \epsilon_t^2 + \left| \log\left(1 + \frac{\sigma_t^2 - \tilde{\sigma}_t^2}{\tilde{\sigma}_t^2}\right) \right| + \frac{|\epsilon_t^2 - \tilde{\epsilon}_t^2|}{\tilde{\sigma}_t^2} \right\}$$

$$\leq \left\{ \sup_{\varphi \in \Phi} \max\left(\frac{1}{\omega^2}, \frac{1}{\omega}\right) \right\} K n^{-1} \sum_{t=1}^{n} \rho^t(\epsilon_t^2 + 1) \sum_{k=-q}^{t} (|\epsilon_k| + 1).$$

This inequality is analogous to inequality (7.31), $\epsilon_t^2 + 1$ being replaced by $\xi_t = (\epsilon_t^2 + 1) \sum_{k=-q}^{t}(|\epsilon_k| + 1)$. Following the lines of the proof of (a) in Theorem 7.1 (see Exercise 7.2), it suffices to show that for all real $r > 0$, $E(\rho^t\xi_t)^r$ is the general term of a finite series. Note that[7]

$$E(\rho^t\xi_t)^{s/2} \leq \rho^{ts/2} \sum_{k=-q}^{t} E(\epsilon_t^2|\epsilon_k| + \epsilon_t^2 + |\epsilon_k| + 1)^{s/2}$$

$$\leq \rho^{ts/2} \sum_{k=-q}^{t} [\{E(\epsilon_t^{2s})E \mid \epsilon_k|^s\}^{1/2} + E|\epsilon_t|^s + E|\epsilon_k|^{s/2} + 1]$$

$$= O(t\rho^{ts/2}),$$

since, by Corollary 2.3, $E(\epsilon_t^{2s}) < \infty$. Statement (a) follows.

(b) Identifiability of the parameter.
If $\epsilon_t(\vartheta) = \epsilon_t(\vartheta_0)$ almost surely, assumptions A8 and A9 imply that there exists a constant linear combination of the variables $X_{t-j}, j \geq 0$. The linear innovation of (X_t), equal to $X_t - E(X_t \mid X_u, u < t) = \eta_t\sigma_t(\varphi_0)$, is zero almost surely only if $\eta_t = 0$ a.s. (since $\sigma_t^2(\varphi_0) \geq \omega_0 > 0$). This is precluded, since $E(\eta_t^2) = 1$. It follows that $\vartheta = \vartheta_0$, and thus that $\theta = \theta_0$ by the argument used in the proof of Theorem 7.1.

[7] We use the fact that if X and Y are positive random variables, $E(X+Y)^r \leq E(X^r) + E(Y^r)$ for all $r \in (0, 1]$, this inequality being trivially obtained from the inequality already used: $(a+b)^r \leq a^r + b^r$ for all positive real numbers a and b.

(c) The limit criterion is minimised at the true value.
By the arguments used in the proof of (c) in Theorem 7.1, it can be shown that, for all φ, $E_{\varphi_0}\mathbf{l}_n(\varphi) = E_{\varphi_0}\ell_t(\varphi)$ is defined in $\mathbb{R}\cup\{+\infty\}$, and in $\mathbb{R}$ at $\varphi=\varphi_0$. We have

$$\begin{aligned}
E_{\varphi_0}\ell_t(\varphi) - E_{\varphi_0}\ell_t(\varphi_0) &= E_{\varphi_0}\log\frac{\sigma_t^2(\varphi)}{\sigma_t^2(\varphi_0)} + E_{\varphi_0}\left[\frac{\epsilon_t^2(\vartheta)}{\sigma_t^2(\varphi)} - \frac{\epsilon_t^2(\vartheta_0)}{\sigma_t^2(\varphi_0)}\right]\\
&= E_{\varphi_0}\left\{\log\frac{\sigma_t^2(\varphi)}{\sigma_t^2(\varphi_0)} + \frac{\sigma_t^2(\varphi_0)}{\sigma_t^2(\varphi)} - 1\right\}\\
&\quad + E_{\varphi_0}\frac{\{\epsilon_t(\vartheta)-\epsilon_t(\vartheta_0)\}^2}{\sigma_t^2(\varphi)}\\
&\quad + E_{\varphi_0}\frac{2\eta_t\sigma_t(\varphi_0)\{\epsilon_t(\vartheta)-\epsilon_t(\vartheta_0)\}}{\sigma_t^2(\varphi)}\\
&\geq 0
\end{aligned}$$

because the last expectation is equal to 0 (noting that $\epsilon_t(\vartheta)-\epsilon_t(\vartheta_0)$ belongs to the past, as well as $\sigma_t(\varphi_0)$ and $\sigma_t(\varphi)$), the other expectations being positive or null by arguments already used. This inequality is strict only if $\epsilon_t(\vartheta)=\epsilon_t(\vartheta_0)$ and if $\sigma_t^2(\varphi)=\sigma_t^2(\varphi_0)$ P_{φ_0} a.s. which, by (b), implies $\varphi=\varphi_0$ and completes the proof of (c).

(d) Use of the compactness of Φ and of the ergodicity of $(\ell_t(\varphi))$.
The end of the proof is the same as that of Theorem 7.1. □

Proof of Theorem 7.5 The proof follows the steps of that of Theorem 7.2. The block-diagonal form of the matrices $\mathcal{I}$ and $\mathcal{J}$ when the distribution of η_t is symmetric is shown in Exercise 7.7. It suffices to establish the following properties.

(a) $E_{\varphi_0}\left\|\frac{\partial\ell_t(\varphi_0)}{\partial\varphi}\frac{\partial\ell_t(\varphi_0)}{\partial\varphi'}\right\| < \infty$, $E_{\varphi_0}\left\|\frac{\partial^2\ell_t(\varphi_0)}{\partial\varphi\partial\varphi'}\right\| < \infty$.

(b) $\mathcal{I}$ and $\mathcal{J}$ are invertible.

(c) $\left\|n^{-1/2}\sum_{t=1}^n\left\{\frac{\partial\ell_t(\varphi_0)}{\partial\varphi} - \frac{\partial\tilde{\ell}_t(\varphi_0)}{\partial\varphi}\right\}\right\|$ and $\sup_{\varphi\in\mathcal{V}(\varphi_0)}\left\|n^{-1}\sum_{t=1}^n\left\{\frac{\partial^2\ell_t(\varphi)}{\partial\varphi\partial\varphi'} - \frac{\partial^2\tilde{\ell}_t(\varphi)}{\partial\varphi\partial\varphi'}\right\}\right\|$ tend in probability to 0 as $n\to\infty$.

(d) $n^{-1/2}\sum_{t=1}^n\frac{\partial\ell_t}{\partial\varphi}(\varphi_0) \Rightarrow \mathcal{N}(0,\mathcal{I})$.

(e) $n^{-1}\sum_{t=1}^n\frac{\partial^2\ell_t}{\partial\varphi_i\partial\varphi_j}(\varphi^*) \to \mathcal{J}(i,j)$ a.s., for all φ^* between $\hat{\varphi}_n$ and φ_0.

Formulas (7.38) and (7.39) giving the derivatives with respect to the GARCH parameters (that is, the vector θ) remain valid in the presence of an ARMA part (writing $\epsilon_t^2=\epsilon_t^2(\vartheta)$). The same is true for all the results established in (a) and (b) of the proof of Theorem 7.2, with obvious changes of notation. The derivatives of $\ell_t(\varphi)=\epsilon_t^2(\vartheta)/\sigma_t^2(\varphi)+\log\sigma_t^2(\varphi)$ with respect to the parameter ϑ, and the cross derivatives with respect to θ and ϑ, are given by

$$\frac{\partial\ell_t(\varphi)}{\partial\vartheta} = \left(1-\frac{\epsilon_t^2}{\sigma_t^2}\right)\frac{1}{\sigma_t^2}\frac{\partial\sigma_t^2}{\partial\vartheta} + \frac{2\epsilon_t}{\sigma_t^2}\frac{\partial\epsilon_t}{\partial\vartheta}, \tag{7.74}$$

$$\begin{aligned}
\frac{\partial^2\ell_t(\varphi)}{\partial\vartheta\,\partial\vartheta'} &= \left(1-\frac{\epsilon_t^2}{\sigma_t^2}\right)\frac{1}{\sigma_t^2}\frac{\partial^2\sigma_t^2}{\partial\vartheta\,\partial\vartheta'} + \left(2\frac{\epsilon_t^2}{\sigma_t^2}-1\right)\frac{1}{\sigma_t^2}\frac{\partial\sigma_t^2}{\partial\vartheta}\frac{1}{\sigma_t^2}\frac{\partial\sigma_t^2}{\partial\vartheta'}\\
&\quad + \frac{2}{\sigma_t^2}\frac{\partial\epsilon_t}{\partial\vartheta}\frac{\partial\epsilon_t}{\partial\vartheta'} + \frac{2\epsilon_t}{\sigma_t^2}\frac{\partial^2\epsilon_t}{\partial\vartheta\,\partial\vartheta'} - \frac{2\epsilon_t}{\sigma_t^2}\left(\frac{\partial\epsilon_t}{\partial\vartheta}\frac{1}{\sigma_t^2}\frac{\partial\sigma_t^2}{\partial\vartheta'} + \frac{1}{\sigma_t^2}\frac{\partial\sigma_t^2}{\partial\vartheta}\frac{\partial\epsilon_t}{\partial\vartheta'}\right),
\end{aligned} \tag{7.75}$$

$$\frac{\partial^2 \ell_t(\varphi)}{\partial\theta\,\partial\theta'} = \left(1-\frac{\epsilon_t^2}{\sigma_t^2}\right)\frac{1}{\sigma_t^2}\frac{\partial^2\sigma_t^2}{\partial\theta\,\partial\theta'} + \left(2\frac{\epsilon_t^2}{\sigma_t^2}-1\right)\frac{1}{\sigma_t^2}\frac{\partial\sigma_t^2}{\partial\theta}\frac{1}{\sigma_t^2}\frac{\partial\sigma_t^2}{\partial\theta'} - \frac{2\epsilon_t}{\sigma_t^2}\frac{\partial\epsilon_t}{\partial\vartheta}\frac{1}{\sigma_t^2}\frac{\partial\sigma_t^2}{\partial\vartheta'}. \tag{7.76}$$

The derivatives of ϵ_t are of the form

$$\frac{\partial\epsilon_t}{\partial\vartheta} = \left(-A_\vartheta(1)B_\vartheta^{-1}(1), v_{t-1}(\vartheta), \dots, v_{t-P}(\vartheta), u_{t-1}(\vartheta), \dots, u_{t-Q}(\vartheta)\right)'$$

where

$$v_t(\vartheta) = -A_\vartheta^{-1}(B)\epsilon_t(\vartheta), \quad u_t(\vartheta) = B_\vartheta^{-1}(B)\epsilon_t(\vartheta) \tag{7.77}$$

and

$$\frac{\partial^2\epsilon_t}{\partial\vartheta\,\partial\vartheta'} = \begin{pmatrix} & & 0_{1,Q} \\ & 0_{P+1,P+1} & \\ & & -A_\vartheta^{-1}(B)B_\vartheta^{-1}(B)H_{P,Q}(t) \\ 0_{Q,1} & -A_\vartheta^{-1}(B)B_\vartheta^{-1}(B)H_{Q,P}(t) & -2B_\vartheta^{-2}(B)H_{Q,Q}(t) \end{pmatrix}, \tag{7.78}$$

where $H_{k,\ell}(t)$ is the $k\times\ell$ (Hankel) matrix of general term ϵ_{t-i-j}, and $0_{k,\ell}$ denotes the null matrix of size $k\times\ell$. Moreover, by relation (7.28),

$$\frac{\partial\sigma_t^2}{\partial\vartheta_j} = \sum_{k=0}^{\infty} B^k(1,1)\sum_{i=1}^{q} 2\alpha_i\epsilon_{t-k-i}\frac{\partial\epsilon_{t-k-i}}{\partial\vartheta_j}, \tag{7.79}$$

where ϑ_j denotes the jth component of ϑ, and

$$\frac{\partial^2\sigma_t^2}{\partial\vartheta_j\partial\vartheta_\ell} = \sum_{k=0}^{\infty} B^k(1,1)\sum_{i=1}^{q} 2\alpha_i\left(\frac{\partial\epsilon_{t-k-i}}{\partial\vartheta_j}\frac{\partial\epsilon_{t-k-i}}{\partial\vartheta_\ell} + \epsilon_{t-k-i}\frac{\partial^2\epsilon_{t-k-i}}{\partial\vartheta_j\partial\vartheta_\ell}\right). \tag{7.80}$$

(a) Integrability of the derivatives of the criterion at ϕ_0.
The existence of the expectations in (7.40) remains true. By (7.74)–(7.76), the independence between $(\epsilon_t/\sigma_t)(\varphi_0) = \eta_t$ and $\sigma_t^2(\varphi_0)$, its derivatives, and the derivatives of $\epsilon_t(\vartheta_0)$, using $E(\eta_t^4) < \infty$ and $\sigma_t^2(\varphi_0) > \omega_0 > 0$, it suffices to show that

$$E_{\varphi_0}\left\|\frac{\partial\epsilon_t}{\partial\vartheta}\frac{\partial\epsilon_t}{\partial\vartheta'}(\theta_0)\right\| < \infty, \quad E_{\varphi_0}\left\|\frac{\partial^2\epsilon_t}{\partial\vartheta\,\partial\vartheta'}(\theta_0)\right\| < \infty, \tag{7.81}$$

$$E_{\varphi_0}\left\|\frac{1}{\sigma_t^4}\frac{\partial\sigma_t^2}{\partial\vartheta}\frac{\partial\sigma_t^2}{\partial\vartheta'}(\varphi_0)\right\| < \infty, \quad E_{\varphi_0}\left\|\frac{\partial^2\sigma_t^2}{\partial\vartheta\,\partial\vartheta'}(\varphi_0)\right\| < \infty, \quad E_{\varphi_0}\left\|\frac{\partial^2\sigma_t^2}{\partial\theta\,\partial\vartheta'}(\varphi_0)\right\| < \infty \tag{7.82}$$

to establish point (a), together with the existence of the matrices $\mathcal{I}$ and $\mathcal{J}$. By the expressions for the derivatives of ϵ_t, (7.77)–(7.78), and using $E\epsilon_t^2(\vartheta_0) < \infty$, we obtain the moment conditions (7.81).

The Cauchy–Schwarz inequality implies that

$$\left|\sum_{i=1}^{q}\alpha_{0i}\epsilon_{t-k-i}(\vartheta_0)\frac{\partial\epsilon_{t-k-i}(\vartheta_0)}{\partial\vartheta_j}\right| \le \left\{\sum_{i=1}^{q}\alpha_{0i}\epsilon_{t-k-i}^2(\vartheta_0)\right\}^{1/2}\left\{\sum_{i=1}^{q}\alpha_{0i}\left(\frac{\partial\epsilon_{t-k-i}(\vartheta_0)}{\partial\vartheta_j}\right)^2\right\}^{1/2}.$$

Thus, in view of relation (7.79) and the positivity of ω_0,

$$\frac{\partial\sigma_t^2}{\partial\vartheta_j}(\varphi_0) \le 2\sum_{k=0}^{\infty} B_0^k(1,1)\left\{\omega_0 + \sum_{i=1}^{q}\alpha_{0i}\epsilon_{t-k-i}^2(\vartheta_0)\right\}^{1/2}\left\{\sum_{i=1}^{q}\alpha_{0i}\left(\frac{\partial\epsilon_{t-k-i}(\vartheta_0)}{\partial\vartheta_j}\right)^2\right\}^{1/2}.$$

Using the triangle inequality and the elementary inequalities $(\sum |x_i|)^{1/2} \leq \sum |x_i|^{1/2}$ and $x/(1+x^2) \leq 1$, it follows that

$$\begin{aligned}
\left\| \frac{1}{\sigma_t^2} \frac{\partial \sigma_t^2}{\partial \vartheta_j}(\varphi_0) \right\|_2 &\leq \left\| 2 \sum_{k=0}^{\infty} \frac{B^{k/2}(1,1) \underline{c}_{t-k}^{1/2}(1) B^{k/2}(1,1) \sum_{i=1}^{q} \alpha_i^{1/2} \left| \frac{\partial \epsilon_{t-k-i}}{\partial \vartheta_j} \right|}{\omega + B^k(1,1) \underline{c}_{t-k}(1)} (\varphi_0) \right\|_2 \\
&\leq \left\| \frac{2}{\sqrt{\omega}} \sum_{k=0}^{\infty} B^{k/2}(1,1) \frac{\frac{B^{k/2}(1,1)\underline{c}_{t-k}^{1/2}(1)}{\sqrt{\omega}}}{1 + \left(\frac{B^{k/2}(1,1)\underline{c}_{t-k}^{1/2}(1)}{\sqrt{\omega}} \right)^2} \sum_{i=1}^{q} \alpha_i^{1/2} \left| \frac{\partial \epsilon_{t-k-i}}{\partial \vartheta_j} \right| (\varphi_0) \right\|_2 \\
&\leq \frac{K}{\sqrt{\omega_0}} \sum_{k=0}^{\infty} \rho^{k/2} \sum_{i=1}^{q} \alpha_{0i}^{1/2} \left\| \frac{\partial \epsilon_{t-k-i}(\vartheta_0)}{\partial \vartheta_j} \right\|_2 < \infty. \qquad (7.83)
\end{aligned}$$

The first inequality of (7.82) follows. The existence of the second expectation in (7.82) is a consequence of (7.80), the Cauchy–Schwarz inequality, and the square integrability of ϵ_t and its derivatives. To handle the second-order partial derivatives of σ_t^2, first note that $(\partial^2 \sigma_t^2 / \partial \vartheta \partial \omega)(\varphi_0) = 0$ by (7.41). Moreover, using relation (7.79),

$$E_{\varphi_0} \left| \frac{\partial^2 \sigma_t^2}{\partial \vartheta_j \partial \alpha_i}(\varphi_0) \right| = E_{\varphi_0} \left| 2 \sum_{k=0}^{\infty} B^k(1,1) \epsilon_{t-k-i} \frac{\partial \epsilon_{t-k-i}}{\partial \vartheta_j}(\varphi_0) \right| < \infty. \qquad (7.84)$$

By the arguments used to show inequality (7.44), we obtain

$$\beta_{0\ell} E_{\varphi_0} \left| \frac{\partial^2 \sigma_t^2}{\partial \vartheta_j \partial \beta_\ell}(\varphi_0) \right| \leq E_{\varphi_0} \left| \sum_{k=1}^{\infty} k B^k(1,1) \sum_{i=1}^{q} 2 \alpha_{0i} \epsilon_{t-k-i} \frac{\partial \epsilon_{t-k-i}}{\partial \vartheta_j}(\varphi_0) \right| < \infty, \qquad (7.85)$$

which entails the existence of the third expectation in (7.82).

(b) Invertibility of $\mathcal{I}$ and $\mathcal{J}$.
Assume that $\mathcal{I}$ is non-invertible. There exists a non-zero vector λ in $\mathbb{R}^{P+Q+p+q+2}$ such that $\lambda' \partial \ell_t(\varphi_0)/\partial \varphi' = 0$ a.s. By relations (7.38) and (7.74), this implies that

$$(1 - \eta_t^2) \frac{1}{\sigma_t^2(\varphi_0)} \lambda' \frac{\partial \sigma_t^2(\varphi_0)}{\partial \varphi} + \frac{2\eta_t}{\sigma_t(\varphi_0)} \lambda' \frac{\partial \epsilon_t(\vartheta_0)}{\partial \varphi} = 0 \quad \text{a.s.} \qquad (7.86)$$

Taking the variance of the left-hand side, conditionally on the σ-field generated by $\{\eta_u, u < t\}$, we obtain a.s., at $\varphi = \varphi_0$,

$$\begin{aligned}
0 &= (\kappa_\eta - 1) \left(\frac{1}{\sigma_t^2} \lambda' \frac{\partial \sigma_t^2}{\partial \varphi} \right)^2 - 2\nu_\eta \frac{1}{\sigma_t^2} \lambda' \frac{\partial \sigma_t^2}{\partial \varphi} \frac{2}{\sigma_t} \lambda' \frac{\partial \epsilon_t}{\partial \varphi} + \left(\frac{2}{\sigma_t} \lambda' \frac{\partial \epsilon_t}{\partial \varphi} \right)^2 \\
&:= (\kappa_\eta - 1) a_t^2 - 2\nu_\eta a_t b_t + b_t^2 \\
&= (\kappa_\eta - 1 - \nu_\eta^2) a_t^2 + (b_t - \nu_\eta a_t)^2,
\end{aligned}$$

where $\nu_\eta = E(\eta_t^3)$. It follows that $\kappa_\eta - 1 - \nu_\eta^2 \leq 0$ and $b_t = a_t\{\nu_\eta \pm (\nu_\eta^2 + 1 - \kappa_\eta)^{1/2}\}$ almost surely. By stationarity, we have either $b_t = a_t\{\nu_\eta - (\nu_\eta^2 + 1 - \kappa_\eta)^{1/2}\}$ a.s. for all t, or $b_t = a_t\{\nu_\eta + (\nu_\eta^2 + 1 - \kappa_\eta)^{1/2}\}$ a.s. for all t. Consider for instance the latter case, the first one being treated similarly. Relation (7.86) implies $a_t[1 - \eta_t^2 + \{\nu_\eta + (\nu_\eta^2 + 1 - \kappa_\eta)^{1/2}\}\eta_t] = 0$ a.s. The term in brackets cannot vanish almost surely, otherwise η_t would take at least two different values, which would be in contradiction to assumption A12. It follows that $a_t = 0$ a.s. and thus

$b_t = 0$ a.s. We have shown that almost surely

$$\lambda' \frac{\partial \epsilon_t(\varphi_0)}{\partial \varphi} = \lambda_1' \frac{\partial \epsilon_t(\vartheta_0)}{\partial \vartheta} = 0 \quad \text{and} \quad \lambda' \frac{\partial \sigma_t^2(\varphi_0)}{\partial \varphi} = 0, \tag{7.87}$$

where λ_1 is the vector of the first $P+Q+1$ components of λ. By stationarity of $(\partial \epsilon_t/\partial \varphi)_t$, the first equality implies that

$$0 = \lambda_1' \begin{pmatrix} -A_{\vartheta_0}(1) \\ c_0 - X_{t-1} \\ \vdots \\ c_0 - X_{t-P} \\ \epsilon_{t-1} \\ \vdots \\ \epsilon_{t-Q} \end{pmatrix} + \sum_{j=1}^{Q} b_{0j} \lambda_1' \frac{\partial \epsilon_{t-j}(\vartheta_0)}{\partial \vartheta} = \lambda_1' \begin{pmatrix} -A_{\vartheta_0}(1) \\ c_0 - X_{t-1} \\ \vdots \\ c_0 - X_{t-P} \\ \epsilon_{t-1} \\ \vdots \\ \epsilon_{t-Q} \end{pmatrix}.$$

We now use assumption A9, that the ARMA representation is minimal, to conclude that $\lambda_1 = 0$. The third equality in (7.87) is then written, with obvious notation, as $\lambda_2' \frac{\partial \sigma_t^2(\varphi_0)}{\partial \theta} = 0$. We have already shown in the proof of Theorem 7.2 that this entails $\lambda_2 = 0$. We are led to a contradiction, which proves that $\mathcal{I}$ is invertible. Using relations (7.39) and (7.75)–(7.76), we obtain

$$\mathcal{J} = E_{\varphi_0} \left(\frac{1}{\sigma_t^4} \frac{\partial \sigma_t^2}{\partial \varphi} \frac{\partial \sigma_t^2}{\partial \varphi'}(\varphi_0) \right) + 2E_{\varphi_0} \left(\frac{1}{\sigma_t^2} \frac{\partial \epsilon_t}{\partial \varphi} \frac{\partial \epsilon_t}{\partial \varphi'}(\varphi_0) \right).$$

We have just shown that the first expectation is a positive definite matrix. The second expectation being a positive semidefinite matrix, $\mathcal{J}$ is positive definite and thus invertible, which completes the proof of (b).

(c) Asymptotic unimportance of the initial values.
The initial values being fixed, the derivatives of $\underline{\tilde{\sigma}}_t^2$, obtained from (7.71), are given by

$$\frac{\partial \underline{\tilde{\sigma}}_t^2}{\partial \omega} = \sum_{k=0}^{t-1} B^k \underline{1}, \quad \frac{\partial \underline{\tilde{\sigma}}_t^2}{\partial \alpha_i} = \sum_{k=0}^{t-1} B^k \underline{\tilde{\epsilon}}_{t-k-i}^2, \quad \frac{\partial \underline{\tilde{\sigma}}_t^2}{\partial \beta_j} = \sum_{k=1}^{t-1} \left\{ \sum_{i=1}^{k} B^{i-1} B^{(j)} B^{k-i} \right\} \underline{\tilde{c}}_{t-k},$$

with the notation introduced in relations (7.41)–(7.42) and (7.55)–(7.56). As for relation (7.79), we obtain

$$\frac{\partial \tilde{\sigma}_t^2}{\partial \vartheta_j} = \sum_{k=0}^{t-1} B^k(1,1) \sum_{i=1}^{q} 2\alpha_i \tilde{\epsilon}_{t-k-i} \frac{\partial \tilde{\epsilon}_{t-k-i}}{\partial \vartheta_j}$$

and, by an obvious extension of inequality (7.72),

$$\sup_{\varphi \in \Phi} \max \left\{ |\epsilon_k - \tilde{\epsilon}_k|, \left| \frac{\partial \epsilon_k}{\partial \vartheta_j} - \frac{\partial \tilde{\epsilon}_k}{\partial \vartheta_j} \right| \right\} \le K\rho^k, \quad \text{almost surely.} \tag{7.88}$$

Thus

$$\left| \frac{\partial \sigma_t^2}{\partial \vartheta_j} - \frac{\partial \tilde{\sigma}_t^2}{\partial \vartheta_j} \right|$$
$$\le \left| \sum_{k=t}^{\infty} B^k(1,1) \sum_{i=1}^{q} 2\alpha_i \epsilon_{t-k-i} \frac{\partial \epsilon_{t-k-i}}{\partial \vartheta_j} \right|$$

$$+\sum_{k=0}^{t-1} B^k(1,1)\sum_{i=1}^{q} 2\alpha_i\left|(\epsilon_{t-k-i}-\tilde{\epsilon}_{t-k-i})\frac{\partial \epsilon_{t-k-i}}{\partial \vartheta_j}+\tilde{\epsilon}_{t-k-i}\left(\frac{\partial \epsilon_{t-k-i}}{\partial \vartheta_j}-\frac{\partial \tilde{\epsilon}_{t-k-i}}{\partial \vartheta_j}\right)\right|$$

$$\leq K\rho^t\sum_{k=0}^{\infty}\rho^k\left|\epsilon_{-k-1}\frac{\partial \epsilon_{-k-1}}{\partial \vartheta_j}\right|+K\sum_{k=0}^{t-1}\rho^k\sum_{i=1}^{q}\rho^{t-k-i}\left\{\left|\frac{\partial \epsilon_{t-k-i}}{\partial \vartheta_j}\right|+|\tilde{\epsilon}_{t-k-i}|\right\}$$

$$\leq K\rho^t\sum_{k=0}^{\infty}\rho^k\left|\epsilon_{-k-1}\frac{\partial \epsilon_{-k-1}}{\partial \vartheta_j}\right|+K\rho^{t/2}\sum_{k=1}^{t-1+q}\rho^{k/2}\rho^{\frac{t-k}{2}}\left\{\left|\frac{\partial \epsilon_{t-k}}{\partial \vartheta_j}\right|+|\tilde{\epsilon}_{t-k}|\right\}$$

$$\leq K\rho^{t/2}\sum_{k=0}^{\infty}\rho^{k/2}\left\{\left|\epsilon_{-k-1}\frac{\partial \epsilon_{-k-1}}{\partial \vartheta_j}\right|+\left|\frac{\partial \epsilon_{k+1-q}}{\partial \vartheta_j}\right|+|\tilde{\epsilon}_{k+1-q}|\right\}.$$

The latter sum converges almost surely because its expectation is finite. We have thus shown that

$$\sup_{\varphi\in\Phi}\left|\frac{\partial \sigma_t^2}{\partial \vartheta_j}-\frac{\partial \tilde{\sigma}_t^2}{\partial \vartheta_j}\right|\leq K\rho^t \quad \text{a.s.}$$

The other derivatives of σ_t^2 are handled similarly, and we obtain

$$\sup_{\varphi\in\Phi}\left\|\frac{\partial \sigma_t^2}{\partial \varphi}-\frac{\partial \tilde{\sigma}_t^2}{\partial \varphi}\right\|<K\rho^t \quad \text{a.s.}$$

We have, in view of inequality (7.73),

$$\left|\frac{1}{\sigma_t^2}-\frac{1}{\tilde{\sigma}_t^2}\right|=\left|\frac{\tilde{\sigma}_t^2-\sigma_t^2}{\sigma_t^2\tilde{\sigma}_t^2}\right|\leq\frac{K}{\sigma_t^2}\rho^t S_{t-1},\qquad \frac{\sigma_t^2}{\tilde{\sigma}_t^2}\leq 1+K\rho^t S_{t-1},$$

where $S_{t-1}=\sum_{k=1-q}^{t-1}(|\epsilon_k|+1)$. It is also easy to check that for $\varphi=\varphi_0$,

$$|\tilde{\epsilon}_t^2-\epsilon_t^2|\leq K\rho^t(1+\sigma_t\eta_t),\qquad \left|1-\frac{\tilde{\epsilon}_t^2}{\tilde{\sigma}_t^2}\right|\leq 1+\eta_t^2+K\rho^t(1+|\eta_t|S_{t-1}+\eta_t^2S_{t-1}).$$

It follows that, using the inequality (7.88),

$$\begin{aligned}\left|\frac{\partial \ell_t(\varphi_0)}{\partial \varphi_i}-\frac{\partial \tilde{\ell}_t(\varphi_0)}{\partial \varphi_i}\right| &= \left|\{\tilde{\epsilon}_t^2-\epsilon_t^2\}\left\{\frac{1}{\tilde{\sigma}_t^2}\right\}\left\{\frac{1}{\sigma_t^2}\frac{\partial \sigma_t^2}{\partial \varphi_i}\right\}+\epsilon_t^2\left\{\frac{1}{\tilde{\sigma}_t^2}-\frac{1}{\sigma_t^2}\right\}\left\{\frac{1}{\sigma_t^2}\frac{\partial \sigma_t^2}{\partial \varphi_i}\right\}\right.\\
&\quad+\left\{1-\frac{\tilde{\epsilon}_t^2}{\tilde{\sigma}_t^2}\right\}\left\{\frac{1}{\sigma_t^2}-\frac{1}{\tilde{\sigma}_t^2}\right\}\left\{\frac{\partial \sigma_t^2}{\partial \varphi_i}\right\}\\
&\quad+\left\{1-\frac{\tilde{\epsilon}_t^2}{\tilde{\sigma}_t^2}\right\}\left\{\frac{1}{\tilde{\sigma}_t^2}\right\}\left\{\frac{\partial \sigma_t^2}{\partial \varphi_i}-\frac{\partial \tilde{\sigma}_t^2}{\partial \varphi_i}\right\}\\
&\quad+2\{\epsilon_t-\tilde{\epsilon}_t\}\left\{\frac{1}{\sigma_t^2}\right\}\left\{\frac{\partial \epsilon_t}{\partial \varphi_i}\right\}+2\tilde{\epsilon}_t\left\{\frac{1}{\sigma_t^2}-\frac{1}{\tilde{\sigma}_t^2}\right\}\left\{\frac{\partial \epsilon_t}{\partial \varphi_i}\right\}\\
&\quad\left.+2\tilde{\epsilon}_t\left\{\frac{1}{\tilde{\sigma}_t^2}\right\}\left\{\frac{\partial \epsilon_t}{\partial \varphi_i}-\frac{\partial \tilde{\epsilon}_t}{\partial \varphi_i}\right\}\right|(\varphi_0)\\
&\leq K\rho^t\{1+S_{t-1}^2(|\eta_t|+\eta_t^2)\}\left\{1+\frac{1}{\sigma_t^2}\frac{\partial \sigma_t^2}{\partial \varphi_i}+\left|\frac{\partial \epsilon_t}{\partial \varphi_i}\right|\right\}(\varphi_0).\end{aligned}$$

Using the independence between η_t and S_{t-1}, (7.40), (7.83), the Cauchy–Schwarz inequality and $E(\epsilon_t^4) < \infty$, we obtain

$$P\left(\left|n^{-1/2}\sum_{t=1}^{n}\left\{\frac{\partial \ell_t(\varphi_0)}{\partial \varphi_i} - \frac{\partial \tilde{\ell}_t(\varphi_0)}{\partial \varphi_i}\right\}\right| > \epsilon\right)$$

$$\leq \frac{K}{\epsilon}\left\|1 + \frac{1}{\sigma_t^2}\frac{\partial \sigma_t^2}{\partial \varphi_i}(\varphi_0) + \left|\frac{\partial \epsilon_t}{\partial \varphi_i}\right|(\varphi_0)\right\|_2 n^{-1/2}\sum_{t=1}^{n}\rho^t\{1 + (\|\eta_t^2\|_2 + \|\eta_t\|_2)\|S_{t-1}^2\|_2\}$$

$$\leq \frac{K}{\epsilon}\left\|1 + \frac{1}{\sigma_t^2}\frac{\partial \sigma_t^2}{\partial \varphi_i}(\varphi_0) + \left|\frac{\partial \epsilon_t}{\partial \varphi_i}\right|(\varphi_0)\right\|_2 n^{-1/2}\sum_{t=1}^{n}\rho^t t^2 \to 0,$$

which shows the first part of (c). The second is established by the same arguments.

(d) Use of a CLT for martingale increments.
The proof of this point is exactly the same as that of the pure GARCH case (see the proof of Theorem 7.2).

(e) Convergence to the matrix $\mathcal{J}$.
This part of the proof differs drastically from that of Theorem 7.2. For pure GARCH, we used a Taylor expansion of the second-order derivatives of the criterion, and showed that the third-order derivatives were uniformly integrable in a neighbourhood of θ_0. Without additional assumptions, this argument fails in the ARMA–GARCH case because variables of the form $\sigma_t^{-2}(\partial\sigma_t^2/\partial\vartheta)$ do not necessarily have moments of all orders, even at the true value of the parameter. First note that, since $\mathcal{J}$ exists, the ergodic theorem implies that

$$\lim_{n\to\infty}\frac{1}{n}\sum_{t=1}^{n}\frac{\partial^2 \ell_t(\varphi_0)}{\partial\varphi\partial\varphi'} = \mathcal{J} \quad \text{a.s.}$$

The consistency of $\hat{\varphi}_n$ having already been established, and it suffices to show that for all $\varepsilon > 0$, there exists a neighbourhood $\mathcal{V}(\varphi_0)$ of φ_0 such that almost surely

$$\lim_{n\to\infty}\frac{1}{n}\sum_{t=1}^{n}\sup_{\varphi\in\mathcal{V}(\varphi_0)}\left\|\frac{\partial^2 \ell_t(\varphi)}{\partial\varphi\partial\varphi'} - \frac{\partial^2 \ell_t(\varphi_0)}{\partial\varphi\partial\varphi'}\right\| \leq \varepsilon \tag{7.89}$$

(see Exercise 7.9). We first show that there exists $\mathcal{V}(\varphi_0)$ such that

$$E\sup_{\varphi\in\mathcal{V}(\varphi_0)}\left\|\frac{\partial^2 \ell_t(\varphi)}{\partial\varphi\partial\varphi'}\right\| < \infty. \tag{7.90}$$

By Hölder's inequality, relations (7.39), (7.75), and (7.76), it suffices to show that for any neighbourhood $\mathcal{V}(\varphi_0) \subset \Phi$ whose elements have their components α_i and β_j bounded above by a positive constant, the quantities

$$\left\|\sup_{\varphi\in\mathcal{V}(\varphi_0)}\epsilon_t^2\right\|_2, \quad \left\|\sup_{\varphi\in\mathcal{V}(\varphi_0)}\left|\frac{\partial\epsilon_t}{\partial\vartheta}\right|\right\|_4, \quad \left\|\sup_{\varphi\in\mathcal{V}(\varphi_0)}\left|\frac{\partial^2\epsilon_t}{\partial\vartheta\partial\vartheta'}\right|\right\|_4, \tag{7.91}$$

$$\left\|\sup_{\varphi\in\mathcal{V}(\varphi_0)}\frac{1}{\sigma_t^2}\right\|_\infty, \quad \left\|\sup_{\varphi\in\mathcal{V}(\varphi_0)}\left|\frac{1}{\sigma_t^2}\frac{\partial\sigma_t^2}{\partial\vartheta}\right|\right\|_4, \quad \left\|\sup_{\varphi\in\mathcal{V}(\varphi_0)}\left|\frac{1}{\sigma_t^2}\frac{\partial\sigma_t^2}{\partial\theta}\right|\right\|_4, \tag{7.92}$$

$$\left\|\sup_{\varphi\in\mathcal{V}(\varphi_0)}\left|\frac{\partial^2\sigma_t^2}{\partial\vartheta\partial\vartheta'}\right|\right\|_2, \quad \left\|\sup_{\varphi\in\mathcal{V}(\varphi_0)}\left|\frac{\partial^2\sigma_t^2}{\partial\theta\partial\vartheta'}\right|\right\|_2, \quad \left\|\sup_{\varphi\in\mathcal{V}(\varphi_0)}\left|\frac{1}{\sigma_t^2}\frac{\partial^2\sigma_t^2}{\partial\theta\partial\theta'}\right|\right\|_2 \tag{7.93}$$

are finite. Using the expansion of the series

$$\epsilon_t(\vartheta) = A_\vartheta(B)B_\vartheta^{-1}(B)A_{\vartheta_0}^{-1}(B)B_{\vartheta_0}(B)\epsilon_t(\vartheta_0),$$

similar expansions for the derivatives, and $\|\epsilon_t(\vartheta_0)\|_4 < \infty$, it can be seen that the norms in (7.91) are finite. In (7.92), the first norm is finite, as an obvious consequence of $\sigma_t^2 \geq \inf_{\phi\in\Phi} \omega$, this latter term being strictly positive by compactness of Φ. An extension of inequality (7.83) leads to

$$\sup_{\varphi\in\Phi} \left\| \frac{1}{\sigma_t^2} \frac{\partial \sigma_t^2}{\partial \vartheta_j}(\varphi) \right\|_4 \leq K \sum_{k=0}^{\infty} \rho^{k/2} \sup_{\varphi\in\Phi} \left\| \frac{\partial \epsilon_{t-k}}{\partial \vartheta_j} \right\|_4 < \infty.$$

Moreover, since relations (7.41)–(7.44) remain valid when ϵ_t is replaced by $\epsilon_t(\vartheta)$, it can be shown that

$$\sup_{\varphi\in\mathcal{V}(\varphi_0)} \left\| \frac{1}{\sigma_t^2} \frac{\partial \sigma_t^2}{\partial \theta}(\varphi) \right\|_d < \infty$$

for any $d > 0$ and any neighbourhood $\mathcal{V}(\varphi_0)$ whose elements have their components α_i and β_j bounded from below by a positive constant. The norms in (7.92) are thus finite. The existence of the first norm of (7.93) follows from (7.80) and (7.91). To handle the second one, we use (7.84), (7.85), (7.91), and the fact that $\sup_{\varphi\in\mathcal{V}(\varphi_0)} \beta_j^{-1} < \infty$. Finally, it can be shown that the third norm is finite by (7.47), (7.48) and by arguments already used. The property (7.90) is thus established. The ergodic theorem shows that the limit in (7.89) is equal almost surely to

$$E \sup_{\varphi\in\mathcal{V}(\varphi_0)} \left\| \frac{\partial^2 \ell_t(\varphi)}{\partial\varphi\partial\varphi'} - \frac{\partial^2 \ell_t(\varphi_0)}{\partial\varphi\partial\varphi'} \right\|.$$

By the dominated convergence theorem, using (7.90), this expectation tends to 0 when the neighbourhood $\mathcal{V}(\varphi_0)$ tends to the singleton $\{\varphi_0\}$. Thus (7.89) hold true, which proves (e). The proof of Theorem 7.5 is now complete.

7.5 Bibliographical Notes

The asymptotic properties of the QMLE of the ARCH models have been established by Weiss (1986) under the condition that the moment of order 4 exists. In the GARCH(1, 1) case, the asymptotic properties have been established by Lumsdaine (1996) (see also Lee and Hansen 1994) for the local QMLE under the strict stationarity assumption. In Lumsdaine (1996), the conditions on the coefficients α_1 and β_1 allow to handle the IGARCH(1, 1) model. They are, however, very restrictive with regard to the iid process: it is assumed that $E|\eta_t|^{32} < \infty$ and that the density of η_t has a unique mode and is bounded in a neighbourhood of 0. In Lee and Hansen (1994), the consistency of the global estimator is obtained under the assumption of second-order stationarity.

Berkes, Horváth, and Kokoszka (2003b) was the first paper to give a rigorous proof of the asymptotic properties of the QMLE in the GARCH(p, q) case under very weak assumptions; see also Berkes and Horváth (2003b, 2004), together with Boussama (1998, 2000). The assumptions given in Berkes, Horváth, and Kokoszka (2003b) were weakened slightly in Francq and Zakoïan (2004). The proofs presented here come from that paper. An extension to non-iid errors was proposed by Escanciano (2009).

Jensen and Rahbek (2004a,2004b) have shown that the parameter α_0 of an ARCH(1) model, or the parameters α_0 and β_0 of a GARCH(1, 1) model, can be consistently estimated, with a standard Gaussian asymptotic distribution and a standard rate of convergence, even if the parameters are outside the strict stationarity region. They considered a constrained version of the QMLE, in which the intercept ω is fixed (see Exercises 7.13 and 7.14). These results were misunderstood by a number of researchers and practitioners, who wrongly claimed that the QMLE of the GARCH parameters is consistent and asymptotically normal without any stationarity constraint. We have seen in Section 3.1.3 that the QMLE of ω_0 is inconsistent in the non-stationary ARCH(1) case. The asymptotic properties of the unconstrained QMLE for non-stationary GARCH(1,1)-type models were studied in Francq and Zakoian (2012b, 2013b), as well as tests of strict stationarity and non-stationarity.

For ARMA–GARCH models, asymptotic results have been established by Ling and Li (1997, 1998), Ling and McAleer (2003a,2003b), and Francq and Zakoïan (2004). A comparison of the assumptions used in these papers can be found in the last reference. We refer the reader to Straumann (2005) for a detailed monograph on the estimation of GARCH models, to Francq and Zakoïan (2009a) for a review of the literature, and to Straumann and Mikosch (2006) and Bardet and Wintenberger (2009) for extensions to other conditionally heteroscedastic models. Li, Ling, and McAleer (2002) reviewed the literature on the estimation of ARMA–GARCH models, including in particular the case of nonstationary models.

The proof of the asymptotic normality of the QMLE of ARMA models under the second-order moment assumption can be found, for instance, in Brockwell and Davis (1991). For ARMA models with infinite variance noise, see Davis, Knight, and Liu (1992), Mikosch et al. (1995), and Kokoszka and Taqqu (1996).

7.6 Exercises

7.1 (The distribution of η_t is symmetric for GARCH models)
The aim of this exercise is to show property (7.24).

1. Show the result for $j < 0$.
2. For $j \geq 0$, explain why $E\{g(\epsilon_t^2, \epsilon_{t-1}^2, \ldots) \mid \epsilon_{t-j}, \epsilon_{t-j-1}, \ldots)\}$ can be written as $h(\epsilon_{t-j}^2, \epsilon_{t-j-1}^2, \ldots)$ for some function h.
3. Complete the proof of (7.24).

7.2 (Almost sure convergence to zero at an exponential rate)
Let (ϵ_t) be a strictly stationary process admitting a moment order $s > 0$. Show that if $\rho \in (0, 1)$, then $\rho^t \epsilon_t^2 \to 0$ a.s.

7.3 (Ergodic theorem for nonintegrable processes)
Prove the following ergodic theorem. If (X_t) is an ergodic and strictly stationary process and if EX_1 exists in $\mathbb{R} \cup \{+\infty\}$, then

$$n^{-1} \sum_{t=1}^{n} X_t \to EX_1 \text{ a.s.,} \quad \text{as } n \to \infty.$$

The result is shown in Billingsley (1995, p. 284) for iid variables.
Hint: Consider the truncated variables $X_t^\kappa = X_t \mathbb{1}_{X_t \leq \kappa}$ where $\kappa > 0$ with κ tending to $+\infty$.

7.4 (Uniform ergodic theorem)
Let $\{X_t(\theta)\}$ be a process of the form

$$X_t(\theta) = f(\theta, \eta_t, \eta_{t-1}, \ldots), \tag{7.94}$$

where (η_t) is strictly stationary and ergodic and f is continuous in $\theta \in \Theta$, Θ being a compact subset of $\mathbb{R}^d$.

1. Show that the process $\{\inf_{\theta \in \Theta} X_t(\theta)\}$ is strictly stationary and ergodic.
2. Does the property still hold true if $X_t(\theta)$ is not of the form (7.94), but it is assumed that $\{X_t(\theta)\}$ is strictly stationary and ergodic and that $X_t(\theta)$ is a continuous function of θ?

7.5 (OLS estimator of a GARCH)
In the framework of the GARCH(p, q) model (7.1), an OLS estimator of θ is defined as any measurable solution $\hat{\theta}_n$ of

$$\hat{\theta}_n = \arg\min_{\theta \in \Theta} Q_n(\theta), \quad \Theta \subset \mathbb{R}^{p+q+1},$$

where

$$\tilde{Q}_n(\theta) = n^{-1} \sum_{t=1}^{n} \tilde{e}_t^2(\theta), \quad \tilde{e}_t(\theta) = \epsilon_t^2 - \tilde{\sigma}_t^2(\theta),$$

and $\tilde{\sigma}_t^2(\theta)$ is defined by (7.4) with, for instance, initial values given by (7.6) or (7.7). Note that the estimator is unconstrained and that the variable $\tilde{\sigma}_t^2(\theta)$ can take negative values. Similarly, a constrained OLS estimator is defined by

$$\hat{\theta}_n^c = \arg\min_{\theta\in\Theta^c} Q_n(\theta), \quad \Theta^c \subset (0,+\infty)\times[0,+\infty)^{p+q}.$$

The aim of this exercise is to show that under the assumptions of Theorem 7.1, and if $E_{\theta_0}\epsilon_t^4 < \infty$, the constrained and unconstrained OLS estimators are strongly consistent. We consider the theoretical criterion

$$Q_n(\theta) = n^{-1}\sum_{t=1}^{n} e_t^2(\theta), \quad e_t(\theta) = \epsilon_t^2 - \sigma_t^2(\theta).$$

1. Show that $\sup_{\theta\in\Theta}|\tilde{Q}_n(\theta) - Q_n(\theta)| \to 0$ almost surely as $n \to \infty$.
2. Show that the asymptotic criterion is minimised at θ_0,

$$\forall \theta \in \Theta, \quad \lim_{n\to\infty} Q(\theta) \geq \lim_{n\to\infty} Q(\theta_0),$$

   and that θ_0 is the unique minimum.
3. Prove that $\hat{\theta}_n \to \theta_0$ almost surely as $n \to \infty$.
4. Show that $\hat{\theta}_n^c \to \theta_0$ almost surely as $n \to \infty$.

7.6 (The mean of the squares of the normalised residuals is equal to 1)
For a GARCH model, estimated by QML with initial values set to zero, the normalized residuals are defined by $\hat{\eta}_t = \epsilon_t/\tilde{\sigma}_t(\hat{\theta}_n)$, $t = 1, \dots, n$. Show that almost surely, for Θ large enough,

$$\frac{1}{n}\sum_{t=1}^{n}\hat{\eta}_t^2 = 1.$$

Hint: Note that for all $c > 0$ and all $\theta \in \Theta$, there exists θ_c such that $\tilde{\sigma}_t^2(\theta_c) = c\tilde{\sigma}_t^2(\theta)$ for all $t \geq 0$, and consider the function $c \mapsto \mathbf{l}_n(\theta)$.

7.7 ($\mathcal{I}$ and $\mathcal{J}$ *block-diagonal*)
Show that $\mathcal{I}$ and $\mathcal{J}$ have the block-diagonal form given in Theorem 7.5 when the distribution of η_t is symmetric.

7.8 (Forms of $\mathcal{I}$ and $\mathcal{J}$ in the AR(1)-ARCH(1) case)
We consider the QML estimation of the AR(1)-ARCH(1) model

$$X_t = a_0X_{t-1} + \epsilon_t, \quad \epsilon_t = \sigma_t\eta_t, \quad \sigma_t^2 = 1 + \alpha_0\epsilon_{t-1}^2,$$

assuming that $\omega_0 = 1$ is known and without specifying the distribution of η_t.

1. Give the explicit form of the matrices $\mathcal{I}$ and $\mathcal{J}$ in Theorem 7.5 (with an obvious adaptation of the notation because the parameter here is (a_0, α_0)).
2. Give the block-diagonal form of these matrices when the distribution of η_t is symmetric, and verify that the asymptotic variance of the estimator of the ARCH parameter
   (i) does not depend on the AR parameter, and
   (ii) is the same as for the estimator of a pure ARCH (without the AR part).
3. Compute Σ when $\alpha_0 = 0$. Is the asymptotic variance of the estimator of a_0 the same as that obtained when estimating an AR(1)? Verify the results obtained by simulation in the corresponding column of Table 7.3.

7.9 (A useful result in showing asymptotic normality)
Let $(J_t(\theta))$ be a sequence of random matrices, which are function of a vector of parameters θ. We consider an estimator $\hat{\theta}_n$ which strongly converges to the vector θ_0. Assume that

$$\frac{1}{n}\sum_{t=1}^{n} J_t(\theta_0) \to J, \quad \text{a.s.},$$

where J is a matrix. Show that if for all $\varepsilon > 0$ there exists a neighbourhood $V(\theta_0)$ of θ_0 such that

$$\lim_{n\to\infty} \frac{1}{n}\sum_{t=1}^{n} \sup_{\theta\in V(\theta_0)} \|J_t(\theta) - J_t(\theta_0)\| \leq \varepsilon, \quad \text{a.s.}, \tag{7.95}$$

where $\|\cdot\|$ denotes a matrix norm, then

$$\frac{1}{n}\sum_{t=1}^{n} J_t(\hat{\theta}_n) \to J, \quad \text{a.s.}$$

Give an example showing that condition (7.95) is not necessary for the latter convergence to hold in probability.

7.10 (A lower bound for the asymptotic variance of the QMLE of an ARCH)
Show that, for the ARCH(q) model, under the assumptions of Theorem 7.2,

$$\text{Var}_{as}\{\sqrt{n}(\hat{\theta}_n - \theta_0)\} \geq (\kappa_\eta - 1)\theta_0\theta_0',$$

in the sense that the difference is a positive semi-definite matrix.
Hint: Compute $\theta_0 \partial\sigma_t^2(\theta_0)/\partial\theta'$ and show that $J - J\theta_0\theta_0' J$ is a variance matrix.

7.11 (A striking property of $\mathcal{J}$)
For a GARCH(p, q) model we have, under the assumptions of Theorem 7.2,

$$J = E(Z_t Z_t'), \quad \text{where } Z_t = \frac{1}{\sigma_t^2(\theta_0)} \frac{\partial\sigma_t^2(\theta_0)}{\partial\theta}.$$

The objective of the exercise is to show that

$$\Omega' J^{-1} \Omega = 1, \quad \text{where } \Omega = E(Z_t). \tag{7.96}$$

1. Show the property in the ARCH case.
   Hint: Compute $\theta_0' Z_t$, $\theta_0' J$ and $\theta_0' J \theta_0$.
2. In the GARCH case, let $\overline{\theta} = (\omega, \alpha_1, \dots, \alpha_q, 0, \dots, 0)'$. Show that

$$\overline{\theta}' \frac{\partial\sigma_t^2(\theta)}{\partial\theta'} = \sigma_t^2(\theta).$$

3. Complete the proof of (7.96).

7.12 (A condition required for the generalised Bartlett formula)
Using (7.24), show that if the distribution of η_t is symmetric and if $E(\epsilon_t^4) < \infty$, then formula (B.13) holds true, that is,

$$E\epsilon_{t_1}\epsilon_{t_2}\epsilon_{t_3}\epsilon_{t_4} = 0 \quad \text{when } t_1 \neq t_2, t_1 \neq t_3 \text{ and } t_1 \neq t_4.$$

7.13 (Constrained QMLE of the parameter α_0 of a nonstationary ARCH(1) process)
Jensen and Rahbek (2004a) consider the ARCH(1) model (7.15), in which the parameter $\omega_0 > 0$ *is assumed to be known* ($\omega_0 = 1$ for instance) and where only α_0 is unknown. They work with the *constrained QMLE* of α_0 defined by

$$\hat{\alpha}_n^c(\omega_0) = \arg\min_{\alpha \in [0,\infty)} \frac{1}{n}\sum_{t=1}^{n} \ell_t(\alpha), \quad \ell_t(\alpha) = \frac{\epsilon_t^2}{\sigma_t^2(\alpha)} + \log \sigma_t^2(\alpha), \tag{7.97}$$

where $\sigma_t^2(\alpha) = \omega_0 + \alpha\epsilon_{t-1}^2$. Assume therefore that $\omega_0 = 1$ and suppose that the nonstationarity condition (7.16) is satisfied.

1. Verify that

$$\frac{1}{\sqrt{n}}\sum_{t=1}^{n} \frac{\partial}{\partial\alpha}\ell_t(\alpha_0) = \frac{1}{\sqrt{n}}\sum_{t=1}^{n}(1-\eta_t^2)\frac{\epsilon_{t-1}^2}{1+\alpha_0\epsilon_{t-1}^2}$$

and that

$$\frac{\epsilon_{t-1}^2}{1+\alpha_0\epsilon_{t-1}^2} \to \frac{1}{\alpha_0} \quad \text{a.s.} \quad \text{as } t \to \infty.$$

2. Prove that

$$\frac{1}{\sqrt{n}}\sum_{t=1}^{n} \frac{\partial}{\partial\alpha}\ell_t(\alpha_0) \xrightarrow{\mathcal{L}} \mathcal{N}\left(0, \frac{\kappa_\eta - 1}{\alpha_0^2}\right).$$

3. Determine the almost sure limit of

$$\frac{1}{n}\sum_{t=1}^{n} \frac{\partial^2}{\partial\alpha^2}\ell_t(\alpha_0).$$

4. Show that for all $\underline{\alpha} > 0$, almost surely

$$\sup_{\alpha \geq \underline{\alpha}} \left| \frac{1}{n}\sum_{t=1}^{n} \frac{\partial^3}{\partial\alpha^3}\ell_t(\alpha) \right| = O(1).$$

5. Prove that if $\hat{\alpha}_n^c = \hat{\alpha}_n^c(\omega_0) \to \alpha_0$ almost surely (see Exercise 7.14) then

$$\sqrt{n}(\hat{\alpha}_n^c - \alpha_0) \xrightarrow{\mathcal{L}} \mathcal{N}\{0, (\kappa_\eta - 1)\alpha_0^2\}.$$

6. Does the result change when $\hat{\alpha}_n^c = \hat{\alpha}_n^c(1)$ and $\omega_0 \neq 1$?
7. Discuss the practical usefulness of this result for estimating ARCH models.

7.14 (Strong consistency of Jensen and Rahbek's estimator)
We consider the framework of Exercise 7.13, and follow the lines of the proof of (7.19)

1. Show that $\hat{\alpha}_n^c(1)$ converges almost surely to α_0 when $\omega_0 = 1$.
2. Does the result change if $\hat{\alpha}_n^c(1)$ is replaced by $\hat{\alpha}_n^c(\omega)$ and if ω and ω_0 are arbitrary positive numbers? Does it entail the convergence result (7.19)?

8

Tests Based on the Likelihood

In the previous chapter, we saw that the asymptotic normality of the QMLE of a GARCH model holds true under general conditions, in particular without any moment assumption on the observed process. An important application of this result concerns testing problems. In particular, we are able to test the IGARCH assumption, or more generally a given GARCH model with infinite variance. This problem is the subject of Section 8.1.

The main aim of this chapter is to derive tests for the nullity of coefficients. These tests are complex in the GARCH case because of the constraints that are imposed on the estimates of the coefficients to guarantee that the estimated conditional variance is positive. Without these constraints, it is impossible to compute the Gaussian log-likelihood of the GARCH model. Moreover, asymptotic normality of the QMLE has been established assuming that the parameter belongs to the interior of the parameter space (assumption A5 in Chapter 7). When some coefficients, α_i or β_j, are null, Theorem 7.2 does not apply. It is easy to see that, in such a situation, the asymptotic distribution of $\sqrt{n}(\hat{\theta}_n - \theta_0)$ cannot be Gaussian. Indeed, the components, $\hat{\theta}_{in}$ of $\hat{\theta}_n$, are constrained to be positive or null. If, for instance $\theta_{0i} = 0$, then $\sqrt{n}(\hat{\theta}_{in} - \theta_{0i}) = \sqrt{n}\hat{\theta}_{in} \geq 0$ for all n and the asymptotic distribution of this variable cannot be Gaussian.

Before considering the significance tests, we shall, therefore, establish in Section 8.2 the asymptotic distribution of the QMLE without assumption A5, at the cost of a moment assumption on the observed process. In Section 8.3, we present the main tests (Wald, score and likelihood ratio) used for testing the nullity of some coefficients. The asymptotic distribution obtained for the QMLE will lead to modification of the standard critical regions. Two cases of particular interest will be examined in detail: the test of nullity of only one coefficient and the test of conditional homoscedasticity, which corresponds to the nullity of all the coefficients α_i and β_j. Section 8.4 is devoted to testing the adequacy of a particular GARCH(p, q) model, using portmanteau tests. The chapter also contains a numerical application in which the pre-eminence of the GARCH(1, 1) model is questioned.

8.1 Test of the Second-Order Stationarity Assumption

For the GARCH(p, q) model defined by Eq. (7.1), testing for second-order stationarity involves testing

$$H_0 : \sum_{i=1}^{q} \alpha_{0i} + \sum_{j=1}^{p} \beta_{0j} < 1 \quad \text{against } H_1 : \sum_{i=1}^{q} \alpha_{0i} + \sum_{j=1}^{p} \beta_{0j} \geq 1.$$

GARCH Models: Structure, Statistical Inference and Financial Applications, Second Edition. Christian Francq and Jean-Michel Zakoian.

Introducing the vector $c = (0, 1, \ldots, 1)' \in \mathbb{R}^{p+q+1}$, the testing problem is

$$H_0 : c'\theta_0 < 1 \quad \text{against} \quad H_1 : c'\theta_0 \geq 1. \tag{8.1}$$

In view of Theorem 7.2, the QMLE $\hat{\theta}_n = (\hat{\omega}_n, \hat{\alpha}_{1n}, \ldots, \hat{\alpha}_{qn}, \hat{\beta}_{1n}, \ldots \hat{\beta}_{pn})$ of θ_0 satisfies

$$\sqrt{n}(\hat{\theta}_n - \theta_0) \xrightarrow{\mathcal{L}} \mathcal{N}(0, (\kappa_\eta - 1)J^{-1}),$$

under assumptions which are compatible with H_0 and H_1. In particular, if $c'\theta_0 = 1$ we have

$$\sqrt{n}(c'\hat{\theta}_n - 1) \xrightarrow{\mathcal{L}} \mathcal{N}(0, (\kappa_\eta - 1)c'J^{-1}c).$$

It is thus natural to consider the Wald statistic

$$\mathbf{T}_n = \frac{\sqrt{n}\left(\sum_{i=1}^{q} \hat{\alpha}_{in} + \sum_{j=1}^{p} \hat{\beta}_{jn} - 1\right)}{\{(\hat{\kappa}_\eta - 1)c'\hat{J}^{-1}c\}^{1/2}},$$

where $\hat{\kappa}_\eta$ and $\hat{J}$ are consistent estimators in probability of κ_η and J. The following result follows immediately from the convergence of $\mathbf{T}_n$ to $\mathcal{N}(0, 1)$ when $c'\theta_0 = 1$.

Proposition 8.1 (Critical region of stationarity test) *Under the assumptions of Theorem 7.2, a test of the form (8.1) at the asymptotic level α is defined by the rejection region*

$$\{\mathbf{T}_n > \Phi^{-1}(1 - \alpha)\},$$

where Φ is the $\mathcal{N}(0, 1)$ cumulative distribution function.

Note that for most real series (see, for instance Table 7.4), the sum of the estimated coefficients $\hat{\alpha}$ and $\hat{\beta}$ is strictly less than 1: second-order stationarity thus cannot be rejected, for any reasonable asymptotic level (when $\mathbf{T}_n < 0$, the p-value of the test is greater than 1/2). Of course, the non-rejection of H_0 does not mean that the stationarity is proved. It is interesting to test the reverse assumption that the data-generating process is an IGARCH, or more generally that it does not have moments of order 2. We thus consider the problem

$$H_0 : c'\theta_0 \geq 1 \quad \text{against } H_1 : c'\theta_0 < 1. \tag{8.2}$$

Proposition 8.2 (Critical region of non-stationarity test) *Under the assumptions of Theorem 7.2, a test of the form (8.2) at the asymptotic level α is defined by the rejection region*

$$\{\mathbf{T}_n < \Phi^{-1}(\alpha)\}.$$

As an application, we take up the data sets of Table 7.4 again, and we give the p-values of the previous test for the 11 series of daily returns. For the FTSE (DAX, Nasdaq, and S&P 500), the assumption of infinite variance cannot be rejected at the 5% (3%, 2%, and 1%) level (see Table 8.1). The other series can be considered as second-order stationary (if one believes in the GARCH(1, 1) model, of course).

8.2 Asymptotic Distribution of the QML When θ_0 is at the Boundary

In view of the form of the parameter space (7.3), the QMLE $\hat{\theta}_n$ is constrained to have a strictly positive first component, while the other components are constrained to be positive or null. A general technique

Table 8.1 Test of the infinite variance assumption for 11 stock market returns.

Index	$\hat{\alpha} + \hat{\beta}$	p-Value
CAC	0.983 (0.007)	0.0089
DAX	0.981 (0.011)	0.0385
DJA	0.982 (0.007)	0.0039
DJI	0.986 (0.006)	0.0061
DJT	0.983 (0.009)	0.0023
DJU	0.983 (0.007)	0.0060
FTSE	0.990 (0.006)	0.0525
Nasdaq	0.993 (0.003)	0.0296
Nikkei	0.980 (0.007)	0.0017
SMI	0.962 (0.015)	0.0050
S&P 500	0.989 (0.005)	0.0157

Estimated standard deviations are in parentheses.

for determining the distribution of a constrained estimator involves expressing it as a function of the unconstrained estimator $\hat{\theta}_n^{nc}$ (see Gouriéroux and Monfort 1995). For the QMLE of a GARCH, this technique does not work because the objective function

$$\tilde{\mathbf{l}}_n(\theta) = n^{-1} \sum_{t=1}^{n} \tilde{\ell}_t, \quad \tilde{\ell}_t = \tilde{\ell}_t(\theta) = \frac{\epsilon_t^2}{\tilde{\sigma}_t^2} + \log \tilde{\sigma}_t^2, \quad \text{where } \tilde{\sigma}_t^2 \text{ is defined by (7.4),}$$

cannot be computed outside Θ (for an ARCH(1), it may happen that $\tilde{\sigma}_t{}^2 := \omega + \alpha_1 \epsilon_t^2$ is negative when $\alpha_1 < 0$). It is thus impossible to define $\hat{\theta}_n^{nc}$.

The technique that we will utilise here (see, in particular, Andrews 1999), involves writing $\hat{\theta}_n$ with the aid of the normalised score vector, evaluated at θ_0:

$$Z_n := -J_n^{-1} n^{1/2} \frac{\partial \mathbf{l}_n(\theta_0)}{\partial \theta}, \quad J_n = \frac{\partial^2 \mathbf{l}_n(\theta_0)}{\partial \theta \partial \theta'}, \tag{8.3}$$

with

$$\mathbf{l}_n(\theta) = \mathbf{l}_n(\theta; \epsilon_n, \epsilon_{n-1}, \ldots) = n^{-1} \sum_{t=1}^{n} \ell_t, \quad \ell_t = \ell_t(\theta) = \frac{\epsilon_t^2}{\sigma_t^2} + \log \sigma_t^2,$$

where the components of $\partial \mathbf{l}_n(\theta_0)/\partial \theta$ and of J_n are right derivatives (see (a) in the proof of Theorem 8.1 below).

In the proof of Theorem 7.2, we showed that

$$n^{1/2}(\hat{\theta}_n - \theta_0) = Z_n + o_P(1), \quad \text{when } \theta_0 \in \overset{\circ}{\Theta}. \tag{8.4}$$

For any value of $\theta_0 \in \Theta$ (even when $\theta_0 \notin \overset{\circ}{\Theta}$), it will be shown that the vector Z_n is well defined and satisfies

$$Z_n \xrightarrow{\mathcal{L}} Z \sim \mathcal{N}\{0, (\kappa_\eta - 1) J^{-1}\}, \quad J = E_{\theta_0} \left\{ \frac{1}{\sigma_t^4} \frac{\partial \sigma_t^2}{\partial \theta} \frac{\partial \sigma_t^2}{\partial \theta'} (\theta_0) \right\}, \tag{8.5}$$

provided J exists. By contrast, when

$$\theta_0 \in \partial\Theta := \{\theta \in \Theta : \theta_i = 0 \text{ for some } i > 1\},$$

Result (8.4) is no longer valid. However, we will show that the asymptotic distribution of $n^{1/2}(\hat{\theta}_n - \theta_0)$ is well approximated by that of the vector $n^{1/2}(\theta - \theta_0)$ which is located at the minimal distance of Z_n, under the constraint $\theta \in \Theta$. Consider thus a random vector $\theta_{J_n}(Z_n)$ (which is not an estimator, of course) solving the minimisation problem

$$\theta_{J_n}(Z_n) = \arg\inf_{\theta\in\Theta} \{Z_n - n^{1/2}(\theta - \theta_0)\}' J_n \{Z_n - n^{1/2}(\theta - \theta_0)\}. \tag{8.6}$$

It will be shown that J_n converges to the positive definite matrix J. For n large enough, we thus have

$$\theta_{J_n}(Z_n) = \arg\inf_{\theta\in\Theta} \text{dist}^2_{J_n}\{Z_n, \sqrt{n}(\theta - \theta_0)\},$$

where $\text{dist}_{J_n}(x, y) := \{(x-y)' J_n (x-y)\}^{1/2}$ is a distance between two points x and y of $\mathbb{R}^{p+q+1}$.

We allow θ_0 to have null components, but we do not consider the (less interesting) case where θ_0 reaches another boundary of Θ. More precisely, we assume that

B1: $\theta_0 \in (\underline{\omega}, \overline{\omega}) \times [0, \overline{\theta}_2) \times \cdots \times [0, \overline{\theta}_{p+q+1}) \subset \Theta$,

where $0 < \underline{\omega} < \overline{\omega}$ and $0 < \min\{\overline{\theta}_2, \ldots, \overline{\theta}_{p+q+1}\}$. In this case $\sqrt{n}(\theta_{J_n}(Z_n) - \theta_0)$ and $\sqrt{n}(\hat{\theta}_n - \theta_0)$ belong to the 'local parameter space'

$$\Lambda := \bigcup_n \sqrt{n}(\Theta - \theta_0) = \Lambda_1 \times \cdots \times \Lambda_{q+q+1}, \tag{8.7}$$

where $\Lambda_1 = \mathbb{R}$, for $i = 2, \ldots, p+q+1$, $\Lambda_i = \mathbb{R}$ if $\theta_{0i} \neq 0$ and $\Lambda_i = [0, \infty)$ if $\theta_{0i} = 0$. With the notation

$$\lambda_n^\Lambda = \arg\inf_{\lambda\in\Lambda} \{\lambda - Z_n\}' J_n \{\lambda - Z_n\},$$

we thus have, with probability 1,

$$\sqrt{n}(\theta_{J_n}(Z_n) - \theta_0) = \lambda_n^\Lambda, \quad \text{for } n \text{ large enough.} \tag{8.8}$$

The vector λ_n^Λ is the projection of Z_n on Λ, with respect to the norm $\|x\|^2_{J_n} := x' J_n x$ (see Figure 8.1). Since Λ is closed and convex, such a projection is unique. We will show that

$$n^{1/2}(\hat{\theta}_n - \theta_0) = \lambda_n^\Lambda + o_P(1). \tag{8.9}$$

Since (Z_n, J_n) tends in law to (Z, J) and λ_n^Λ is a function of (Z_n, J_n) which is continuous everywhere except at the points where J_n is singular (that is, almost everywhere with respect to the distribution of (Z, J) because J is invertible), we have $\lambda_n^\Lambda \xrightarrow{\mathcal{L}} \lambda^\Lambda$, where λ^Λ is the solution of limiting problem

$$\lambda^\Lambda := \arg\inf_{\lambda\in\Lambda} \{\lambda - Z\}' J \{\lambda - Z\}, \quad Z \sim \mathcal{N}\{0, (\kappa_\eta - 1) J^{-1}\}. \tag{8.10}$$

In addition to B1, we retain most of the assumptions of Theorem 7.2:

B2: $\theta_0 \in \Theta$ and Θ is a compact set.

B3: $\gamma(\mathbf{A}_0) < 0$ and for all $\theta \in \Theta$, $\sum_{j=1}^p \beta_j < 1$.

B4: η_t^2 has a non-degenerate distribution with $E\eta_t^2 = 1$.

B5: If $p > 0$, $\mathcal{A}_{\theta_0}(z)$, and $\mathcal{B}_{\theta_0}(z)$ do not have common roots, $\mathcal{A}_{\theta_0}(1) \neq 0$, and $\alpha_{0q} + \beta_{0p} \neq 0$.

B6: $\kappa_\eta = E\eta_t^4 < \infty$.

We also need the following moment assumption:

B7: $E\epsilon_t^6 < \infty$.

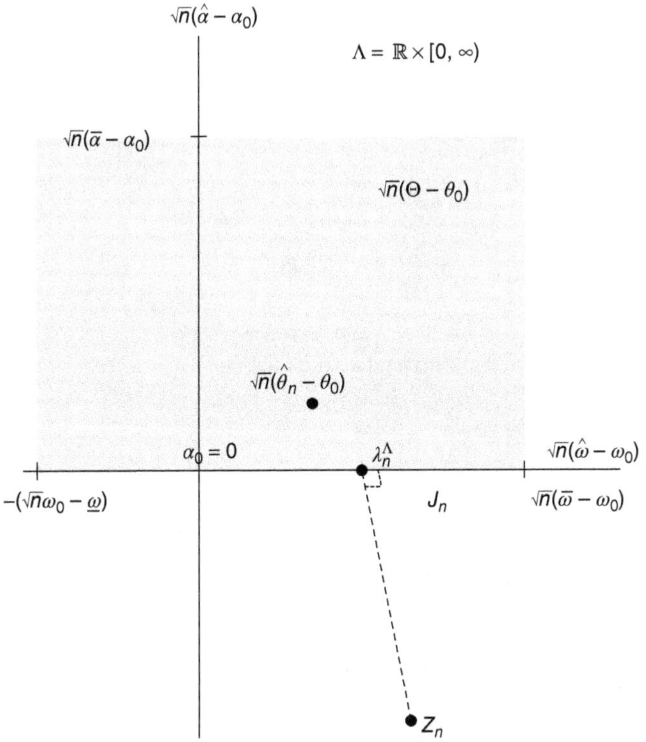

Figure 8.1 ARCH(1) model with $\theta_0 = (\omega_0, 0)$ and $\Theta = [\underline{\omega}, \overline{\omega}] \times [0, \overline{\alpha}]$: $\sqrt{n}(\Theta - \theta_0) = [-\sqrt{n}(\omega_0 - \underline{\omega}), \sqrt{n}(\overline{\omega} - \omega_0)] \times [0, \sqrt{n}(\overline{\alpha} - \alpha_0)]$ is the grey area; $Z_n \xrightarrow{\mathcal{L}} \mathcal{N}\{0, (\kappa_\eta - 1)J^{-1}\}$; $\sqrt{n}(\hat{\theta}_n - \theta_0)$ and λ_n^Λ have the same asymptotic distribution.

When $\theta_0 \in \overset{\circ}{\Theta}$, we can show the existence of the information matrix

$$J = E_{\theta_0}\left\{\frac{1}{\sigma_t^4}\frac{\partial \sigma_t^2}{\partial \theta}\frac{\partial \sigma_t^2}{\partial \theta'}(\theta_0)\right\}$$

without moment assumptions similar to B7. The following example shows that, in the ARCH case, this is no longer possible when we allow $\theta_0 \in \partial\Theta$.

Example 8.1 (The existence of *J* may require a moment of order 4) Consider the ARCH(2) model

$$\epsilon_t = \sigma_t \eta_t, \quad \sigma_t^2 = \omega_0 + \alpha_{01}\epsilon_{t-1}^2 + \alpha_{02}\epsilon_{t-2}^2, \tag{8.11}$$

where the true values of the parameters are such that $\omega_0 > 0$, $\alpha_{01} \geq 0$, $\alpha_{02} = 0$, and the distribution of the iid sequence (η_t) is defined, for $a > 1$, by

$$\mathbb{P}(\eta_t = a) = \mathbb{P}(\eta_t = -a) = \frac{1}{2a^2}, \quad \mathbb{P}(\eta_t = 0) = 1 - \frac{1}{a^2}.$$

The process (ϵ_t) is always stationary, for any value of α_{01} (since $\exp\{-E(\log \eta_t^2)\} = +\infty$, the strict stationarity constraint (2.10) holds true). By contrast, ϵ_t does not possess moments of order 2 when $\alpha_{01} \geq 1$ (see Proposition 2.2).

We have

$$\frac{1}{\sigma_t^2}\frac{\partial\sigma_t^2}{\partial\alpha_2}(\theta_0)=\frac{\epsilon_{t-2}^2}{\omega_0+\alpha_{01}\epsilon_{t-1}^2},$$

so that

$$E\left\{\frac{1}{\sigma_t^2}\frac{\partial\sigma_t^2}{\partial\alpha_2}(\theta_0)\right\}^2\geq E\left[\left\{\frac{\epsilon_{t-2}^2}{\omega_0+\alpha_{01}\epsilon_{t-1}^2}\right\}^2\mid\eta_{t-1}=0\right]\mathbb{P}(\eta_{t-1}=0)$$
$$=\frac{1}{\omega_0^2}\left(1-\frac{1}{a^2}\right)E(\epsilon_{t-2}^4)$$

Because, on the one hand, $\eta_{t-1}=0$ entails $\epsilon_{t-1}=0$, and on the other hand, η_{t-1} and ϵ_{t-2} are independent. Consequently, if $E\epsilon_t^4=\infty$, then the matrix J does not exist.

We then have the following result.

Theorem 8.1 (QML asymptotic distribution at the boundary) *Under assumptions B1–B7, the asymptotic distribution of $\sqrt{n}(\hat{\theta}_n-\theta_0)$ is that of λ^Λ satisfying (8.10), where Λ is given by (8.7).*

Remark 8.1 (We retrieve the standard results in $\overset{\circ}{\Theta}$) For $\theta_0\in\overset{\circ}{\Theta}$, the result is shown in Theorem 7.2. Indeed, in this case $\Lambda=\mathbb{R}^{p+q+1}$ and

$$\lambda^\Lambda=Z\sim\mathcal{N}\{0,(\kappa_\eta-1)J^{-1}\}.$$

Theorem 8.1 is thus only of interest when θ_0 is at the boundary $\partial\Theta$ of the parameter space.

Remark 8.2 (The moment condition B7 can sometimes be relaxed) Apart from the ARCH(q) case, it is sometimes possible to get rid of the moment assumption B7. Note that under the condition $\gamma(\mathbf{A}_0)<0$, we have $\sigma_t^2(\theta_0)=b_{00}+\sum_{j=1}^{\infty}b_{0j}\epsilon_{t-j}^2$ with $b_{00}>0$, $b_{0j}\geq 0$. The derivatives $\partial\sigma_t^2/\partial\theta_k$ have the form of similar series. It can be shown that the ratio $\{\partial\sigma_t^2/\partial\theta\}/\sigma_t^2$ admits moments of all orders whenever any term ϵ_{t-j}^2 which appears in the numerator is also present in the denominator. This allows us to show (see the references at the end of the chapter) that, in the theorem, assumption B7 can be replaced by

B7:′ $b_{0j}>0$ for all $j\geq 1$, where $\sigma_t^2(\theta_0)=b_{00}+\sum_{j=1}^{\infty}b_{0j}\epsilon_{t-j}^2$.

Note that a sufficient condition for B7′ is $\alpha_{01}>0$ and $\beta_{01}>0$ (because $b_{0j}\geq\alpha_{01}\beta_{01}^{j-1}$). A necessary condition is obviously that $\alpha_{01}>0$ (because $b_{01}=\alpha_{01}$). Finally, a necessary and sufficient condition for B7′ is

$$\{j\mid\beta_{0,j}>0\}\neq\emptyset\quad\text{and}\quad\prod_{i=1}^{j_0}\alpha_{0i}>0\quad\text{for}\quad j_0=\min\{j\mid\beta_{0,j}>0\}.$$

Obviously, according to Example 8.1, assumption B7′ is not satisfied in the ARCH case.

Computation of the Asymptotic Distribution

In this section, we will show how to compute the solutions of the optimization problem (8.10). Switching the components of θ, if necessary, it can be assumed without loss of generality that the vector $\theta_0^{(1)}$ of the first d_1 components of θ_0 has strictly positive elements and that the vector $\theta_0^{(2)}$ of the last $d_2=p+q+1-d_1$ components of θ_0 is null. This can be written as

$$K\theta_0=0_{d_2\times 1},\quad K=(0_{d_2\times d_1},I_{d_2}).\tag{8.12}$$

More generally, it will be useful to consider all the subsets of these constraints. Let

$$\mathcal{K} = \{K_1, \dots, K_{2^{d_2}-1}\}$$

be the set of the matrices obtained by deleting no, one, or several (but not all) rows of K. Note that the solution of the constrained minimisation problem (8.10) is the unconstrained solution $\lambda = Z$ when the latter satisfies the constraint, that is, when

$$Z \in \Lambda = \mathbb{R}^{d_1} \times [0, \infty)^{d_2} = \{\lambda \in \mathbb{R}^{p+q+1} \mid K\lambda \geq 0\}.$$

When $Z \notin \Lambda$, the solution λ^Λ coincides with that of an equality constrained problem of the form

$$\lambda_{K_i} = \arg\min_{\lambda : K_i\lambda = 0} (\lambda - Z)' J(\lambda - Z), \quad K_i \in \mathcal{K}.$$

An important difference, compared to the initial minimisation program (8.10), is that the minimisation is done here on a vectorial space. The solution is given by a projection (non-orthogonal when J is not the identity matrix). We thus obtain (see Exercise 8.1)

$$\lambda_{K_i} = P_i Z, \quad \text{where } P_i = I_{p+q+1} - J^{-1}K_i'(K_i J^{-1} K_i')^{-1} K_i \tag{8.13}$$

is the projection matrix (orthogonal for the metric defined by J) on the orthogonal subspace of the space generated by the rows of K_i. Note that λ_{K_i} does not necessarily belong to Λ because $K_i\lambda = 0$ does not imply that $K\lambda \geq 0$. Let $C = \{\lambda_{K_i} : K_i \in \mathcal{K} \text{ and } K\lambda_{K_i} \geq 0\}$ be the class of the admissible solutions. It follows that the solution that we are looking for is

$$\lambda^\Lambda = Z \mathbb{1}_\Lambda(Z) + \mathbb{1}_{\Lambda^c}(Z) \times \arg\min_{\lambda \in C} Q(\lambda) \quad \text{where } Q(\lambda) = (\lambda - Z)' J(\lambda - Z).$$

This formula can be used in practice to obtain realisations of λ^Λ from realisations of Z. The $Q(\lambda_{K_i})$ can be obtained by writing

$$Q(P_i Z) = Z' K_i'(K_i J^{-1} K_i')^{-1} K_i Z. \tag{8.14}$$

Another expression (of theoretical interest) for λ^Λ is

$$\lambda^\Lambda = Z\, \mathbb{1}_{\mathcal{D}_0}(Z) + \sum_{i=1}^{2^{d_2}-1} P_i Z \mathbb{1}_{\mathcal{D}_i}(Z)$$

where $\mathcal{D}_0 = \Lambda$ and the $\mathcal{D}_i$ form a partition of $\mathbb{R}^{p+q+1}$. Indeed, according to the zone to which Z belongs, a solution $\lambda^\Lambda = \lambda_{K_i}$ is obtained. We will make explicit these formulas in a few examples. Let $d = p + q + 1$, $z^+ = z\mathbb{1}_{(0,+\infty)}(z)$ and $z^- = z\mathbb{1}_{(-\infty,0)}(z)$.

Example 8.2 (Law when only one component is at the boundary) When $d_2 = 1$, that is, when only the last component of θ_0 is zero, we have

$$\Lambda = \mathbb{R}^{d_1} \times [0, \infty), \qquad K = (0, \dots, 0, 1), \qquad \mathcal{K} = \{K\},$$

and

$$\lambda^\Lambda = Z\mathbb{1}_{\{Z_d \geq 0\}} + PZ\mathbb{1}_{\{Z_d < 0\}}, \quad P = I_d - J^{-1}K'(KJ^{-1}K')^{-1}K.$$

We finally obtain

$$\lambda^\Lambda = Z - Z_d^- c,$$

where c is the last column of J^{-1} divided by the (d, d)th element of this matrix. Note that the last component of λ^Λ is Z_d^+. Noting that J^{-1} is, up to a multiplicative factor, the variance of Z, it can also be seen that

$$\lambda^\Lambda = \begin{pmatrix} Z_1 - \gamma_1 Z_d^- \\ \vdots \\ Z_{p+q} - \gamma_{p+q} Z_d^- \\ Z_d^+ \end{pmatrix}, \quad \gamma_i = \frac{E(Z_d Z_i)}{\mathrm{Var}(Z_d)}. \tag{8.15}$$

Thus $\lambda_i^\Lambda = Z_i$ if and only if $\mathrm{Cov}(Z_i, Z_d) = 0$.

Example 8.3 (ARCH(2) model when the data-generating process is a white noise) Consider an ARCH(2) model with $\theta_0 = (\omega_0, 0, 0)$. We thus have $d_2 = 2$, $d_1 = 1$ and

$$\Lambda = \mathbb{R} \times [0, \infty)^2, \quad K = \begin{pmatrix} 0 & 1 & 0 \\ 0 & 0 & 1 \end{pmatrix}, \quad \mathcal{K} = \{K_1, K_2, K_3\}$$

with $K_1 = K$, $K_2 = (0, 1, 0)$ and $K_3 = (0, 0, 1)$. Exercise 8.6 shows that

$$Z = \begin{pmatrix} Z_1 \\ Z_2 \\ Z_3 \end{pmatrix} \sim \mathcal{N}\left\{0, \Sigma = (\kappa_\eta - 1)J^{-1} = \begin{pmatrix} (\kappa_\eta + 1)\omega_0^2 & -\omega_0 & -\omega_0 \\ -\omega_0 & 1 & 0 \\ -\omega_0 & 0 & 1 \end{pmatrix}\right\}.$$

Using $K\Sigma K' = I_2$ and $K_i \Sigma K_i' = 1$ for $i = 2, 3$ in particular, we thus obtain

$$P_1 Z = (Z_1 + \omega_0(Z_2 + Z_3), 0, 0)',$$
$$P_2 Z = (Z_1 + \omega_0 Z_2, 0, Z_3)',$$
$$P_3 Z = (Z_1 + \omega_0 Z_3, Z_2, 0)'.$$

Let $P_0 = I_3$ and $K_0 = 0$. Using relation (8.14), we have

$$Q(P_i Z) = (\kappa_\eta - 1) Z' K_i' K_i Z$$
$$= (\kappa_\eta - 1) \begin{cases} 0 & \text{for } i = 0 \\ Z_2^2 + Z_3^2 & \text{for } i = 1 \\ Z_2^2 & \text{for } i = 2 \\ Z_3^2 & \text{for } i = 3. \end{cases}$$

This shows that

$$\lambda^\Lambda = \begin{pmatrix} Z_1 + \omega Z_2^- + \omega Z_3^- \\ Z_2^+ \\ Z_3^+ \end{pmatrix}. \tag{8.16}$$

In order to obtain a slightly simpler expression for the projections (8.13), note that the constraint (8.12) can also be written as

$$\theta_0 = H\theta_0^{(1)}, \quad H = \begin{pmatrix} I_{d_1} \\ 0_{d_2 \times d_1} \end{pmatrix}. \tag{8.17}$$

We define a dual of $\mathcal{K}$, by

$$\mathcal{H} = \{H_0, \ldots, H_{2^{d_2}-1}\},$$

the set of the matrices obtained by deleting from 0 to d_2 of the last d_2 columns of the matrix I_d. Note that the elements of $\mathcal{H}$ can always be numbered in such a way that $H_0 = I_d$ corresponds to the absence of constraint on θ_0 and that, for $i = 1, \dots, 2^{d_2} - 1$, the constraint $K_i\theta_0 = 0$ corresponds to the constraint $\theta_0 = H_i H_i' \theta_0$. Exercise 8.2 then shows that

$$P_i Z = H_i(H_i' J H_i)^{-1} H_i' J Z, \tag{8.18}$$

for $i = 0, \dots, 2^{d_2} - 1$ (with $P_0 = I_d$). Note that projection (8.18) requires the inversion of only one matrix of size $(d-k) \times (d-k)$ ($d-k$ being the number of columns of H_i), whereas projection (8.13) requires the inversion of one matrix of size $d \times d$ and another matrix of size $k \times k$. To illustrate this new formula, we return to our previous examples.

Example 8.4 (Example 8.2 continued) We have

$$H = \begin{pmatrix} I_{d_1} \\ 0_{1\times d_1} \end{pmatrix}, \quad \mathcal{H} = \{I_d, H\}$$

and

$$P_1 Z = H(H'JH)^{-1}H'JZ = \begin{pmatrix} Z^{(1)} + J_{11}^{-1} J_{12} Z^{(2)} \\ 0 \end{pmatrix},$$

using the notation

$$J = \begin{pmatrix} J_{11} & J_{12} \\ J_{21} & J_{22} \end{pmatrix}, \quad Z = \begin{pmatrix} Z^{(1)} \\ Z^{(2)} \end{pmatrix},$$

where the matrix J_{11} is of size $d_1 \times d_1$, the vectors $J_{12} = J_{21}'$ and $Z^{(1)}$ are of size $d_1 \times 1$, and J_{22} and $Z^{(2)} = Z_d$ are scalars. We finally obtain

$$\lambda^\Lambda = Z - Z_d^- \begin{pmatrix} -J_{11}^{-1} J_{12} \\ 1 \end{pmatrix},$$

which can be shown using (8.15) and

$$H' J^{-1} K (K J^{-1} K')^{-1} = -J_{11}^{-1} J_{12}.$$

Example 8.5 (Example 8.3 continued) We have

$$H = \begin{pmatrix} 1 \\ 0 \\ 0 \end{pmatrix}, \quad \mathcal{H} = \{I_3, H_1, H_2, H_3\},$$

with $H_1 = H$,

$$H_2 = \begin{pmatrix} 1 & 0 \\ 0 & 0 \\ 0 & 1 \end{pmatrix} \quad \text{and} \quad H_3 = \begin{pmatrix} 1 & 0 \\ 0 & 1 \\ 0 & 0 \end{pmatrix}.$$

In view of Exercise 8.6, we have

$$J = \frac{1}{\omega_0^2} \begin{pmatrix} 1 & \omega_0 & \omega_0 \\ \omega_0 & \omega_0^2 \kappa_\eta & \omega_0^2 \\ \omega_0 & \omega_0^2 & \omega_0^2 \kappa_\eta \end{pmatrix},$$

which allows us to obtain

$$P_i Z = H_i(H_i' J H_i)^{-1} H_i' J Z = \begin{cases} Z & \text{for} \quad i = 0 \\ (Z_1 + \omega_0(Z_2 + Z_3), 0, 0)' & \text{for} \quad i = 1 \\ (Z_1 + \omega_0 Z_2, 0, Z_3)' & \text{for} \quad i = 2 \\ (Z_1 + \omega_0 Z_3, Z_2, 0)' & \text{for} \quad i = 3, \end{cases}$$

and to retrieve relation (8.16). Note that, in this example, the calculation is simpler with projection (8.13) than with projection (8.18), because J^{-1} has a simpler expression than J.

8.3 Significance of the GARCH Coefficients

We make the assumptions of Theorem 8.1 and use the notation of Section 8.2. Assume $\theta_0^{(1)} > 0$, and consider the testing problem

$$H_0 : \theta_0^{(2)} = 0 \quad \text{against} \quad H_1 : \theta_0^{(2)} \neq 0. \tag{8.19}$$

Recall that under H_0, we have

$$\sqrt{n}\hat{\theta}_n^{(2)} \xrightarrow{\mathcal{L}} K\lambda^\Lambda, \quad K = (0_{d_2 \times d_1}, I_{d_2}),$$

where the distribution of λ^Λ is defined by

$$\lambda^\Lambda := \arg\inf_{\lambda \in \Lambda} \{\lambda - Z\}' J \{\lambda - Z\}, \quad Z \sim \mathcal{N}\{0, (\kappa_\eta - 1)J^{-1}\}, \tag{8.20}$$

with $\Lambda = \mathbb{R}^{d_1} \times [0, \infty)^{d_2}$.

8.3.1 Tests and Rejection Regions

For parametric assumptions of the form (8.19), the most popular tests are the Wald, score, and likelihood ratio tests.

Wald Statistic

The Wald test looks at whether $\hat{\theta}_n^{(2)}$ is close to 0. The usual Wald statistic is defined by

$$\mathbf{W}_n = n\hat{\theta}_n^{(2)'} \{K\hat{\Sigma}K'\}^{-1} \hat{\theta}_n^{(2)},$$

where $\hat{\Sigma}$ is a consistent estimator of $\Sigma = (\kappa_\eta - 1)J^{-1}$.

Score (or Lagrange Multiplier or Rao) Statistic

Let

$$\hat{\theta}_{n|2} = \begin{pmatrix} \hat{\theta}_{n|2}^{(1)} \\ 0 \end{pmatrix}$$

denote the QMLE of θ constrained by $\theta^{(2)} = 0$. The score test aims to determine whether $\partial \mathbf{l}_n(\hat{\theta}_{n|2})/\partial\theta$ is not too far from 0, using a statistic of the form

$$\mathbf{R}_n = \frac{n}{\hat{\kappa}_{\eta|2} - 1} \frac{\partial \tilde{\mathbf{l}}_n(\hat{\theta}_{n|2})}{\partial \theta'} \hat{J}_{n|2}^{-1} \frac{\partial \tilde{\mathbf{l}}_n(\hat{\theta}_{n|2})}{\partial \theta},$$

where $\hat{\kappa}_{\eta|2}$ and $\hat{J}_{n|2}$ denote consistent estimators of κ_η and J.

Likelihood Ratio Statistic

The likelihood ratio test is based on the fact that under $H_0 : \theta^{(2)} = 0$, the constrained (quasi) log-likelihood $\log L_n(\hat{\theta}_{n|2}) = -(n/2)\tilde{\mathbf{l}}_n(\hat{\theta}_{n|2})$ should not be much smaller than the unconstrained log-likelihood $-(n/2)\tilde{\mathbf{l}}_n(\hat{\theta}_n)$. The test employs the statistic

$$\mathbf{L}_n = n\{\tilde{\mathbf{l}}_n(\hat{\theta}_{n|2}) - \tilde{\mathbf{l}}_n(\hat{\theta}_n)\}.$$

Usual Rejection Regions

From the practical viewpoint, the score statistic presents the advantage of only requiring constrained estimation, which is sometimes much simpler than the unconstrained estimation required by the two other tests. The likelihood ratio statistic does not require estimation of the information matrix J, nor the kurtosis coefficient κ_η. For each test, it is clear that the null hypothesis must be rejected for large values of the statistic. For standard statistical problems, the three statistics asymptotically follow the same $\chi^2_{d_2}$ distribution under the null. At the asymptotic level α, the standard rejection regions are thus

$$\{\mathbf{W}_n > \chi^2_{d_2}(1-\alpha)\}, \quad \{\mathbf{R}_n > \chi^2_{d_2}(1-\alpha)\}, \quad \{\mathbf{L}_n > \chi^2_{d_2}(1-\alpha)\}$$

where $\chi^2_{d_2}(1-\alpha)$ is the $(1-\alpha)$-quantile of the χ^2 distribution with d_2 degrees of freedom. In the case $d_2 = 1$, for testing the significance of only one coefficient, the most widely used test is Student's t test, defined by the rejection region

$$\{|\mathbf{t}_n| > \Phi^{-1}(1-\alpha/2)\}, \tag{8.21}$$

where $\mathbf{t}_n = \sqrt{n}\hat{\theta}_n^{(2)}\{K\hat{\Sigma}K'\}^{-1/2}$. This test is equivalent to the standard Wald test because $\mathbf{t}_n = \sqrt{\mathbf{W}_n}$ ($\mathbf{t}_n$ being here always positive or null, because of the positivity constraints of the QML estimates) and

$$\{\Phi^{-1}(1-\alpha/2)\}^2 = \chi^2_1(1-\alpha).$$

Our testing problem is not standard because, by Theorem 8.1, the asymptotic distribution of $\hat{\theta}_n$ is not normal. We will see that, among the previous rejection regions, only that of the score test asymptotically has the level α.

8.3.2 Modification of the Standard Tests

The following proposition shows that for the Wald and likelihood ratio tests, the asymptotic distribution is not the usual $\chi^2_{d_2}$ under the null hypothesis. The proposition also shows that the asymptotic distribution of the score test remains the $\chi^2_{d_2}$ distribution. The asymptotic distribution of $\mathbf{R}_n$ is not affected by the fact that, under the null hypothesis, the parameter is at the boundary of the parameter space. These results are not very surprising. Take the example of an ARCH(1) with the hypothesis $H_0: \alpha_0 = 0$ of absence of ARCH effect. As illustrated by Figure 8.2, there is a non-zero probability that $\hat{\alpha}$ be at the boundary, that is, that $\hat{\alpha} = 0$. Consequently $\mathbf{W}_n = n\hat{\alpha}^2$ admits a mass at 0 and does not follow, even asymptotically,

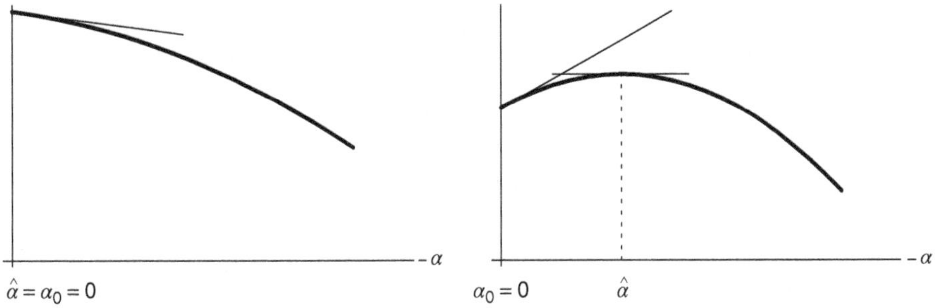

Figure 8.2 Concentrated log-likelihood (solid line) $\alpha \mapsto \log L_n(\hat{\omega}, \alpha)$ for an ARCH(1) model. Assume there is no ARCH effect: the true value of the ARCH parameter is $\alpha_0 = 0$. In the configuration on the right, the likelihood maximum does not lie at the boundary and the three statistics $\mathbf{W}_n$, $\mathbf{R}_n$, and $\mathbf{L}_n$ take strictly positive values. In the configuration on the left, we have $\mathbf{W}_n = n\hat{\alpha}^2 = 0$ and $\mathbf{L}_n = 2\{\log L_n(\hat{\omega}, \hat{\alpha}) - \log L_n(\hat{\omega}, 0)\} = 0$, whereas $\mathbf{R}_n = \{\partial \log L_n(\hat{\omega}, 0)/\partial\alpha\}^2$ continues to take a strictly positive value.

the χ^2_1 law. The same conclusion can be drawn for the likelihood ratio test. On the contrary, the score $n^{1/2}\partial \mathbf{l}_n(\theta_0)/\partial\theta$ can take as well positive or negative values, and does not seem to have a specific behaviour when θ_0 is at the boundary.

Proposition 8.3 (Asymptotic distribution of the three statistics under H_0) *Under H_0 and the assumptions of Theorem 8.1,*

$$\mathbf{W}_n \xrightarrow{\mathcal{L}} \mathbf{W} = \lambda^{\Lambda'}\Omega\lambda^{\Lambda}, \tag{8.22}$$

$$\mathbf{R}_n \xrightarrow{\mathcal{L}} \mathbf{R} \sim \chi^2_{d_2}, \tag{8.23}$$

$$\begin{aligned}\mathbf{L}_n \xrightarrow{\mathcal{L}} \mathbf{L} &= -\frac{1}{2}(\lambda^{\Lambda} - Z)'J(\lambda^{\Lambda} - Z) + \frac{1}{2}Z'K'\{KJ^{-1}K'\}^{-1}KZ \\ &= -\frac{1}{2}\left\{\inf_{K\lambda\geq 0}\|Z-\lambda\|^2_J - \inf_{K\lambda=0}\|Z-\lambda\|^2_J\right\},\end{aligned} \tag{8.24}$$

where $\Omega = K'\{(\kappa_\eta - 1)KJ^{-1}K'\}^{-1}K$ and λ^{Λ} satisfies (8.20).

Remark 8.3 (Equivalence of the statistics $\mathbf{W}_n$ and $\mathbf{L}_n$) Let $\hat{\kappa}_\eta$ be an estimator which converges in probability to κ_η. We can show that

$$\mathbf{W}_n = \frac{2}{\hat{\kappa}_\eta - 1}\mathbf{L}_n + o_P(1)$$

under the null hypothesis. The Wald and likelihood ratio tests will thus have the same asymptotic critical values and will have the same local asymptotic powers (see Exercises 8.8 and 8.9, and Section 8.3.4). They may, however, have different asymptotic behaviours under non-local alternatives.

Remark 8.4 (Assumptions on the tests) In order for the Wald statistic to be well defined, Ω must exist, that is, $J = J(\theta_0)$ must exist and must be invertible. This is not the case, in particular, for a GARCH(p, q) at $\theta_0 = (\omega_0, 0, \ldots, 0)$, when $p \neq 0$. It is thus impossible to carry out a Wald test on the simultaneous nullity of all the α_i and β_j coefficients in a GARCH(p, q), $p \neq 0$. The assumptions of Theorem 8.1 are actually required, in particular the identifiability assumptions. It is thus impossible to test, for instance, an ARCH(1) against a GARCH(2, 1), but we can test, for instance, an ARCH(1) against an ARCH(3).

A priori, the asymptotic distributions **W** and **L** depend on J, and thus on nuisance parameters. We will consider two particular cases: the case where we test the nullity of only one GARCH coefficient and the case where we test the nullity of all the coefficients of an ARCH. In the two cases, the asymptotic laws of the test statistics are simpler and do not depend on nuisance parameters. In the second case, both the test statistics and their asymptotic laws are simplified.

8.3.3 Test for the Nullity of One Coefficient

Consider the case $d_2 = 1$, which is perhaps the most interesting case and corresponds to testing the nullity of only one coefficient. In view of relations (8.15), the last component of λ^{Λ} is equal to Z_d^+. We thus have

$$\mathbf{W}_n \xrightarrow{\mathcal{L}} \mathbf{W} = \frac{(K\lambda^{\Lambda})^2}{K\Sigma K'} = \frac{(KZ)^2}{\mathrm{Var}KZ}\mathbb{1}_{\{KZ\geq 0\}} = (Z^*)^2\mathbb{1}_{\{Z^*\geq 0\}}$$

where $Z^* \sim \mathcal{N}(0,1)$. Using the symmetry of the Gaussian distribution, and the independence between Z^{*2} and $\mathbb{1}_{\{Z^*>0\}}$ when Z^* follows the real normal law, we obtain

$$\mathbf{W}_n \xrightarrow{\mathcal{L}} \mathbf{W} \sim \frac{1}{2}\delta_0 + \frac{1}{2}\chi^2_1 \quad \text{(where } \delta_0 \text{ denotes the Dirac measure at 0).}$$

Testing

$$H_0 : \theta_0^{(2)} := \theta_0(p+q+1) = 0 \quad \text{against} \quad H_1 : \theta_0(p+q+1) > 0,$$

can thus be achieved by using the critical region $\{\mathbf{W}_n > \chi_1^2(1-2\alpha)\}$ at the asymptotic level $\alpha \leq 1/2$. In view of Remark 8.3, we can define a modified likelihood ratio test of critical region $\{2\mathbf{L}_n/(\hat{\kappa}_\eta - 1) > \chi_1^2(1-2\alpha)\}$. Note that the standard Wald test $\{\mathbf{W}_n > \chi_1^2(1-\alpha)\}$ has the asymptotic level $\alpha/2$, and that the asymptotic level of the standard likelihood ratio test $\{\mathbf{L}_n > \chi_1^2(1-\alpha)\}$ is much larger than α when the kurtosis coefficient κ_η is large. A modified version of the Student t test is defined by the rejection region

$$\{\mathbf{t}_n > \Phi^{-1}(1-\alpha)\}, \tag{8.25}$$

We observe that commercial software – such as GAUSS, R, RATS, SAS, and SPSS – do not use the modified version (8.25), but the standard version (8.21). This standard test is not of asymptotic level α but only $\alpha/2$. To obtain a t test of asymptotic level α, it then suffices to use a test of nominal level 2α.

Example 8.6 (Empirical behaviour of the tests under the null) We simulated 5000 independent samples of length $n = 100$ and $n = 1000$ of a strong $\mathcal{N}(0,1)$ white noise. On each realisation, we fitted an ARCH(1) model $\epsilon_t = \{\omega + \alpha\epsilon_{t-1}^2\}^{1/2}\eta_t$, by QML, and carried out tests of $H_0 : \alpha = 0$ against $H_1 : \alpha > 0$.

We began with the modified Wald test with rejection region

$$\{\mathbf{W}_n = n\hat{\alpha}^2 > \chi_1^2(0.90) = 2.71\}.$$

This test is of asymptotic level 5%. For the sample size $n = 100$, we observed a relative rejection frequency of 6.22%. For $n = 1000$, we observe a relative rejection frequency of 5.38%, which is not significantly different from the theoretical 5%. Indeed, an elementary calculation shows that, on 5000 independent replications of a same experiment with success probability 5%, the success percentage should vary between 4.4% and 5.6% with a probability of approximately 95%. Figure 8.3 shows that the empirical distribution of $\mathbf{W}_n$ is quite close to the asymptotic distribution $\delta_0/2 + \chi_1^2/2$, even for the small sample size $n = 100$.

We then carried out the score test defined by the rejection region

$$\{\mathbf{R}_n = nR^2 > \chi_1^2(0.95) = 3.84\},$$

where R^2 is the determination coefficient of the regression of ϵ_t^2 on 1 and ϵ_{t-1}^2. This test is also of asymptotic level 5%. For the sample size $n = 100$, we observed a relative rejection frequency of 3.40%. For $n = 1000$, we observed a relative frequency of 4.32%.

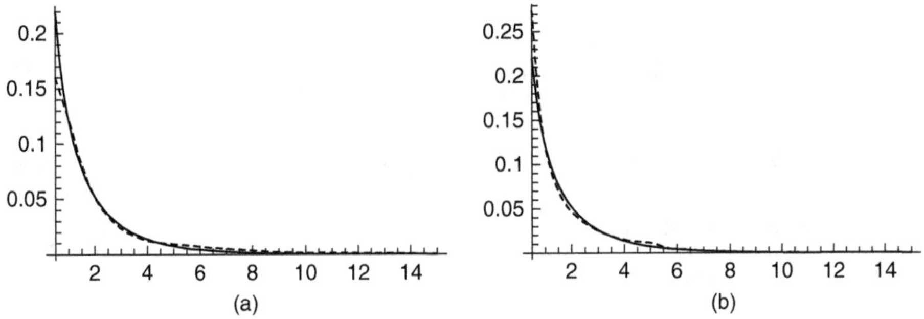

Figure 8.3 Comparison between a kernel density estimator of the Wald statistic (dotted line) and the $\chi_1^2/2$ density on $[0.5, \infty)$ (solid line) on 5000 simulations of an ARCH(1) process with $\alpha_{01} = 0$: (a) for sample size $n = 100$; (b) for $n = 1000$.

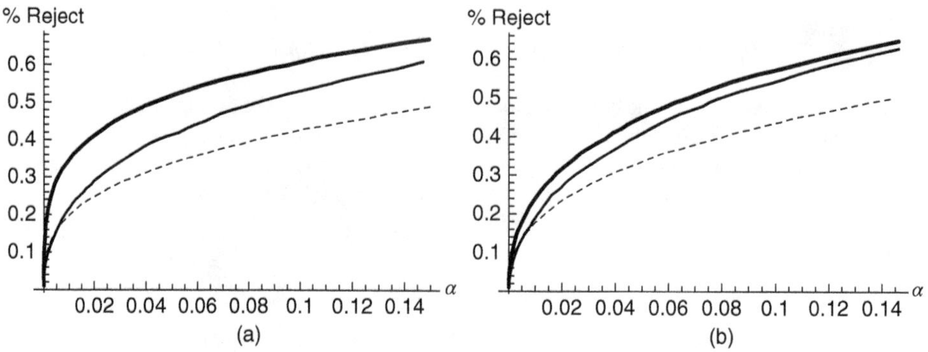

Figure 8.4 Comparison of the observed powers of the Wald test (thick line), of the score test (dotted line) and of the likelihood ratio test (thin line), as function of the nominal level α, on 5000 simulations of an ARCH(1) process: (a) for $n = 100$ and $\alpha_{01} = 0.2$; (b) for $n = 1000$ and $\alpha_{01} = 0.05$.

We also used the modified likelihood ratio test. For the sample size $n = 100$, we observed a relative rejection frequency of 3.20%, and for $n = 1000$ we observed 4.14%.

On these simulation experiments, the Type I error is thus slightly better controlled by the modified Wald test than by the score and modified likelihood ratio tests.

Example 8.7 (Comparison of the tests under the alternative hypothesis) We implemented the $\mathbf{W}_n$, $\mathbf{R}_n$, and $\mathbf{L}_n$ tests of the null hypothesis $H_0 : \alpha_{01} = 0$ in the ARCH(1) model $\epsilon_t = \{\omega + \alpha_{01}\epsilon_{t-1}^2\}^{1/2}\eta_t$, $\eta_t \sim \mathcal{N}(0, 1)$. Figure 8.4 compares the observed powers of the three tests, that is, the relative frequency of rejection of the hypothesis H_0 that there is no ARCH effect, on 5000 independent realisations of length $n = 100$ and $n = 1000$ of an ARCH(1) model with $\alpha_{01} = 0.2$ when $n = 100$, and $\alpha_{01} = 0.05$ when $n = 1000$. On these simulated series, the modified Wald test turns out to be the most powerful.

8.3.4 Conditional Homoscedasticity Tests with ARCH Models

Another interesting case is that obtained with $d_1 = 1$, $\theta^{(1)} = \omega$, $p = 0$, and $d_2 = q$. This case corresponds to the test of the conditional homoscedasticity null hypothesis

$$H_0 : \alpha_{01} = \cdots = \alpha_{0q} = 0 \tag{8.26}$$

in an ARCH(q) model

$$\begin{cases} \epsilon_t = \sigma_t \eta_t, & (\eta_t) \text{ iid } (0, 1) \\ \sigma_t^2 = \omega + \sum_{i=1}^{q} \alpha_i \epsilon_{t-i}^2, & \omega > 0, \quad \alpha_i \geq 0. \end{cases} \tag{8.27}$$

We will see that for testing problem (8.26), there exist very simple forms of the Wald and score statistics.

Simplified Form of the Wald Statistic

Using Exercise 8.6, we have

$$\Sigma(\theta_0) = (\kappa_\eta - 1)J(\theta_0)^{-1} = \begin{pmatrix} (\kappa_\eta + q - 1)\omega^2 & -\omega \cdots -\omega \\ -\omega & \\ \vdots & I_q \\ -\omega & \end{pmatrix}.$$

Since $K\Sigma K' = I_q$, we obtain a very simple form for the Wald statistic:

$$\mathbf{W}_n = n \sum_{i=1}^{q} \hat{\alpha}_i^2. \tag{8.28}$$

Asymptotic Distribution W and L

A trivial extension of Example 8.3 yields

$$\lambda^\Lambda = \begin{pmatrix} Z_1 + \omega \sum_{i=2}^{d} Z_i^- \\ Z_2^+ \\ \vdots \\ Z_d^+ \end{pmatrix}. \tag{8.29}$$

The asymptotic distribution of $n \sum_{i=2}^{d} \hat{\alpha}_i^2$ is thus that of

$$\sum_{i=2}^{d} (Z_i^+)^2,$$

where the Z_i's are independent and $\mathcal{N}(0,1)$-distributed. Thus, in the case where an ARCH(q) is fitted to a white noise we have

$$\mathbf{W}_n = n \sum_{i=1}^{q} \hat{\alpha}_i^2 \xrightarrow{\mathcal{L}} \frac{1}{2^q}\delta_0 + \sum_{i=1}^{q} C_q^i \frac{1}{2^q}\chi_i^2. \tag{8.30}$$

This asymptotic distribution is tabulated and the critical values are given in Table 8.2. In view of Remark 8.3, Table 8.2 also yields the asymptotic critical values of the modified likelihood ratio statistic $2\mathbf{L}_n/(\hat{\kappa}_\eta - 1)$. Table 8.3 shows that the use of the standard $\chi_q^2(1-\alpha)$-based critical values of the Wald test would lead to large discrepancies between the asymptotic levels and the nominal level α.

Score Test

For the hypothesis (8.26) that all the α coefficients of an ARCH(q) model are equal to zero, the score statistic $\mathbf{R}_n$ can be simplified. To work within the linear regression framework, write

$$\frac{\partial \tilde{\mathbf{l}}_n(\hat{\theta}_{n|2}^{(1)}, 0)}{\partial \theta} = n^{-1} X'Y,$$

where Y is the vector of length n of the 'dependent' variable $1 - \epsilon_t^2/\hat{\omega}$, where X is the $n \times (q+1)$ matrix of the constant $\hat{\omega}^{-1}$ (in the first column) and of the 'explanatory' variables $\epsilon_{t-i}^2 \hat{\omega}^{-1}$ (in column $i+1$, with

Table 8.2 Asymptotic critical value $c_{q,\alpha}$, at level α, of the Wald test of rejection region $\{n \sum_{i=1}^{q} \hat{\alpha}_i^2 > c_{q,\alpha}\}$ for the conditional homoscedasticity hypothesis $H_0 : \alpha_1 = \cdots = \alpha_q = 0$ in an ARCH(q) model.

q	α (%)					
	0.1	1	2.5	5	10	15
1	9.5493	5.4119	3.8414	2.7055	1.6424	1.0742
2	11.7625	7.2895	5.5369	4.2306	2.9524	2.2260
3	13.4740	8.7464	6.8610	5.4345	4.0102	3.1802
4	14.9619	10.0186	8.0230	6.4979	4.9553	4.0428
5	16.3168	11.1828	9.0906	7.4797	5.8351	4.8519

Table 8.3 Exact asymptotic level (%) of erroneous Wald tests, of rejection region $\{n\sum_{i=1}^{q}\hat{\alpha}_i^2 > \chi_q^2(1-\alpha)\}$, under the conditional homoscedasticity assumption $H_0: \alpha_1 = \cdots = \alpha_q = 0$ in an ARCH(q) model.

q	α (%)					
	0.1	1	2.5	5	10	15
1	0.05	0.5	1.25	2.5	5	7.5
2	0.04	0.4	0.96	1.97	4.09	6.32
3	0.02	0.28	0.75	1.57	3.36	5.29
4	0.02	0.22	0.59	1.28	2.79	4.47
5	0.01	0.17	0.48	1.05	2.34	3.81

the convention $\epsilon_t = 0$ for $t \leq 0$), and $\hat{\omega} = \hat{\theta}_{n|2}^{(1)} = n^{-1}\sum_{t=1}^{n}\epsilon_t^2$. Estimating $J(\theta_0)$ by $n^{-1}X'X$, and $\kappa_\eta - 1$ by $n^{-1}Y'Y$, we obtain

$$\mathbf{R}_n = n\frac{Y'X(X'X)^{-1}X'Y}{Y'Y},$$

and one recognizes n times the coefficient of determination in the linear regression of Y on the columns of X. Since this coefficient is not changed by linear transformation of the variables (see Exercise 5.11), we simply have $\mathbf{R}_n = nR^2$, where R^2 is the coefficient of determination in the regression of ϵ_t^2 on a constant and q lagged values $\epsilon_{t-1}^2, \dots, \epsilon_{t-q}^2$. Under the null hypothesis of conditional homoscedasticity, $\mathbf{R}_n$ asymptotically follows the χ_q^2 law.

The previous simple forms of the Wald and score tests are obtained with estimators of J which exploit the particular form of the matrix under the null. Note that there exist other versions of these tests, obtained with other consistent estimators of J. The different versions are equivalent under the null, but can have different asymptotic behaviours under the alternative.

8.3.5 Asymptotic Comparison of the Tests

The Wald and score tests that we have just defined are in general consistent, that is, their powers converge to 1 when they are applied to a wide class of conditionally heteroscedastic processes. An asymptotic study will be conducted via two different approaches: Bahadur's approach compares the rates of convergence to zero of the p-values under fixed alternatives, whereas Pitman's approach compares the asymptotic powers under a sequence of local alternatives, that is, a sequence of alternatives tending to the null as the sample size increases.

Bahadur's Approach

Let $S_\mathbf{W}(t) = \mathbb{P}(\mathbf{W} > t)$ and $S_\mathbf{R}(t) = \mathbb{P}(\mathbf{R} > t)$ be the asymptotic survival functions of the two test statistics, under the null hypothesis H_0 defined by (8.26). Consider, for instance, the Wald test. Under the alternative of an ARCH(q) which does not satisfy H_0, the p-value of the Wald test $S_\mathbf{W}(\mathbf{W}_n)$ converges almost surely to zero as $n \to \infty$ because

$$\frac{\mathbf{W}_n}{n} \to \sum_{i=1}^{q}\alpha_i^2 \neq 0.$$

The p-value of a test is typically equivalent to $\exp\{-nc/2\}$, where c is a positive constant called the Bahadur slope. Using the fact that

$$\log S_\mathbf{W}(x) \sim \log P(\chi_q^2 > x) \quad x \to \infty, \tag{8.31}$$

and that $\lim_{x\to\infty} \log P(\chi^2_q > x) \sim -x/2$, the (approximate[1]) Bahadur slope of the Wald test is thus

$$\lim_{n\to\infty} -\frac{2}{n} \log S_{\mathbf{W}}(\mathbf{W}_n) = \sum_{i=1}^{q} \alpha_i^2, \quad \text{a.s.}$$

To compute the Bahadur slope of the score test, note that we have the linear regression model $\epsilon_t^2 = \omega + \mathcal{A}(B)\epsilon_t^2 + \nu_t$, where $\nu_t = (\eta_t^2 - 1)\sigma_t^2(\theta_0)$ is the linear innovation of ϵ_t^2. We then have

$$\frac{\mathbf{R}_n}{n} = R^2 \to 1 - \frac{\text{Var}(\nu_t)}{\text{Var}(\epsilon_t^2)}.$$

The previous limit is thus equal to the Bahadur slope of the score test. The comparison of the two slopes favours the score test over the Wald test.

Proposition 8.4 *Let (ϵ_t) be a strictly stationary and non-anticipative solution of the ARCH(q) model (8.27), with $E(\epsilon_t^4) < \infty$ and $\sum_{i=1}^{q} \alpha_{0i} > 0$. The score test is considered as more efficient than the Wald test in Bahadur's sense because its slope is always greater or equal to that of the Wald test, with equality when $q = 1$.*

Example 8.8 (Slopes in the ARCH(1) and ARCH(2) cases) The slopes are the same in the ARCH(1) case because

$$\lim_{n\to\infty} \frac{\mathbf{W}_n}{n} = \lim_{n\to\infty} \frac{R_n}{n} = \alpha_1^2.$$

In the ARCH(2) case with fourth-order moment, we have

$$\lim_{n\to\infty} \frac{\mathbf{W}_n}{n} = \alpha_1^2 + \alpha_2^2, \quad \lim_{n\to\infty} \frac{\mathbf{R}_n}{n} = \alpha_1^2 + \alpha_2^2 + \frac{2\alpha_1^2\alpha_2}{1-\alpha_2}.$$

We see that the second limit is always larger than the first. Consequently, in Bahadur's sense, the Wald and Rao tests have the same asymptotic efficiency in the ARCH(1) case. In the ARCH(2) case, the score test is, still in Bahadur's sense, asymptotically more efficient than the Wald test for testing the conditional homoscedasticity (that is, $\alpha_1 = \alpha_2 = 0$).

Bahadur's approach is sometimes criticised for not taking account of the critical value of test, and thus for not really comparing the powers. This approach only takes into account the (asymptotic) distribution of the statistic under the null and the rate of divergence of the statistic under the alternative. It is unable to distinguish a two-sided test from its one-sided counterpart (see Exercise 8.8). In this sense, the result of Proposition 8.4 must be put into perspective.

Pitman's Approach

In the ARCH(1) case, consider a sequence of local alternatives $H_n(\tau) : \alpha_1 = \tau/\sqrt{n}$. We can show that under this sequence of alternatives,

$$\mathbf{W}_n = n\hat{\alpha}_1^2 \xrightarrow{\mathcal{L}} (U+\tau)^2 \, \mathbb{1}_{\{U+\tau>0\}}, \quad U \sim \mathcal{N}(0,1).$$

Consequently, the local asymptotic power of the Wald test is

$$\mathbb{P}(U + \tau > c_1) = 1 - \Phi(c_1 - \tau), \quad c_1 = \Phi^{-1}(1-\alpha). \tag{8.32}$$

[1] The term 'approximate' is used by Bahadur (1960) to emphasise that the exact survival function $S_{\mathbf{W}_n}(t)$ is approximated by the asymptotic survival function $S_{\mathbf{W}}(t)$. See also Bahadur (1967) for a discussion on the exact and approximate slopes.

The score test has the local asymptotic power

$$\mathbb{P}\{(U+\tau)^2 > c_2^2\} = 1 - \Phi(c_2 - \tau) + \Phi(-c_2 - \tau), \quad c_2 = \Phi^{-1}(1 - \alpha/2) \tag{8.33}$$

Note that the probability in (8.32) is the power of the test of the assumption $H_0 : \theta = 0$ against the assumption $H_1 : \theta = \tau > 0$, based on the rejection region of $\{X > c_1\}$ with only one observation $X \sim \mathcal{N}(\theta, 1)$. The power (8.33) is that of the two-sided test $\{|X| > c_2\}$. The tests $\{X > c_1\}$ and $\{|X| > c_2\}$ have the same level, but the first test is uniformly more powerful than the second (by the Neyman–Pearson lemma, $\{X > c_1\}$ is even uniformly more powerful than any test of level less than or equal to α, for any one-sided alternative of the form H_1). The local asymptotic power of the Wald test is thus uniformly strictly greater than that of Rao's test for testing for conditional homoscedasticity in an ARCH(1) model.

Consider the ARCH(2) case, and a sequence of local alternatives $H_n(\tau) : \alpha_1 = \tau_1/\sqrt{n}, \alpha_2 = \tau_2/\sqrt{n}$. Under this sequence of alternatives

$$\mathbf{W}_n = n(\hat{\alpha}_1^2 + \hat{\alpha}_1^2) \xrightarrow{\mathcal{L}} (U_1 + \tau_1)^2 \, \mathbb{1}_{\{U_1+\tau_1>0\}} + (U_2 + \tau_2)^2 \mathbb{1}_{\{U_2+\tau_2>0\}},$$

with $(U_1, U_2)' \sim \mathcal{N}(0, I_2)$. Let c_1 be the critical value of the Wald test of level α. The local asymptotic power of the Wald test is

$$\begin{aligned}
&\mathbb{P}(U_1 + \tau_1 > \sqrt{c_1})\mathbb{P}(U_2 + \tau_2 < 0) \\
&\quad + \int_{-\tau_2}^{\infty} \mathbb{P}\{(U_1 + \tau_1)^2 \mathbb{1}_{\{U_1+\tau_1>0\}} > c_1 - (x + \tau_2)^2\}\phi(x)dx \\
&= \{1 - \Phi(\sqrt{c_1} - \tau_1)\}\Phi(-\tau_2) + 1 - \Phi(-\tau_2 + \sqrt{c_1}) \\
&\quad + \int_{-\tau_2}^{-\tau_2+\sqrt{c_1}} \left\{1 - \Phi\left(\sqrt{c_1 - (x + \tau_2)^2} - \tau_1\right)\right\} \phi(x)dx
\end{aligned}$$

Let c_2 be the critical value of the Rao test of level α. The local asymptotic power of the Rao test is

$$\mathbb{P}\{(U_1 + \tau_1)^2 + (U_1 + \tau_2)^2 > c_2\},$$

where $(U_1 + \tau_1)^2 + (U_2 + \tau_2)^2$ follows a non-central χ^2 distribution, with two degrees of freedom and non-centrality parameter $\tau_1^2 + \tau_2^2$. Figure 8.5 compares the powers of the two tests when $\tau_1 = \tau_2$.

Thus, the comparison of the local asymptotic powers clearly favours the Wald test over the score test, counter-balancing the result of Proposition 8.4.

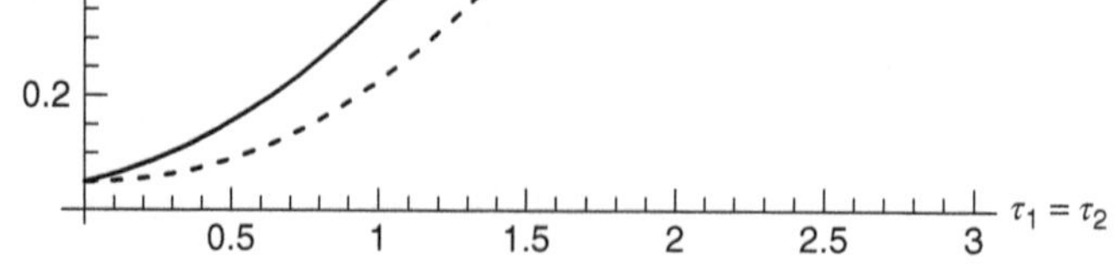

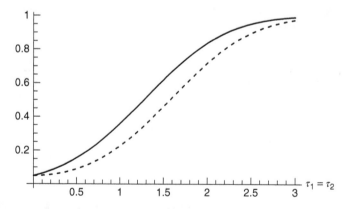

Figure 8.5 Local asymptotic power of the Wald test (solid line) and of the Rao test (dotted line) for testing for conditional homoscedasticity in an ARCH(2) model.

8.4 Diagnostic Checking with Portmanteau Tests

To check the adequacy of a given time series model, for instance an ARMA(p, q) model, it is common practice to test the significance of the residual autocorrelations. In the GARCH framework this approach is not relevant because the process $\tilde{\eta}_t = \epsilon_t/\tilde{\sigma}_t$ is always a white noise (more precisely a martingale difference) even when the volatility is misspecified, that is, when $\epsilon_t = \sqrt{h_t}\eta_t$ with $h_t \neq \tilde{\sigma}_t^2$. To check the adequacy of a volatility model, for instance a GARCH(p, q) of the form (7.1), it is much more fruitful to look at the squared residual autocovariances

$$\hat{r}(h) = \frac{1}{n}\sum_{t=|h|+1}^{n}(\hat{\eta}_t^2 - 1)(\hat{\eta}_{t-|h|}^2 - 1), \quad \hat{\eta}_t^2 = \frac{\epsilon_t^2}{\hat{\sigma}_t^2},$$

where $|h| < n$, $\hat{\sigma}_t = \tilde{\sigma}_t(\hat{\theta}_n)$, $\tilde{\sigma}_t$ is defined by Eq. (7.4) and $\hat{\theta}_n$ is the QMLE given by Eq. (7.9).

For any fixed integer m, $1 \leq m < n$, consider the statistic $\hat{\mathbf{r}}_m = (\hat{r}(1), \dots, \hat{r}(m))'$. Let $\hat{\kappa}_\eta$ and $\hat{J}$ be weakly consistent estimators of κ_η and J. For instance, one can take

$$\hat{\kappa}_\eta = \frac{1}{n}\sum_{t=1}^{n}\frac{\epsilon_t^4}{\tilde{\sigma}_t^4(\hat{\theta}_n)}, \quad \hat{J} = \frac{1}{n}\sum_{t=1}^{n}\frac{1}{\tilde{\sigma}_t^4(\hat{\theta}_n)}\frac{\partial\tilde{\sigma}_t^2(\hat{\theta}_n)}{\partial\theta}\frac{\partial\tilde{\sigma}_t^2(\hat{\theta}_n)}{\partial\theta'}.$$

Define also the $m \times (p+q+1)$ matrix $\hat{C}_m$ whose (h, k)th element, for $1 \leq h \leq m$ and $1 \leq k \leq p+q+1$, is given by

$$\hat{C}_m(h,k) = -\frac{1}{n}\sum_{t=h+1}^{n}(\hat{\eta}_{t-h}^2 - 1)\frac{1}{\tilde{\sigma}_t^2(\hat{\theta}_n)}\frac{\partial\tilde{\sigma}_t^2(\hat{\theta}_n)}{\partial\theta_k}.$$

Theorem 8.2 (Asymptotic distribution of a portmanteau test statistic) *Under the assumptions of Theorem 7.2 ensuring the consistency and asymptotic normality of the QMLE,*

$$n\hat{\mathbf{r}}_m'\hat{D}^{-1}\hat{\mathbf{r}}_m \xrightarrow{\mathcal{L}} \chi_m^2,$$

with $\hat{D} = (\hat{\kappa}_\eta - 1)^2 I_m - (\hat{\kappa}_\eta - 1)\hat{C}_m\hat{J}^{-1}\hat{C}_m'$.

The adequacy of the GARCH(p, q) model is rejected at the asymptotic level α when

$$\{n\hat{\mathbf{r}}_m'\hat{D}^{-1}\hat{\mathbf{r}}_m > \chi_m^2(1-\alpha)\}.$$

8.5 Application: Is the GARCH(1,1) Model Overrepresented?

The GARCH(1,1) model is by far the most widely used by practitioners who wish to estimate the volatility of daily returns. In general, this model is chosen *a priori*, without implementing any statistical identification procedure. This practice is motivated by the common belief that the GARCH(1,1) (or its simplest asymmetric extensions) is sufficient to capture the properties of financial series and that higher-order models may be unnecessarily complicated.

We will show that, for a large number of series, this practice is not always statistically justified. We consider daily and weekly series of 11 returns (CAC, DAX, DJA, DJI, DJT, DJU, FTSE, Nasdaq, Nikkei, SMI and S&P 500) and five exchange rates. The observations cover the period from 2 January 1990 to 22 January 2009 for the daily returns and exchange rates, and from 2 January 1990 to 20 January 2009 for the weekly returns (except for the indices for which the first observations are after 1990). We begin with the portmanteau tests defined in Section 8.4. Table 8.4 shows that the ARCH models (even with large order q) are generally rejected, whereas the GARCH(1,1) is only occasionally rejected. This table

Table 8.4 Portmanteau test p-values for adequacy of the ARCH(5) and GARCH(1,1) models for daily returns of stock market indices, based on m squared residual autocovariances.

	M											
	1	2	3	4	5	6	7	8	9	10	11	12
Portmanteau tests for adequacy of the ARCH(5)												
CAC	0.194	**0.010**	**0.001**	**0.000**	**0.000**	**0.000**	**0.000**	**0.000**	**0.000**	**0.000**	**0.000**	**0.000**
DAX	0.506	0.157	0.140	**0.049**	**0.044**	0.061	0.080	0.119	0.140	0.196	0.185	0.237
DJA	0.441	0.34	0.139	0.002	**0.000**	**0.000**	**0.000**	**0.000**	**0.000**	**0.000**	**0.000**	**0.000**
DJI	0.451	0.374	**0.015**	**0.000**	**0.000**	**0.000**	**0.000**	**0.000**	**0.000**	**0.000**	**0.000**	**0.000**
DJT	0.255	0.514	0.356	**0.044**	**0.025**	**0.013**	**0.020**	**0.023**	**0.000**	**0.000**	**0.000**	**0.000**
DJU	0.477	0.341	**0.002**	**0.000**	**0.000**	**0.000**	**0.000**	**0.000**	**0.000**	**0.000**	**0.000**	**0.000**
FTSE	0.139	**0.001**	**0.000**	**0.000**	**0.000**	**0.000**	**0.000**	**0.000**	**0.000**	**0.000**	**0.000**	**0.000**
Nasdaq	0.025	**0.031**	**0.001**	**0.000**	**0.000**	**0.000**	**0.000**	**0.000**	**0.000**	**0.000**	**0.000**	**0.000**
Nikkei	**0.004**	**0.000**	**0.001**	**0.001**	**0.000**	**0.000**	**0.000**	**0.000**	**0.000**	**0.000**	**0.000**	**0.000**
SMI	0.502	0.692	0.407	0.370	0.211	0.264	0.351	0.374	0.463	0.533	0.623	0.700
S&P 500	0.647	0.540	**0.012**	**0.000**	**0.000**	**0.000**	**0.000**	**0.000**	**0.000**	**0.000**	**0.000**	**0.000**
Portmanteau tests for adequacy of the GARCH(1,1)												
CAC	0.312	0.379	0.523	0.229	0.301	0.396	0.495	0.578	0.672	0.660	0.704	0.743
DAX	0.302	0.583	0.574	0.704	0.823	0.901	0.938	0.968	0.983	0.989	0.994	0.995
DJA	0.376	0.424	0.634	0.740	0.837	0.908	0.838	0.886	0.909	0.916	0.938	0.959
DJI	0.202	0.208	0.363	0.505	0.632	0.742	0.770	0.812	0.871	0.729	0.748	0.811
DJT	0.750	0.100	0.203	0.276	0.398	0.518	0.635	0.721	0.804	0.834	0.885	0.925
DJU	**0.000**	**0.000**	**0.000**	**0.000**	**0.000**	**0.000**	**0.000**	**0.000**	**0.000**	**0.000**	**0.000**	**0.000**
FTSE	0.733	0.940	0.934	0.980	0.919	0.964	0.328	0.424	0.465	0.448	0.083	0.108
Nasdaq	0.523	**0.024**	0.061	**0.019**	**0.001**	**0.001**	**0.002**	**0.001**	**0.002**	**0.001**	**0.001**	**0.002**
Nikkei	0.049	0.146	0.246	0.386	0.356	0.475	0.567	0.624	0.703	0.775	0.718	0.764
SMI	0.586	0.758	0.908	0.959	0.986	0.995	0.996	0.999	0.999	0.999	0.999	0.999
S&P 500	0.598	0.364	0.528	0.643	0.673	0.394	0.512	0.535	0.639	0.432	0.496	0.594

p-values less than 5% are in bold, those less than 1% are underlined.

only concerns the daily returns, but similar conclusions hold for the weekly returns and exchange rates. The portmanteau tests are known to be omnibus tests, powerful for a broad spectrum of alternatives. As we will now see, for the specific alternatives for which they are built, the tests defined in Section 8.3 (Wald, score, and likelihood ratio) may be much more powerful.

The GARCH(1,1) model is chosen as the benchmark model and is successively tested against the GARCH(1,2), GARCH(1,3), GARCH(1,4), and GARCH(2,1) models. In each case, the three tests (Wald, score, and likelihood ratio) are applied. The empirical p-values are displayed in Table 8.5. This table shows that (i) the results of the tests strongly depend on the alternative; (ii) the p-values of the three tests can be quite different; (iii) for most of the series, the GARCH(1,1) model is clearly rejected. Point (ii) is not surprising because the asymptotic equivalence between the three tests is only shown under the null hypothesis or under local alternatives. Moreover, because of the positivity constraints, it is possible (see, for instance, the DJU) that the estimated GARCH(1,2) model satisfies $\hat{\alpha}_2 = 0$ with $\partial \tilde{\mathbf{l}}_n(\hat{\theta}_{n|2})/\partial \alpha_2 \ll 0$. In this case, when the estimators lie at the boundary of the parameter space and the score is strongly positive, the Wald and LR tests do not reject the GARCH(1,1) model, whereas the score does reject it. In other situations, the Wald or LR test rejects the GARCH(1,1), whereas the score does not (see, for instance the DAX for the GARCH(1,4) alternative). This study shows that it is often relevant

Table 8.5 p-values for tests of the null of a GARCH(1,1) model against the GARCH(1,2), GARCH(1,3), GARCH(1,4), and GARCH(2,1) alternatives, for returns of stock market indices and exchange rates.

	Alternative											
	GARCH(1,2)			GARCH(1,3)			GARCH(1,4)			GARCH(2,1)		
	$\mathbf{W}_n$	$\mathbf{R}_n$	$\mathbf{L}_n$	$\mathbf{W}_n$	$\mathbf{R}_n$	$\mathbf{L}_n$	$\mathbf{W}_n$	$\mathbf{R}_n$	$\mathbf{L}_n$	$\mathbf{W}_n$	$\mathbf{R}_n$	$\mathbf{L}_n$
Daily returns of indices												
CAC	**0.007**	**0.033**	**0.013**	**0.005**	**0.000**	**0.001**	**0.024**	0.188	**0.040**	0.500	0.280	0.500
DAX	**0.002**	**0.001**	**0.003**	**0.001**	**0.000**	**0.000**	**0.001**	0.162	**0.014**	0.350	**0.031**	0.143
DJA	0.158	0.337	0.166	0.259	0.285	0.269	0.081	0.134	0.064	0.500	0.189	0.500
DJI	**0.044**	0.100	**0.049**	0.088	0.071	0.094	0.107	0.143	0.114	0.500	**0.012**	0.500
DJT	0.469	0.942	0.470	0.648	**0.009**	0.648	0.519	0.116	0.517	0.369	0.261	0.262
DJU	0.500	**0.000**	0.500	0.643	**0.000**	0.643	0.725	**0.001**	0.725	**0.017**	**0.000**	**0.005**
FTSE	0.080	0.122	0.071	0.093	0.223	0.083	0.213	0.423	0.205	0.458	0.843	0.442
Nasdaq	0.469	0.922	0.468	0.579	0.983	0.578	0.683	0.995	0.702	0.500	0.928	0.500
Nikkei	**0.004**	**0.002**	**0.004**	**0.042**	0.332	0.081	0.052	0.526	0.108	0.238	**0.000**	**0.027**
SMI	0.224	0.530	0.245	0.058	0.202	0.063	0.086	0.431	0.108	0.500	0.932	0.500
SP 500	0.053	0.079	**0.047**	0.089	**0.035**	0.078	0.055	0.052	**0.043**	0.500	**0.045**	0.500
Weekly returns of indices												
CAC	**0.017**	0.143	**0.049**	**0.028**	0.272	0.068	0.061	0.478	0.142	0.500	0.575	0.500
DAX	0.154	**0.000**	**0.004**	0.674	0.798	0.674	0.667	0.892	0.661	**0.043**	**0.000**	**0.000**
DJA	0.194	**0.001**	0.052	0.692	0.607	0.692	0.679	0.899	0.597	**0.003**	**0.000**	**0.000**
DJI	0.173	**0.000**	**0.030**	0.682	0.482	0.682	0.788	0.358	0.788	**0.000**	**0.000**	**0.000**
DJT	0.428	0.623	0.385	0.628	0.456	0.628	0.693	0.552	0.693	**0.002**	**0.000**	**0.004**
DJU	0.500	0.747	0.500	0.646	**0.011**	0.646	0.747	**0.038**	0.747	0.071	**0.003**	**0.017**
FTSE	0.188	0.484	0.222	0.183	0.534	0.214	0.242	0.472	0.272	0.500	0.532	0.500
Nasdaq	0.441	0.905	0.448	0.387	0.868	0.412	0.199	0.927	0.266	0.069	0.961	0.344
Nikkei	0.500	0.140	0.500	0.310	0.154	0.260	0.330	0.316	0.462	**0.030**	0.138	0.053
SMI	0.500	0.720	0.500	0.217	0.144	0.150	0.796	0.754	0.796	0.314	0.769	0.360
SP 500	0.117	**0.000**	**0.001**	0.659	0.114	0.659	0.724	0.051	0.724	**0.000**	**0.000**	**0.000**
Daily exchange rates												
\$/€	0.452	0.904	0.452	0.194	0.423	0.181	0.066	**0.000**	**0.015**	0.500	**0.002**	0.500
¥/€	**0.037**	**0.000**	**0.002**	0.616	0.090	0.618	0.304	**0.000**	0.227	0.136	**0.000**	**0.000**
£/€	0.439	0.879	0.440	0.471	0.905	0.464	0.677	0.981	0.677	0.258	0.493	0.248
CHF/€	0.141	**0.000**	**0.012**	0.641	0.152	0.641	0.520	0.154	0.562	**0.012**	**0.000**	**0.000**
C\$/€	0.500	0.268	0.500	0.631	0.714	0.631	**0.032**	**0.000**	**0.002**	**0.045**	**0.045**	**0.029**

p-values less than 5% are in bold, those less than 1% are underlined.

to employ several tests and several alternatives. The conservative approach of Bonferroni (rejecting if the minimal p-value multiplied by the number of tests is less than a given level α), leads to rejection of the GARCH(1,1) model for 16 out of the 24 series in Table 8.5. Other procedures, less conservative than that of Bonferroni, could also be applied (see Wright 1992) without changing the general conclusion.

In conclusion, this study shows that the GARCH(1,1) model is certainly overrepresented in empirical studies. The tests presented in this chapter are easily implemented and lead to selection of GARCH models that are more elaborate than the GARCH(1,1).

8.6 Proofs of the Main Results*

Proof of Theorem 8.1 We will split the proof into seven parts.

(a) *Asymptotic normality of score vector.* When $\theta_0 \in \partial\Theta$, the function $\sigma_t^2(\theta)$ can take negative values in a neighbourhood of θ_0, and $\ell_t(\theta) = \epsilon_t^2/\sigma_t^2(\theta) + \log \sigma_t^2(\theta)$ is then undefined in this neighbourhood. Thus the derivative of $\ell_t(\cdot)$ does not exist at θ_0. By contrast, the right derivatives exist, and the vector $\partial\ell_t(\theta_0)/\partial\theta$ of the right partial derivatives is written as an ordinary derivative. The same convention is used for the higher-order derivatives, as well as for the right derivatives of $\mathbf{l}_n$, $\tilde{\ell}_t$ and $\tilde{\mathbf{l}}_n$ at θ_0. With these conventions, the formulas for the derivative of criterion remain valid:

$$\frac{\partial\ell_t(\theta)}{\partial\theta} = \left\{1 - \frac{\epsilon_t^2}{\sigma_t^2}\right\}\left\{\frac{1}{\sigma_t^2}\frac{\partial\sigma_t^2}{\partial\theta}\right\},$$

$$\frac{\partial^2\ell_t(\theta)}{\partial\theta\partial\theta'} = \left\{1 - \frac{\epsilon_t^2}{\sigma_t^2}\right\}\left\{\frac{1}{\sigma_t^2}\frac{\partial^2\sigma_t^2}{\partial\theta\partial\theta'}\right\} + \left\{2\frac{\epsilon_t^2}{\sigma_t^2} - 1\right\}\left\{\frac{1}{\sigma_t^2}\frac{\partial\sigma_t^2}{\partial\theta}\right\}\left\{\frac{1}{\sigma_t^2}\frac{\partial\sigma_t^2}{\partial\theta'}\right\}. \tag{8.34}$$

It is then easy to see that $J = E\partial^2\ell_t(\theta_0)/\partial\theta\partial\theta'$ exists under the moment assumption B7. The ergodic theorem immediately yields

$$J_n \to J, \quad \text{almost surely,} \tag{8.35}$$

where J is invertible, by assumptions B4 and B5 (cf. Proof of Theorem 7.2). The convergence (8.5) then directly follows from Slutsky's lemma and the central limit theorem given in Corollary A.1.

(b) *Uniform integrability and continuity of the second-order derivatives.* It will be shown that, for all $\varepsilon > 0$, there exists a neighbourhood $\mathcal{V}(\theta_0)$ of θ_0 such that, almost surely,

$$E_{\theta_0} \sup_{\theta\in\mathcal{V}(\theta_0)\cap\Theta} \left\| \frac{\partial^2\ell_t(\theta)}{\partial\theta\partial\theta'} \right\| < \infty \tag{8.36}$$

and

$$\lim_{n\to\infty} \frac{1}{n}\sum_{t=1}^{n} \sup_{\theta\in\mathcal{V}(\theta_0)\cap\Theta} \left\| \frac{\partial^2\ell_t(\theta)}{\partial\theta\partial\theta'} - \frac{\partial^2\ell_t(\theta_0)}{\partial\theta\partial\theta'} \right\| \le \varepsilon. \tag{8.37}$$

Using elementary derivative calculations and the compactness of Θ, it can be seen that

$$\frac{\partial\sigma_t^2(\theta)}{\partial\theta_i} = b_0^{(i)}(\theta) + \sum_{k=1}^{\infty} b_k^{(i)}(\theta)\epsilon_{t-k}^2 \quad \text{and} \quad \frac{\partial^2\sigma_t^2(\theta)}{\partial\theta_i\partial\theta_j} = b_0^{(i,j)}(\theta) + \sum_{k=1}^{\infty} b_k^{(i,j)}(\theta)\epsilon_{t-k}^2,$$

with

$$\sup_{\theta\in\Theta} | b_k^{(i)}(\theta) | \le K\rho^k, \quad \sup_{\theta\in\Theta} | b_k^{(i,j)}(\theta) | \le K\rho^k,$$

where $K > 0$ and $0 < \rho < 1$. Since $\sup_{\theta\in\Theta} 1/\sigma_t^2(\theta) \le K$, assumption B7 then entails that

$$\left\| \sup_{\theta\in\Theta} \frac{1}{\sigma_t^2}\frac{\partial\sigma_t^2(\theta)}{\partial\theta} \right\|_3 < \infty, \quad \left\| \sup_{\theta\in\Theta} \frac{1}{\sigma_t^2}\frac{\partial^2\sigma_t^2(\theta)}{\partial\theta\partial\theta'} \right\|_3 < \infty, \quad \left\| \sup_{\theta\in\Theta} \frac{\epsilon_t^2}{\sigma_t^2} \right\|_3 < \infty.$$

In view of derivatives (8.34), the Hölder and Minkowski inequalities then show (8.36) for all neighbourhood of θ_0. The ergodic theorem entails that

$$\lim_{n\to\infty} \frac{1}{n}\sum_{t=1}^{n} \sup_{\theta\in\mathcal{V}(\theta_0)\cap\Theta} \left\| \frac{\partial^2\ell_t(\theta)}{\partial\theta\partial\theta'} - \frac{\partial^2\ell_t(\theta_0)}{\partial\theta\partial\theta'} \right\| = E_{\theta_0} \sup_{\theta\in\mathcal{V}(\theta_0)\cap\Theta} \left\| \frac{\partial^2\ell_t(\theta)}{\partial\theta\partial\theta'} - \frac{\partial^2\ell_t(\theta_0)}{\partial\theta\partial\theta'} \right\|.$$

This expectation decreases to 0 when the neighbourhood $\mathcal{V}(\theta_0)$ decreases to the singleton $\{\theta_0\}$, which shows inequality (8.37).

(c) *Convergence in probability of $\theta_{J_n}(Z_n)$ to θ_0 at rate $\sqrt{n}$.* In view of (8.35), for n large enough, $\|x\|_{J_n} = (x'J_nx)^{1/2}$ defines a norm. The definition (8.6) of $\theta_{J_n}(Z_n)$ entails that $\|\sqrt{n}(\theta_{J_n}(Z_n) - \theta_0) - Z_n\|_{J_n} \le \|Z_n\|_{J_n}$. The triangular inequality then implies that

$$\begin{aligned}\|\sqrt{n}(\theta_{J_n}(Z_n) - \theta_0)\|_{J_n} &= \|\sqrt{n}(\theta_{J_n}(Z_n) - \theta_0) - Z_n + Z_n\|_{J_n} \\ &\le \|\sqrt{n}(\theta_{J_n}(Z_n) - \theta_0) - Z_n\|_{J_n} + \|Z_n\|_{J_n} \\ &\le 2(Z_n'J_nZ_n)^{1/2} = O_P(1),\end{aligned}$$

where the last equality comes from the convergence in law of (Z_n, J_n) to (Z, J). This entails that $\theta_{J_n}(Z_n) - \theta_0 = O_P(n^{-1/2})$.

(d) *Quadratic approximation of the objective function.* A Taylor expansion yields

$$\begin{aligned}\tilde{\mathbf{l}}_n(\theta) &= \tilde{\mathbf{l}}_n(\theta_0) + \frac{\partial \tilde{\mathbf{l}}_n(\theta_0)}{\partial \theta'}(\theta - \theta_0) + \frac{1}{2}(\theta - \theta_0)'\left[\frac{\partial^2 \tilde{\mathbf{l}}_n(\theta_{ij}^*)}{\partial \theta \partial \theta'}\right](\theta - \theta_0) \\ &= \tilde{\mathbf{l}}_n(\theta_0) + \frac{\partial \tilde{\mathbf{l}}_n(\theta_0)}{\partial \theta'}(\theta - \theta_0) + \frac{1}{2}(\theta - \theta_0)'\left[\frac{\partial^2 \tilde{\mathbf{l}}_n(\theta_0)}{\partial \theta \partial \theta'}\right](\theta - \theta_0) + R_n(\theta),\end{aligned}$$

where

$$R_n(\theta) = \frac{1}{2}(\theta - \theta_0)'\left\{\left[\frac{\partial^2 \tilde{\mathbf{l}}_n(\theta_{ij}^*)}{\partial \theta \partial \theta'}\right] - \frac{\partial^2 \tilde{\mathbf{l}}_n(\theta_0)}{\partial \theta \partial \theta'}\right\}(\theta - \theta_0)$$

and θ_{ij}^* is between θ and θ_0. Note that inequality (8.37) implies that, for any sequence (θ_n) such that $\theta_n - \theta_0 = O_P(1)$, we have $R_n(\theta_n) = o_P(\|\theta_n - \theta_0\|^2)$. In particular, in view of (c), we have $R_n\{\theta_{J_n}(Z_n)\} = o_P(n^{-1})$. Introducing the vector Z_n defined by (8.3), we can write

$$\frac{\partial \mathbf{l}_n(\theta_0)}{\partial \theta'}(\theta - \theta_0) = -\frac{1}{n}Z_n'J_n\sqrt{n}(\theta - \theta_0)$$

and

$$\begin{aligned}\tilde{\mathbf{l}}_n(\theta) - \tilde{\mathbf{l}}_n(\theta_0) &= -\frac{1}{2n}Z_n'J_n\sqrt{n}(\theta - \theta_0) - \frac{1}{2n}\sqrt{n}(\theta - \theta_0)'J_nZ_n \\ &\quad + \frac{1}{2}(\theta - \theta_0)'J_n(\theta - \theta_0) + R_n(\theta) + R_n^*(\theta) \\ &= \frac{1}{2n}\|Z_n - \sqrt{n}(\theta - \theta_0)\|_{J_n}^2 \\ &\quad - \frac{1}{2n}Z_n'J_nZ_n + R_n(\theta) + R_n^*(\theta),\end{aligned} \tag{8.38}$$

where

$$R_n^*(\theta) = \left\{\frac{\partial \tilde{\mathbf{l}}_n(\theta_0)}{\partial \theta} - \frac{\partial \mathbf{l}_n(\theta_0)}{\partial \theta}\right\}(\theta - \theta_0) + \frac{1}{2}(\theta - \theta_0)'\left\{\frac{\partial^2 \tilde{\mathbf{l}}_n(\theta_0)}{\partial \theta \partial \theta'} - J_n\right\}(\theta - \theta_0).$$

The initial conditions are asymptotically negligible, even when the parameter stays at the boundary. Result (d) in the proof of Theorem 7.2 remaining valid, we have $R_n^*(\theta_n) = o_P(n^{-1/2}\|\theta_n - \theta_0\|)$ for any sequence (θ_n) such that $\theta_n \to \theta_0$ in probability.

(e) *Convergence in probability of $\hat{\theta}_n$ to θ_0 at rate $\sqrt{n}$.* We know that

$$\sqrt{n}(\hat{\theta}_n - \theta_0) = O_P(1)$$

when $\theta_0 \in \overset{\circ}{\Theta}$. We will show that this result remains valid when $\theta_0 \in \partial\Theta$. Theorem 7.1 applies. In view of (d), the almost sure convergence of $\hat{\theta}_n$ to θ_0 and of J_n to the non-singular matrix J, we have

$$R_n(\hat{\theta}_n) = o_P(\|\hat{\theta}_n - \theta_0\|^2) = o_P(\|\hat{\theta}_n - \theta_0\|_{J_n}^2)$$

and

$$R_n^*(\hat{\theta}_n) = o_P(n^{-1/2}\|\hat{\theta}_n - \theta_0\|_{J_n}).$$

Since $\tilde{\mathbf{l}}_n(\cdot)$ is minimised at $\hat{\theta}_n$, we have

$$\begin{aligned}\tilde{\mathbf{l}}_n(\hat{\theta}_n) - \tilde{\mathbf{l}}_n(\theta_0) &= \frac{1}{2n}\{\|Z_n - \sqrt{n}(\hat{\theta}_n - \theta_0)\|_{J_n}^2 - \|Z_n\|_{J_n}^2 \\ &\quad + o_P(\|\sqrt{n}(\hat{\theta}_n - \theta_0)\|_{J_n}^2) + o_P(\|\sqrt{n}(\hat{\theta}_n - \theta_0)\|_{J_n})\} \\ &\leq 0.\end{aligned}$$

It follows that

$$\begin{aligned}\|Z_n - \sqrt{n}(\hat{\theta}_n - \theta_0)\|_{J_n}^2 &\leq \|Z_n\|_{J_n}^2 + o_P(\|\sqrt{n}(\hat{\theta}_n - \theta_0)\|_{J_n}) \\ &\quad + o_P(\|\sqrt{n}(\hat{\theta}_n - \theta_0)\|_{J_n}^2) \\ &\leq \{\|Z_n\|_{J_n} + o_P(\|\sqrt{n}(\hat{\theta}_n - \theta_0)\|_{J_n})\}^2,\end{aligned}$$

where the last inequality follows from $\|Z_n\|_{J_n} = O_P(1)$. The triangular inequality then yields

$$\begin{aligned}\|\sqrt{n}(\hat{\theta}_n - \theta_0)\|_{J_n} &\leq \|\sqrt{n}(\hat{\theta}_n - \theta_0) - Z_n\|_{J_n} + \|Z_n\|_{J_n} \\ &\leq 2\|Z_n\|_{J_n} + o_P(\|\sqrt{n}(\hat{\theta}_n - \theta_0)\|_{J_n}).\end{aligned}$$

Thus $\|\sqrt{n}(\hat{\theta}_n - \theta_0)\|_{J_n}\{1 + o_P(1)\} \leq 2\|Z_n\|_{J_n} = O_P(1)$.

(f) *Approximation of* $\|Z_n - \sqrt{n}(\hat{\theta}_n - \theta_0)\|_{J_n}^2$ *by* $\|Z_n - \lambda_n^\Lambda\|_{J_n}^2$. We have

$$\begin{aligned}0 &\leq \|Z_n - \sqrt{n}(\hat{\theta}_n - \theta_0)\|_{J_n}^2 - \|Z_n - \sqrt{n}(\theta_{J_n}(Z_n) - \theta_0)\|_{J_n}^2 \\ &= 2n\{\tilde{\mathbf{l}}_n(\hat{\theta}_n) - \tilde{\mathbf{l}}_n(\theta_{J_n}(Z_n))\} - 2n\{(R_n + R_n^*)(\hat{\theta}_n) - (R_n + R_n^*)(\theta_{J_n}(Z_n))\} \\ &\leq -2n\{(R_n + R_n^*)(\hat{\theta}_n) - (R_n + R_n^*)(\theta_{J_n}(Z_n))\} = o_P(1),\end{aligned}$$

where the first line comes from the definition of $\theta_{J_n}(Z_n)$, the second line comes from relation (8.38), and the inequality in third line follows from the fact that $\tilde{\mathbf{l}}_n(\cdot)$ is minimised at $\hat{\theta}_n$, the final equality having been shown in (d). In view of equality (8.8), we conclude that

$$\|Z_n - \sqrt{n}(\hat{\theta}_n - \theta_0)\|_{J_n}^2 - \|Z_n - \lambda_n^\Lambda\|_{J_n}^2 = o_P(1). \tag{8.39}$$

(g) *Approximation of* $\sqrt{n}(\hat{\theta}_n - \theta_0)$ *by* λ_n^Λ. The vector λ_n^Λ, which is the projection of Z_n on Λ with respect to the scalar product $<x, y>_{J_n} := x'J_n y$, is characterised (see Lemma 1.1 in Zarantonello 1971) by

$$\lambda_n^\Lambda \in \Lambda, \quad \langle Z_n - \lambda_n^\Lambda, \lambda_n^\Lambda - \lambda\rangle_{J_n} \geq 0, \quad \forall \lambda \in \Lambda$$

(see Figure 8.1). Since $\hat{\theta}_n \in \Theta$ and $\Lambda = \lim \uparrow \sqrt{n}(\Theta - \theta_0)$, we have almost surely $\sqrt{n}(\hat{\theta}_n - \theta_0) \in \Lambda$ for n large enough. The characterisation then entails

$$\begin{aligned}\|\sqrt{n}(\hat{\theta}_n - \theta_0) - Z_n\|^2_{J_n} &= \|\sqrt{n}(\hat{\theta}_n - \theta_0) - \lambda^{\Lambda}_n\|^2_{J_n} + \|\lambda^{\Lambda}_n - Z_n\|^2_{J_n} \\ &\quad + 2\langle\sqrt{n}(\hat{\theta}_n - \theta_0) - \lambda^{\Lambda}_n, \lambda^{\Lambda}_n - Z_n\rangle_{J_n} \\ &\geq \|\sqrt{n}(\hat{\theta}_n - \theta_0) - \lambda^{\Lambda}_n\|^2_{J_n} + \|\lambda^{\Lambda}_n - Z_n\|^2_{J_n}.\end{aligned}$$

Using the convergence (8.39), this yields

$$\|\sqrt{n}(\hat{\theta}_n - \theta_0) - \lambda^{\Lambda}_n\|^2_{J_n} \leq \|\sqrt{n}(\hat{\theta}_n - \theta_0) - Z_n\|^2_{J_n} - \|\lambda^{\Lambda}_n - Z_n\|^2_{J_n} = o_P(1),$$

which entails relation (8.9), and completes the proof.

Proof of Proposition 8.3 The first result is an immediate consequence of Slutsky's lemma and of the fact that under H_0,

$$\sqrt{n}\hat{\theta}^{(2)\prime}_n = K\sqrt{n}(\hat{\theta}_n - \theta_0) \xrightarrow{\mathcal{L}} K\lambda^{\Lambda}, \quad K'\{(\hat{\kappa}_\eta - 1)K\hat{J}^{-1}K'\}^{-1}K \xrightarrow{P} \Omega.$$

To show convergence (8.23) in the standard case where $\theta_0 \in \overset{\circ}{\Theta}$, the asymptotic $\chi^2_{d_2}$ distribution is established by showing that $\mathbf{R}_n - \mathbf{W}_n = o_P(1)$. This equation does not hold true in our testing problem $H_0 : \theta = \theta_0$, where θ_0 is on the boundary of Θ. Moreover, the asymptotic distribution of $\mathbf{W}_n$ is not $\chi^2_{d_2}$. A more direct proof is thus necessary.

Since $\hat{\theta}^{(1)}_{n|2}$ is a consistent estimator of $\theta^{(1)}_0 > 0$, we have, for n large enough, $\hat{\theta}^{(1)}_{n|2} > 0$ and

$$\frac{\partial\tilde{\mathbf{l}}_n(\hat{\theta}_{n|2})}{\partial\theta_i} = 0 \quad \text{for } i = 1, \dots, d_1.$$

Let

$$\tilde{K}\frac{\partial\tilde{\mathbf{l}}_n(\hat{\theta}_{n|2})}{\partial\theta} = 0,$$

where $\tilde{K} = (I_{d_1}, 0_{d_1 \times d_2})$. We again obtain

$$\frac{\partial\tilde{\mathbf{l}}_n(\hat{\theta}_{n|2})}{\partial\theta} = K'\frac{\partial\tilde{\mathbf{l}}_n(\hat{\theta}_{n|2})}{\partial\theta^{(2)}}, \quad K = (0_{d_2 \times d_1}, I_{d_2}). \tag{8.40}$$

Since

$$\frac{\partial^2\mathbf{l}_n(\theta_0)}{\partial\theta\partial\theta'} \to J \quad \text{almost surely}, \tag{8.41}$$

a Taylor expansion shows that

$$\sqrt{n}\frac{\partial\tilde{\mathbf{l}}_n(\hat{\theta}_{n|2})}{\partial\theta} \overset{o_P(1)}{=} \sqrt{n}\frac{\partial\mathbf{l}_n(\theta_0)}{\partial\theta} + J\sqrt{n}(\hat{\theta}_{n|2} - \theta_0),$$

where $a \overset{c}{=} b$ means $a = b + c$. The last d_2 components of this vectorial relation yield

$$\sqrt{n}\frac{\partial\tilde{\mathbf{l}}_n(\hat{\theta}_{n|2})}{\partial\theta^{(2)}} \overset{o_P(1)}{=} \sqrt{n}\frac{\partial\mathbf{l}_n(\theta_0)}{\partial\theta^{(2)}} + KJ\sqrt{n}(\hat{\theta}_{n|2} - \theta_0), \tag{8.42}$$

and the first d_1 components yield

$$0 \overset{o_P(1)}{=} \sqrt{n}\frac{\partial\mathbf{l}_n(\theta_0)}{\partial\theta^{(1)}} + \tilde{K}J\tilde{K}'\sqrt{n}(\hat{\theta}^{(1)}_{n|2} - \theta^{(1)}_0),$$

using

$$(\hat{\theta}_{n|2} - \theta_0) = \tilde{K}'(\hat{\theta}^{(1)}_{n|2} - \theta^{(1)}_0). \tag{8.43}$$

We thus have

$$\sqrt{n}(\hat{\theta}^{(1)}_{n|2} - \theta^{(1)}_0) \overset{o_P(1)}{=} -(\tilde{K}\hat{J}\tilde{K}')^{-1}\sqrt{n}\frac{\partial \mathbf{l}_n(\theta_0)}{\partial \theta^{(1)}}. \tag{8.44}$$

Using successively (8.40), (8.42) and (8.43), we obtain

$$\begin{aligned}
\mathbf{R}_n &= \frac{n}{\hat{\kappa}_\eta - 1}\frac{\partial \tilde{\mathbf{l}}_n(\hat{\theta}_{n|2})}{\partial \theta^{(2)'}}K\hat{J}^{-1}K'\frac{\partial \tilde{\mathbf{l}}_n(\hat{\theta}_{n|2})}{\partial \theta^{(2)}} \\
&\overset{o_P(1)}{=} \frac{n}{\hat{\kappa}_\eta - 1}\left\{\frac{\partial \mathbf{l}_n(\theta_0)}{\partial \theta^{(2)'}} + (\hat{\theta}^{(1)'}_{n|2} - \theta^{(1)'}_0)\tilde{K}\hat{J}K'\right\}K\hat{J}^{-1}K' \\
&\quad \times \left\{\frac{\partial \mathbf{l}_n(\theta_0)}{\partial \theta^{(2)}} + K\hat{J}\tilde{K}'(\hat{\theta}^{(1)}_{n|2} - \theta^{(1)}_0)\right\}.
\end{aligned}$$

Let

$$W = \begin{pmatrix} W_1 \\ W_2 \end{pmatrix} \sim \mathcal{N}\left\{0, J = \begin{pmatrix} J_{11} & J_{12} \\ J_{21} & J_{22} \end{pmatrix}\right\},$$

where W_1 and W_2 are vectors of respective sizes d_1 and d_2, and J_{11} is of size $d_1 \times d_1$. Thus

$$KJ\tilde{K}' = J_{21}, \; \tilde{K}JK' = J_{12}, \; \tilde{K}J\tilde{K}' = J_{11}, \; KJ^{-1}K' = (J_{22} - J_{21}J_{11}^{-1}J_{12})^{-1},$$

where the last equality comes from Exercise 6.7. Using convergence (8.44), the asymptotic distribution of $\mathbf{R}_n$ is thus that of

$$(W_2 - J_{21}J_{11}^{-1}W_1)'(J_{22} - J_{21}J_{11}^{-1}J_{12})^{-1}(W_2 - J_{21}J_{11}^{-1}W_1)$$

which follows a $\chi^2_{d_2}$ because it is easy to check that

$$W_2 - J_{21}J_{11}^{-1}W_1 \sim \mathcal{N}(0, J_{22} - J_{21}J_{11}^{-1}J_{12}).$$

We have thus shown (8.23).

Now we show (8.24). Using (8.43) and (8.44), several Taylor expansions yield

$$\begin{aligned}
n\tilde{\mathbf{l}}_n(\hat{\theta}_{n|2}) &\overset{o_P(1)}{=} n\mathbf{l}_n(\theta_0) + n\frac{\partial \mathbf{l}_n(\theta_0)}{\partial \theta'}(\hat{\theta}_{n|2} - \theta_0) + \frac{n}{2}(\hat{\theta}_{n|2} - \theta_0)'J(\hat{\theta}_{n|2} - \theta_0) \\
&\overset{o_P(1)}{=} n\mathbf{l}_n(\theta_0) - \frac{n}{2}\frac{\partial \mathbf{l}_n(\theta_0)}{\partial \theta^{(1)'}}(\tilde{K}J\tilde{K}')^{-1}\frac{\partial \mathbf{l}_n(\theta_0)}{\partial \theta^{(1)}}
\end{aligned}$$

and

$$n\tilde{\mathbf{l}}_n(\hat{\theta}_n) \overset{o_P(1)}{=} n\mathbf{l}_n(\theta_0) + n\frac{\partial \mathbf{l}_n(\theta_0)}{\partial \theta'}(\hat{\theta}_n - \theta_0) + \frac{n}{2}(\hat{\theta}_n - \theta_0)'J(\hat{\theta}_n - \theta_0).$$

By subtraction,

$$\mathbf{L}_n \overset{o_P(1)}{=} -n\left\{\frac{1}{2}\frac{\partial \mathbf{l}_n(\theta_0)}{\partial \theta^{(1)'}}(\tilde{K}J\tilde{K}')^{-1}\frac{\partial \mathbf{l}_n(\theta_0)}{\partial \theta^{(1)}} + \frac{\partial \mathbf{l}_n(\theta_0)}{\partial \theta'}(\hat{\theta}_n - \theta_0) + \frac{1}{2}(\hat{\theta}_n - \theta_0)'J(\hat{\theta}_n - \theta_0)\right\}.$$

It can be checked that

$$\sqrt{n}\begin{pmatrix} \frac{\partial \mathbf{l}_n(\theta_0)}{\partial \theta} \\ \hat{\theta}_n - \theta_0 \end{pmatrix} \overset{\mathcal{L}}{\to} \begin{pmatrix} -JZ \\ \lambda^\Lambda \end{pmatrix}.$$

Thus, the asymptotic distribution of $\mathbf{L}_n$ is that of

$$\mathbf{L} = -\frac{1}{2}Z'J'\tilde{K}'J_{11}^{-1}\tilde{K}JZ + Z'J'\lambda^{\Lambda} - \frac{1}{2}\lambda^{\Lambda'}J\lambda^{\Lambda}.$$

Moreover, it can easily be verified that

$$J'\tilde{K}'J_{11}^{-1}\tilde{K}J = J - (\kappa_\eta - 1)\Omega \quad \text{where } (\kappa_\eta - 1)\Omega = \begin{pmatrix} 0 & 0 \\ 0 & J_{22} - J_{21}J_{11}^{-1}J_{12} \end{pmatrix}.$$

It follows that

$$\begin{aligned} \mathbf{L} &= -\frac{1}{2}Z'JZ + \frac{\kappa_\eta - 1}{2}Z'\Omega Z + Z'J'\lambda^{\Lambda} - \frac{1}{2}\lambda^{\Lambda'}J\lambda^{\Lambda} \\ &= -\frac{1}{2}(\lambda^{\Lambda} - Z)'J(\lambda^{\Lambda} - Z) + \frac{\kappa_\eta - 1}{2}Z'\Omega Z, \end{aligned}$$

which gives the first equality of (8.24). The second equality follows using Exercise 8.2.

Proof of Proposition 8.4 Since $\text{Cov}(\nu_t, \sigma_t^2) = 0$, we have

$$\begin{aligned} 1 - \frac{\text{Var}(\nu_t)}{\text{Var}(\epsilon_t^2)} = \frac{\text{Var}(\sigma_t^2)}{\text{Var}(\epsilon_t^2)} &= \frac{\text{Var}\left(\sum_{i=1}^q \alpha_i \epsilon_{t-i}^2\right)}{\text{Var}(\epsilon_t^2)} \\ &= \sum_{i=1}^q \alpha_i^2 + 2\sum_{i<j} \alpha_i \alpha_j \frac{\text{Cov}(\epsilon_{t-i}^2, \epsilon_{t-j}^2)}{\text{Var}(\epsilon_t^2)} \end{aligned}$$

and the result follows from $\rho_{\epsilon^2}(i) \geq 0$ (see Proposition 2.2).

Proof of Theorem 8.2 We first study the asymptotic impact of the unknown initial values on the statistic $\hat{\boldsymbol{r}}_m$. Introduce the vector $\boldsymbol{r}_m = (r(1), \dots, r(m))'$, where

$$r(h) = n^{-1}\sum_{t=h+1}^{n} s_t s_{t-h}, \quad \text{with } s_t = \eta_t^2 - 1 \text{ and } 0 < h < n.$$

Let $s_t(\theta)$ $(\tilde{s}_t(\theta))$ be the random variable obtained by replacing η_t by $\eta_t(\theta) = \epsilon_t/\sigma_t(\theta)$ $(\tilde{\eta}_t(\theta) = \epsilon_t/\tilde{\sigma}_t(\theta))$ in s_t. The vectors $\boldsymbol{r}_m(\theta)$ and $\tilde{\boldsymbol{r}}_m(\theta)$ are defined similarly so that $\boldsymbol{r}_m = \boldsymbol{r}_m(\theta_0)$ and $\hat{\boldsymbol{r}}_m = \tilde{\boldsymbol{r}}_m(\hat{\theta}_n)$. Write $a \stackrel{c}{=} b$ when $a = b + c$. Using inequality (7.30) and the arguments used to show (d) in the proof of Theorem 7.2, it can be shown that, as $n \to \infty$,

$$\sqrt{n}\|\boldsymbol{r}_m - \tilde{\boldsymbol{r}}_m(\theta_0)\| = o_P(1), \quad \sup_{\theta\in\Theta}\left\|\frac{\partial \boldsymbol{r}_m(\theta)}{\partial \theta'} - \frac{\partial \tilde{\boldsymbol{r}}_m(\theta)}{\partial \theta'}\right\| = o_P(1). \tag{8.45}$$

We now show that the asymptotic distribution of $\sqrt{n}\hat{\boldsymbol{r}}_m$ is a function of the joint asymptotic distribution of $\sqrt{n}\boldsymbol{r}_m$ and of the QMLE. By the arguments used to show (c) in the proof of Theorem 7.2, it can be shown that there exists a neighbourhood $\mathcal{V}(\theta_0)$ of θ_0 such that

$$\lim_{n\to\infty} E \sup_{\theta\in\mathcal{V}(\theta_0)}\left\|\frac{\partial^2 \boldsymbol{r}_m(\theta)}{\partial\theta_i \partial\theta_j}\right\| < \infty \quad \text{for all } i,j \in \{1, \dots, p+q+1\}. \tag{8.46}$$

Using the convergences (8.45) and the fact that $\sqrt{n}(\hat{\theta}_n - \theta_0) = O_P(1)$, a Taylor expansion of $\boldsymbol{r}_m(\cdot)$ around $\hat{\theta}_n$ and θ_0 shows that

$$\begin{aligned} \sqrt{n}\hat{\boldsymbol{r}}_m &= \sqrt{n}\tilde{\boldsymbol{r}}_m(\theta_0) + \frac{\partial \tilde{\boldsymbol{r}}_m(\theta^*)}{\partial\theta'}\sqrt{n}(\hat{\theta}_n - \theta_0) \\ &\stackrel{o_P(1)}{=} \sqrt{n}\boldsymbol{r}_m + \frac{\partial \boldsymbol{r}_m(\theta^*)}{\partial\theta'}\sqrt{n}(\hat{\theta}_n - \theta_0) \end{aligned}$$

for some θ^* between $\hat{\theta}_n$ and θ_0. Using result (8.46), the ergodic theorem, the strong consistency of the QMLE, and a second Taylor expansion, we obtain

$$\frac{\partial \boldsymbol{r}_m(\theta^*)}{\partial \theta'} \overset{o_P(1)}{=} \frac{\partial \boldsymbol{r}_m(\theta_0)}{\partial \theta'} \to C_m := \begin{pmatrix} c_1' \\ \vdots \\ c_m' \end{pmatrix},$$

where

$$c_h = E\left\{ s_{t-h} \frac{\partial s_t(\theta_0)}{\partial \theta} \right\} = -E\left\{ s_{t-h} \frac{1}{\sigma_t^2(\theta_0)} \frac{\partial \sigma_t^2(\theta_0)}{\partial \theta} \right\}.$$

For the next to last equality, we use the fact that $E\{s_t \partial s_{t-h}(\theta_0)/\partial \theta\} = 0$. It follows that

$$\sqrt{n}\hat{\boldsymbol{r}}_m \overset{o_P(1)}{=} \sqrt{n}\boldsymbol{r}_m + C_m\sqrt{n}(\hat{\theta}_n - \theta_0). \tag{8.47}$$

We now derive the asymptotic distribution of $\sqrt{n}(\boldsymbol{r}_m, \hat{\theta}_n - \theta_0)$. In the proof of Theorem 7.2, it is shown that

$$\sqrt{n}(\hat{\theta}_n - \theta_0) \overset{o_P(1)}{=} -J^{-1}\frac{1}{\sqrt{n}}\sum_{t=1}^{n} \frac{\partial}{\partial \theta}\ell_t(\theta_0) \overset{\mathcal{L}}{\to} \mathcal{N}\{0, (\kappa_\eta - 1)J^{-1}\} \tag{8.48}$$

as $n \to \infty$, where

$$\frac{\partial}{\partial \theta}\ell_t(\theta_0) = -s_t \frac{1}{\sigma_t^2}\frac{\partial \sigma_t^2(\theta_0)}{\partial \theta}.$$

Note that $\boldsymbol{r}_m \overset{o_P(1)}{=} n^{-1}\sum_{t=1}^{n} s_t \boldsymbol{s}_{t-1:t-m}$, where $\boldsymbol{s}_{t-1:t-m} = (s_{t-1}, \ldots, s_{t-m})'$. In view of convergence (8.48), the central limit theorem applied to the martingale difference $\left\{ \left(\frac{\partial}{\partial \theta'}\ell_t(\theta_0), s_t \boldsymbol{s}_{t-1:t-m}' \right)'; \sigma(\eta_u, u \le t) \right\}$ shows that

$$\sqrt{n}\begin{pmatrix} \hat{\theta}_n - \theta_0 \\ \boldsymbol{r}_m \end{pmatrix} \overset{o_P(1)}{=} \frac{1}{\sqrt{n}}\sum_{t=1}^{n} s_t \begin{pmatrix} J^{-1}\frac{1}{\sigma_t^2}\frac{\partial \sigma_t^2(\theta_0)}{\partial \theta} \\ \boldsymbol{s}_{t-1:t-m} \end{pmatrix}$$

$$\overset{\mathcal{L}}{\to} \mathcal{N}\left\{ 0, \begin{pmatrix} (\kappa_\eta - 1)J^{-1} & \Sigma_{\hat{\theta}_n \boldsymbol{r}_m} \\ \Sigma_{\hat{\theta}_n \boldsymbol{r}_m}' & (\kappa_\eta - 1)^2 I_m \end{pmatrix} \right\}, \tag{8.49}$$

where

$$\Sigma_{\hat{\theta}_n \boldsymbol{r}_m} = (\kappa_\eta - 1)J^{-1}E\frac{1}{\sigma_t^2}\frac{\partial \sigma_t^2(\theta_0)}{\partial \theta}\boldsymbol{s}_{t-1:t-m}' = -(\kappa_\eta - 1)J^{-1}C_m'.$$

Using convergences (8.47) and (8.49) together, we obtain

$$\sqrt{n}\hat{\boldsymbol{r}}_m \overset{\mathcal{L}}{\to} \mathcal{N}(0, D), \quad D = (\kappa_\eta - 1)^2 I_m - (\kappa_\eta - 1)C_m J^{-1} C_m'.$$

We now show that D is invertible. Because the law of η_t^2 is non-degenerate, we have $\kappa_\eta > 1$. We thus have to show the invertibility of

$$(\kappa_\eta - 1)I_m - C_m J^{-1} C_m' = E\mathbf{V}\mathbf{V}', \quad \mathbf{V} = \boldsymbol{s}_{-1:-m} + C_m J^{-1}\frac{1}{\sigma_0^2}\frac{\partial \sigma_0^2(\theta_0)}{\partial \theta}.$$

If the previous matrix is singular, then there exists $\lambda = (\lambda_1, \ldots, \lambda_m)'$ such that $\lambda \neq 0$ and

$$\lambda'\mathbf{V} = \lambda'\boldsymbol{s}_{-1:-m} + \mu'\frac{1}{\sigma_0^2}\frac{\partial \sigma_0^2(\theta_0)}{\partial \theta} = 0 \quad \text{a.s.,} \tag{8.50}$$

with $\mu = \lambda' C_m J^{-1}$. Note that $\mu = (\mu_1, \dots, \mu_{p+q+1})' \neq 0$. Otherwise $\lambda' s_{-1:-m} = 0$ almost surely, which implies that there exists $j \in \{1, \dots, m\}$ such that s_{-j} is measurable with respect to $\sigma\{s_t, t \neq -j\}$. This is impossible because s_t is independent and non-degenerate by assumption A3 in Section 7.1.1 (see Exercise 10.3). Denoting by R_t *any* random variable measurable with respect to $\sigma\{\eta_u, u \leq t\}$, we have

$$\mu' \frac{\partial \sigma_0^2(\theta_0)}{\partial \theta} = \mu_2 \sigma_{-1}^2 \eta_{-1}^2 + R_{-2}$$

and

$$\sigma_0^2 \lambda' s_{-1:-m} = (\alpha_1 \sigma_{-1}^2 \eta_{-1}^2 + R_{-2})(\lambda_1 \eta_{-1}^2 + R_{-2}) = \lambda_1 \alpha_1 \sigma_{-1}^2 \eta_{-1}^4 + R_{-2} \eta_{-1}^2 + R_{-2}.$$

Thus equality (8.50) entails that

$$\lambda_1 \alpha_1 \sigma_{-1}^2 \eta_{-1}^4 + R_{-2} \eta_{-1}^2 + R_{-2} = 0 \quad \text{a.s.}$$

Solving this quadratic equation in η_{-1}^2 shows that either $\eta_{-1}^2 = R_{-2}$, which is impossible by arguments already given, or $\lambda_1 \alpha_1 = 0$. Let $\lambda'_{2:m} = (\lambda_2, \dots, \lambda_m)'$. If $\lambda_1 = 0$, then equality (8.50) implies that

$$\alpha_1 \lambda'_{2:m} \mathbf{s}_{-2:-m} \eta_{-1}^2 = \mu_2 \eta_{-1}^2 + R_{-2} \quad \text{a.s.}$$

Taking the expectation with respect to $\sigma\{\eta_t, t \leq -2\}$, it can be seen that $R_{-2} = \alpha_1 \lambda'_{2:m} \mathbf{s}_{-2:-m} - \mu_2$ in the previous equality. Thus we have

$$(\alpha_1 \lambda'_{2:m} \mathbf{s}_{-2:-m} - \mu_2)(\eta_{-1}^2 - 1) = 0 \quad \text{a.s.}$$

which entails $\alpha_1 = \mu_2 = 0$ because $P\{(\lambda_2, \dots, \lambda_m)' \mathbf{s}_{-2:-m} = 0\} < 1$ (see Exercise 8.12). For GARCH(p, 1) models, it is impossible to have $\alpha_1 = 0$ by assumption A4. The invertibility of D is thus shown in this case. In the general case, we show by induction that equality (8.50) entails $\alpha_1 = \dots \alpha_p = 0$, which contradicts A4.

It is easy to show that $\hat{D} \to D$ in probability (and even almost surely) as $n \to \infty$. The conclusion follows.

8.7 Bibliographical Notes

It is well known that when the parameter is at the boundary of the parameter space, the maximum likelihood estimator does not necessarily satisfy the first-order conditions and, in general, does not admit a limiting normal distribution. The technique, employed in particular by Chernoff (1954) and Andrews (1997) in a general framework, involves approximating the quasi-likelihood by a quadratic function, and defining the asymptotic distribution of the QML as that of the projection of a Gaussian vector on a convex cone. Particular GARCH models are considered by Andrews (1997, 1999) and Jordan (2003). The general GARCH(p, q) case is considered by Francq and Zakoïan (2007). A proof of Theorem 8.1, when the moment assumption B7 is replaced by assumption B7′ of Remark 8.2, can be found in the latter reference. When the nullity of GARCH coefficients is tested, the parameter is at the boundary of the parameter space under the null, and the alternative is one-sided. Numerous works deal with testing problems where, under the null hypothesis, the parameter is at the boundary of the parameter space. Such problems have been considered by Chernoff (1954), Bartholomew (1959), Perlman (1969), and Gouriéroux, Holly, and Monfort (1982), among many others. General one-sided tests have been studied by, for instance Rogers (1986), Wolak (1989), Silvapulle and Silvapulle (1995), and King and Wu (1997). Papers dealing more specifically with ARCH and GARCH models are Lee and King (1993), Hong (1997), Demos and Sentana (1998), Andrews (2001), Hong and Lee (2001), Dufour et al. (2004), Francq and Zakoïan (2009b) and Pedersen and Rahbek (2018).

The portmanteau tests based on the squared residual autocovariances were proposed by McLeod and Li (1983), Li and Mak (1994), and Ling and Li (1997). The results presented here closely follow

Berkes, Horváth, and Kokoszka (2003a). Problems of interest that are not studied in this book are the tests on the distribution of the iid process (see Horváth, Kokoszka, and Teyssiére 2004; Horváth and Zitikis 2006).

Concerning the overrepresentation of the GARCH(1, 1) model in financial studies, we mention Stărică (2006). This paper highlights, on a very long S&P 500 series, the poor performance of the GARCH(1, 1) in terms of prediction and modelling, and suggests a non-stationary dynamics of the returns.

8.8 Exercises

8.1 *(Minimisation of a distance under a linear constraint)*
Let J be an $n\times n$ invertible matrix, let x_0 be a vector of $\mathbb{R}^n$, and let K be a full-rank $p\times n$ matrix, $p\leq n$. Solve the problem of the minimisation of $Q(x)=(x-x_0)'J(x-x_0)$ under the constraint $Kx=0$.

8.2 *(Minimisation of a distance when some components are equal to zero)*
Let J be an $n\times n$ invertible matrix, x_0 a vector of $\mathbb{R}^n$, and $p<n$. Minimise $Q(x)=(x-x_0)'J(x-x_0)$ under the constraints $x_{i_1}=\cdots=x_{i_p}=0$ (x_i denoting the ith component of x, and assuming that $1\leq i_1<\cdots<i_p\leq n$).

8.3 *(Lagrangian or method of substitution for optimisation with constraints)*
Compare the solutions of the optimisation problem of Exercise 8.2, with

$$J=\begin{pmatrix}2&-1&0\\-1&2&1\\0&1&2\end{pmatrix}$$

and the constraints

(a) $x_3=0$,
(b) $x_2=x_3=0$.

8.4 *(Minimisation of a distance under inequality constraints)*
Find the minimum of the function

$$\lambda=\begin{pmatrix}\lambda_1\\\lambda_2\\\lambda_3\end{pmatrix}\mapsto Q(\lambda)=(\lambda-Z)'J(\lambda-Z),\quad J=\begin{pmatrix}2&-1&0\\-1&2&1\\0&1&2\end{pmatrix},$$

under the constraints $\lambda_2\geq 0$ and $\lambda_3\geq 0$, when

i. $Z=(-2,1,2)'$,

ii. $Z=(-2,-1,2)'$,

iii. $Z=(-2,1,-2)'$,

iv. $Z=(-2,-1,-2)'$.

8.5 *(Influence of the positivity constraints on the moments of the QMLE)*
Compute the mean and variance of the vector λ^{Λ} defined by equality (8.16). Compare these moments with the corresponding moments of $Z=(Z_1,Z_2,Z_3)'$.

8.6 *(Asymptotic distribution of the QMLE of an ARCH in the conditionally homoscedastic case)*
For an ARCH(q) model, compute the matrix Σ involved in the asymptotic distribution of the QMLE in the case where all the α_{0i} are equal to zero.

8.7 *(Asymptotic distribution of the QMLE when an ARCH(1) is fitted to a strong white noise)*
Let $\hat{\theta} = (\hat{\omega}, \hat{\alpha})$ be the QMLE in the ARCH(1) model $\epsilon_t = \sqrt{\omega + \alpha\epsilon_{t-1}^2}\eta_t$ when the true parameter is equal to $(\omega_0, \alpha_0) = (\omega_0, 0)$ and when $\kappa_\eta := E\eta_t^4$. Give an expression for the asymptotic distribution of $\sqrt{n}(\hat{\theta} - \theta_0)$ with the aid of

$$Z = (Z_1, Z_2)' \sim \mathcal{N}\left\{0, \begin{pmatrix} \omega_0^2\kappa_\eta & -\omega_0 \\ -\omega_0 & 1 \end{pmatrix}\right\}.$$

Compute the mean vector and the variance matrix of this asymptotic distribution. Determine the density of the asymptotic distribution of $\sqrt{n}(\hat{\omega} - \omega_0)$. Give an expression for the kurtosis coefficient of this distribution as function of κ_η.

8.8 *(One-sided and two-sided tests have the same Bahadur slopes)*
Let $X_1, \dots, X_n$ be a sample from the $\mathcal{N}(\theta, 1)$ distribution. Consider the null hypothesis $H_0: \theta = 0$. Denote by Φ the $\mathcal{N}(0, 1)$ cumulative distribution function. By the Neyman–Pearson lemma, we know that, for alternatives of the form $H_1: \theta > 0$, the one-sided test of rejection region

$$C = \left\{ n^{-1/2} \sum_{i=1}^{n} X_i > \Phi^{-1}(1 - \alpha) \right\}$$

is uniformly more powerful than the two-sided test of rejection region

$$C^* = \left\{ \left| n^{-1/2} \sum_{i=1}^{n} X_i \right| > \Phi^{-1}(1 - \alpha/2) \right\}$$

(moreover, C is uniformly more powerful than any other test of level α or less). Although the previous argument shows that the test C is superior to the test C^* in finite samples, we will conduct an asymptotic comparison of the two tests, using the Bahadur and Pitman approaches.

- The asymptotic Bahadur slope $c(\theta)$ is defined as the almost sure limit of $-2/n$ times the logarithm of the p-value under P_θ, when the limit exists. Compare the Bahadur slopes of the two tests.
- In the Pitman approach, we define a local power around $\theta = 0$ as being the power at $\tau/\sqrt{n}$. Compare the local powers of C and C^*. Compare also the local asymptotic powers of the two tests for non-Gaussian samples.

8.9 *(The local asymptotic approach cannot distinguish the Wald, score and likelihood ratio tests)*
Let $X_1, \dots, X_n$ be a sample of the $\mathcal{N}(\theta, \sigma^2)$ distribution, where θ and σ^2 are unknown. Consider the null hypothesis $H_0: \theta = 0$ against the alternative $H_1: \theta > 0$. Consider the following three tests:

- $C_1 = \{\mathbf{W}_n \geq \chi_1^2(1 - \alpha)\}$, where

$$\mathbf{W}_n = \frac{n\overline{X}_n^2}{S_n^2} \quad \left(S_n^2 = \frac{1}{n} \sum_{i=1}^{n} (X_i - \overline{X}_n)^2 \text{ and } \overline{X}_n = \frac{1}{n} \sum_{i=1}^{n} X_i \right)$$

  is the Wald statistic;

- $C_2 = \{\mathbf{R}_n \geq \chi_1^2(1 - \alpha)\}$, where

$$\mathbf{R}_n = \frac{n\overline{X}_n^2}{n^{-1} \sum_{i=1}^{n} X_i^2}$$

  is the Rao score statistic;

- $C_3 = \{\mathbf{L}_n \geq \chi_1^2(1-\alpha)\}$, where

$$\mathbf{L}_n = n \log \frac{n^{-1}\sum_{i=1}^n X_i^2}{S_n^2}$$

is the likelihood ratio statistic.

Give a justification for these three tests. Compare their local asymptotic powers and their Bahadur slopes.

8.10 *(The Wald and likelihood ratio statistics have the same asymptotic distribution)*
Consider the case $d_2 = 1$, that is, the framework of Section 8.3.3 where only one coefficient is equal to zero. Without using Remark 8.3, show that the asymptotic laws $\mathbf{W}$ and $\mathbf{L}$ defined by (8.22) and (8.24) are such that

$$\mathbf{W} = \frac{2}{\kappa_\eta - 1}\mathbf{L}.$$

8.11 *(For testing conditional homoscedasticity, the Wald and likelihood ratio statistics have the same asymptotic distribution)*
Repeat Exercise 8.10 for the conditional homoscedasticity test (8.26) in the ARCH(q) case.

8.12 *(The product of two independent random variables is null if and only if one of the two variables is null)*
Let X and Y be two independent random variables such that $XY = 0$ almost surely. Show that either $X = 0$ almost surely or $Y = 0$ almost surely.

9

Optimal Inference and Alternatives to the QMLE*

The most commonly used estimation method for GARCH models is the QML method studied in Chapter 7. One of the attractive features of this method is that the asymptotic properties of the QMLE are valid under mild assumptions. In particular, no moment assumption is required on the observed process in the pure GARCH case. However, the QML method has several drawbacks, motivating the introduction of alternative approaches. These drawbacks are the following: (i) the estimator is not explicit and requires a numerical optimisation algorithm; (ii) the asymptotic normality of the estimator requires the existence of a moment of order 4 for the noise η_t; (iii) the QMLE is inefficient in general; (iv) the asymptotic normality requires the existence of moments for ϵ_t in the general ARMA–GARCH case; and (v) a complete parametric specification is required.

In the ARCH case, the QLS estimator defined in Section 6.2 addresses point (i) satisfactorily, at the cost of additional moment conditions. The maximum likelihood (ML) estimator studied in Section 9.1 of this chapter provides an answer to points (ii) and (iii), but it requires knowledge of the density f of η_t. Indeed, it will be seen that adaptive estimators for the set of all the parameters do not exist in general semi-parametric GARCH models. Concerning point (iii), it will be seen that the QML can sometimes be optimal outside of trivial case where f is Gaussian. In Section 9.2, the ML estimator will be studied in the (quite realistic) situation where f is mis-specified. It will also be seen that the so-called local asymptotic normality (LAN) property allows us to show the local asymptotic optimality of test procedures based on the ML. In Section 9.3, less standard estimators are presented in order to address to some of the points (i)–(v).

In this chapter, we focus on the main principles of the estimation methods and do not give all the mathematical details. Precise regularity conditions justifying the arguments used can be found in the references that are given throughout the chapter or in Section 9.4.

9.1 Maximum Likelihood Estimator

In this section, the density f of the strong white noise (η_t) is assumed known. This assumption is obviously very strong and the effect of the mis-specification of f will be examined in Section 9.2.

GARCH Models: Structure, Statistical Inference and Financial Applications, Second Edition. Christian Francq and Jean-Michel Zakoian.

Conditionally on the σ-field $\mathcal{F}_{t-1}$ generated by $\{\epsilon_u : u < t\}$, the variable ϵ_t has the density $x \to \sigma_t^{-1} f(x/\sigma_t)$. It follows that, given the observations $\epsilon_1, \dots, \epsilon_n$, and the initial values $\epsilon_0, \dots, \epsilon_{1-q}, \tilde{\sigma}_0^2, \dots, \tilde{\sigma}_{1-p}^2$, the conditional likelihood is defined by

$$L_{n,f}(\theta) = L_{n,f}(\theta; \epsilon_1, \dots, \epsilon_n) = \prod_{t=1}^{n} \frac{1}{\tilde{\sigma}_t} f\left(\frac{\epsilon_t}{\tilde{\sigma}_t}\right),$$

where $\tilde{\sigma}_t^2$ is recursively defined, for $t \geq 1$, by

$$\tilde{\sigma}_t^2 = \tilde{\sigma}_t^2(\theta) = \omega + \sum_{i=1}^{q} \alpha_i \epsilon_{t-i}^2 + \sum_{j=1}^{p} \beta_j \tilde{\sigma}_{t-j}^2. \tag{9.1}$$

A maximum likelihood estimator (MLE) is obtained by maximising the likelihood on a compact subset Θ^* of the parameter space. Such an estimator is denoted by $\hat{\theta}_{n,f}$.

9.1.1 Asymptotic Behaviour

Under the above-mentioned regularity assumptions, the initial conditions are asymptotically negligible and, using the ergodic theorem, we have almost surely

$$\log \frac{L_{n,f}(\theta)}{L_{n,f}(\theta_0)} \to E_{\theta_0} \log \frac{\sigma_t(\theta_0)}{\sigma_t(\theta)} \frac{f\left(\frac{\epsilon_t}{\sigma_t(\theta)}\right)}{f\left(\frac{\epsilon_t}{\sigma_t(\theta_0)}\right)} \leq \log E_{\theta_0} \frac{\sigma_t(\theta_0)}{\sigma_t(\theta)} \frac{f\left(\frac{\epsilon_t}{\sigma_t(\theta)}\right)}{f\left(\frac{\epsilon_t}{\sigma_t(\theta_0)}\right)} = 0,$$

using Jensen's inequality and the fact that

$$E_{\theta_0}\left\{ \frac{\sigma_t(\theta_0)}{\sigma_t(\theta)} \frac{f\left(\frac{\epsilon_t}{\sigma_t(\theta)}\right)}{f\left(\frac{\epsilon_t}{\sigma_t(\theta_0)}\right)} \middle| \mathcal{F}_{t-1} \right\} = \int \frac{1}{\sigma_t(\theta)} f\left(\frac{x}{\sigma_t(\theta)}\right) dx = 1.$$

Adapting the proof of the consistency of the QMLE, it can be shown that $\hat{\theta}_{n,f} \to \theta_0$ almost surely as $n \to \infty$.

Assuming in particular that θ_0 belongs to the interior of the parameter space, a Taylor expansion yields

$$0 = \frac{1}{\sqrt{n}} \frac{\partial}{\partial \theta} \log L_{n,f}(\theta_0) + \frac{1}{n} \frac{\partial^2}{\partial \theta \, \partial \theta'} \log L_{n,f}(\theta_0) \sqrt{n}(\hat{\theta}_{n,f} - \theta_0) + o_P(1). \tag{9.2}$$

We have

$$\frac{\partial}{\partial \theta} \log L_{n,f}(\theta_0) = -\sum_{t=1}^{n} \frac{1}{2\sigma_t^2} \left\{ 1 + \frac{f'(\eta_t)}{f(\eta_t)} \eta_t \right\} \frac{\partial \sigma_t^2}{\partial \theta}(\theta_0) := \sum_{t=1}^{n} \nu_t. \tag{9.3}$$

It is easy to see that $(\nu_t, \mathcal{F}_t)$ is a martingale difference (using, for instance, the computations of Exercise 9.1). It follows that

$$S_{n,f}(\theta_0) := n^{-1/2} \frac{\partial}{\partial \theta} \log L_{n,f}(\theta_0) \xrightarrow{\mathcal{L}} \mathcal{N}(0, \mathfrak{I}), \tag{9.4}$$

where $\mathfrak{I}$ is the Fisher information matrix, defined by

$$\mathfrak{I} = E \nu_1 \nu_1' = \frac{\zeta_f}{4} E \frac{1}{\sigma_t^4} \frac{\partial \sigma_t^2}{\partial \theta} \frac{\partial \sigma_t^2}{\partial \theta'}(\theta_0), \quad \zeta_f = \int \left\{ 1 + \frac{f'(x)}{f(x)} x \right\}^2 f(x)\, dx.$$

Note that ζ_f is equal to σ^2 times the Fisher information for the scale parameter $\sigma > 0$ of the densities $\sigma^{-1} f(\cdot/\sigma)$. When f is the $\mathcal{N}(0,1)$ density, we thus have $\zeta_f = \sigma^2 \times 2/\sigma^2 = 2$.

We now turn to the other terms of the Taylor expansion (9.2). Let

$$\hat{\mathfrak{I}}(\theta) = -\frac{1}{n}\frac{\partial^2}{\partial\theta\,\partial\theta'} \log L_{n,f}(\theta).$$

We have

$$\hat{\mathfrak{I}}(\theta_0) = \mathfrak{I} + o_P(1), \tag{9.5}$$

thus, using the invertibility of $\mathfrak{I}$,

$$n^{1/2}(\hat{\theta}_{n,f} - \theta_0) \xrightarrow{\mathcal{L}} \mathcal{N}\{0, \mathfrak{I}^{-1}\}. \tag{9.6}$$

Note that

$$\hat{\theta}_{n,f} = \theta_0 + \mathfrak{I}^{-1} S_{n,f}(\theta_0)/\sqrt{n} + o_P(n^{-1/2}). \tag{9.7}$$

With the previous notation, the QMLE has the asymptotic variance

$$\begin{aligned}\mathfrak{I}_{\mathcal{N}}^{-1} &:= (E\eta_t^4 - 1)\left\{E\frac{1}{\sigma_t^4}\frac{\partial\sigma_t^2}{\partial\theta}\frac{\partial\sigma_t^2}{\partial\theta'}(\theta_0)\right\}^{-1}\\ &= \frac{\left(\int x^4 f(x)\,dx - 1\right)\zeta_f}{4}\mathfrak{I}^{-1}.\end{aligned} \tag{9.8}$$

The following proposition shows that the QMLE is not only optimal in the Gaussian case.

Proposition 9.1 (Densities Ensuring the Optimality of the QMLE) *Under the previous assumptions, the QMLE has the same asymptotic variance as the MLE if and only if the density of η_t is of the form*

$$f(y) = \frac{a^a}{\Gamma(a)}\exp(-ay^2)|y|^{2a-1}, \qquad a > 0, \quad \Gamma(a) = \int_0^\infty t^{a-1}\exp(-t)\,dt. \tag{9.9}$$

Proof. Given the asymptotic variances of the ML and QML estimators, it suffices to show that

$$(E\eta_t^4 - 1)\zeta_f \geq 4, \tag{9.10}$$

with equality if and only if f satisfies (9.9). In view of Exercise 9.2, we have

$$\int (y^2 - 1)\left(1 + \frac{f'(y)}{f(y)}y\right) f(y) dy = -2.$$

The Cauchy–Schwarz inequality then entails that

$$4 \leq \int (y^2 - 1)^2 f(y)\,dy \int \left(1 + \frac{f'(y)}{f(y)}y\right)^2 f(y)\,dy = (E\eta_t^4 - 1)\zeta_f$$

with equality if and only if there exists $a \neq 0$ such that $1 + \eta_t f'(\eta_t)/f(\eta_t) = -2a(\eta_t^2 - 1)$ almost surely. This occurs if and only if $f'(y)/f(y) = -2ay + (2a-1)/y$ almost everywhere. Under the constraints $f \geq 0$ and $\int f(y)\,dy = 1$, the solution of this differential equation is (9.9). □

Note that when f is of the form (9.9) then we have

$$\log L_{n,f}(\theta) = -a\sum_{t=1}^{n}\left(\frac{\epsilon_t^2}{\tilde{\sigma}_t^2(\theta)} + \log\tilde{\sigma}_t^2(\theta)\right)$$

up to a constant which does not depend on θ. It follows that in this case the MLE coincides with the QMLE, which entails the sufficient part of Proposition 9.1.

9.1.2 One-Step Efficient Estimator

Figure 9.1 shows the graph of the family of densities for which the QMLE and MLE coincide (and thus for which the QML is efficient). When the density f does not belong to this family of distributions, we have $\zeta_f(\int x^4 f(x)dx - 1) > 4$, and the QMLE is asymptotically inefficient in the sense that

$$\mathrm{Var}_{as}\sqrt{n}\{\hat{\theta}_n - \theta_0\} - \mathrm{Var}_{as}\sqrt{n}\{\hat{\theta}_{n,f} - \theta_0\} = \left(E\eta_1^4 - 1 - \frac{4}{\zeta_f}\right) J^{-1}$$

is positive definite. Table 9.1 shows that the efficiency loss can be important.

An efficient estimator can be obtained from a simple transformation of the QMLE, using the following result (which is intuitively true by (9.7)).

Proposition 9.2 (One-step Efficient Estimator) *Let $\tilde{\theta}_n$ be a preliminary estimator of θ_0 such that $\sqrt{n}(\tilde{\theta}_n - \theta_0) = O_P(1)$. Under the previous assumptions, the estimator defined by*

$$\overline{\theta}_{n,f} = \tilde{\theta}_n + \hat{\mathfrak{J}}^{-1}(\tilde{\theta}_n) S_{n,f}(\tilde{\theta}_n)/\sqrt{n}$$

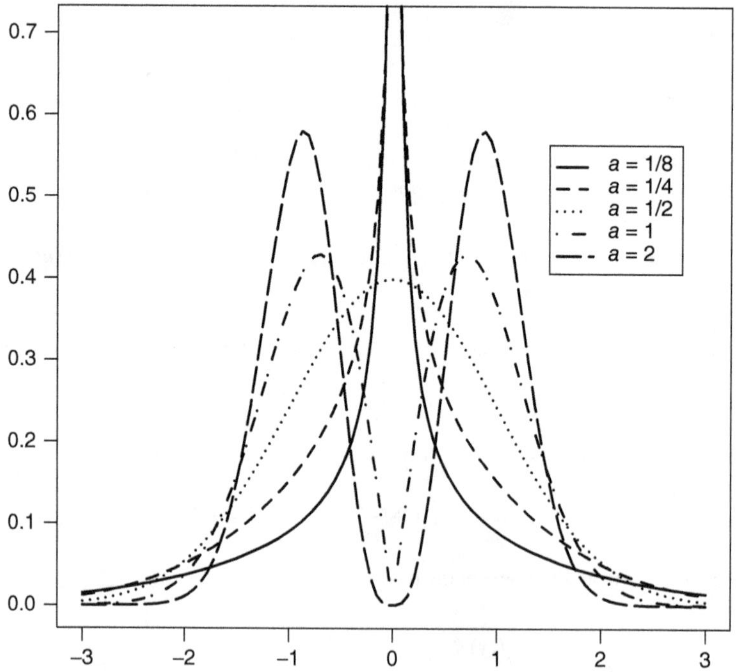

Figure 9.1 Density (9.9) for different values of $a > 0$. When η_t has this density, the QMLE and MLE have the same asymptotic variance.

Table 9.1 Asymptotic relative efficiency (ARE) of the MLE with respect to the QMLE, $\mathrm{Var}_{as}\hat{\theta}_n/\mathrm{Var}_{as}\hat{\theta}_{n,f}$, when $f(y) = \sqrt{\nu/\nu-2}f_\nu(y\sqrt{\nu/\nu-2})$, where f_ν denotes the Student t density with ν degrees of freedom.

ν	5	6	7	8	9	10	20	30	∞
ARE	5/2	5/3	7/5	14/11	6/5	15/13	95/92	145/143	1

is asymptotically equivalent to the MLE:

$$\sqrt{n}(\overline{\theta}_{n,f} - \theta_0) \xrightarrow{\mathcal{L}} \mathcal{N}\{0, \mathfrak{I}^{-1}\}.$$

Proof. A Taylor expansion of $S_{n,f}(\cdot)$ around θ_0 yields

$$S_{n,f}(\tilde{\theta}_n) = S_{n,f}(\theta_0) - \mathfrak{I}\sqrt{n}(\tilde{\theta}_n - \theta_0) + o_P(1).$$

We thus have

$$\begin{aligned}\sqrt{n}(\overline{\theta}_{n,f} - \theta_0) &= \sqrt{n}(\overline{\theta}_{n,f} - \tilde{\theta}_n) + \sqrt{n}(\tilde{\theta}_n - \theta_0)\\ &= \mathfrak{I}^{-1}S_{n,f}(\tilde{\theta}_n) + \mathfrak{I}^{-1}\{S_{n,f}(\theta_0) - S_{n,f}(\tilde{\theta}_n)\} + o_p(1)\\ &= \mathfrak{I}^{-1}S_{n,f}(\theta_0) + o_p(1) \xrightarrow{\mathcal{L}} \mathcal{N}\{0, \mathfrak{I}^{-1}\},\end{aligned}$$

using (9.4). □

In practice, one can take the QMLE as a preliminary estimator: $\tilde{\theta} = \hat{\theta}_n$.

Example 9.1 (QMLE and one-step MLE) $N = 1000$ independent samples of length $n = 100$ and 1000 were simulated for an ARCH(1) model with parameter $\omega = 0.2$ and $\alpha = 0.9$, where the distribution of the noise η_t is the standardised Student t given by $f(y) = \sqrt{\nu/\nu - 2}f_\nu(y\sqrt{\nu/\nu - 2})$ (f_ν denoting the Student density with ν degrees of freedom). Table 9.2 summarises the estimation results of the QMLE $\hat{\theta}_n$ and of the efficient estimator $\overline{\theta}_{n,f}$. This table shows that the one-step estimator $\overline{\theta}_{n,f}$ is, for this example, always more accurate than the QMLE. The observed relative efficiency is close to the theoretical asymptotic relative efficiency (ARE) computed in Table 9.1.

Table 9.2 QMLE and efficient estimator $\overline{\theta}_{n,f}$, on $N = 1000$ realisations of the ARCH(1) model $\epsilon_t = \sigma_t\eta_t$, $\sigma_t^2 = \omega + \alpha\epsilon_{t-1}^2$, $\omega = 0.2$, $\alpha = 0.9$, $\eta_t \sim f(y) = \sqrt{\nu/\nu - 2}f_\nu(y\sqrt{\nu/\nu - 2})$.

			QMLE $\hat{\theta}_n$		$\overline{\theta}_{n,f}$		
ν	n	θ_0	Mean	RMSE	Mean	RMSE	$\widehat{ARE}$
5	100	$\omega = 0.2$	0.202	0.0794	0.211	0.0646	1.51
		$\alpha = 0.9$	0.861	0.5045	0.857	0.3645	1.92
	1000	$\omega = 0.2$	0.201	0.0263	0.201	0.0190	1.91
		$\alpha = 0.9$	0.897	0.1894	0.886	0.1160	2.67
6	100	$\omega = 0.2$	0.212	0.0816	0.215	0.0670	1.48
		$\alpha = 0.9$	0.837	0.3852	0.845	0.3389	1.29
	1000	$\omega = 0.2$	0.202	0.0235	0.202	0.0186	1.61
		$\alpha = 0.9$	0.889	0.1384	0.888	0.1060	1.70
20	100	$\omega = 0.2$	0.207	0.0620	0.209	0.0619	1.00
		$\alpha = 0.9$	0.847	0.2899	0.845	0.2798	1.07
	1000	$\omega = 0.2$	0.199	0.0170	0.199	0.0165	1.06
		$\alpha = 0.9$	0.899	0.0905	0.898	0.0885	1.05

The last column gives the estimated ARE obtained from the ratio of the MSE of the two estimators on the N realisations.

For $\nu = 5$, 6 and 20 the theoretical AREs are, respectively, 2.5, 1.67, and 1.03 (for α and ω).

9.1.3 Semiparametric Models and Adaptive Estimators

In general, the density f of the noise is unknown, but f and f' can be estimated from the normalised residuals $\hat{\eta}_t = \epsilon_t/\sigma_t(\hat{\theta}_n)$, $t = 1, \ldots, n$ (for instance, using a kernel non-parametric estimator). The estimator $\hat{\theta}_{n,\hat{f}}$ (or the one-step estimator $\overline{\theta}_{n,\hat{f}}$) can then be utilised. This estimator is said to be *adaptive* if it inherits the efficiency property of $\hat{\theta}_{n,f}$ for any value of f. In general, it is not possible to estimate all the GARCH parameters adaptively.

Take the ARCH(1) example

$$\begin{cases} \epsilon_t = \sigma_t \eta_t, & \eta_t \sim f_\lambda \\ \sigma_t^2 = \omega + \alpha \epsilon_{t-1}^2, \end{cases} \tag{9.11}$$

where η_t has the double Weibull density

$$f_\lambda(x) = \frac{\lambda}{2}|x|^{\lambda-1}\exp(-|x|^\lambda), \quad \lambda > 0.$$

The subscript 0 is added to signify the true values of the parameters. The parameter $\vartheta_0 = (\theta_0', \lambda_0)'$, where $\theta_0 = (\omega_0, \alpha_0)'$, is estimated by maximising the likelihood of the observations $\epsilon_1, \ldots, \epsilon_n$ conditionally on the initial value ϵ_0. In view of (9.3), the first two components of the score are given by

$$\frac{\partial}{\partial\theta}\log L_{n,f_\lambda}(\vartheta_0) = -\sum_{t=1}^{n}\frac{1}{2\sigma_t^2}\left\{1 + \frac{f_\lambda'(\eta_t)}{f_\lambda(\eta_t)}\eta_t\right\}\frac{\partial\sigma_t^2}{\partial\theta}(\vartheta_0),$$

with

$$\left\{1 + \frac{f_\lambda'(\eta_t)}{f_\lambda(\eta_t)}\eta_t\right\} = \lambda_0(1 - |\eta_t|^{\lambda_0}), \qquad \frac{\partial\sigma_t^2}{\partial\theta} = \begin{pmatrix}1\\ \epsilon_{t-1}^2\end{pmatrix}.$$

The last component of the score is

$$\frac{\partial}{\partial\lambda}\log L_{n,f_\lambda}(\vartheta_0) = \sum_{t=1}^{n}\left\{\frac{1}{\lambda_0} + (1 - |\eta_t|^{\lambda_0})\log|\eta_t|\right\}.$$

Note that

$$\tilde{I}_{f_{\lambda_0}} = E\{\lambda_0(1 - |\eta_t|^{\lambda_0})\}^2 = \lambda_0^2,$$

$$E\left\{\frac{1}{\lambda_0} + (1 - |\eta_t|^{\lambda_0})\log|\eta_t|\right\}^2 = \frac{1 - 2\gamma + \gamma^2 + \pi^2/3}{\lambda_0^2}$$

$$E\left[\{\lambda_0(1 - |\eta_t|^{\lambda_0})\}\left\{\frac{1}{\lambda_0} + (1 - |\eta_t|^{\lambda_0})\log|\eta_t|\right\}\right] = 1 - \gamma,$$

where $\gamma = 0.577\ldots$ is the Euler constant. It follows that the score satisfies

$$n^{-1/2}\frac{\partial}{\partial\vartheta}\log L_{n,f_\lambda}(\vartheta_0) \xrightarrow{\mathcal{L}} \mathcal{N}\left\{0, \mathfrak{J} = \begin{pmatrix}\mathfrak{J}_{11} & \mathfrak{J}_{12}\\ \mathfrak{J}_{12}' & \mathfrak{J}_{22}\end{pmatrix}\right\},$$

where

$$\mathfrak{J}_{11} = E\frac{\lambda_0^2}{4(\omega_0 + \alpha_0\epsilon_{t-1}^2)^2}\begin{pmatrix}1 & \epsilon_{t-1}^2\\ \epsilon_{t-1}^2 & \epsilon_{t-1}^4\end{pmatrix}, \quad \mathfrak{J}_{12} = \frac{\gamma - 1}{2}E\frac{1}{\omega_0 + \alpha_0\epsilon_{t-1}^2}\begin{pmatrix}1\\ \epsilon_{t-1}^2\end{pmatrix},$$

and $\mathfrak{J}_{22} = \lambda_0^{-2}(1 - 2\gamma + \gamma^2 + \pi^2/6)$. By the general properties of an information matrix (see Exercise 9.4 for a direct verification), we also have

$$n^{-1}\frac{\partial^2}{\partial\vartheta\,\partial\vartheta'}\log L_{n,f_\lambda}(\vartheta_0) \to -\mathfrak{J} \quad \text{a.s. as } n \to \infty.$$

The information matrix $\mathfrak{I}$ being such that $\mathfrak{I}_{12} \neq 0$, the necessary Stein's condition (see Bickel 1982) for the existence of an adaptive estimator is not satisfied. The intuition behind this condition is the following. In view of the previous discussion, the asymptotic variance of the MLE of ϑ_0 should be of the form

$$\mathfrak{I}^{-1} = \begin{pmatrix} \mathfrak{I}^{11} & \mathfrak{I}^{12} \\ \mathfrak{I}^{21} & \mathfrak{I}^{22} \end{pmatrix}.$$

When λ_0 is unknown, the optimal asymptotic variance of a regular estimator of θ_0 is thus $\mathfrak{I}^{11}$. Knowing λ_0, the asymptotic variance of the MLE of θ_0 is $\mathfrak{I}_{11}^{-1}$. If there exists an adaptive estimator for the class of the densities of the form f_λ (or for a larger class of densities), then we have $\mathfrak{I}^{11} = \mathfrak{I}_{11}^{-1}$. Since $\mathfrak{I}^{11} = (\mathfrak{I}_{11} - \mathfrak{I}_{12}\mathfrak{I}_{22}^{-1}\mathfrak{I}_{21})^{-1}$ (see Exercise 6.7), this is possible only if $\mathfrak{I}_{12} = 0$, which is not the case here.

Reparameterising the model, Drost and Klaassen (1997) showed that it is, however, possible to obtain adaptive estimates of certain parameters. To illustrate this point, return to the ARCH(1) example with the parameterisation

$$\begin{cases} \epsilon_t = c\sigma_t\eta_t, & \eta_t \sim f_\lambda \\ \sigma_t^2 = 1 + \alpha\epsilon_{t-1}^2. \end{cases} \tag{9.12}$$

Let $\vartheta = (\alpha, c, \lambda)$ be an element of the parameter space. The score now satisfies

$$\frac{\partial}{\partial\alpha}\log L_{n,f_\lambda}(\vartheta_0) = -\sum_{t=1}^{n}\frac{1}{2\sigma_t^2}\{\lambda_0(1 - |\eta_t|^{\lambda_0})\}\epsilon_{t-1}^2,$$

$$\frac{\partial}{\partial c}\log L_{n,f_\lambda}(\vartheta_0) = -\sum_{t=1}^{n}\frac{1}{c_0}\{\lambda_0(1 - |\eta_t|^{\lambda_0})\},$$

$$\frac{\partial}{\partial\lambda}\log L_{n,f_\lambda}(\vartheta_0) = \sum_{t=1}^{n}\left\{\frac{1}{\lambda_0} + (1 - |\eta_t|^{\lambda_0})\log|\eta_t|\right\}.$$

Thus $n^{-1/2}\partial\log L_{n,f_{\lambda_0}}(\vartheta_0)/\partial\vartheta \xrightarrow{\mathcal{L}} \mathcal{N}(0, \mathfrak{I})$ with

$$\mathfrak{I} = \begin{pmatrix} \frac{\lambda_0^2}{4}A & \frac{\lambda_0^2}{2c_0}B & -\frac{1-\gamma}{2}B \\ \frac{\lambda_0^2}{2c_0}B & \frac{\lambda_0^2}{c_0^2} & -\frac{1-\gamma}{c_0} \\ -\frac{1-\gamma}{2}B & -\frac{1-\gamma}{c_0} & \frac{1-2\gamma+\gamma^2+\pi^2/6}{\lambda_0^2} \end{pmatrix},$$

where

$$A = E\frac{\epsilon_{t-1}^4}{(1 + \alpha_0\epsilon_{t-1}^2)^2}, \quad B = E\frac{\epsilon_{t-1}^2}{1 + \alpha_0\epsilon_{t-1}^2}.$$

It can be seen that this matrix is invertible because its determinant is equal to $\pi^2\lambda_0^2(A - B^2)/24c_0^2 > 0$. Moreover,

$$\mathfrak{I}^{-1} = \begin{pmatrix} \frac{4}{\lambda_0^2(A-B^2)} & -\frac{2c_0B}{\lambda_0^2(A-B^2)} & 0 \\ -\frac{2c_0B}{\lambda_0^2(A-B^2)} & \frac{c_0^2[A\{\pi^2+6(1-\gamma)^2\}-6B^2(1-\gamma)^2]}{\pi^2\lambda_0^2(A-B^2)} & \frac{6c_0(1-\gamma)}{\pi^2} \\ 0 & \frac{6c_0(1-\gamma)}{\pi^2} & \frac{6\lambda_0^2}{\pi^2} \end{pmatrix}.$$

The MLE enjoying optimality properties in general, when λ_0 is unknown, the optimal variance of an estimator of (α_0, c_0) should be equal to

$$\Sigma_{ML} = \begin{pmatrix} \frac{4}{\lambda_0^2(A-B^2)} & -\frac{2c_0B}{\lambda_0^2(A-B^2)} \\ -\frac{2c_0B}{\lambda_0^2(A-B^2)} & \frac{c_0^2[A\{\pi^2+6(1-\gamma)^2\}-6B^2(1-\gamma)^2]}{\pi^2\lambda_0^2(A-B^2)} \end{pmatrix}.$$

When λ_0 is known, a similar calculation shows that the MLE of (α_0, c_0) should have the asymptotic variance

$$\Sigma_{ML|\lambda_0} = \begin{pmatrix} \frac{\lambda_0^2}{4}A & \frac{\lambda_0^2}{2c_0}B \\ \frac{\lambda_0^2}{2c_0}B & \frac{\lambda_0^2}{c_0^2} \end{pmatrix}^{-1} = \begin{pmatrix} \frac{4}{\lambda_0^2(A-B^2)} & -\frac{2c_0B}{\lambda_0^2(A-B^2)} \\ -\frac{2c_0B}{\lambda_0^2(A-B^2)} & \frac{c_0^2A}{\lambda_0^2(A-B^2)} \end{pmatrix}.$$

We note that $\Sigma_{ML|\lambda_0}(1,1) = \Sigma_{ML}(1,1)$. Thus, in presence of the unknown parameter c, the MLE of α_0 is equally accurate when λ_0 is known or unknown. This is not particular to the chosen form of the density of the noise, which leads us to think that there might exist an estimator of α_0 that adapts to the density f of the noise (in presence of the nuisance parameter c). Drost and Klaassen (1997) showed the actual existence of adaptive estimators for some parameters of an extension of (9.12).

9.1.4   Local Asymptotic Normality

In this section, we will see that the GARCH model satisfies the so-called LAN property, which has interesting consequences for the local asymptotic properties of estimators and tests. Let $\theta_n = \overset{\circ}{\theta} + h_n/\sqrt{n}$ be a sequence of local parameters around the parameter $\theta \in \overset{\circ}{\Theta}$, where (h_n) is a bounded sequence of $\mathbb{R}^{p+q+1}$. Consider the local log-likelihood ratio function

$$h_n \rightarrow \Lambda_{n,f}(\theta_n, \theta) := \log \frac{L_{n,f}(\theta_n)}{L_{n,f}(\theta)}.$$

The Taylor expansion of this function around 0 leads to

$$\Lambda_{n,f}(\theta + h_n/\sqrt{n}, \theta) = h_n' S_{n,f}(\theta) - \frac{1}{2}h_n' \mathfrak{I}(\theta)h_n + o_{P_\theta}(1), \tag{9.13}$$

where, as we have already seen,

$$S_{n,f}(\theta) \overset{\mathcal{L}}{\rightarrow} \mathcal{N}\{0, \mathfrak{I}(\theta)\} \quad \text{under } P_\theta. \tag{9.14}$$

It follows that

$$\Lambda_{n,f}(\theta + h_n/\sqrt{n}, \theta) \overset{o_{P_\theta}(1)}{\sim} \mathcal{N}\left(-\frac{1}{2}\tau_n, \tau_n\right), \quad \tau_n = h_n' \mathfrak{I}(\theta)h_n.$$

Remark 9.1 (Limiting Gaussian Experiments)   Denoting by $L(h)$ the $\mathcal{N}\{h, \mathfrak{I}^{-1}(\theta)\}$ density evaluated at the point X, we have

$$\Lambda(h, 0) := \log \frac{L(h)}{L(0)} = -\frac{1}{2}(X-h)' \mathfrak{I}(\theta)(X-h) + \frac{1}{2}X' \mathfrak{I}X$$

$$= -\frac{1}{2}h' \mathfrak{I}(\theta)h + h' \mathfrak{I}X \sim \mathcal{N}\left(-\frac{1}{2}h' \mathfrak{I}(\theta)h, h' \mathfrak{I}(\theta)h\right)$$

under the null hypothesis that $X \sim \mathcal{N}\{0, \mathfrak{I}^{-1}(\theta)\}$. It follows that the local log-likelihood ratio $\Lambda_{n,f}(\theta + h/\sqrt{n}, \theta)$ of n observations converges in law to the likelihood ratio $\Lambda(h, 0)$ of one Gaussian observation $X \sim \mathcal{N}\{h, \mathfrak{I}^{-1}(\theta)\}$. Using Le Cam's terminology, we say that the 'sequence of local experiments' $\{L_{n,f}(\theta + h/\sqrt{n}), h \in \mathbb{R}^{p+q+1}\}$ converges to the Gaussian experiment $\{\mathcal{N}(h, \mathfrak{I}^{-1}(\theta)), h \in \mathbb{R}^{p+q+1}\}$.

The properties (9.13)–(9.14) are called LAN. It entails that the MLE is locally asymptotically optimal (in the minimax sense and in various other senses; see van der Vaart 1998). The LAN property also makes it very easy to compute the local asymptotic distributions of statistics, or the asymptotic local powers of tests. As an example, consider tests of the null hypothesis

$$H_0 : \alpha_q = \alpha_0 > 0$$

against the sequence of local alternatives

$$H_n : \alpha_q = \alpha_0 + c/\sqrt{n}.$$

The performance of the Wald, score, and of likelihood ratio tests will be compared.

Wald Test Based on the MLE

Let $\hat{\alpha}_{q,f}$ be the $(q+1)$th component of the MLE $\hat{\theta}_{n,f}$. In view of (9.7) and (9.13)–(9.14), we have under H_0 that

$$\begin{pmatrix} \sqrt{n}(\hat{\alpha}_{q,f} - \alpha_0) \\ \Lambda_{n,f}(\theta_0 + h/\sqrt{n}, \theta_0) \end{pmatrix} = \begin{pmatrix} e'_{q+1}\mathfrak{I}^{-1/2}X \\ h'\mathfrak{I}^{1/2}X - \frac{h'\mathfrak{I}h}{2} \end{pmatrix}' + o_p(1), \tag{9.15}$$

where $X \sim \mathcal{N}(0, I_{1+p+q})$, and e_i denotes the ith vector of the canonical basis of $\mathbb{R}^{p+q+1}$, noting that the $(q+1)$th component of $\theta_0 \in \overset{\circ}{\Theta}$ is equal to α_0. Consequently, the asymptotic distribution of the vector defined in (9.15) is

$$\mathcal{N}\left\{\begin{pmatrix} 0 \\ -\frac{h'\mathfrak{I}h}{2} \end{pmatrix}, \begin{pmatrix} e'_{q+1}\mathfrak{I}^{-1}e_{q+1} & e'_{q+1}h \\ h'e_{q+1} & h'\mathfrak{I}h \end{pmatrix}\right\} \quad \text{under } H_0. \tag{9.16}$$

Le Cam's third lemma (see van der Vaart 1998, p. 90; see also Exercise 9.3) and the contiguity of the probabilities P_{θ_0} and $P_{\theta_0+h/\sqrt{n}}$ (implied by the LAN properties (9.13)–(9.14)) show that, for $e'_{q+1}h = c$,

$$\sqrt{n}(\hat{\alpha}_{q,f} - \alpha_0) \xrightarrow{\mathcal{L}} \mathcal{N}\{c, e'_{q+1}\mathfrak{I}^{-1}e_{q+1}\} \quad \text{under } H_n.$$

The Wald test (and also the t test) is defined by the rejection region $\{\mathbf{W}_{n,f} > \chi^2_1(1-\alpha)\}$ where

$$\mathbf{W}_{n,f} = n(\hat{\alpha}_{q,f} - \alpha_0)^2 / \{e'_{q+1}\hat{\mathfrak{I}}^{-1}(\hat{\theta}_{n,f})e_{q+1}\}$$

and $\chi^2_1(1-\alpha)$ denotes the $(1-\alpha)$-quantile of a chi-square distribution with 1 degree of freedom. This test has asymptotic level α and local asymptotic power $c \mapsto 1 - \Phi_c\{\chi^2_1(1-\alpha)\}$, where $\Phi_c(\cdot)$ denotes the cumulative distribution function of a non-central chi-square with 1 degree of freedom and non-centrality parameter[1]

$$\frac{c^2}{e'_{q+1}\mathfrak{I}^{-1}e_{q+1}}.$$

This test is locally asymptotically uniformly most powerful among the asymptotically unbiased tests.

Score Test Based on the MLE

The score (or Lagrange multiplier) test is based on the statistic

$$\mathbf{R}_{n,f} = \frac{1}{n}\frac{\partial \log L_{n,f}(\hat{\theta}^c_{n,f})}{\partial \theta'}\hat{\mathfrak{I}}^{-1}(\hat{\theta}^c_{n,f})\frac{\partial \log L_{n,f}(\hat{\theta}^c_{n,f})}{\partial \theta}, \tag{9.17}$$

where $\hat{\theta}^c_{n,f}$ is the MLE under H_0, that is, constrained by the condition that the $(q+1)$th component of the estimator is equal to α_0. By the definition of $\hat{\theta}^c_{n,f}$, we have

$$\frac{\partial \log L_{n,f}(\hat{\theta}^c_{n,f})}{\partial \theta_i} = 0, \quad i \neq q+1. \tag{9.18}$$

---

[1] By definition, if $Z_1, \dots, Z_k$ are independently and normally distributed with variance 1 and means $m_1, \dots, m_k$, then $\sum_{i=1}^k Z_i^2$ follows a noncentral chi-square with k degrees of freedom and noncentrality parameter $\sum_{i=1}^k m_i^2$.

In view of (9.17) and (9.18), the test statistic can be written as

$$\mathbf{R}_{n,f} = \frac{1}{n}\left\{\frac{\partial \log L_{n,f}(\hat{\theta}^c_{n,f})}{\partial \theta_{q+1}}\right\}^2 e'_{q+1}\hat{\mathfrak{F}}^{-1}(\hat{\theta}^c_{n,f})e_{q+1}. \tag{9.19}$$

Under H_0, almost surely $\hat{\theta}^c_{n,f} \to \theta_0$ and $\hat{\theta}_{n,f} \to \theta_0$. Consequently,

$$0 = \frac{1}{\sqrt{n}}\frac{\partial \log L_{n,f}(\hat{\theta}_{n,f})}{\partial \theta} \overset{o_P(1)}{=} \frac{1}{\sqrt{n}}\frac{\partial \log L_{n,f}(\theta_0)}{\partial \theta} - \mathfrak{F}\sqrt{n}(\hat{\theta}_{n,f} - \theta_0)$$

and

$$\frac{1}{\sqrt{n}}\frac{\partial \log L_{n,f}(\hat{\theta}^c_{n,f})}{\partial \theta} \overset{o_P(1)}{=} \frac{1}{\sqrt{n}}\frac{\partial \log L_{n,f}(\theta_0)}{\partial \theta} - \mathfrak{F}\sqrt{n}(\hat{\theta}^c_{n,f} - \theta_0).$$

Taking the difference, we obtain

$$\begin{aligned}\frac{1}{\sqrt{n}}\frac{\partial \log L_{n,f}(\hat{\theta}^c_{n,f})}{\partial \theta} &\overset{o_P(1)}{=} \mathfrak{F}\sqrt{n}(\hat{\theta}_{n,f} - \hat{\theta}^c_{n,f}) \\ &\overset{o_P(1)}{=} \hat{\mathfrak{F}}(\hat{\theta}^c_{n,f})\sqrt{n}(\hat{\theta}_{n,f} - \hat{\theta}^c_{n,f}),\end{aligned} \tag{9.20}$$

which, using (9.17), gives

$$\mathbf{R}_{n,f} \overset{o_P(1)}{=} n(\hat{\theta}_{n,f} - \hat{\theta}^c_{n,f})'\hat{\mathfrak{F}}(\hat{\theta}^c_{n,f})(\hat{\theta}_{n,f} - \hat{\theta}^c_{n,f}). \tag{9.21}$$

From (9.20), we obtain

$$\sqrt{n}(\hat{\theta}_{n,f} - \hat{\theta}^c_{n,f}) \overset{o_P(1)}{=} \mathfrak{F}^{-1}\frac{1}{\sqrt{n}}\frac{\partial \log L_{n,f}(\hat{\theta}^c_{n,f})}{\partial \theta}.$$

Using this relation, $e'_{q+1}\hat{\theta}^c_{n,f} = \alpha_0$ and (9.18), it follows that

$$\sqrt{n}(\hat{\alpha}_{n,f} - \alpha_0) \overset{o_P(1)}{=} \{e'_{q+1}\hat{\mathfrak{F}}^{-1}(\theta_0)e_{q+1}\}\frac{1}{\sqrt{n}}\frac{\partial \log L_{n,f}(\hat{\theta}^c_{n,f})}{\partial \theta_{q+1}}.$$

Using (9.19), we have

$$\mathbf{R}_{n,f} \overset{o_P(1)}{=} n\frac{(\hat{\alpha}_{n,f} - \alpha_0)^2}{e'_{q+1}\hat{\mathfrak{F}}^{-1}(\hat{\theta}^c_{n,f})e_{q+1}} \overset{o_P(1)}{=} \mathbf{W}_{n,f} \quad \text{under } H_0. \tag{9.22}$$

By Le Cam's third lemma, the score test thus inherits the local asymptotic optimality properties of the Wald test.

Likelihood Ratio Test Based on the MLE

The likelihood ratio test is based on the statistic $\mathbf{L}_{n,f} = 2\Lambda_{n,f}(\hat{\theta}_{n,f}, \hat{\theta}^c_{n,f})$. The Taylor expansion of the log-likelihood around $\hat{\theta}_{n,f}$ leads to

$$\begin{aligned}\log L_{n,f}(\hat{\theta}^c_{n,f}) &\overset{o_P(1)}{=} \log L_{n,f}(\hat{\theta}_{n,f}) + (\hat{\theta}^c_{n,f} - \hat{\theta}_{n,f})'\frac{\partial \log L_{n,f}(\hat{\theta}_{n,f})}{\partial \theta} \\ &+ \frac{1}{2}(\hat{\theta}^c_{n,f} - \hat{\theta}_{n,f})'\frac{\partial^2 \log L_{n,f}(\hat{\theta}_{n,f})}{\partial \theta\, \partial \theta'}(\hat{\theta}^c_{n,f} - \hat{\theta}_{n,f}),\end{aligned}$$

thus, using $\partial \log L_{n,f}(\hat{\theta}_{n,f})/\partial\theta = 0$, (9.5) and (9.21),

$$\mathbf{L}_{n,f} \overset{o_P(1)}{=} n(\hat{\theta}^c_{n,f} - \hat{\theta}_{n,f})'\mathfrak{I}(\hat{\theta}^c_{n,f} - \hat{\theta}_{n,f}) \overset{o_P(1)}{=} \mathbf{R}_{n,f},$$

under H_0 and H_n. It follows that the three tests exhibit the same asymptotic behaviour, both under the null hypothesis and under local alternatives.

Tests Based on the QML

We have seen that the $\mathbf{W}_{n,f}$, $\mathbf{R}_{n,f}$, and $\mathbf{L}_{n,f}$ tests based on the MLE are all asymptotically equivalent under H_0 and H_n (in particular, they are all asymptotically locally optimal). We now compare these tests to those based on the QMLE, focusing on the QML Wald whose statistic is

$$\mathbf{W}_n = n(\hat{\alpha}_q - \alpha_0)^2/e'_{q+1}\hat{\mathfrak{I}}^{-1}_{\mathcal{N}}(\hat{\theta}_n)e_{q+1},$$

where $\hat{\alpha}_q$ is the $(q+1)$th component of the QML $\hat{\theta}_n$, and $\hat{\mathfrak{I}}^{-1}_{N}(\theta)$ is

$$\hat{\mathfrak{I}}^{-1}_{\mathcal{N}}(\theta) = \frac{1}{n}\sum_{t=1}^{n}\left\{\frac{\epsilon_t^2}{\sigma_t^2(\theta)} - 1\right\}^2\left\{\frac{1}{n}\sum_{t=1}^{n}\frac{1}{\sigma_t^4}\frac{\partial\sigma_t^2}{\partial\theta}\frac{\partial\sigma_t^2}{\partial\theta'}(\theta)\right\}^{-1}$$

or an asymptotically equivalent estimator.

Remark 9.2 Obviously, $\mathfrak{I}_{\mathcal{N}}$ should not be estimated by the analog of (9.5),

$$\hat{\mathfrak{I}}_1 := -\frac{1}{n}\sum_{t=1}^{n}\frac{\partial^2 \log L_n(\hat{\theta}_n)}{\partial\theta\,\partial\theta'} \overset{o_P(1)}{=} \frac{1}{2n}\sum_{t=1}^{n}\frac{1}{\sigma_t^4}\frac{\partial\sigma_t^2}{\partial\theta}\frac{\partial\sigma_t^2}{\partial\theta'}(\hat{\theta}_n), \tag{9.23}$$

nor by the empirical variance of the (pseudo-)score vector

$$\begin{aligned}\hat{\mathfrak{I}}_2 &:= \frac{1}{n}\sum_{t=1}^{n}\frac{\partial \log L_n(\hat{\theta}_n)}{\partial\theta}\frac{\partial \log L_n(\hat{\theta}_n)}{\partial\theta'}\\ &\overset{o_P(1)}{=} \frac{1}{4n}\sum_{t=1}^{n}\left(1 - \frac{\epsilon_t^2}{\sigma_t^2(\hat{\theta}_n)}\right)^2\frac{1}{\sigma_t^4}\frac{\partial\sigma_t^2}{\partial\theta}\frac{\partial\sigma_t^2}{\partial\theta'}(\hat{\theta}_n),\end{aligned} \tag{9.24}$$

which do not always converge to $\mathfrak{I}_{\mathcal{N}}$, when the observations are not Gaussian.

Remark 9.3 The score test based on the QML can be defined by means of the statistic

$$\mathbf{R}_n = \frac{1}{n}\left\{\frac{\partial \log L_n(\hat{\theta}^c_n)}{\partial\theta_{q+1}}\right\}^2 e'_{q+1}\hat{\mathfrak{I}}^{-1}_{\mathcal{N}}(\hat{\theta}^c_n)e_{q+1},$$

denoting by $\hat{\theta}^c_n$ the QML constrained by $\hat{\alpha}_q = \alpha_0$. Similarly, we define the likelihood ratio test statistic $\mathbf{L}_n$ based on the QML. Taylor expansions similar to those previously used show that $\mathbf{W}_n \overset{o_P(1)}{=} \mathbf{R}_n \overset{o_P(1)}{=} \mathbf{L}_n$ under H_0 and under H_n.

Recall that

$$\begin{aligned}\sqrt{n}(\hat{\theta}_n - \theta_0) &\overset{o_P(1)}{=} 2\left\{E\frac{1}{\sigma_t^4}\frac{\partial\sigma_t^2}{\partial\theta}\frac{\partial\sigma_t^2}{\partial\theta'}(\theta_0)\right\}^{-1}\frac{-1}{\sqrt{n}}\sum_{t=1}^{n}\frac{1}{2\sigma_t^2}\left\{1 - \frac{\epsilon_t^2}{\sigma_t^2}\right\}\frac{\partial\sigma_t^2}{\partial\theta}(\theta_0)\\ &\overset{o_P(1)}{=} \frac{\zeta_f}{2}\mathfrak{I}^{-1}\frac{-1}{\sqrt{n}}\sum_{t=1}^{n}\frac{1}{2\sigma_t^2}\left\{1 - \frac{\epsilon_t^2}{\sigma_t^2}\right\}\frac{\partial\sigma_t^2}{\partial\theta}(\theta_0).\end{aligned}$$

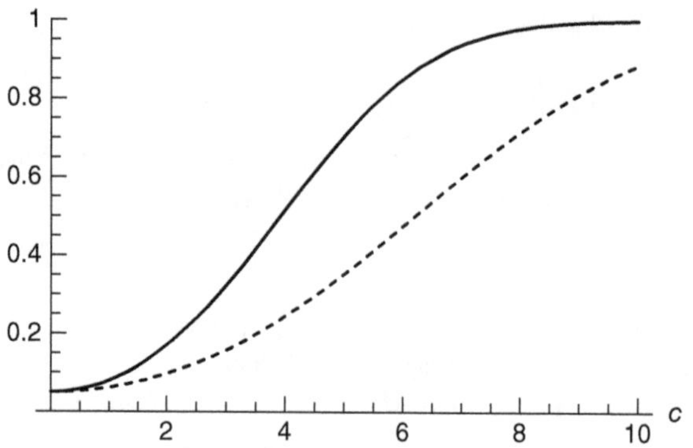

Figure 9.2 Local asymptotic power of the optimal Wald test $\{\mathbf{W}_{n,f} > \chi_1^2(0.95)\}$ (solid line) and of the standard Wald test $\{\mathbf{W}_n > \chi_1^2(0.95)\}$ (dotted line), when $f(y) = \sqrt{\nu/\nu-2}f_\nu(y\sqrt{\nu/\nu-2})$ and $\nu = 5$.

Using (9.13)–(9.14), (9.8), and Exercise 9.2, we obtain

$$\text{Cov}_{as}\{\sqrt{n}(\hat{\theta}_n - \theta_0), \Lambda_{n,f}\left(\theta_0 + h/\sqrt{n}, \theta_0\right)\}$$
$$\overset{o_P(1)}{=} \frac{\zeta_f}{2}\mathfrak{J}^{-1}E\left\{(1-\eta_t^2)\left(1+\eta_t\frac{f'(\eta_t)}{f(\eta_t)}\right)\right\}E_{\theta_0}\left\{\frac{1}{4\sigma_t^4}\frac{\partial\sigma_t^2}{\partial\theta}\frac{\partial\sigma_t^2}{\partial\theta'}\right\}h = h$$

under H_0. The previous arguments, in particular Le Cam's third lemma, show that

$$\sqrt{n}(\hat{\alpha}_q - \alpha_0) \xrightarrow{\mathcal{L}} \mathcal{N}\{c, e'_{q+1}\mathfrak{J}_{\mathcal{N}}^{-1}(\theta_0)e_{q+1}\} \quad \text{under } H_n.$$

The local asymptotic power of the $\{\mathbf{W}_n > \chi_1^2(1-\alpha)\}$ test is thus $c \mapsto 1 - \Phi_{\tilde{c}}\{\chi_1^2(1-\alpha)\}$, where the non-centrality parameter is

$$\tilde{c} = \frac{c^2}{e'_{q+1}\mathfrak{J}_{\mathcal{N}}^{-1}e_{q+1}} = \frac{4}{\left(\int x^4 f(x)dx - 1\right)\zeta_f} \times \frac{c^2}{e'_{q+1}\mathfrak{J}^{-1}e_{q+1}}.$$

Figure 9.2 displays the local asymptotic powers of the two tests, $c \mapsto 1 - \Phi_c\{\chi_1^2(0.95)\}$ (solid line) and $c \mapsto 1 - \Phi_{\tilde{c}}\{\chi_1^2(0.95)\}$ (dashed line), when f is the normalised Student t density with 5 degrees of freedom and when θ_0 is such that $e'_{q+1}\mathfrak{J}_{\mathcal{N}}^{-1}e_{q+1} = 4$. Note that the local asymptotic power of the optimal Wald test is sometimes twice as large as that of score test.

9.2 Maximum Likelihood Estimator with Mis-specified Density

The MLE requires the (unrealistic) assumption that f is known. What happens when f is mis-specified, that is, when we use $\hat{\theta}_{n,h}$ with $h \neq f$?

In this section, the usual assumption $E\eta_t^2 = 1$ will be replaced by alternative moment assumptions that will be more relevant for the estimators considered. Under some regularity assumptions, the ergodic theorem entails that

$$\hat{\theta}_{n,h} = \arg\max_\theta Q_n(\theta), \quad \text{where } Q_n(\theta) \to Q(\theta) = E_f \log\frac{\sigma_t(\theta_0)}{\sigma_t(\theta)}h\left(\eta_t\frac{\sigma_t(\theta_0)}{\sigma_t(\theta)}\right) \quad \text{a.s.}$$

Here, the subscript f is added to the expectation symbol in order to emphasise the fact that the random variable η_0 follows the distribution f, which does not necessarily coincide with the 'instrumental' density h. This allows us to show that

$$\hat{\theta}_{n,h} \to \theta^* = \arg\max_{\theta} Q(\theta) \quad \text{a.s.}$$

Note that the estimator $\hat{\theta}_{n,h}$ can be seen as a non-Gaussian QMLE.

9.2.1 Condition for the Convergence of $\hat{\theta}_{n,h}$ to θ_0

Note that under suitable identifiability conditions, $\sigma_t(\theta_0)/\sigma_t(\theta) = 1$ if and only if $\theta = \theta_0$. For the consistency of the estimator (that is, for $\theta^* = \theta_0$), it is thus necessary for the function $\sigma \to E_f g(\eta_0, \sigma)$, where $g(x, \sigma) = \log \sigma h(x\sigma)$, to have a unique maximum at 1:

$$E_f g(\eta_0, \sigma) < E_f g(\eta_0, 1) \qquad \forall \sigma > 0, \quad \sigma \neq 1. \tag{9.25}$$

Remark 9.4 (Interpretation of the Condition) If the distribution of X has a density f, and if h denotes any density, the quantity $-2E_f \log h(X)$ is sometimes called the Kullback–Leibler contrast of h with respect to f. The Jensen inequality shows that the contrast is minimal for $h = f$. Note that $h_\sigma(x) = \sigma h(x\sigma)$ is the density of Y/σ, where Y has density h. The condition thus signifies that the density h minimises the Kullback–Leibler contrast of any density of the family h_σ, $\sigma > 0$, with respect to the density f. In other words, the condition says that it is impossible to get closer to f by scaling h.

It is sometimes useful to replace condition (9.25) by one of its consequences that is easier to handle. Assume the existence of

$$g_1(x, \sigma) = \frac{\partial g(x, \sigma)}{\partial \sigma} = \frac{1}{\sigma} + \frac{h'(\sigma x)}{h(\sigma x)} x.$$

If there exists a neighborhood $V(1)$ of 1 such that $E_f \sup_{\sigma \in V(1)} |g_1(\eta_0, \sigma)| < \infty$, the dominated convergence theorem shows that (9.25) implies the moment condition

$$\int \frac{h'(x)}{h(x)} x f(x) dx = -1. \tag{9.26}$$

Obviously, condition (9.25) is satisfied for the ML, that is, when $h = f$ (see Exercise 9.5), and also for the QML, as the following example shows.

Example 9.2 (QML) When h is the $\mathcal{N}(0, 1)$ density, the estimator $\hat{\theta}_{n,h}$ corresponds to the QMLE. In this case, if $E\eta_t^2 = 1$, we have $E_f g(\eta_0, \sigma) = -\sigma^2/2 + \log \sigma - \log \sqrt{2\pi}$, and this function possesses a unique maximum at $\sigma = 1$. We recall the fact that the QMLE is consistent even when f is not the $\mathcal{N}(0, 1)$ density.

The following example shows that for condition (9.25) to be satisfied, it is sometimes necessary to reparameterise the model and to change the identifiability constraint $E\eta^2 = 1$.

Example 9.3 (Laplace QML) Consider the Laplace density

$$h(y) = \frac{1}{2} \exp(- \mid y \mid), \quad \lambda > 0.$$

Then $E_f g(\eta_0, \sigma) = -\sigma E \mid \eta_0 \mid + \log \sigma - \log 2$. This function possesses a unique maximum at $\sigma = 1/E \mid \eta_0 \mid$. In order to have consistency for a large class of density f, it thus suffices to replace to usual constraint $E\eta_t^2 = 1$ in the GARCH model $\epsilon_t = \sigma_t \eta_t$ by the new identifiability constraint $E \mid \eta_t \mid = 1$. Of course σ_t^2 no longer corresponds to the conditional variance, but to the conditional moment $\sigma_t = E(|\epsilon_t| \mid \epsilon_u,\ u < t)$.

Table 9.3 Identifiability constraint under which $\hat{\theta}_{n,h}$ is consistent.

Law	Instrumental density h	Constraint
Gaussian	$\frac{1}{\sqrt{2\pi}\sigma}\exp\left\{-\frac{(x-m)^2}{2\sigma^2}\right\}$	$\frac{E\eta_t^2}{\sigma^2}-\frac{mE\eta_t}{\sigma^2}=1$
Double gamma	$\frac{b^p}{2\Gamma(p)}\lvert x\rvert^{p-1}\exp\{-b\mid x\mid\},\quad b,p>0$	$E\lvert\eta_t\rvert=\frac{p}{b}$
Laplace	$\frac{1}{2}\exp\{-\mid x\mid\}$	$E\lvert\eta_t\rvert=1$
Gamma	$\frac{b^p}{\Gamma(p)}\lvert x\rvert^{p-1}\exp\{-b\mid x\mid\}\mathbb{1}_{(0,\infty)}(x)$	$E\eta_t=\frac{p}{b}$
Double inverse gamma	$\frac{b^p}{2\Gamma(p)}\lvert x\rvert^{-p-1}\exp\{-b/\mid x\mid\}$	$E\frac{1}{\lvert\eta_t\rvert}=\frac{p}{b}$
Double inverse χ^2	$\frac{(\sigma^2\nu/2)^{\nu/2}}{2\Gamma(\nu/2)}\lvert x\rvert^{-\nu/2-1}\exp\{-\nu\sigma^2/2\mid x\mid\}$	$E\frac{1}{\lvert\eta_t\rvert}=\frac{1}{\sigma^2}$
Double Weibull	$\frac{\lambda}{2}\lvert x\rvert^{\lambda-1}\exp(-\lvert x\rvert^{\lambda}),\quad \lambda>0$	$E\lvert\eta_t\rvert^{\lambda}=1$
Gaussian generalised	$\frac{\lambda^{1-1/\lambda}}{2\Gamma(1/\lambda)}\exp(-\lvert x\rvert^{\lambda}/\lambda)$	$E\lvert\eta_t\rvert^{\lambda}=1$
Inverse Weibull	$\frac{\lambda}{2}\lvert x\rvert^{-\lambda-1}\exp(-\lvert x\rvert^{-\lambda}),\quad \lambda>0$	$E\lvert\eta_t\rvert^{-\lambda}=1$
Double log-normal	$\frac{1}{2\mid x\mid\sqrt{2\pi}\sigma}\exp\left\{-\frac{(\log\mid x\mid-m)^2}{2\sigma^2}\right\}$	$E\log\lvert\eta_t\rvert=m$

The previous examples show that a particular choice of h corresponds to a natural identifiability constraint. This constraint applies to a moment of η_t ($E\eta_t^2=1$ when h is $\mathcal{N}(0,1)$, and $E\mid\eta_t\mid=1$ when h is Laplace). Table 9.3 gives the natural identifiability constraints associated with various choices of h. When these natural identifiability constraints are imposed on the GARCH model, the estimator $\hat{\theta}_{n,h}$ can be interpreted as a non-Gaussian QMLE, and converges to θ_0, even when $h\neq f$.

9.2.2 Convergence of $\hat{\theta}_{n,h}$ and Interpretation of the Limit

The following examples show that the estimator $\hat{\theta}_{n,h}$ based on the mis-specified density $h\neq f$ generally converges to a value $\theta^*\neq\theta_0$ which can be interpreted in a reparameterised model.

Example 9.4 (Laplace QML for a Usual GARCH Model) Take h to be the Laplace density and assume that the GARCH model is of the form

$$\begin{cases}\epsilon_t=\sigma_t\eta_t\\ \sigma_t^2=\omega_0+\sum_{i=1}^{q}\alpha_{0i}\epsilon_{t-i}^2+\sum_{j=1}^{p}\beta_{0j}\sigma_{t-j}^2,\end{cases}$$

with the usual constraint $E\eta_t^2=1$. The estimator $\hat{\theta}_{n,h}$ does not converge to the parameter

$$\theta_0=(\omega_0,\alpha_{01},\dots,\alpha_{0q},\beta_{01},\dots,\beta_{0p})'.$$

The model can, however, always be rewritten as

$$\begin{cases} \epsilon_t = \sigma_t^* \eta_t^* \\ \sigma_t^{*2} = \omega^* + \sum_{i=1}^{q} \alpha_i^* \epsilon_{t-i}^2 + \sum_{j=1}^{p} \beta_j^* \sigma_{t-j}^{*2}, \end{cases}$$

with $\eta_t^* = \eta_t / \varrho$, $\sigma_t^* = \varrho\sigma_t$ and $\varrho = \int |x| f(x)dx$. Since $E \mid \eta_t^* \mid = 1$, the estimator $\hat{\theta}_{n,h}$ converges to

$$\theta^* = (\varrho^2\omega_0, \varrho^2\alpha_{01}, \dots, \varrho^2\alpha_{0q}, \beta_{01}, \dots, \beta_{0p})'.$$

Example 9.5 (QMLE of GARCH under the Constraint $E \mid \eta_t \mid = 1$) Assume now that the GARCH model is of the form

$$\begin{cases} \epsilon_t = \sigma_t \eta_t \\ \sigma_t^2 = \omega_0 + \sum_{i=1}^{q} \alpha_{0i} \epsilon_{t-i}^2 + \sum_{j=1}^{p} \beta_{0j} \sigma_{t-j}^2, \end{cases}$$

with the constraint $E \mid \eta_t \mid = 1$. If h is the Laplace density, the estimator $\hat{\theta}_{n,h}$ converges to the parameter θ_0, regardless of the density f of η_t (satisfying mild regularity conditions). The model can be written as

$$\begin{cases} \epsilon_t = \sigma_t^* \eta_t^* \\ \sigma_t^{*2} = \omega^* + \sum_{i=1}^{q} \alpha_i^* \epsilon_{t-i}^2 + \sum_{j=1}^{p} \beta_j^* \sigma_{t-j}^{*2}, \end{cases}$$

with the usual constraint $E\eta_t^{*2} = 1$ when $\eta_t^* = \eta_t / \varrho$, $\sigma_t^* = \varrho\sigma_t$ and $\varrho = \sqrt{\int x^2 f(x)dx}$. The standard QMLE does not converge to θ_0, but to $\theta^* = (\varrho^2\omega_0, \varrho^2\alpha_{01}, \dots, \varrho^2\alpha_{0q}, \beta_{01}, \dots, \beta_{0p})'$.

9.2.3 Choice of Instrumental Density *h*

We have seen that, for any fixed h, there exists an identifiability constraint implying the convergence of $\hat{\theta}_{n,h}$ to θ_0 (see Table 9.3). In practice, we do not choose the parameterisation for which $\hat{\theta}_{n,h}$ converges, but the estimator that guarantees a consistent estimation of the model of interest. The instrumental function h is chosen to estimate the model under a given constraint, corresponding to a given problem. As an example, suppose that we wish to estimate the conditional moment $E_{t-1}(\epsilon_t^4) := E(\epsilon_t^4 \mid \epsilon_u,\ u < t)$ of a GARCH (p, q) process. It will be convenient to consider the parameterisation $\epsilon_t = \sigma_t \eta_t$ under the constraint $E\eta_t^4 = 1$. The volatility σ_t will then be directly related to the conditional moment of interest, by the relation $\sigma_t^4 = E_{t-1}(\epsilon_t^4)$. In this particular case, the Gaussian QMLE is inconsistent (because, in particular, the QMLE of α_i converges to $\alpha_i E\eta_t^2$). In view of (9.26), to find relevant instrumental functions h, one can solve

$$1 + \frac{h'(x)}{h(x)} x = \lambda - \lambda x^4, \qquad \lambda \neq 0,$$

since $E(\lambda - \lambda\eta_t^4) = 0$ and $E\{1 + h'(x)/h(\eta_t)\eta_t\} = 0$. The densities that solve this differential equation are of the form

$$h(x) = c|x|^{\lambda - 1} \exp(-\lambda x^4/4), \qquad \lambda > 0.$$

For $\lambda = 1$, we obtain the double Weibull, and for $\lambda = 4$ a generalised Gaussian, which is in accordance with the results given in Table 9.3.

For the more general problem of estimating conditional moments of $|\epsilon_t|$ or $\log |\epsilon_t|$, Table 9.4 gives the parameterisation (that is, the moment constraint on η_t) and the type of estimator (that is, the choice of h) for the solution to be only a function of the volatility σ_t (a solution which is thus independent of the distribution f of η_t). It is easy to see that for the instrumental function h of Table 9.4,

Table 9.4 Choice of h as function of the prediction problem.

Problem	Constraint	Solution	Instrumental density, h
$E_{t-1}\|\epsilon_t\|^r, r>0$	$E\|\eta_t\|^r=1$	σ_t^r	$c\|x\|^{\lambda-1}\exp(-\lambda\|x\|^r/r), \lambda>0$
$E_{t-1}\|\epsilon_t\|^{-r}$	$E\|\eta_t\|^{-r}=1$	σ_t^{-r}	$c\|x\|^{-\lambda-1}\exp(-\lambda\|x\|^{-r}/r)$
$E_{t-1}\log\|\epsilon_t\|$	$E\log\|\eta_t\|=0$	$\log\sigma_t$	$\sqrt{\lambda/\pi}\|2x\|^{-1}\exp\{-\lambda(\log\|x\|)^2\}$

the estimator $\hat{\theta}_{n,h}$ depends only on r and not on c and λ. Indeed, taking the case $r>0$, up to some constant we have

$$\log L_{n,h}(\theta) = -\frac{\lambda}{r}\sum_{t=1}^{n}\left(\log\tilde{\sigma}_t^r(\theta) + \left|\frac{\epsilon_t}{\tilde{\sigma}_t(\theta)}\right|^r\right),$$

which shows that $\hat{\theta}_{n,h}$ does not depend on c and λ. In practice, one can thus choose the simplest constants in the instrumental function, for instance $c=\lambda=1$.

9.2.4 Asymptotic Distribution of $\hat{\theta}_{n,h}$

Using arguments similar to those of Section 7.4, a Taylor expansion shows that, under (9.25),

$$\begin{aligned}0 &= \frac{1}{\sqrt{n}}\frac{\partial}{\partial\theta}\log L_{n,h}(\hat{\theta}_{n,h})\\ &= \frac{1}{\sqrt{n}}\frac{\partial}{\partial\theta}\log L_{n,h}(\theta_0) + \frac{1}{n}\frac{\partial^2}{\partial\theta\partial\theta'}\log L_{n,h}(\theta_0)\sqrt{n}(\hat{\theta}_{n,h}-\theta_0) + o_P(1),\end{aligned}$$

where

$$\begin{aligned}\frac{1}{\sqrt{n}}\frac{\partial}{\partial\theta}\log L_{n,h}(\theta) &= \frac{1}{\sqrt{n}}\sum_{t=1}^{n}\frac{\partial}{\partial\theta}\log\frac{1}{\sigma_t(\theta)}h\left(\frac{\sigma_t(\theta_0)}{\sigma_t(\theta)}\eta_t\right)\\ &= \frac{1}{\sqrt{n}}\sum_{t=1}^{n}g_1\left(\eta_t,\frac{\sigma_t(\theta_0)}{\sigma_t(\theta)}\right)\frac{-\sigma_t(\theta_0)}{2\sigma_t^3(\theta)}\frac{\partial\sigma_t^2(\theta)}{\partial\theta}\end{aligned}$$

and

$$\frac{1}{n}\frac{\partial^2}{\partial\theta\,\partial\theta'}\log L_{n,h}(\theta_0) = \frac{1}{n}\sum_{t=1}^{n}g_2(\eta_t,1)\frac{1}{4\sigma_t^4(\theta_0)}\frac{\partial\sigma_t^2(\theta_0)}{\partial\theta}\frac{\partial\sigma_t^2(\theta_0)}{\partial\theta'} + o_P(1).$$

The ergodic theorem and the CLT for martingale increments (see Section A.2) then entail that

$$\begin{aligned}\sqrt{n}(\hat{\theta}_{n,h}-\theta_0) &= \frac{2}{Eg_2(\eta_0,1)}J^{-1}\frac{1}{\sqrt{n}}\sum_{t=1}^{n}g_1(\eta_t,1)\frac{1}{\sigma_t^2(\theta_0)}\frac{\partial\sigma_t^2(\theta_0)}{\partial\theta} + o_p(1)\\ &\xrightarrow{\mathcal{L}} \mathcal{N}(0, 4\tau_{h,f}^2 J^{-1})\end{aligned} \tag{9.27}$$

where

$$J = E\frac{1}{\sigma_t^4}\frac{\partial\sigma_t^2}{\partial\theta}\frac{\partial\sigma_t^2}{\partial\theta'}(\theta_0) \quad\text{and}\quad \tau_{h,f}^2 = \frac{E_f g_1^2(\eta_0,1)}{\{E_f g_2(\eta_0,1)\}^2}, \tag{9.28}$$

with $g_1(x,\sigma)=\partial g(x,\sigma)/\partial\sigma$ and $g_2(x,\sigma)=\partial g_1(x,\sigma)/\partial\sigma$.

Example 9.6 (Asymptotic distribution of the MLE) When $h=f$ (that is, for the MLE), we have $g(x,\sigma)=\log\sigma f(x\sigma)$ and $g_1(x,\sigma)=\sigma^{-1}+xf'(x\sigma)/f(x\sigma)$. Thus $E_f g_1^2(\eta_0,1)=\zeta_f$, as defined after (9.4).

Table 9.5 Asymptotic relative efficiency of the MLE with respect to the QMLE and to the Laplace QMLE: $\tau^2_{\phi,f}/\tau^2_{f,f}$ and $\tau^2_{l,f}/\tau^2_{f,f}$, where ϕ denotes the $\mathcal{N}(0,1)$ density, and $\ell(x) = \frac{1}{2}\exp(-\mid x\mid)$ the Laplace density.

$\tau^2_{h,f}/\tau^2_{f,f}$	ν								
	5	6	7	8	9	10	20	30	100
MLE – QMLE	2.5	1.667	1.4	1.273	1.2	1.154	1.033	1.014	1.001
MLE – Laplace	1.063	1.037	1.029	1.028	1.030	1.034	1.070	1.089	1.124

For the QMLE, the Student t density f_ν with ν degrees of freedom is normalised so that $E\eta_t^2 = 1$, that is, the density of η_t is $f(y) = \sqrt{\nu/\nu-2}f_\nu(y\sqrt{\nu/\nu-2})$. For the Laplace QMLE, η_t has the density $f(y) = E\mid t_\nu\mid f_\nu(yE\mid t_\nu\mid)$, so that $E\mid\eta_t\mid = 1$.

This Fisher information can also be expressed as $\zeta_f = -E_f g_2(\eta_0, 1)$. This shows that $\tau^2_{f,f} = \zeta_f^{-1}$, and we obtain (9.6).

Example 9.7 (Asymptotic distribution of the QMLE) If we choose for h the density $\phi(x) = (2\pi)^{-1/2}\exp(-x^2/2)$, we have $g(x,\sigma) = \log\sigma - \sigma^2x^2/2 - \log\sqrt{2\pi}$, $g_1(x,\sigma) = \sigma^{-1} - \sigma x^2$ and $g_2(x,\sigma) = -\sigma^{-2} - x^2$. Thus $E_f g_1^2(\eta_0, 1) = (\kappa_\eta - 1)$, with $\kappa_\eta = \int x^4 f(x)dx$, and $E_f g_2(\eta_0, 1) = -2$. Therefore, $\tau^2_{\phi,f} = (\kappa_\eta - 1)/4$, and we obtain the usual expression for the asymptotic variance of the QMLE.

Example 9.8 (Asymptotic Distribution of the Laplace QMLE) Write the GARCH model as $\epsilon_t = \sigma_t\eta_t$, with the constraint $E_f\mid\eta_t\mid = 1$. Let $\ell(x) = \frac{1}{2}\exp(-\mid x\mid)$ be the Laplace density. For $h = \ell$ we have $g(x,\sigma) = \log\sigma - \sigma\mid x\mid - \log 2$, $g_1(x,\sigma) = \sigma^{-1} - |x|$ and $g_2(x,\sigma) = -\sigma^{-2}$. We thus have $\tau^2_{\ell,f} = E_f\eta_t^2 - 1$.

Table 9.5 completes Table 9.1. Using the previous examples, this table gives the ARE of the QMLE and Laplace QMLE with respect to the MLE, in the case where f follows the Student t distribution. The table does not allow us to obtain the ARE of the QMLE with respect to Laplace QMLE, because the noise η_t has a different normalisation with the standard QMLE or the Laplace QMLE (in other words, the two estimators do not converge to the same parameter).

9.3 Alternative Estimation Methods

The estimation methods presented in this section are less popular among practitioners than the QML and LS methods, but each has specific features of interest.

9.3.1 Weighted LSE for the ARMA Parameters

Consider the estimation of the ARMA part of the ARMA(P, Q)-GARCH (p, q) model

$$\begin{cases} X_t - c_0 = \sum_{i=1}^{P} a_{0i}(X_{t-i} - c_0) + e_t - \sum_{j=1}^{Q} b_{0j}e_{t-j} \\ e_t = \sqrt{h_t}\eta_t \\ h_t = \omega_0 + \sum_{i=1}^{q}\alpha_{0i}e_{t-i}^2 + \sum_{j=1}^{p}\beta_{0j}h_{t-j}, \end{cases} \tag{9.29}$$

where (η_t) is an iid(0,1) sequence and the coefficients ω_0, α_{0i} and β_{0j} satisfy the usual positivity constraints. The orders P, Q, p, and q are assumed known. The parameter vector is

$$\vartheta = (c, a_1, \dots, a_P, b_1, \dots, b_Q)',$$

the true value of which is denoted by ϑ_0, and the parameter space $\Psi \subset \mathbb{R}^{P+Q+1}$. Given observations $X_1, \ldots, X_n$ and initial values, the sequence $\tilde{\epsilon}_t$ is defined recursively by (7.22). The weighted LSE is defined as a measurable solution of

$$\hat{\vartheta}_n = \arg\min_{\vartheta} n^{-1} \sum_{t=1}^{n} \omega_t^2 \tilde{\epsilon}_t^2(\vartheta),$$

where the weights ω_t are known positive measurable functions of $X_{t-1}, X_{t-2}, \ldots$. One can take, for instance,

$$\omega_t^{-1} = 1 + \sum_{k=1}^{t-1} k^{-1-1/s} \mid X_{t-k} \mid$$

with $E|X_1|^{2s} < \infty$ and $s \in (0, 1)$. It can be shown that there exist constants $K > 0$ and $\rho \in (0, 1)$ such that

$$\mid \tilde{\epsilon}_t \mid \leq K(1 + \mid \eta_t \mid) \left(1 + \sum_{k=1}^{t-1} \rho^k \mid X_{t-k} \mid \right) \quad \text{and} \quad \left| \frac{\partial \tilde{\epsilon}_t}{\partial \vartheta_i} \right| \leq K \sum_{k=1}^{t-1} \rho^k \mid X_{t-k} \mid .$$

This entails that

$$\mid \omega_t \tilde{\epsilon}_t \mid \leq K(1 + \mid \eta_t \mid) \left(1 + \sum_{k=1}^{\infty} k^{1+1/s} \rho^k \right), \quad \left| \omega_t \frac{\partial \tilde{\epsilon}_t}{\partial \vartheta_i} \right| \leq K \left(1 + \sum_{k=1}^{\infty} k^{1+1/s} \rho^k \right).$$

Thus

$$E \left| \omega_t^2 \tilde{\epsilon}_t \frac{\partial \tilde{\epsilon}_t}{\partial \vartheta_i} \right|^2 \leq K^4 E(1 + \mid \eta_1 \mid)^2 \left(\sum_{k=1}^{\infty} k^{1+1/s} \rho^k \right)^4 < \infty,$$

which implies a finite variance for the score vector $\omega_t^2 \tilde{\epsilon}_t \partial \tilde{\epsilon}_t / \partial \vartheta$. Ling (2005) deduces the asymptotic normality of $\sqrt{n}(\hat{\vartheta}_n - \vartheta_0)$, even in the case $EX_1^2 = \infty$.

9.3.2 Self-Weighted QMLE

Recall that, for the ARMA-GARCH models, the asymptotic normality of the QMLE has been established under the condition $EX_1^4 < \infty$ (see Theorem 7.5). To obtain an asymptotically normal estimator of the parameter $\varphi_0 = (\vartheta_0', \theta_0')'$ of the ARMA-GARCH model (9.29) with weaker moment assumptions on the observed process, Ling (2007) proposed a self-weighted QMLE of the form

$$\hat{\varphi}_n = \arg\min_{\varphi \in \Phi} n^{-1} \sum_{t=1}^{n} \omega_t \tilde{\ell}_t(\varphi),$$

where $\tilde{\ell}_t(\varphi) = \tilde{\epsilon}_t^2(\vartheta)/\tilde{\sigma}_t^2(\varphi) + \log \tilde{\sigma}_t^2(\varphi)$, using standard notation. To understand the principle of this estimator, note that the minimised criterion converges to the limit criterion $\mathbf{l}(\varphi) = E_\varphi \omega_t \ell_t(\varphi)$ satisfying

$$\begin{aligned} \mathbf{l}(\varphi) - \mathbf{l}(\varphi_0) =& E_{\varphi_0} \omega_t \left\{ \log \frac{\sigma_t^2(\varphi)}{\sigma_t^2(\varphi_0)} + \frac{\sigma_t^2(\varphi_0)}{\sigma_t^2(\varphi)} - 1 \right\} + E_{\varphi_0} \omega_t \frac{\{\epsilon_t(\vartheta) - \epsilon_t(\vartheta_0)\}^2}{\sigma_t^2(\varphi)} \\ &+ E_{\varphi_0} \omega_t \frac{2\eta_t \sigma_t(\varphi_0) \{\epsilon_t(\vartheta) - \epsilon_t(\vartheta_0)\}}{\sigma_t^2(\varphi)}. \end{aligned}$$

The last expectation (when it exists) is null, because η_t is centred and independent of the other variables. The inequality $x - 1 \geq \log x$ entails that

$$E_{\varphi_0} \omega_t \left\{ \log \frac{\sigma_t^2(\varphi)}{\sigma_t^2(\varphi_0)} + \frac{\sigma_t^2(\varphi_0)}{\sigma_t^2(\varphi)} - 1 \right\} \geq E_{\varphi_0} \omega_t \left\{ \log \frac{\sigma_t^2(\varphi)}{\sigma_t^2(\varphi_0)} + \log \frac{\sigma_t^2(\varphi_0)}{\sigma_t^2(\varphi)} \right\} = 0.$$

Thus, under the usual identifiability conditions, we have $\mathbf{l}(\varphi) \geq \mathbf{l}(\varphi_0)$, with equality if and only if $\varphi = \varphi_0$. Note that the orthogonality between η_t and the weights ω_t is essential. Ling (2007) showed the convergence and asymptotic normality of $\hat{\varphi}_n$ under the assumption $E|X_1|^s < \infty$ for some $s > 0$.

9.3.3 L_p Estimators

The previous weighted estimator requires the assumption $E\eta_1^4 < \infty$. Practitioners often claim that financial series admit few moments. A GARCH process with infinite variance is obtained either by taking large values of the parameters, or by taking an infinite variance for η_t. Indeed, for a GARCH(1, 1) process, each of the two sets of assumptions

(i) $\alpha_{01} + \beta_{01} \geq 1, E\eta_1^2 = 1,$

(ii) $E\eta_1^2 = \infty$

implies an infinite variance for ϵ_t. Under (i), and strict stationarity, the asymptotic distribution of the QMLE is generally Gaussian (see Section 7.1.1), whereas the usual estimators have non-standard asymptotic distributions under (ii) (see Berkes and Horváth 2003b; Hall and Yao 2003; Mikosch and Straumann 2002), which causes difficulties for inference. As an alternative to the QMLE, it is thus interesting to define estimators having an asymptotic normal distribution under (ii), or even in the more general situation where both $\alpha_{01} + \beta_{01} > 1$ and $E\eta_1^2 = \infty$ are allowed. A GARCH model is usually defined under the normalisation constraint $E\eta_1^2 = 1$. When the assumption that $E\eta_1^2$ exists is relaxed, the GARCH coefficients can be identified by imposing, for instance, that the median of η_1^2 is $\tau = 1$. In the framework of ARCH(q) models, Horváth and Liese (2004) consider L_p estimators, including the L_1 estimator

$$\arg\min_{\theta} n^{-1} \sum_{t=1}^{n} \omega_t \left| \epsilon_t^2 - \omega - \sum_{i=1}^{q} \alpha_i \epsilon_{t-1}^2 \right|,$$

where, for instance, $\omega_t^{-1} = 1 + \sum_{i=1}^{p} \epsilon_{t-i}^2 + \epsilon_{t-i}^4$. When η_t^2 admits a density, continuous and positive around its median $\tau = 1$, the consistency and asymptotic normality of these estimators are shown in Horváth and Liese (2004), without any moment assumption. An alternative to L_p-estimators, which only requires $E\eta_t^2 < \infty$ and can be applied to ARMA-GARCH, is the self-weighted quasi-maximum exponential likelihood estimator studied by Zhu and Ling (2011).

9.3.4 Least Absolute Value Estimation

For ARCH and GARCH models, Peng and Yao (2003) studied several least absolute deviations estimators. An interesting specification is

$$\arg\min_{\theta} n^{-1} \sum_{t=1}^{n} |\log \epsilon_t^2 - \log \tilde{\sigma}_t^2(\theta)|. \tag{9.30}$$

With this estimator, it is convenient to define the GARCH parameters under the constraint that the median of η_1^2 is 1. A reparameterisation of the standard GARCH models is thus necessary. Consider, for instance, a GARCH(1, 1) with parameters ω, α_1 and β_1, and a Gaussian noise η_t. Since the median of η_1^2 is $\tau = 0.4549\ldots$, the median of the square of $\eta_t^* = \eta_t/\sqrt{\tau}$ is 1, and the model is rewritten as

$$\epsilon_t = \sigma_t \eta_t^*, \quad \sigma_t^2 = \tau\omega + \tau\alpha_1\epsilon_{t-1}^2 + \beta_1\sigma_{t-1}^2.$$

It is interesting to note that the error terms $\log \eta_t^{*2} = \log \epsilon_t^2 - \log \tilde{\sigma}_t^2(\theta)$ are iid with median 0 when $\theta = \theta_0$. Intuitively, this is the reason why it is pointless to introduce weights in the sum (9.30). Under the moment assumption $E\epsilon_1^2 < \infty$ and some regularity assumptions, Peng and Yao (2003) show that there exists a local solution of (9.30) which is weakly consistent and asymptotically normal, with rate of convergence $n^{1/2}$. This convergence holds even when the distribution of the errors has a fat tail: only the moment condition $E\eta_t^2 = 1$ is required.

9.3.5 Whittle Estimator

In Chapter 2, we have seen that, under the condition that the fourth-order moments exist, the square of a GARCH(p, q) satisfies the ARMA(max(p, q), q) representation

$$\phi_{\theta_0}(L)\epsilon_t^2 = \omega_0 + \psi_{\theta_0}(L)u_t, \tag{9.31}$$

where

$$\phi_{\theta_0}(z) = 1 - \sum_{i=1}^{\max(p,q)} (\alpha_{0i} + \beta_{0i})z^i, \quad \psi_{\theta_0}(z) = 1 - \sum_{i=1}^{p} \beta_{0i}z^i, \quad u_t = (\eta_t^2 - 1)\sigma_t^2.$$

The spectral density of ϵ_t^2 is

$$f_{\theta_0}(\lambda) = \frac{\sigma_u^2}{2\pi}\frac{|\psi_{\theta_0}(e^{-i\lambda})|^2}{|\phi_{\theta_0}(e^{-i\lambda})|^2}, \quad \sigma_u^2 = Eu_t^2.$$

Let $\hat{\gamma}_{\epsilon^2}(h)$ be the empirical autocovariance of ϵ_t^2 at lag h. At Fourier frequencies $\lambda_j = 2\pi j/n \in (-\pi, \pi]$, the periodogram

$$I_n(\lambda_j) = \sum_{|h|<n} \hat{\gamma}_{\epsilon^2}(h)e^{-ih\lambda_j}, \quad j \in \mathfrak{J} = \left\{\left[-\frac{n}{2}\right] + 1, \dots, \left[\frac{n}{2}\right]\right\},$$

can be considered as a non-parametric estimator of $2\pi f_{\theta_0}(\lambda_j)$. Let

$$u_t(\theta) = \frac{\phi_\theta(L)}{\psi_\theta(L)}\{\epsilon_t^2 - \omega\phi_\theta^{-1}(1)\}.$$

It can be shown that

$$\sigma_u^2(\theta) := Eu_t^2(\theta) = \frac{\sigma_u^2(\theta_0)}{2\pi}\int_{-\pi}^{\pi}\frac{f_{\theta_0}(\lambda)}{f_\theta(\lambda)}d\lambda \geq \sigma_u^2(\theta_0),$$

with equality if and only if $\theta = \theta_0$ (see Proposition 10.8.1 in Brockwell and Davis 1991). In view of this inequality, it is natural to consider the so-called Whittle estimator

$$\arg\min_\theta \frac{1}{n}\sum_{j\in\mathfrak{J}}\frac{I_n(\lambda_j)}{f_\theta(\lambda_j)}.$$

For ARMA with iid innovations, the Whittle estimator has the same asymptotic behaviour as the QMLE and LSE (which coincide in that case). For GARCH models, the Whittle estimator still exhibits the same asymptotic behaviour as the LSE, but it is generally less accurate than the QMLE. Moreover, Giraitis and Robinson (2001), Mikosch and Straumann (2002), and Straumann (2005) have shown that the consistency requires the existence of $E\epsilon_t^4$, and that the asymptotic normality requires $E\epsilon_t^8 < \infty$.

9.4 Bibliographical Notes

The central reference of Sections 9.1 and 9.2 is Berkes and Horváth (2004), who give precise conditions for the consistency and asymptotic normality of the estimators $\hat{\theta}_{n,h}$. Slightly different conditions implying consistency and asymptotic normality of the MLE can be found in Francq and Zakoïan (2006c). These results were extended to a general conditionally heteroskedastic model and used for prediction purposes by Francq and Zakoian (2013b). Additional results, in particular concerning the interesting situation where the density f of the iid noise is known up to a nuisance parameter, are available in Straumann (2005). Newey and Steigerwald (1997) show that, in general conditional heteroscedastic models, a suitable parameterisation allows to consistently estimate part of the volatility parameter by non-Gaussian QML estimation. Fan, Qi, and Xiu (2014) and Francq, Lepage, and Zakoian (2011) propose multi-steps consistent estimators of the GARCH volatility parameters based on non-Gaussian QML estimators. The adaptative estimation of the GARCH models is studied in Drost and Klaassen (1997) and also in Engle and González-Rivera (1991), Linton (1993), González-Rivera and Drost (1999), and Ling and McAleer (2003b). Drost and Klaassen (1997), Drost, Klaassen and Werker (1997), Ling and McAleer (2003a), and Lee and Taniguchi (2005) give mild regularity conditions ensuring the LAN property of GARCH.

Several estimation methods for GARCH models have not been discussed here, among them Bayesian methods (see Geweke 1989), the generalised method of moments (see Rich, Raymond, and Butler 1991), variance targeting and robust methods (see Muler and Yohai 2008, Hill 2015). Rank-based estimators for GARCH coefficients (except the intercept) were recently proposed by Andrews (2012). These estimators are shown to be asymptotically normal under assumptions which do not include the existence of a finite fourth moment for the iid noise.

9.5 Exercises

9.1 *(The score of a scale parameter is centred)*
Show that if f is a differentiable density such that $\int |x| f(x)dx < \infty$, then

$$\int \left\{1 + \frac{f'(x/\sigma_t)}{f(x/\sigma_t)} \frac{x}{\sigma_t}\right\} \frac{1}{\sigma_t} f\left(\frac{x}{\sigma_t}\right) dx = 0.$$

Deduce that the score vector defined by (9.3) is centred.

9.2 *(Covariance between the square and the score of the scale parameter)*
Show that if f is a differentiable density such that $\int |x|^3 f(x)dx < \infty$, then

$$\int (1 - x^2) \left(1 + x\frac{f'(x)}{f(x)}\right) f(x)dx = 2.$$

9.3 *(Intuition behind Le Cam's third lemma)*
Let $\phi_\theta(x) = (2\pi\sigma^2)^{-1/2} \exp(-(x-\theta)^2/2\sigma^2)$ be the $\mathcal{N}(\theta, \sigma^2)$ density and let the log-likelihood ratio

$$\Lambda(\theta, \theta_0, x) = \log \frac{\phi_\theta(x)}{\phi_{\theta_0}(x)}.$$

Determine the distribution of

$$\begin{pmatrix} aX + b \\ \Lambda(\theta, \theta_0, X) \end{pmatrix}$$

when $X \sim \mathcal{N}(\theta_0, \sigma^2)$, and then when $X \sim \mathcal{N}(\theta, \sigma^2)$.

9.4 *(Fisher information)*
For the parametrisation (9.11), verify that

$$-n^{-1} \frac{\partial^2}{\partial \vartheta \, \partial \vartheta'} \log L_{n,f_\lambda}(\vartheta_0) \to \mathfrak{I} \quad \text{a.s. as } n \to \infty.$$

9.5 *(Condition for the consistency of the MLE)*
Let η be a random variable with density f such that $E|\eta|^r < \infty$ for some $r \neq 0$. Show that

$$E \log \sigma f(\eta\sigma) < E \log f(\eta) \qquad \forall \sigma \neq 1.$$

9.6 *(Case where the Laplace QMLE is optimal)*
Consider a GARCH model whose noise has the $\Gamma(b, b)$ distribution with density

$$f_b(x) = \frac{b^b}{2\Gamma(b)} |x|^{b-1} \exp(-b \mid x \mid), \qquad \Gamma(b) = \int_0^\infty x^{b-1} \exp(-x)dx,$$

where $b > 0$. Show that the Laplace QMLE is optimal.

9.7 *(Comparison of the MLE, QMLE and Laplace QMLE)*
Give a table similar to Table 9.5, but replace the Student t distribution f_ν by the double $\Gamma(b, p)$ distribution

$$f_{b,p}(x) = \frac{b^p}{2\Gamma(p)}|x|^{p-1}\exp(-b \mid x \mid), \qquad \Gamma(p) = \int_0^\infty x^{p-1}\exp(-x)dx,$$

where $b, p > 0$.

9.8 *(Asymptotic comparison of the estimators $\hat{\theta}_{n,h}$)*
Compute the coefficient $\tau_{h,f}^2$ defined by (9.28) for each of the instrumental densities h of Table 9.4. Compare the asymptotic behaviour of the estimators $\hat{\theta}_{n,h}$.

9.9 *(Fisher information at a pseudo-true value)*
Consider a GARCH (p, q) model with parameter

$$\theta_0 = (\omega_0, \alpha_{01}, \dots, \alpha_{0q}, \beta_{01}, \dots, \beta_{0p})'.$$

1. Give an example of an estimator which does not converge to θ_0, but which converges to a vector of the form

$$\theta^* = (\varrho^2\omega_0, \varrho^2\alpha_{01}, \dots, \varrho^2\alpha_{0q}, \beta_{01}, \dots, \beta_{0p})',$$

where ϱ is a constant.
2. What is the relationship between $\sigma_t^2(\theta_0)$ and $\sigma_t^2(\theta^*)$?
3. Let $\Lambda_\varrho = \mathrm{diag}(\varrho^{-2}I_{q+1}, I_p)$ and

$$J(\theta) = E\frac{1}{\sigma_t^4}\frac{\partial\sigma_t^2}{\partial\theta}\frac{\partial\sigma_t^2}{\partial\theta'}(\theta).$$

Give an expression for $J(\theta^*)$ as a function of $J(\theta_0)$ and Λ_ϱ.

9.10 *(Asymptotic distribution of the Laplace QMLE)*
Determine the asymptotic distribution of the Laplace QMLE when the GARCH model does not satisfy the natural identifiability constraint $E \mid \eta_t \mid = 1$, but the usual constraint $E\eta_t^2 = 1$.

Part III

Extensions and Applications

10

Multivariate GARCH Processes

While the volatility of univariate series has been the focus of the previous chapters, modelling the co-movements of several series is of great practical importance. When several series displaying temporal or contemporaneous dependencies are available, it is useful to analyse them jointly, by viewing them as the components of a vector-valued (multivariate) process. The standard linear modelling of real-time series has a natural multivariate extension through the framework of the vector ARMA (VARMA) models. In particular, the subclass of vector autoregressive (VAR) models has been widely studied in the econometric literature. This extension entails numerous specific problems and has given rise to new research areas (such as co-integration).

Similarly, it is important to introduce the concept of multivariate GARCH (MGARCH) model. For instance, asset pricing and risk management crucially depend on the conditional covariance structure of the assets of a portfolio. Unlike the ARMA models, however, the GARCH model specification does not suggest a *natural* extension to the multivariate framework. Indeed, the (conditional) expectation of a vector of size m is a vector of size m, but the (conditional) variance is an $m \times m$ matrix. A general extension of the univariate GARCH processes would involve specifying each of the $m(m+1)/2$ entries of this matrix as a function of its past values and the past values of the other entries. Given the excessive number of parameters that this approach would entail, it is not feasible from a statistical point of view. An alternative approach is to introduce some specification constraints which, while preserving a certain generality, make these models operational.

We start by reviewing the main concepts for the analysis of the multivariate time series.

10.1 Multivariate Stationary Processes

In this section, we consider a vector process $(X_t)_{t\in\mathbb{Z}}$ of dimension m, $X_t = (X_{1t}, \dots, X_{mt})'$. The definition of strict stationarity (see Chapter 1, Definition 1.1) remains valid for vector processes, while second-order stationarity is defined as follows.

Definition 10.1 (Second-order stationarity) *The process (X_t) is said to be second-order stationary if:*

GARCH Models: Structure, Statistical Inference and Financial Applications, Second Edition. Christian Francq and Jean-Michel Zakoian.

(i) $EX_{it}^2 < \infty, \forall t \in \mathbb{Z}, \ i = 1, \ldots, m;$

(ii) $EX_t = \mu, \forall t \in \mathbb{Z};$

(iii) $Cov(X_t, X_{t+h}) = E[(X_t - \mu)(X_{t+h} - \mu)'] = \Gamma(h), \forall t, h \in \mathbb{Z}.$
The function $\Gamma(\cdot)$*, taking values in the space of* $m \times m$ *matrices, is called the autocovariance function of* (X_t).

Obviously, $\Gamma_X(h) = \Gamma_X(-h)'$. In particular, $\Gamma_X(0) = \mathrm{Var}(X_t)$ is a symmetric matrix.

The simplest example of a multivariate stationary process is white noise, defined as a sequence of centred and uncorrelated variables whose covariance matrix is time-independent.

The following property can be used to construct a stationary process by linear transformation of another stationary process.

Theorem 10.1 (Stationary linear filter) *Let* (Z_t) *denote a stationary process,* $Z_t \in \mathbb{R}^m$*. Let* $(C_k)_{k\in\mathbb{Z}}$ *denote a sequence of non-random* $n \times m$ *matrices, such that for all* $i = 1, \ldots, n$*, for all* $j = 1, \ldots, m$*,* $\sum_{k\in\mathbb{Z}} | c_{ij}^{(k)} | < \infty$*, where* $C_k = (c_{ij}^{(k)})$*. Then the* $\mathbb{R}^n$*-valued process defined by* $X_t = \sum_{k\in\mathbb{Z}} C_k Z_{t-k}$ *is stationary, and we have, in obvious notation,*

$$\mu_X = \sum_{k\in\mathbb{Z}} C_k \mu_Z, \quad \Gamma_X(h) = \sum_{k,l\in\mathbb{Z}} C_l \Gamma_Z(h + k - l) C_k'.$$

The proof of an analogous result is given by Brockwell and Davis (1991, pp. 83–84) and the arguments used extend straightforwardly to the multivariate setting. When, in this theorem, (Z_t) is a white noise and $C_k = 0$ for all $k < 0$, (X_t) is called a vector moving average process of infinite order, VMA(∞). A multivariate extension of Wold's representation theorem (see Hannan 1970, pp. 157–158) states that if (X_t) is a stationary and purely non-deterministic process, it can be represented as an infinite-order moving average,

$$X_t = \sum_{k=0}^{\infty} C_k \epsilon_{t-k} := C(B)\epsilon_t, \quad C_0 = I_m, \tag{10.1}$$

where (ϵ_t) is a $(m \times 1)$ white noise, B is the lag operator, $C(B) = \sum_{k=0}^{\infty} C_k B^k$, and the matrices C_k are not necessarily absolutely summable but satisfy the (weaker) condition $\sum_{k=0}^{\infty} \|C_k\|^2 < \infty$, for any matrix norm $\|\cdot\|$. The following definition generalises the notion of a scalar ARMA process to the multivariate case.

Definition 10.2 (VARMA(p, q) process) *An* $\mathbb{R}^m$*-valued process* $(X_t)_{t\in\mathbb{Z}}$ *is called a vector ARMA process of orders* p *and* q *(VARMA(p, q)) if* $(X_t)_{t\in\mathbb{Z}}$ *is a stationary solution to the difference equation*

$$\Phi(B)X_t = c + \Psi(B)\epsilon_t, \tag{10.2}$$

where (ϵ_t) *is a* $(m \times 1)$ *white noise with covariance matrix* Ω*,* c *is a* $(m \times 1)$ *vector, and* $\Phi(z) = I_m - \Phi_1 z - \cdots - \Phi_p z^p$ *and* $\Psi(z) = I_m - \Psi_1 z - \cdots - \Psi_q z^q$ *are matrix-valued polynomials.*

Denote by $\det(A)$, or more simply $|A|$ when there is no ambiguity, the determinant of a square matrix A. A sufficient condition for the existence of a stationary and invertible solution to the preceding equation is

$$|\Phi(z)||\Psi(z)| \neq 0, \quad \text{for all } z \in \mathbb{C} \text{ such that } | z | \leq 1$$

(see Brockwell and Davis 1991, Theorems 11.3.1 and 11.3.2).

When $p = 0$, the process is called vector moving average of order q (VMA(q)); when $q = 0$, the process is called VAR of order p (VAR(p)).

Note that the determinant $|\Phi(z)|$ is a polynomial admitting a finite number of roots $z_1, \dots, z_{mp}$. Let $\delta = \min_i |z_i| > 1$. The power series expansion

$$\Phi(z)^{-1} := |\Phi(z)|^{-1}\Phi^*(z) := \sum_{k=0}^{\infty} C_k z^k, \tag{10.3}$$

where A^* denotes the adjoint of the matrix A (that is, the transpose of the matrix of the cofactors of A), is well defined for $|z| < \delta$, and is such that $\Phi(z)^{-1}\Phi(z) = I$. The matrices C_k are recursively obtained by

$$C_0 = I, \quad \text{and for } k \geq 1, \qquad C_k = \sum_{\ell=1}^{\min(k,p)} C_{k-\ell}\Phi_\ell. \tag{10.4}$$

10.2 Multivariate GARCH Models

As in the univariate case, we can define MGARCH models by specifying their first two conditional moments. An $\mathbb{R}^m$-valued GARCH process (ϵ_t), with $\epsilon_t = (\epsilon_{1t}, \dots, \epsilon_{mt})'$, must then satisfy, for all $t \in \mathbb{Z}$,

$$E(\epsilon_t \mid \epsilon_u, u < t) = 0, \quad \text{Var}(\epsilon_t \mid \epsilon_u, u < t) = E(\epsilon_t\epsilon_t' \mid \epsilon_u, u < t) = H_t. \tag{10.5}$$

The multivariate extension of the notion of the strong GARCH process is based on an equation of the form

$$\epsilon_t = H_t^{1/2}\eta_t, \tag{10.6}$$

where (η_t) is a sequence of iid $\mathbb{R}^m$-valued variables with zero mean and identity covariance matrix. The square root has to be understood in the sense of the Cholesky factorization, that is, $H_t^{1/2}(H_t^{1/2})' = H_t$. The matrix $H_t^{1/2}$ can be chosen to be symmetric and positive definite,[1] but it can also be chosen to be triangular, with positive diagonal elements (see, for instance, Harville 1997, Theorem 14.5.11). The latter choice may be of interest because if, for instance, $H_t^{1/2}$ is chosen to be lower triangular, the first component of ϵ_t only depends on the first component of η_t. When $m = 2$, we can thus set

$$\begin{cases} \epsilon_{1t} = h_{11,t}^{1/2}\eta_{1t}, \\ \epsilon_{2t} = \frac{h_{12,t}}{h_{11,t}^{1/2}}\eta_{1t} + \left(\frac{h_{11,t}h_{22,t} - h_{12,t}^2}{h_{11,t}}\right)^{1/2}\eta_{2t}, \end{cases} \tag{10.7}$$

where η_{it} and $h_{ij,t}$ denote the generic elements of η_t and H_t.

Note that any square integral solution (ϵ_t) of (10.6) is a martingale difference satisfying (10.5).

Choosing a specification for H_t is obviously more delicate than in the univariate framework because (i) H_t should be (almost surely) symmetric, and positive definite for all t; (ii) the specification should be simple enough to be amenable to probabilistic study (existence of solutions, stationarity, etc.), while being of sufficient generality; (iii) the specification should be parsimonious enough to enable feasible estimation. However, the model should not be too simple to be able to capture the – possibly sophisticated – dynamics in the covariance structure.

Moreover, it may be useful to have the so-called *stability by aggregation* property. If ϵ_t satisfies (10.5), the process $(\tilde{\epsilon}_t)$ defined by $\tilde{\epsilon}_t = P\epsilon_t$, where P is an invertible square matrix, is such that

$$E(\tilde{\epsilon}_t \mid \tilde{\epsilon}_u, u < t) = 0, \quad \text{Var}(\tilde{\epsilon}_t \mid \tilde{\epsilon}_u, u < t) = \tilde{H}_t = PH_tP'. \tag{10.8}$$

The stability by aggregation of a class of specifications for H_t requires that the conditional variance matrices $\tilde{H}_t$ belong to the same class for any choice of P. This property is particularly relevant in finance

[1] The choice is then unique because to any positive definite matrix A, one can associate a unique positive definite matrix R such that $A = R^2$ (see Harville 1997, Theorem 21.9.1). We have $R = P\Lambda^{1/2}P'$, where $\Lambda^{1/2}$ is a diagonal matrix, with diagonal elements equal to the square roots of the eigenvalues of A, and P is the orthogonal matrix of the corresponding eigenvectors.

because if the components of the vector ϵ_t are asset returns, $\tilde{\epsilon}_t$ is a vector of *portfolios* of the same assets, each of its components consisting of amounts (coefficients of the corresponding row of P) of the initial assets.

10.2.1 Diagonal Model

A popular specification, known as the *diagonal representation*, is obtained by assuming that each element $h_{k\ell,t}$ of the covariance matrix H_t is formulated in terms only of the product of the prior k and ℓ returns. Specifically,

$$h_{k\ell,t} = \omega_{k\ell} + \sum_{i=1}^{q} a_{k\ell}^{(i)} \epsilon_{k,t-i}\epsilon_{\ell,t-i} + \sum_{j=1}^{p} b_{k\ell}^{(j)} h_{k\ell,t-j},$$

with $\omega_{k\ell} = \omega_{\ell k}$, $a_{k\ell}^{(i)} = a_{\ell k}^{(i)}$, $b_{k\ell}^{(j)} = b_{\ell k}^{(j)}$ for all (k, ℓ). For $m = 1$, this model coincides with the usual univariate formulation. When $m > 1$, the model obviously has a large number of parameters and will not in general produce positive definite covariance matrices H_t. We have

$$H_t = \begin{bmatrix} \omega_{11} & \dots & \omega_{1m} \\ \vdots & & \vdots \\ \omega_{1m} & \dots & \omega_{mm} \end{bmatrix} + \sum_{i=1}^{q} \begin{bmatrix} a_{11}^{(i)}\epsilon_{1,t-i}^2 & \dots & a_{1m}^{(i)}\epsilon_{1,t-i}\epsilon_{m,t-i} \\ \vdots & & \vdots \\ a_{1m}^{(i)}\epsilon_{1,t-i}\epsilon_{m,t-i} & \dots & a_{mm}^{(i)}\epsilon_{m,t-i}^2 \end{bmatrix}$$

$$+ \sum_{j=1}^{p} \begin{bmatrix} b_{11}^{(j)} h_{11,t-j} & \dots & b_{1m}^{(j)} h_{1m,t-j} \\ \vdots & & \vdots \\ b_{1m}^{(j)} h_{1m,t-j} & \dots & b_{mm}^{(j)} h_{mm,t-j}^2 \end{bmatrix}$$

$$:= \Omega + \sum_{i=1}^{q} \mathrm{diag}(\epsilon_{t-i}) A^{(i)} \mathrm{diag}(\epsilon_{t-i}) + \sum_{j=1}^{p} B^{(j)} \odot H_{t-j}$$

where $\odot$ denotes the Hadamard product, that is, the element by element product.[2] Thus, in the ARCH case ($p = 0$), sufficient positivity conditions are that Ω is positive definite and the $A^{(i)}$ are positive semi-definite, but these constraints do not easily generalise to the GARCH case. We shall give further positivity conditions obtained by expressing the model in a different way, viewing it as a particular case of a more general class.

It is easy to see that the model is not stable by aggregation: for instance, the conditional variance of $\epsilon_{1,t} + \epsilon_{2,t}$ can in general be expressed as a function of the $\epsilon_{1,t-i}^2$ and $\epsilon_{2,t-i}^2$, but not of the $(\epsilon_{1,t-i} + \epsilon_{2,t-i})^2$. A final drawback of this model is that there is no interaction between the different components of the conditional covariance, which appears unrealistic for applications to financial series.

In what follows, we present the main specifications introduced in the literature, before turning to the existence of solutions. Let η denote a probability distribution on $\mathbb{R}^m$, with zero mean and unit covariance matrix.

10.2.2 Vector GARCH Model

The vector GARCH (VEC-GARCH) model is the most direct generalisation of univariate GARCH: every conditional covariance is a function of lagged conditional variances as well as lagged cross-products of all components. In some sense, everything is explained by everything, which makes this model not only very general but also not very parsimonious.

Denote by vech($\cdot$) the operator that stacks the columns of the lower triangular part of its argument square matrix (if $A = (a_{ij})$, then $\mathrm{vech}(A) = (a_{11}, a_{21}, \dots, a_{m1}, a_{22}, \dots, a_{m2}, \dots, a_{mm})'$). The next definition is a natural extension of the standard GARCH(p, q) specification.

[2] For two matrices $A = (a_{ij})$ and $B = (b_{ij})$ of the same dimension, $A \odot B = (a_{ij}b_{ij})$.

Definition 10.3 (VEC-GARCH(p,q) process) *Let (η_t) be a sequence of iid variables with distribution η. The process (ϵ_t) is said to admit a VEC-GARCH(p, q) representation (relative to the sequence (η_t)) if it satisfies*

$$\begin{cases} \epsilon_t = H_t^{1/2}\eta_t, & \text{where } H_t \text{ is positive definite such that} \\ \text{vech}(H_t) = \omega + \sum_{i=1}^{q} A^{(i)}\text{vech}(\epsilon_{t-i}\epsilon'_{t-i}) + \sum_{j=1}^{p} B^{(j)}\text{vech}(H_{t-j}) \end{cases} \tag{10.9}$$

where ω is a vector of size $m(m+1)/2 \times 1$, and the $A^{(i)}$ and $B^{(j)}$ are matrices of dimension $m(m+1)/2 \times m(m+1)/2$.

Remark 10.1 (The diagonal model is a special case of the VEC-GARCH model) The diagonal model admits a vector representation, obtained for diagonal matrices $A^{(i)}$ and $B^{(j)}$.

We will show that the class of VEC-GARCH models is stable by aggregation. Recall that the vec($\cdot$) operator converts any matrix to a vector by stacking all the columns of the matrix into one vector. It is related to the vech operator by the formulas

$$\text{vec}(A) = D_m\text{vech}(A), \qquad \text{vech}(A) = D_m^+\text{vec}(A), \tag{10.10}$$

where A is any $m \times m$ symmetric matrix, D_m is a full-rank $m^2 \times m(m+1)/2$ matrix (the so-called 'duplication matrix'), whose entries are only 0 and 1, $D_m^+ = (D'_m D_m)^{-1} D'_m$.[3] We also have the relation

$$\text{vec}(ABC) = (C' \otimes A)\text{vec}(B), \tag{10.11}$$

where $\otimes$ denotes the Kronecker matrix product,[4] provided the product ABC is well defined.

Theorem 10.2 (The VEC-GARCH is stable by aggregation) *Let (ϵ_t) be a VEC-GARCH(p, q) process. Then, for any invertible $m \times m$ matrix P, the process $\tilde{\epsilon}_t = P\epsilon_t$ is a VEC-GARCH(p, q) process.*

Proof. Setting $\tilde{H}_t = PH_tP'$, we have $\tilde{\epsilon}_t = \tilde{H}_t^{1/2}\eta_t$ and

$$\begin{aligned} \text{vech}(\tilde{H}_t) &= D_m^+(P \otimes P)D_m\text{vech}(H_t) \\ &= \tilde{\omega} + \sum_{i=1}^{q} \tilde{A}^{(i)}\text{vech}(\tilde{\epsilon}_{t-i}\tilde{\epsilon}'_{t-i}) + \sum_{j=1}^{p} \tilde{B}^{(j)}\text{vech}(\tilde{H}_{t-j}), \end{aligned}$$

where

$$\begin{aligned} \tilde{\omega} &= D_m^+(P \otimes P)D_m\omega, \\ \tilde{A}^{(i)} &= D_m^+(P \otimes P)D_m A^{(i)} D_m^+(P^{-1} \otimes P^{-1})D_m, \\ \tilde{B}^{(i)} &= D_m^+(P \otimes P)D_m B^{(i)} D_m^+(P^{-1} \otimes P^{-1})D_m. \end{aligned}$$

---

[3] For instance,

$$D_1 = (1), \quad D_2 = \begin{pmatrix} 1 & 0 & 0 \\ 0 & 1 & 0 \\ 0 & 1 & 0 \\ 0 & 0 & 1 \end{pmatrix}, \quad D_2^+ = \begin{pmatrix} 1 & 0 & 0 & 0 \\ 0 & 1/2 & 1/2 & 0 \\ 0 & 0 & 0 & 1 \end{pmatrix}.$$

More generally, for $i \geq j$, the $[(j-1)m+i]$th and $[(i-1)m+j]$th rows of D_m equal the $m(m+1)/2$-dimensional row vector all of whose entries are null, with the exception of the $[(j-1)(m-j/2)+i]$th, equal to 1.

[4] If $A = (a_{ij})$ is an $m \times n$ matrix and B is an $m' \times n'$ matrix, $A \otimes B$ is the $mm' \times nn'$ matrix admitting the block elements $a_{ij}B$.

To derive the form of $\tilde{A}^{(i)}$ we use

$$\mathrm{vec}(\epsilon_t\epsilon_t') = \mathrm{vec}(P^{-1}\tilde{\epsilon}_t\tilde{\epsilon}_t'P^{-1'}) = (P^{-1}\otimes P^{-1})\mathrm{D}_m\mathrm{vech}(\tilde{\epsilon}_t\tilde{\epsilon}_t'),$$

and for $\tilde{B}^{(i)}$ we use

$$\mathrm{vec}(H_t) = \mathrm{vec}(P^{-1}\tilde{H}_tP^{-1'}) = (P^{-1}\otimes P^{-1})\mathrm{D}_m\mathrm{vech}(\tilde{H}_t).$$

□

Positivity Conditions

To ensure the positive semi-definiteness of H_t, the initial values and Ω (where $\omega = \mathrm{vech}(\Omega)$) have to be positive semi-definite, and the matrices $A^{(i)}$ and $B^{(j)}$ need to map the vectorized positive semi-definite matrices into themselves. To be more specific, a generic element of

$$h_t = \mathrm{vech}(H_t)$$

is denoted by $h_{k\ell,t}$ $(k \geq \ell)$, and we will denote by $a^{(i)}_{k\ell,k'\ell'}$ $(b^{(j)}_{k\ell,k'\ell'})$ the entry of $A^{(i)}$ $(B^{(j)})$ located on the same row as $h_{k\ell,t}$ and belonging to the same column as the element $h_{k'\ell',t}$ of h_t'. We thus have an expression of the form

$$\begin{aligned} h_{k\ell,t} &= \mathrm{Cov}(\epsilon_{kt}, \epsilon_{\ell t} \mid \epsilon_u, u < t) \\ &= \omega_{k\ell} + \sum_{i=1}^{q}\sum_{\substack{k',\ell'=1\\k'\geq\ell'}}^{m} a^{(i)}_{k\ell,k'\ell'}\epsilon_{k',t-i}\epsilon_{\ell',t-i} + \sum_{j=1}^{p}\sum_{\substack{k',\ell'=1\\k'\geq\ell'}}^{m} b^{(j)}_{k\ell,k'\ell'}h_{k'\ell',t-j}. \end{aligned}$$

Denoting by $A^{(i)}_{k\ell}$ the $m\times m$ symmetric matrix with (k', ℓ')th entry $a^{(i)}_{k\ell,k'\ell'}/2$, for $k' \neq \ell'$, and the elements $a^{(i)}_{k\ell,k'k'}$ on the diagonal, the preceding equality is written as

$$h_{k\ell,t} = \omega_{k\ell} + \sum_{i=1}^{q}\epsilon_{t-i}'A^{(i)}_{k\ell}\epsilon_{t-i} + \sum_{j=1}^{p}\sum_{\substack{k',\ell'=1\\k'\geq\ell'}}^{m} b^{(j)}_{k\ell,k'\ell'}h_{k'\ell',t-j}. \tag{10.12}$$

In order to obtain a more compact form for the last part of this expression, let us introduce the spectral decomposition of the symmetric matrices H_t, assumed to be positive semi-definite. We have $H_t = \sum_{r=1}^{m}\lambda_t^{(r)}v_t^{(r)}v_t^{(r)'}$, where $V_t = (v_t^{(1)}, \dots, v_t^{(m)})$ is an orthogonal matrix of eigenvectors $v_t^{(r)}$ associated with the (positive) eigenvalues $\lambda_t^{(r)}$ of H_t. Defining the matrices $B^{(j)}_{k\ell}$ by analogy with the $A^{(i)}_{k\ell}$, we get

$$h_{k\ell,t} = \omega_{k\ell} + \sum_{i=1}^{q}\epsilon_{t-i}'A^{(i)}_{k\ell}\epsilon_{t-i} + \sum_{j=1}^{p}\sum_{r=1}^{m}\lambda_{t-j}^{(r)}v_{t-j}^{(r)'}B^{(j)}_{k\ell}v_{t-j}^{(r)}. \tag{10.13}$$

Finally, consider the $m^2\times m^2$ matrix admitting the block form $A_i = (A^{(i)}_{k\ell})$, and let $B_j = (B^{(j)}_{k\ell})$. The preceding expressions are equivalent to

$$\begin{aligned} \mathrm{H}_t &= \sum_{r=1}^{m}\lambda_t^{(r)}v_t^{(r)}v_t^{(r)'} \\ &= \Omega + \sum_{i=1}^{q}(I_m\otimes\epsilon_{t-i}')A_i(I_m\otimes\epsilon_{t-i}) \\ &\quad + \sum_{j=1}^{p}\sum_{r=1}^{m}\lambda_{t-j}^{(r)}(I_m\otimes v_{t-j}^{(r)'})B_j(I_m\otimes v_{t-j}^{(r)}), \end{aligned} \tag{10.14}$$

where Ω is the symmetric matrix such that $\mathrm{vech}(\Omega) = \omega$.

In this form, it is evident that the assumption

$$A_i \text{ and } B_j \text{ are positive semi-definite and } \Omega \text{ is positive definite} \tag{10.15}$$

ensures that if the H_{t-j} are almost surely positive definite, then so is H_t.

Example 10.1 (Three representations of a vector ARCH(1) model) For $p = 0$, $q = 1$, and $m = 2$, the conditional variance is written, in the form (10.9), as

$$\text{vech}(H_t) = \begin{bmatrix} h_{11,t} \\ h_{12,t} \\ h_{22,t} \end{bmatrix} = \begin{bmatrix} \omega_{11} \\ \omega_{12} \\ \omega_{22} \end{bmatrix} + \begin{bmatrix} a_{11,11} & a_{11,12} & a_{11,22} \\ a_{12,11} & a_{12,12} & a_{12,22} \\ a_{22,11} & a_{22,12} & a_{22,22} \end{bmatrix} \begin{bmatrix} \epsilon^2_{1,t-1} \\ \epsilon_{1,t-1}\epsilon_{2,t-1} \\ \epsilon^2_{2,t-1} \end{bmatrix},$$

in the form (10.12) as

$$h_{k\ell,t} = \omega_{k\ell} + (\epsilon_{1,t-1}, \epsilon_{2,t-1}) \begin{bmatrix} a_{k\ell,11} & \frac{a_{k\ell,12}}{2} \\ \frac{a_{k\ell,12}}{2} & a_{k\ell,22} \end{bmatrix} \begin{bmatrix} \epsilon_{1,t-1} \\ \epsilon_{2,t-1} \end{bmatrix}, \quad k, \ell = 1, 2,$$

and in the form (10.14) as

$$H_t = \begin{bmatrix} \omega_{11} & \omega_{12} \\ \omega_{12} & \omega_{22} \end{bmatrix} + (I_2 \otimes \epsilon'_{t-1}) \begin{bmatrix} a_{11,11} & \frac{a_{11,12}}{2} & a_{12,11} & \frac{a_{12,12}}{2} \\ \frac{a_{11,12}}{2} & a_{11,22} & \frac{a_{12,12}}{2} & a_{12,22} \\ a_{12,11} & \frac{a_{12,12}}{2} & a_{22,11} & \frac{a_{22,12}}{2} \\ \frac{a_{12,12}}{2} & a_{12,22} & \frac{a_{22,12}}{2} & a_{22,22} \end{bmatrix} (I_2 \otimes \epsilon_{t-1}).$$

This example shows that, even for small orders, the VEC model potentially has an enormous number of parameters, which can make estimation of the parameters computationally demanding. Moreover, the positivity conditions are not directly obtained from (10.9) but from (10.14), involving the spectral decomposition of the matrices H_{t-j}.

The following classes provide more parsimonious and tractable models.

10.2.3 Constant Conditional Correlations Models

Suppose that, for a MGARCH process of the form (10.6), all the past information on ϵ_{kt}, involving all the variables $\epsilon_{\ell,t-i}$, is summarised in the variable $h_{kk,t}$, with $Eh_{kk,t} = E\epsilon^2_{kt}$. Then, letting $\tilde{\eta}_{kt} = h^{-1/2}_{kk,t}\epsilon_{kt}$, we define for all k a sequence of iid variables with zero mean and unit variance. The variables $\tilde{\eta}_{kt}$ are generally correlated, so let $R = \text{Var}(\tilde{\eta}_t) = (\rho_{k\ell})$, where $\tilde{\eta}_t = (\tilde{\eta}_{1t}, \dots, \tilde{\eta}_{mt})'$. The conditional variance of

$$\epsilon_t = \text{diag}(h^{1/2}_{11,t}, \dots, h^{1/2}_{mm,t})\tilde{\eta}_t$$

is then written as

$$H_t = \text{diag}(h^{1/2}_{11,t}, \dots, h^{1/2}_{mm,t})R\ \text{diag}(h^{1/2}_{11,t}, \dots, h^{1/2}_{mm,t}). \tag{10.16}$$

By construction, the conditional correlations between the components of ϵ_t are time-invariant:

$$\frac{h_{k\ell,t}}{h^{1/2}_{kk,t}h^{1/2}_{\ell\ell,t}} = \frac{E(\epsilon_{kt}\epsilon_{\ell t} \mid \epsilon_u, u < t)}{\{E(\epsilon^2_{kt} \mid \epsilon_u, u < t)E(\epsilon^2_{\ell t} \mid \epsilon_u, u < t)\}^{1/2}} = \rho_{kl}.$$

To complete the specification, the dynamics of the conditional variances $h_{kk,t}$ has to be defined. The simplest constant conditional correlations (CCC) model relies on the following univariate GARCH specifications:

$$h_{kk,t} = \omega_k + \sum_{i=1}^{q} a_{k,i}\epsilon^2_{k,t-i} + \sum_{j=1}^{p} b_{k,j}h_{kk,t-j}, \quad k = 1, \dots, m, \tag{10.17}$$

where $\omega_k > 0$, $a_{k,i} \geq 0$, $b_{k,j} \geq 0$, $-1 \leq \rho_{k\ell} \leq 1$, $\rho_{kk} = 1$, and R is symmetric and positive semi-definite. Observe that the conditional variances are specified as in the diagonal model. The conditional covariances clearly are not linear in the squares and cross products of the returns.

In a multivariate framework, it seems natural to extend the specification (10.17) by allowing $h_{kk,t}$ to depend not only on its own past, but also on the past of all the variables $\epsilon_{\ell,t}$. Set

$$\underline{h}_t = \begin{pmatrix} h_{11,t} \\ \vdots \\ h_{mm,t} \end{pmatrix}, \quad D_t = \begin{pmatrix} \sqrt{h_{11,t}} & 0 & \dots & 0 \\ 0 & \ddots & & \vdots \\ \vdots & & \ddots & \vdots \\ 0 & \dots & & \sqrt{h_{mm,t}} \end{pmatrix}, \quad \underline{\epsilon}_t = \begin{pmatrix} \epsilon_{1t}^2 \\ \vdots \\ \epsilon_{mt}^2 \end{pmatrix}.$$

Definition 10.4 (CCC-GARCH(p, q) process) *Let (η_t) be a sequence of iid variables with distribution η. A process (ϵ_t) is called CCC-GARCH(p, q) if it satisfies*

$$\begin{cases} \epsilon_t = H_t^{1/2}\eta_t, \\ H_t = D_t R D_t, \\ \underline{h}_t = \underline{\omega} + \sum_{i=1}^{q} \mathbf{A}_i \underline{\epsilon}_{t-i} + \sum_{j=1}^{p} \mathbf{B}_j \underline{h}_{t-j}, \end{cases} \tag{10.18}$$

where R is a correlation matrix, $\underline{\omega}$ is a $m \times 1$ vector with positive coefficients, and the $\mathbf{A}_i$ and $\mathbf{B}_j$ are $m \times m$ matrices with non-negative coefficients.

We have $\epsilon_t = D_t\tilde{\eta}_t$, where $\tilde{\eta}_t = R^{1/2}\eta_t$ is a centred vector with covariance matrix R. The components of ϵ_t thus have the usual expression, $\epsilon_{kt} = h_{kk,t}^{1/2}\tilde{\eta}_{kt}$, but the conditional variance $h_{kk,t}$ depends on the past of all the components of ϵ_t.

Note that the conditional covariances are generally non-linear functions of the components of $\epsilon_{t-i}\epsilon'_{t-i}$ and of past values of the components of H_t. Model (10.18) is thus not a VEC-GARCH model, defined by (10.9), except when R is the identity matrix.

One advantage of this specification is that a simple condition ensuring the positive definiteness of H_t is obtained through the positive coefficients for the matrices $\mathbf{A}_i$ and $\mathbf{B}_j$ and the choice of a positive definite matrix for R. Conrad and Karanasos (2010) showed that less restrictive assumptions ensuring the positive definiteness of H_t can be found. Moreover there exists a representation of the CCC model in which the matrices B_j are diagonal. Another advantage of the CCC specification is that the study of the stationarity is remarkably simple.

Two limitations of the CCC model are, however, (i) its non-stability by aggregation and (ii) the arbitrary nature of the assumption of constant conditional correlations.

10.2.4 Dynamic Conditional Correlations Models

Dynamic conditional correlations GARCH (DCC-GARCH) models are an extension of CCC-GARCH, obtained by introducing a dynamic for the conditional correlation. Hence, the constant matrix R in Definition 10.4 is replaced by a matrix R_t which is measurable with respect to the past variables $\{\epsilon_u, u < t\}$. For reasons of parsimony, it seems reasonable to choose diagonal matrices $\mathbf{A}_i$ and $\mathbf{B}_i$ in (10.18), corresponding to univariate GARCH models for each component as in (10.17). Different DCC models are obtained depending on the specification of R_t. A simple formulation is

$$R_t = \theta_1 R + \theta_2 \Psi_{t-1} + \theta_3 R_{t-1}, \tag{10.19}$$

where the θ_i are positive weights summing to 1, R is a constant correlation matrix, and Ψ_{t-1} is the empirical correlation matrix of $\epsilon_{t-1}, \dots, \epsilon_{t-M}$. The matrix R_t is thus a correlation matrix (see Exercise 10.9). Equation (10.19) is reminiscent of the GARCH(1, 1) specification, $\theta_1 R$ playing the role of the parameter ω, θ_2 that of α, and θ_3 that of β.

Another way of specifying the dynamics of R_t is by setting

$$R_t = (\text{diag}\, Q_t)^{-1/2} Q_t (\text{diag}\, Q_t)^{-1/2},$$

where diag Q_t is the diagonal matrix constructed with the diagonal elements of Q_t, and Q_t is a sequence of covariance matrices which is measurable with respect to $\sigma(\epsilon_u, u < t)$. In the original DCC model of Engle (2002), the dynamics of Q_t is given by

$$Q_t = (1 - \alpha - \beta)S + \alpha \eta^*_{t-1}\eta^{*\prime}_{t-1} + \beta Q_{t-1}, \tag{10.20}$$

where $\eta^*_t = (\epsilon_{1t}/\sqrt{h_{11,t}}, \dots, \epsilon_{mt}/\sqrt{h_{mm,t}})'$ denotes the vector of standardized returns, $\alpha, \beta \geq 0, \alpha + \beta < 1$, and S is a positive-definite matrix.

Matrix S turns out to be difficult to interpret, and Aielli (2013) pointed out that $S \neq E(\eta^*_t \eta^{*'}_t)$ in general. Thus, the commonly used estimator of S defined as the sample second moment of the standardized returns is not consistent in this formulation.

In the so-called corrected DCC (cDCC) model of Aielli (2013), the dynamics of Q_t is reformulated as

$$Q_t = (1 - \alpha - \beta)S + \alpha(\text{diag}Q_{t-1})^{1/2}\eta^*_{t-1}\eta^{*\prime}_{t-1}(\text{diag}Q_{t-1})^{1/2} + \beta Q_{t-1},$$

under the same constraints on the coefficients. In this model, under stationarity conditions, $S = E(Q_t) = E\{(\text{diag}Q_t)^{1/2}\eta^*_t\eta^{*'}_t(\text{diag}Q_t)^{1/2}\}$.

Multiplying the left-hand side and the right-hand side of Q_t by an arbitrary positive-definite matrix yields the same conditional correlation matrix R_t. It is thus necessary to introduce an identifiability condition, as for instance imposing that S be a correlation matrix.

10.2.5 BEKK-GARCH Model

The BEKK acronym refers to a specific parameterisation of the MGARCH model developed by Baba, Engle, Kraft, and Kroner, in a preliminary version of Engle and Kroner (1995).

Definition 10.5 (BEKK-GARCH(p, q) process) *Let (η_t) denote an iid sequence with common distribution η. The process (ϵ_t) is called a strong BEKK-GARCH(p, q), with respect to the sequence (η_t), if it satisfies*

$$\begin{cases} \epsilon_t = H_t^{1/2}\eta_t \\ H_t = \Omega + \sum_{i=1}^{q}\sum_{k=1}^{K} A_{ik}\epsilon_{t-i}\epsilon'_{t-i}A'_{ik} + \sum_{j=1}^{p}\sum_{k=1}^{K} B_{jk}H_{t-j}B'_{jk}, \end{cases}$$

where K is an integer, Ω, A_{ik} and B_{jk} are square $m \times m$ matrices, and Ω is positive definite.

The specification obviously ensures that if the matrices, H_{t-i}, $i = 1, \dots, p$, are almost surely positive definite, then so is H_t.

To compare this model with the representation (10.9), let us derive the vector form of the equation for H_t. Using the relations (10.10) and (10.11), we get

$$\text{vech}(H_t) = \text{vech}(\Omega) + \sum_{i=1}^{q} \text{D}_m^+ \sum_{k=1}^{K}(A_{ik} \otimes A_{ik})\text{D}_m \text{vech}(\epsilon_{t-i}\epsilon'_{t-i})$$

$$+ \sum_{j=1}^{p} \text{D}_m^+ \sum_{k=1}^{K}(B_{jk} \otimes B_{jk})\text{D}_m \text{vech}H_{t-j}.$$

The model can thus be written in the VEC-GARCH(p,q) form (10.9), with

$$A^{(i)} = \text{D}_m^+ \sum_{k=1}^{K}(A_{ik} \otimes A_{ik})\text{D}_m, \quad B^{(j)} = \text{D}_m^+ \sum_{k=1}^{K}(B_{jk} \otimes B_{jk})\text{D}_m, \tag{10.21}$$

for $i = 1, \dots, q$ and $j = 1, \dots, p$. In particular, it can be seen that the number of coefficients of a matrix $A^{(i)}$ in (10.9) is $[m(m+1)/2]^2$, whereas it is Km^2 in this particular case. However, the converse is not true.

Stelzer (2008) showed that, for $m \geq 3$, there exist VEC-GARCH models that cannot be represented in the BEKK form.

The BEKK class contains (Exercise 10.13) the diagonal models obtained by choosing diagonal matrices A_{ik} and B_{jk}. The following theorem establishes a converse to this property.

Theorem 10.3 *For the model defined by the diagonal vector representation (10.9) with*

$$A^{(i)} = \mathrm{diag}\{\mathrm{vech}(\tilde{A}^{(i)})\}, \quad B^{(j)} = \mathrm{diag}\{\mathrm{vech}(\tilde{B}^{(j)})\},$$

where $\tilde{A}^{(i)} = (\tilde{a}^{(i)}_{\ell\ell'})$ *and* $\tilde{B}^{(j)} = (\tilde{b}^{(j)}_{\ell\ell'})$ *are* $m \times m$ *symmetric positive semi-definite matrices, there exist matrices* A_{ik} *and* B_{jk} *such that (10.21) holds, for* $K = m$.

Proof. There exists an upper triangular matrix

$$\tilde{D}^{(i)} = \begin{bmatrix} d^{(i)}_{11,m} & d^{(i)}_{11,m-1} & \dots & d^{(i)}_{11,1} \\ 0 & d^{(i)}_{22,m-1} & \dots & d^{(i)}_{22,1} \\ \vdots & & & \vdots \\ 0 & 0 & \dots & d^{(i)}_{mm,1} \end{bmatrix}$$

such that $\tilde{A}^{(i)} = \tilde{D}^{(i)}\{\tilde{D}^{(i)}\}'$. Let $A_{ik} = \mathrm{diag}(d^{(i)}_{11,k}, d^{(i)}_{22,k}, \dots, d^{(i)}_{rr,k}, 0, \dots, 0)$, where $r = m-k+1$, for $k = 1, \dots, m$. It is easy to show that the first equality in (10.21) is satisfied with $K = m$. The second equality is obtained similarly. □

Example 10.2 By way of illustration, consider the particular case where $m = 2$, $q = K = 1$ and $p = 0$. If $A = (a_{ij})$ is a 2×2 matrix, it is easy to see that

$$\mathrm{D}_2^+(A \otimes A)\mathrm{D}_2 = \begin{bmatrix} a_{11}^2 & 2a_{11}a_{12} & a_{12}^2 \\ a_{11}a_{21} & a_{11}a_{22} + a_{12}a_{21} & a_{12}a_{22} \\ a_{21}^2 & 2a_{21}a_{22} & a_{22}^2 \end{bmatrix}.$$

Hence, cancelling out the unnecessary indices,

$$\begin{cases} h_{11,t} = \omega_{11} + a_{11}^2\epsilon_{1,t-1}^2 + 2a_{11}a_{12}\epsilon_{1,t-1}\epsilon_{2,t-1} + a_{12}^2\epsilon_{2,t-1}^2 \\ h_{12,t} = \omega_{12} + a_{11}a_{21}\epsilon_{1,t-1}^2 + (a_{11}a_{22} + a_{12}a_{21})\epsilon_{1,t-1}\epsilon_{2,t-1} + a_{12}a_{22}\epsilon_{2,t-1}^2 \\ h_{22,t} = \omega_{22} + a_{21}^2\epsilon_{1,t-1}^2 + 2a_{21}a_{22}\epsilon_{1,t-1}\epsilon_{2,t-1} + a_{22}^2\epsilon_{2,t-1}^2. \end{cases}$$

In particular, the diagonal models belonging to this class are of the form

$$\begin{cases} h_{11,t} = \omega_{11} + a_{11}^2\epsilon_{1,t-1}^2 \\ h_{12,t} = \omega_{12} + a_{11}a_{22}\epsilon_{1,t-1}\epsilon_{2,t-1} \\ h_{22,t} = \omega_{22} + a_{22}^2\epsilon_{2,t-1}^2. \end{cases}$$

Remark 10.2 (Interpretation of the BEKK coefficients) Example 10.2 shows that the BEKK specification imposes highly artificial constraints on the volatilities and co-volatilities of the components. As a consequence, the coefficients of a BEKK representation are difficult to interpret.

Remark 10.3 (Identifiability) Identifiability of a BEKK representation requires additional constraints. Indeed, the same representation holds if A_{ik} is replaced by $-A_{ik}$, or if the matrices $A_{1k}, \dots, A_{qk}$ and $A_{1k'}, \dots, A_{qk'}$ are permuted for $k \neq k'$.

Example 10.3 (A general and identifiable BEKK representation) Consider the case $m = 2$, $q = 1$, and $p = 0$. Suppose that the distribution η is non-degenerate, so that there exists no non-trivial constant linear combination of a finite number of the $\epsilon_{k,t-i}\epsilon_{\ell,t-i}$. Let

$$H_t = \Omega + \sum_{k=1}^{m^2} A_k \epsilon_{t-1}\epsilon'_{t-1} A'_k,$$

where Ω is a symmetric positive definite matrix,

$$A_1 = \begin{pmatrix} a_{11,1} & a_{12,1} \\ a_{21,1} & a_{22,1} \end{pmatrix}, \quad A_2 = \begin{pmatrix} 0 & 0 \\ a_{21,2} & a_{22,2} \end{pmatrix},$$

$$A_3 = \begin{pmatrix} 0 & a_{12,3} \\ 0 & a_{22,3} \end{pmatrix}, \quad A_4 = \begin{pmatrix} 0 & 0 \\ 0 & a_{22,4} \end{pmatrix},$$

with $a_{11,1} \geq 0$, $a_{12,3} \geq 0$, $a_{21,2} \geq 0$ and $a_{22,4} \geq 0$.

Let us show that this BEKK representation is both identifiable and quite general. Easy, but tedious, computation shows that an expression of the form (10.9) holds with

$$A^{(1)} = \sum_{k=1}^{4} \mathrm{D}_2^+ (A_k \otimes A_k) \mathrm{D}_2$$

$$= \begin{pmatrix} a_{11,1}^2 & 2a_{11,1}a_{12,1} & a_{12,1}^2 + a_{12,3}^2 \\ a_{11,1}a_{21,1} & a_{12,1}a_{21,1} + a_{11,1}a_{22,1} & a_{12,1}a_{22,1} + a_{12,3}a_{22,3} \\ a_{21,1}^2 + a_{21,2}^2 & 2a_{21,1}a_{22,1} + 2a_{21,2}a_{22,2} & a_{22,1}^2 + a_{22,2}^2 + a_{22,3}^2 + a_{22,4}^2 \end{pmatrix}.$$

In view of the sign constraint, the (1, 1)th element of $A^{(1)}$ allows us to identify $a_{11,1}$. The (1, 2)th and (2, 1)th elements then allow us to find $a_{12,1}$ and $a_{21,1}$, whence the (2, 2)th element yields $a_{22,1}$. The two elements of A_3 are deduced from the (1, 3)th and (2, 3)th elements of $A^{(1)}$, and from the constraint $a_{12,3} \geq 0$ (which could be replaced by a constraint on the sign of $a_{22,3}$). A_2 is identified similarly, and the non-zero element of A_4 is finally identified by considering the (3, 3)th element of $A^{(1)}$.

In this example, the BEKK representation contains the same number of parameters as the corresponding VEC representation, but has the advantage of automatically providing a positive definite solution H_t.

It is interesting to consider the stability by aggregation of the BEKK class.

Theorem 10.4 (Stability of the BEKK model by aggregation) *Let (ϵ_t) be a BEKK-GARCH(p, q) process. Then, for any invertible $m \times m$ matrix P, the process $\tilde{\epsilon}_t = P\epsilon_t$ is a BEKK-GARCH(p, q) process.*

Proof. Letting $\tilde{H}_t = PH_tP'$, $\tilde{\Omega} = P\Omega P'$, $\tilde{A}_{ik} = PA_{ik}P^{-1}$, and $\tilde{B}_{jk} = PB_{jk}P^{-1}$, we get

$$\tilde{\epsilon}_t = \tilde{H}_t^{1/2}\eta_t, \quad \tilde{H}_t = \tilde{\Omega} + \sum_{i=1}^{q}\sum_{k=1}^{K} \tilde{A}_{ik}\tilde{\epsilon}_{t-i}\tilde{\epsilon}'_{t-i}\tilde{A}'_{ik} + \sum_{j=1}^{p}\sum_{k=1}^{K} \tilde{B}_{jk}\tilde{H}_{t-i}\tilde{B}'_{jk}$$

and, $\tilde{\Omega}$ being a positive definite matrix, the result is proved. □

As in the univariate case, the 'square' of the (ϵ_t) process is the solution of an ARMA model. Indeed, define the innovation of the process $\mathrm{vec}(\epsilon_t\epsilon'_t)$:

$$\nu_t = \mathrm{vec}(\epsilon_t\epsilon'_t) - \mathrm{vec}[E(\epsilon_t\epsilon'_t \mid \epsilon_u, u < t)] = \mathrm{vec}(\epsilon_t\epsilon'_t) - \mathrm{vec}(H_t). \tag{10.22}$$

Applying the vec operator, and substituting the variables $\mathrm{vec}(H_{t-j})$ in the model of Definition 10.5 by $\mathrm{vec}(\epsilon_{t-j}\epsilon'_{t-j}) - \nu_{t-j}$, we get the representation

$$\mathrm{vec}(\epsilon_t\epsilon'_t) = \mathrm{vec}(\Omega) + \sum_{i=1}^{r}\sum_{k=1}^{K}\{(A_{ik} + B_{ik}) \otimes (A_{ik} + B_{ik})\}\mathrm{vec}(\epsilon_{t-i}\epsilon'_{t-i})$$

$$+ \nu_t - \sum_{j=1}^{p}\sum_{k=1}^{K}(B_{jk} \otimes B_{jk})\nu_{t-j}, \quad t \in \mathbb{Z}, \tag{10.23}$$

where $r = \max(p, q)$, with the convention $A_{ik} = 0$ ($B_{jk} = 0$) if $i > q$ ($j > p$). This representation cannot be used to obtain stationarity conditions because the process (v_t) is not iid in general. However, it can be used to derive the second-order moment, when it exists, of the process ϵ_t as

$$E\{\text{vec}(\epsilon_t\epsilon_t')\} = \text{vec}(\Omega) + \sum_{i=1}^{r}\sum_{k=1}^{K}\{(A_{ik} + B_{ik}) \otimes (A_{ik} + B_{ik})\}E\{\text{vec}(\epsilon_t\epsilon_t')\},$$

that is,

$$E\{\text{vec}(\epsilon_t\epsilon_t')\} = \left\{I - \sum_{i=1}^{r}\sum_{k=1}^{K}(A_{ik} + B_{ik}) \otimes (A_{ik} + B_{ik})\right\}^{-1} \text{vec}(\Omega),$$

provided that the matrix in braces is non-singular.

10.2.6 Factor GARCH Models

In these models, it is assumed that a non-singular linear combination f_t of the m components of ϵ_t, or an exogenous variable summarising the co-movements of the components, has a GARCH structure.

Factor Models with Idiosyncratic Noise

A very popular factor model links individual returns ϵ_{it} to the market return f_t through a regression model

$$\epsilon_{it} = \beta_i f_t + \eta_{it}, \quad i = 1, \dots, m. \tag{10.24}$$

The parameter β_i can be interpreted as a sensitivity to the factor, and the noise η_{it} as a specific risk (often called idiosyncratic risk) which is conditionally uncorrelated with f_t. It follows that $H_t = \Omega + \lambda_t \beta\beta'$, where $\beta = (\beta_1, \dots, \beta_m)'$ is the vector of sensitivities, λ_t is the conditional variance of f_t, and Ω is the covariance matrix of the idiosyncratic terms. More generally, assuming the existence of r conditionally uncorrelated factors, we obtain the decomposition

$$H_t = \Omega + \sum_{j=1}^{r} \lambda_{jt} \beta^{(j)} \beta^{(j)\prime}. \tag{10.25}$$

It is not restrictive to assume that the factors are linear combinations of the components of ϵ_t (Exercise 10.10). If, in addition, the conditional variances λ_{jt} are specified as univariate GARCH, the model remains parsimonious in terms of unknown parameters and (10.25) reduces to a particular BEKK model (Exercise 10.11). If Ω is chosen to be positive definite and if the univariate series $(\lambda_{jt})_t$, $j = 1, \dots, r$ are independent, strictly and second-order stationary, then it is clear that (10.25) defines a sequence of positive definite matrices (H_t) that are strictly and second-order stationary.

Principal Components GARCH Model

The concept of factor is central to principal components analysis (PCA) and to other methods of exploratory data analysis. PCA relies on decomposing the covariance matrix V of m quantitative variables as $V = P\Lambda P'$, where Λ is a diagonal matrix whose elements are the eigenvalues $\lambda_1 \geq \lambda_2 \geq \cdots \geq \lambda_m$ of V, and where P is the orthonormal matrix of the corresponding eigenvectors. The first principal component is the linear combination of the m variables, with weights given by the first column of P, which, in some sense, is the factor which best summarises the set of m variables (Exercise 10.12). There exist m principal components, which are uncorrelated and whose variances $\lambda_1, \dots, \lambda_m$ (and hence whose explanatory powers) are in decreasing order. It is natural to consider this method for extracting the key factors of the volatilities of the m components of ϵ_t.

We obtain a principal component GARCH (PC-GARCH) or orthogonal GARCH (O-GARCH) model by assuming that

$$H_t = P\Lambda_t P', \tag{10.26}$$

where P is an orthogonal matrix ($P' = P^{-1}$) and $\Lambda_t = \text{diag}(\lambda_{1t}, \dots, \lambda_{mt})$, where the λ_{it} are the volatilities, which can be obtained from univariate GARCH-type models. This is equivalent to assuming

$$\epsilon_t = Pf_t, \tag{10.27}$$

where $\mathbf{f}_t = P'\epsilon_t$ is the principal component vector, whose components are orthogonal factors. If univariate GARCH(1, 1) models are used for the factors $f_{it} = \sum_{j=1}^{m} P(j, i)\epsilon_{jt}$, then

$$\lambda_{it} = \omega_i + \alpha_i f_{it-1}^2 + \beta_i \lambda_{it-1}. \tag{10.28}$$

Remark 10.4 (Interpretation, factor estimation, and extensions)

1. Model (10.26) can also be interpreted as a full-factor GARCH (FF-GARCH) model, that is, a model with as many factors as components and no idiosyncratic terms. Let $P(\cdot, j)$ be the jth column of P (an eigenvector of H_t associated with the eigenvalue λ_{jt}). We get a spectral expression for the conditional variance,

$$H_t = \sum_{j=1}^{m} \lambda_{jt} P(\cdot, j)P'(\cdot, j),$$

which is of the form (10.25) with an idiosyncratic variance $\Omega = 0$.

2. A PCA of the conditional variance H_t should, in full generality, give $H_t = P_t \Lambda_t P_t'$ with factors (that is, principal components) $\mathbf{f}_t = P_t'\epsilon_t$. Model (10.26) thus assumes that all factors are linear combinations, with fixed coefficients, of the same returns ϵ_{it}. For instance, the first factor f_{1t} is the conditionally most risky factor (with the largest conditional variance λ_{1t}, see Exercise 10.12). But since it is assumed that the direction of f_{1t} is fixed, in the subspace of $\mathbb{R}^m$ generated by the components of ϵ_{it}, the first factor is also the most risky unconditionally. This can be seen through the PCA of the unconditional variance $H = EH_t = P\Lambda P'$, which is assumed to exist.

3. It is easy to estimate P by applying PCA to the empirical variance $\hat{H} = n^{-1} \sum_{t=1}^{n} (\epsilon_t - \overline{\epsilon})(\epsilon_t - \overline{\epsilon})'$, where $\overline{\epsilon} = n^{-1} \sum_{t=1}^{n} \epsilon_t$. The components of $\hat{P}'\epsilon_t$ are specified as GARCH-type univariate models. Estimation of the conditional variance $\hat{H}_t = \hat{P}\hat{\Lambda}_t\hat{P}'$ thus reduces to estimating m univariate models.

4. It is common practice to apply PCA on centred and standardised data, in order to remove the influence of the units of the various variables. For returns ϵ_{it}, standardisation does not seem appropriate if one wishes to retain a size effect, that is, if one expects an asset with a relatively large variance to have more weight in the riskier factors.

5. In the spirit of the standard PCA, it is possible to only consider the first r principal components, which are the key factors of the system. The variance H_t is thus approximated by

$$\hat{P}\text{diag}(\hat{\lambda}_{1t}, \dots, \hat{\lambda}_{rt}, 0'_{m-r})\hat{P}', \tag{10.29}$$

where the $\hat{\lambda}_{it}$ is estimated from simple univariate models, such as GARCH(1, 1) models of the form (10.28), the matrix $\hat{P}$ is obtained from PCA of the empirical covariance matrix $\hat{H} = \hat{P}\text{diag}(\hat{\lambda}_1, \dots, \hat{\lambda}_m)\hat{P}'$, and the factors are approximated by $\hat{\mathbf{f}}_t = \hat{P}'\epsilon_t$. Instead of the approximation (10.29), one can use

$$\hat{H}_t = \hat{P}\text{diag}(\hat{\lambda}_{1t}, \dots, \hat{\lambda}_{rt}, \hat{\lambda}_{r+1}, \dots, \hat{\lambda}_m)\hat{P}'. \tag{10.30}$$

The approximation in (10.30) is as simple as (10.29) and does not require additional computations (in particular, the r GARCH equations are retained) but has the advantage of providing an almost surely invertible estimation of H_t (for fixed n), which is required in the computation of certain statistics (such as the AIC-type information criteria based on the Gaussian log-likelihood).

6. Note that the assumption that P is orthogonal can be restrictive. The class of generalised orthogonal GARCH (GO-GARCH) processes assumes only that P is any non-singular matrix.

10.2.7 Cholesky GARCH

Suppose that the conditional covariance matrix H_t of ϵ_t is positive-definite, i.e. that the components of ϵ_t are not multicolinear. Given the information $\mathcal{F}_{t-1}$ generated by the past values of ϵ_t, let $\ell_{21,t}$ be the *conditional beta* in the regression of ϵ_{2t} on $v_{1t} := \epsilon_{1t}$. One can write

$$\epsilon_{2t} = \ell_{21,t}v_{1t} + v_{2t} = \beta_{21,t}\epsilon_{1t} + v_{2t},$$

with $\beta_{21,t} = \ell_{21,t} \in \mathcal{F}_{t-1}$, and v_{2t} is orthogonal to ϵ_{1t} conditionally on $\mathcal{F}_{t-1}$. More generally, we have

$$\epsilon_{it} = \sum_{j=1}^{i-1} \ell_{ij,t}v_{jt} + v_{it} = \sum_{j=1}^{i-1} \beta_{ij,t}\epsilon_{jt} + v_{it}, \quad \text{for } i = 2, \dots, m, \tag{10.31}$$

where v_{it} is uncorrelated with $v_{1t}, \dots, v_{i-1,t}$, and thus uncorrelated with $\epsilon_{1t}, \dots, \epsilon_{i-1,t}$, conditionally on $\mathcal{F}_{t-1}$. In matrix form, Eq. (10.31) is written as

$$\epsilon_t = L_t v_t \quad \text{and} \quad B_t\epsilon_t = v_t,$$

where L_t and $B_t = L_t^{-1}$ are lower unitriangular (i.e. triangular with 1 on the diagonal) matrices, with $\ell_{ij,t}$ (respectively $-\beta_{ij,t}$) at row i and column j of L_t (respectively B_t) for $i > j$. For $m = 3$, we have

$$L_t = \begin{pmatrix} 1 & 0 & 0 \\ \ell_{21,t} & 1 & 0 \\ \ell_{31,t} & \ell_{32,t} & 1 \end{pmatrix}, \quad B_t = \begin{pmatrix} 1 & 0 & 0 \\ -\beta_{21,t} & 1 & 0 \\ -\beta_{31,t} & -\beta_{32,t} & 1 \end{pmatrix}.$$

The vector $\boldsymbol{v}_t$ of the error terms in the linear regressions (10.31) can be interpreted as a vector of orthogonal factors, whose covariance matrix is $G_t = \text{diag}(g_{1t}, \dots, g_{mt})$ with $g_{it} > 0$ for $i = 1, \dots, m$. We then obtain the so-called Cholesky decomposition of the covariance matrix of ϵ_t:

$$H_t = L_t G_t L_t'. \tag{10.32}$$

Note that the Cholesky decomposition also extends to positive semi-definite matrices.

Taking $H_t^{1/2} = L_t G_t^{1/2}$, we obtain $\boldsymbol{v}_t = \boldsymbol{G}_t^{1/2}\eta_t$. A simple parametric form of the volatility $\sqrt{g_{it}}$ of the ith factor v_{it} could be of the GARCH(1,1)-type

$$g_{it} = \omega_i + \alpha_i v_{i,t-1}^2 + \beta_i g_{i,t-1}. \tag{10.33}$$

In view of the results of the Chapter 2, a necessary and sufficient condition for the existence of a strictly stationary process (v_t) in this parameterisation is

$$E \log(\alpha_i\eta_{it}^2 + \beta_i) < 0, \quad \text{for } i = 1, \dots, m. \tag{10.34}$$

Of course, Eq. (10.33) could also include explanatory variables $v_{j,t-1}^2$ with $j \neq i$, or other variables belonging $\mathcal{F}_{t-1}^{\epsilon}$.

To obtain a complete parametric specification for H_t, it now suffices to specify a time series model for the conditional betas, i.e. for the elements of L_t (or alternatively for those of B_t). Note that this model can be quite general because, to ensure the positive-definiteness of H_t, the conditional betas are not a priori subject to any constraint.

For instance, one can assume a dynamics of the form

$$\beta_{ij,t} = f_{ij}(v_{t-1}; \theta) + c_{ij}\beta_{ij,t-1} \tag{10.35}$$

where f_{ij} is a real-valued function, depending on v_{t-1} and on some parameter θ, and c_{ij} is a real coefficient. Under the conditions $|c_{ij}| < 1$, and the condition (10.34) ensuring the existence of the stationary sequence (v_t), there exists a stationary solution to Eq. (10.35). The Cholesky decomposition (10.32) then provides a model for the conditional covariance matrix H_t.

The absence of strong constraints on the coefficients constitutes an attractive feature of the Cholesky decomposition (10.32), compared in particular to the DCC decomposition $H_t = D_t R_t D_t'$ for which the unexplicit constraints of positive definiteness of R_t render the determination of stationarity conditions challenging. Another obvious interest of the Cholesky approach is to provide a direct way to predict the conditional betas, which appear to be of primary interest for particular financial applications (see Engle 2016).

Now, let us briefly explore the relationships between the Cholesky decomposition and the factor models. Suppose that the first $k<m$ columns of the matrix L_t are constant so that $L_t = [P:P_t]$ with P a non-random matrix of size $m\times k$ and P_t of size $m\times(m-r)$. The Cholesky GARCH can then be interpreted as a factor model:

$$\epsilon_t = Pf_t + \mathrm{u}_t,$$

where $f_t = (v_{1t},\dots, v_{kt})'$ is a vector of k orthogonal factors, P is called the loading matrix and $\mathrm{u}_t = P_t(v_{k+1,t},\dots, v_{mt})'$ is a so-called idiosyncratic disturbance (independent of f_t). If ϵ_t is a vector of excess returns (i.e. each component is the return of a risky asset minus the return of a risk free asset), and if the first component is the excess return of the market portfolio, in the one-factor case ($k = 1$) the previous factor model corresponds to the Capital Asset Pricing Model (CAPM) of Sharpe (1964), Lintner (1965), and Merton (1973). More precisely, we have $P = (1, \beta')'$, where β represents the 'sensitivity of returns to market returns', and $\mathrm{u}_t = (0, e_t')'$. Denoting by $r_t = (\epsilon_{2t},\dots,\epsilon_{mt})'$ the vector of excess returns, the last $(m-1)$ lines of the one-factor model gives the CAPM equation

$$r_t = \beta v_{1t} + e_t,$$

where v_{1t} is the market excess return.

10.3 Stationarity

In this section, we will first discuss the difficulty of establishing stationarity conditions, or the existence of moments, for MGARCH models. For the general vector model (10.9), and in particular for the BEKK model, there exist sufficient stationarity conditions. The stationary solution being non-explicit, we propose an algorithm that converges, under certain assumptions, to the stationary solution. We will then see that the problem is much simpler for the CCC model (10.18).

10.3.1 Stationarity of VEC and BEKK Models

It is not possible to provide stationary solutions, in explicit form, for the general VEC model (10.9). To illustrate the difficulty, recall that a univariate ARCH(1) model admits a solution $\epsilon_t = \sigma_t\eta_t$ with σ_t explicitly given as a function of $\{\eta_{t-u}, u>0\}$ as the square root of

$$\sigma_t^2 = \omega + \alpha\eta_{t-1}^2\sigma_{t-1}^2 = \omega\{1 + \alpha\eta_{t-1}^2 + \alpha^2\eta_{t-1}^2\eta_{t-2}^2 + \cdots\},$$

provided that the series converges almost surely. Now consider a bivariate model of the form (10.6) with $H_t = I_2 + \alpha\epsilon_{t-1}\epsilon_{t-1}'$, where α is assumed, for the sake of simplicity, to be scalar and positive. Also choose $H_t^{1/2}$ to be a lower triangular so as to have Eq. (10.7). Then

$$\begin{cases} h_{11,t} = 1 + \alpha h_{11,t-1}\eta_{1,t-1}^2 \\ h_{12,t} = \alpha h_{12,t-1}\eta_{1,t-1}^2 + \alpha(h_{11,t-1}h_{22,t-1} - h_{12,t-1}^2)^{1/2}\eta_{1,t-1}\eta_{2,t-1} \\ h_{22,t} = 1 + \alpha\dfrac{h_{12,t-1}^2}{h_{11,t-1}}\eta_{1,t-1}^2 + \alpha\dfrac{h_{11,t-1}h_{22,t-1}-h_{12,t-1}^2}{h_{11,t-1}}\eta_{2,t-1}^2 \\ \qquad +2\alpha\dfrac{h_{12,t-1}(h_{11,t-1}h_{22,t-1}-h_{12,t-1}^2)^{1/2}}{h_{11,t-1}}\eta_{1,t-1}\eta_{2,t-1}. \end{cases}$$

It can be seen that given η_{t-1}, the relationship between $h_{11,t}$ and $h_{11,t-1}$ is linear, and can be iterated to yield

$$h_{11,t} = 1 + \sum_{i=1}^{\infty} \alpha^i \prod_{j=1}^{i} \eta_{1,t-j}^2$$

under the constraint $\alpha < \exp(-E \log \eta_{1t}^2)$. In contrast, the relationships between $h_{12,t}$, or $h_{22,t}$, and the components of H_{t-1} are not linear, which makes it impossible to express $h_{12,t}$ and $h_{22,t}$ as a simple function of α, $\{\eta_{t-1}, \eta_{t-2}, \dots, \eta_{t-k}\}$ and H_{t-k} for $k \geq 1$. This constitutes a major obstacle for determining sufficient stationarity conditions.

Remark 10.5 (Stationarity does not follow from the ARMA model) Similar to (10.22), letting $\nu_t = \text{vech}(\epsilon_t\epsilon_t') - \text{vech}(H_t)$, we obtain the ARMA representation

$$\text{vech}(\epsilon_t\epsilon_t') = \omega + \sum_{i=1}^{r} C^{(i)}\text{vech}(\epsilon_{t-i}\epsilon_{t-i}') + \nu_t - \sum_{j=1}^{p} B^{(j)}\nu_{t-j},$$

by setting $C^{(i)} = A^{(i)} + B^{(i)}$ and by using the usual notation and conventions. In the literature, one may encounter the argument that the model is weakly stationary if the polynomial $z \mapsto \det\left(I_s - \sum_{i=1}^{r} C^{(i)}z^i\right)$ has all its roots outside the unit circle ($s = m(m+1)/2$). Although the result is certainly true with additional assumptions on the noise density (see Theorem 10.5 and the subsequent discussion), the argument is not correct since

$$\text{vech}(\epsilon_t\epsilon_t') = \left(\sum_{i=1}^{r} C^{(i)}B^i\right)^{-1}\left(\omega + \nu_t - \sum_{j=1}^{p} B^{(j)}\nu_{t-j}\right)$$

constitutes a solution only if $\nu_t = \text{vech}(\epsilon_t\epsilon_t') - \text{vech}(H_t)$ can be expressed as a function of $\{\eta_{t-u}, u > 0\}$.

Boussama, Fuchs and Stelzer (2011) (see also Boussama (2006)) obtained the following stationarity condition. Recall that $\rho(A)$ denotes the spectral radius of a square matrix A.

Theorem 10.5 (Stationarity and ergodicity) *There exists a strictly stationary and non-anticipative solution of the vector GARCH model (10.9), if:*

(i) the positivity condition (10.15) is satisfied;

(ii) the distribution of η has a density, positive on a neighbourhood of 0, with respect to the Lebesgue measure on $\mathbb{R}^m$;

(iii) $\rho\left(\sum_{i=1}^{r} C^{(i)}\right) < 1$.

This solution is unique, β-mixing, second-order stationary and ergodic.

In the particular case of the BEKK model of Definition 10.5, condition (iii) takes the form

$$\rho\left(\text{D}_m^+ \sum_{k=1}^{K}\sum_{i=1}^{r}(A_{ik} \otimes A_{ik} + B_{ik} \otimes B_{ik})\text{D}_m\right) < 1.$$

The proof of Theorem 10.5 relies on sophisticated algebraic tools. Assumption (ii) is a standard technical condition for showing the β-mixing property (but is of no use for stationarity). Note that condition (iii), written as $\sum_{i=1}^{r} \alpha_i + \beta_i < 1$ in the univariate case, is generally not necessary for the strict stationarity.

This theorem does not provide explicit stationary solutions, that is, a relationship between ϵ_t and the η_{t-i}. However, it is possible to construct an algorithm which, when it converges, allows a stationary solution to the vector GARCH model (10.9) to be defined.

Construction of a Stationary Solution

For any $t, k \in \mathbb{Z}$, we define

$$\epsilon_t^{(k)} = H_t^{(k)} = 0, \quad \text{when } k < 0,$$

and, recursively on $k \geq 0$,

$$\text{vech}(H_t^{(k)}) = \omega + \sum_{i=1}^{q} A^{(i)} \text{vech}(\epsilon_{t-i}^{(k-i)} \epsilon_{t-i}^{(k-i)\prime}) + \sum_{j=1}^{p} B^{(j)} \text{vech}(H_{t-j}^{(k-j)}), \tag{10.36}$$

with $\epsilon_t^{(k)} = H_t^{(k)\,1/2} \eta_t$.

Observe that, for $k \geq 1$,

$$H_t^{(k)} = f_k(\eta_{t-1}, \dots, \eta_{t-k}) \quad \text{and} \quad EH_t^{(k)} = H^{(k)},$$

where f_k is a measurable function and $H^{(k)}$ is a square matrix. The processes $(H_t^{(k)\,1/2})_t$ and $(\epsilon_t^{(k)})_t$ are thus stationary with components in the Banach space L^2 of the (equivalence classes of) square integral random variables. It is then clear that Eq. (10.9) admits a strictly stationary solution, which is non-anticipative and ergodic, if, for all t,

$$H_t^{(k)} \text{ converges almost surely when } k \to \infty. \tag{10.37}$$

Indeed, letting $H_t^{1/2} = \lim_{k\to\infty} H_t^{(k)\,1/2}$ and $\epsilon_t = H_t^{1/2}\eta_t$, and taking the limit of each side of (10.36), we note that (10.9) is satisfied. Moreover, (ϵ_t) constitutes a strictly stationary and non-anticipative solution, because ϵ_t is a measurable function of $\{\eta_u, u \leq t\}$. In view of Theorem A.1, such a process is also ergodic. Note also that if H_t exists, it is symmetric and positive definite because the matrices $H_t^{(k)}$ are symmetric and satisfy

$$\lambda' H_t^{(k)} \lambda \geq \lambda' \Omega \lambda > 0, \quad \text{for } \lambda \neq 0.$$

This solution (ϵ_t) is also second-order stationary if

$$H_t^{(k)} \text{ converges in } L^1 \text{ when } k \to \infty. \tag{10.38}$$

Let

$$\Delta_t^{(k)} = \text{vech}|H_t^{(k)} - H_t^{(k-1)}|.$$

From Exercise 10.8 and its proof, we obtain (10.37), and hence the existence of strictly stationary solution to the vector GARCH (10.9), if there exists $\rho \in (0, 1)$ such that $\|\Delta_t^{(k)}\| = O(\rho^k)$ almost surely as $k \to \infty$, which is equivalent to

$$\lim_{k\to\infty} \frac{1}{k} \log \|\Delta_t^{(k)}\| < 0, \quad \text{a.s.} \tag{10.39}$$

Similarly, we obtain (10.38) if $\|E\Delta_t^{(k)}\| = O(\rho^k)$. The criterion in (10.39) is not very explicit but the left-hand side of the inequality can be evaluated by simulation, just as for a Lyapunov coefficient.

10.3.2 Stationarity of the CCC Model

In model (10.18), letting $\tilde{\eta}_t = R^{1/2}\eta_t$, we get

$$\underline{\epsilon}_t = \Upsilon_t \underline{h}_t, \quad \text{where } \Upsilon_t = \begin{pmatrix} \tilde{\eta}_{1t}^2 & 0 & \dots & 0 \\ 0 & \ddots & & \\ \vdots & & \ddots & \\ 0 & \dots & & \tilde{\eta}_{mt}^2 \end{pmatrix}.$$

Multiplying by Υ_t the equation for $\underline{h}_t$, we thus have

$$\underline{\epsilon}_t = \Upsilon_t\underline{\omega} + \sum_{i=1}^{q} \Upsilon_t \mathbf{A}_i \underline{\epsilon}_{t-i} + \sum_{j=1}^{p} \Upsilon_t \mathbf{B}_j \underline{h}_{t-j},$$

which can be written

$$\underline{z}_t = \underline{b}_t + A_t \underline{z}_{t-1}, \tag{10.40}$$

where

$$\underline{b}_t = \underline{b}(\eta_t) = \begin{pmatrix} \Upsilon_t\underline{\omega} \\ 0 \\ \vdots \\ \underline{\omega} \\ 0 \\ \vdots \\ 0 \end{pmatrix} \in \mathbb{R}^{m(p+q)}, \quad \underline{z}_t = \begin{pmatrix} \underline{\epsilon}_t \\ \vdots \\ \underline{\epsilon}_{t-q+1} \\ \underline{h}_t \\ \vdots \\ \underline{h}_{t-p+1} \end{pmatrix} \in \mathbb{R}^{m(p+q)},$$

and

$$A_t = \begin{pmatrix}
\Upsilon_t\mathbf{A}_1 & & \cdots & & \Upsilon_t\mathbf{A}_q & \Upsilon_t\mathbf{B}_1 & & \cdots & & \Upsilon_t\mathbf{B}_p \\
I_m & 0 & \cdots & & 0 & 0 & & \cdots & & 0 \\
0 & I_m & \cdots & & 0 & 0 & & \cdots & & 0 \\
\vdots & \ddots & \ddots & & \vdots & \vdots & \ddots & \ddots & & \vdots \\
0 & & \cdots & I_m & 0 & 0 & & \cdots & 0 & 0 \\
\mathbf{A}_1 & & \cdots & & \mathbf{A}_q & \mathbf{B}_1 & & \cdots & & \mathbf{B}_p \\
0 & & \cdots & & 0 & I_m & 0 & \cdots & & 0 \\
0 & & \cdots & & 0 & 0 & I_m & \cdots & & 0 \\
\vdots & \ddots & \ddots & & \vdots & \vdots & \ddots & \ddots & & \vdots \\
0 & & \cdots & 0 & 0 & 0 & & \cdots & I_m & 0
\end{pmatrix} \tag{10.41}$$

is a $(p+q)m \times (p+q)m$ matrix.

We obtain a vector representation, analogous to (2.16) obtained in the univariate case. This allows us to state the following result.

Theorem 10.6 (Strict stationarity of the CCC model) *A necessary and sufficient condition for the existence of a strictly stationary and non-anticipative solution process for model (10.18) is $\gamma < 0$, where γ is the top Lyapunov exponent of the sequence $\{A_t, t \in \mathbb{Z}\}$ defined in (10.41). This stationary and non-anticipative solution, when $\gamma < 0$, is unique and ergodic.*

Proof. The proof is similar to that of Theorem 2.4. The variables η_t admitting a variance, the condition $E \log^+ \|A_t\| < \infty$ is satisfied.

It follows that when $\gamma < 0$, the series

$$\underline{\tilde{z}}_t = \underline{b}_t + \sum_{n=0}^{\infty} A_t A_{t-1} \dots A_{t-n} \underline{b}_{t-n-1} \tag{10.42}$$

converges almost surely for all t. A strictly stationary solution to model (10.18) is obtained as $\epsilon_t = \{\mathrm{diag}(\underline{\tilde{z}}_{q+1,t})\}^{1/2} R^{1/2} \eta_t$, where $\underline{\tilde{z}}_{q+1,t}$ denotes the $(q+1)$th sub-vector of size m of $\underline{\tilde{z}}_t$. This solution is thus non-anticipative and ergodic. The proof of the uniqueness is exactly the same as in the univariate case.

The proof of the necessary part can also be easily adapted. From Lemma 2.1, it is sufficient to prove that $\lim_{t\to\infty} \|A_0 \dots A_{-t}\| = 0$. It suffices to show that, for $1 \le i \le p+q$,

$$\lim_{t\to\infty} A_0 \dots A_{-t}\underline{e}_i = 0, \quad \text{a.s.,} \tag{10.43}$$

where $\underline{e}_i = e_i \otimes I_m$ and e_i is the ith element of the canonical basis of $\mathbb{R}^{p+q}$, since any vector x of $\mathbb{R}^{m(p+q)}$ can be uniquely decomposed as $x = \sum_{i=1}^{p+q} \underline{e}_i x_i$, where $x_i \in \mathbb{R}^m$. As in the univariate case, the existence of a strictly stationary solution implies that $A_0 \dots A_{-k}\underline{b}_{-k-1}$ tends to 0, almost surely, as $k \to \infty$. It follows that, using the relation $\underline{b}_{-k-1} = \underline{e}_1 \Upsilon_{-k-1}\underline{\omega} + \underline{e}_{q+1}\underline{\omega}$, we have

$$\lim_{k\to\infty} A_0 \cdots A_{-k}\underline{e}_1 \Upsilon_{-k-1}\underline{\omega} = 0, \quad \lim_{k\to\infty} A_0 \dots A_{-k}\underline{e}_{q+1}\underline{\omega} = 0, \quad \text{a.s.} \tag{10.44}$$

Since the components of $\underline{\omega}$ are strictly positive, condition (10.43) thus holds for $i = q+1$. Using

$$A_{-k}\underline{e}_{q+i} = \Upsilon_{-k}\mathbf{B}_i\underline{e}_1 + \mathbf{B}_i\underline{e}_{q+1} + \underline{e}_{q+i+1}, \quad i = 1, \dots, p, \tag{10.45}$$

with the convention that $\underline{e}_{p+q+1} = 0$, for $i = 1$ we obtain

$$0 = \lim_{k\to\infty} A_0 \dots A_{-k}\underline{e}_{q+1} \ge \lim_{k\to\infty} A_0 \dots A_{-k+1}\underline{e}_{q+2} \ge 0,$$

where the inequalities are taken componentwise. Therefore, condition (10.43) holds true for $i = q+2$, and by induction, for $i = q+j$, $j = 1, \dots, p$ in view of Eq. (10.45). Since $\Upsilon_{k-1}\underline{\omega} > 0$ with positive probability, the first convergence in (10.44) implies that (10.43) holds for $i = 1$. Since $A_{-k}\underline{e}_i \ge \underline{e}_{i+1}$ for $i = 1, \dots, q-2$, we can show by an ascendent recursion that (10.43) holds for $i = 1, \dots, q-1$. Finally, since $A_{-k}\underline{e}_q = \Upsilon_{-k}\mathbf{A}_q\underline{e}_1 + \mathbf{A}_q\underline{e}_{q+1}$, (10.43) also holds for $i = q$. □

The following result provides a necessary strict stationarity condition which is simple to check.

Corollary 10.1 (Consequence of the strict stationarity) *Let γ denote the top Lyapunov exponent of the sequence $\{A_t, t \in \mathbb{Z}\}$ defined in condition (10.41). Consider the matrix polynomial defined by: $B(z) = I_m - z\boldsymbol{B}_1 - \cdots - z^p\boldsymbol{B}_p$, $z \in \mathbb{C}$. Let*

$$\mathbb{B} = \begin{pmatrix} \mathbf{B}_1 & \mathbf{B}_2 & \cdots & & \mathbf{B}_p \\ I_m & 0 & \cdots & & 0 \\ 0 & I_m & \cdots & & 0 \\ \vdots & \ddots & \ddots & & \vdots \\ 0 & & \cdots & I_m & 0 \end{pmatrix}.$$

Then, if $\gamma < 0$ the following equivalent properties hold:

1. *The roots of $\det B(z)$ are outside the unit disk.*
2. *$\rho(\mathbb{B}) < 1$.*

Proof. Because all the entries of the matrices A_t are positive, it is clear that γ is larger than the top Lyapunov exponent of the sequence (A_t^*) obtained by replacing the matrices $\mathbf{A}_i$ by 0 in A_t. It is easily seen that the top Lyapunov coefficient of (A_t^*) coincides with that of the constant sequence equal to $\mathbb{B}$, that is, with $\rho(\mathbb{B})$. It follows that $\gamma \ge \log \rho(\mathbb{B})$. Hence, $\gamma < 0$ entails that all the eigenvalues of $\mathbb{B}$ are outside the unit disk. Finally, in view of Exercise 10.14, the equivalence between the two properties follows from

$$\begin{aligned}\det(\mathbb{B} - \lambda I_{mp}) &= (-1)^{mp} \det\{\lambda^p I_m - \lambda^{p-1}\mathbf{B}_1 - \cdots - \lambda\mathbf{B}_{p-1} - \mathbf{B}_p\} \\ &= (-\lambda)^{mp} \det B\left(\frac{1}{\lambda}\right), \quad \lambda \ne 0.\end{aligned}$$

□

Corollary 10.2 *Suppose that $\gamma < 0$. Let ϵ_t be the strictly stationary and non-anticipative solution of model (10.18). There exists $s > 0$ such that $E\|\underline{h}_t\|^s < \infty$ and $E\|\epsilon_t\|^{2s} < \infty$.*

Proof. It is shown in the proof of Corollary 2.3 that the strictly stationary solution defined by condition (10.42) satisfies $E\|\underline{z}_t\|^s < \infty$ for some $s > 0$. The conclusion follows from $\|\underline{\epsilon}_t\| \leq \|\underline{z}_t\|$ and $\|\underline{h}_t\| \leq \|\underline{z}_t\|$. □

10.3.3 Stationarity of DCC models

Stationarity conditions for the DCC model, with the specification (10.19) for R_t, have been established by Fermanian and Malongo (2016). However, except in very specific cases, such conditions are non explicit. The corrected DCC model allows for more tractable conditions. Sufficient stationarity conditions – based on the results by Boussama, Fuchs and Stelzer (2011) – are provided for the cDDD model by Aielli (2013).

10.4 QML Estimation of General MGARCH

In this section, we define the quasi-maximum likelihood estimator (QMLE) of a general MGARCH model, and we provide high-level assumptions entailing its strong consistency and asymptotic normality (CAN). In the next section, these assumptions will be made more explicit for the particular case of the CCC-GARCH model.

Assume a general parametric conditional variance $H_t = H_t(\theta_0)$, with an unknown d-dimensional parameter θ_0. Let Θ be a compact parameter space which contains θ_0. For all $\theta = (\theta_1, \dots, \theta_d)' \in \Theta$, assume that

$$H_t(\theta) = H(\epsilon_{t-1}, \epsilon_{t-2}, \dots ; \theta) \text{ is a well-defined symmetric positive-definite matrix.} \tag{10.46}$$

For particular MGARCH models, it is possible to write H_t as a measurable function of $\{\eta_u, u < t\}$, which entails that (ϵ_t) is stationary and ergodic (see the ergodic theorem, Theorem A.1). In (10.46), the matrix $H_t(\theta)$ is written as a function of the past observations. A model satisfying this requirement is said to be invertible at θ. For prediction purposes, it is obviously necessary that a model of parameter θ_0 be invertible at θ_0. For estimating θ_0, it seems also crucial that the model possesses this property at any $\theta \in \Theta$, since the estimation methods are typically based on comparisons between $\epsilon_t \epsilon_t'$ and $H_t(\theta)$, for all observations ϵ_t and all values of $\theta \in \Theta$.

Given observations $\epsilon_1, \dots, \epsilon_n$, and arbitrary fixed initial values $\tilde{\epsilon}_i$ for $i \leq 0$, let the statistics

$$\tilde{H}_t(\theta) = H(\epsilon_{t-1}, \dots, \epsilon_1, \tilde{\epsilon}_0, \tilde{\epsilon}_{-1}, \dots ; \theta).$$

A QMLE of θ_0 is defined as any measurable solution $\hat{\theta}_n$ of

$$\hat{\theta}_n = \arg\min_{\theta \in \Theta} \sum_{n}^{t=1} \tilde{\ell}_t(\theta), \qquad \tilde{\ell}_t(\theta) = \epsilon_t' \tilde{H}_t^{-1}(\theta) \epsilon_t + \log|\tilde{H}_t(\theta)|. \tag{10.47}$$

Note that the QMLE would be simply the maximum likelihood estimation if the conditional distribution of ϵ_t was Gaussian with mean zero and variance H_t.

Asymptotic Properties of the QMLE

Let $\ell_t(\theta) = \epsilon_t' H_t^{-1}(\theta) \epsilon_t + \log|H_t(\theta)|$,

$$\mathbf{l}_n(\theta) = n^{-1} \sum_{n}^{t=1} \ell_t(\theta) \quad \text{and} \quad \tilde{\mathbf{l}}_n(\theta) = n^{-1} \sum_{n}^{t=1} \tilde{\ell}_t(\theta).$$

Recall that ρ denotes a generic constant belonging to [0, 1), and K denotes a positive constant or a positive random variable measurable with respect to $\{\epsilon_u, u<0\}$ (and thus which does not depend on n). The regularity conditions (which do not depend on the choice of the norms) are the following.

A1: $\sup_{\theta\in\Theta}\|\tilde{H}_t^{-1}(\theta)\| \leq K, \quad \sup_{\theta\in\Theta}\|H_t^{-1}(\theta)\| \leq K$, a.s.

A2: $\sup_{\theta\in\Theta} \| H_t(\theta) - H_t(\theta) \| \leq K\rho^t$, a.s.

A3: $E\|\epsilon_t{}^s\| < \infty$ and $E\|H_t(\theta_0)\|^s < \infty$ for some $s > 0$.

A4: For $\theta \in \Theta$, $H_t(\theta) = H_t(\theta_0)$ a.s. implies $\theta = \theta_0$.

A5: For any sequence $x_1, x_2, \ldots$ of vectors of $\mathbb{R}^m$, the function $\theta \mapsto H(x_1, x_2, \ldots; \theta)$ is continuous on Θ.

A6: θ_0 belongs to the interior $\overset{\circ}{\Theta}$ of Θ.

A7: For any sequence $x_1, x_2, \ldots$ of vectors of $\mathbb{R}^m$, the function $\theta \mapsto H(x_1, x_2, \ldots; \theta)$ admits continuous second-order derivatives.

A8: For some neighbourhood $V(\theta_0)$ of θ_0,

$$\sup_{\theta\in V(\theta_0)} \sqrt{n} \left| \frac{\partial \mathbf{l}_n(\theta)}{\partial\theta} - \frac{\partial \tilde{\mathbf{l}}_n(\theta)}{\partial\theta} \right| = o_P(1).$$

A9: For some neighbourhood $V(\theta_0)$ of θ_0, for all $i, j \in \{1, \ldots, m\}$ and $p > 1$, $q > 2$ and $r > 2$ such that $2q^{-1} + 2r^{-1} = 1$ and $p^{-1} + 2r^{-1} = 1$, we have[5]

$$E \sup_{\theta\in V(\theta_0)} \left\| H_t^{-1/2}(\theta) \frac{\partial^2 H_t(\theta)}{\partial\theta_i\partial\theta_j} H_t^{-1/2'}(\theta) \right\|^p < \infty,$$

$$E \sup_{\theta\in V(\theta_0)} \left\| H_t^{-1/2}(\theta) \frac{\partial H_t(\theta)}{\partial\theta_i} H_t^{-1/2'}(\theta) \right\|^q < \infty,$$

$$E \sup_{\theta\in V(\theta_0)} \|H_t^{-1/2}(\theta)H_t^{1/2}(\theta_0)\|^r < \infty.$$

A10: $E \| \eta_t\|^4 < \infty$.

A11: The matrices $\{\partial H_t(\theta_0)/\partial\theta_i, i = 1, \ldots, d\}$ are linearly independent with non-zero probability.

In the case $m = 1$, Assumption A1 requires that the volatility be bounded away from zero uniformly on Θ. For a standard GARCH satisfying $\omega > 0$ and Θ compact, the assumption is satisfied. Assumption A2 is related to the invertibility and entails that $H_t(\theta)$ is well estimated by the statistic $\tilde{H}_t(\theta)$ when t is large. For some models it will be useful to replace A2 by the weaker, but more complicated, assumption

A2': $\sup_{\theta\in\Theta} \| H_t(\theta) - \tilde{H}_t(\theta) \| \leq K\rho_t$, a.s. where ρ_t is a random variable satisfying $\sum_{t=1}^{\infty} (E\rho_t^{s(p-1)/2})^{1/p} < \infty$, for some $p > 1$ and $s \in (0, 1)$ satisfying A3.

This assumption, as well as A8, will be used to show that the choice of the initial values $\tilde{\epsilon}_i$ is asymptotically unimportant. Corollary 10.2 shows that, for some particular models, the strict stationarity implies the existence of marginal moments, as required in A3. Assumption A4 is an identifiability condition. As shown in Section 8.2, A6 is necessary to obtain the asymptotic normality (AN) of the QMLE. Assumptions A9 and A10 are used to show the existence of the information matrices $\boldsymbol{I}$ and $\boldsymbol{J}$

[5] We use the convention that $H_t^{-1/2'} = (H_t^{1/2})^{-1}$. It follows that $H_t^{-1/2}H_t^{1/2} \neq I_m$ in general.

involved in the sandwich form $\boldsymbol{J}^{-1}\boldsymbol{I}\boldsymbol{J}^{-1}$ of the asymptotic variance of the QMLE. Assumption A11 is used to show the invertibility of $\boldsymbol{J}$.

Theorem 10.7 (CAN of the QMLE) *Let (ϵ_t) be a stationary MGARCH process satisfying (10.6). Assume that Eq. (10.46) holds for all $\theta \in \Theta$ and that $H_t = H_t(\theta_0)$ is a measurable function of $\{\boldsymbol{\eta}_u, u < t\}$. Let $(\hat{\theta}_n)$ be a sequence of QML estimators satisfying condition (10.47). Under A1, A2 or A2′, and A3-A5 we have*

$$\hat{\theta}_n \to \theta_0, \quad \textit{almost surely} \text{ as } n \to \infty.$$

Under the additional Assumptions A6–A10, we have the existence of the $d \times d$ matrix

$$\boldsymbol{J} = E\boldsymbol{\Delta}_t'\{H_t^{-1}(\theta_0) \otimes H_t^{-1}(\theta_0)\}\boldsymbol{\Delta}_t, \quad \boldsymbol{\Delta}_t = \frac{\partial \text{vec} H_t(\theta_0)}{\partial \theta'},$$

and of the $d \times d$ matrix $\boldsymbol{I}$, whose generic term is

$$\boldsymbol{I}(i,j) = \text{Tr}\{\boldsymbol{K}E\boldsymbol{C}_{i,t}\boldsymbol{C}_{j,t}'\},$$

with $\boldsymbol{K} = E\,\text{vec}(\boldsymbol{I}_m - \boldsymbol{\eta}_t\boldsymbol{\eta}_t')\text{vec}'(\boldsymbol{I}_m - \boldsymbol{\eta}_t\boldsymbol{\eta}_t')$ and

$$\boldsymbol{C}_{i,t} = \{H_t^{-1/2}(\theta_0) \otimes H_t^{-1/2}(\theta_0)\}\text{vec}\frac{\partial H_t(\theta_0)}{\partial \theta_i}.$$

Moreover, under the additional Assumption A11, $\boldsymbol{J}$ is invertible and

$$\sqrt{n}(\hat{\theta}_n - \theta_0) \xrightarrow{\mathcal{L}} \mathcal{N}\{0, \boldsymbol{J}^{-1}\boldsymbol{I}\boldsymbol{J}^{-1}\} \text{ as } n \to \infty. \tag{10.48}$$

We also have the Bahadur representation

$$\hat{\theta}_n - \theta_0 = \boldsymbol{J}^{-1}\frac{1}{n}\sum_{t=1}^{n} \boldsymbol{\Lambda}_t \text{vec}(\boldsymbol{\eta}_t\boldsymbol{\eta}_t' - \boldsymbol{I}_m) + o_P(n^{-1/2}) \tag{10.49}$$

where $\boldsymbol{\Lambda}_t = \boldsymbol{\Delta}_t'\{H_t^{-1/2'}(\theta_0) \otimes H_t^{-1/2'}(\theta_0)\}$.

Remark 10.6 (Relative usefulness of (too) general CAN results) The interest of the previous theorem is to highlight that, under a set of regularity conditions that seem reasonable, the QMLE of a GARCH model is CAN. The section devoted to the CCC model will reveal, however, that the regularity conditions are not always straightforwardly verifiable on particular specifications. Moreover, one has to be aware that the MLE/QMLE is not always consistent. The statistical literature provides numerous examples of inconsistent MLE (see e.g. Le Cam 1990). In the framework of GARCH models, checking the irrelevance of the initial value is an important point that should not be neglected. The point is related to the invertibility. As seen in Chapter 4, for some stationary EGARCH processes, Assumption A2 (or even A2′) may not hold and the MLE may fail.

10.5 Estimation of the CCC Model

We now turn to the estimation of the m-dimensional CCC-GARCH(p, q) model by the quasi-maximum likelihood method. Recall that (ϵ_t) is called a CCC-GARCH(p, q) if it satisfies

$$\begin{cases} \epsilon_t = H_t^{1/2}\eta_t, \\ H_t = D_t R D_t, \quad D_t^2 = \text{diag}(\underline{h}_t), \\ \underline{h}_t = \underline{\omega} + \sum_{i=1}^{q} \mathbf{A}_i \underline{\epsilon}_{t-i} + \sum_{j=1}^{p} \mathbf{B}_j \underline{h}_{t-j}, \quad \underline{\epsilon}_t = (\epsilon_{1t}^2, \dots, \epsilon_{mt}^2)', \end{cases} \tag{10.50}$$

where R is a correlation matrix, $\underline{\omega}$ is a vector of size $m \times 1$ with strictly positive coefficients, the $\mathbf{A}_i$ and $\mathbf{B}_j$ are matrices of size $m \times m$ with positive coefficients, and (η_t) is a sequence of iid centred variables in $\mathbb{R}^m$ with identity covariance matrix.

As in the univariate case, the criterion is written as if the iid process were Gaussian.

The parameters are the coefficients of the matrices $\underline{\omega}$, $\mathbf{A}_i$ and $\mathbf{B}_j$, and the coefficients of the lower triangular part (excluding the diagonal) of the correlation matrix $R = (\rho_{ij})$. The number of unknown parameters is thus

$$s_0 = m + m^2(p+q) + \frac{m(m-1)}{2}.$$

The parameter vector is denoted by

$$\theta = (\theta_1, \dots, \theta_{s_0})' = (\underline{\omega}', \alpha_1', \dots, \alpha_q', \beta_1', \dots, \beta_p', \rho')' := (\underline{\omega}', \alpha', \beta', \rho')',$$

where $\rho' = (\rho_{21}, \dots, \rho_{m1}, \rho_{32}, \dots, \rho_{m2}, \dots, \rho_{m,m-1})$, $\alpha_i = \mathrm{vec}(\mathbf{A}_i)$, $i = 1, \dots, q$, and $\beta_j = \mathrm{vec}(\mathbf{B}_j)$, $j = 1, \dots, p$. The parameter space is a subspace Θ of

$$(0, +\infty)^m \times (0, \infty)^{m^2(p+q)} \times (-1, 1)^{m(m-1)/2}.$$

The true parameter value is denoted by

$$\theta_0 = (\underline{\omega}_0', \alpha_{01}', \dots, \alpha_{0q}', \beta_{01}', \dots, \beta_{0p}', \rho_0')' = (\underline{\omega}_0', \alpha_0', \beta_0', \rho_0')'.$$

Before detailing the estimation procedure and its properties, we discuss the conditions that need to be imposed on the matrices $\mathbf{A}_i$ and $\mathbf{B}_j$ in order to ensure the uniqueness of the parameterisation.

10.5.1 Identifiability Conditions

Let $\mathcal{A}_\theta(z) = \sum_{i=1}^q \mathbf{A}_i z^i$ and $\mathcal{B}_\theta(z) = I_m - \sum_{j=1}^p \mathbf{B}_j z^j$. By convention, $\mathcal{A}_\theta(z) = 0$ if $q = 0$ and $\mathcal{B}_\theta(z) = I_m$ if $p = 0$.

If $\mathcal{B}_\theta(z)$ is non-singular, that is, if the roots of $\det(\mathcal{B}_\theta(z)) = 0$ are outside the unit disk, we deduce from $\mathcal{B}_\theta(B)\underline{h}_t = \underline{\omega} + \mathcal{A}_\theta(B)\underline{\epsilon}_t$ the representation

$$\underline{h}_t = \mathcal{B}_\theta(1)^{-1}\underline{\omega} + \mathcal{B}_\theta(B)^{-1}\mathcal{A}_\theta(B)\underline{\epsilon}_t. \tag{10.51}$$

In the vector case, assuming that the polynomials $\mathcal{A}_{\theta_0}$ and $\mathcal{B}_{\theta_0}$ have no common root is insufficient to ensure that there exists no other pair $(\mathcal{A}_\theta, \mathcal{B}_\theta)$, with the same degrees (p, q), such that

$$\mathcal{B}_\theta(B)^{-1}\mathcal{A}_\theta(B) = \mathcal{B}_{\theta_0}(B)^{-1}\mathcal{A}_{\theta_0}(B). \tag{10.52}$$

This condition is equivalent to the existence of an operator $U(B)$ such that

$$\mathcal{A}_\theta(B) = U(B)\mathcal{A}_{\theta_0}(B) \quad \text{and} \quad \mathcal{B}_\theta(B) = U(B)\mathcal{B}_{\theta_0}(B), \tag{10.53}$$

this common factor vanishing in $\mathcal{B}_\theta(B)^{-1}\mathcal{A}_\theta(B)$ (Exercise 10.2).

The polynomial $U(B)$ is called *unimodular* if $\det\{U(B)\}$ is a non-zero constant. When the only common factors of the polynomials $P(B)$ and $Q(B)$ are unimodular, that is, when

$$P(B) = U(B)P_1(B), \quad Q(B) = U(B)Q_1(B) \Rightarrow \det\{U(B)\} = \text{constant},$$

then $P(B)$ and $Q(B)$ are called *left coprime*.

The following example shows that, in the vector case, assuming that $\mathcal{A}_{\theta_0}(B)$ and $\mathcal{B}_{\theta_0}(B)$ are left coprime is insufficient to ensure that condition (10.52) has no solution $\theta \neq \theta_0$ (in the univariate case this is sufficient because the condition $\mathcal{B}_\theta(0) = \mathcal{B}_{\theta_0}(0) = 1$ imposes $U(B) = U(0) = 1$).

Example 10.4 (Non-identifiable bivariate model) For $m = 2$, let

$$\mathcal{A}_{\theta_0}(B) = \begin{pmatrix} a_{11}(B) & a_{12}(B) \\ a_{21}(B) & a_{22}(B) \end{pmatrix}, \quad \mathcal{B}_{\theta_0}(B) = \begin{pmatrix} b_{11}(B) & b_{12}(B) \\ b_{21}(B) & b_{22}(B) \end{pmatrix},$$

$$U(B) = \begin{pmatrix} 1 & 0 \\ B & 1 \end{pmatrix},$$

with

$$\deg(a_{21}) = \deg(a_{22}) = q, \quad \deg(a_{11}) < q, \quad \deg(a_{12}) < q$$

and

$$\deg(b_{21}) = \deg(b_{22}) = p, \quad \deg(b_{11}) < p, \quad \deg(b_{12}) < p.$$

The polynomial $\mathcal{A}(B) = U(B)\mathcal{A}_{\theta_0}(B)$ has the same degree q as $\mathcal{A}_{\theta_0}(B)$, and $\mathcal{B}(B) = U(B)\mathcal{B}_{\theta_0}(B)$ is a polynomial of the same degree p as $\mathcal{B}_{\theta_0}(B)$. On the other hand, $U(B)$ has a nonzero determinant which is independent of B, hence it is unimodular. Moreover, $\mathcal{B}(0) = \mathcal{B}_{\theta_0}(0) = I_m$ and $\mathcal{A}(0) = \mathcal{A}_{\theta_0}(0) = 0$. It is thus possible to find θ such that $\mathcal{B}(B) = \mathcal{B}_\theta(B)$, $\mathcal{A}(B) = \mathcal{A}_\theta(B)$ and $\underline{\omega} = U(1)\underline{\omega_0}$. The model is thus non-identifiable, θ and θ_0 corresponding to the same representation (10.51).

Identifiability can be ensured by several types of conditions; see Reinsel (1997, pp. 37–40), Hannan (1976) or Hannan and Deistler (1988, Section 2.7). To obtain a mild condition define, for any column i of the matrix operators $\mathcal{A}_\theta(B)$ and $\mathcal{B}_\theta(B)$, the maximal degrees $q_i(\theta)$ and $p_i(\theta)$, respectively. Suppose that maximal values are imposed for these orders, that is,

$$\forall \theta \in \Theta, \forall i = 1, \dots, m, \quad q_i(\theta) \le q_i \quad \text{and} \quad p_i(\theta) \le p_i, \tag{10.54}$$

where $q_i \le q$ and $p_i \le p$ are fixed integers. Denote by $a_{q_i}(i)$ (resp. $b_{p_i}(i)$) the column vector of the coefficients of B^{q_i} (resp. B^{p_i}) in the ith column of $\mathcal{A}_{\theta_0}(B)$ (resp. $\mathcal{B}_{\theta_0}(B)$).

Example 10.5 (Illustration of the notation) For

$$\mathcal{A}_{\theta_0}(B) = \begin{pmatrix} 1 + a_{11}B^2 & a_{12}B \\ a_{21}B^2 + a_{21}^*B & 1 + a_{22}B \end{pmatrix}, \quad \mathcal{B}_{\theta_0}(B) = \begin{pmatrix} 1 + b_{11}B^4 & b_{12}B \\ b_{21}B^4 & 1 + b_{22}B \end{pmatrix},$$

with $a_{11}a_{21}a_{12}a_{22}b_{11}b_{21}b_{12}b_{22} \neq 0$, we have

$$q_1(\theta_0) = 2, \quad q_2(\theta_0) = 1, \quad p_1(\theta_0) = 4, \quad p_2(\theta_0) = 1$$

and

$$a_2(1) = \begin{pmatrix} a_{11} \\ a_{21} \end{pmatrix}, \quad a_1(2) = \begin{pmatrix} a_{12} \\ a_{22} \end{pmatrix}, \quad b_4(1) = \begin{pmatrix} b_{11} \\ b_{21} \end{pmatrix}, \quad b_1(2) = \begin{pmatrix} b_{12} \\ b_{22} \end{pmatrix}.$$

Proposition 10.1 (A simple identifiability condition) *If the matrix*

$$M(\mathcal{A}_{\theta_0}, \mathcal{B}_{\theta_0}) = [a_{q_1}(1) \cdots a_{q_m}(m) b_{p_1}(1) \cdots b_{p_m}(m)] \tag{10.55}$$

has full rank m, the parameters α_0 and β_0 are identified by the constraints (10.54) with $q_i = q_i(\theta_0)$ and $p_i = p_i(\theta_0)$ for any value of i.

Proof. From the proof of the theorem in Hannan (1969), $U(B)$ satisfying condition (10.54) is a unimodular matrix of the form $U(B) = U_0 + U_1B + \cdots + U_kB^k$. Since the term of highest degree (column by column) of $\mathcal{A}_{\theta_0}(B)$ is $[a_{q_1}(1)B^{q_1} \cdots a_{q_m}(m)B^{q_m}]$, the ith column of $\mathcal{A}_\theta(B) = U(B)\mathcal{A}_{\theta_0}(B)$ is a polynomial in B of degree less than q_i if and only if $U_j a_{q_i}(i) = 0$, for $j = 1, \dots, k$. Similarly, we must have $U_j b_{p_i}(i) = 0$, for $j = 1, \dots, k$ and $i = 1, \dots m$. It follows that $U_j M(\mathcal{A}_{\theta_0}, \mathcal{B}_{\theta_0}) = 0$, which implies that $U_j = 0$

for $j = 1, \dots, k$ thanks to condition (10.55). Consequently $U(B) = U_0$ and, since, for all θ, $\mathcal{B}_\theta(0) = I_m$, we have $U(B) = I_m$. □

Example 10.6 (Illustration of the identifiability condition) In Example 10.4,

$$M(\mathcal{A}_{\theta_0}, \mathcal{B}_{\theta_0}) = [a_q(1)a_q(2)b_p(1)b_p(2)] = \begin{bmatrix} 0 & 0 & 0 & 0 \\ \times & \times & \times & \times \end{bmatrix}$$

is not a full-rank matrix. Hence, the identifiability condition of Proposition 10.1 is not satisfied. Indeed, the model is not identifiable.

A simpler, but more restrictive, condition is obtained by imposing the requirement that

$$M_1(\mathcal{A}_{\theta_0}, \mathcal{B}_{\theta_0}) = \begin{bmatrix} \mathbf{A}_q & \mathbf{B}_p \end{bmatrix}$$

has full rank m. This entails uniqueness under the constraint that the degrees of $\mathcal{A}_\theta$ and $\mathcal{B}_\theta$ are less than q and p, respectively.

Example 10.7 (Another illustration of the identifiability condition) Turning again to Example 10.5 with $a_{12}b_{21} = a_{22}b_{11}$ and, for instance, $a_{21} = 0$ and $a_{22} \neq 0$, observe that the matrix

$$M_1(\mathcal{A}_{\theta_0}, \mathcal{B}_{\theta_0}) = \begin{bmatrix} 0 & a_{12} & b_{11} & 0 \\ 0 & a_{22} & b_{21} & 0 \end{bmatrix}$$

does not have full rank, but the matrix

$$M(\mathcal{A}_{\theta_0}, \mathcal{B}_{\theta_0}) = \begin{bmatrix} a_{11} & a_{12} & b_{11} & b_{12} \\ 0 & a_{22} & b_{21} & b_{22} \end{bmatrix}$$

does have full rank.

More restrictive forms, such as the echelon form, are sometimes required to ensure identifiability.

10.5.2 Asymptotic Properties of the QMLE of the CCC-GARCH model

For particular classes of MGARCH models, the assumptions of the previous section can be made more explicit. Let us consider the CCC-GARCH model. Let $(\epsilon_1, \dots, \epsilon_n)$ be an observation of length n of the unique non-anticipative and strictly stationary solution (ϵ_t) of model (10.50). Conditionally on non-negative initial values $\epsilon_0, \dots, \epsilon_{1-q}, \underline{\tilde{h}}_0, \dots, \underline{\tilde{h}}_{1-p}$, the Gaussian quasi-likelihood is written as

$$L_n(\theta) = L_n(\theta; \epsilon_1, \dots, \epsilon_n) = \prod_{t=1}^{n} \frac{1}{(2\pi)^{m/2}|\tilde{\boldsymbol{H}}_t|^{1/2}} \exp\left(-\frac{1}{2}\epsilon_t' \tilde{\boldsymbol{H}}_t^{-1} \epsilon_t\right),$$

where the $\tilde{\boldsymbol{H}}_t$ are recursively defined, for $t \geq 1$, by

$$\begin{cases} \tilde{\boldsymbol{H}}_t = \tilde{\boldsymbol{D}}_t \boldsymbol{R} \tilde{\boldsymbol{D}}_t, \quad \tilde{\boldsymbol{D}}_t = \{\mathrm{diag}(\underline{\tilde{h}}_t)\}^{1/2} \\ \underline{\tilde{h}}_t = \underline{\tilde{h}}_t(\theta) = \underline{\omega} + \sum_{i=1}^{q} \mathbf{A}_i \underline{\epsilon}_{t-i} + \sum_{j=1}^{p} \mathbf{B}_j \underline{\tilde{h}}_{t-j}. \end{cases}$$

A QMLE of $\boldsymbol{\theta}$ is defined as in (10.47) by:

$$\hat{\theta}_n = \mathrm{argmax}_{\theta \in \Theta} L_n(\theta) = \mathrm{argmin}_{\theta \in \Theta} \tilde{\mathbf{l}}_n(\theta). \tag{10.56}$$

Remark 10.7 (Choice of initial values) It will be shown later that, as in the univariate case, the initial values have no influence on the asymptotic properties of the estimator. These initial values can be fixed, for instance, so that

$$\underline{\epsilon}_0 = \cdots = \underline{\epsilon}_{1-q} = \underline{\tilde{h}}_1 = \cdots = \underline{\tilde{h}}_{1-p} = 0.$$

They can also be chosen as functions of $\boldsymbol{\theta}$, such as

$$\underline{\epsilon}_0 = \cdots = \underline{\epsilon}_{1-q} = \underline{\tilde{h}}_1 = \cdots = \underline{\tilde{h}}_{1-p} = \underline{\omega},$$

or as random variable functions of the observations, such as

$$\underline{\tilde{h}}_t = \underline{\epsilon}_t = \begin{pmatrix} \epsilon_{1t}^2 \\ \vdots \\ \epsilon_{mt}^2 \end{pmatrix}, \quad t = 0, -1, \ldots, 1-r,$$

where the first $r = \max\{p, q\}$ observations are denoted by $\epsilon_{1-r}, \ldots, \epsilon_0$.

Let $\gamma(\mathbf{A}_0)$ denote the top Lyapunov coefficient of the sequence of matrices $\mathbf{A}_0 = (A_{0t})$ defined as in (10.41), at $\boldsymbol{\theta} = \boldsymbol{\theta}_0$. The following assumptions will be used to establish the strong consistency of the QMLE.

CC1: $\theta_0 \in \boldsymbol{\Theta}$ and $\boldsymbol{\Theta}$ is compact.

CC2: $\gamma(\mathbf{A}_0) < 0$ and, for all $\theta \in \boldsymbol{\Theta}$, $\det \mathcal{B}(z) = 0 \Rightarrow |z| > 1$.

CC3: The components of $\boldsymbol{\eta}_t$ are independent and their squares have non-degenerate distributions.

CC4: If $p > 0$, then $\mathcal{A}_{\theta_0}(z)$ and $\mathcal{B}_{\theta_0}(z)$ are left coprime and $M_1(\mathcal{A}_{\theta_0}, \mathcal{B}_{\theta_0})$ has full rank m.

CC5: $\boldsymbol{R}$ is a positive definite correlation matrix for all $\theta \in \boldsymbol{\Theta}$.

If the space $\boldsymbol{\Theta}$ is constrained by (10.54), that is, if maximal orders are imposed for each component of $\underline{\epsilon}_t$ and $\underline{h}_t$ in each equation, then Assumption CC4 can be replaced by the following more general condition:

CC4′: If $p > 0$, then $\mathcal{A}_{\theta_0}(z)$ and $\mathcal{B}_{\theta_0}(z)$ are left coprime and $M(\mathcal{A}_{\theta_0}, \mathcal{B}_{\theta_0})$ has full rank m.

It will be useful to approximate the sequence $(\tilde{\ell}_t(\theta))$ by an ergodic and stationary sequence. Assumption CC2 implies that, for all $\theta \in \boldsymbol{\Theta}$, the roots of $\mathcal{B}_\theta(z)$ are outside the unit disk. Denote by $(\underline{h}_t)_t = \{\underline{h}_t(\theta)\}_t$ the strictly stationary, non-anticipative and ergodic solution of

$$\underline{h}_t = \underline{\omega} + \sum_{i=1}^{q} \mathbf{A}_i \underline{\epsilon}_{t-i} + \sum_{j=1}^{p} \mathbf{B}_j \underline{h}_{t-j}, \quad \forall t. \tag{10.57}$$

Now, letting $\boldsymbol{D}_t = \{\mathrm{diag}(\underline{h}_t)\}^{1/2}$ and $\boldsymbol{H}_t = \boldsymbol{D}_t \boldsymbol{R} \boldsymbol{D}_t$, we define

$$\mathbf{l}_n(\theta) = \mathbf{l}_n(\theta; \epsilon_n, \epsilon_{n-1} \ldots,) = n^{-1} \sum_{t=1}^{n} \ell_t, \quad \ell_t = \ell_t(\theta) = \epsilon_t' \boldsymbol{H}_t^{-1} \epsilon_t + \log |\boldsymbol{H}_t|.$$

Let $\boldsymbol{H}_t^{1/2} = \boldsymbol{D}_t \boldsymbol{R}^{1/2}$ and $\tilde{\boldsymbol{H}}_t^{1/2} = \tilde{\boldsymbol{D}}_t \boldsymbol{R}^{1/2}$, where $\boldsymbol{R}^{1/2}$ is the symmetric positive definite square root of $\boldsymbol{R}$. We also set $\boldsymbol{H}_t^{-1/2} = \boldsymbol{R}^{-1/2} \boldsymbol{D}_t^{-1}$ and $\tilde{\boldsymbol{H}}_t^{-1/2} = \boldsymbol{R}^{-1/2} \tilde{\boldsymbol{D}}_t^{-1}$. We are now in a position to state the following consistency theorem.

Theorem 10.8 (Strong consistency) *Let $(\hat{\theta}_n)$ be a sequence of QMLEs satisfying (10.56). Then, under CC1–CC5 (or CC1–CC3, CC4′ and CC5),*

$$\hat{\theta}_n \to \theta_0, \quad \textit{almost surely when } n \to \infty.$$

To establish the AN we require the following additional assumptions:

CC6: $\theta_0 \in \overset{\circ}{\boldsymbol{\Theta}}$, where $\overset{\circ}{\boldsymbol{\Theta}}$ is the interior of $\boldsymbol{\Theta}$.

CC7: $E \| \eta_t \eta_t' \|^2 < \infty$.

Theorem 10.9 (Asymptotic normality) *Under the assumptions of Theorem 10.8 and CC6–CC7, $\sqrt{n}(\hat{\theta}_n - \theta_0)$ converges in distribution to $\mathcal{N}(\mathbf{0}, \boldsymbol{J}^{-1}\boldsymbol{I}\boldsymbol{J}^{-1})$, where $\boldsymbol{J}$ is a positive definite matrix and $\boldsymbol{I}$ is a positive semi-definite matrix, defined by*

$$\boldsymbol{I} = E\left(\frac{\partial \ell_t(\theta_0)}{\partial \theta}\frac{\partial \ell_t(\theta_0)}{\partial \theta'}\right), \quad J = E\left(\frac{\partial^2 \ell_t(\theta_0)}{\partial \theta \partial \theta'}\right).$$

10.6 Looking for Numerically Feasible Estimation Methods

Despite the wide range of applicability of the QML estimation method, this approach may entail formidable numerical difficulties when the dimension of the vector of financial returns under study is large. In asset pricing applications or portfolio management, practitioners may have to handle cross sections of hundreds – even thousands – of stocks. Whatever the class of MGARCH used, the number of unknown parameters increases dramatically as the dimension of the cross section increases. This is particularly problematic when the Gaussian QML estimation method is used, because the high-dimensional conditional variance matrix has to be inverted at every step of the optimisation procedure.

In this section, we present two methods aiming at alleviating the dimensionality curse.

10.6.1 Variance Targeting Estimation

Variance targeting (VT) estimation is a two-step procedure in which the unconditional variance–covariance matrix of the returns vector process is estimated by a moment estimator in a first step. VT is based on a reparameterisation of the MGARCH model, in which the matrix of intercepts in the volatility equation is replaced by the unconditional covariance matrix.

To be more specific, consider the CCC-GARCH(p, q) model

$$\begin{cases} \epsilon_t = \boldsymbol{H}_t^{1/2}\boldsymbol{\eta}_t, \\ \boldsymbol{H}_t = \boldsymbol{D}_t\boldsymbol{R}_0\boldsymbol{D}_t, \quad \boldsymbol{D}_t^2 = \mathrm{diag}(\underline{\boldsymbol{h}}_t), \\ \underline{\boldsymbol{h}}_t = \omega_0 + \sum_{i=1}^{q} \boldsymbol{A}_{0i}\underline{\epsilon}_{t-i} + \sum_{j=1}^{p} \boldsymbol{B}_{0j}\underline{\boldsymbol{h}}_{t-j}, \end{cases} \tag{10.58}$$

where $\underline{\epsilon}_t = (\epsilon_{1t}^2, \dots, \epsilon_{mt}^2)'$ and $\boldsymbol{R}_0$ is a correlation matrix, $\boldsymbol{A}_{0i}$ and $\boldsymbol{B}_{0j}$ are $m \times m$ matrices with positive coefficients. Model (10.58) admits a strict and second-order non-anticipative stationary solution (ϵ_t) when

A: the spectral radius of $\sum_{i=1}^{q} \boldsymbol{A}_{0i} + \sum_{j=1}^{p} \boldsymbol{B}_{0j}$ is strictly less than 1.

Moreover, under this assumption, we have that

$$E\underline{\epsilon}_t = E\underline{\boldsymbol{h}}_t = \left\{\boldsymbol{I}_m - \sum_{i=1}^{r}(\boldsymbol{A}_{0i} + \boldsymbol{B}_{0i})\right\}^{-1} \omega_0 := \boldsymbol{h}_0. \tag{10.59}$$

It follows that the last equation in model (10.58) can be equivalently written

$$\underline{\boldsymbol{h}}_t - \boldsymbol{h}_0 = \sum_{i=1}^{q} \boldsymbol{A}_{0i}(\underline{\epsilon}_{t-i} - \boldsymbol{h}_0) + \sum_{j=1}^{p} \boldsymbol{B}_{0j}(\underline{\boldsymbol{h}}_{t-j} - \boldsymbol{h}_0). \tag{10.60}$$

With this new formulation, the generic parameter value consists of the coefficients of the vector $\boldsymbol{h}$ and the matrices $\boldsymbol{A}_i$ and $\boldsymbol{B}_j$ (corresponding to the true values $\boldsymbol{h}_0$, $\boldsymbol{A}_{0i}$ and $\boldsymbol{B}_{0j}$, respectively), and the coefficients of the lower triangular part (excluding the diagonal) of the correlation matrix $\boldsymbol{R} = (\rho_{ij})$. One advantage of this parameterisation over the initial one is that the parameter $\boldsymbol{h}_0$ can be straightforwardly estimated by the empirical mean

$$\hat{\boldsymbol{h}}_n = \frac{1}{n}\sum_{t=1}^{n} \underline{\epsilon}_t.$$

Table 10.1 Seconds of CPU time for computing the VTE and QMLE (average of 100 replications).

	$n = 500$			$n = 5000$		
	$m = 2$	$m = 3$	$m = 4$	$m = 2$	$m = 3$	$m = 4$
VTE	3.44	7.98	17.29	41.40	94.43	197.53
QMLE	5.48	13.82	25.17	65.22	145.41	284.85

Source: Francq, Horváth, and Zakoian (2016).

In the VT estimation method, the components of $\boldsymbol{h}$ are thus estimated empirically in a first step, and the other parameters are estimated in a second step, via a QML optimisation. Under appropriate assumptions, the resulting estimator is shown to be consistent and asymptotically normal (see Francq, Horváth, and Zakoian 2016).

One interest of this procedure is computational. Table 10.1 shows the reduction of computational time compared to the full quasi-maximum likelihood (FQML), when both the dimension m of the series and the sample size n vary. For both methods, the computation time increases rapidly with m, but the relative time-computation gain does not depend much on m, nor on n.

One issue is whether the time-computation gain is paid in terms of accuracy of the estimator. In the univariate case, $m = 1$ (that is, for a standard GARCH(p, q) model), it can be shown that the VTE is never asymptotically more efficient than the QMLE, regardless of the values of the GARCH parameters and the distribution of the iid process (see Francq, Horváth, and Zakoian 2011). In the multivariate setting, the asymptotic distributions of the two estimators are difficult to compare but, on simulation experiments, the accuracy loss entailed by the two-step VT procedure is often barely visible. In finite sample, the VT estimator may even perform much better than the QML estimator.

A nice feature of the VT estimator is that it ensures robust estimation of the marginal variance, provided that it exists. Indeed, the variance of a model estimated by VT converges to the theoretical variance, even if the model is mis-specified. For the convergence to hold true, it suffices that the observed process be stationary and ergodic with a finite second-order moment. This is generally not the case when the misspecified model is estimated by QML. For some specific purposes such as long-horizon prediction or long-term value-at-risk evaluation, the essential point is to well estimate the marginal distribution, in particular the marginal moments. The fact that the VTE guarantees a consistent estimation of the marginal variance may then be a crucial advantage over the QMLE.

10.6.2 Equation-by-Equation Estimation

Another approach for alleviating the dimensionality curse in the estimation of MGARCH models consists in estimating the individual volatilities in a first step – 'equation-by-equation' (EbE) – and to estimate the remaining parameters in a second step.

More precisely, any $\mathbb{R}^m$-valued process (ϵ_t) with zero-conditional mean and positive-definite conditional variance matrix $\boldsymbol{H}_t$ can be represented as

$$\begin{cases} \epsilon_t = \boldsymbol{H}_t^{1/2}\boldsymbol{\eta}_t, \quad E(\boldsymbol{\eta}_t|\mathcal{F}_{t-1}^{\epsilon}) = \boldsymbol{0}, \quad \mathrm{Var}(\boldsymbol{\eta}_t|\mathcal{F}_{t-1}^{\epsilon}) = \boldsymbol{I}_m, \\ \boldsymbol{H}_t = \boldsymbol{H}(\epsilon_{t-1}, \epsilon_{t-2}, \dots) = \boldsymbol{D}_t\boldsymbol{R}_t\boldsymbol{D}_t, \end{cases} \tag{10.61}$$

where $\mathcal{F}_{t-1}^{\epsilon}$ is the σ-field generated by $\{\epsilon_u, u < t\}$, $\boldsymbol{D}_t = \{\mathrm{diag}(\boldsymbol{H}_t)\}^{1/2}$ and $\boldsymbol{R}_t = \mathrm{Corr}(\epsilon_t, \epsilon_t|\mathcal{F}_{t-1}^{\epsilon})$. Let σ_{kt}^2 the kth diagonal element of $\boldsymbol{H}_t$, that is the variance of ϵ_{kt} conditional on $\mathcal{F}_{t-1}^{\epsilon}$. Assuming that σ_{kt}^2 is parameterised by some parameter $\boldsymbol{\theta}_0^{(k)} \in \mathbb{R}^{d_k}$, belonging to a compact parameter space $\boldsymbol{\Theta}_k$, we can write

$$\begin{cases} \epsilon_{kt} = \sigma_{kt}\eta_{kt}^*, \\ \sigma_{kt} = \sigma_k(\epsilon_{t-1}, \epsilon_{t-2}, \dots; \boldsymbol{\theta}_0^{(k)}), \end{cases} \tag{10.62}$$

where σ_k is a positive function. The vector $\boldsymbol{\eta}_t^* = (\eta_{1t}^*, \dots, \eta_{mt}^*) = \boldsymbol{D}_t^{-1}\epsilon_t = (\epsilon_{1t}/\sigma_{1t}, \dots, \epsilon_{mt}/\sigma_{mt})'$ satisfies $E(\boldsymbol{\eta}_t^* | \mathcal{F}_{t-1}^{\epsilon}) = \mathbf{0}$ and $\mathrm{Var}(\boldsymbol{\eta}_t^* | \mathcal{F}_{t-1}^{\epsilon}) = \boldsymbol{R}_t$. Because $E(\eta_{kt}^* | \mathcal{F}_{t-1}^{\epsilon}) = 0$ and $\mathrm{Var}(\eta_{kt}^* | \mathcal{F}_{t-1}^{\epsilon}) = 1$, $\boldsymbol{\eta}_t^*$ can be called the vector of EbE innovations of (ϵ_t).

It is important to note that model (10.62), satisfied by the components of (ϵ_t), is not a standard univariate GARCH. First, because the volatility σ_{kt} depends on the past of all components of ϵ_t – not only the past of ϵ_{kt}. And second, because the innovation sequence (η_{kt}^*) is not iid (except when the conditional correlation matrix is constant, $\boldsymbol{R}_t = \boldsymbol{R}$). For instance, a parametric specification of σ_{kt} mimicking the classical GARCH(1,1) is given by

$$\sigma_{kt}^2 = \omega_k + \sum_{\ell=1}^{m} \alpha_{k,\ell}\epsilon_{\ell,t-1}^2 + \beta_k \sigma_{k,t-1}^2, \quad \omega_k > 0, \ \alpha_{k,\ell} \geq 0, \ \beta_k \geq 0. \tag{10.63}$$

Of course, other formulations – for instance including leverage effects – could be considered as well.

Having parameterised the individual conditional variances of (ϵ_t), the model can be completed by specifying the dynamics of the conditional correlation matrix $\boldsymbol{R}_t$. For instance, in the DCC model of Engle (2002a), the conditional correlation matrix is modelled as a function of the past standardized returns, as follows

$$\boldsymbol{R}_t = \boldsymbol{Q}_t^{*-1/2}\boldsymbol{Q}_t\boldsymbol{Q}_t^{*-1/2}, \quad \boldsymbol{Q}_t = (1-\alpha-\beta)\boldsymbol{S} + \alpha\boldsymbol{\eta}_{t-1}^*\boldsymbol{\eta}_{t-1}^{*\prime} + \beta\boldsymbol{Q}_{t-1},$$

where $\alpha, \beta \geq 0$, $\alpha + \beta < 1$, $\boldsymbol{S}$ is a positive-definite matrix, and $\boldsymbol{Q}_t^*$ is the diagonal matrix with the same diagonal elements as $\boldsymbol{Q}_t$.

More generally, suppose that matrix $\boldsymbol{R}_t$ is parameterised by some parameter $\rho_0 \in \mathbb{R}^r$, together with the volatility parameter $\boldsymbol{\theta}_0$, as

$$\boldsymbol{R}_t = \boldsymbol{R}(\epsilon_{t-1}, \epsilon_{t-2}, \dots; \boldsymbol{\theta}_0, \rho_0) = \mathcal{R}(\boldsymbol{\eta}_{t-1}^*, \boldsymbol{\eta}_{t-2}^*, \dots; \rho_0).$$

Given observations $\epsilon_1, \dots, \epsilon_n$, and arbitrary initial values $\tilde{\epsilon}_i$ for $i \leq 0$, we define $\tilde{\sigma}_{kt}(\cdot) = \sigma_k(\epsilon_{t-1}, \epsilon_{t-2}, \dots, \epsilon_1, \tilde{\epsilon}_0, \tilde{\epsilon}_{-1}, \dots; \cdot)$ for $k = 1, \dots, m$. A two-step estimation procedure can be developed as follows.

1. *First step*: EbE estimation of the volatility parameters $\boldsymbol{\theta}_0^{(k)}$, by

$$\hat{\boldsymbol{\theta}}_n^{(k)} = \arg\min_{\boldsymbol{\theta}^{(k)} \in \boldsymbol{\Theta}_k} \tilde{Q}_n^{(k)}(\boldsymbol{\theta}^{(k)}), \quad \tilde{Q}_n^{(k)}(\boldsymbol{\theta}^{(k)}) = \frac{1}{n}\sum_{t=1}^{n} \log \tilde{\sigma}_{kt}^2(\boldsymbol{\theta}^{(k)}) + \frac{\epsilon_{kt}^2}{\tilde{\sigma}_{kt}^2(\boldsymbol{\theta}^{(k)})},$$

and extraction of the vectors of residuals $\hat{\boldsymbol{\eta}}_t^* = (\hat{\eta}_{1t}^*, \dots, \hat{\eta}_{mt}^*)'$ where $\hat{\eta}_{kt}^* = \tilde{\sigma}_{kt}^{-1}(\hat{\boldsymbol{\theta}}^{(k)})\epsilon_{kt}$;

2. *Second step*: QML estimation of the conditional correlation matrix ρ_0, as a solution of

$$\arg\min_{\rho \in \Lambda} n^{-1}\sum_{t=1}^{n} \hat{\boldsymbol{\eta}}_t^{*\prime}\hat{\boldsymbol{R}}_t^{-1}\hat{\boldsymbol{\eta}}_t^* + \log|\hat{\boldsymbol{R}}_t|, \tag{10.64}$$

where $\Lambda \subset \mathbb{R}^r$ is a compact set, $\hat{\boldsymbol{R}}_t = \mathcal{R}(\hat{\boldsymbol{\eta}}_{t-1}^*, \hat{\boldsymbol{\eta}}_{t-2}^*, \dots, \hat{\boldsymbol{\eta}}_1^*, \tilde{\boldsymbol{\eta}}_0^*, \tilde{\boldsymbol{\eta}}_{-1}^*, \dots; \rho)$, and the $\tilde{\boldsymbol{\eta}}_{-i}^*$'s are initial values.

Asymptotic properties of the procedure can be established for particular models (BEKK, CCC, etc.).

Example 10.8 (Estimating semi-diagonal BEKK models) Full BEKK models are generally impossible to estimate for large cross-sectional dimensions (see for instance Laurent, Rombouts, and Violante 2012) and practitioners generally only consider diagonal or even scalar models. The EbE approach can be applied to estimate a BEKK-GARCH(p, q) model given by

$$\epsilon_t = \boldsymbol{H}_t^{1/2}\boldsymbol{\eta}_t,$$

$$\boldsymbol{H}_t = \boldsymbol{\Omega}_0 + \sum_{i=1}^{q} \boldsymbol{A}_{0i}\epsilon_{t-i}\epsilon_{t-i}'\boldsymbol{A}_{0i}' + \sum_{j=1}^{p} \boldsymbol{B}_{0j}\boldsymbol{H}_{t-j}\boldsymbol{B}_{0j}, \tag{10.65}$$

where (η_t) is an iid $\mathbb{R}^m$-valued centred sequence with $E\eta_t\eta_t' = I_m$, $A_{0i} = (a_{ik\ell})_{1\le k,\ell\le m}$, $B_{0j} = \text{diag}(b_{j1}, \dots, b_{jm})$, and $\Omega_0 = (\omega_{k\ell})_{1\le k,\ell\le m}$ is a positive definite $m \times m$ matrix. In this model, the conditional variance $h_{kk,t}$ of the kth return may depend on the past of all returns; however, the equation of $h_{kk,t}$ does not involve past values of the other conditional variances $h_{jj,t}$. The model can thus be called *semi-diagonal* (as opposed to the diagonal BEKK in which both the B_{0j} and A_{0i} are diagonal matrices).

The dynamics of the kth diagonal entry of H_t is given by

$$h_{kk,t} = \omega_{kk} + \sum_{i=1}^{q}\left(\sum_{\ell=1}^{m} a_{ik\ell}\epsilon_{\ell,t-i}\right)^2 + \sum_{j=1}^{p} b_{jk}^2 h_{kk,t-j}. \tag{10.66}$$

Let $\theta_0^{(k)} = (\omega_{kk}, a_{1k}', \dots, a_{qk}', b_k)'$ for $k = 1, \dots, m$, where a_{ik}' denotes the kth row of the matrix A_{0i}, and $b_k = (b_{1k}^2, \dots, b_{pk}^2)$. Note that $h_{kk,t}$ is invariant to a change of sign of the kth row of any matrix A_i. For identifiability, we therefore impose $a_{ik1} > 0$ for $i = 1, \dots, q$. Let $\theta^{(k)} = (\theta_i^{(k)}) \in \mathbb{R}^{1+mq+p}$ denote a generic parameter value. The parameter space Θ_k is any compact subset of

$$\left\{\theta^{(k)} \mid \theta_1^{(k)}, \theta_2^{(k)}, \theta_{m+2}^{(k)}, \dots, \theta_{(q-1)m+2}^{(k)} > 0, \theta_{1+mq+1}^{(k)}, \dots, \theta_{1+mq+p}^{(k)} \ge 0, \sum_{j=1}^{p}\theta_{1+mq+j}^{(k)} < 1\right\}.$$

Under the strict stationarity condition (see Section 10.3.1), the existence of a positive density around 0 for η_1 and the moment condition $E|\eta_{kt}|^{4(1+\delta)} < \infty$, the equation-by-equation estimator (EbEE) of $\theta_0^{(k)}$ is shown to be strongly consistent and asymptotically normal (provided that $\theta_0^{(k)}$ belongs to the interior of Θ_k).

The diagonal elements of Ω_0 and the matrices A_{0i} and B_{0j} can thus be consistently estimated by successively applying the EbEE to each equation. Note that this is possible because each parameter of the model appears in one, and only one, equation.

Can the semi-diagonal BEKK-GARCH(p, q) model (10.65) be fully estimated by this approach? The answer is positive using a moment estimator. More precisely, let

$$A_0 = \sum_{i=1}^{q} D_m^+(A_{0i} \otimes A_{0i})D_m, \quad B_0 = \sum_{j=1}^{p} D_m^+(B_{0j} \otimes B_{0j})D_m.$$

Under the second-order stationarity assumption, we have

$$\text{vech}(\Omega_0) = (I_{m(m+1)/2} - A_0 - B_0)\text{vech}E(\epsilon_t\epsilon_t').$$

It follows that, letting $\hat{A}$ and $\hat{B}$ denote the EbE estimators of A_0 and B_0, respectively, a consistent estimator of Ω_0 is obtained from

$$\text{vech}(\hat{\Omega}) = (I_{m(m+1)/2} - \hat{A} - \hat{B})\text{vech}\left(\frac{1}{n}\sum_{t=1}^{n}\epsilon_t\epsilon_t'\right).$$

In this example, a complex MGARCH model is consistently estimated without relying on a cumbersome full-likelihood optimisation. When the number m of assets increases, the number of optimisations involved in Step 1 increases accordingly. However, the estimation may remain feasible provided that the number of parameters involved in each volatility equation does not explode.

Numerical experiments confirm that the gains in computation time can be huge compared with the FQML estimator in which all the parameters are estimated in one step. Table 10.2 compares the effective computation times required by the two estimators as a function of the dimension m, for the CCC-GARCH(1, 1) model with $A_1 = 0.05I_m$, $B_1 = 0.9I_m$, $R = I_m$, and $\eta_t \sim \mathcal{N}(\mathbf{0}, I_m)$. The $m(m-1)/2$ sub-diagonal terms of R were estimated, together with the $3m$ other parameters of the model. The two estimators were fitted on simulations of length $n = 2000$. The comparison of the CPUs is clearly in

Table 10.2 Computation time (CPU time in seconds) and relative efficiency (RE) of the EbE with respect to the FQMLE (NA = not available, due to the impossibility to compute the FQMLE), for m-dimensional CCC-GARCH(1,1) models.

Dim. m	2	4	6	8	10
No. of parameters	7	18	33	52	75
CPU for EbEE	0.57	1.18	1.52	2.04	2.82
CPU for FQMLE	32.49	123.33	317.85	876.52	1 292.34
ratio of CPU	57.00	104.52	209.11	429.67	458.28
RE	0.96	0.99	0.99	0.97	102.42
Dim. m	50	100	200	400	800
No. of parameters	1 375	5 250	20 500	81 000	322 000
CPU for EbEE	13.67	27.89	56.58	110.00	226.32
CPU for FQMLE	NA	NA	NA	NA	NA
Ratio of CPU	NA	NA	NA	NA	NA
RE	NA	NA	NA	NA	NA

Source: Francq and Zakoïan (2016).

favour of the EbEE, the FQML being in failure for $m \geq 10$. When m increases, the computation time of the FQML estimator becomes prohibitive, and more importantly, the optimisation fails to provide a reasonable value for $\hat{\vartheta}_{\text{QML}}$. Table 10.2 also compares the relative efficiencies (RE) of the two approaches. To this aim, we first computed the approximated information matrix $\boldsymbol{J}_n = -\frac{1}{2n}\frac{\partial^2}{\partial\vartheta\partial\vartheta'}\sum_{t=1}^{n}\epsilon_t'\boldsymbol{H}_t^{-1}\epsilon_t + \log|\boldsymbol{H}_t|$. The quadratic form $n(\hat{\vartheta}_n - \vartheta_0)'\boldsymbol{J}_n(\hat{\vartheta}_n - \vartheta_0)$ is used as a measure of accuracy of an estimator $\hat{\vartheta}_n$ (the Euclidean distance, obtained by replacing $\boldsymbol{J}_n$ by the identity matrix, has the drawback of being scale dependent). The relative efficiency displayed in Table 10.2 is defined by

$$\text{RE} = \frac{(\hat{\vartheta}_{\text{QML}} - \vartheta_0)'\boldsymbol{J}_n(\hat{\vartheta}_{\text{QML}} - \vartheta_0)}{(\hat{\vartheta}_{\text{EbE}} - \vartheta_0)'\boldsymbol{J}_n(\hat{\vartheta}_{\text{EbE}} - \vartheta_0)}$$

where $\hat{\vartheta}_{\text{EbE}}$ and $\hat{\vartheta}_{\text{QML}}$ denote, respectively, the EbE and FQML estimators. The computation time of the FQML estimator being huge when m is large, the RE and CPU times are only computed on 1 simulation, but they are representative of what is generally observed. When $m \leq 9$, the accuracies are very similar, with a slight advantage for the FQML (which corresponds here to the ML). When the number of parameters becomes too large, the optimisation fails to give a reasonable value of $\hat{\vartheta}_{\text{QMLE}}$, and the RE clearly indicates the superiority of the EbEE over the QMLE for $m \geq 10$.

10.7 Proofs of the Asymptotic Results

10.7.1 Proof of the CAN in Theorem 10.7

We shall use the multiplicative norm (see Exercises 10.5 and 10.6) defined by

$$\|A\| := \sup_{\|x\| \leq 1} \|Ax\| = \rho^{1/2}(A'A), \tag{10.67}$$

where A is a $d_1 \times d_2$ matrix, $\|x\|$ is the Euclidean norm of vector $x \in \mathbb{R}^{d_2}$, and $\rho(\cdot)$ denotes the spectral radius. This norm satisfies, for any $d_2 \times d_1$ matrix B,

$$\|A\|^2 \leq \sum_{i,j} a_{ij}^2 = \text{Tr}(A'A) \leq d_2\|A\|^2, \quad |A'A| \leq \|A\|^{2d_2}, \tag{10.68}$$

$$|\mathrm{Tr}(AB)| \leq \left(\sum_{i,j} a_{i,j}^2\right)^{1/2}\left(\sum_{i,j} b_{i,j}^2\right)^{1/2} \leq \{d_2 d_1\}^{1/2}\|A\|\|B\|. \tag{10.69}$$

When A is a square $d \times d$ matrix, the last inequality of (10.68) yields

$$|A| \leq \|A\|^d. \tag{10.70}$$

We also often use the elementary relation

$$\mathrm{Tr}(AB) = \mathrm{Tr}(BA) = \mathrm{vec}'(A')\mathrm{vec}(B) \tag{10.71}$$

when A is a $d_1 \times d_2$ matrix and B is $d_2 \times d_1$ matrix.

To show the AN, we will use the following elementary results on the differentiation of expressions involving matrices. If $f(A)$ is a real-valued function of a matrix A whose entries a_{ij} are functions of some variable x, the chain rule for differentiation of compositions of functions states that

$$\frac{\partial f(A)}{\partial x} = \sum_{i,j} \frac{\partial f(A)}{\partial a_{ij}} \frac{\partial a_{ij}}{\partial x} = \mathrm{Tr}\left\{\frac{\partial f(A)}{\partial A'} \frac{\partial A}{\partial x}\right\}. \tag{10.72}$$

Moreover, for A invertible we have

$$\frac{\partial c'Ac}{\partial A'} = cc', \tag{10.73}$$

$$\frac{\partial Tr(CA'BA')}{\partial A'} = C'AB' + B'AC', \tag{10.74}$$

$$\frac{\partial \log |\det(A)|}{\partial A'} = A^{-1}, \tag{10.75}$$

$$\frac{\partial A^{-1}}{\partial x} = -A^{-1}\frac{\partial A}{\partial x}A^{-1}, \tag{10.76}$$

$$\frac{\partial Tr(CA^{-1}B)}{\partial A'} = -A^{-1}BCA^{-1}, \tag{10.77}$$

$$\frac{\partial Tr(CAB)}{\partial A'} = BC. \tag{10.78}$$

Proof of the Consistency

We shall establish the intermediate results (a), (c), and (d) which are stated as in the univariate case (see the proof of Theorem 7.1 in Section 7.4), the result (b) being satisfied by Assumption A4.

Proof of (a): initial values are forgotten asymptotically. We have

$$\sup_{\theta\in\Theta} |\mathbf{l}_n(\theta) - \tilde{\mathbf{l}}_n(\theta)| \leq n^{-1}\sum_{t=1}^{n} \sup_{\theta\in\Theta}|\epsilon_t'(\boldsymbol{H}_t^{-1}(\theta) - \tilde{\boldsymbol{H}}_t^{-1}(\theta))\epsilon_t|$$
$$+ n^{-1}\sum_{t=1}^{n} \sup_{\theta\in\Theta}|\log|\boldsymbol{H}_t(\theta)| - \log|\tilde{\boldsymbol{H}}_t(\theta)||. \tag{10.79}$$

Using (10.71), (10.69), A1–A2, and omitting the subscript '(θ)', the first sum of (10.79) satisfies

$$n^{-1}\sum_{t=1}^{n} \sup_{\theta\in\Theta}|\epsilon_t'\tilde{\boldsymbol{H}}_t^{-1}(\boldsymbol{H}_t - \tilde{\boldsymbol{H}}_t)\boldsymbol{H}_t^{-1}\epsilon_t|$$
$$= n^{-1}\sum_{t=1}^{n} \sup_{\theta\in\Theta}|\mathrm{Tr}\{\tilde{\boldsymbol{H}}_t^{-1}(\boldsymbol{H}_t - \tilde{\boldsymbol{H}}_t)\boldsymbol{H}_t^{-1}\epsilon_t\epsilon_t'\}|$$

$$\le Kn^{-1}\sum_{t=1}^{n}\sup_{\theta\in\Theta}\|\tilde{\boldsymbol{H}}_t^{-1}\|\|\boldsymbol{H}_t-\tilde{\boldsymbol{H}}_t\|\|\boldsymbol{H}_t^{-1}\|\|\epsilon_t\epsilon_t'\|$$

$$\le Kn^{-1}\sum_{t=1}^{n}\rho^t\|\epsilon_t\epsilon_t'\|=O(n^{-1})\text{ a.s.}$$

as $n\to\infty$. Indeed $\sum_{t=1}^{\infty}\rho^t\|\epsilon_t\epsilon_t'\|$ is finite a.s. since, for some $s<1$,

$$E\left(\sum_{t=1}^{\infty}\rho^t\epsilon_t\epsilon_t'\right)^s\le\sum_{t=1}^{\infty}\rho^{st}E\|\epsilon_t\|^{2s}<\infty\quad\text{a.s.}$$

by A3. If A2 is replaced by A2′, the previous result follows from the Hölder inequality. Now, consider the second term of the right-hand side of (10.79). The relation (10.70), the Minkowski inequality, the elementary inequality $\log(x)\le x+1$ and the multiplicativity of the spectral matrix norm imply

$$\begin{aligned}\log|\boldsymbol{H}_t|-\log|\tilde{\boldsymbol{H}}_t|&=\log|I_m+(\boldsymbol{H}_t-\tilde{\boldsymbol{H}}_t)\tilde{\boldsymbol{H}}_t^{-1}|\\&\le m\log\|I_m+(\boldsymbol{H}_t-\tilde{\boldsymbol{H}}_t)\tilde{\boldsymbol{H}}_t^{-1}\|\\&\le m\|\boldsymbol{H}_t-\tilde{\boldsymbol{H}}_t\|\|\tilde{\boldsymbol{H}}_t^{-1}\|,\end{aligned}$$

and, by symmetry,

$$\log|\tilde{\boldsymbol{H}}_t|-\log|\boldsymbol{H}_t|\le m\|\boldsymbol{H}_t-\tilde{\boldsymbol{H}}_t\|\|\boldsymbol{H}_t^{-1}\|.$$

Using again A1, A2, or A2′, we thus have shown that

$$\sup_{\theta\in\Theta}|\mathbf{l}_n(\theta)-\tilde{\mathbf{l}}_n(\theta)|=O(n^{-1})\quad\text{a.s.}\tag{10.80}$$

Proof of (c): the limit criterion is minimised at the true value. As in the univariate case, we first show that $E\ell_t(\theta)$ is well defined in $\mathbb{R}\cup\{+\infty\}$ for all θ, and in $\mathbb{R}$ for $\theta=\theta_0$. By A1 and (10.70) we have $|\boldsymbol{H}_t(\theta)|^{-1}=|\boldsymbol{H}_t^{-1}(\theta)|<\infty$ and thus

$$E\ell_t^-(\theta)\le E\log^-|\boldsymbol{H}_t(\theta)|<\infty.$$

At θ_0, Jensen's inequality and A3 entail

$$E\ell_t(\theta_0)=E\eta_t'\eta_t+\frac{1}{s}E\log|H_t(\theta_0)|^s<\infty.$$

Now we show that $E\ell_t(\theta)$ is minimised at θ_0. Without loss of generality, assume that $E\ell_t^+(\theta)<\infty$. Let $\lambda_{1,t},\dots,\lambda_{m,t}$ be the eigenvalues of $H_t(\theta_0)\boldsymbol{H}_t^{-1}(\theta)$, which are positive (see Exercise 10.15). We have

$$\begin{aligned}&E\ell_t(\theta)-E\ell_t(\theta_0)\\&=E\log\{|\boldsymbol{H}_t(\theta)\boldsymbol{H}_t^{-1}(\theta_0)|\}+E(\mathrm{Tr}\{[\boldsymbol{H}_t(\theta_0)\boldsymbol{H}_t^{-1}(\theta)-I_m]\})\\&=E\left\{\sum_{i=1}^{m}(\lambda_{it}-1-\log\lambda_{it})\right\}\ge 0,\end{aligned}$$

where the inequality is strict unless if $\lambda_{it}=1$ a.s. for all i, that is iff $\boldsymbol{H}_t(\theta)=\boldsymbol{H}_t(\theta_0)$ a.s., which is equivalent to $\theta=\theta_0$ under A4.

Proof of (d). The previous results and the ergodic theorem then entail that

$$\lim_{n\to\infty}\tilde{\mathbf{l}}_n(\theta_0)=\lim_{n\to\infty}\mathbf{l}_n(\theta_0)=E\ell_t(\theta_0).$$

Similarly, (10.80) and the ergodic theorem applied to the stationary process (X_t) with $X_t = \inf_{\theta^* \in V_m(\theta)} \ell_t(\theta^*)$ show that

$$\liminf_{n\to\infty} \inf_{\theta^*\in V_m(\theta)} \tilde{\mathrm{l}}_n(\theta^*) \geq \lim_{n\to\infty} \frac{1}{n}\sum_{t=1}^{n} \inf_{\theta^*\in V_m(\theta)} \ell_t(\theta^*) = E \inf_{\theta^*\in V_m(\theta)} \ell_t(\theta^*)$$

where $V_m(\theta)$ denotes the ball of centre θ and radius $1/m$. By A1 we have $K_0 := \inf_{\theta\in\Theta} |\tilde{H}_t(\theta)| > -\infty$. Subtracting the constant K_0 to $\ell_t(\theta^*)$ if necessary, one can always assume that $\inf_{\theta^*\in V_m(\theta)} \ell_t(\theta^*)$ is positive. If $E|\ell_t(\theta)| < \infty$, by Fatou's lemma and A5, for any $\varepsilon > 0$ there exists m sufficiently large such that

$$E \inf_{\theta^*\in V_m(\theta)} \ell_t(\theta^*) > E\ell_t(\theta) - \varepsilon.$$

If $E\ell_t^+(\theta) = \infty$, then the left-hand side of the previous inequality can be made arbitrarily large. The consistency follows.

Proof of the Asymptotic Normality (AN)

The matrix derivative rules (10.72), (10.77), and (10.75) yield

$$\frac{\partial}{\partial\theta_i}\ell_t(\theta) = \mathrm{Tr}\left\{-H_t^{-1}(\theta)\epsilon_t\epsilon_t' H_t^{-1}(\theta)\frac{\partial H_t(\theta)}{\partial\theta_i} + H_t^{-1}(\theta)\frac{\partial H_t(\theta)}{\partial\theta_i}\right\}. \tag{10.81}$$

We thus have

$$\begin{aligned}\frac{\partial}{\partial\theta_i}\ell_t(\theta_0) &= \mathrm{Tr}\left\{(I_m - \eta_t\eta_t')H_t^{-1/2}(\theta_0)\frac{\partial H_t(\theta_0)}{\partial\theta_i}H_t^{-1/2'}(\theta_0)\right\}\\ &= \frac{\partial \mathrm{vec}' H_t(\theta_0)}{\partial\theta_i}\{H_t^{-1/2'}(\theta_0)\otimes H_t^{-1/2'}(\theta_0)\}\mathrm{vec}(I_m - \eta_t\eta_t').\end{aligned} \tag{10.82}$$

Under A10, the matrix K is well defined, and the second moment condition in A9 entails $E\|C_{i,t}\|^2 < \infty$. The existence of the matrix I is thus guaranteed by A9 and A10. It is then clear from (10.82) that $\left\{\frac{\partial}{\partial\theta}\ell_t(\theta_0)\right\}_t$ is a square integrable stationary and ergodic martingale difference. The central limit theorem in Corollary A.1 and the Cramér–Wold device then entail that

$$\frac{1}{\sqrt{n}}\sum_{t=1}^{n}\frac{\partial\ell_t(\theta_0)}{\partial\theta} \xrightarrow{\mathcal{L}} \mathcal{N}(\mathbf{0}, I) \quad \text{as } n\to\infty. \tag{10.83}$$

Differentiating (10.81), we also have

$$\frac{\partial^2}{\partial\theta_i\partial\theta_j}\ell_t(\theta) = \sum_{i=1}^{5} c_i(\theta),$$

with

$$c_1(\theta) = \epsilon_t' H_t^{-1}(\theta)\frac{\partial H_t(\theta)}{\partial\theta_i}H_t^{-1}(\theta)\frac{\partial H_t(\theta)}{\partial\theta_j}H_t^{-1}(\theta)\epsilon_t,$$

$$c_2(\theta) = \epsilon_t' H_t^{-1}(\theta)\frac{\partial H_t(\theta)}{\partial\theta_j}H_t^{-1}(\theta)\frac{\partial H_t(\theta)}{\partial\theta_i}H_t^{-1}(\theta)\epsilon_t$$

$$c_3(\theta) = -\epsilon_t' H_t^{-1}(\theta)\frac{\partial^2 H_t(\theta)}{\partial\theta_i\partial\theta_j}H_t^{-1}(\theta)\epsilon_t.$$

$$c_4(\theta) = -\mathrm{Tr}\left(\frac{\partial H_t(\theta)}{\partial\theta_i}H_t^{-1}(\theta)\frac{\partial H_t(\theta)}{\partial\theta_j}H_t^{-1}(\theta)\right)$$

$$c_5(\theta) = \mathrm{Tr}\left(H_t^{-1}(\theta)\frac{\partial^2 H_t(\theta)}{\partial\theta_i\partial\theta_j}\right).$$

Under A9, noting that $Ec_1(\theta_0) = -Ec_4(\theta_0)$ and $Ec_3(\theta_0) = -Ec_5(\theta_0)$, we thus have

$$E\left\{\frac{\partial^2}{\partial\theta_i\partial\theta_j}\ell_t(\theta_0)\right\} = E\,\mathrm{Tr}\left(H_t^{-1/2}(\theta_0)\frac{\partial H_t(\theta_0)}{\partial\theta_i}H_t^{-1}(\theta_0)\frac{\partial H_t(\theta_0)}{\partial\theta_j}H_t^{-1/2'}(\theta_0)\right)$$

$$= E\,\mathrm{vec}\left(\frac{\partial H_t(\theta_0)}{\partial\theta_i}\right)' H_t^{-1}(\theta_0)\otimes H_t^{-1}(\theta_0)\mathrm{vec}\left(\frac{\partial H_t(\theta_0)}{\partial\theta_i}\right),$$

using elementary properties of the vec and Kronecker operators. By the consistency and A6, we have $\hat{\theta}_n \to \theta_0 \in \mathring{\Theta}$, and thus almost surely $\partial\sum_{t=1}^n \tilde{\ell}_t(\hat{\theta}_n)/\partial\theta = 0$ for n large enough. Taylor expansions and A7–A8 thus show that almost surely

$$0 = \frac{1}{\sqrt{n}}\sum_{t=1}^n \frac{\partial\ell_t(\theta_0)}{\partial\theta} + \left(\frac{1}{n}\sum_{t=1}^n \frac{\partial^2}{\partial\theta_i\partial\theta_j}\ell_t(\theta_{ij}^*)\right)\sqrt{n}(\hat{\theta}_n - \theta_0)$$

where the θ_{ij}^*'s are between $\hat{\theta}_n$ and θ_0 component-wise. To show that the previous matrix into brackets converges almost surely to J, it suffices to use the ergodic theorem, the continuity of the derivatives, and to show that

$$E\sup_{\theta\in V(\theta_0)}\left|\frac{\partial^2\ell_t(\theta)}{\partial\theta_i\partial\theta_j}\right| < \infty$$

for some neighbourhood $V(\theta_0)$ of θ_0, which follows from A7 and A9.

If J was singular, there would exist some non-zero vector $\lambda\in\mathbb{R}^d$ such that $\lambda' J\lambda = 0$. Since $H_t^{-1}(\theta_0)\otimes H_t^{-1}(\theta_0)$ is almost surely positive definite, this entails that $\mathbf{\Delta}_t\lambda = 0$ with probability one, which is excluded by A10. The AN, as well as the Bahadur linearisation (10.49), easily follow from (10.83).

10.7.2 Proof of the CAN in Theorems 10.8 and 10.9

The proof consists in showing that the regularity conditions of Theorem 10.7 are satisfied.

Proof of Theorem 10.8

Note that CC2 implies that the invertibility condition (10.46) holds for all $\theta\in\Theta$, and that $H_t = H_t(\theta_0)$ is a measurable function of $\{\eta_u, u < t\}$. Corollary 10.2 shows that, under the stationarity condition CC2, the moment conditions A3 are satisfied. The smoothness conditions A5 and A7 are obviously satisfied for the CCC model.

Proof that A1 and A2′ are satisfied: Rewrite Eq. (10.57) in matrix form as

$$\mathbf{H}_t = \underline{c}_t + \mathbb{B}\mathbf{H}_{t-1}, \tag{10.84}$$

where $\mathbb{B} = \mathbb{B}(\theta)$ is defined in Corollary 10.1 and

$$\mathbf{H}_t = \begin{pmatrix}\underline{h}_t\\ \underline{h}_{t-1}\\ \vdots\\ \underline{h}_{t-p+1}\end{pmatrix},\quad \underline{c}_t = \begin{pmatrix}\underline{\omega} + \sum_{i=1}^q \mathbf{A}_i\underline{\epsilon}_{t-i}\\ 0\\ \vdots\\ 0\end{pmatrix}. \tag{10.85}$$

In view of assumption CC2 and Corollary 10.1, we have $\rho(\mathbb{B}) < 1$. By the compactness of Θ, we even have

$$\sup_{\theta\in\Theta}\rho(\mathbb{B}) < 1. \tag{10.86}$$

Corollary 10.2 and (10.86) thus entail that

$$E \sup_{\theta\in\Theta} \|\mathbf{H}_t\|^s < \infty, \quad E \sup_{\theta\in\Theta} \|\boldsymbol{D}_t\|^s < \infty \tag{10.87}$$

for some $s > 0$. Using Eq. (10.84) iteratively, as in the univariate case, we deduce that almost surely

$$\sup_{\theta\in\Theta} \| \mathbf{H}_t - \tilde{\mathbf{H}}_t \| \leq K\rho^t, \quad \forall t,$$

where $\tilde{\mathbf{H}}_t$ denotes the vector obtained by replacing the variables $\underline{h}_{t-i}$ by $\underline{\tilde{h}}_{t-i}$ in $\mathbf{H}_t$. Observe that K is a random variable that depends on the past values $\{\epsilon_t, t \leq 0\}$. Since K does not depend on n, it can be considered as a constant, like ρ. We deduce that

$$\sup_{\theta\in\Theta} \| \boldsymbol{D}_t^2 - \tilde{\boldsymbol{D}}_t^2 \| \leq K\rho^t, \quad \forall t, \qquad \sup_{\theta\in\Theta} \| \boldsymbol{D}_t - \tilde{\boldsymbol{D}}_t \| \leq K\rho^t, \quad \forall t, \tag{10.88}$$

where the second inequality is obtained by arguing that the elements of $\underline{\omega}$ are strictly positive uniformly in $\boldsymbol{\Theta}$. We then have

$$\sup_{\theta\in\Theta} \|\boldsymbol{H}_t - \tilde{\boldsymbol{H}}_t\| = \sup_{\theta\in\Theta} \|(\boldsymbol{D}_t - \tilde{\boldsymbol{D}}_t)\boldsymbol{R}\boldsymbol{D}_t + \tilde{\boldsymbol{D}}_t\boldsymbol{R}(\boldsymbol{D}_t - \tilde{\boldsymbol{D}}_t)\| \leq K\rho_t, \tag{10.89}$$

with $\rho_t = \rho^t \sup_{\theta\in\Theta} \|\boldsymbol{D}_t\|$. Therefore, the second moment condition of (10.87) shows that A2′ holds true, for any p and for s sufficiently small.

Noting that $\|\boldsymbol{R}^{-1}\|$ is the inverse of the eigenvalue of smallest modulus of $\boldsymbol{R}$, and that $\| \tilde{\boldsymbol{D}}_t^{-1} \| = \{\min_i (h_{ii,t})\}^{-1}$, we have

$$\sup_{\theta\in\Theta} \| \tilde{\boldsymbol{H}}_t^{-1} \| \leq \sup_{\theta\in\Theta} \| \tilde{\boldsymbol{D}}_t^{-1} \|^2 \| \boldsymbol{R}^{-1} \| \leq \sup_{\theta\in\Theta} \{\min_i \underline{\omega}(i)\}^{-2} \| \boldsymbol{R}^{-1} \| \leq K, \tag{10.90}$$

using CC5, the compactness of $\boldsymbol{\Theta}$ and the strict positivity of the components of $\underline{\omega}$. Similarly, we have

$$\sup_{\theta\in\Theta} \| \boldsymbol{H}_t^{-1} \| \leq K. \tag{10.91}$$

We thus have shown that A1 and A2′ hold true, meaning that the initial values are asymptotically irrelevant, in the sense of (10.80).

Proof of A4: Suppose that, for some $\boldsymbol{\theta} \neq \boldsymbol{\theta}_0$,

$$\underline{h}_t(\boldsymbol{\theta}) = \underline{h}_t(\boldsymbol{\theta}_0), \quad P_{\theta_0}\text{–a.s. and } \boldsymbol{R}(\boldsymbol{\theta}) = \boldsymbol{R}(\boldsymbol{\theta}_0).$$

Then it readily follows that $\rho = \rho_0$ and, using the invertibility of the polynomial $\mathcal{B}_\theta(B)$ under assumption CC2, by (10.87),

$$\mathcal{B}_\theta(1)^{-1}\underline{\omega} + \mathcal{B}_\theta(B)^{-1}\mathcal{A}_\theta(B)\underline{\epsilon}_t = \mathcal{B}_{\theta_0}(1)^{-1}\underline{\omega}_0 + \mathcal{B}_{\theta_0}(B)^{-1}\mathcal{A}_{\theta_0}(B)\underline{\epsilon}_t$$

that is,

$$\begin{aligned}\mathcal{B}_\theta(1)^{-1}\underline{\omega} - \mathcal{B}_{\theta_0}(1)^{-1}\underline{\omega}_0 &= \{\mathcal{B}_{\theta_0}(B)^{-1}\mathcal{A}_{\theta_0}(B) - \mathcal{B}_\theta(B)^{-1}\mathcal{A}_\theta(B)\}\underline{\epsilon}_t \\ &:= \mathcal{P}(B)\underline{\epsilon}_t \quad \text{a.s.} \quad \forall t.\end{aligned}$$

Let $\mathcal{P}(B) = \sum_{i=0}^{\infty} \mathcal{P}_i B^i$. Noting that $\mathcal{P}_0 = \mathcal{P}(0) = 0$ and isolating the terms that are functions of η_{t-1},

$$\mathcal{P}_1(h_{11,t-1}\eta_{1,t-1}^2, \ldots, h_{mm,t-1}\eta_{m,t-1}^2)' = Z_{t-2}, \quad \text{a.s.},$$

where Z_{t-2} belongs to the σ-field generated by $\{\eta_{t-2}, \eta_{t-3}, \ldots\}$. Since η_{t-1} is independent of this σ-field, Exercise 10.3 shows that the latter equality contradicts CC3 unless when $p_{ij}h_{jj,t} = 0$ almost surely, where the p_{ij} are the entries of $\mathcal{P}_1$ for $i, j = 1, \ldots, m$. Because $h_{jj,t} > 0$ for all j, we thus have $\mathcal{P}_1 = 0$. Similarly, we show that $\mathcal{P}(B) = 0$ by successively considering the past values of η_{t-1}. Therefore, in view of CC4 (or CC4′), we have $\alpha = \alpha_0$ and $\beta = \beta_0$ by arguments already used. It readily follows that $\underline{\omega} = \underline{\omega}_0$. Hence $\boldsymbol{\theta} = \boldsymbol{\theta}_0$. We have thus established the identifiability condition A4, and the proof of Theorem 10.8 follows from Theorem 10.7. □

Proof of Theorem 10.9

It remains to show that A8, A9 and A11 hold.

Proof of A9: Denoting by $\underline{h}_{0t}(i_1)$ the i_1th component of $\underline{h}_t(\theta_0)$,

$$\underline{h}_{0t}(i_1) = c_0 + \sum_{k=0}^{\infty}\sum_{j_1=1}^{m}\sum_{j_2=1}^{m}\sum_{i=1}^{q} \mathbf{A}_{0i}(j_1,j_2)\mathbb{B}_0^k(i_1,j_1)\epsilon_{j_2,t-k-i}^2,$$

where c_0 is a strictly positive constant and, by the usual convention, the index 0 corresponds to quantities evaluated at $\theta = \theta_0$. Under CC6, for a sufficiently small neighbourhood $\mathcal{V}(\theta_0)$ of θ_0, we have

$$\sup_{\theta\in\mathcal{V}(\theta_0)} \frac{\mathbf{A}_{0i}(j_1,j_2)}{\mathbf{A}_i(j_1,j_2)} < K, \qquad \sup_{\theta\in\mathcal{V}(\theta_0)} \frac{\mathbb{B}_0^k(i_1,j_1)}{\mathbb{B}^k(i_1,j_1)} < 1+\delta,$$

for all $i_1,j_1,j_2 \in \{1,\dots,m\}$ and all $\delta > 0$. Moreover, in $\underline{h}_t(i_1)$, the coefficient of $\mathbb{B}^k(i_1,j_1)\epsilon_{j_2,t-k-i}^2$ is bounded below by a constant $c > 0$ uniformly on $\theta \in \mathcal{V}(\theta_0)$. We thus have

$$\begin{aligned}\frac{\underline{h}_{0t}(i_1)}{\underline{h}_t(i_1)} &\le K + K\sum_{k=0}^{\infty}\sum_{j_1=1}^{m}\sum_{j_2=1}^{m}\sum_{i=1}^{q} \frac{(1+\delta)^k\mathbb{B}^k(i_1,j_1)\epsilon_{j_2,t-k-i}^2}{\omega + c\mathbb{B}^k(i_1,j_1)\epsilon_{j_2,t-k-i}^2}\\ &\le K + K\sum_{j_2=1}^{m}\sum_{i=1}^{q}\sum_{k=0}^{\infty}(1+\delta)^k\rho^{ks}\epsilon_{j_2,t-k-i}^{2s},\end{aligned}$$

for some $\rho \in [0, 1)$, all $\delta > 0$ and all $s \in [0, 1]$. Corollary 10.2 then implies that, for all $r_0 \ge 0$, there exists $\mathcal{V}(\theta_0)$ such that

$$E\sup_{\theta\in\mathcal{V}(\theta_0)}\left|\frac{\underline{h}_{0t}(i_1)}{\underline{h}_t(i_1)}\right|^{r_0} < \infty.$$

Since

$$\boldsymbol{H}_t^{-1/2}(\theta)\boldsymbol{H}_t^{1/2}(\theta_0) = \boldsymbol{R}^{-1/2}\boldsymbol{D}_t^{-1}(\theta)\boldsymbol{D}_t(\theta_0)\boldsymbol{R}_0^{1/2},$$

where $\boldsymbol{R}_0$ denotes the true value of $\boldsymbol{R}$, the last moment condition of A9 is satisfied (for all r).

By (10.84)–(10.86),

$$\sup_{\theta\in\Theta}\left\|\frac{\partial \mathbf{H}_t}{\partial\theta_i}\right\| < \infty, \quad i = 1,\dots,m,$$

and, setting $s_2 = m + qm^2$,

$$\theta_i\frac{\partial \mathbf{H}_t}{\partial\theta_i} \le \mathbf{H}_t, \quad i = m+1,\dots,s_2.$$

Setting $s_1 = m + (p+q)m^2$, we have

$$\frac{\partial \mathbf{H}_t}{\partial\theta_i} = \sum_{k=1}^{\infty}\left\{\sum_{j=1}^{k}\mathbb{B}^{j-1}\mathbb{B}^{(i)}\mathbb{B}^{k-j}\right\}\underline{c}_{t-k}, \quad i = s_2+1,\dots,s_1,$$

where $\mathbb{B}^{(i)} = \partial\mathbb{B}/\partial\theta_i$ is a matrix whose entries are all 0, apart from a 1 located at the same place as θ_i in $\mathbb{B}$. By abuse of notation, we denote by $\mathbf{H}_t(i_1)$ and $\underline{h}_{0t}(i_1)$ the i_1th components of $\mathbf{H}_t$ and $\underline{h}_t(\theta_0)$. Note that, by the arguments used to show (10.87), the previous expressions show that

$$E\sup_{\theta\in\Theta}\left\|\frac{\partial \boldsymbol{D}_t}{\partial\theta_i}\right\|^s < \infty, \quad E\sup_{\theta\in\Theta}\left\|\frac{\partial \boldsymbol{H}_t}{\partial\theta_i}\right\|^s < \infty, \qquad i = 1,\dots,s_1, \tag{10.92}$$

for some $s > 0$. With arguments similar to those used in the univariate case, that is, the inequality $x/(1+x) \leq x^s$ for all $x \geq 0$ and $s \in [0, 1]$, and the inequalities

$$\theta_i \frac{\partial \mathbf{H}_t}{\partial \theta_i} \leq \sum_{k=1}^{\infty} k \mathbb{B}^k \underline{c}_{t-k}, \quad \theta_i \frac{\partial \mathbf{H}_t(i_1)}{\partial \theta_i} \leq \sum_{k=1}^{\infty} k \sum_{j_1=1}^{m} \mathbb{B}^k(i_1, j_1) \underline{c}_{t-k}(j_1)$$

and, setting $\omega = \inf_{1 \leq i \leq m} \underline{\omega}(i)$,

$$\mathbf{H}_t(i_1) \geq \omega + \sum_{j_1=1}^{m} \mathbb{B}^k(i_1, j_1) \underline{c}_{t-k}(j_1), \qquad \forall k,$$

we obtain

$$\frac{\theta_i}{\mathbf{H}_t(i_1)} \frac{\partial \mathbf{H}_t(i_1)}{\partial \theta_i} \leq \sum_{j_1=1}^{m} \sum_{k=1}^{\infty} k \left\{ \frac{\mathbb{B}^k(i_1, j_1) \underline{c}_{t-k}(j_1)}{\omega} \right\}^{s/r_0} \leq K \sum_{j_1=1}^{m} \sum_{k=1}^{\infty} k \rho_{j_1}^k \underline{c}_{t-k}^{s/r_0}(j_1),$$

where the constants ρ_{j_1} (which also depend on i_1, s and r_0) belong to the interval $[0, 1)$. Noting that these inequalities are uniform on a neighbourhood of $\boldsymbol{\theta}_0 \in \overset{\circ}{\boldsymbol{\Theta}}$, that they can be extended to higher-order derivatives, and that Corollary 10.2 implies that $\| \underline{c}_t \|_s < \infty$, we can show, as in the univariate case, that for all $i_1 = 1, \ldots, m$, all $i, j, k = 1, \ldots, s_1$ and all $r_0 \geq 0$, there exists a neighbourhood $\mathcal{V}(\boldsymbol{\theta}_0)$ of $\boldsymbol{\theta}_0$ such that

$$E \sup_{\theta \in \mathcal{V}(\theta_0)} \left| \frac{1}{\underline{h}_t(i_1)} \frac{\partial \underline{h}_t(i_1)}{\partial \theta_i}(\boldsymbol{\theta}) \right|^{r_0} < \infty, \tag{10.93}$$

and

$$E \sup_{\theta \in \mathcal{V}(\theta_0)} \left| \frac{1}{\underline{h}_t(i_1)} \frac{\partial^2 \underline{h}_t(i_1)}{\partial \theta_i \partial \theta_j}(\boldsymbol{\theta}) \right|^{r_0} < \infty. \tag{10.94}$$

Omitting '$(\boldsymbol{\theta})$' to lighten the notation, note that

$$\boldsymbol{H}_t^{-1/2} \frac{\partial \boldsymbol{H}_t}{\partial \theta_i} \boldsymbol{H}_t^{-1/2'} = \boldsymbol{R}^{-1/2} \boldsymbol{D}_t^{-1} \left\{ \boldsymbol{D}_t \boldsymbol{R} \frac{\partial \boldsymbol{D}_t}{\partial \theta_i} + \frac{\partial \boldsymbol{D}_t}{\partial \theta_i} \boldsymbol{R} \boldsymbol{D}_t \right\} \boldsymbol{D}_t^{-1} \boldsymbol{R}^{-1/2},$$

for $i = 1, \ldots, s_1$, and

$$\boldsymbol{H}_t^{-1/2} \frac{\partial \boldsymbol{H}_t}{\partial \theta_i} \boldsymbol{H}_t^{-1/2'} = \boldsymbol{R}^{-1/2} \boldsymbol{D}_t^{-1} \boldsymbol{D}_t \frac{\partial \boldsymbol{R}}{\partial \theta_i} \boldsymbol{D}_t \boldsymbol{D}_t^{-1} \boldsymbol{R}^{-1/2} = \boldsymbol{R}^{-1/2} \frac{\partial \boldsymbol{R}}{\partial \theta_i} \boldsymbol{R}^{-1/2}$$

for $i = s_1 + 1, \ldots, s_0$. It follows that (10.93) entails the second moment condition of A9 (for any q). Similarly, (10.94) entails the first moment condition of A9 (for any p).

Proof of A8: First note that by

$$\boldsymbol{H}_t^{-1} - \tilde{\boldsymbol{H}}_t^{-1} = \boldsymbol{H}_t^{-1} (\tilde{\boldsymbol{H}}_t - \boldsymbol{H}_t) \tilde{\boldsymbol{H}}_t^{-1}$$

and (10.88)–(10.91), we have

$$\sup_{\theta \in \Theta} \| \boldsymbol{H}_t^{-1} - \tilde{\boldsymbol{H}}_t^{-1} \| \leq K \rho^t \sup_{\theta \in \Theta} \| \boldsymbol{D}_t \| . \tag{10.95}$$

From Eq. (10.84), we have

$$\mathbf{H}_t = \sum_{k=0}^{t-r-1} \mathbb{B}^k \underline{c}_{t-k} + \mathbb{B}^{t-r} \mathbf{H}_r, \qquad \tilde{\mathbf{H}}_t = \sum_{k=0}^{t-r-1} \mathbb{B}^k \tilde{\underline{c}}_{t-k} + \mathbb{B}^{t-r} \tilde{\mathbf{H}}_r,$$

where $r = \max\{p, q\}$ and the tilde means that initial values are taken into account. Since $\tilde{\underline{c}}_t = \underline{c}_t$ for all $t > r$, we have $\mathbf{H}_t - \tilde{\mathbf{H}}_t = \mathbb{B}^{t-r}(\mathbf{H}_r - \tilde{\mathbf{H}}_r)$ and

$$\frac{\partial}{\partial\theta_i}(\mathbf{H}_t - \tilde{\mathbf{H}}_t) = \mathbb{B}^{t-r}\frac{\partial}{\partial\theta_i}(\mathbf{H}_r - \tilde{\mathbf{H}}_r) + \sum_{j=1}^{t-r}\mathbb{B}^{j-1}\mathbb{B}^{(i)}\mathbb{B}^{t-r-j}(\mathbf{H}_r - \tilde{\mathbf{H}}_r).$$

Thus condition (10.86) entails that

$$\sup_{\theta\in\Theta}\left\|\frac{\partial}{\partial\theta_i}(\boldsymbol{D}_t - \tilde{\boldsymbol{D}}_t)\right\| \le K\rho^t. \tag{10.96}$$

Because

$$\frac{\partial}{\partial\theta_i}(\boldsymbol{H}_t - \tilde{\boldsymbol{H}}_t) = \frac{\partial}{\partial\theta_i}\{(\boldsymbol{D}_t - \tilde{\boldsymbol{D}}_t)\boldsymbol{R}\boldsymbol{D}_t + \tilde{\boldsymbol{D}}_t R(\boldsymbol{D}_t - \tilde{\boldsymbol{D}}_t)\},$$

the second inequality of (10.88) and (10.96) imply that

$$\sup_{\theta\in\Theta}\left\|\frac{\partial\boldsymbol{H}_t}{\partial\theta_i} - \frac{\partial\tilde{\boldsymbol{H}}_t}{\partial\theta_i}\right\| \le K\rho^t\left\{\sup_{\theta\in\Theta}\|\boldsymbol{D}_t\| + \sup_{\theta\in\Theta}\left\|\frac{\partial\boldsymbol{D}_t}{\partial\theta_i}\right\| + 1\right\}. \tag{10.97}$$

Note that (10.81) continues to hold when $\ell_t(\theta)$ and $\boldsymbol{H}_t(\theta)$ are replaced by $\tilde{\ell}_t(\theta)$ and $\tilde{\boldsymbol{H}}_t(\theta)$. Therefore,

$$\frac{\partial\ell_t(\theta)}{\partial\theta_i} - \frac{\partial\tilde{\ell}_t(\theta)}{\partial\theta_i} = \mathrm{Tr}(\boldsymbol{C}_1 + \boldsymbol{C}_2),$$

where

$$\boldsymbol{C}_1 = (\boldsymbol{I}_m - \boldsymbol{H}_t^{-1}\epsilon_t\epsilon_t')(\boldsymbol{H}_t^{-1} - \tilde{\boldsymbol{H}}_t^{-1})\frac{\partial\boldsymbol{H}_t}{\partial\theta_i} - (\boldsymbol{H}_t^{-1} - \tilde{\boldsymbol{H}}_t^{-1})\epsilon_t\epsilon_t'\tilde{\boldsymbol{H}}_t^{-1}\frac{\partial\tilde{\boldsymbol{H}}_t}{\partial\theta_i},$$

$$\boldsymbol{C}_2 = (\boldsymbol{I}_m - \boldsymbol{H}_t^{-1}\epsilon_t\epsilon_t')\tilde{\boldsymbol{H}}_t^{-1}\left(\frac{\partial\boldsymbol{H}_t}{\partial\theta_i} - \frac{\partial\tilde{\boldsymbol{H}}_t}{\partial\theta_i}\right).$$

Using conditions (10.87), (10.90), (10.91), (10.92), (10.95), and (10.97), it can be shown that $\mathrm{Tr}(\boldsymbol{C}_1 + \boldsymbol{C}_2) \le K\rho^t \mathrm{u}_t$, where u_t is a random variable such that $\sup_t E|\mathrm{u}_t|^s < \infty$ for some small $s \in (0, 1)$. Arguing that $\sum_{t=1}^{\infty}\rho^t\mathrm{u}_t$ is almost surely finite because its moment of order s is finite, the conclusion follows.

Proof of A11: Recall that we applied Theorem 10.7 with $d = s_0$. We thus have to show that it cannot exist as a non-zero vector $\mathbf{c} \in \mathbb{R}^{s_0}$, such that $\boldsymbol{\Delta}_t\mathbf{c} = 0_{m^2}$ with probability 1. Decompose $\mathbf{c}$ into $\mathbf{c} = (\mathbf{c}_1', \mathbf{c}_2')'$ with $\mathbf{c}_1 \in \mathbb{R}^{s_1}$ and $\mathbf{c}_2 \in \mathbb{R}^{s_3}$, where $s_3 = s_0 - s_1 = m(m-1)/2$. Rows $1, m+1, \dots, m^2$ of the equations

$$\boldsymbol{\Delta}_t\mathbf{c} = \sum_{i=1}^{s_0} c_i\frac{\partial}{\partial\theta_i}\mathrm{vec}\,\boldsymbol{H}_{0t} = \sum_{i=1}^{s_0} c_i\frac{\partial}{\partial\theta_i}(\boldsymbol{D}_{0t}\otimes\boldsymbol{D}_{0t})\mathrm{vec}\,\boldsymbol{R}_0 = 0_{m^2}, \quad \text{a.s.} \tag{10.98}$$

give

$$\sum_{i=1}^{s_1} c_i\frac{\partial}{\partial\theta_i}\underline{h}_t(\theta_0) = 0_m, \quad \text{a.s.} \tag{10.99}$$

Differentiating Eq. (10.57) yields

$$\sum_{i=1}^{s_1} c_i\frac{\partial}{\partial\theta_i}\underline{h}_t = \underline{\omega}^* + \sum_{j=1}^{q}\mathbf{A}_j^*\underline{\epsilon}_{t-j} + \sum_{j=1}^{p}\mathbf{B}_j^*\underline{h}_{t-j} + \sum_{j=1}^{p}\mathbf{B}_j\sum_{i=1}^{s_1} c_i\frac{\partial}{\partial\theta_i}\underline{h}_{t-j},$$

where

$$\underline{\omega}^* = \sum_{i=1_1}^{s} c_i\frac{\partial}{\partial\theta_i}\underline{\omega}, \quad \mathbf{A}_j^* = \sum_{i=1}^{s_1} c_i\frac{\partial}{\partial\theta_i}\mathbf{A}_j, \quad \mathbf{B}_j^* = \sum_{i=1}^{s_1} c_i\frac{\partial}{\partial\theta_i}\mathbf{B}_j.$$

Because (10.99) is satisfied for all t, we have

$$\underline{\omega}_0^* + \sum_{j=1}^{q} \mathbf{A}_{0j}^* \underline{\epsilon}_{t-j} + \sum_{j=1}^{p} \mathbf{B}_{0j}^* \underline{h}_{t-j}(\theta_0) = 0,$$

where quantities evaluated at $\theta = \theta_0$ are indexed by 0. This entails that

$$\underline{h}_t(\theta_0) = \underline{\omega}_0 - \underline{\omega}_0^* + \sum_{j=1}^{q} (\mathbf{A}_{0j} - \mathbf{A}_{0j}^*) \underline{\epsilon}_{t-j} + \sum_{j=1}^{p} (\mathbf{B}_{0j} - \mathbf{B}_{0j}^*) \underline{h}_{t-j}(\theta_0),$$

and finally, introducing a vector θ_1 whose s_1 first components are

$$\text{vec}(\underline{\omega}_0 - \underline{\omega}_0^* | \mathbf{A}_{01} - \mathbf{A}_{01}^* | \cdots | \mathbf{B}_{0p} - \mathbf{B}_{0p}^*),$$

we have

$$\underline{h}_t(\theta_0) = \underline{h}_t(\theta_1)$$

by choosing $\mathbf{c}_1$ small enough so that $\theta_1 \in \Theta$. If $\mathbf{c}_1 \neq 0$ then $\theta_1 \neq \theta_0$. This is in contradiction to the identifiability of the parameter (see the proof of A4), hence $\mathbf{c}_1 = 0$. Equations (10.98) thus become

$$(\boldsymbol{D}_{0t} \otimes \boldsymbol{D}_{0t}) \sum_{i=s_1+1}^{s_0} c_i \frac{\partial}{\partial \theta_i} \text{vec} \boldsymbol{R}_0 = 0_{m^2}, \quad \text{a.s.}.$$

Therefore,

$$\sum_{i=s_1+1}^{s_0} c_i \frac{\partial}{\partial \theta_i} \text{vec}\, \boldsymbol{R}_0 = 0_{m^2}.$$

Because the vectors, $\partial \text{vec}\, \boldsymbol{R} / \partial \theta_i$, $i = s_1 + 1, \ldots, s_0$, are linearly independent, the vector $\mathbf{c}_2 = (c_{s_1+1}, \ldots, c_{s_0})'$ is null, and thus $\mathbf{c} = 0$, which completes the proof. □

10.8 Bibliographical Notes

Multivariate ARCH models were first considered by Engle, Granger, and Kraft (1984) in the guise of the diagonal model. This model was extended and studied by Bollerslev, Engle, and Woolridge (1988). The reader may refer to Hafner and Preminger (2009a), Lanne and Saikkonen (2007), van der Weide (2002), and Vrontos, Dellaportas, and Politis (2003) for the definition and study of FF-GARCH models of the form (10.26) where P is not assumed to be orthonormal. The CCC-GARCH model based on (10.17) was introduced by Bollerslev (1990) and extended to (10.18) by Jeantheau (1998). A sufficient condition for strict stationarity and the existence of fourth-order moments of the CCC-GARCH(p, q) is established by Aue et al. (2009). The DCC formulations based on (10.19) and (10.20) were proposed, respectively, by Tse and Tsui (2002), and Engle (2002a). The cDCC model (10.1) is extended by allowing for a clustering structure of the univariate GARCH parameters in Aielli and Caporin (2014). The single-factor model (10.24), which can be viewed as a dynamic version of the capital asset pricing model of Sharpe (1964), was proposed by Engle, Ng, and Rothschild (1990). The main references on the O-GARCH or PC-GARCH models are Alexander (2002) and Ding and Engle (2001). See van der Weide (2002) and Boswijk and van der Weide (2006) for references on the GO-GARCH model. More details on BEKK models can be found in McAleer et al. (2009). . Caporin and McAleer (2012) compared the BEKK and DCC specifications. Hafner (2003) and He and Teräsvirta (2004) studied the fourth-order moments of MGARCH models. Dynamic conditional correlations models were introduced by Engle (2002a) and Tse and Tsui (2002). Pourahmadi (1999) showed that an unconstrained parametrisation of a covariance matrix can be conveniently obtained from its Cholesky decomposition (see Chapter 10 in Tsay 2010 and Dellaportas and Pourahmadi 2012 for modelling the Cholesky decomposition of a conditional variance and for applications in finance). These references, and those given in the text, can be complemented by the surveys by Bauwens, Laurent, and Rombouts (2006), Silvennoinen and Teräsvirta (2009), Bauwens, Hafner, and Laurent (2012), Almeida, Hotta, and Ruiz (2018), and by the book by Engle (2009). Stationarity conditions for DCC models have been established by Fermanian and Malongo (2016).

Jeantheau (1998) gave general conditions for the strong consistency of the QMLE for MGARCH models. Comte and Lieberman (2003) showed the CAN of the QMLE for the BEKK formulation under some high-level assumptions, in particular the existence of eight-order moments for the returns process. On the other hand, Avarucci, Beutner, and Zaffaroni (2013) showed that for the BEKK, the finiteness of the variance of the scores requires at least the existence of second-order moments of the observable process. Asymptotic results were established by Ling and McAleer (2003a) for the CCC formulation of an ARMA-GARCH, and by Hafner and Preminger (2009a) for a factor GARCH model of the FF-GARCH form. Theorem 10.1 provides high-level assumptions implying the CAN of the QMLE of a general MGARCH model. Theorems 10.2 and 10.3 are concerned with the CCC formulation, and allow us to study a subclass of the models considered by Ling and McAleer (2003a), but do not cover the models studied by Comte and Lieberman (2003) or those studied by Hafner and Preminger (2009b). Theorems 10.2 and 10.3 are mainly of interest because they do not require any moment on the observed process and do not use high-level assumptions. An extension of Theorems 10.2 and 10.3 to asymmetric CCC-GARCH can be found in Francq and Zakoian (2012a). For additional information on identifiability, in particular on the echelon form, one may for instance refer to Hannan (1976), Hannan and Deistler (1988), Lütkepohl (1991), and Reinsel (1997). Pedersen (2017) recently studied inference and testing in extended CCC GARCH models in the case where the true parameter vector is a boundary point of the parameter space.

Portmanteau tests on the normalised residuals of MGARCH processes were proposed, in particular, by Tse (2002) and Duchesne and Lalancette (2003).

Bardet and Wintenberger (2009) gave regularity conditions for the strong CAN of the QMLE, for a general class of multidimensional causal processes. Their framework is more general than that of Theorem 10.1, since it also incorporates a conditional mean.

Among models not studied in this book are the spline GARCH models in which the volatility is written as a product of a slowly varying deterministic component and a GARCH-type component. These models were introduced by Engle and Rangel (2008), and their multivariate generalization is due to Hafner and Linton (2010).

The VT approach is not limited to the standard CCC model. Pedersen and Rahbek (2014) showed that the BEKK-GARCH model can be reparameterized in such a way that the variance of the observed process appears explicitly in the conditional variance equation. They established that the VT estimator is consistent and asymptotically normal when the process has finite sixth-order moments. See Hill and Renault (2012), Vaynman and Beare (2014), and Pedersen (2016) for results with infinite fourth moments. Bauwens, Braione, and Storti (2016) use the VT approach for a class of dynamic models for realized covariance matrices. Models incorporating leverage effects can also be estimated in a similar way: instead of targeting the variances, moments related to the signs of the returns are targeted (see Francq, Horváth, and Zakoian 2016). The idea of variance targeting has been extended to covariance targeting by Noureldin, Shephard and Sheppard (2014).

The EbE approach was initially proposed by Engle and Sheppard (2001) and Engle (2002a) in the context of DCC models. It was also suggested by Pelletier (2006) for regime-switching dynamic correlation models, by Aielli (2013) for DCC models, and it was used in several empirical studies (see for instance Hafner and Reznikova 2012; Sucarrat Grønneberg, and Escribano 2016). The statistical properties of such two-step estimators have been established by Francq and Zakoïan (2016).

10.9 Exercises

10.1 *(More or less parsimonious representations)*
Compare the number of parameters of the various GARCH(p, q) representations, as a function of the dimension m.

10.2 *(Identifiability of a matrix rational fraction)*
Let $\mathcal{A}_\theta(z)$, $\mathcal{B}_\theta(z)$, $\mathcal{A}_{\theta_0}(z)$, and $\mathcal{B}_{\theta_0}(z)$ denote square matrices of polynomials. Show that

$$\mathcal{B}_\theta(z)^{-1}\mathcal{A}_\theta(z) = \mathcal{B}_{\theta_0}(z)^{-1}\mathcal{A}_{\theta_0}(z) \qquad (10.100)$$

for all z such that $\det \mathcal{B}_\theta(z)\mathcal{B}_{\theta_0}(z) \neq 0$ if and only if there exists an operator $U(z)$ such that

$$\mathcal{A}_\theta(z) = U(z)\mathcal{A}_{\theta_0}(z) \quad \text{and} \quad \mathcal{B}_\theta(z) = U(z)\mathcal{B}_{\theta_0}(z). \qquad (10.101)$$

10.3 *(Two independent non-degenerate random variables cannot be equal)*
Let X and Y be two independent real random variables such that $Y = X$ almost surely. We aim to prove that X and Y are almost surely constant.

1. Suppose that Var(X) exists. Compute Var(X) and show the stated result in this case.
2. Suppose that X is discrete and $P(X = x_1)P(X = x_2) \neq 0$. Show that necessarily $x_1 = x_2$ and show the result in this case.
3. Prove the result in the general case.

10.4 *(Duplication and elimination)*
Consider the duplication matrix D_m and the elimination matrix D_m^+ defined by

$$\mathrm{vec}(A) = \mathrm{D}_m \mathrm{vech}(A), \quad \mathrm{vech}(A) = \mathrm{D}_m^+ \mathrm{vec}(A),$$

where A is any symmetric $m \times m$ matrix. Show that

$$\mathrm{D}_m^+\mathrm{D}_m = I_{m(m+1)/2}.$$

10.5 *(Norm and spectral radius)*
Show that

$$\|A\| := \sup_{\|x\|\leq 1} \|Ax\| = \rho^{1/2}(A'A).$$

10.6 *(Elementary results on matrix norms)*
Show the equalities and inequalities of (10.68) and (10.69).

10.7 *(Scalar GARCH)*
The scalar GARCH model has a volatility of the form

$$H_t = \Omega + \sum_{i=1}^{q} \alpha_i \epsilon_{t-i}\epsilon'_{t-i} + \sum_{j=1}^{p} \beta_j H_{t-j},$$

where the α_i and β_j are positive numbers. Give the positivity and second-order stationarity conditions.

10.8 *(Condition for the L^p and almost sure convergence)*
Let $p \in [1, \infty)$ and let (u_n) be a sequence of real random variables of L^p such that

$$E|u_n - u_{n-1}|^p \leq C\rho^n,$$

for some positive constant C, and some constant ρ in (0, 1). Prove that

$$u_n \text{ converges almost surely and in } L^p$$

to some random variable u of L^p.

10.9 *(An average of correlation matrices is a correlation matrix)*
Let R and Q be two correlation matrices of the same size and let $p \in [0, 1]$. Show that $pR + (1-p)Q$ is a correlation matrix.

10.10 *(Factors as linear combinations of individual returns)*
Consider the factor model

$$\mathrm{Var}(\epsilon_t \epsilon_t' \mid \epsilon_u, u < t) = \Omega + \sum_{j=1}^{r} \lambda_{jt} \boldsymbol{\beta}_j \boldsymbol{\beta}_j',$$

where the $\boldsymbol{\beta}_j$ are linearly independent. Show there exist vectors $\boldsymbol{\alpha}_j$ such that

$$\mathrm{Var}(\epsilon_t \epsilon_t' \mid \epsilon_u, u < t) = \Omega^* + \sum_{j=1}^{r} \lambda_{jt}^* \boldsymbol{\beta}_j \boldsymbol{\beta}_j',$$

where the λ_{jt}^* are conditional variances of the portfolios $\boldsymbol{\alpha}_j' \epsilon_t$. Compute the conditional covariance between these factors.

10.11 *(BEKK representation of factor models)*
Consider the factor model

$$H_t = \Omega + \sum_{j=1}^{r} \lambda_{jt} \boldsymbol{\beta}_j \boldsymbol{\beta}_j', \quad \lambda_{jt} = \omega_j + a_j \epsilon_{jt-1}^2 + b_j \lambda_{jt-1},$$

where the $\boldsymbol{\beta}_j$ are linearly independent, $\omega_j > 0$, $a_j \geq 0$, and $0 \leq b_j < 1$ for $j = 1, \dots, r$. Show that a BEKK representation holds, of the form

$$H_t = \Omega^* + \sum_{k=1}^{K} A_k \epsilon_{t-1} \epsilon_{t-1}' A_k' + \sum_{k=1}^{K} B_k H_{t-1} B_k'.$$

10.12 *(PCA of a covariance matrix)*
Let X be a random vector of $\mathbb{R}^m$ with variance matrix Σ.

1. Find the (or a) first principal component of X, that is a random variable $C^1 = u_1' X$ of maximal variance, where $u_1' u_1 = 1$. Is C^1 unique?
2. Find the second principal component, that is, a random variable $C^2 = u_2' X$ of maximal variance, where $u_2' u_2 = 1$ and $\mathrm{Cov}(C^1, C^2) = 0$.
3. Find all the principal components.

10.13 *(BEKK-GARCH models with a diagonal representation)*
Show that the matrices $A^{(i)}$ and $B^{(j)}$ defined in (10.21) are diagonal when the matrices A_{ik} and B_{jk} are diagonal.

10.14 *(Determinant of a block companion matrix)*
If A and D are square matrices, with D invertible, we have

$$\det \begin{pmatrix} A & B \\ C & D \end{pmatrix} = \det(D) \det(A - BD^{-1}C).$$

Use this property to show that matrix $\mathbb{B}$ in Corollary 10.1 satisfies

$$\det(\mathbb{B} - \lambda I_{mp}) = (-1)^{mp} \det\{\lambda^p I_m - \lambda^{p-1} \mathbf{B}_1 - \cdots - \lambda \mathbf{B}_{p-1} - \mathbf{B}_p\}.$$

10.15 *(Eigenvalues of a product of positive definite matrices)*
Let A and B denote symmetric positive definite matrices of the same size. Show that AB is diagonalizable and that its eigenvalues are positive.

10.16 *(Positive definiteness of a sum of positive semi-definite matrices)*
Consider two matrices of the same size, symmetric, and positive semi-definite, of the form

$$A = \begin{pmatrix} A_{11} & A_{12} \\ A_{21} & A_{22} \end{pmatrix} \quad \text{and} \quad B = \begin{pmatrix} B_{11} & 0 \\ 0 & 0 \end{pmatrix},$$

where A_{11} and B_{11} are also square matrices of the same size. Show that if A_{22} and B_{11} are positive definite, then so is $A + B$.

10.17 *(Positive definite matrix and almost surely positive definite matrix)*
Let A by a symmetric random matrix such that for all real vectors $c \neq 0$,

$$c'Ac > 0 \quad \text{almost surely.}$$

Show that this does not entail that A is almost surely positive definite.

11

Financial Applications

In this chapter, we discuss several financial applications of GARCH models. In connecting these models with those frequently used in mathematical finance, one is faced with the problem that the latter are generally written in continuous time. We start by studying the relation between GARCH and continuous-time processes. We present sufficient conditions for a sequence of stochastic difference equations to converge in distribution to a stochastic differential equation (SDE) as the length of the discrete time intervals between observations goes to zero. We then apply these results to GARCH(1,1)-type models. The second part of this chapter is devoted to the pricing of derivatives. We introduce the notion of stochastic discount factor (SDF) and show how it can be used in the GARCH framework. The final part of the chapter is devoted to risk measurement.

11.1 Relation Between GARCH and Continuous-Time Models

Continuous-time models are central to mathematical finance. Most theoretical results on derivative pricing rely on continuous-time processes, obtained as solutions of diffusion equations. However, discrete-time models are the most widely used in applications. The literature on discrete-time models and that on continuous-time models was developed independently, but it is possible to establish connections between the two approaches.

11.1.1 Some Properties of Stochastic Differential Equations

This first section reviews the basic material from diffusion processes, which will be known to many readers. In some probability space $(\Omega, \mathcal{A}, P)$, a d-dimensional process $\{W_t; 0 \le t < \infty\}$ is called *standard Brownian motion* if $W_0 = 0$ almost surely, for $s \le t$, the increment $W_t - W_s$ is independent of $\sigma\{W_u; u \le s\}$ and is $\mathcal{N}(0, (t-s)I_d)$ distributed, where I_d is the $d \times d$ identity matrix. Brownian motion is a Gaussian process and admits a version with continuous paths.

A SDE in $\mathbb{R}^p$ is an equation of the form

$$dX_t = \mu(X_t)dt + \sigma(X_t)dW_t, \quad 0 \le t < \infty, \quad X_0 = x_0, \tag{11.1}$$

GARCH Models: Structure, Statistical Inference and Financial Applications, Second Edition. Christian Francq and Jean-Michel Zakoian.

where $x_0 \in \mathbb{R}^p$, μ and σ are measurable functions, defined on $\mathbb{R}^p$ and, respectively, taking values in $\mathbb{R}^p$ and $\mathcal{M}_{p\times d}$, the space of $p \times d$ matrices. Here, we only consider time-homogeneous SDEs, in which the functions μ and σ do not depend on t. A process $(X_t)_{t\in[0,T]}$ is a solution of this equation, and is called a *diffusion process*, if it satisfies

$$X_t = x_0 + \int_0^t \mu(X_s)ds + \int_0^t \sigma(X_s)dW_s.$$

Existence and uniqueness of a solution require additional conditions on the functions μ and σ. The simplest conditions require Lipschitz and sublinearity properties:

$$\|\mu(x) - \mu(y)\| + \|\sigma(x) - \sigma(y)\| \leq K\|x - y\|,$$
$$\|\mu(x)\|^2 + \|\sigma(x)\|^2 \leq K^2(1 + \|x\|^2),$$

where $t \in [0, +\infty[$, $x, y \in \mathbb{R}^p$, and K is a positive constant. In these inequalities, $\|\cdot\|$ denotes a norm on either $\mathbb{R}^p$ or $\mathcal{M}_{p\times d}$. These hypotheses also ensure the 'non-explosion' of the solution on every time interval of the form $[0, T]$ with $T > 0$ (see Karatzas and Shreve 1988, Theorem 5.2.9). They can be considerably weakened, in particular when $p = d = 1$. The term $\mu(X_t)$ is called the *drift* of the diffusion, and the term $\sigma(X_t)$ is called the *volatility*. They have the following interpretation:

$$\mu(x) = \lim_{\tau\to 0} \tau^{-1} E(X_{t+\tau} - X_t \mid X_t = x), \tag{11.2}$$

$$\sigma(x)\sigma(x)' = \lim_{\tau\to 0} \tau^{-1} \mathrm{Var}(X_{t+\tau} - X_t \mid X_t = x). \tag{11.3}$$

These relations can be generalised using the *second-order differential operator* defined, in the case $p = d = 1$, by

$$L = \mu\frac{\partial}{\partial x} + \frac{1}{2}\sigma^2\frac{\partial^2}{\partial x^2}.$$

Indeed, for a class of twice continuously differentiable functions f, we have

$$Lf(x) = \lim_{\tau\to 0} \tau^{-1}\{E(f(X_{t+\tau}) \mid X_t = x) - f(x)\}.$$

Moreover, the following property holds: if ϕ is a twice continuously differentiable function with compact support, then the process

$$Y_t = \phi(X_t) - \phi(X_0) - \int_0^t L\phi(X_s)ds$$

is a martingale with respect to the filtration (F_t), where F_t is the σ-field generated by $\{W_s, s \leq t\}$. This result admits a reciprocal which provides a useful characterisation of diffusions. Indeed, it can be shown that if, for a process (X_t), the process (Y_t) just defined is a F_t-martingale, for a class of sufficiently smooth functions ϕ, then (X_t) is a diffusion and solves Eq. (11.1).

Stationary Distribution

In certain cases, the solution of an SDE admits a stationary distribution, but in general, this distribution is not available in explicit form. Wong (1964) showed that, for model (11.1) in the univariate case ($p = d = 1$) with $\sigma(\cdot) \geq 0$, if there exists a function f that solves the equation

$$f(x)\mu(x) = \frac{1}{2}\frac{d}{dx}\{f(x)\sigma^2(x)\}, \tag{11.4}$$

and belongs to the Pearson family of distributions, that is, of the form

$$f(x) = \lambda x^a e^{b/x}, \quad x > 0, \tag{11.5}$$

where $a < -1$ and $b < 0$, then Eq. (11.1) admits a stationary solution with density f.

Example 11.1 (Linear model) A linear SDE is an equation of the form

$$dX_t = (\omega + \mu X_t)dt + \sigma X_t dW_t, \tag{11.6}$$

where ω, μ, and σ are constants. For any initial value x_0, this equation admits a strictly positive solution if $\omega \geq 0$, $x_0 \geq 0$ and $(\omega, x_0) \neq (0, 0)$ (Exercise 11.1). If f is assumed to be of the form (11.5), solving Eq. (11.4) leads to

$$a = -2\left(1 - \frac{\mu}{\sigma^2}\right), \quad b = \frac{-2\omega}{\sigma^2}.$$

Under the constraints

$$\omega > 0, \qquad \sigma \neq 0, \qquad \zeta := 1 - \frac{2\mu}{\sigma^2} > 0,$$

we obtain the stationary density

$$f(x) = \frac{1}{\Gamma(\zeta)}\left(\frac{2\omega}{\sigma^2}\right)^{\zeta} \exp\left(\frac{-2\omega}{\sigma^2 x}\right) x^{-1-\zeta}, \quad x > 0,$$

where Γ denotes the gamma function. If this distribution is chosen for the initial distribution (i.e. the law of X_0), then the process (X_t) is stationary and its inverse follows a gamma distribution,[1]

$$\frac{1}{X_t} \sim \Gamma\left(\frac{2\omega}{\sigma^2}, \zeta\right). \tag{11.7}$$

11.1.2 Convergence of Markov Chains to Diffusions

Consider a Markov chain $Z^{(\tau)} = (Z^{(\tau)}_{k\tau})_{k\in\mathbb{N}}$ with values in $\mathbb{R}^d$, indexed by the time unit $\tau > 0$. We transform $Z^{(\tau)}$ into a continuous-time process, $(Z^{(\tau)}_t)_{t\in\mathbb{R}^+}$, by means of the time interpolation

$$Z^{(\tau)}_t = Z^{(\tau)}_{k\tau} \quad \text{if } k\tau \leq t < (k+1)\tau.$$

Under conditions given in the next theorem, the process $(Z^{(\tau)}_t)$ converges in distribution to a diffusion. Denote by $\|\cdot\|$ the Euclidean norm on $\mathbb{R}^d$.

Theorem 11.1 (Convergence of $(Z^{(\tau)}_t)$ to a diffusion) *Suppose there exist continuous applications μ and σ from $\mathbb{R}^d$ to $\mathbb{R}^d$ and $\mathcal{M}_{p\times d}$, respectively, such that for all $r > 0$ and for some $\delta > 0$,*

$$\lim_{\tau\to 0} \sup_{\|z\|\leq r} \left| \tau^{-1} E\left(Z^{(\tau)}_{(k+1)\tau} - Z^{(\tau)}_{k\tau} \mid Z^{(\tau)}_{k\tau} = z\right) - \mu(z)\right| = 0, \tag{11.8}$$

$$\lim_{\tau\to 0} \sup_{\|z\|\leq r} \left| \tau^{-1} \mathrm{Var}\left(Z^{(\tau)}_{(k+1)\tau} - Z^{(\tau)}_{k\tau} \mid Z^{(\tau)}_{k\tau} = z\right) - \sigma(z)\sigma(z)'\right| = 0, \tag{11.9}$$

$$\overline{\lim}_{\tau\to 0} \sup_{\|z\|\leq r} \tau^{-(2+\delta)/2} E\left(\|Z^{(\tau)}_{(k+1)\tau} - Z^{(\tau)}_{k\tau}\|^{2+\delta} \mid Z^{(\tau)}_{k\tau} = z\right) < \infty. \tag{11.10}$$

Then, if the equation

$$dZ_t = \mu(Z_t)dt + \sigma(Z_t)dW_t, \quad 0 \leq t < \infty, \quad Z_0 = z_0, \tag{11.11}$$

admits a solution (Z_t) which is unique in distribution, and if $Z^{(\tau)}_0$ converges in distribution to z_0, then the process $(Z^{(\tau)}_t)$ converges in distribution to (Z_t).

---

[1] The $\Gamma(a, b)$ density, for $a, b > 0$, is defined on $\mathbb{R}^+$ by

$$f(x) = \frac{a^b}{\Gamma(b)} e^{-ax} x^{b-1}.$$

Remark 11.1 Condition (11.10) ensures, in particular, that by applying the Markov inequality, for all $\epsilon > 0$,

$$\lim_{\tau \to 0} \tau^{-1} P(\| Z^{(\tau)}_{(k+1)\tau} - Z^{(\tau)}_{k\tau} \| > \epsilon \mid Z^{(\tau)}_{k\tau} = z) = 0.$$

As a consequence, the limiting process has continuous paths.

Euler Discretisation of a Diffusion

Diffusion processes do not admit an exact discretisation in general. An exception is the *geometric Brownian motion*, defined as a solution of the real SDE

$$dX_t = \mu X_t dt + \sigma X_t dW_t \tag{11.12}$$

where μ and σ are constants. It can be shown that if the initial value x_0 is strictly positive, then $X_t \in (0, \infty)$ for any $t > 0$. By Itô's lemma,[2] we obtain

$$d\log(X_t) = \left(\mu - \frac{\sigma^2}{2}\right) dt + \sigma dW_t \tag{11.13}$$

and then, by integration of this equation between times $k\tau$ and $(k+1)\tau$, we get the discretised version of model (11.12),

$$\log X_{(k+1)\tau} = \log X_{k\tau} + \left(\mu - \frac{\sigma^2}{2}\right)\tau + \sqrt{\tau}\sigma\epsilon_{(k+1)\tau}, \quad (\epsilon_{k\tau}) \overset{iid}{\sim} \mathcal{N}(0, 1). \tag{11.14}$$

For general diffusions, an explicit discretised model does not exist, but a natural approximation, called the *Euler discretisation*, is obtained by replacing the differential elements by increments. The Euler discretisation of the SDE (11.1) is then given, for the time unit τ, by

$$X_{(k+1)\tau} = X_{k\tau} + \tau\mu(X_{k\tau}) + \sqrt{\tau}\sigma(X_{k\tau})\epsilon_{(k+1)\tau}, \qquad (\epsilon_{k\tau}) \overset{iid}{\sim} \mathcal{N}(0, 1). \tag{11.15}$$

The Euler discretisation of a diffusion converges in distribution to this diffusion (Exercise 11.2).

Convergence of GARCH-M Processes to Diffusions

It is natural to assume that the return of a financial asset increases with risk. Economic agents who are risk-averse must receive compensation when they own risky assets. ARMA-GARCH type time series models do not take this requirement into account because the conditional mean and variance are modelled separately. A simple way to model the dependence between the average return and risk is to specify the conditional mean of the returns in the form

$$\mu_t = \xi + \lambda\sigma_t,$$

where ξ and λ are parameters. By doing so we obtain, when σ_t^2 is specified as an ARCH, a particular case of the *ARCH in mean* (ARCH-M) model, introduced by Engle, Lilien, and Robins (1987). The parameter λ can be interpreted as the price of risk and can thus be assumed to be positive. Other specifications of

---

[2] For $Y_t = f(t, X_t)$ where $f: (t, x) \in [0, T] \times \mathbb{R} \mapsto f(t, x) \in \mathbb{R}$ is continuous, continuously differentiable with respect to the first component and twice continuously differentiable with respect to the second component, if (X_t) satisfies $dX_t = \mu_t dt + \sigma_t dW_t$ where μ_t and σ_t are adapted processes such that almost surely $\int_0^T |\mu_t| \, dt < \infty$ and $\int_0^T \sigma_t^2 dt < \infty$, we have

$$dY_t = \frac{\partial f}{\partial t}(t, X_t)dt + \frac{\partial f}{\partial x}(t, X_t)dX_t + \frac{1}{2}\frac{\partial^2 f}{\partial x^2}(t, X_t)\sigma_t^2 dt.$$

the conditional mean are obviously possible. In this section, we focus on a GARCH(1, 1)-M model of the form

$$\begin{cases} X_t = X_{t-1} + f(\sigma_t) + \sigma_t \eta_t, & (\eta_t)\ iid\ (0,1), \\ g(\sigma_t) = \omega + a(\eta_{t-1}) g(\sigma_{t-1}), \end{cases} \tag{11.16}$$

where $\omega > 0$, f is a continuous function from $\mathbb{R}^+$ to $\mathbb{R}$, g is a continuous one-to-one map from $\mathbb{R}^+$ to itself and a is a positive function. The previous interpretation implies that f is increasing, but at this point, it is not necessary to make this assumption. When $g(x) = x^2$ and $a(x) = \alpha x^2 + \beta$ with $\alpha \geq 0$, $\beta \geq 0$, we get the classical GARCH(1, 1) model. Asymmetric effects can be introduced, for instance by taking $g(x) = x$ and $a(x) = \alpha_+ x^+ - \alpha_- x^- + \beta$, with $x^+ = \max(x, 0)$, $x^- = \min(x, 0)$, $\alpha_+ \geq 0$, $\alpha_- \geq 0$, $\beta \geq 0$.

Observe that the constraint for the existence of a strictly stationary and non-anticipative solution (Y_t), with $Y_t = X_t - X_{t-1}$, is written as

$$E \log\{a(\eta_t)\} < 0,$$

by the techniques studied in Chapter 2.

Now, in view of the Euler discretisation (11.15), we introduce the sequence of models indexed by the time unit τ, defined by

$$\begin{cases} X_{k\tau}^{(\tau)} = X_{(k-1)\tau}^{(\tau)} + f(\sigma_{k\tau}^{(\tau)})\tau + \sqrt{\tau}\sigma_{k\tau}\eta_{k\tau}^{(\tau)}, & (\eta_{k\tau}^{(\tau)})\ iid\ (0,1) \\ g(\sigma_{(k+1)\tau}^{(\tau)}) = \omega_\tau + a_\tau(\eta_{k\tau}^{(\tau)}) g(\sigma_{k\tau}^{(\tau)}), \end{cases} \tag{11.17}$$

for $k > 0$, with initial values $X_0^{(\tau)} = x_0$, $\sigma_0^{(\tau)} = \sigma_0 > 0$ and assuming $E(\eta_{k\tau}^{(\tau)})^4 < \infty$. The introduction of a delay k in the second equation is due to the fact that $\sigma_{(k+1)\tau}$ belongs to the σ-field generated by $\eta_{k\tau}$ and its past values.

Noting that the pair $Z_{k\tau} = (X_{(k-1)\tau}, g(\sigma_{k\tau}))$ defines a Markov chain, we obtain its limiting distribution by application of Theorem 11.1. We have, for $z = (x, g(\sigma))$,

$$\tau^{-1} E(X_{k\tau}^{(\tau)} - X_{(k-1)\tau}^{(\tau)} \mid Z_{(k-1)\tau}^{(\tau)} = z) = f(\sigma), \tag{11.18}$$

$$\tau^{-1} E(g(\sigma_{(k+1)\tau}^{(\tau)}) - g(\sigma_{k\tau}^{(\tau)}) \mid Z_{(k-1)\tau}^{(\tau)} = z) = \tau^{-1}\omega_\tau + \tau^{-1}\{E a_\tau(\eta_{k\tau}^{(\tau)}) - 1\} g(\sigma). \tag{11.19}$$

This latter quantity converges if

$$\lim_{\tau\to 0} \tau^{-1}\omega_\tau = \omega, \quad \lim_{\tau\to 0} \tau^{-1}\{E a_\tau(\eta_{k\tau}^{(\tau)}) - 1\} = -\delta, \tag{11.20}$$

where ω and δ are constants.

Similarly,

$$\tau^{-1} \mathrm{Var}[X_{k\tau}^{(\tau)} - X_{(k-1)\tau}^{(\tau)} \mid Z_{(k-1)\tau}^{(\tau)} = z] = \sigma^2, \tag{11.21}$$

$$\tau^{-1} \mathrm{Var}[g(\sigma_{(k+1)\tau}^{(\tau)}) - g(\sigma_{k\tau}^{(\tau)}) \mid Z_{(k-1)\tau}^{(\tau)} = z] = \tau^{-1} \mathrm{Var}\{a_\tau(\eta_{k\tau}^{(\tau)})\} g^2(\sigma), \tag{11.22}$$

which converges if and only if

$$\lim_{\tau\to 0} \tau^{-1} \mathrm{Var}\{a_\tau(\eta_{k\tau}^{(\tau)})\} = \zeta, \tag{11.23}$$

where ζ is a positive constant. Finally,

$$\tau^{-1} \mathrm{Cov}[X_{k\tau}^{(\tau)} - X_{(k-1)\tau}^{(\tau)}, g(\sigma_{(k+1)\tau}^{(\tau)}) - g(\sigma_{k\tau}^{(\tau)}) \mid Z_{(k-1)\tau}^{(\tau)} = z]$$
$$= \tau^{-1/2} \mathrm{Cov}\{\eta_{k\tau}^{(\tau)}, a_\tau(\eta_{k\tau}^{(\tau)})\} \sigma g(\sigma), \tag{11.24}$$

converges if and only if

$$\lim_{\tau\to 0} \tau^{-1/2} \mathrm{Cov}\{\eta_{k\tau}^{(\tau)}, a_\tau(\eta_{k\tau}^{(\tau)})\} = \rho, \tag{11.25}$$

where ρ is a constant such that $\rho^2 \leq \zeta$.

Under these conditions, we thus have

$$\lim_{\tau\to 0} \tau^{-1}\mathrm{Var}[Z_{k\tau}^{(\tau)} - Z_{(k-1)\tau}^{(\tau)} \mid Z_{(k-1)\tau}^{(\tau)} = z] = A(\sigma) = \begin{pmatrix} \sigma^2 & \rho\sigma g(\sigma) \\ \rho\sigma g(\sigma) & \zeta g^2(\sigma) \end{pmatrix}.$$

Moreover, we have

$$A(\sigma) = B(\sigma)B'(\sigma), \quad \text{where } B(\sigma) = \begin{pmatrix} \sigma & 0 \\ \rho g(\sigma) & \sqrt{\zeta - \rho^2}g(\sigma) \end{pmatrix}.$$

We are now in a position to state our next result.

Theorem 11.2 (Convergence of $(X_t^{(\tau)}, g(\sigma_t^{(\tau)}))$ to a diffusion) *Under Conditions (11.20), (11.23), and (11.25), and if, for $\delta > 0$,*

$$\overline{\lim}_{\tau\to 0}\, \tau^{-(1+\delta)} E\{a_\tau(\eta_{k\tau}^{(\tau)}) - 1\}^{2(1+\delta)} < \infty, \tag{11.26}$$

the limiting process when $\tau \to 0$, in the sense of Theorem 11.1, of the sequence of solutions of models (11.17) is the bivariate diffusion

$$\begin{cases} dX_t &= f(\sigma_t)dt + \sigma_t dW_t^1 \\ dg(\sigma_t) &= \{\omega - \delta g(\sigma_t)\}dt + g(\sigma_t)(\rho dW_t^1 + \sqrt{\zeta - \rho^2}dW_t^2), \end{cases} \tag{11.27}$$

where (W_t^1) and (W_t^2) are independent Brownian motions, with initial values x_0 and σ_0.

Proof. It suffices to verify the conditions of Theorem 11.1. It is immediate from (11.18), (11.19), (11.21), (11.22), (11.24) and the hypotheses on f and g, that, in (11.8) and (11.9), the limits are uniform on every ball of $\mathbb{R}^2$. Moreover, for $\tau \le \tau_0$ and $\delta < 2$, we have

$$\begin{aligned} \tau^{-\frac{2+\delta}{2}} E(|X_{k\tau}^{(\tau)} - X_{(k-1)\tau}^{(\tau)}|^{2+\delta} \mid Z_{k\tau}^{(\tau)} = z) &= \tau^{-\frac{2+\delta}{2}} E(|f(\sigma)\tau + \sqrt{\tau}\sigma\eta_0^{(\tau)}|^{2+\delta}) \\ &\le E(| f(\sigma) \mid \sqrt{\tau_0} + \sigma \mid \eta_0^{(\tau)} \mid)^{2+\delta}, \end{aligned}$$

which is bounded uniformly in σ on every compact. On the other hand, introducing the $L^{2+\delta}$ norm and using the triangle and Hölder's inequalities,

$$\begin{aligned} &\tau^{-\frac{2+\delta}{2}} E(|g(\sigma_{(k+1)\tau}^{(\tau)}) - g(\sigma_{k\tau}^{(\tau)})|^{2+\delta} \mid Z_{k\tau}^{(\tau)} = z) \\ &= E|\tau^{-1/2}\omega_\tau + \tau^{-1/2}\{a_\tau(\eta_{k\tau}^{(\tau)}) - 1\}g(\sigma)|^{2+\delta} \\ &= \|\tau^{-1/2}\omega_\tau + \tau^{-1/2}\{a_\tau(\eta_{k\tau}^{(\tau)}) - 1\}g(\sigma)\|_{2+\delta}^{2+\delta} \\ &\le (\tau^{-1/2}\omega_\tau + \tau^{-1/2}\|a_\tau(\eta_{k\tau}^{(\tau)}) - 1\|_{2+\delta} g(\sigma))^{2+\delta} \\ &\le \left\{\tau^{-1/2}\omega_\tau + \tau^{-1/2}(E|a_\tau(\eta_{k\tau}^{(\tau)}) - 1|^2 E|a_\tau(\eta_{k\tau}^{(\tau)}) - 1|^{2(1+\delta)})^{\frac{1}{2(2+\delta)}} g(\sigma)\right\}^{2+\delta}. \end{aligned}$$

Since the limit superior of this quantity, when $\tau \to 0$, is bounded uniformly in σ on every compact, we can conclude that condition (11.10) is satisfied.

It remains to show that the SDE (11.27) admits a unique solution. Note that $g(\sigma_t)$ satisfies a linear SDE given by

$$dg(\sigma_t) = \{\omega - \delta g(\sigma_t)\}dt + \sqrt{\zeta}g(\sigma_t)dW_t^3, \tag{11.28}$$

where (W_t^3) is the Brownian motion $W_t^3 = (\rho W_t^1 + \sqrt{\zeta - \rho^2}W_t^2)/\sqrt{\zeta}$. This equation admits a unique solution (Exercise 11.1)

$$g(\sigma_t) = Y_t\left(g(\sigma_0) + \omega\int_0^t \frac{1}{Y_s}ds\right), \quad \text{where } Y_t = \exp\{-(\delta + \zeta/2)t + \sqrt{\zeta}W_t^3\}.$$

The function g being one-to-one, we deduce σ_t and the solution (X_t), uniquely obtained as

$$X_t = x_0 + \int_0^t f(\sigma_s)ds + \int_0^t \sigma_s dW_s^1.$$

□

Remark 11.2

1. It is interesting to note that the limiting diffusion involves two Brownian motions, whereas GARCH processes involve only one noise. This can be explained by the fact that, to obtain a Markov chain, it is necessary to consider the pair $(X_{(k-1)\tau}, g(\sigma_{k\tau}))$. The Brownian motions involved in the equations of X_t and $g(\sigma_t)$ are independent if and only if $\rho = 0$. This is, for instance, the case when the function a_τ is even and the distribution of the iid process is symmetric.

2. Equation (11.28) shows that $g(\sigma_t)$ is the solution of a linear model of the form (11.6). From the study of this model, we know that under the constraints

$$\omega > 0, \qquad \zeta > 0, \qquad 1 + \frac{2\delta}{\zeta} > 0,$$

there exists a stationary distribution for $g(\sigma_t)$. If the process is initialised with this distribution, then

$$\frac{1}{g(\sigma_t)} \sim \Gamma\left(\frac{2\omega}{\zeta}, 1 + \frac{2\delta}{\zeta}\right). \tag{11.29}$$

Example 11.2 (GARCH(1, 1)) The volatility of the GARCH(1, 1) model is obtained for $g(x) = x^2$, $a(x) = \alpha x^2 + \beta$. Suppose for simplicity that the distribution of the process $(\eta_{k\tau}^{(\tau)})$ does not depend on τ and admits moments of order $4(1+\delta)$, for $\delta > 0$. Denote by μ_r the rth-order moment of this process. Conditions (11.20) and (11.23) take the form

$$\lim_{\tau\to 0} \tau^{-1}\omega_\tau = \omega, \quad \lim_{\tau\to 0} \tau^{-1}(\mu_4 - 1)\alpha_\tau^2 = \zeta, \quad \lim_{\tau\to 0} \tau^{-1}(\alpha_\tau + \beta_\tau - 1) = -\delta.$$

A choice of parameters satisfying these constraints is, for instance,

$$\omega_\tau = \omega\tau, \quad \alpha_\tau = \sqrt{\frac{\zeta\tau}{\mu_4 - 1}}, \quad \beta_\tau = 1 - \alpha_\tau - \delta\tau.$$

Condition (11.25) is then automatically satisfied with

$$\rho = \sqrt{\zeta}\frac{\mu_3}{\sqrt{\mu_4 - 1}},$$

as well as Condition (11.26). The limiting diffusion takes the form

$$\begin{cases} dX_t = f(\sigma_t)dt + \sigma_t dW_t^1 \\ d\sigma_t^2 = \{\omega - \delta\sigma_t^2\}dt + \sqrt{\frac{\zeta}{\mu_4-1}}\sigma_t^2(\mu_3 dW_t^1 + \sqrt{\mu_4 - 1 - \mu_3^2}dW_t^2) \end{cases}$$

and, if the law of $\eta_{k\tau}^{(\tau)}$ is symmetric,

$$\begin{cases} dX_t = f(\sigma_t)dt + \sigma_t dW_t^1 \\ d\sigma_t^2 = \{\omega - \delta\sigma_t^2\}dt + \sqrt{\zeta}\sigma_t^2 dW_t^2. \end{cases} \tag{11.30}$$

Note that, with other choices of the rates of convergence of the parameters, we can obtain a limiting process involving only one Brownian motion but with a degenerate volatility equation, in the sense that it is an ordinary differential equation (Exercise 11.3).

Example 11.3 (TGARCH(1, 1)) For $g(x) = x$, $a(x) = \alpha_+ x^+ - \alpha_- x^- + \beta$, we have the volatility of the threshold GARCH model. Under the assumptions of the previous example, let $\mu_{r+} = E(\eta_0^+)^r$ and $\mu_{r-} = E(-\eta_0^-)^r$. Conditions (11.20) and (11.23) take the form

$$\lim_{\tau\to 0} \tau^{-1}\omega_\tau = \omega, \quad \lim_{\tau\to 0} \tau^{-1}\{\alpha_{\tau+}^2\mu_{2+} + \alpha_{\tau-}^2\mu_{2-} - (\alpha_{\tau+}\mu_{1+} + \alpha_{\tau-}\mu_{1-})^2\} = \zeta,$$

$$\lim_{\tau\to 0} \tau^{-1}(\alpha_{\tau+}\mu_{1+} + \alpha_{\tau-}\mu_{1-} + \beta_\tau - 1) = -\delta.$$

These constraints are satisfied by taking, for instance, $\alpha_{\tau+} = \sqrt{\tau}\alpha_+$, $\alpha_{\tau-} = \sqrt{\tau}\alpha_-$ and $\omega_\tau = \omega\tau$, $\alpha_+^2\mu_{2+} + \alpha_-^2\mu_{2-} - (\alpha_+\mu_{1+} + \alpha_-\mu_{1-})^2 = \zeta$, $\beta_\tau = 1 - \alpha_{\tau+}\mu_{1+} - \alpha_{\tau-}\mu_{1-} - \delta\tau$. Condition (11.25) is then satisfied with $\rho = \alpha_+\mu_{2+} - \alpha_-\mu_{2-}$, as well as condition (11.26). The limiting diffusion takes the form

$$\begin{cases} dX_t = f(\sigma_t)dt + \sigma_t dW_t^1 \\ d\sigma_t = \{\omega - \delta\sigma_t\}dt + \sigma_t(\rho dW_t^1 + \sqrt{\zeta - \rho^2}dW_t^2). \end{cases}$$

In particular, if the law of $\eta_{k\tau}^{(\tau)}$ is symmetric and if $\alpha_{\tau+} = \alpha_{\tau-}$, the correlation between the Brownian motions of the two equations vanishes, and we get a limiting diffusion of the form

$$\begin{cases} dX_t = f(\sigma_t)dt + \sigma_t dW_t^1 \\ d\sigma_t = \{\omega - \delta\sigma_t\}dt + \sqrt{\zeta}\sigma_t dW_t^2. \end{cases}$$

By applying the Itô lemma to this equation, we obtain

$$d\sigma_t^2 = 2\{\omega - \delta\sigma_t\}\sigma_t dt + 2\sqrt{\zeta}\sigma_t^2 dW_t^2,$$

which shows that, even in the symmetric case, the limiting diffusion does not coincide with that obtained in Eq. (11.30) for the classical GARCH model. When the law of $\eta_{k\tau}^{(\tau)}$ is symmetric and $\alpha_{\tau+} \neq \alpha_{\tau-}$, the asymmetry in the discrete-time model results in a correlation between the two Brownian motions of the limiting diffusion.

11.2 Option Pricing

Classical option pricing models rely on independent Gaussian returns processes. These assumptions are incompatible with the empirical properties of prices, as we saw, in particular, in Chapter 1. It is thus natural to consider pricing models founded on more realistic, GARCH-type, or stochastic volatility, price dynamics.

We start by briefly recalling the terminology and basic concepts related to the Black–Scholes model. Appropriate financial references are provided at the end of this chapter.

11.2.1 Derivatives and Options

The need to hedge against several types of risks gave rise to a number of financial assets called *derivatives*. A derivative (*derivative security* or *contingent claim*) is a financial asset whose *payoff* depends on the price process of an *underlying asset*: action, portfolio, stock index, currency, etc. The definition of this payoff is settled in a contract.

There are two basic types of option. A *call option* (*put option*) or more simply a call (put) is a derivative giving to the holder the right, but not the obligation, to buy (sell) an agreed quantity of the underlying asset S, from the seller of the option on (or before) the *expiration date* T, for a specified price K, *the strike price* or *exercise price*. The seller (or 'writer') of a call is obliged to sell the underlying asset should the buyer so decide. The buyer pays a fee, called a *premium*, for this right. The most common options, since their introduction in 1973, are the *European options*, which can be exercised only at the option expiry date, and the *American options*, which can be exercised at any time during the life of the option. For a European call option, the buyer receives, at the expiry date, the amount $\max(S_T - K, 0) = (S_T - K)^+$ since the option will not be exercised unless it is 'in the money'. Similarly, for a put, the payoff at time T is $(K - S_T)^+$. *Asset pricing* involves determining the option price at time t. In what follows, we shall only consider European options.

11.2.2 The Black–Scholes Approach

Consider a market with two assets, an underlying asset, and a risk-free asset. The Black and Scholes (1973) model assumes that the price of the underlying asset is driven by a geometric Brownian motion

$$dS_t = \mu S_t dt + \sigma S_t dW_t, \tag{11.31}$$

where μ and σ are constants and (W_t) is a standard Brownian motion. The risk-free interest rate r is assumed to be constant. By Itô's lemma, we obtain

$$d\log(S_t) = \left(\mu - \frac{\sigma^2}{2}\right) dt + \sigma dW_t, \tag{11.32}$$

showing that the logarithm of the price follows a generalised Brownian motion, with drift $\mu - \sigma^2/2$ and constant volatility. Integrating the two sides of Equality (11.32) between times $t-1$ and t yields the discretised version

$$\log\left(\frac{S_t}{S_{t-1}}\right) = \mu - \frac{\sigma^2}{2} + \sigma\epsilon_t, \qquad \epsilon_t \overset{iid}{\sim} \mathcal{N}(0, 1). \tag{11.33}$$

The assumption of constant volatility is obviously unrealistic. However, this model allows for explicit formulas for option prices, or more generally for any derivative based on the underlying asset S, with payoff $g(S_T)$ at the expiry date T. The price of this product at time t is unique under certain regularity conditions[3] and is denoted by $C(S, t)$ for simplicity. The set of conditions ensuring the uniqueness of the derivative price is referred to as the *complete market hypothesis*. In particular, these conditions imply the *absence of arbitrage opportunities*, that is, that there is no 'free lunch'. It can be shown[4] that the derivative price is

$$C(S, t) = e^{-r(T-t)} E^{\pi}[g(S_T) \mid S_t], \tag{11.34}$$

where the expectation is computed under the probability π corresponding to the equation

$$dS_t = rS_t dt + \sigma S_t dW_t^*, \tag{11.35}$$

where (W_t^*) denotes a standard Brownian motion. The probability π is called the *risk-neutral* probability, because under π the expected return of the underlying asset is the risk-free interest rate r. It is important to distinguish this from the *historic probability*, that is, the law under which the data are generated (here defined by model (11.31)). Under the risk-neutral probability, the price process is still a geometric Brownian motion, with the same volatility σ but with drift r. Note that the initial drift term, μ, does not play a role in the pricing formula (11.34). Moreover, the actualised price $X_t = e^{-rt}S_t$ satisfies $dX_t = \sigma X_t dW_t^*$. This implies that the actualised price is a martingale for the risk-neutral probability: $e^{-r(T-t)}E^{\pi}[S_T \mid S_t] = S_t$. Note that this formula is obvious in view of Relation (11.34), by considering the underlying asset as a product with payoff S_T at time T.

---

[3] These conditions are the absence of transaction costs, the possibility of constructing a portfolio with any allocations (sign and size) of the two assets, the possibility of continuously adjusting the composition of the portfolio and the existence of a price for the derivative depending only on the present and past values of S_t.

[4] The three classical methods for proving this formula are the method based on the binomial model (Cox, Ross, and Rubinstein 1979), the method based on the resolution of equations with partial derivatives, and the method based on the martingale theory.

The Black–Scholes formula is an explicit version of (11.34) when the derivative is a call, that is, when $g(S_T) = (K - S_T)^+$, given by

$$C(S,t) = S_t\Phi(x_t + \sigma\sqrt{\tau}) - e^{-r\tau}K\Phi(x_t), \tag{11.36}$$

where Φ is the conditional distribution function (cdf) of the $\mathcal{N}(0,1)$ distribution and

$$\tau = T - t, \quad x_t = \frac{\log(S_t/e^{-r\tau}K)}{\sigma\sqrt{\tau}} - \frac{1}{2}\sigma\sqrt{\tau}.$$

In particular, it can be seen that if S_t is large compared to K, we have $\Phi(x_t + \sigma\sqrt{\tau}) \approx \Phi(x_t) \approx 1$, and the call price is approximately given by $S_t - e^{-r\tau}K$, that is, the current underlying price minus the actualised exercise price. The price of a put $P(S,t)$ follows from the *put–call parity* relationship (Exercise 11.4): $C(S,t) = P(S,t) + S_t - e^{-r\tau}K$.

A simple computation (Exercise 11.5) shows that the European call option price is an increasing function of S_t, which is intuitive. The derivative of $C(S,t)$ with respect to S_t, called *delta*, is used in the construction of a riskless hedge, a portfolio obtained from the risk-free and risky assets allowing the seller of a call to cover the risk of a loss when the option is exercised. The construction of a riskless hedge is often referred to as *delta hedging*.

The previous approach can be extended to other price processes, in particular if (S_t) is solution of a SDE of the form

$$dS_t = \mu(t, S_t)dt + \sigma(t, S_t)dW_t$$

under regularity assumptions on μ and σ. When the geometric Brownian motion for S_t is replaced by another dynamics, the complete market property is generally lost.[5]

11.2.3 Historic Volatility and Implied Volatilities

Note that, from a statistical point of view, the sole unknown parameter in the Black–Scholes pricing formula (11.36) is the volatility of the underlying asset. Assuming that the prices follow a geometric Brownian motion, application of this formula thus requires estimating σ. Any estimate of σ based on a history of prices $S_0, \dots, S_n$ is referred to as *historic volatility*. For geometric Brownian motion, the log-returns $\log(S_t/S_{t-1})$ are, by (11.32), iid $\mathcal{N}(\mu - \sigma^2/2, \sigma^2)$ distributed variables. Several estimation methods for σ can be considered, such as the method of moments and the maximum likelihood method (Exercise 11.7). An estimator of $C(S,t)$ is then obtained by replacing σ by its estimate.

Another approach involves using option prices. In practice, traders usually work with the so-called *implied volatilities*. These are the volatilities implied by option prices observed in the market. Consider a European call option whose price at time t is $\tilde{C}_t$. If $\tilde{S}_t$ denotes the price of the underlying asset at time t, an implied volatility σ_t^I is defined by solving the equation

$$\tilde{C}_t = \tilde{S}_t\Phi(x_t + \sigma_t^I\sqrt{\tau}) - e^{-r\tau}K\Phi(x_t), \quad x_t = \frac{\log(\tilde{S}_t/e^{-r\tau}K)}{\sigma_t^I\sqrt{\tau}} - \frac{1}{2}\sigma_t^I\sqrt{\tau}.$$

This equation cannot be solved analytically, and numerical procedures are called for. Note that the solution is unique because the call price is an increasing function of σ (Exercise 11.8).

If the assumptions of the Black–Scholes model, that is, the geometric Brownian motion, are satisfied, implied volatilities calculated from options with different characteristics, but the same underlying asset

[5] This is the case for the stochastic volatility model of Hull and White (1987), defined by

$$\begin{cases} dS_t = \phi S_t dt + \sigma_t S_t dW_t \\ d\sigma_t^2 = \mu\sigma_t^2 dt + \xi\sigma_t^2 dW_t^*, \end{cases}$$

where W_t, W_t^* are independent Brownian motions.

should coincide with the theoretical volatility σ. In practice, implied volatilities calculated with different strikes or expiration dates are very unstable, which is not surprising since we know that the geometric Brownian motion is a misspecified model.

11.2.4 Option Pricing when the Underlying Process is a GARCH

In discrete time, with time unit δ, the binomial model (in which, given S_t, $S_{t+\delta}$ can only take two values) allows us to define a unique risk-neutral probability, under which the actualised price is a martingale. This model is used, in the Cox, Ross, and Rubinstein (1979) approach, as an analog in discrete time of the geometric Brownian motion. Intuitively, the assumption of a complete market is satisfied (in the binomial model as well as in the Black–Scholes model) because the number of assets, two, coincides with the number of states of the world at each date. Apart from this simple situation, the complete market property is generally lost in discrete time. It follows that a multiplicity of probability measures may exist, under which the prices are martingales, and consequently, a multiplicity of pricing formulas such as (11.34). Roughly speaking, there is too much variability in prices between consecutive dates.

To determine options prices in incomplete markets, additional assumptions can be made on the risk premium and/or the preferences of the agents. A modern alternative relies on the concept of SDF, which allows pricing formulas in discrete time similar to those in continuous time to be obtained.

Stochastic Discount Factor

We start by considering a general setting. Suppose that we observe a vector process $Z = (Z_t)$ and let I_t denote the information available at time t, that is, the σ-field generated by $\{Z_s, s \le t\}$. We are interested in the pricing of a derivative whose payoff is $g = g(Z_T)$ at time T. Suppose that there exists, at time $t < T$, a price $C_t(Z, g, T)$ for this asset. It can be shown that, under mild assumptions on the function $g \mapsto C_t(Z, g, T)$,[6] we have the representation

$$C(g,t,T) = E[g(Z_T)M_{t,T} \mid I_t], \qquad \text{where } M_{t,T} > 0, \quad M_{t,T} \in I_T. \tag{11.37}$$

The variable $M_{t,T}$ is called the SDF for the period $[t, T]$. The SDF introduced in representation (11.37) is not unique and can be parameterised. The formula applies, in particular, to the *zero-coupon bond* of expiry date T, defined as the asset with payoff 1 at time T. Its price at t is the conditional expectation of the SDF,

$$B(t,T) = C(1,t,T) = E[M_{t,T} \mid I_t].$$

It follows that Relation (11.37) can be written as

$$C(g,t,T) = B(t,T)E\left[g(Z_T)\frac{M_{t,T}}{B(t,T)} \mid I_t\right]. \tag{11.38}$$

Forward Risk-Neutral Probability

Observe that the ratio $M_{t,T}/B(t, T)$ is positive and that its mean, conditional on I_t, is 1. Consequently, a probability change removing this factor in formula (11.38) can be done.[7] Denoting by $\pi_{t,T}$, the new probability and by $E^{\pi_{t,T}}$ the expectation under this probability, we obtain the pricing formula

$$C(g,t,T) = B(t,T)E^{\pi_{t,T}}[g(Z_T) \mid I_t]. \tag{11.39}$$

[6] In particular, linearity and positivity (see Hansen and Richard 1987; Gouriéroux and Tiomo 2007).

[7] If X is a random variable with distribution P_X and G is a function with real positive values such that $E\{G(X)\} = 1$, we have, interpreting G as the density of some measure P_X^* with respect to P_X,

$$E\{XG(X)\} = \int xG(x)dP_X(x) = \int xdP_X^*(x) =: E^*(X).$$

The probability law $\pi_{t,T}$ is called *forward risk-neutral* probability. Note the analogy between this formula and Relation (11.34), the latter corresponding to a particular form of $B(t, T)$. To make this formula operational, it remains to specify the SDF.

Risk-Neutral Probability

As mentioned earlier, the SDF is not unique in incomplete markets. A natural restriction, referred to as a temporal coherence restriction, is given by

$$M_{t,T} = M_{t,t+1}M_{t+1,t+2} \cdots M_{T-1,T}. \tag{11.40}$$

On the other hand, the one-step SDFs are constrained by

$$B(t, t+1) = E[M_{t,t+1} \mid I_t], \quad S_t = E[S_{t+1}M_{t,t+1} \mid I_t], \tag{11.41}$$

where $S_t \in I_t$ is the price of an underlying asset (or a vector of assets). We have

$$C(g, t, T) = B(t, t+1)E\left[g(Z_T)\prod_{i=t+1}^{T-1} B(i, i+1)\prod_{i=t}^{T-1}\frac{M_{i,i+1}}{B(i, i+1)} \mid I_t\right]. \tag{11.42}$$

Noting that

$$E\left[\prod_{i=t}^{T-1}\frac{M_{i,i+1}}{B(i, i+1)} \mid I_t\right] = E\left[\prod_{i=t}^{T-2}\frac{M_{i,i+1}}{B(i, i+1)}E\left(\frac{M_{T-1,T}}{B(T-1, T)} \mid I_{T-1}\right) \mid I_t\right] = 1,$$

we can make a change of probability such that the SDF vanishes. Under a probability law $\pi^*_{t,T}$, called *risk-neutral* probability, we thus have

$$C(g, t, T) = B(t, t+1)E^{\pi^*_{t,T}}\left[g(Z_T)\prod_{i=t+1}^{T-1} B(i, i+1) \mid I_t\right]. \tag{11.43}$$

The risk-neutral probability satisfies the temporal coherence property: $\pi^*_{t,T+1}$ is related to $\pi^*_{t,T}$ through the factor $M_{T,T+1}/B(T, T+1)$. Without the restriction (11.40), the risk-neutral *forward* probability does not satisfy this property.

Pricing Formulas

One approach to deriving pricing formulas is to specify, parametrically, the dynamics of Z_t and of $M_{t,t+1}$, taking the constraint (11.41) into account.

Example 11.4 (Black–Scholes model) Consider model (11.33) with $Z_t = \log(S_t/S_{t-1})$ and suppose that $B(t, t+1) = e^{-r}$. A simple specification of the one-step SDF is given by the affine model $M_{t,t+1} = \exp(a + bZ_{t+1})$, where a and b are constants. The constraints (11.41) are written as

$$e^{-r} = E\exp(a + bZ_{t+1}), \quad 1 = E\exp\{a + (b+1)Z_{t+1}\},$$

that is, in view of the $\mathcal{N}(\mu - \sigma^2/2, \sigma^2)$ distribution of Z_t,

$$0 = a + r + b\left(\mu - \frac{\sigma^2}{2}\right) + \frac{b^2\sigma^2}{2} = a + (b+1)\left(\mu - \frac{\sigma^2}{2}\right) + \frac{(b+1)^2\sigma^2}{2}.$$

These equations provide a unique solution (a, b). We then obtain the risk-neutral probability $\pi_{t,t+1} = \pi$ through the characteristic function

$$E^{\pi}(e^{uZ_{t+1}}) = E\left(e^{uZ_{t+1}}\frac{M_{t,t+1}}{B(t, t+1)}\right) = E(e^{a+r+(b+u)Z_{t+1}}) = e^{u(r-\sigma^2/2)+u^2\sigma^2/2}.$$

Because the latter is the characteristic function of the $\mathcal{N}(r-\sigma^2/2,\sigma^2)$ distribution, we retrieve the geometric Brownian motion model (11.35) for (S_t). The Black–Scholes formula is thus obtained by specifying an affine exponential SDF with constant coefficients.

Now consider a general GARCH-type model of the form

$$\begin{cases} Z_t = \log(S_t/S_{t-1}) = \mu_t + \epsilon_t, \\ \epsilon_t = \sigma_t\eta_t, \qquad (\eta_t) \overset{iid}{\sim} \mathcal{N}(0,1), \end{cases} \tag{11.44}$$

where μ_t and σ_t belong to the σ-field generated by the past of Z_t, with $\sigma_t > 0$. Suppose that the σ-fields generated by the past of ϵ_t, Z_t, and η_t are the same, and denote this σ-field by I_{t-1}. Suppose again that $B(t, t+1) = e^{-r}$. Consider for the SDF an affine exponential specification with random coefficients, given by

$$M_{t,t+1} = \exp(a_t + b_t\eta_{t+1}), \tag{11.45}$$

where $a_t, b_t \in I_t$. The constraints (11.41) are written as

$$\begin{aligned} e^{-r} &= E\exp(a_t + b_t\eta_{t+1} \mid I_t), \\ 1 &= E\exp\{a_t + b_t\eta_{t+1} + Z_{t+1} \mid I_t\} = E\exp\{a_t + \mu_{t+1} + (b_t + \sigma_{t+1})\eta_{t+1} \mid I_t\}, \end{aligned}$$

that is, after simplification,

$$a_t = -r - \frac{b_t^2}{2}, \quad b_t\sigma_{t+1} = r - \mu_{t+1} - \frac{\sigma_{t+1}^2}{2}. \tag{11.46}$$

As before, these equations provide a unique solution (a_t, b_t). The risk-neutral probability $\pi_{t,t+1}$ is defined through the characteristic function

$$\begin{aligned} E^{\pi_{t,t+1}}(e^{uZ_{t+1}} \mid I_t) &= E(e^{uZ_{t+1}}M_{t,t+1}/B(t,t+1) \mid I_t) \\ &= E(e^{a_t + r + u\mu_{t+1} + (b_t + u\sigma_{t+1})\eta_{t+1}} \mid I_t) \\ &= \exp\left(u(\mu_{t+1} + b_t\sigma_{t+1}) + u^2\frac{\sigma_{t+1}^2}{2} \right) \\ &= \exp\left(u\left(r - \frac{\sigma_{t+1}^2}{2}\right) + u^2\frac{\sigma_{t+1}^2}{2} \right). \end{aligned}$$

The last two equalities are obtained by taking into account the constraints on a_t and b_t. Thus, under the probability $\pi_{t,t+1}$, the law of the process (Z_t) is given by the model

$$\begin{cases} Z_t = r - \frac{\sigma_t^2}{2} + \epsilon_t^*, \\ \epsilon_t^* = \sigma_t\eta_t^*, \qquad (\eta_t^*) \overset{iid}{\sim} \mathcal{N}(0,1). \end{cases} \tag{11.47}$$

The independence of the η_t^* follows from the independence between η_{t+1}^* and I_t (because η_{t+1}^* has a fixed distribution conditional on I_t) and from the fact that $\eta_t^* = \sigma_t^{-1}(Z_t - r + \sigma_t^2/2) \in I_t$.

The model under the risk-neutral probability is then a GARCH-type model if the variable σ_t^2 is a measurable function of the past of ϵ_t^*. This generally does not hold because the relation

$$\epsilon_t^* = \mu_t - r + \frac{\sigma_t^2}{2} + \epsilon_t \tag{11.48}$$

entails that the past of ϵ_t^* is included in the past of ϵ_t, but not the reverse.

If the relation (11.48) is invertible, in the sense that there exists a measurable function f such that $\epsilon_t = f(\epsilon_t^*, \epsilon_{t-1}^*, \dots)$, model (11.47) is of the GARCH type, but the volatility σ_t^2 can take a very complicated form as a function of the ϵ^*_{t-j}. Specifically, if the volatility under the historic probability is that of a classical GARCH(1, 1), we have

$$\sigma_t^2 = \omega + \alpha\left(r - \frac{\sigma_{t-1}^2}{2} - \mu_{t-1} + \epsilon^*_{t-1}\right)^2 + \beta\sigma_{t-1}^2.$$

Finally, using $\pi_{t,T}$, the forward risk-neutral probability for the expiry date T, the price of a derivative is given by the formula

$$C_t(Z, g, T) = e^{-r(T-t)}E^{\pi_{t,T}}[g(Z_T) \mid S_t] \tag{11.49}$$

or, under the historic probability, in view of the temporal coherence restriction (11.40),

$$C_t(Z, g, T) = E[g(Z_T)M_{t,t+1}M_{t+1,t+2}\dots M_{T-1,T} \mid S_t]. \tag{11.50}$$

It is important to note that, with the affine exponential specification of the SDF, the volatilities coincide with the two probability measures. This will not be the case for other SDFs (Exercise 11.11).

Example 11.5 (Constant coefficient SDF) We have seen that the Black–Scholes formula relies on (i) a Gaussian marginal distribution for the log-returns, and (ii) an affine exponential SDF with constant coefficients. In the framework of model (11.44), it is thus natural to look for a SDF of the same type, with a_t and b_t independent of t. It is immediate from the relations (11.46) that it is necessary and sufficient to take μ_t of the form

$$\mu_t = \mu + \lambda\sigma_t - \frac{\sigma_t^2}{2},$$

where μ and λ are constants. We thus obtain a model of GARCH in mean type, because the conditional mean is a function of the conditional variance. The volatility in the risk-neutral model is thus written as

$$\sigma_t^2 = \omega + \alpha\{(r - \mu) - \lambda\sigma_{t-1} + \epsilon^*_{t-1}\}^2 + \beta\sigma_{t-1}^2.$$

If, moreover, $r = \mu$, then under the historic probability, the model is expressed as

$$\begin{cases} \log(S_t/S_{t-1}) &= r + \lambda\sigma_t - \frac{\sigma_t^2}{2} + \epsilon_t, \\ \epsilon_t &= \sigma_t\eta_t, \qquad (\eta_t) \overset{iid}{\sim} \mathcal{N}(0, 1), \\ \sigma_t^2 &= \omega + \alpha\epsilon_{t-1}^2 + \beta\sigma_{t-1}^2, \qquad \omega > 0, \alpha, \beta \geq 0, \end{cases} \tag{11.51}$$

and under the risk-neutral probability,

$$\begin{cases} \log(S_t/S_{t-1}) &= r - \frac{\sigma_t^2}{2} + \epsilon_t^*, \\ \epsilon_t^* &= \sigma_t\eta_t^*, \qquad (\eta_t^*) \overset{iid}{\sim} \mathcal{N}(0, 1), \\ \sigma_t^2 &= \omega + \alpha(\epsilon^*_{t-1} - \lambda\sigma_{t-1})^2 + \beta\sigma_{t-1}^2. \end{cases} \tag{11.52}$$

Under the latter probability, the actualised price $e^{-rt}S_t$ is a martingale (Exercise 11.9). Note that in Model (11.52) the coefficient λ appears in the conditional variance, but the risk has been neutralised (the conditional mean no longer depends on σ_t^2). This risk-neutral probability was obtained by Duan (1995) using a different approach based on the utility of the representative agent.

Note that $\sigma_t^2 = \omega + \sigma_{t-1}^2\{\alpha(\lambda - \eta^*_{t-1})^2 + \beta\}$. Under the strict stationarity constraint

$$E\log\{\alpha(\lambda - \eta_t^*)^2 + \beta\} < 0,$$

the variable σ_t^2 is a function of the past of η_t^* and can be interpreted as the volatility of Z_t under the risk-neutral probability. Under this probability, the model is not a classical GARCH(1, 1) unless $\lambda = 0$, but in this case, the risk-neutral probability coincides with the historic one, which is not the case in practice.

Numerical Pricing of Option Prices

Explicit computation of the expectation involved in Formula (11.49) is not possible, but the expectation can be evaluated by simulation. Note that, under $\pi_{t,T}$, S_T, and S_t are linked by the formula

$$S_T = S_t \exp\left\{ (T-t)r - \frac{1}{2}\sum_{s=t+1}^{T} h_s + \sum_{s=t+1}^{T} \sqrt{h_s} v_s \right\},$$

where $h_s = \sigma_s^2$. At time t, suppose that an estimate $\hat{\theta} = (\hat{\lambda}, \hat{\omega}, \hat{\alpha}, \hat{\beta})$ of the coefficients of model (11.51) is available, obtained from observations $S_1, \dots, S_t$ of the underlying asset. Simulated values $S_T^{(i)}$ of S_T, and thus simulated values $Z_T^{(i)}$ of Z_T, for $i = 1, \dots, N$, are obtained by simulating, at step i, $T-t$ independent realisations $v_s^{(i)}$ of the $\mathcal{N}(0,1)$ and by setting

$$S_T^{(i)} = S_t \exp\left\{ (T-t)r - \frac{1}{2}\sum_{s=t+1}^{T} h_s^{(i)} + \sum_{s=t+1}^{T} \sqrt{h_s^{(i)}} v_s^{(i)} \right\},$$

where the $h_s^{(i)}$, $s = t+1, \dots, T$, are recursively computed from

$$h_s^{(i)} = \hat{\omega} + \{\hat{\alpha}(v_{s-1}^{(i)} - \hat{\lambda})^2 + \hat{\beta}\} h_{s-1}^{(i)}$$

taking, for instance, as initial value $h_t^{(i)} = \hat{\sigma}_t^2$, the volatility estimated from the initial GARCH model (this volatility being computed recursively, and the effect of the initialisation being negligible for t large enough under stationarity assumptions). This choice can be justified by noting that for SDFs of the form (11.45), the volatilities coincide under the historic and risk-neutral probabilities. Finally, a simulation of the derivative price is obtained by taking

$$\hat{C}_t(Z, g, T) = e^{-r(T-t)} \frac{1}{N} \sum_{i=1}^{N} g(Z_T^{(i)}).$$

The previous approach can obviously be extended to more general GARCH models, with larger orders and/or different volatility specifications. It can also be extended to other SDFs.

Empirical studies show that, comparing the computed prices with actual prices observed on the market, GARCH option pricing provides much better results than the classical Black–Scholes approach (see, for instance, Sabbatini and Linton 1998; Härdle and Hafner 2000).

To conclude this section, we observe that the formula providing the theoretical option prices can also be used to estimate the parameters of the underlying process, using observed options (see, for instance, Hsieh and Ritchken 2005).

11.3 Value at Risk and Other Risk Measures

Risk measurement is becoming more and more important in the financial risk management of banks and other institutions involved in financial markets. The need to quantify risk typically arises when a financial institution has to determine the amount of capital to hold as a protection against unexpected losses. In fact, risk measurement is concerned with all types of risks encountered in finance. *Market risk*, the best-known type of risk, is the risk of change in the value of a financial position. *Credit risk*, also a very important type of risk, is the risk of not receiving repayments on outstanding loans, as a result of borrower's default. *Operational risk*, which has received more and more attention in recent years, is the risk of losses resulting from failed internal processes, people, and systems, or external events. *Liquidity risk* occurs when, due to a lack of marketability, an investment cannot be bought or sold quickly enough to prevent a loss. *Model risk* can be defined as the risk due to the use of a misspecified model for the risk measurement.

The need for risk measurement has increased dramatically, in the last two decades, due to the introduction of new regulation procedures. In 1996, the Basel Committee on Banking Supervision (a committee established by the central bank governors in 1974) prescribed a so-called *standardised model* for market risk. At the same time, the Committee allowed the larger financial institutions to develop their own *internal model*. The second Basel Accord (Basel II), initiated in 2001, considers operational risk as a new risk class and prescribes the use of finer approaches to assess the risk of credit portfolios. By using sophisticated approaches, the banks may reduce the amount of regulatory capital (the capital required to support the risk), but in the event of frequent losses, a larger amount may be imposed by the regulator. Parallel developments took place in the insurance sector, giving rise to the Solvency projects. In response to the financial crisis of 2008, the third Basel accord, whose first measures started in 2013, strengthened the regulation and risk managment of banks, in particular by introducing a liquidity coverage ratio.

A risk measure that is used for specifying capital requirements can be thought of as the amount of capital that must be added to a position to make its risk acceptable to regulators. Value at risk (VaR) is arguably the most widely used risk measure in financial institutions. In 1993, the business bank JP Morgan publicised its estimation method, *RiskMetrics*, for the VaR of a portfolio. VaR is now an indispensable tool for banks, regulators, and portfolio managers. We start by defining VaR and discussing its properties.

11.3.1 Value at Risk

Definition

VaR is concerned with the possible loss of a portfolio in a given time horizon. A natural risk measure is the maximum possible loss. However, in most models, the support of the loss distribution is unbounded so that the maximum loss is infinite. The concept of VaR replaces the maximum loss by a *maximum loss which is not exceeded with a given (high) probability.*

VaR should be computed using the *predictive distribution* of future losses, that is the conditional distribution of future losses using the current information. However, for horizons $h > 1$, this conditional distribution may be hard to obtain.

To be more specific, consider a portfolio whose value at time t is a random variable denoted V_t. At horizon h, the *loss* is denoted

$$L_{t,t+h} = -(V_{t+h} - V_t).$$

The distribution of $L_{t,t+h}$ is called the *loss distribution* (conditional or not). This distribution is used to compute the regulatory capital which allows certain risks to be covered, but not all of them. In general, V_t is specified as a function of d unobservable *risk factors*.

Example 11.6 Suppose, for instance, that the portfolio is composed of d stocks. Denote by $S_{i,t}$ the price of stock i at time t and by $r_{i,t,t+h} = \log S_{i,t+h} - \log S_{i,t}$ the log-return. If a_i is the number of stocks i in the portfolio, we have

$$V_t = \sum_{i=1}^{d} a_i S_{i,t}.$$

Assuming that the composition of the portfolio remains fixed between the dates t and $t+h$, we have

$$L_{t,t+h} = -\sum_{i=1}^{d} a_i S_{i,t}(e^{r_{i,t,t+h}} - 1).$$

The distribution of $-L_{t,t+h}$ conditional on the available information at time t is called the profit and loss (P&L) distribution.

The determination of reserves depends on

- the portfolio,
- the available information at time t and the horizon h,[8]
- a level $\alpha \in (0, 1)$ characterising the acceptable risk.[9]

Denote by $R_{t,h}(\alpha)$ the level of the reserves. Including these reserves, which are not subject to remuneration, the value of the portfolio at time $t+h$ becomes $V_{t+h} + R_{t,h}(\alpha)$. The capital used to support risk, the VaR, also includes the current portfolio value,

$$\mathrm{VaR}_{t,h}(\alpha) = V_t + R_{t,h}(\alpha),$$

and satisfies

$$P_t[V_{t+h} - V_t < -\mathrm{VaR}_{t,h}(\alpha)] < \alpha,$$

where P_t is the probability conditional on the information available at time t.[10] VaR can thus be interpreted as the capital exposed to risk in the event of bankruptcy. Equivalently,

$$P_t[\mathrm{VaR}_{t,h}(\alpha) < L_{t,t+h}] < \alpha, \quad \text{i.e. } P_t\,[L_{t,t+h} \leq \mathrm{VaR}_{t,h}(\alpha)] \geq 1 - \alpha. \tag{11.53}$$

In probabilistic terms, $\mathrm{VaR}_{t,h}(\alpha)$ is thus simply the $(1-\alpha)$-quantile of the conditional loss distribution. If, for instance, for a confidence level 99% and a horizon of 10 days, the VaR of a portfolio is €5000, this means that, if the composition of the portfolio does not change, there is a probability of 1% that the potential loss over 10 days will be larger than €5000.

Definition 11.1 *The $(1-\alpha)$-quantile of the conditional loss distribution is called the VaR at the level α:*

$$\mathrm{VaR}_{t,h}(\alpha) := \inf\{x \in \mathbb{R} \mid P_t[L_{t,t+h} \leq x] \geq 1 - \alpha\},$$

when this quantile is positive. By convention $\mathrm{VaR}_{t,h}(\alpha) = 0$ *otherwise.*

In particular, it is obvious that $\mathrm{VaR}_{t,h}(\alpha)$ is a decreasing function of α.

From Definition (11.53), computing a VaR simply reduces to determining a quantile of the conditional loss distribution. Figure 11.1 compares the VaR of three distributions, with the same variance but different tail thicknesses. The thickest tail, proportional to $1/x^4$, is that of the Student t distribution with 3 degrees of freedom, here denoted as $\mathcal{S}$; the thinnest tail, proportional to $e^{-x^2/2}$, is that of the Gaussian $\mathcal{N}$; and the double exponential $\mathcal{E}$ possesses a tail of intermediate size, proportional to $e^{-\sqrt{2}|x|}$. For some very small level α, the VaRs are ranked in the order suggested by the thickness of the tails: $\mathrm{VaR}(\mathcal{N}) < \mathrm{VaR}(\mathcal{E}) < \mathrm{VaR}(\mathcal{S})$. However, Figure 11.1b shows that this ranking does not hold for the standard levels $\alpha = 1\%$ or 5%.

[8] For market risk management, h is typically 1 or 10 days. For the regulator (concerned with credit or operational risk) h is 1 year.

[9] $1-\alpha$ is often referred to as the confidence level. Typical values of α are 5% or 3%.

[10] In the 'standard' approach, the conditional distribution is replaced by the unconditional.

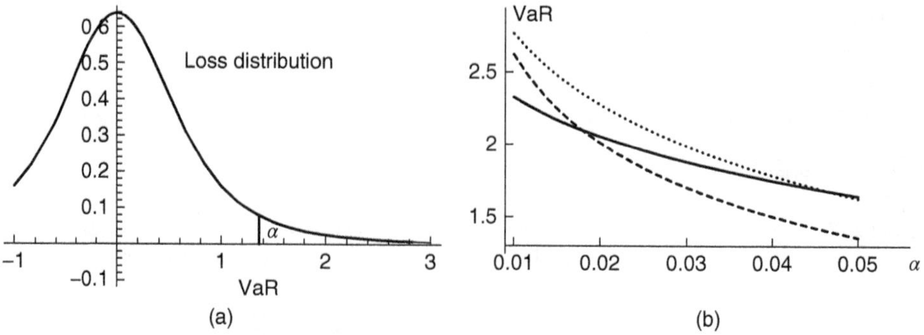

Figure 11.1 (a) VaR is the $(1-\alpha)$-quantile of the conditional loss distribution. (b) VaR as a function of $\alpha \in [1\%, 5\%]$ for a Gaussian distribution $\mathcal{N}$ (solid line), a Student t distribution with 3 degrees of freedom $\mathcal{S}$ (dashed line) and a double exponential distribution $\mathcal{E}$ (thin dotted line). The three laws are standardised so as to have unit variances. For $\alpha = 1\%$, we have $\mathrm{VaR}(\mathcal{N}) < \mathrm{VaR}(\mathcal{S}) < \mathrm{VaR}(\mathcal{E})$, whereas for $\alpha = 5\%$, we have $\mathrm{VaR}(\mathcal{S}) < \mathrm{VaR}(\mathcal{E}) < \mathrm{VaR}(\mathcal{N})$.

VaR and Conditional Moments

Let us introduce the first two moments of $L_{t,t+h}$ conditional on the information available at time t:

$$m_{t,t+h} = E_t(L_{t,t+h}), \quad \sigma^2_{t,t+h} = \mathrm{Var}_t(L_{t,t+h}).$$

Suppose that

$$L_{t,t+h} = m_{t,t+h} + \sigma_{t,t+h} L_h^*, \tag{11.54}$$

where L_h^* is a random variable with cumulative distribution function F_h. Denote by $F_h^{\leftarrow}$ the quantile function of the variable L_h^*, defined as the generalised inverse of F_h:

$$F_h^{\leftarrow}(\alpha) = \inf\{x \in \mathbb{R} \mid F_h(x) \geq \alpha\}, \qquad 0 < \alpha < 1.$$

If F_h is continuous and strictly increasing, we simply have $F_h^{\leftarrow}(\alpha) = F_h^{-1}(\alpha)$, where F_h^{-1} is the ordinary inverse of F_h. In view of the relations (11.53) and (11.54), it follows that

$$1 - \alpha = P_t[\mathrm{VaR}_{t,h}(\alpha) \geq m_{t,t+h} + \sigma_{t,t+h} L^*] = F_h\left(\frac{\mathrm{VaR}_{t,h}(\alpha) - m_{t,t+h}}{\sigma_{t,t+h}}\right).$$

Consequently,

$$\mathrm{VaR}_{t,h}(\alpha) = m_{t,t+h} + \sigma_{t,t+h} F_h^{\leftarrow}(1-\alpha). \tag{11.55}$$

VaR can thus be decomposed into an 'expected loss' $m_{t,t+h}$, the conditional mean of the loss, and an 'unexpected loss' $\sigma_{t,t+h} F^{\leftarrow}(1-\alpha)$, also called *economic capital*.

The apparent simplicity of formula (11.55) masks difficulties in (i) deriving the first conditional moments for a given model, and (ii) determining the law F_h, supposedly independent of t, of the standardised loss at horizon h.

Consider the price of a portfolio, defined as a combination of the prices of d assets, $p_t = a' P_t$, where $a, P_t \in \mathbb{R}^d$. Introducing the price variations $\Delta P_t = P_t - P_{t-1}$, we have

$$L_{t,t+h} = -(p_{t+h} - p_t) = -a'(P_{t+h} - P_t) = -a' \sum_{i=1}^{h} \Delta P_{t+i}.$$

The *term structure* of the VaR, that is its evolution as a function of the horizon, can be analysed in different cases.

Example 11.7 (Independent and identically distributed price variations) If the ΔP_{t+i} are iid $\mathcal{N}(m, \Sigma)$ distributed, the law of $L_{t,t+h}$ is $\mathcal{N}(-a'mh, a'\Sigma ah)$. In view of Relation (11.55), it follows that

$$\mathrm{VaR}_{t,h}(\alpha) = -a'mh + \sqrt{a'\Sigma a}\sqrt{h}\Phi^{-1}(1-\alpha). \tag{11.56}$$

In particular, if $m = 0$, we have

$$\mathrm{VaR}_{t,h}(\alpha) = \sqrt{h}\mathrm{VaR}_{t,1}(\alpha). \tag{11.57}$$

The rule that one multiplies the VaR at horizon 1 by $\sqrt{h}$ to obtain the VaR at horizon h is often erroneously used when the prices variations are not iid, centred and Gaussian (Exercise 11.12).

Example 11.8 (AR(1) price variations) Suppose now that

$$\Delta P_t - m = A(\Delta P_{t-1} - m) + U_t, \quad (U_t) \ iid \ \mathcal{N}(0, \Sigma),$$

where A is a matrix whose eigenvalues have a modulus strictly less than 1. The process (ΔP_t) is then stationary with expectation m. It can be verified (Exercise 11.13) that

$$\mathrm{VaR}_{t,h}(\alpha) = a'\mu_{t,h} + \sqrt{a'\Sigma_h a}\Phi^{-1}(1-\alpha), \tag{11.58}$$

where, letting $A_i = (I - A^i)(I - A)^{-1}$,

$$\mu_{t,h} = -mh - AA_h(\Delta P_t - m), \quad \Sigma_h = \sum_{j=1}^{h} A_{h-j+1}\Sigma A'_{h-j+1}.$$

If $A = 0$, Relation (11.58) reduces to Relation (11.56). Apart from this case, the term multiplying $\Phi^{-1}(1-\alpha)$ is not proportional to $\sqrt{h}$.

Example 11.9 (ARCH(1) price variations) For simplicity, let $d = 1$, $a = 1$ and

$$\Delta P_t = \sqrt{\omega + \alpha_1 \Delta P_{t-1}^2} U_t, \quad \omega > 0, \alpha_1 \geq 0, \quad (U_t) \ iid \ \mathcal{N}(0, 1). \tag{11.59}$$

The conditional law of $L_{t,t+1}$ is $\mathcal{N}(0, \omega + \alpha_1 \Delta P_t^2)$. Therefore,

$$\mathrm{VaR}_{t,1}(\alpha) = \sqrt{\omega + \alpha_1 \Delta P_t^2}\Phi^{-1}(1-\alpha).$$

Here VaR computation at horizons larger than 1 is problematic. Indeed, the conditional distribution of $L_{t,t+h}$ is no longer Gaussian when $h > 1$ (Exercise 11.14).

It is often more convenient to work with the log-returns $r_t = \Delta_1 \log p_t$, assumed to be stationary, than with the price variations. Letting $q_t(h, \alpha)$ be the α-quantile of the conditional distribution of the future returns $r_{t+1} + \dots + r_{t+h}$, we obtain (Exercise 11.15)

$$\mathrm{VaR}_{t,h}(\alpha) = \{1 - e^{q_t(h,\alpha)}\}p_t. \tag{11.60}$$

Non Subadditivity of VaR

VaR is often criticised for not satisfying, for *any* distribution of the price variations, the 'subadditivity' property. Subadditivity means that the VaR for two portfolios after they have been merged should be no greater than the sum of their VaRs before they were merged. However, this property does not hold: if L_1 and L_2 are two loss variables, we do not necessarily have, in obvious notation,

$$\mathrm{VaR}_{t,h}^{L_1+L_2}(\alpha) \leq \mathrm{VaR}_{t,h}^{L_1}(\alpha) + \mathrm{VaR}_{t,h}^{L_2}(\alpha), \quad \forall \alpha, t, h. \tag{11.61}$$

Example 11.10 (Pareto distribution) Let L_1 and L_2 be two independent variables, Pareto distributed, with density $f(x) = (2+x)^{-2}\mathbb{1}_{x>-1}$. The cdf of this distribution is $F(x) = (1-(2+x)^{-1})\mathbb{1}_{x>-1}$, whence the VaR at level α is $\mathrm{Var}(\alpha) = \alpha^{-1} - 2$. It can be verified, for instance using Mathematica, that

$$P[L_1 + L_2 \le x] = 1 - \frac{2}{4+x} - \frac{2\log(3+x)}{(4+x)^2}, \qquad x > -2.$$

Thus

$$P[L_1 + L_2 \le 2\mathrm{VaR}(\alpha)] = 1 - \alpha - \frac{\alpha^2}{2}\log\left(\frac{2-\alpha}{\alpha}\right) < 1 - \alpha.$$

It follows that $\mathrm{VaR}_{L_1+L_2}(\alpha) > \mathrm{VaR}_{L_1}(\alpha) + \mathrm{VaR}_{L_2}(\alpha) = 2\mathrm{VaR}_{L_1}(\alpha), \forall\alpha \in]0,1[$. If, for instance, $\alpha = 0.01$, we find $\mathrm{VaR}_{L_1}(0.01) = 98$ and, numerically, $\mathrm{VaR}_{L_1+L_2}(0.01) \approx 203.2$.

The lack of subadditivity of the VaR is often interpreted negatively as this does not encourage diversification. Note however that, as underlined by the previous example, the distribution of an average of non integrable iid variables can be more dispersed than the common distribution of the individual variables. In such an extreme situation, diversification does not reduce risk, and any good risk measure should be overadditive, as it is the case for the VaR.

11.3.2 Other Risk Measures

Even if VaR is the most widely used risk measure, the choice of an adequate risk measure is an open issue. As already seen, the convexity property, with respect to the portfolio composition, is not satisfied for VaR with some distributions of the loss variable. In what follows, we present several alternatives to VaR, together with a conceptualisation of the 'expected' properties of risks measures.

Volatility and Moments

In the Markowitz (1952) portfolio theory, the variance is used as a risk measure. It might then seem natural, in a dynamic framework, to use the volatility as a risk measure. However, volatility does not take into account the signs of the differences from the conditional mean. More importantly, this measure does not satisfy some 'coherency' properties, as will be seen later (translation invariance, subadditivity).

Expected Shortfall

The expected shortfall (ES), or anticipated loss, is the standard risk measure used in insurance since Solvency II. This risk measure is closely related to VaR, but avoids certain of its conceptual difficulties. It is more sensitive than VaR to the shape of the conditional loss distribution in the tail of the distribution. In contrast to VaR, it is informative about the expected loss when a big loss occurs.

Let $L_{t,t+h}$ be such that $EL^+_{t,t+h} < \infty$. In this section, the conditional distribution of $L_{t,t+h}$ is assumed to have a continuous and strictly increasing cdf. The ES at level α, also referred to as Tailvar, is defined as the conditional expectation of the loss given that the loss exceeds the VaR:

$$\mathrm{ES}_{t,h}(\alpha) := E_t[L_{t,t+h} \mid L_{t,t+h} > \mathrm{VaR}_{t,h}(\alpha)]. \tag{11.62}$$

We have

$$E_t[L_{t,t+h}\mathbb{1}_{L_{t,t+h}>\mathrm{VaR}_{t,h}(\alpha)}] = E_t[L_{t,t+h} \mid L_{t,t+h} > \mathrm{VaR}_{t,h}(\alpha)]P_t[L_{t,t+h} > \mathrm{VaR}_{t,h}(\alpha)].$$

Now $P_t[L_{t,t+h} > \mathrm{VaR}_{t,h}(\alpha)] = 1 - P_t[L_{t,t+h} \le \mathrm{VaR}_{t,h}(\alpha)] = 1 - (1-\alpha) = \alpha$, where the last but one equality follows from the continuity of the cdf at $\mathrm{VaR}_{t,h}(\alpha)$. Thus

$$\mathrm{ES}_{t,h}(\alpha) = \frac{1}{\alpha}E_t[L_{t,t+h}\mathbb{1}_{L_{t,t+h}>\mathrm{VaR}_{t,h}(\alpha)}]. \tag{11.63}$$

The following characterisation also holds (Exercise 11.16):

$$\mathrm{ES}_{t,h}(\alpha) = \frac{1}{\alpha}\int_0^\alpha \mathrm{VaR}_{t,h}(u)du. \tag{11.64}$$

ES thus can be interpreted, for a given level α, as the mean of the VaR over all levels $u \leq \alpha$. Obviously, $\mathrm{ES}_{t,h}(\alpha) \geq \mathrm{VaR}_{t,h}(\alpha)$.

Note that the integral representation makes $\mathrm{ES}_{t,h}(\alpha)$ a continuous function of α, whatever the distribution (continuous or not) of the loss variable. VaR does not satisfy this property (for loss variables which have a zero mass over certain intervals).

Example 11.11 (The Gaussian case) If the conditional loss distribution is $\mathcal{N}(m_{t,t+h}, \sigma^2_{t,t+h})$ then, by Relation (11.55), $\mathrm{VaR}_{t,h}(\alpha) = m_{t,t+h} + \sigma_{t,t+h}\Phi^{-1}(1-\alpha)$, where Φ is the $\mathcal{N}(0,1)$ cdf. Using Definition (11.62) and introducing L^*, a variable of law $\mathcal{N}(0,1)$, we have

$$\begin{aligned}\mathrm{ES}_{t,h}(\alpha) &= m_{t,t+h} + \sigma_{t,t+h}E[L^* \mid L^* \geq \Phi^{-1}(1-\alpha)]\\ &= m_{t,t+h} + \sigma_{t,t+h}\frac{1}{\alpha}E[L^* 1 1_{L^*\geq\Phi^{-1}(1-\alpha)}]\\ &= m_{t,t+h} + \sigma_{t,t+h}\frac{1}{\alpha}\phi\{\Phi^{-1}(1-\alpha)\},\end{aligned}$$

where ϕ is the density of the standard Gaussian. For instance, if $\alpha = 0.05$, the conditional standard deviation is multiplied by 1.65 in the VaR formula, and by 2.06 in the ES formula.

More generally, we have under assumption (11.54), in view of Expressions (11.64) and (11.55),

$$\mathrm{ES}_{t,h}(\alpha) = m_{t,t+h} + \sigma_{t,t+h}\frac{1}{\alpha}\int_0^\alpha F_h^{\leftarrow}(1-u)du. \tag{11.65}$$

Distortion Risk Measures

Continue to assume that the cdf F of the loss distribution is continuous and strictly increasing. For notational simplicity, we omit the indices t and h. From Definition (11.64), the ES is written as

$$\mathrm{ES}(\alpha) = \int_0^1 F^{-1}(1-u)\,\mathbb{1}_{[0,\alpha]}(u)\frac{1}{\alpha}du,$$

where the term $\mathbb{1}_{[0,\alpha]}\frac{1}{\alpha}$ can be interpreted as the density of the uniform distribution over $[0, \alpha]$. More generally, a *distortion risk measure* (DRM) is defined as a number

$$r(F;G) = \int_0^1 F^{-1}(1-u)dG(u),$$

where G is a cdf on [0,1], called *distortion function*, and F is the loss distribution. The introduction of a probability distribution on the confidence levels is often interpreted in terms of optimism or pessimism. If G admits a density g which is increasing on [0,1], that is, if G is convex, the weight of the quantile $F^{-1}(1-u)$ increases with u: large risks receive small weights with this choice of G. Conversely, if g decreases, those large risks receive the bigger weights.

VaR at level α is a DRM, obtained by taking for G the Dirac mass at α. As we have seen, the ES corresponds to the constant density g on $[0, \alpha]$: it is simply an average over all levels below α.

A family of DRMs is obtained by parameterising the distortion measure as

$$r_p(F;G) = \int_0^1 F^{-1}(1-u)dG_p(u),$$

where the parameter p reflects the confidence level, that is the degree of optimism in the face of risk.

Example 11.12 (Exponential DRM) Let

$$G_p(u) = \frac{1 - e^{-pu}}{1 - e^{-p}},$$

where $p \in]0, +\infty[$. We have.

$$r_p(F; G) = \int_0^1 F^{-1}(1-u) \frac{pe^{-pu}}{1 - e^{-p}} du.$$

The density function g is decreasing whatever the value of p, which corresponds to an over-weighting of the larger risks.

Coherent Risk Measures

In response to criticisms of VaR, several notions of *coherent* risk measures have been introduced. One of the proposed definitions is the following.

Definition 11.2 *Let $\mathcal{L}$ denote a set of real random loss variables defined on a measurable space $(\Omega, \mathcal{A})$. Suppose that $\mathcal{L}$ contains all the variables that are almost surely constant and is closed under addition and multiplication by scalars. An application $\rho : \mathcal{L} \mapsto \mathbb{R}$ is called a coherent risk measure if it has the following properties:*

1. *Monotonicity: $\forall L_1, L_2 \in \mathcal{L}, L_1 \leq L_2 \Rightarrow \rho(L_1) \leq \rho(L_2)$.*
2. *Subadditivity: $\forall L_1, L_2 \in \mathcal{L}, L_1 + L_2 \in \mathcal{L} \Rightarrow \rho(L_1 + L_2) \leq \rho(L_1) + \rho(L_2)$.*
3. *Positive homogeneity: $\forall L \in \mathcal{L}, \forall \lambda \geq 0, \rho(\lambda L) = \lambda \rho(L)$.*
4. *Translation invariance: $\forall L \in \mathcal{L}, \forall c \in \mathbb{R}, \rho(L + c) = \rho(L) + c$.*

This definition has the following immediate consequences:

1. $\rho(0) = 0$, using the homogeneity property with $L = 0$. More generally, $\rho(c) = c$ for all constants c (if a loss of amount c is certain, a cash amount c should be added to the portfolio).
2. If $L \geq 0$, then $\rho(L) \geq 0$. If a loss is certain, an amount of capital must be added to the position.
3. $\rho(L - \rho(L)) = 0$, that is the deterministic amount $\rho(L)$ cancels the risk of L.

These requirements, which are controversial (see Example 11.10), are not satisfied for most risk measures used in finance. The variance, or more generally any risk measure based on the centred moments of the loss distribution, does not satisfy the monotonicity property, for instance. The expectation can be seen as a coherent, but uninteresting, risk measure. VaR satisfies all conditions except subadditivity: we have seen that this property holds for (dependent or independent) Gaussian variables, but not for general variables. ES is a coherent risk measure in the sense of Definition 11.2 (Exercise 11.17). It can be shown (see Wang and Dhaene 1998) that DRMs with G concave satisfy the subadditivity requirement. Note that, because finite expectation is required, ES and DRMs do not apply to fat tailed distributions such as the Pareto distributions of Example 11.10, for which overaddidivity is desirable.

11.3.3 Estimation Methods

Unconditional VaR

The simplest estimation method is based on the K last returns at horizon h, that is, $r_{t+h-i}(h) = \log(p_{t+h-i}/p_{t-i})$, for $i = h \dots, h + K - 1$. These K returns are viewed as scenarios for future returns. The non-parametric *historical* VaR is simply obtained by replacing $q_t(h, \alpha)$ in Relation (11.60) by the empirical α-quantile of the last K returns. Typical values are $K = 250$ and $\alpha = 1\%$, which means that the third worst return is used as the empirical quantile. A parametric version is obtained by fitting a particular distribution to the returns, for example a Gaussian $\mathcal{N}(\mu, \sigma^2)$ which amounts to replacing $q_t(h, \alpha)$ by $\hat{\mu} + \hat{\sigma}\, \Phi^{-1}(\alpha)$, where $\hat{\mu}$ and $\hat{\sigma}$ are the estimated mean and standard deviation. Apart from the (somewhat unrealistic) case where the returns are iid, these methods have little theoretical justification.

RiskMetrics Model

A popular estimation method for the conditional VaR relies on the RiskMetrics model. This model is defined by the equations

$$\begin{cases} r_t := \log(p_t/p_{t-1}) = \epsilon_t = \sigma_t \eta_t, & (\eta_t) \; iid \; \mathcal{N}(0,1), \\ \sigma_t^2 = \lambda \sigma_{t-1}^2 + (1-\lambda)\epsilon_{t-1}^2, & \end{cases} \tag{11.66}$$

where $\lambda \in]0, 1[$ is a smoothing parameter, for which, according to RiskMetrics (Longerstaey 1996), a reasonable choice is $\lambda = 0.94$ for daily series. Thus, σ_t^2 is simply the prediction of ϵ_t^2 obtained by simple exponential smoothing. This model can also be viewed as an IGARCH(1, 1) without intercept. It is worth noting that no non-degenerate solution $(r_t)_{t\in\mathbb{Z}}$ to Model (11.66) exists (Exercise 11.18). Thus Model (11.66) is not a realistic data generating process for any usual financial series. This model can, however, be used as a simple tool for VaR computation. From Relation (11.60), we get

$$\mathrm{VaR}_{t,1}(\alpha) = \{1 - e^{\sigma_{t+1}\Phi^{-1}(\alpha)}\} p_t \simeq_t \sigma_{t+1} \Phi^{-1}(1-\alpha).$$

Let Ω_t denote the information generated by $\epsilon_t, \epsilon_{t-1}, \ldots, \epsilon_1$. Choosing an arbitrary initial value to σ_1^2, we obtain $\sigma_{t+1}^2 \in \Omega_t$ and

$$E(\sigma_{t+i}^2 \mid \Omega_t) = E(\lambda\sigma_{t+i-1}^2 + (1-\lambda)\sigma_{t+i-1}^2 \mid \Omega_t) = E(\sigma_{t+i-1}^2 \mid \Omega_t) = \sigma_{t+1}^2$$

for $i \geq 2$. It follows that $\mathrm{Var}(r_{t+1} + \cdots + r_{t+h} \mid \Omega_t) = h\sigma_{t+1}^2$. Note however that the conditional distribution of $r_{t+1} + \cdots + r_{t+h}$ is not exactly $\mathcal{N}(0, h\sigma_{t+1}^2)$ (Exercise 11.19). Many practitioners, however, systematically use the erroneous formula

$$\mathrm{VaR}_{t,h}(\alpha) = \sqrt{h}\; \mathrm{VaR}_t(1, \alpha). \tag{11.67}$$

GARCH-Based Estimation

Of course, one can use more sophisticated GARCH-type models, rather than the degenerate version of RiskMetrics. To estimate $\mathrm{VaR}_t(1, \alpha)$, it suffices to use Relation (11.60) and to estimate $q_t(1, \alpha)$ by $\hat{\sigma}_{t+1}\hat{F}^{-1}(\alpha)$, where $\hat{\sigma}_t^2$ is the conditional variance estimated by a GARCH-type model (for instance, an EGARCH or TGARCH to account for the leverage effect; see Chapter 4), and $\hat{F}$ is an estimate of the distribution of the normalised residuals. It is, however, important to note that, even for a simple Gaussian GARCH(1, 1), there is no explicit available formula for computing $q_t(h, \alpha)$ when $h > 1$. Apart from the case $h = 1$, simulations are required to evaluate this quantile (but, as can be seen from Exercise 11.19, this should also be the case with the RiskMetrics method). The following procedure may then be suggested:

(a) Fit a model, for instance a GARCH(1, 1), on the observed returns $r_t = \epsilon_t$, $t = 1, \ldots, n$, and deduce the estimated volatility $\hat{\sigma}_t^2$ for $t = 1, \ldots, n+1$.

(b) Simulate a large number N of scenarios for $\epsilon_{n+1}, \ldots, \epsilon_{n+h}$ by iterating, independently for $i = 1, \ldots, N$, the following three steps:

(b1) simulate the values $\eta_{n+1}^{(i)}, \ldots, \eta_{n+h}^{(i)}$ iid with the distribution $\hat{F}$;

(b2) set $\sigma_{n+1}^{(i)} = \hat{\sigma}_{n+1}$ and $\epsilon_{n+1}^{(i)} = \sigma_{n+1}^{(i)}\eta_{n+1}^{(i)}$;

(b3) for $k = 2, \ldots, h$, set $(\sigma_{n+k}^{(i)})^2 = \hat{\omega} + \hat{\alpha}(\epsilon_{n+k-1}^{(i)})^2 + \hat{\beta}(\sigma_{n+k-1}^{(i)})^2$ and $\epsilon_{n+k}^{(i)} = \sigma_{n+k}^{(i)}\eta_{n+k}^{(i)}$.

(c) Determine the empirical quantile of simulations $\epsilon_{t+h}^{(i)}$, $i = 1, \ldots, N$.

The distribution $\hat{F}$ can be obtained parametrically or non-parametrically. A simple non-parametric method involves taking for $\hat{F}$ the empirical distribution of the standardised residuals $r_t/\hat{\sigma}_t$, which amounts to taking, in step (b1), a bootstrap sample of the standardised residuals.

Table 11.1 Comparison of the four VaR estimation methods for the CAC 40.

	Historic	Risk metrics	GARCH-$\mathcal{N}$	GARCH-NP
Average estimated VaR (€)	38 323	32 235	31 950	35 059
Number of losses > VaR	29	37	37	21

On the 2120 values, the VaR at the 1% level should only be violated $2120 \times 1\% = 21.2$ times on average.

Assessment of the Estimated VaR (Backtesting)

The Basel accords allow financial institutions to develop their own internal procedures to evaluate their techniques for risk measurement. The term 'backtesting' refers to procedures comparing, on a test (out-of-sample) period, the observed violations of the VaR (or any other risk measure), the latter being computed from a model estimated on an earlier period (in-sample).

To fix ideas, define the variables corresponding to the violations of VaR ('hit variables')

$$I_{t+1}(\alpha) = \mathbb{1}_{\{L_{t,t+1} > \mathrm{VaR}_{t,1}(\alpha)\}}.$$

Ideally, we should have

$$\frac{1}{n}\sum_{t=1}^{n} I_{t+1}(\alpha) \simeq \alpha \qquad \text{and} \qquad \frac{1}{n}\sum_{t=1}^{n} \mathrm{Var}_{t,1}(\alpha) \text{ minimal},$$

that is, a correct proportion of effective losses which violate the estimated VaRs, with a minimal average cost.

Numerical Illustration

Consider a portfolio constituted solely by the CAC 40 index, over the period from 1 March 1990 to 23 April 2007. We use the first 2202 daily returns, corresponding to the period from 2 March 1990 to 30 December 1998, to estimate the volatility using different methods. To fix ideas, suppose that on 30 December 1998, the value of the portfolio was, in French Francs, the equivalent of €1 million. For the second period, from 4 January 1999 to 23 April 2007 (2120 values), we estimated VaR at horizon $h = 1$ and level $\alpha = 1\%$ using four methods.

The first method (historical) is based on the empirical quantiles of the last 250 returns. The second method is RiskMetrics. The initial value for σ_1^2 was chosen equal to the average of the squared last 250 returns of the period from 2 March 1990 to 30 December 1998, and we took $\lambda = 0.94$. The third method (GARCH-$\mathcal{N}$) relies on a GARCH(1, 1) model with Gaussian $\mathcal{N}(0, 1)$ innovations. With this method we set $\hat{F}^{-1}(0.01) = -2.32635$, the 1% quantile of the $\mathcal{N}(0, 1)$ distribution. The last method (GARCH-NP) estimates volatility using a GARCH(1, 1) model, and approximates $\hat{F}^{-1}(0.01)$ by the empirical 1% quantile of the standardised residuals. For the last two methods, we estimated a GARCH(1, 1) on the first period, and kept this GARCH model for all VaR estimations of the second period. The estimated VaR and the effective losses were compared for the 2120 data of the second period.

Table 11.1 and Figure 11.2 do not allow us to draw definitive conclusions, but the historical method appears to be outperformed by the NP-GARCH method. On this example, the only method which adequately controls the level 1% is the NP-GARCH, which is not surprising since the empirical distribution of the standardised residuals is very far from Gaussian.

11.4 Bibliographical Notes

A detailed presentation of the financial concepts introduced in this chapter is provided in the books by Gouriéroux and Jasiak (2001) and Franke, Härdle, and Hafner (2004). A classical reference on the stochastic calculus is the book by Karatzas and Shreve (1988).

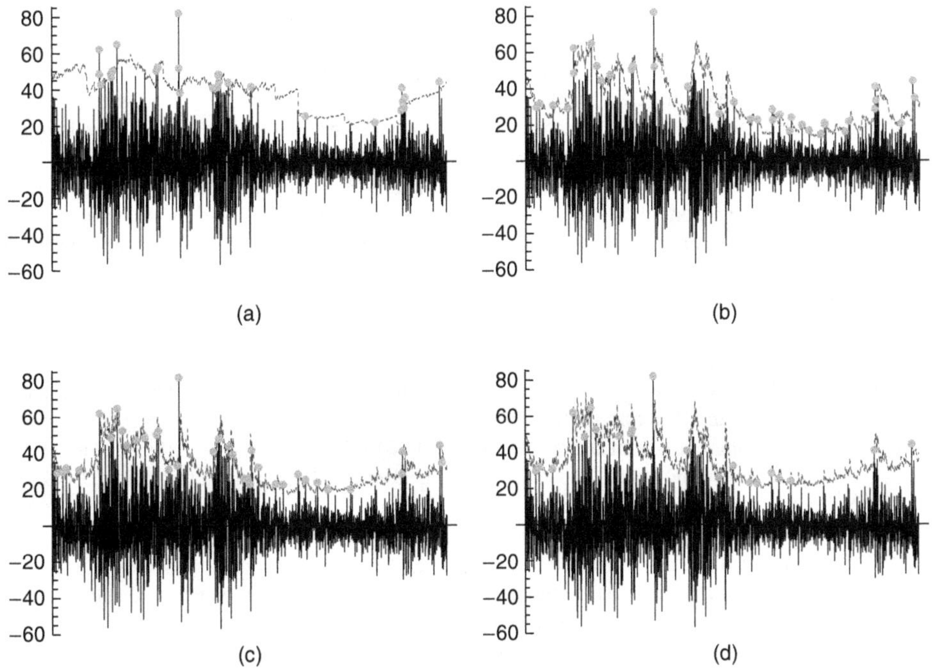

Figure 11.2 Effective losses of the CAC 40 (solid lines) and estimated VaRs (dotted lines) in thousands of euros for the historical method (a), RiskMetrics (b), GARCH-$\mathcal{N}$ (c), and GARCH-NP (d).

The relation between continuous-time processes and GARCH processes was established by Nelson (1990b) (see also Nelson 1992; Nelson and Foster 1994, 1995). The results obtained by Nelson rely on concepts presented in the monograph by Stroock and Varadhan (1979). A synthesis of these results is presented in Elie (1994). An application of these techniques to the TARCH model with contemporaneous asymmetry is developed in El Babsiri and Zakoïan (1990).

When applied to high-frequency (intraday) data, diffusion processes obtained as GARCH limits when the time unit tends to zero are often found inadequate, in particular because they do not allow for daily periodicity. There is a vast literature on the so-called *realised volatility*, which is a daily measure of daily return variability. See Barndorff-Nielsen and Shephard (2002), Andersen et al. (2003) and Andersen et al. (2011) for econometric approaches to realised volatility. In Andersen et al. (2003), it is argued that 'standard volatility models used for forecasting at the daily level cannot readily accommodate the information in intraday data, and models specified directly for the intraday data generally fail to capture the longer interdaily volatility movements sufficiently well'. Another point of view is defended in the thesis by Visser (2009) in which it is shown that intraday price movements can be incorporated into daily GARCH models.

Concerning the pricing of derivatives, we have purposely limited our presentation to the elementary definitions. Specialised monographs on this topic are those of Dana and Jeanblanc-Picqué (1994) and Duffie (1994). Many continuous-time models have been proposed to extend the Black and Scholes (1973) formula to the case of a non-constant volatility. The Hull and White (1987) approach introduces a SDE for the volatility but is not compatible with the assumption of a complete market. To overcome this difficulty, Hobson and Rogers (1998) developed a stochastic volatility model in which no additional Brownian motion is introduced. A discrete-time version of this model was proposed and studied by Jeantheau (2004).

The characterisation of the risk-neutral measure in the GARCH case is due to Duan (1995). Numerical methods for computing option prices were developed by Engle and Mustafa (1992) and Heston and Nandi (2000), among many others. Problems of option hedging with pricing models based on GARCH or stochastic volatility are discussed in Garcia, Ghysels, and Renault (1998). The empirical performance of pricing models in the GARCH framework is studied by Härdle and Hafner (2000), Christoffersen and Jacobs (2004) and the references therein. Valuation of American options in the GARCH framework is studied in Duan and Simonato (2001) and Stentoft (2005). The use of the realised volatility, based on high-frequency data is considered in Stentoft (2008). Statistical properties of the realised volatility in stochastic volatility models are studied by Barndorff-Nielsen and Shephard (2002).

Introduced by Engle, Lilien, and Robins (1987), ARCH-M models are characterised by a linear relationship between the conditional mean and variance of the returns. These models were used to test the validity of the intertemporal capital asset pricing model of Merton (1973) which postulates such a relationship (see, for instance, Lanne and Saikkonen 2006). The asymptotic properties of the quasi-maximum likelihood estimator (QMLE) of GARCH-in-Mean models have been studied by Conrad and Mammen (2016).

The concept of the SDF was developed by Hansen and Richard (1987) and, more recently, by Cochrane (2001). Our presentation follows that of Gouriéroux and Tiomo (2007). This method is used in Bertholon, Monfort, and Pegoraro (2008).

The concept of coherent risk measures (Definition 11.2) was introduced by Artzner et al. (1999), initially on a finite probability space, and extended by Delbaen (2002). In the latter article it is shown that, for the existence of coherent risk measures, the set $\mathcal{L}$ cannot be too large, for instance the set of all absolutely continuous random variables. Alternative axioms were introduced by Wang, Young, and Panjer (1997), initially for risk analysis in insurance. Dynamic VaR models were proposed by Koenker and Xiao (2006) (quantile autoregressive models), Engle and Manganelli (2004) (conditional autoregressive VaR), Gouriéroux and Jasiak (2008) (dynamic additive quantile). The issue of assessing risk measures was considered by Christoffersen (1998), Christoffersen and Pelletier (2004), Engle and Manganelli (2004) and Hurlin and Tokpavi (2006), among others. The article by Escanciano and Olmo (2010) considers the impact of parameter estimation in risk measure assessment. Evaluation of VaR at horizons longer than 1, under GARCH dynamics, is discussed by Ardia (2008).

11.5 Exercises

11.1 *(Linear SDE)*
Consider the linear SDE (11.6). Letting X_t^0 denote the solution obtained for $\omega = 0$, what is the equation satisfied by $Y_t = X_t / X_t^0$?
Hint: the following result, which is a consequence of the multidimensional Itô formula, can be used. If $X = (X_t^1, X_t^2)$ is a two-dimensional process such that, for a real Brownian motion (W_t),

$$\begin{cases} dX_t^1 &= \mu_t^1 dt + \sigma_t^1 dW_t \\ dX_t^2 &= \mu_t^2 dt + \sigma_t^2 dW_t \end{cases}$$

under standard assumptions, then

$$d(X_t^1 X_t^2) = X_t^1 dX_t^2 + X_t^2 dX_t^1 + \sigma_t^1 \sigma_t^2 dt.$$

Deduce the solution of Eq. (11.6) and verify that if $\omega \geq 0$, $x_0 \geq 0$ and $(\omega, x_0) \neq (0, 0)$, then this solution will remain strictly positive.

11.2 *(Convergence of the Euler discretisation)*
Show that the Euler discretisation (11.15), with μ and σ continuous, converges in distribution to the solution of the SDE (11.1), assuming that this equation admits a unique (in distribution) solution.

11.3 *(Another limiting process for the GARCH(1, 1) (Corradi 2000))*
Instead of the rates of convergence of Example 11.2 for the parameters of a GARCH(1, 1), consider

$$\lim_{\tau \to 0} \tau^{-1} \omega_\tau = \omega, \quad \lim_{\tau \to 0} \tau^{-s} \alpha_\tau = 0, \quad \forall s < 1, \quad \lim_{\tau \to 0} \tau^{-1} (\alpha_\tau + \beta_\tau - 1) = -\delta.$$

Give an example of the sequence $(\omega_\tau, \alpha_\tau, \beta_\tau)$ compatible with these conditions. Determine the limiting process of $(X_t^{(\tau)}, (\sigma_t^{(\tau)})^2)$ when $\tau \to 0$. Show that, in this model, the volatility σ_t^2 has a non-stochastic limit when $t \to \infty$.

11.4 *(Put–call parity)*
Using the martingale property for the actualised price under the risk-neutral probability, deduce the European put option price from the European call option price.

11.5 *(Delta of a European call)*
Compute the derivative with respect to S_t of the European call option price and check that it is positive.

11.6 *(Volatility of an option price)*
Show that the European call option price $C_t = C(S, t)$ is solution of an SDE of the form

$$dC_t = \mu_t C_t dt + \sigma_t C_t dW_t$$

with $\sigma_t > \sigma$.

11.7 *(Estimation of the drift and volatility)*
Compute the maximum likelihood estimators of μ and σ^2 based on observations $S_1, \dots, S_n$ of the geometric Brownian motion.

11.8 *(Vega of a European call)*
A measure of the sensitivity of an option to the volatility of the underlying asset is the so-called *vega* coefficient defined by $\partial C_t / \partial \sigma$. Compute this coefficient for a European call and verify that it is positive. Is this intuitive?

11.9 *(Martingale property under the risk-neutral probability)*
Verify that under the measure π defined in Model (11.52), the actualised price $e^{-rt}S_t$ is a martingale.

11.10 *(Risk-neutral probability for a non-linear GARCH model)*
Duan (1995) considered the model

$$\begin{cases} \log(S_t/S_{t-1}) & = r + \lambda\sigma_t - \frac{\sigma_t^2}{2} + \epsilon_t, \\ \epsilon_t & = \sigma_t \eta_t, \\ \sigma_t^2 & = \omega + \{\alpha(\eta_{t-1} - \gamma)^2 + \beta\}\sigma_{t-1}^2, \end{cases}$$

where $\omega > 0$, $\alpha, \beta \geq 0$, and $(\eta_t) \overset{iid}{\sim} \mathcal{N}(0, 1)$. Establish the strict and second-order stationarity conditions for the process (ϵ_t). Determine the risk-neutral probability using SDFs, chosen to be affine exponential with time-dependent coefficients.

11.11 *(A non-affine exponential SDF)*
Consider an SDF of the form

$$M_{t,t+1} = \exp(a_t + b_t \eta_{t+1} + c_t \eta_{t+1}^2).$$

Show that, by an appropriate choice of the coefficients a_t, b_t, and c_t, with $c_t \neq 0$, a risk-neutral probability can be obtained for model (11.51). Derive the risk-neutral version of the model and verify that the volatility differs from that of the initial model.

11.12 *(An erroneous computation of the VaR at horizon h)*
The aim of this exercise is to show that Relation (11.57) may be wrong if the price variations are iid but non-Gaussian. Suppose that $(a'\Delta P_t)$ is iid, with a double exponential density with parameter λ, given by $f(x) = 0.5\lambda \exp\{-\lambda |x|\}$. Calculate $\text{VaR}_{t,1}(\alpha)$. What is the density of $L_{t,t+2}$? Deduce the equation for VaR at horizon 2. Show, for instance for $\lambda = 0.1$, that VaR is overevaluated if Relation (11.57) is applied with $\alpha = 0.01$, but is underevaluated with $\alpha = 0.05$.

11.13 *(VaR for AR(1) prices variations)*
Check that formula (11.58) is satisfied.

11.14 *(VaR for ARCH(1) prices variations)*
Suppose that the price variations follow an ARCH(1) model (11.59). Show that the distribution of ΔP_{t+2} conditional on the information at time t is not Gaussian if $\alpha_1 > 0$. Deduce that VaR at horizon 2 is not easily computable.

11.15 *(VaR and conditional quantile)*
Derive formula (11.60), giving the relationship between VaR and the returns conditional quantile.

11.16 *(Integral formula for the ES)*
Using the fact that $L_{t,t+h}$ and $F^{-1}(U)$ have the same distribution, where U denotes a variable uniformly distributed on [0,1] and F the cdf of $L_{t,t+h}$, derive formula (11.64).

11.17 *(Coherence of the ES)*
Prove that the ES is a coherent risk measure.
Hint for proving subadditivity: For L_i such that $EL_i^+ < \infty$, $i = 1, 2, 3$, denote the VaR at level α by $\mathrm{VaR}_i(\alpha)$ and the ES by $\mathrm{ES}_i(\alpha) = \alpha^{-1}E[L_i \mathbb{1}_{L_i \geq \mathrm{VaR}_i(\alpha)}]$. For $L_3 = L_1 + L_2$, compute $\alpha\{\mathrm{ES}_1(\alpha) + \mathrm{ES}_2(\alpha) - \mathrm{ES}_3(\alpha)\}$ using expectations and observe that

$$(L_1 - \mathrm{VaR}_1(\alpha))(\mathbb{1}_{L_1 \geq \mathrm{VaR}_1(\alpha)} - \mathbb{1}_{L_3 \geq \mathrm{VaR}_3(\alpha)}) \geq 0.$$

11.18 *(RiskMetrics is a prediction method, not really a model)*
For any initial value σ_0^2 let $(\epsilon_t)_{t\geq 1}$ be a sequence of random variables satisfying the Risk Metrics model (11.66) for any $t \geq 1$. Show that $\epsilon_t \to 0$ almost surely as $t \to \infty$.

11.19 *(At horizon* h > *1 the conditional distribution of the future returns is not Gaussian with RiskMetrics)*
Prove that in the RiskMetrics model, the conditional distribution of the returns at horizon 2, $r_{t+1} + r_{t+2}$, is not Gaussian. Conclude that formula (11.67) is incorrect.

12

Parameter-Driven Volatility Models

In this chapter, we consider volatility models in which the volatility no longer coincides with the conditional variance. In other words, volatility would remain unobservable even if the parameters of the data-generating process were known. Such models are naturally related to GARCH and can be seen as their competitors/alternatives in financial applications.

The difference between the different volatility modellings can be perceived through the concepts of *observation-driven* and *parameter-driven* models introduced by Cox (1981) and recently revisited by Koopman, Lucas and Scharth (2016). Suppose that ϕ_t is a 'parameter' at time t. Let ϵ_t be an observed random variable at time t. In an observation-driven model,

$$\phi_t = \Phi(\epsilon_{t-1}, \epsilon_{t-2}, \ldots),$$

where Φ is a measurable function. In a parameter-driven model,

$$\phi_t = \Phi^*(\phi_{t-1}, \phi_{t-2}, \ldots, \eta_t).$$

where η_t is an idiosyncratic innovation at time t and Φ^* is a measurable function. In the latter case, there is a latent structure in the model which, in general, cannot be directly related to the observations.

Let us a consider a GARCH model, choosing the volatility for the parameter: $\phi_t = \sigma_t$. Under stationarity and invertibility conditions, σ_t can be expressed as a function of the past observations, making the model observation-driven. This enables to write the likelihood function easily, for any purported distribution of the innovation η_t. Most GARCH models can also be considered as parameter-driven since, for instance, in the GARCH(1,1) case,

$$\sigma_t^2 = \omega + \alpha\epsilon_{t-1}^2 + \beta\sigma_{t-1}^2 = \omega + (\alpha\eta_{t-1}^2 + \beta)\sigma_{t-1}^2 := \Phi(\sigma_{t-1}^2, \eta_{t-1}).$$

We will focus in this chapter on two classes of volatility models – the stochastic volatility (SV) and the Markov switching GARCH models – which do not have an observation-driven form, but can be given a parameter-driven representation.

GARCH Models: Structure, Statistical Inference and Financial Applications, Second Edition. Christian Francq and Jean-Michel Zakoian.

12.1 Stochastic Volatility Models

In GARCH models, the concepts of conditional variance and volatility coincide. By making volatility a function of the past observations, these specifications are particularly convenient for statistical purposes and are amenable to multivariate extensions. However, they may be difficult to study from a probability point of view and may imply important dynamic restrictions.[1]

By contrast, in the so-called *SV* models, volatility is a latent variable which cannot be expressed as a function of observable variables and satisfies a dynamic model. The observed process, ϵ_t, and its volatility, σ_t, are still related by the equation

$$\epsilon_t = \sigma_t \eta_t$$

where (η_t) is a strong white noise sequence, which may or may not be independent from the process (σ_t). The model is completed by specifying a dynamic for σ_t, which only has to be compatible with the positivity of σ_t.

Contrary to GARCH models – for which a plethora of formulations have been developed, and which can be used with different orders like ARMA models – a prominent part of the literature on SV models focused on a simple first-order specification, often considered sufficient to capture the main characteristics of financial series. This specification – called the canonical SV model – postulates an AR(1) model for the dynamics of $\log \sigma_t^2$. While the probability structure is much simpler, the statistical inference is much more complex than in GARCH models, which has refrained practitioners from considering SV as a benchmark model for financial data.

12.1.1 Definition of the Canonical SV Model

The *canonical* SV model is defined by

$$\begin{cases} \epsilon_t = \sqrt{h_t}\eta_t \\ \log h_t = \omega + \beta \log h_{t-1} + \sigma v_t, \end{cases} \tag{12.1}$$

where (η_t) and (v_t) are two independent sequences of iid zero-mean and unit-variance random variables, ω, β and σ are parameters.

Under conditions ensuring the existence of a solution, the variable $\sqrt{h_t}$ is called the volatility of ϵ_t. As in GARCH models, the *magnitude* of ε_t is proportional to $\sqrt{h_t}$, and the *sign* of ε_t is independent from $\sqrt{h_t}$ and from any past variable. Dependence between ϵ_t and its past values is introduced indirectly, through the log-volatility dynamics. However, contrary to what occurs with GARCH-type models, the volatility at time t cannot be interpreted as the conditional variance of ε_t.[2] Another difference with GARCH specifications is that, by specifying an AR(1) dynamics on the log-volatility rather than on the volatility, the parameters ω, β, and σ are *a priori* free of positivity constraints. Such constraints have nevertheless to be introduced to ensure either identifiability of the parameters, or sensible interpretation and compatibility with the data.

Parameter β plays the role of a persistence coefficient. Indeed, when β is close to 1, a positive shock on the log-volatility (due to a large positive value of v_t) has a persistent effect on h_t: the volatility will remain high over several dates. A negative shock on the log-volatility, implying a small level for the volatility h_t, is also persistent when β is close to 1. If β is close to 0, the effect of large shocks vanishes rapidly. Now if β is close to −1, the instantaneous effect of a positive shock is a large volatility, but at the next period, this volatility takes a small value, and subsequently oscillates between large and small values provided no other big shock occurs. A negative shock produces the same kind of alternate effects. Such

[1] For instance, in standard GARCH(p, q) models, the autocorrelation function of the squared returns, ϵ_t^2, is positive, whatever the GARCH coefficients (provided $E\epsilon_t^4 < \infty$).

[2] The conditional variance is non-explicit in the SV model.

behaviours being rarely observed on financial series, negative values of β can be considered unrealistic for applications.

The interpretation of the other coefficients is more straightforward. Parameter ω is a scale factor for the volatility. The canonical SV model is clearly stable by scaling: if the returns ϵ_t are multiplied by some constant c, then the process $\tilde{\epsilon}_t = c\epsilon_t$ satisfies the same SV model, except for the coefficient ω which is replaced by $\tilde{\omega} = \omega + \log c^2(1-\beta)$. Finally, parameter σ can be interpreted as the volatility of the log-volatility. Without generality loss, it can be assumed that $\sigma \geq 0$.

One motivation for considering the SV model is that it can be interpreted as the discretised version of diffusion models used in finance. For instance, a popular continuous-time model for the asset price S_t is given by

$$\begin{cases} d\log S_t = \mu dt + \sigma_t dW_{1t}, \\ d\log \sigma_t^2 = \{\omega + (\beta - 1)\log \sigma_{t-1}^2\}dt + \sigma dW_{2t}, \end{cases} \tag{12.2}$$

where (W_{1t}) and (W_{2t}) are independent Brownian motions. In this model, the log-volatility follows an Ornstein–Uhlenbeck process. The canonical SV model – with an intercept μ added in the equation of $\epsilon_t = \log(S_t/S_{t-1})$ – can be seen as a discrete-time version of the previous continuous-time model.

12.1.2 Stationarity

The probabilistic structure of the SV model is much simpler than that of GARCH models. Indeed, the stationarity of the volatility sequence (h_t) in Model (12.1) can be studied independently from the iid sequence (η_t). Call non-anticipative any solution (ϵ_t) of Model (12.1) such that ϵ_t belongs to $\sigma\{(\eta_u, v_u): u \leq t\}$. The proof of the following result is immediate.

Theorem 12.1 (Strict stationarity condition) *If $|\beta| < 1$, Model (12.1) admits a unique strictly stationary solution, which is non-anticipative and given by*

$$\epsilon_t = \exp\left\{\frac{\omega}{2(1-\beta)} + \frac{\sigma}{2}\sum_{i=0}^{\infty}\beta^i v_{t-i}\right\}\eta_t, \quad t \in \mathbb{Z}. \tag{12.3}$$

If $|\beta| \geq 1$ and $\sigma > 0$, there exists no strictly stationary and non-anticipative solution.

The second-order properties of the strictly stationary solution can be summarised as follows (see Exercise 12.1 for a proof). Let

$$\alpha_i = E\{\exp(\sigma\beta^i v_t)\} \in (0, \infty], \quad i \geq 0.$$

Theorem 12.2 (Second-order stationarity condition) *Suppose $|\beta| < 1$ and $\prod_{i=0}^{\infty} E\{e^{\sigma|\beta^i v_t|}\} < \infty$. Then, the process (ϵ_t) defined by (12.3) is a white noise with variance*

$$Var(\epsilon_t) = e^{\frac{\omega}{1-\beta}}\prod_{i=0}^{\infty}\alpha_i.$$

Remark 12.1 (The Gaussian case) When $v_t \sim \mathcal{N}(0,1)$, more explicit results can be given. We have $E(e^{\rho v_t}) = e^{\rho^2/2}$ and $E(e^{|\rho v_t|}) = 2e^{\rho^2/2}\Phi(|\rho|)$ for any constant ρ. Thus $\alpha_i = \exp\left(\frac{\sigma^2}{2}\beta^{2i}\right)$ and

$$0 < \prod_{i=0}^{\infty} E\{e^{\sigma|\beta^i v_t|}\} < \infty \quad \Longleftrightarrow \quad |\beta| < 1.$$

Moreover, $\mathrm{Var}(\epsilon_t) = \exp\left(\frac{\omega}{1-\beta} + \frac{\sigma^2}{2(1-\beta^2)}\right)$.

Kurtosis of the Marginal Distribution

In view of (12.3), the existence of a fourth-order moment for ε_t requires the conditions

$$|\beta|<1 \quad \text{and} \quad E(\eta_t^4)\prod_{i=0}^{\infty} E\{e^{2\sigma|\beta^i v_t|}\} < \infty. \tag{12.4}$$

We then have

$$E(\varepsilon_t^4) = e^{\frac{2\omega}{1-\beta}} \prod_{i=0}^{\infty} E\{e^{2\sigma\beta^i v_t}\}E(\eta_t^4),$$

and the kurtosis coefficient of the marginal distribution of ϵ_t is given by

$$\kappa_\varepsilon = E(\varepsilon_t^4)/\{E(\varepsilon_t^2)\}^2 = \kappa_\eta \prod_{i=0}^{\infty} \kappa^{(i)},$$

where κ_η is the kurtosis coefficient of η_t and $\kappa^{(i)}$ is the kurtosis coefficient of $e^{\sigma/2\beta^i v_t}$. Because the law of v_t is non-degenerate, $\kappa^{(i)} > 1$. Thus, the kurtosis of the marginal distribution is strictly larger than that of the conditional distribution of ϵ_t. When $v_t \sim \mathcal{N}(0, 1)$, we have $\kappa_\varepsilon = \kappa_\eta e^{\frac{\sigma^2}{1-\beta^2}}$. In particular if we also have $\eta_t \sim \mathcal{N}(0, 1)$, then the distribution of ε_t is leptokurtic ($\kappa_\varepsilon > 3$).

Remark 12.2 (A difference with GARCH models) It was shown in Chapter 2 that a strictly stationary GARCH process admits (at least) a moment of small order (see Corollary 2.3). This property is, in particular, useful to establish the asymptotic properties of the QML estimator without any additional moment assumption on the observed process. This property does not hold for the strictly stationary solution (12.3) of Model (12.1). Indeed, for any $\delta > 0$,

$$Eh_t^\delta = e^{\delta\omega} E e^{\delta\sigma v_t} E h_{t-1}^{\delta\beta}$$

may be infinite if, for instance, the law of v_t is sub-exponential.[3] For such laws, we have $E(e^{\rho v_t}) = \infty$ for any $\rho > 0$. Thus, to obtain a stationary SV process admitting small order moments, the tails of the process (v_t) must not be too heavy.

Tails of the Marginal Distribution

When stable laws are estimated on series of financial returns, one generally obtains heavy-tailed marginal distributions.[4] It is, therefore, useful to study in more detail the tail properties of the marginal distribution of SV processes. We relax in this section the assumption of finite variance for η_t.

The distribution of a non-negative random variable X is said to be *regularly varying* with index α, which can be denoted $X \in RV(\alpha)$ if

$$P(X > x) = x^{-\alpha} L(x), \quad x > 0, \tag{12.5}$$

with $\alpha \in (0, 2)$ and $L(x)$ a slowly varying function at infinity, which is defined by the asymptotic relation $L(tx)/L(x) \to 1$ as $x \to \infty$, for all $t > 0$.[5] If X satisfies (12.5), then $E(X^r) = \infty$ for $r > \alpha$ and $E(X^r) < \infty$ for $r < \alpha$ (for $r = \alpha$, the expectation may be finite or infinite, depending on the slowly varying function L). To

---

[3] A distribution F is called sub-exponential if for any independent copies $X_1, \dots, X_n$ of $X \sim F$, we have $P[X_1 + \cdots + X_n > x] \sim P[\max(X_1, \dots, X_n) > x]$ as $x \to \infty$. The sub-exponential class includes commonly used heavy-tailed distributions such as the Student and Pareto, but also moderately heavy-tailed distributions such as the log-normal and Weibull (see Embrechts, Klüppelberg, and Mikosch 1997 for an overview).

[4] For instance Francq and Zakoïan (2013c) found indexes $\alpha < 2$ on nine major series of stock returns.

[5] Examples of slowly varying functions are functions converging to a positive constants, logarithms and iterated logarithms.

handle products, the following result by Breiman (1965, Eq. (3.1)) can be useful: for any non-negative independent random variables X and Y, where $X \in \mathrm{RV}(\alpha)$ and $E(Y^{\alpha+\delta}) < \infty$ for some $\delta > 0$, we have

$$P(XY > x) \sim E(Y^{\alpha})P(X > x), \quad \text{as } x \to \infty.$$

It follows that if, in Model (12.1) without the assumption that $E\eta_t^2 = 1$, $|\eta_t| \in \mathrm{RV}(\alpha)$, and if the noise (v_t) is such that $Eh_t^{(\alpha+\tau)/2} < \infty$, for some $\tau > 0$, then ϵ_t inherits the heavy-tailed property of the noise (η_t). This is in particular the case if (v_t) is Gaussian, since h_t then admits moments at any order.

12.1.3 Autocovariance Structures

Under the conditions of Theorem 12.2, the SV process (ε_t) is uncorrelated yet dependent. In other words, like GARCH processes, it is not white noise in the strong sense. As for GARCH models, the dependence structure of (ε_t) can be studied through the autocorrelation function of appropriate transformations of the SV process.

ARMA Representation for log ε_t^2

Assuming $P(\eta_t = 0) = 0$, and by taking the logarithm of ε_t^2, the first equation of (12.1) writes as follows:

$$\log \varepsilon_t^2 = \log h_t + \log \eta_t^2. \tag{12.6}$$

This formulation allows us to derive the autocovariance function of the process $(\log \varepsilon_t^2)$. Let $X_t = \log \varepsilon_t^2$, and $Z_t = \log \eta_t^2$. Assume that Z_t admits fourth-order moments, and let $\mu_Z = E(Z_t)$, $\sigma_Z^2 = \mathrm{Var}(Z_t)$, and $\sigma_{Z^2}^2 = \mathrm{Var}(Z_t^2)$.[6] The next result (see Exercise 12.3 for a proof) shows that (X_t) admits an ARMA representation.

Theorem 12.3 (ARMA representation for the log-squared returns) *If $|\beta| < 1$ and $\sigma_{Z^2}^2 < \infty$, the process $(X_t) = (\log \varepsilon_t^2)$ admits an ARMA(1,1) representation of the form*

$$X_t - \mu_X = \beta(X_{t-1} - \mu_X) + u_t - \alpha u_{t-1}, \tag{12.7}$$

where (u_t) is a white noise, $\mu_X = \frac{\omega}{1-\beta} + \mu_Z$, and α is a coefficient depending on β and σ/σ_Z. If $\beta \neq 0$ and $\sigma \neq 0$, the orders of the ARMA(1,1) representation are exact.

Note that, in Eq. (12.7), the noise (u_t) is not strong. This sequence is not even a martingale difference sequence (unlike the noise of the ARMA representation for the square of a GARCH process). This can be seen by computing, for instance,

$$E\{u_t(X_{t-1} - \mu_X)^2\} = \frac{\sigma^3\alpha\{1 + \beta^2(1 - \alpha)\}}{(1 - \beta^3)(1 - \alpha\beta^2)}E(v_t^3) + (\alpha - \beta)E(Z_t - \mu_Z)^3.$$

Even when (v_t) is symmetrically distributed, this quantity is generally non-zero. This shows that u_t is correlated with some (non-linear) functions of its past: thus $E(u_t \mid u_{t-1}, u_{t-2}, \ldots) \neq 0$ (a. s.).

Example 12.1 If η_t and v_t are $\mathcal{N}(0,1)$ distributed, and if $\omega = -1$, $\beta = 0.9$, $\sigma = 0.4$, (X_t) admits the following ARMA(1,1) representation:

$$(X_t - 11.27) - 0.9(X_{t-1} - 11.27) = u_t - 0.81u_{t-1}, \tag{12.8}$$

where (u_t) is a white noise with variance $\sigma_u^2 = 5.52$.

[6] When $\eta_t \sim \mathcal{N}(0,1)$ we have $\mu_Z = -1.270$, $\sigma_Z^2 = \pi^2/2 = 4.935$, $\sigma_{Z^2}^2 = 263.484$ (*cf* Abramowitz and Stegun 1965, pp. 260 and 943).

Autocovariance of the Process (ϵ_t^2)

When the process (ϵ_t^2) is second-order stationary, that is under Eq. (12.4), deriving its autocovariance function has interest for comparison with GARCH models and for statistical applications.

Using Eq. (12.3) we get, for all $k \geq 0$

$$E(\epsilon_t^2 \epsilon_{t-k}^2) = e^{\frac{2\omega}{1-\beta}} E\left\{ \prod_{i=0}^{\infty} e^{\sigma \beta^i v_{t-i}} \prod_{i=0}^{\infty} e^{\sigma \beta^i v_{t-k-i}} \right\} E(\eta_t^2 \eta_{t-k}^2).$$

For $k > 0$, since η_t^2 and η_{t-k}^2 are independent and have unit expectation,

$$E(\epsilon_t^2 \epsilon_{t-k}^2) = e^{\frac{2\omega}{1-\beta}} E\left\{ \prod_{i=0}^{k-1} e^{\sigma \beta^i v_{t-i}} \prod_{i=0}^{\infty} e^{\sigma \beta^i (1+\beta^k) v_{t-k-i}} \right\} = e^{\frac{2\omega}{1-\beta}} \prod_{i=0}^{k-1} \alpha_i \prod_{i=0}^{\infty} \alpha_{k,i},$$

where, for $i \geq k$, $\alpha_{k,i} = E[\exp\{\sigma \beta^i (1+\beta^k) v_t\}]$. Thus, we have, for all $k > 0$

$$\text{Cov}(\epsilon_t^2, \epsilon_{t-k}^2) = e^{\frac{2\omega}{1-\beta}} \left(\prod_{i=0}^{k-1} \alpha_i \prod_{i=0}^{\infty} \alpha_{k,i} - \prod_{i=0}^{\infty} \alpha_i^2 \right).$$

Now suppose that $v_t \sim \mathcal{N}(0,1)$. It follows that $\alpha_{k,i} = \exp\left\{ \frac{\sigma^2}{2} \beta^{2i} (1+\beta^k)^2 \right\}$, and thus the second-order stationarity condition reduces to $E(\eta_t^4) < \infty$ and $|\beta| < 1$ in this case. Therefore, for any $k > 0$

$$\begin{aligned}\text{Cov}(\epsilon_t^2, \epsilon_{t-k}^2) &= e^{\frac{2\omega}{1-\beta}} \left(\prod_{i=0}^{k-1} e^{\frac{\sigma^2}{2}\beta^{2i}} \prod_{i=0}^{\infty} e^{\frac{\sigma^2}{2}\beta^{2i}(1+\beta^k)^2} - \prod_{i=0}^{\infty} e^{\sigma^2 \beta^{2i}} \right) \\ &= e^{\frac{2\omega}{1-\beta} + \frac{\sigma^2}{1-\beta^2}} \left(e^{\frac{\sigma^2}{1-\beta^2}\beta^k} - 1 \right).\end{aligned}$$

It can be noted that the autocovariance function of (ϵ_t^2) tends to zero when k goes to infinity, but that the decreasing is not compatible with a linear recurrence equation on the autocovariances. In other words, the process (ϵ_t^2) – though second-order stationary and admitting a Wold representation – does not have an ARMA representation. However, we have an equivalent of the form $\text{Cov}(\epsilon_t^2, \epsilon_{t-k}^2) \sim cste \times \beta^k$ as $k \to \infty$, which shows that the asymptotic (exponential) rate of decrease of the autocovariances is the same as for an ARMA process. Recall that, by contrast, squared fourth-order stationary GARCH processes satisfy an ARMA model.

12.1.4 Extensions of the Canonical SV Model

Many other specifications could be considered for the volatility dynamics. Obviously, the memory of the process could be extended by introducing more lags in the AR model, that is an AR(p) instead of an AR(1). An insightful extension of model (12.1) is obtained by allowing a form of dependence between the two noise sequences. We consider two natural extensions of the basic model.

Contemporaneous Correlation

A contemporaneous correlation coefficient ρ between the disturbances of model (12.1) is introduced by assuming that (η_t, v_t) forms an iid centred normal sequence with $\text{Var}(\eta_t) = \text{Var}(v_t) = 1$ and $\text{Cov}(\eta_t, v_t) = \rho$. The model can thus be equivalently written as

$$\begin{cases} \epsilon_t = \sqrt{h_t}\eta_t \\ \ln h_t = \omega_t + \beta \ln h_{t-1} + \sigma^* v_t^* \end{cases} \tag{12.9}$$

where $\omega_t = \omega + \sigma\rho\eta_t$, $\sigma^* = \sigma\sqrt{1-\rho^2}$, (η_t) and (v_t^*) are two independent sequences of iid normal variables with zero mean and variance equal to 1. We still assume that $|\beta| < 1$, which ensures stationarity of the solution (ϵ_t) defined in (12.3). If the correlation ρ is negative, a negative shock on the price – corresponding to a negative η_t – is likely to be associated with an increase of the volatility at the current date and – if $\beta > 0$ – at the subsequent dates. Conversely, a positive shock on the price – $\eta_t > 0$ – is likely to be compensated by a decrease of the volatility. The contemporaneous correlation between the iid processes of the SV model thus induces asymmetry effects, similar to the leverage effect described in Chapter 1. This effect, characterised by a negative correlation between the current price and the future volatility, has been often detected in stock prices. However, the correlation coefficient ρ also entails a non-zero mean

$$E(\varepsilon_t) = e^{\omega/2} E h_{t-1}^{\beta/2} E\left(e^{\frac{\sigma}{2}v_t}\eta_t\right) \neq 0,$$

and non-zero autocorrelations for the process (ε_t) which, in financial applications, is often inappropriate (see Exercise 12.4). Therefore, it appears that the correlation coefficient ρ has several roles in this model, which makes it difficult to interpret.

Leverage Effect

A non-contemporaneous correlation between the disturbances of model (12.1) allows to pick up the leverage effect, without removing the zero-mean property. Therefore, we assume that (η_{t-1}, v_t) forms an iid centred normal sequence with $\text{Var}(\eta_t) = \text{Var}(v_t) = 1$ and $\text{Cov}(\eta_{t-1}, v_t) = \rho$. The model is given by

$$\begin{cases} \varepsilon_t = \sqrt{h_t}\eta_t \\ \ln h_t = \omega_{t-1} + \beta \ln h_{t-1} + \sigma^* v_t^* \end{cases} \tag{12.10}$$

with the same notations as in the last section. By (12.3), we have $E(\varepsilon_t) = 0$ and $\text{Cov}(\varepsilon_t, \varepsilon_{t-k}) = 0$ for $k > 0$. Thus, the white noise property of the stationary solution is preserved with this correlation structure of the two disturbances sequences.

Intuitively, when $\rho < 0$, a large negative shock on the noise η at time $t-1$ – that is, a sudden price fall – is likely to produce (through the correlation between v_t and η_{t-1}) a large volatility at the next period (and at the subsequent ones if $\beta > 0$). The following result shows that the model is indeed able to capture the leverage effect and provides a quantification of this effect.

Proposition 12.1 *For the SV model with Gaussian iid zero-mean and unit-variance disturbances (η_{t-1}, v_t) such that $\text{Cov}(\eta_{t-1}, v_t) = \rho$, we have*

$$\text{Cov}(\varepsilon_{t-1}, \sqrt{h_t}) = \frac{\sigma\rho}{2} e^{\frac{\sigma^2+4\omega}{4(1-\beta)}}.$$

It follows that the leverage effect $\text{Cov}(-\varepsilon_{t-1}^-, \sqrt{h_t}) > \text{Cov}(\varepsilon_{t-1}^+, \sqrt{h_t})$ *holds if and only if* $\rho < 0$.

At horizon $k > 0$, we have

$$\text{Cov}(\varepsilon_{t-k}, \sqrt{h_t}) = \frac{\sigma\rho}{2}\beta^{k-1} e^{\frac{\omega}{1-\beta} + \frac{\sigma^2}{8(1-\beta^2)}\{1-\beta^{2(k-1)} + (1+\beta^k)^2\}}.$$

Proof. Using expansion (12.3), we obtain

$$E(\varepsilon_{t-1}\sqrt{h_t}) = E\left(\exp\left\{\frac{\omega}{(1-\beta)} + \frac{\sigma}{2}\sum_{i=0}^{\infty}\beta^i v_{t-i-1} + \frac{\sigma}{2}\sum_{i=0}^{\infty}\beta^i v_{t-i}\right\}\eta_{t-1}\right)$$

$$= e^{\frac{\omega}{1-\beta}} E\left(e^{\frac{\sigma}{2}v_t}\eta_{t-1}\right)\prod_{i=1}^{\infty} E\left(e^{\frac{\sigma}{2}\beta^{i-1}(1+\beta)v_t}\right).$$

Write $v_t = \rho\eta_{t-1} + \sqrt{1-\rho^2}u_t$, where $V(u_t) = 1$, and u_t and η_{t-1} are non-correlated – thus independent under the normality assumption. Thus

$$E\left(e^{\frac{\sigma}{2}v_t}\eta_{t-1}\right) = E\left(e^{\frac{\sigma}{2}\rho\eta_{t-1}}\eta_{t-1}\right)E\left(e^{\frac{\sigma}{2}\sqrt{1-\rho^2}u_t}\right) = \frac{\sigma\rho}{2}e^{\frac{\sigma^2}{8}},$$

and the conclusion follows for $k = 1$. Similarly, for $k > 1$ we have

$$E(\epsilon_{t-k}\sqrt{h_t}) = e^{\frac{\omega}{1-\beta}}E\left(e^{\frac{\sigma}{2}\beta^{k-1}v_{t-k+1}}\eta_{t-k}\right)\prod_{i=0}^{k-2}E\left(e^{\frac{\sigma}{2}\beta^i v_{t-i}}\right)\prod_{i=0}^{\infty}E\left(e^{\frac{\sigma}{2}\beta^i(1+\beta^k)v_{t-i-k}}\right),$$

and the conclusion follows. □

To summarise, the SV model where η_{t-1} and v_t are negatively correlated is able to capture the leverage effect, while preserving the white noise property of (ϵ_t). As the horizon increases, however, this effect is maintained only if $\beta > 0$ and it vanishes exponentially.

12.1.5 Quasi-Maximum Likelihood Estimation

Estimating parameters in a SV model – even in the simple canonical SV model – is a difficult task because the volatility process is *latent*. In GARCH-type models, the period-t volatility is also non-observed, but it only depends on (i) unknown parameters, and (ii) the sample path of the (observed) returns process up to time $t-1$. It follows that the log-likelihood can be written in closed-form for all invertible GARCH models. Because the log-likelihood cannot be explicitly written for SV models, a plethora of alternative methods have been suggested. To mention just a few, the following approaches can be used: the Kalman filter, Generalised Moments Methods (GMM), simulations-based methods (in particular Bayesian Markov chain Monte Carlo (MCMC) methods). We will limit ourselves to the Kalman filter method (see Appendix D for a general presentation) of evaluating the likelihood. Although this method may be numerically less accurate than more sophisticated ones, it is easy to implement, computationally fast, and amenable to extensions – in particular multivariate extensions – of the SV model.

The state-space form of model (12.1) is straightforwardly obtained by taking logarithms:

$$\begin{cases} \log\epsilon_t^2 = \log h_t + \mu_Z + u_t \\ \log h_t = \beta\log h_{t-1} + \omega + \sigma v_t \end{cases}$$

using the notations of Section 12.1.3.

Applying the Kalman filter is not without difficulty in this framework, because it requires assuming a Gaussian distribution for $\log\eta_t^2$. For financial returns, the marginal distributions are known to be heavy tailed. Even if the relation between such marginal distributions and the distribution of η_t^2 is complex, it seems unrealistic to assume a normal distribution for $\log\eta_t^2$. As mentioned in Appendix D, when the normal assumption is in failure, the Kalman filter does not provide the exact conditional variance of the observed process, but rather an approximation.

Let

$$\alpha_{t|t-1} = E(\log h_t \mid \epsilon_1^2, \dots, \epsilon_{t-1}^2), \quad P_{t|t-1} = \text{Var}(\log h_t \mid \epsilon_1^2, \dots, \epsilon_{t-1}^2),$$

The algorithm described in Appendix D takes here a simplified form:

$$\alpha_{t|t-1} = \beta\alpha_{t-1|t-2} + K_t(y_{t-1} - \alpha_{t-1|t-2} - \mu_Z) + \omega \tag{12.11}$$

$$P_{t|t-1} = \beta^2 P_{t-1|t-2} - K_t^2 F_{t-1|t-2} + \sigma^2 \tag{12.12}$$

$$F_{t-1|t-2} = P_{t-1|t-2} + \sigma_Z^2, \quad K_t = \beta P_{t-1|t-2}F_{t-1|t-2}^{-1} \tag{12.13}$$

$$\alpha_{1|0} = \beta_0 a_0 + \omega, \quad P_{1|0} = \beta^2 P_0 + \sigma^2. \tag{12.14}$$

In view of Proposition D.1 in Appendix D, if $|\beta| < 1$ the sequence $(P_{t|t-1})$ converges. The limit P^* is obtained as the unique positive solution of the Ricatti equation (D.22), that is

$$P^* = 0.5[\sigma^2 - (1-\beta^2)\sigma_Z^2 + \{(\sigma^2 - (1-\beta^2)\sigma_Z^2)^2 + 4\sigma^2\sigma_Z^2\}^{1/2}].$$

One can easily check that $|P_{t|t-1} - P^*| < \beta^2 |P_{t-1|t-2} - P^*|$, the rate of convergence is thus exponential.

Finally, the log-likelihood writes:

$$\log \ell(\epsilon_1, \dots, \epsilon_n; \theta) = -\frac{n}{2}\log 2\pi - \frac{1}{2}\sum_{t=1}^{n} \log F_{t|t-1} + \frac{(\log(\epsilon_t^2) - \alpha_{t|t-1} - \mu_Z)^2}{F_{t|t-1}}.$$

For fixed values of μ_Z and σ_Z^2, parameter $\theta = (\omega, \beta, \sigma)$ can thus be estimated by maximising this function, which is built by applying the previous algorithm. It can be noted that the conditional variance $F_{t|t-1}$ obtained by this algorithm does not depend on the observations.

12.2 Markov Switching Volatility Models

When GARCH models are fitted over sub-periods of long time series of returns, one often find important differences in estimated parameters (see Exercise 12.5). This discrepancy can be interpreted as the indication that the dynamics is changing over time. Such dynamic changes are not surprising given the existence of business cycles in the economy, characterised by well-known phases/regimes of recession, depression, recovery, and expansion. A natural way to model changes in regime is to postulate the existence of an underlying latent[7] variable of regime change, taking values in a finite set and satisfying – for simplicity – the Markov property.

Models constructed with a latent Markov chain have found widespread applications in various domains including economics, biology, finance, and speech recognition. We start by presenting a class of Hidden Markov Models (HMM), which can be seen as dynamic extensions of finite mixture models.

12.2.1 Hidden Markov Models

A process $(\epsilon_t)_{t\geq 0}$ satisfies a HMM if

1. conditional on some unobserved Markov chain (Δ_t), the (observed) variables $\epsilon_0, \epsilon_1, \dots$ are independent;
2. the conditional law of ϵ_s given (Δ_t) only depends on Δ_s.

For instance, the observed and latent processes can be related through the multiplicative model

$$\begin{cases} \epsilon_t = \sigma_t \eta_t, \\ \sigma_t^2 = \omega(\Delta_t), \end{cases} \tag{12.15}$$

where

$$0 < \omega(1) < \cdots < \omega(d), \tag{12.16}$$

(η_t) is a sequence of iid centred variables with unit variance, and Δ_t is a Markov chain on $\mathcal{E} = \{1, \dots, d\}$. Set $\mathcal{E}$ is called the set of *states*, or *regimes*, of the Markov chain (see Appendix C for basic elements of the Markov Chain theory). The transition matrix is denoted by $\mathbb{P}$. It is assumed that:

---

[7] Or hidden, which clearly indicates that the variable is not observed.

A1 The Markov chain (Δ_t) is stationary, irreducible and aperiodic.

A2 The processes (η_t) and (Δ_t) are independent.

Note that the condition (12.16) is not restrictive because if $\omega(i) = \omega(j)$, the states i and j can be merged by changing the Markov chain. This model is pertinent for financial returns, as it belongs to the general class of random conditional variance models defined in (1.6), where $\mathcal{F}_{t-1}$ is the σ-field generated by the random variable Δ_t. Although very simple, this model presents strong similarities – but also noticeable differences – with standard GARCH models.

Comparison with GARCH Models

As in GARCH or SV models, the observed process (ϵ_t) is centred and its sample paths fluctuate around zero. The magnitude of the fluctuations will be different according to the state of the Markov chain Δ_t. Regime 1 is characterised by small fluctuations, while regime d is associated with the most turbulent periods. The length of stay in a given regime, as well as the number of transitions from one regime to another only depend on the transition probabilities of the Markov chain. An example of sample path (100 observations) is displayed in Figure 12.1. The path is simulated from a three-regime model with $\eta_t \sim \mathcal{N}(0, 1)$ and

$$\omega(1) = 1, \quad \omega(2) = 9, \quad \omega(3) = 81, \quad \mathbb{P} = \begin{pmatrix} 0.85 & 0.1 & 0.05 \\ 0.3 & 0.7 & 0 \\ 0.3 & 0 & 0.7 \end{pmatrix}.$$

This very simple model presents similarities with GARCH-type models. By construction, the clustering property [8] is satisfied.

The assumptions made on the Markov chain and on (η_t) entail that any solution of model (12.15) is obviously strictly stationary. Moments of ϵ_t exist at any order, provided that the same moments exist for

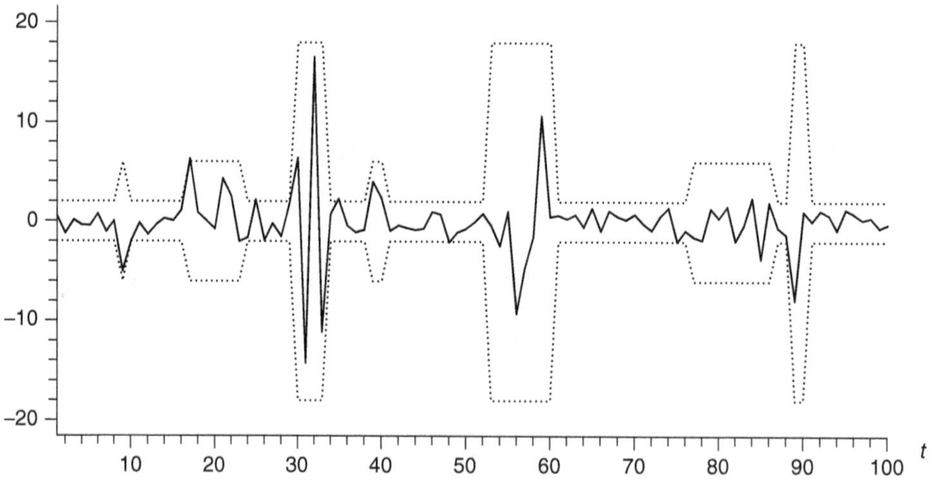

Figure 12.1 Simulation of length 100 of a three-regime HMM. The full line corresponds to the simulated values of ϵ_t and the dotted lines correspond to $\pm 2\sqrt{\omega(\Delta_t)}$. Source: Francq and Roussignol (1997).

[8] It is however important to note that, contrary to the volatility GARCH models, σ_t^2 does not represent here the variance de ϵ_t conditional on some available information at time t: $\omega(\Delta_t)$ is not a function of the past values of ϵ_t. We will nevertheless refer to (σ_t) as the volatility process.

η_t. By the independence between η_t and Δ_t, we have for any real positive number r,

$$E(\epsilon_t^r) = E\omega(\Delta_t)^{r/2}E(\eta_t^r) = \sum_{i=1}^{d} \omega(i)^{r/2}\pi(i)E(\eta_t^r).$$

In particular, ϵ_t is centred. Recall that for GARCH processes, moments cannot exist at any order. Although the marginal distribution of ϵ_t does not possess heavy-tails, the leptokurticity property is satisfied in the HMM: the law of ϵ_t has fatter tails than that of η_t. If a Gaussian distribution is assumed for η_t, the Kurtosis coefficient of ϵ_t is strictly greater than 3 (the exact value depends on both the transition probabilities of the chain and the values of the variances $\omega(i)$ in the different regimes).

An important difference with GARCH model pops up when we compute the conditional Kurtosis coefficient. Recall that this coefficient is constant for GARCH-type models. For model (12.15) we have

$$\frac{E(\epsilon_t^4 \mid \epsilon_{t-1}, \epsilon_{t-2}, \dots)}{[E(\epsilon_t^2 \mid \epsilon_{t-1}, \epsilon_{t-2}, \dots)]^2} = \frac{E(\sigma_t^4 \mid \epsilon_{t-1}, \epsilon_{t-2}, \dots)}{[E(\sigma_t^2 \mid \epsilon_{t-1}, \epsilon_{t-2}, \dots)]^2}\kappa_\eta$$

and no simplification occurs here, because σ_t^2 is not a function of past values of ϵ_t. Thus, the shape of the conditional distribution – not only the variance – may also fluctuate over time.

Less straightforward similarities with standard GARCH models appear by studying the autocorrelation functions of (ϵ_t) and (ϵ_t^2).

Autocorrelations, Moments

We have

$$\text{Corr}(\epsilon_t, \epsilon_{t-k}) = 0, \quad \text{for any } k > 0$$

hence (ϵ_t) is a white noise with variance

$$E\epsilon_t^2 = \sum_{i=1}^{d} \omega(i)\pi(i).$$

We now study the autocorrelation of the squared process. For ease of presentation, let us first consider the case where $d = 2$. The eigenvalues of the transition matrix are 1 and $\lambda := p(1,1) + p(2,2) - 1$, since the Markov chain is, by assumption A1, irreducible and aperiodic. Note that $-1 < \lambda < 1$. By diagonalising $\mathbb{P}$, it can easily be seen that the entries of $\mathbb{P}^k$ have the form $p^{(k)}(i,j) = a_1(i,j) + a_2(i,j)\lambda^k$, for $k \geq 0$. By letting k go to infinity, we get $a_1(i,j) = \pi(j)$ and, using the value $k = 0$: $a_1(i,j) + a_2(i,j) = \mathbb{1}_{\{i=j\}}$. It follows that for $j = 1, 2$ and $i \neq j$

$$p^{(k)}(i,j) = \pi(j)(1-\lambda^k), \quad p^{(k)}(j,j) = \pi(j) + \lambda^k(1-\pi(j)),$$

hence, for $i, j = 1, 2$

$$p^{(k)}(i,j) - \pi(j) = \lambda^k[\{1-\pi(j)\}\mathbb{1}_{\{i=j\}} - \pi(j)\mathbb{1}_{\{i\neq j\}}]. \tag{12.17}$$

We have, for $k > 0$,

$$\begin{aligned}\text{Cov}(\epsilon_t^2, \epsilon_{t-k}^2) &= \text{Cov}\{\omega(\Delta_t), \omega(\Delta_{t-k})\} = E\{\omega(\Delta_t)\omega(\Delta_{t-k})\} - \{E\omega(\Delta_t)\}^2\\ &= \sum_{i,j=1}^{2} p^{(k)}(i,j)\pi(i)\omega(i)\omega(j) - \left\{\sum_i \pi(i)\omega(i)\right\}^2\\ &= \sum_{i,j=1}^{2} \{p^{(k)}(i,j) - \pi(j)\}\pi(i)\omega(i)\omega(j),\end{aligned} \tag{12.18}$$

and thus, by using Eq. (12.17),

$$\begin{aligned}\text{Cov}(\epsilon_t^2, \epsilon_{t-k}^2) &= \lambda^k \left\{ \sum_{j=1}^{2} (1-\pi(j))\pi(j)\omega^2(j) - \sum_{i\neq j} \pi(i)\pi(j)\omega(i)\omega(j) \right\} \\ &= \lambda^k \{\omega(1)-\omega(2)\}^2 \pi(1)\pi(2), \quad k>0. \end{aligned} \tag{12.19}$$

It is worth noting that the autocorrelations of the squares decrease at exponential rate like for a second-order stationary GARCH process. An important difference, however, is that the rate of this convergence is not related to the existence of moments. Note also that the larger $|\lambda| = |1-p(1,1)-p(2,2)|$, the slower the decrease of the autocorrelations (in module). In other words, the autocorrelations decrease slowly when the transition probabilities from a regime to the other are either both very small or both close to 1. Obviously, $\text{Cov}(\epsilon_t^2, \epsilon_{t-k}^2) = 0$ for any $k > 0$ when $\omega(1) = \omega(2)$ because ϵ_t is an iid white noise in this case. A similar computation shows that

$$\text{Var}(\epsilon_t^2) = \{\omega(1)-\omega(2)\}^2\pi(1)\pi(2) + \{\omega^2(1)\pi(1) + \omega^2(2)\pi(2)\}\text{Var}(\eta_t^2). \tag{12.20}$$

We deduce from Eqs. (12.19) and (12.20) that ϵ_t^2 admits an ARMA(1,1) representation, with autoregressive coefficient λ.

In the general case, matrix $\mathbb{P}$ may not be diagonalisable but has an eigenvalue equal to 1, with corresponding eigenspace of dimension 1. Denote by $\lambda_1, \dots, \lambda_m$ the other eigenvalues and $n_1, \dots, n_m$ the dimensions of the corresponding eigenspaces ($n_1 + \cdots + n_m = d-1$). The Jordan form of matrix $\mathbb{P}$ is $\mathbb{P} = SJS^{-1}$, where S is non-singular and

$$J = \begin{pmatrix} J_{n_1}(\lambda_1) & 0 & \dots & & 0 \\ 0 & J_{n_2}(\lambda_2) & 0 & & \vdots \\ \vdots & & \ddots & \ddots & \vdots \\ \vdots & & \ddots & J_{n_m}(\lambda_m) & 0 \\ 0 & \dots & & 0 & 1 \end{pmatrix}$$

with $J_l(\lambda) = \lambda \mathbb{I}_l + N_l(1)$, denoting by $N_l(i)$ the square matrix of dimension l whose entries are equal to zero, except for the ith uperdiagonal whose elements are equal to 1. Using $N_l^{k'}(1) = N_l(k')$ for $k' \le l-1$ and $N_l^{k'}(1) = 0$ for $k' > l-1$, we get

$$J_l^k(\lambda) = \sum_{k'=0}^{k} \binom{k}{k'} \lambda^{k-k'} N_l^{k'}(1) = \sum_{k'=0}^{l-1} \binom{k}{k'} \lambda^{k-k'} N_l(k') := \lambda^k P^{(l)}(k)$$

where $P^{(l)}$ is a polynomial of degree $l-1$. It follows that

$$\mathbb{P}^k = S \begin{pmatrix} \lambda_1^k P^{(n_1)}(k) & 0 & \dots & & 0 \\ 0 & \lambda_2^k P^{(n_2)}(k) & 0 & & \vdots \\ \vdots & & \ddots & \ddots & \vdots \\ \vdots & & \ddots & \lambda_m^k P^{(n_m)}(k) & 0 \\ 0 & \dots & & 0 & 1 \end{pmatrix} S^{-1}.$$

We deduce that

$$p^{(k)}(i,j) = \pi(j) + \sum_{l=1}^{m} \lambda_l^k p_{i,j}^{(n_l)}(k)$$

where the $p_{i,j}^{(n_l)}$ are polynomials of degree $n_l - 1$. The first term in the right-hand side is identified by the convergence $p^{(k)}(i,j) \to \pi(j)$ as $k \to \infty$. Note that $|\lambda_l| < 1$ for $l = 1, \dots, m$. Finally, we have, using (12.18),

$$\text{Cov}(\epsilon_t^2, \epsilon_{t-k}^2) = \sum_{i,j=1}^{d} \sum_{l=1}^{m} \lambda_l^k p_{i,j}^{(n_l)}(k) := \sum_{l=1}^{m} \lambda_l^k q^{(n_l)}(k), \quad k>0 \tag{12.21}$$

where the $q^{(n_l)}$'s are polynomials of degree $n_l - 1$.

The computation of $E(\sigma^2_{t+k} \mid \epsilon_{t-1}, \epsilon_{t-2}, \ldots)$ is more tedious, but it can be shown that this expectation depends of k through the λ_l^k's (similarly to (12.21)). Thus, the Markov chain (Δ_t) introduces a source of persistence of shocks on the volatility.

Estimation

The vector of parameters of model (12.15) is

$$\theta = (p(1,1), \ldots, p(1,d-1), \ldots, p(d,d-1), \omega(1), \ldots, \omega(d))' \in [0,1]^{d(d-1)} \times (0,\infty)^d.$$

The likelihood can be written by conditioning on all possible paths $(e_1, \ldots, e_n)$ of the Markov chain, where the e_i belong to $\mathcal{E} = \{1, \ldots, d\}$. For ease of presentation, we assume a standard Gaussian distribution for η_t but the estimation procedure can be adapted to other distributions. The probability of such paths is given by

$$\mathbb{P}(e_1, \ldots, e_n) = \mathbb{P}(\Delta_1 = e_1, \ldots, \Delta_n = e_n) = \pi(e_1)p(e_1, e_2) \ldots p(e_{n-1}, e_n).$$

For each path, we get a conditional likelihood of the form

$$L_\omega^{(e_1,\ldots,e_n)}(\epsilon_1, \ldots, \epsilon_n) = \prod_{t=1}^{n} \phi_{e_t}(\epsilon_t),$$

where $\phi_i(\cdot)$ denotes the density of the law $\mathcal{N}\{0, \omega(i)\}$.

Finally, the likelihood writes

$$L_\theta(\epsilon_1, \ldots, \epsilon_n) = \sum_{(e_1,\ldots,e_n)\in\mathcal{E}^n} L_\omega^{(e_1,\ldots,e_n)}(\epsilon_1, \ldots, \epsilon_n)\mathbb{P}(e_1, \ldots, e_n).$$

Unfortunately, this formula cannot be used in practice because the number of summands is d^n, which is huge even for small samples and two regimes. For instance, 2^{300} is generally considered to be an upper bound for the number of atoms in the universe.[9] Several solutions to this numerical problem – computing the likelihood – exist.

Computing the Likelihood

Matrix Representation Let

$$F_k(i) = g_k(\epsilon_1, \ldots, \epsilon_k \mid \Delta_k = i)\pi(i),$$

where $g_k(\cdot \mid \Delta_k = i)$ is the law of $(\epsilon_1, \ldots, \epsilon_k)$ given $\{\Delta_k = i\}$. One can easily check that

$$F_1(i) = \pi(i)\phi_i(\epsilon_1), \tag{12.22}$$

$$F_k(i) = \phi_i(\epsilon_k) \sum_{j=1}^{d} F_{k-1}(j)p(j,i), \tag{12.23}$$

and

$$L_\theta(\epsilon_1, \ldots, \epsilon_n) = \sum_{i=1}^{d} F_n(i). \tag{12.24}$$

In matrix form we get

$$F_k := (F_k(1), \ldots, F_k(d))' = M(\epsilon_k)F_{k-1},$$

[9] The well-known story of the wheat and chessboard problem (see for instance http://britton.disted.camosun.bc.ca/jbchessgrain.htm) allows to realize that 2^{64} is already a huge number.

where

$$M(x) = \begin{pmatrix} p(1,1)\phi_1(x) & \cdots & p(d,1)\phi_1(x) \\ \vdots & & \\ p(1,d)\phi_d(x) & \cdots & p(d,d)\phi_d(x) \end{pmatrix}.$$

Hence, letting $\mathbf{1}' = (1, \dots, 1)$, we have

$$L_\theta(\epsilon_1, \dots, \epsilon_n) = \mathbf{1}' M(\epsilon_n) M(\epsilon_{n-1}) \cdots M(\epsilon_2) F_1, \tag{12.25}$$

which is computationally usable (with an order of $d^2 n$ multiplications).

Forward–Backward Algorithm Let $B_k(i) = B_k(\epsilon_{k+1}, \dots, \epsilon_n \mid \Delta_k = i)$ the law of $(\epsilon_{k+1}, \dots, \epsilon_n)$ given $\{\Delta_k = i\}$. By the Markov property, we have

$$L_\theta(\epsilon_1, \dots, \epsilon_n \mid \Delta_k = i)\pi(i) = F_k(i) B_k(i).$$

The Forward formulas, allowing us to compute $F_k(i)$ for $k = 1, 2, \dots$, are given by (12.22) and (12.23). The Backward formulas, allowing us to compute $B_k(i)$ for $k = n-1, n-2, \dots$ are

$$B_n(i) = 1 \tag{12.26}$$

$$B_k(i) = \sum_{j=1}^{d} B_{k+1}(j) p(i,j) \phi_j(\epsilon_{k+1}). \tag{12.27}$$

We thus obtain

$$L_\theta(\epsilon_1, \dots, \epsilon_n) = \sum_{i=1}^{d} F_k(i) B_k(i), \tag{12.28}$$

for any $k \in \{1, \dots, n\}$. For $k = n$, we retrieve Eq. (12.24).

Hamilton Filter The Forward–Backward algorithm was developed in the statistical literature (Baum 1972). Models involving a latent Markov chain were put forward in the econometric literature by Hamilton (1989). Let

$$\pi_{t|t} = \begin{pmatrix} \mathbb{P}(\Delta_t = 1 \mid \epsilon_t, \dots, \epsilon_1) \\ \vdots \\ \mathbb{P}(\Delta_t = d \mid \epsilon_t, \dots, \epsilon_1) \end{pmatrix}, \quad \pi_{t|t-1} = \begin{pmatrix} \mathbb{P}(\Delta_t = 1 \mid \epsilon_{t-1}, \dots, \epsilon_1) \\ \vdots \\ \mathbb{P}(\Delta_t = d \mid \epsilon_{t-1}, \dots, \epsilon_1) \end{pmatrix},$$

$\phi(\epsilon_t) = (\phi_1(\epsilon_t), \dots, \phi_d(\epsilon_t))'$, and denote by $\odot$ the element-by-element Hadamard product of matrices.

With obvious notations, we get

$$\pi_{t|t}(i) = \mathbb{P}(\Delta_t = i \mid \epsilon_t, \dots, \epsilon_1) = \frac{\phi_i(\epsilon_t)\mathbb{P}(\Delta_t = i \mid \epsilon_{t-1}, \dots, \epsilon_1)}{f_t(\epsilon_t \mid \epsilon_{t-1}, \dots, \epsilon_1)},$$

where

$$f_t(\epsilon_t \mid \epsilon_{t-1}, \dots, \epsilon_1) = \sum_{i=1}^{d} \phi_i(\epsilon_t) \pi_{t|t-1}(i)$$

Starting from an initial value $\pi_{1|0} = \pi$ (the stationary law) or $\pi_{1|0} = \pi_0$ (a given initial law), we can compute

$$\pi_{t|t} = \frac{\pi_{t|t-1} \odot \phi(\epsilon_t)}{\mathbf{1}'\{\pi_{t|t-1} \odot \phi(\epsilon_t)\}}, \quad \pi_{t+1|t} = \mathbb{P}' \pi_{t|t} \tag{12.29}$$

for $t = 1, \dots, n$, and we deduce the conditional log-likelihood

$$\log L_\theta(\epsilon_1, \dots, \epsilon_n) = \sum_{t=1}^{n} \log f_t(\epsilon_t \mid \epsilon_{t-1}, \dots, \epsilon_1), \tag{12.30}$$

where

$$f_t(\epsilon_t \mid \epsilon_{t-1}, \dots, \epsilon_1) = \mathbf{1}'\{\pi_{t|t-1} \odot \phi(\epsilon_t)\}. \tag{12.31}$$

The Hamilton algorithm (12.29)–(12.31) seems numerically more efficient than the Forward–Backward algorithm in (12.22)–(12.23) and (12.26)–(12.28) which, under this form, produces underflows. Note, however, that conditional versions of the Forward–Backward algorithm have been developed to avoid this problem (Devijver 1985). Note also that the matrix formulation (12.25) is very convenient for establishing asymptotic properties of the ML estimator (see Francq and Roussignol 1997).

Maximising the Likelihood

Maximisation of the (log-)likelihood can be achieved by a classical numerical procedure, or using the EM (Expectation–Maximisation) algorithm, which can be described as follows.

For ease of exposition, we consider that the initial distribution π_0 (the law of Δ_1) is not necessarily the stationary distribution π. In the EM algorithm, π_0 is an additional parameter to be estimated. If – in addition to $(\epsilon_1, \dots, \epsilon_n)$ – one could also observe $(\Delta_1, \dots, \Delta_n)$, estimating θ and π_0 by ML would be an easy task. Indeed,

$$\begin{aligned}
&\log L_{\theta,\pi_0}(\epsilon_1, \dots, \epsilon_n, \Delta_1, \dots, \Delta_n)\\
&= \sum_{t=1}^{n} \log \phi_{\Delta_t}(\epsilon_t) + \log \pi_0(\Delta_1) + \sum_{t=2}^{n} \log p(\Delta_{t-1}, \Delta_t)\\
&= a_1 + a_2 + a_3
\end{aligned}$$

where

$$a_1 = a_1(\omega) = \sum_{i=1}^{d} \sum_{t=1}^{n} \log \phi_i(\epsilon_t) \mathbb{1}_{\{\Delta_t = i\}}, \tag{12.32}$$

$$a_2 = a_2(\pi_0) = \log \pi_0(\Delta_1), \tag{12.33}$$

$$a_3 = a_3(\mathbb{P}) = \sum_{i=1}^{d} \sum_{j=1}^{d} \log p(i,j) \sum_{t=2}^{n} \mathbb{1}_{\{\Delta_{t-1}=i, \Delta_t = j\}}. \tag{12.34}$$

From (12.32), we need to maximise the terms $\sum_{t=1}^{n} \log \phi_i(\epsilon_t) \mathbb{1}_{\{\Delta_t = i\}}$ with respect to $\omega(i)$, for $i = 1, \dots, d$. This yields the estimators

$$\tilde{\omega}(i) = \frac{1}{\sum_{t=1}^{n} \mathbb{1}_{\{\Delta_t = i\}}} \sum_{t=1}^{n} \epsilon_t^2 \mathbb{1}_{\{\Delta_t = i\}}. \tag{12.35}$$

Maximisation of (12.33) with respect to $\pi_0(1), \dots, \pi_0(d)$, under the constraint $\sum_{i=1}^{d} \pi_0(i) = 1$, yields

$$\tilde{\pi}_0(i) = \mathbb{1}_{\{\Delta_1 = i\}}. \tag{12.36}$$

From (12.34), for $i = 1, \dots, d$, one has to maximise with respect to $p(i,1), \dots, p(i,d)$, under the constraint $\sum_{j=1}^{d} p(i,j) = 1$, the sum

$$\sum_{j=1}^{d} \log p(i,j) \frac{\sum_{t=2}^{n} \mathbb{1}_{\{\Delta_{t-1}=i, \Delta_t = j\}}}{\sum_{t=2}^{n} \mathbb{1}_{\{\Delta_{t-1}=i\}}}.$$

We obtain[10]

$$\tilde{p}(i,j) = \frac{1}{\sum_{t=2}^{n} \mathbb{1}_{\{\Delta_t = i\}}} \sum_{t=2}^{n} \mathbb{1}_{\{\Delta_{t-1}=i, \Delta_t = j\}}. \tag{12.37}$$

In practice, formulas (12.35), (12.36), and (12.37) cannot be used, because (Δ_t) is unobserved. Generally speaking, the EM algorithm alternates between steps E – for expectation – where the expectation of the likelihood is evaluated using the current value of the parameter, and steps M – for maximisation – where the objective function computed in step E is maximised. In the present setting, the EM algorithm only requires steps M, together with the computation of the predicted and filtered probabilities – $\pi_{t|t-1}$ and $\pi_{t|t}$, respectively – of the Hamilton filter (12.29), and also an additional step for the computation of the smoothed probabilities.

Step E Suppose that an estimator $(\theta^{(k)}, \pi_0^{(k)})$ de (θ, π_0) is available. It seems sensible to approximate the unknown log-likelihood by its expectation given the observations $(\epsilon_1, \dots, \epsilon_n)$, evaluated under the law parameterised by $(\theta^{(k)}, \pi_0^{(k)})$. We get the criterion

$$\begin{aligned} Q(\theta, \pi_0 \mid \theta^{(k)}, \pi_0^{(k)}) &= E_{\theta^{(k)}, \pi_0^{(k)}} \{\log L_{\theta, \pi_0}(\epsilon_1, \dots, \epsilon_n, \Delta_1, \dots, \Delta_n) \mid \epsilon_1, \dots, \epsilon_n\} \\ &= A_1(\omega) + A_2(\pi_0) + A_3(\mathbb{P}), \end{aligned}$$

where

$$A_1(\omega) = \sum_{i=1}^{d} \sum_{t=1}^{n} \log \phi_i(\epsilon_t) P_{\theta^{(k)}, \pi_0^{(k)}} \{\Delta_t = i \mid \epsilon_1, \dots, \epsilon_n\}, \tag{12.38}$$

$$A_2(\pi_0) = \sum_{i=1}^{d} \log \pi_0(i) P_{\theta^{(k)}, \pi_0^{(k)}} \{\Delta_1 = i \mid \epsilon_1, \dots, \epsilon_n\}, \tag{12.39}$$

$$A_3(\mathbb{P}) = \sum_{i,j} \log p(i,j) \sum_{t=2}^{n} P_{\theta^{(k)}, \pi_0^{(k)}} \{\Delta_{t-1} = i, \Delta_t = j \mid \epsilon_1, \dots, \epsilon_n\}. \tag{12.40}$$

Step M We aim at maximising, with respect to (θ, π_0), the estimated log-likelihood $Q(\theta, \pi_0 \mid \theta^{(k)}, \pi_0^{(k)})$. Maximisation of (12.38) yields the solution

$$\hat{\omega}(i) = \frac{\sum_{t=1}^{n} \epsilon_t^2 P_{\theta^{(k)}, \pi_0^{(k)}} \{\Delta_t = i \mid \epsilon_1, \dots, \epsilon_n\}}{\sum_{t=1}^{n} P_{\theta^{(k)}, \pi_0^{(k)}} \{\Delta_t = i \mid \epsilon_1, \dots, \epsilon_n\}}. \tag{12.41}$$

The estimated variance of regime i is thus a weighted mean of the ϵ_t^2, where the weights are the conditional probabilities that the chain lie in state i at time t. Similarly, (12.39) yields

$$\hat{\pi}_0(i) = P_{\theta^{(k)}, \pi_0^{(k)}} \{\Delta_1 = i \mid \epsilon_1, \dots, \epsilon_n\}, \tag{12.42}$$

and (12.40) yields

$$\hat{p}(i,j) = \frac{\sum_{t=2}^{n} P_{\theta^{(k)}, \pi_0^{(k)}} \{\Delta_{t-1} = i, \Delta_t = j \mid \epsilon_1, \dots, \epsilon_n\}}{\sum_{t=2}^{n} P_{\theta^{(k)}, \pi_0^{(k)}} \{\Delta_{t-1} = i \mid \epsilon_1, \dots, \epsilon_n\}}. \tag{12.43}$$

[10] Let $p_1, \dots, p_n$ be positive numbers such that $\sum_i p_i = 1$. By substitution or by the Lagrange multiplier method, it is easily seen that the global maximum of the function $(\pi_1, \dots, \pi_d) \to \sum_i p_i \log \pi_i$ under the constraint $\sum_{i=1}^{d} \pi_i = 1$ is reached at $(\pi_1, \dots, \pi_d) = (p_1, \dots, p_d)$.

Formulas (12.41), (12.42), and (12.43) require computation of the smoothed probabilities

$$\pi_{t|n} = (P\{\Delta_t = i \mid \epsilon_1, \dots, \epsilon_n\})'_{1\le i\le d} \in \mathbb{R}^d$$

and

$$\pi_{t-1,t|n} = (P\{\Delta_{t-1} = i, \Delta_t = j \mid \epsilon_1, \dots, \epsilon_n\})'_{1\le i,j\le d} \in \mathbb{R}^d \times \mathbb{R}^d.$$

Computation of Smoothed Probabilities The Markov property entails that, given Δ_t, the observations $\epsilon_t, \epsilon_{t+1}, \dots$ do not convey information on Δ_{t-1}. We hence have

$$\mathbb{P}(\Delta_{t-1} = i \mid \Delta_t = j, \epsilon_1, \dots, \epsilon_n) = \mathbb{P}(\Delta_{t-1} = i \mid \Delta_t = j, \epsilon_1, \dots, \epsilon_{t-1})$$

and

$$\begin{aligned}\pi_{t-1,t|n}(i,j) &= \mathbb{P}(\Delta_{t-1} = i \mid \Delta_t = j, \epsilon_1, \dots, \epsilon_n)\pi_{t|n}(j)\\ &= \frac{p(i,j)\pi_{t-1|t-1}(i)\pi_{t|n}(j)}{\pi_{t|t-1}(j)}.\end{aligned}$$

It remains to compute the smoothed probabilities $\pi_{t|n}$, given by

$$\pi_{t-1|n}(i) = \sum_{j=1}^{d} \pi_{t-1,t|n}(i,j) = \sum_{j=1}^{d} \frac{p(i,j)\pi_{t-1|t-1}(i)\pi_{t|n}(j)}{\pi_{t|t-1}(j)}$$

for $t = n, n-1, \dots, 2$. Starting from an initial value $(\theta^{(0)}, \pi_0^{(0)})$, formulas (12.41), (12.42), and (12.43) allow us to obtain a sequence of estimators $(\theta^{(k)}, \pi_0^{(k)})_k$ which increase the likelihood (see Exercise 12.13).

Summary of the Algorithm Starting from initial values for the parameters

$$\begin{aligned}\pi_0 &= \{P(\Delta_1 = 1), \dots, P(\Delta_1 = d)\}',\\ p(i,j) &= P(\Delta_t = j \mid \Delta_{t-1} = i)\\ \omega &= \{\omega(1), \dots, \omega(d)\}',\end{aligned}$$

the algorithm consists in
repeating the following steps until consistency:

1. Set $\pi_{1|0} = \pi_0$ and

$$\pi_{t|t} = \frac{\pi_{t|t-1} \odot \phi(\epsilon_t)}{\mathbb{1}'\{\pi_{t|t-1} \odot \phi(\epsilon_t)\}}, \quad \pi_{t+1|t} = \mathbb{P}'\pi_{t|t}, \qquad \text{for } t = 1, \dots, n.$$

2. Compute the smoothed probabilities $\pi_{t|n}(i) = P(\Delta_t = i \mid \epsilon_1, \dots, \epsilon_n)$ using

$$\pi_{t-1|n}(i) = \sum_{j=1}^{d} \frac{p(i,j)\pi_{t-1|t-1}(i)\pi_{t|n}(j)}{\pi_{t|t-1}(j)} \quad \text{for } t = n, n-1, \dots, 2,$$

and $\pi_{t-1,t|n}(i,j) = P(\Delta_{t-1} = i, \Delta_t = j \mid \epsilon_1, \dots, \epsilon_n)$ from

$$\pi_{t-1,t|n}(i,j) = \frac{p(i,j)\pi_{t-1|t-1}(i)\pi_{t|n}(j)}{\pi_{t|t-1}(j)}.$$

3. Replace the previous values of the parameters by $\pi_0 = \pi_{1|n}$,

$$p(i,j) = \frac{\sum_{t=2}^{n} \pi_{t-1,t|n}(i,j)}{\sum_{t=2}^{n} \pi_{t-1|n}(i)} \quad \text{and} \quad \omega(i) = \frac{\sum_{t=1}^{n} \epsilon_t^2 \pi_{t|n}(i)}{\sum_{t=1}^{n} \pi_{t|n}(i)}.$$

In practice, the sequence converges rather rapidly to the ML estimator (see the illustration below) if an appropriate starting value $\theta^{(0)} > 0$ is chosen (see Exercise 12.9).

Model (12.15) is certainly too simple to take into account, in their complexity, the dynamic properties of real-time series (for instance of financial returns): whenever the series remain in the same regime, the observations have to be independent, and this assumption is quite unrealistic. A natural extension of model (12.15) – and also of the standard GARCH model – is obtained by assuming that, in each regime, the dynamics is governed by a GARCH model.

12.2.2 MS-GARCH(p, q) Process

Consider the Markov–Switching GARCH (MS-GARCH) process:

$$\begin{cases} \epsilon_t = \sigma_t \eta_t \\ \sigma_t^2 = \omega(\Delta_t) + \sum_{i=1}^{q} \alpha_i(\Delta_t)\epsilon_{t-i}^2 + \sum_{j=1}^{p} \beta_j(\Delta_t)\sigma_{t-j}^2 \end{cases} \tag{12.44}$$

with the positivity constraints:

$$\text{for } k = 1, \dots, d, \quad \omega(k) > 0, \quad \alpha_i(k) \geq 0, \quad 1 \leq i \leq q, \quad \beta_j(k) \geq 0, \quad 1 \leq j \leq p.$$

The model has d different GARCH regimes, which entails great flexibility. The standard GARCH – obtained for $d = 1$ – and the HMM – obtained when the $\alpha_i(k)$'s and $\beta_j(k)$s are equal to zero – are particular cases of model (12.44).

Many properties highlighted for model (12.15) also hold for this general formulation. An important difference, however, is that the present model displays two distinct sources of persistence: one is conveyed by the chain (Δ_t), and the other one by the coefficients $\alpha_i(\cdot)$ and $\beta_j(\cdot)$. For instance, regimes where shocks on past variables are very persistent may coexist with regimes where shocks are not persistent.

Another difference concerns the second-order stationarity properties. To ensure that ϵ_t admits a finite, time-independent, variance in model (12.44), constraints on the coefficients $\alpha_i(\cdot)$ and $\beta_j(\cdot)$, as well as on the transition probabilities, have to be imposed. Without entering into details,[11] note that the *local stationarity* – i.e. stationarity within each GARCH regime – is sufficient but not necessary for the *global stationarity*: a stationary solution of model (12.44) may have explosive GARCH regimes, provided that the transition probabilities towards such regimes are small enough. It is also worth noting that when only the intercept $\omega(\cdot)$ is switching (i.e. when the α_i's and β_j's do not depend on Δ_t), we retrieve the usual second-order stationarity condition ($\sum_i \alpha_i + \sum_j \beta_j < 1$). Finally, note that as in standard GARCH, once the existence $Var(\epsilon_t)$ is ensured, (ϵ_t) is a white noise.

Applications of MS-GARCH models are often limited to MS-ARCH. This restriction enables ML estimation with the Hamilton filter, as described in the previous section.

Illustration

For the sake of illustration, consider the CAC 40 and SP 500 stock market indices. Observations cover the period from March 1, 1990, to December 29, 2006. On the daily returns (in %), we fitted the HMM (12.15) and (12.16) with $d = 4$ regimes, using the R code given in Exercise 12.8. Taking initial values with

[11] See exercises 12.10–12.12 and Francq, Roussignol, and Zakoïan (2001).

non-zero transition probabilities (see Exercise 12.9), after roughly 60 iterations of the EM algorithm, we obtain the following estimated model for the SP 500 series

$$\hat{\omega}_{\rm SP} = \begin{pmatrix} 0.26 \\ 0.62 \\ 1.28 \\ 4.8 \end{pmatrix}, \quad \hat{\mathbb{P}}_{\rm SP} = \begin{pmatrix} 0.981 & 0.019 & 0.000 & 0.000 \\ 0.018 & 0.979 & 0.003 & 0.000 \\ 0.000 & 0.003 & 0.986 & 0.011 \\ 0.000 & 0.000 & 0.055 & 0.945 \end{pmatrix}$$

and for the CAC 40

$$\hat{\omega}_{\rm CAC} = \begin{pmatrix} 0.51 \\ 1.19 \\ 2.45 \\ 8.4 \end{pmatrix}, \quad \hat{\mathbb{P}}_{\rm CAC} = \begin{pmatrix} 0.993 & 0.003 & 0.002 & 0.002 \\ 0.003 & 0.991 & 0.003 & 0.003 \\ 0.000 & 0.020 & 0.977 & 0.003 \\ 0.004 & 0.000 & 0.032 & 0.963 \end{pmatrix}.$$

The estimated probabilities of the regimes are

$$\hat{\pi}_{\rm SP} = (0.30, 0.32, 0.32, 0.06)', \quad \hat{\pi}_{\rm CAC} = (0.26, 0.49, 0.19, 0.06)',$$

and the average durations of the four regimes – given by $1/\{1 - p(i,i)\}$ – are approximately

$$\mathrm{D}_{\rm SP} = (53, 48, 71, 18)', \quad \mathrm{D}_{\rm CAC} = (140, 107, 43, 27)'.$$

Thus, the CAC stays in average 27 days in the more volatile regime, namely Regime 4. Figure 12.2 confirms that for the two series, the more volatile regime is also the less-persistent regime, with however a long period of high volatility of 81 days (from June 27, 2002 to October 21, 2002) for the SP, and of 113 days (from June 4, 2002 to November 8, 2002) for the CAC. It is interesting to note that for the model of the SP 500, the possible transitions are always from one regime to an adjacent regime. Being, for instance in Regime 2, the chain can stay in Regime 2 or move to Regime 1 or 3, but the probability $p(2,4)$ of moving directly to Regime 4 is approximately zero. On the contrary, the CAC displays sudden moves from Regime 2 to Regime 4.

Conclusion

Estimation of GARCH-type models over long time series, as those encountered in finance (i.e. typically several thousands of observations), generally produces strong volatility persistence. This effect may be spurious and may be explained by the need to obtain heavy-tailed marginal distributions.

MS-GARCH models enable to disentangle important properties of financial time series: persistence of shocks, decreasing autocorrelations of the squared returns, heavy tailed marginal distributions, and time-varying conditional densities. These models are appropriate for series over a long time period, with successions of phases corresponding to the different regimes. Of course, such models – despite their flexibility – remain merely approximations of the real data generation mechanisms.[12]

12.3 Bibliographical Notes

A recent presentation and statistical analysis of parameter-driven and observation-driven models can be found in Koopman, Lucas, and Scharth (2016). Continuous-time models with SV have been widely studied in the option pricing literature: see, for instance Hull and White (1987), Melino and Turnbull (1990) and, more recently, Aït-Sahalia, Amengual, and Manresa (2015). Harvey, Ruiz, and Shephard (1994) pointed out that the canonical SV model can be studied using a state-space form allowing to apply

[12] For instance, the switching regime mechanism could explicitly depend on the past observations (i.e. not only on the past regimes).

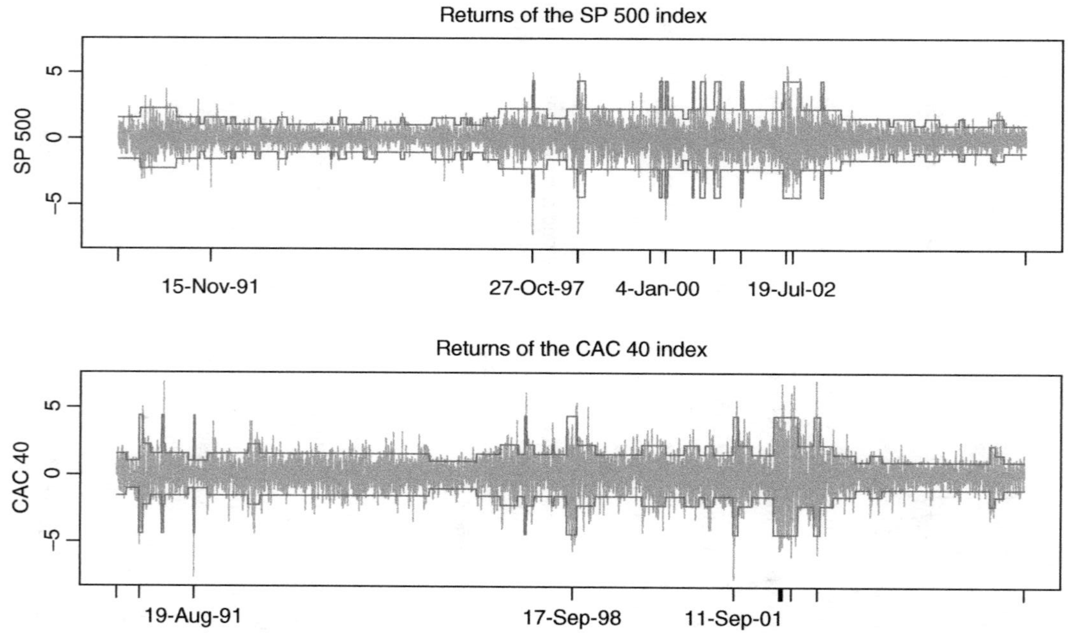

Figure 12.2 CAC 40 and SP 500 from March 1, 1990 to December 29, 2006, with ±2 times the standard deviation of the regime with the highest smoothed probability.

the Kalman filter. The existence of the ARMA(1,1) representation for the log-squared return process was mentioned by Ruiz (1994), and it was used by Breidt and Carriquiry (1996) to obtain QML estimates. See Davis and Mikosch (2009b) for mixing and tail properties of general SV models.

Various approaches have been developed for estimating SV models. Generalised method of moments (GMM) type of estimators have been proposed, for instance by Taylor (1986), Andersen (1994) and Andersen and Sørensen (1996). In the same spirit, Francq and Zakoian (2006a) proposed an estimation method based on the existence of ARMA representations for powers of the logarithm of the observed process. The QML method has been advocated by Harvey, Ruiz, and Sentana (1992), Harvey, Ruiz, and Shephard (1994) and Ruiz (1994), among others. An approach which is closely related to the QML uses a mixture of Gaussian distributions to approximate the non-Gaussian error in the observation equation of the state-space form (see, Mahieu and Schotman 1998). Bayesian MCMC methods of inference were applied to SV models, in particular by Jacquier, Polson, and Rossi (1994, 2004) and Kim, Shephard, and Chib (1998) and by Chib, Nardia and Shepard (2002). The simulated maximum likelihood (SML) method, relying on simulation of the latent volatility conditional on available information, was proposed by Danielsson (1994); the likelihood can also be calculated directly using recursive numerical integration procedures, as suggested by Fridman and Harris (1998). Estimation of SV models with leverage has been studied by Harvey and Shephard (1996), Yu (2005), Omori et al. (2007), among others. See Taylor (1994), Shephard and Andersen (2009), Renault (2009), and Jungbacker and Koopman (2009) for overviews on SV models.

Estimation of HMM was initially studied by Baum and Petrie (1966) and Petrie (1969). They proved consistency and asymptotic normality of the ML estimator when the observations are valued in a finite set. Leroux (1992) showed the consistency, and Bickel, Ritov, and Ryden (1998) established asymptotic normality for general observations. A detailed presentation of the algorithms required for the practical implementation can be found in Rabiner and Juang (1986).

Markov switching models were introduced by Hamilton (1989). In this paper, Hamilton developed a filter which can be applied when the regimes are Markovian (as the MS-AR or the MS-ARCH). MS-ARCH models were initially studied by Hamilton and Susmel (1994), Cai (1994) and Dueker (1997). Stationarity conditions for MS-GARCH, as well as estimation results for the MS-ARCH can be found in Francq, Roussignol, and Zakoïan (2001), and Francq and Zakoïan (2005). Papers dealing with Bayesian estimation of MS-GARCH are Kaufmann and Frühwirth-Schnatter (2002), Das and Yoo (2004), Henneke et al. (2011), and Bauwens, Preminger, and Rombouts (2010). See the books by Cappé, Moulines, and Ryden (2005), Frühwirth-Schnatter (2006), and Douc, Moulines, and Stoffer (2014) for an exhaustive overview on MS models.

12.4 Exercises

12.1 *(Second-order stationarity of the SV model)*
Prove Theorem 12.2.

12.2 *(Logarithmic transformation of the squared returns in the SV model)*
Show that the logarithmic transformation in (12.6) entails no information loss when the distribution of η_t *is symmetric.*

12.3 *(ARMA representation for the log-squared returns in the SV model)*
Prove Theorem 12.3.

12.4 *(Contemporaneously correlated SV model)*
Show that the contemporaneously correlated SV process is not centred and is autocorrelated.

12.5 *(Fitting GARCH on sub-periods of the CAC index)*
Consider a long period of a stock market index (for instance the CAC 40 from March 1, 1990 to December 29, 2006). Fit a GARCH(1,1) model on the returns of the first half of the series, and another GARCH(1,1) on the last part of the series. Compare the two estimated GARCH models, for instance by bootstrapping the estimated models.

12.6 *(Invariant law of Ehrenfest's model)*
This model has been introduced in physics to describe the heat exchanges between two systems. The model considers balls numbered 1 to d in two urns A and B. At the initial step 0, the balls are randomly distributed between the two urns (each ball has probability 1/2 to be in urn A). At step $n \geq 1$, a number i is picked randomly in $\{1, \dots, d\}$, and the ball numbered i is transferred to the other urn. Let Δ_n the number of balls in urn A at step n. Show that Δ_n is a Markov chain on $\mathcal{E} = \{0, 1, \dots, d\}$ which is irreducible and periodic. Provide the transition matrix. Show that the initial law is invariant. If, instead, the initial law is the Dirac mass at 0, show that $\lim_{n\to\infty}\pi_n$ does not exist where π_n denotes the distribution of Δ_n.

12.7 *(Period of an irreducible Markov chain)*
Show that the states of an irreducible Markov chain have the same period.

12.8 *(Implementation of the EM algorithm)*
Implement the EM algorithm of Section 12.2.1 (in R for example).

12.9 *(Choice of the initial value for the EM algorithm)*
Suppose that in the EM algorithm, the initial values are such that $p(i_0, j_0) = 0$ for some i_0, $j_0 \in \{1, \dots, d\}$. What are the next updated values of $p(i_0, j_0)$ in the algorithm?

12.10 *(Strict stationarity of the MS-GARCH)*
Give a condition for strict stationarity of the Markov-switching GARCH(p, q) model (12.44).

12.11 *(Strict stationarity of the MS-GARCH(1,1))*
Give an explicit condition for strict stationarity of the Markov-switching GARCH(1, 1) model. Consider the ARCH(1) case.

12.12 *(Stationarity of the GARCH(1,1) with independent regime changes)*
Consider a GARCH(1,1) model with independent regime changes, namely a model of the form (12.44), where (Δ_t) is an iid sequence. Give a condition for the existence of a second-order stationary solution.

12.13 *(Consistency of the EM algorithm)*
Let $(\theta^{(k)}, \pi_0^{(k)})_k$ be a sequence of estimators obtained from the EM algorithm section. With some abuse of notation, let $L_{\theta,\pi_0}(\epsilon_1, \dots, \epsilon_n)$ be the likelihood and $L_{\theta,\pi_0}(\epsilon_1, \dots, \epsilon_n, \Delta_1, \dots, \Delta_n)$ the joint distribution of the observations and of $(\Delta_1, \dots, \Delta_n)$. Show that $\{L_{\theta^{(k)},\pi_0^{(k)}}(\epsilon_1, \dots, \epsilon_n)\}_k$ is an increasing sequence.

12.14 *(Likelihood of a MS-ARCH)*
Consider an ARCH(q) with Markov-switching regimes, i.e. (12.44) with $p = 0$. Show that the likelihood admits a matrix form similar to (12.25), and that the forward–backward algorithm (12.22)–(12.23) and (12.26)–(12.28) can be applied, as well as the Hamilton filter (12.29)–(12.31). Can the EM algorithm be adapted?

12.15 *(An alternative MS-GARCH(1,1) model)*
The following model has been proposed Haas, Mittnik, and Paolella (2004) and studied by Liu (2006). Each regime $k \in \{1, \dots, d\}$ has the volatility

$$\sigma_t^2(k) = \omega(k) + \alpha(k)\epsilon_{t-1}^2 + \beta(k)\sigma_{t-1}^2(k),$$

and we set $\epsilon_t = \sigma_t(\Delta_t)\eta_t$. Explain the difference between this model and (12.44) with $p = q = 1$.

Appendix A

Ergodicity, Martingales, Mixing

A.1 Ergodicity

A stationary sequence is said to be ergodic if it satisfies the strong law of large numbers.

Definition A.1 (Ergodic stationary processes) *A real valued strictly stationary process $(Z_t)_{t\in\mathbb{Z}}$, is said to be ergodic if and only if, for any Borel set B and any integer k,*

$$n^{-1}\sum_{t=1}^{n}\mathbb{1}_B(Z_t, Z_{t+1}, \dots, Z_{t+k}) \to \mathbb{P}\{(Z_1, \dots, Z_{1+k}) \in B\}$$

with probability 1.[1]

General transformations of ergodic sequences remain ergodic. The proof of the following result can be found, for instance, in Billingsley (1995, Theorem 36.4).

Theorem A.1 *If $(Z_t)_{t\in\mathbb{Z}}$ is an ergodic strictly stationary sequence and if $(Y_t)_{t\in\mathbb{Z}}$ is defined by*

$$Y_t = f(\dots, Z_{t-1}, Z_t, Z_{t+1}, \dots),$$

where f is a measurable function from $\mathbb{R}^\infty$ to $\mathbb{R}$, then $(Y_t)_{t\in\mathbb{Z}}$ is also an ergodic strictly stationary sequence.

In particular, if $(X_t)_{t\in\mathbb{Z}}$ is the non-anticipative stationary solution of the AR(1) equation

$$X_t = aX_{t-1} + \eta_t, \quad |a| < 1, \quad \eta_t \text{ iid } (0, \sigma^2),, \tag{A.1}$$

then the theorem shows that $(X_t)_{t\in\mathbb{Z}}$, $(X_{t-1}\eta_t)_{t\in\mathbb{Z}}$ and $(X_{t-1}^2)_{t\in\mathbb{Z}}$ are also ergodic stationary sequences.

Theorem A.2 (The ergodic theorem for stationary sequences) *If $(Z_t)_{t\in\mathbb{Z}}$ is strictly stationary and ergodic, if f is measurable and if*

$$E\,|f(\dots, Z_{t-1}, Z_t, Z_{t+1}, \dots)| < \infty,$$

[1] The ergodicity concept is much more general and can be extended to non-stationary sequences (see, for instance, Billingsley 1995).

GARCH Models: Structure, Statistical Inference and Financial Applications, Second Edition. Christian Francq and Jean-Michel Zakoian.

then

$$n^{-1}\sum_{t=1}^{n} f(\ldots, Z_{t-1}, Z_t, Z_{t+1}, \ldots) \to Ef(\ldots, Z_{t-1}, Z_t, Z_{t+1}, \ldots) \quad a.s.$$

As an example, consider the least-squares estimator $\hat{a}_n$ of the parameter a in (A.1). By definition

$$\hat{a}_n = \arg\min_a Q_n(a), \quad Q_n(a) = \sum_{t=2}^{n} (X_t - aX_{t-1})^2.$$

From the first-order condition, we obtain

$$\hat{a}_n = \frac{n^{-1}\sum_{t=2}^{n} X_t X_{t-1}}{n^{-1}\sum_{t=2}^{n} X_{t-1}^2}.$$

The ergodic theorem shows that the numerator tends almost surely to $\gamma(1) = \text{Cov}(X_t, X_{t-1}) = a\gamma(0)$ and that the denominator tends to $\gamma(0)$. It follows that $\hat{a}_n \to a$ almost surely as $n \to \infty$. Note that this result still holds true when the assumption that η_t is a strong white noise is replaced by the assumption that η_t is a semi-strong white noise, or even by the assumption that η_t is an ergodic and stationary weak white noise.

A.2 Martingale Increments

In a purely random fair game (for instance, A and B play 'heads or tails', A gives one dollar to B if the coin falls tails, and B gives one dollar to A if the coin falls heads), the winnings of a player constitute a martingale.

Definition A.2 (Martingale) *Let $(Y_t)_{t\in\mathbb{N}}$ be a sequence of real random variables and $(\mathcal{F}_t)_{t\in\mathbb{N}}$ be a sequence of σ-fields. The sequence $(Y_t, \mathcal{F}_t)_{t\in\mathbb{N}}$ is said to be a martingale if and only if*

1. $\mathcal{F}_t \subset \mathcal{F}_{t+1}$;
2. Y_t *is* $\mathcal{F}_t$*-measurable;*
3. $E|Y_t| < \infty$;
4. $E(Y_{t+1} \mid \mathcal{F}_t) = Y_t$.

When $(Y_t)_{t\in\mathbb{N}}$ is said to be a martingale, it is implicitly assumed that $\mathcal{F}_t = \sigma(Y_u, u \le t)$, that is, the σ-field generated by the past and present values.

Definition A.3 (Martingale difference) *Let $(\eta_t)_{t\in\mathbb{N}}$ be a sequence of real random variables and $(\mathcal{F}_t)_{t\in\mathbb{N}}$ be a sequence of σ-fields. The sequence $(\eta_t, \mathcal{F}_t)_{t\in\mathbb{N}}$ is said to be a martingale difference (or a sequence of martingale increments) if and only if*

1. $\mathcal{F}_t \subset \mathcal{F}_{t+1}$;
2. η_t *is* $\mathcal{F}_t$*-measurable;*
3. $E|\eta_t| < \infty$;
4. $E(\eta_{t+1} \mid \mathcal{F}_t) = 0$.

Remark A.1 If $(Y_t, \mathcal{F}_t)_{t\in\mathbb{N}}$ is a martingale and $\eta_0 = Y_0$, $\eta_t = Y_t - Y_{t-1}$, then $(\eta_t, \mathcal{F}_t)_{t\in\mathbb{N}}$ is a martingale difference: $E(\eta_{t+1} \mid \mathcal{F}_t) = E(Y_{t+1} \mid \mathcal{F}_t) - E(Y_t \mid \mathcal{F}_t) = 0$.

Remark A.2 If $(\eta_t, \mathcal{F}_t)_{t\in\mathbb{N}}$ is a martingale difference and $Y_t = \eta_0 + \eta_1 + \cdots + \eta_t$, then $(Y_t, \mathcal{F}_t)_{t\in\mathbb{N}}$ is a martingale: $E(Y_{t+1} \mid \mathcal{F}_t) = E(Y_t + \eta_{t+1} \mid \mathcal{F}_t) = Y_t$.

Remark A.3 In example (A.1),

$$\left\{ \sum_{i=0}^{k} a^i \eta_{t-i}, \sigma(\eta_u, t-k \le u \le t) \right\}_{k\in\mathbb{N}}$$

is a martingale, and $\{\eta_t, \sigma(\eta_u, u \le t)\}_{t\in\mathbb{N}}$, $\{\eta_t X_{t-1}, \sigma(\eta_u, u \le t)\}_{t\in\mathbb{N}}$ are martingale differences.

There exists a central limit theorem (CLT) for triangular sequences of martingale differences (see Billingsley 1995, p. 476).

Theorem A.3 (Lindeberg's CLT) *Assume that, for each $n > 0$, $(\eta_{nk}, \mathcal{F}_{nk})_{k\in\mathbb{N}}$ is a sequence of square integrable martingale increments. Let $\sigma_{nk}^2 = E(\eta_{nk}^2 \mid \mathcal{F}_{n(k-1)})$. If*

$$\sum_{k=1}^{n} \sigma_{nk}^2 \to \sigma_0^2 \text{ in probability as } n \to \infty, \tag{A.2}$$

where σ_0 is a strictly positive constant, and

$$\sum_{k=1}^{n} E\eta_{nk}^2 \mathbb{1}_{\{|\eta_{nk}|\ge\epsilon\}} \to 0 \text{ as } n \to \infty, \tag{A.3}$$

for any positive real ϵ, then $\sum_{k=1}^{n} \eta_{nk} \xrightarrow{\mathcal{L}} \mathcal{N}(0, \sigma_0^2)$

Remark A.4 In numerous applications, η_{nk} and $\mathcal{F}_{nk}$ are only defined for $1 \le k \le n$ and can be displayed as a triangular array

$$\begin{array}{llll} \eta_{11} & & & \\ \eta_{21} & \eta_{22} & & \\ \eta_{31} & \eta_{32} & \eta_{33} & \\ \vdots & & & \\ \eta_{n1} & \eta_{n2} & \cdots & \eta_{nn} \\ \vdots & & & \end{array}$$

One can define η_{nk} and $\mathcal{F}_{nk}$ for all $k \ge 0$, with $\eta_{n0} = 0$, $\mathcal{F}_{n0} = \{\emptyset, \Omega\}$ and $\eta_{nk} = 0$, $\mathcal{F}_{nk} = \mathcal{F}_{nn}$ for all $k > n$. In the theorem, each row of the triangular array is assumed to be a martingale difference.

Remark A.5 The previous theorem encompasses the usual CLT. Let $Z_1, \ldots, Z_n$ be an iid sequence with a finite variance. It suffices to take

$$\eta_{nk} = \frac{Z_k - EZ_k}{\sqrt{n}} \quad \text{and} \quad \mathcal{F}_{nk} = \sigma(Z_1, \ldots, Z_k).$$

It is clear that $(\eta_{nk}, \mathcal{F}_{nk})_{k\in\mathbb{N}}$ is a square integrable martingale difference. We have $\sigma_{nk}^2 = E\eta_{nk}^2 = n^{-1}\mathrm{Var}(Z_0)$. Consequently, the normalisation condition (A.2) is satisfied. Moreover,

$$\begin{aligned} \sum_{k=1}^{n} E\eta_{nk}^2 \mathbb{1}_{\{|\eta_{nk}|\ge\epsilon\}} &= \sum_{k=1}^{n} n^{-1} \int_{\{|Z_k - E Z_k| \ge \sqrt{n}\epsilon\}} |Z_k - EZ_k|^2 dP \\ &= \int_{\{|Z_1 - E Z_1| \ge \sqrt{n}\epsilon\}} |Z_1 - EZ_1|^2 dP \to 0 \end{aligned}$$

because $\{|Z_1 - EZ_1| \geq \sqrt{n}\epsilon\} \downarrow \emptyset$ and $\int_\Omega |Z_1 - EZ_1|^2 dP < \infty$. The Lindeberg condition (A.3) is thus satisfied. The theorem entails the standard CLT since:

$$\sum_{k=1}^{n} \eta_{nk} = \frac{1}{\sqrt{n}} \sum_{k=1}^{n} (Z_k - EZ_k).$$

Remark A.6 In example (A.1), take

$$\eta_{nk} = \frac{\eta_k X_{k-1}}{\sqrt{n}} \quad \text{and} \quad \mathcal{F}_{nk} = \sigma(\eta_u, u \leq k).$$

The sequence $(\eta_{nk}, \mathcal{F}_{nk})_{k\in\mathbb{N}}$ is a square integrable martingale difference. We have $\sigma_{nk}^2 = n^{-1}\sigma^2 X_{k-1}^2$. The ergodic theorem entails (A.2) with $\sigma_0^2 = \sigma^4/(1-a^2)$. We obtain

$$\sum_{k=1}^{n} E\eta_{nk}^2 \mathbb{1}_{\{|\eta_{nk}|\geq\epsilon\}} = \sum_{k=1}^{n} n^{-1} \int_{\{|\eta_k X_{k-1}|\geq\sqrt{n}\epsilon\}} |\eta_k X_{k-1}|^2 dP$$
$$= \int_{\{|\eta_1 X_0|\geq\sqrt{n}\epsilon\}} |\eta_1 X_0|^2 dP \to 0$$

because $\{|\eta_1 X_0| \geq \sqrt{n}\epsilon\} \downarrow \emptyset$ and $\int_\Omega |\eta_1 X_0|^2 dP < \infty$. This shows (A.3). The Lindeberg CLT entails that

$$n^{-1/2} \sum_{k=1}^{n} \eta_k X_{k-1} \xrightarrow{\mathcal{L}} \mathcal{N}(0, \sigma^4/(1-a^2)).$$

It follows that

$$n^{1/2}(\hat{a}_n - a) = \frac{n^{-1/2}\sum_{k=1}^{n} \eta_k X_{k-1}}{n^{-1}\sum_{k=1}^{n} X_{k-1}^2} \xrightarrow{\mathcal{L}} \mathcal{N}(0, 1-a^2), \tag{A.4}$$

because $n^{-1}\sum_{k=1}^{n} X_{k-1}^2 \to \sigma^2/(1-a^2)$.[2]

Remark A.7 The previous result can be used to obtain an asymptotic confidence interval or for testing the nullity of the coefficient a:

1. $[\hat{a}_n \pm 1.96 n^{-1/2}(1-\hat{a}_n^2)^{1/2}]$ is a confidence interval for a at the asymptotic 95% confidence level.
2. The null assumption $H_0 : a = 0$ is rejected at the asymptotic 5% level if $|t_n| > 1.96$, where $t_n = \sqrt{n}\hat{a}_n/\sqrt{1-\hat{a}_n^2}$ is the t-statistic.

In the previous statistics, $1-\hat{a}_n^2$ is often replaced by $\hat{\sigma}^2/\hat{\gamma}(0)$, where $\hat{\sigma}^2 = \sum_{t=1}^{n} (X_t - \hat{a}_n X_{t-1})^2/(n-1)$ and $\hat{\gamma}(0) = n^{-1}\sum_{t=1}^{n} X_{t-1}^2$. Asymptotically, there is no difference:

$$\frac{n-1}{n}\frac{\hat{\sigma}^2}{\hat{\gamma}(0)} = \frac{\sum_{t=1}^{n} (X_t - \hat{a}_n X_{t-1})^2}{\sum_{t=1}^{n} X_{t-1}^2}$$
$$= \frac{\sum_{t=1}^{n} X_t^2 + \hat{a}_n^2 \sum_{t=1}^{n} X_{t-1}^2 - 2\hat{a}_n \sum_{t=1}^{n} X_t X_{t-1}}{\sum_{t=1}^{n} X_{t-1}^2}$$
$$= \frac{\sum_{t=1}^{n} X_t^2}{\sum_{t=1}^{n} X_{t-1}^2} - \hat{a}_n^2.$$

However, it is preferable to use $\hat{\sigma}^2/\hat{\gamma}(0)$, which is always positive, rather than $1-\hat{a}_n^2$ because, in finite samples, one can have $\hat{a}_n^2 > 1$.

---

[2] We have used Slutsky's lemma: if $Y_n \xrightarrow{\mathcal{L}} Y$ and $T_n \to T$ in probability, then $T_n Y_n \xrightarrow{\mathcal{L}} YT$.

The following corollary applies to GARCH processes, which are stationary and ergodic martingale differences.

Corollary A.1 (Billingsley 1961) *If $(v_t, \mathcal{F}_t)_t$ is a stationary and ergodic sequence of square integrable martingale increments such that $\sigma_v^2 = \mathrm{Var}(v_t) \neq 0$, then*

$$n^{-1/2} \sum_{t=1}^{n} v_t \xrightarrow{\mathcal{L}} \mathcal{N}(0, \sigma_v^2).$$

Proof Let $\eta_{nk} = v_k/\sqrt{n}$ and $\mathcal{F}_{nk} = \mathcal{F}_k$. For all n, the sequence $(\eta_{nk}, \mathcal{F}_k)_k$ is a square integrable martingale difference. With the notation of Theorem A.3, we have $\sigma_{nk}^2 = E(\eta_{nk}^2 \mid \mathcal{F}_{k-1})$, and $(n\sigma_{nk}^2)_k = \{E(v_k^2 \mid \mathcal{F}_{k-1})\}_k$ is a stationary and ergodic sequence. We thus have almost surely

$$\sum_{k=1}^{n} \sigma_{nk}^2 = \frac{1}{n} \sum_{k=1}^{n} E(v_k^2 \mid \mathcal{F}_{k-1}) \to E\{E(v_k^2 \mid \mathcal{F}_{k-1})\} = \sigma_v^2 > 0,$$

which shows the normalisation condition (A.2). Moreover,

$$\sum_{k=1}^{n} E\eta_{nk}^2 \mathbb{1}_{\{|\eta_{nk}| \geq \epsilon\}} = \sum_{k=1}^{n} n^{-1} \int_{\{|v_k| \geq \sqrt{n}\epsilon\}} v_k^2 dP = \int_{\{|v_1| \geq \sqrt{n}\epsilon\}} v_1^2 dP \to 0,$$

using stationarity and Lebesgue's theorem. This shows (A.3). The corollary is thus a consequence of Theorem A.3. □

A.3 Mixing

Numerous probabilistic tools have been developed for measuring the dependence between variables. For a finite-variance process, elementary measures of dependence are the autocovariances and autocorrelations. When there is no linear dependence between X_t and X_{t+h}, as is the case for a GARCH process, the autocorrelation is not the right tool, and more elaborate concepts are required. Mixing assumptions, introduced by Rosenblatt (1956), are used to convey different ideas of asymptotic independence between the past and future of a process. We present here two of the most popular mixing coefficients.

A.3.1 α-Mixing and β-Mixing Coefficients

The strong mixing (or α-mixing) coefficient between two σ-fields $\mathcal{A}$ and $\mathcal{B}$[3] is defined by

$$\alpha(\mathcal{A}, \mathcal{B}) = \sup_{A \in \mathcal{A}, B \in \mathcal{B}} |\mathbb{P}(A \cap B) - \mathbb{P}(A)\mathbb{P}(B)|.$$

It is clear that

(i) if $\mathcal{A}$ and $\mathcal{B}$ are independent then $\alpha(\mathcal{A}, \mathcal{B}) = 0$;

(ii) $0 \leq \alpha(\mathcal{A}, \mathcal{B}) \leq 1/4$;[4]

[3] Obviously, we are working with a probability space $(\Omega, \mathcal{A}_0, \mathbb{P})$, and $\mathcal{A}$ and $\mathcal{B}$ are sub-σ-fields of $\mathcal{A}_0$.

[4] It suffices to note that

$$\begin{aligned} |\mathbb{P}(A \cap B) - \mathbb{P}(A)\mathbb{P}(B)| &= |\mathbb{P}(A \cap B) - \mathbb{P}(A \cap B)\mathbb{P}(B) - \mathbb{P}(A \cap B^c)\mathbb{P}(B)| \\ &= |\mathbb{P}(A \cap B)\mathbb{P}(B^c) - \mathbb{P}(A \cap B^c)\mathbb{P}(B)| \\ &\leq \mathbb{P}(B^c)\mathbb{P}(B) \leq 1/4. \end{aligned}$$

(iii) $\alpha(\mathcal{A},\mathcal{A}) = 1/4$ provided that $\mathcal{A}$ contains an event of probability 1/2;

(iv) $\alpha(\mathcal{A},\mathcal{A}) > 0$ provided that $\mathcal{A}$ is non-trivial;[5]

(v) $\alpha(\mathcal{A}',\mathcal{B}') \leq \alpha(\mathcal{A},\mathcal{B})$ provided that $\mathcal{A}' \subset \mathcal{A}$ and $\mathcal{B}' \subset \mathcal{B}$.

The strong mixing coefficients of a process $X = (X_t)$ are defined by

$$\alpha_X(h) = \sup_t \alpha\{\sigma(X_u, u \leq t), \sigma(X_u, u \geq t+h)\}.$$

If X is stationary, the term $\sup_t$ can be omitted. In this case, we have

$$\begin{aligned}\alpha_X(h) &= \sup_{A,B} \mid \mathbb{P}(A \cap B) - \mathbb{P}(A)\mathbb{P}(B) \mid \\ &= \sup_{f,g} \mid \mathrm{Cov}(f(\ldots, X_{-1}, X_0), g(X_h, X_{h+1}, \ldots)) \mid \end{aligned} \tag{A.5}$$

where the first supremum is taken on $A \in \sigma(X_s, s \leq 0)$ and $B \in \sigma(X_s, s \geq h)$ and the second is taken on the set of the measurable functions f and g such that $|f| \leq 1$, $|g| \leq 1$. X is said to be *strongly mixing*, or *α-mixing*, if $\alpha_X(h) \to 0$ as $h \to \infty$. If $\alpha_X(h)$ tends to zero at an exponential rate, then X is said to be *geometrically strongly mixing*.

The β-mixing coefficients of a stationary process X are defined by

$$\begin{aligned}\beta_X(k) &= E \sup_{B \in \sigma(X_s, s \geq k)} \mid \mathbb{P}(B \mid \sigma(X_s, s \leq 0)) - \mathbb{P}(B) \mid \\ &= \frac{1}{2} \sup \sum_{i=1}^{I} \sum_{j=1}^{J} \mid \mathbb{P}(A_i \cap B_j) - \mathbb{P}(A_i)\mathbb{P}(B_j) \mid, \end{aligned} \tag{A.6}$$

where in the last equality, the sup is taken among all the pairs of partitions $\{A_1, \ldots, A_I\}$ and $\{B_1, \ldots, B_J\}$ of Ω such that $A_i \in \sigma(X_s, s \leq 0)$ for all i and $B_j \in \sigma(X_s, s \geq k)$ for all j. The process is said to be β-mixing if $\lim_{k\to\infty} \beta_X(k) = 0$. We have

$$\alpha_X(k) \leq \beta_X(k),$$

so that β-mixing implies α-mixing. If $Y = (Y_t)$ is a process such that $Y_t = f(X_t, \ldots, X_{t-r})$ for a measurable function f and an integer $r \geq 0$, then $\sigma(Y_t, t \leq s) \subset \sigma(X_t, t \leq s)$ and $\sigma(Y_t, t \geq s) \subset \sigma(X_{t-r}, t \geq s)$. In view of point (v) above, this entails that

$$\alpha_Y(k) \leq \alpha_X(k-r) \quad \text{and} \quad \beta_Y(k) \leq \beta_X(k-r) \quad \text{for all } k \geq r. \tag{A.7}$$

Example A.1 It is clear that a q-dependent process such that

$$X_t \in \sigma(\epsilon_t, \epsilon_{t-1}, \ldots, \epsilon_{t-q}), \quad \text{where the } \epsilon_t \text{ are independent,}$$

is strongly mixing, because

$$\alpha_X(h) \leq \alpha_\epsilon(h-q) = 0, \quad \forall h \geq q.$$

Example A.2 Consider the process defined by

$$X_t = Y \cos(\lambda t),$$

For the first inequality, we use $||a| - |b|| \leq \max\{|a|, |b|\}$. Alternatively, one can note that $\mathbb{P}(A \cap B) - \mathbb{P}(A)\mathbb{P}(B) = \mathrm{Cov}(\mathbb{1}_A, \mathbb{1}_B)$ and use the Cauchy–Schwarz inequality.

[5] That is, $\mathcal{A}$ contains an event of probability different from 0 or 1.

where $\lambda \in (0, \pi)$ and Y is a non-trivial random variable. Note that when $\cos(\lambda t) \neq 0$, which occurs for an infinite number of t, we have $\sigma(X_t) = \sigma(Y)$. We thus have, for any t and any h, $\sigma(X_u, u \leq t) = \sigma(X_u, u \geq t+h) = \sigma(Y)$, and $\alpha_X(h) = \alpha\{\sigma(Y), \sigma(Y)\} > 0$ by (iv), which shows that X is not strongly mixing.

Example A.3 Let (u_t) be an iid sequence uniformly distributed on $\{1, \ldots, 9\}$. Let

$$X_t = \sum_{i=0}^{\infty} 10^{-i-1} u_{t-i}.$$

The sequence $u_t, u_{t-1}, \ldots$ constitutes the decimals of X_t; we can write $X_t = 0.\, u_t u_{t-1} \ldots$. The process $X = (X_t)$ is stationary and satisfies a strong AR(1) representation of the form

$$X_t = \frac{1}{10} X_{t-1} + \frac{1}{10} u_t = \frac{1}{10} X_{t-1} + \frac{1}{2} + \epsilon_t$$

where $\epsilon_t = (u_t/10) - (1/2)$ is a strong white noise. The process X is not α-mixing because $\sigma(X_t) \subset \sigma(X_{t+h})$ for all $h \geq 0$,[6] and by (iv) and (v),

$$\alpha_X(h) \geq \alpha\{\sigma(X_t), \sigma(X_t)\} > 0, \quad \forall h \geq 0.$$

A.3.2 Covariance Inequality

Let p, q, and r be three positive numbers such that $p^{-1} + q^{-1} + r^{-1} = 1$. Davydov (1968) showed the covariance inequality

$$|\mathrm{Cov}(X, Y)| \leq K_0 \|X\|_p \|Y\|_q [\alpha\{\sigma(X), \sigma(Y)\}]^{1/r}, \tag{A.8}$$

where $\|X\|_p^p = EX^p$ and K_0 is a universal constant. Davydov initially proposed $K_0 = 12$. Rio (1993) obtained a sharper inequality, involving the quantile functions of X and Y. The latter inequality also shows that one can take $K_0 = 4$ in (A.8). Note that (A.8) entails that the autocovariance function of an α-mixing stationary process (with enough moments) tends to zero.

Example A.4 Consider the process $X = (X_t)$ defined by

$$X_t = Y \cos(\omega t) + Z \sin(\omega t),$$

where $\omega \in (0, \pi)$, and Y and Z are iid $\mathcal{N}(0, 1)$. Then X is Gaussian, centred and stationary, with autocovariance function $\gamma_X(h) = \cos(\omega h)$. Since $\gamma_X(h) \nrightarrow 0$ as $|h| \to \infty$, X is not mixing.

From inequality (A.8), we obtain, for instance, the following results.

Corollary A.2 *Let $X = (X_t)$ be a centred process such that*

$$\sup_t \|X_t\|_{2+\nu} < \infty, \qquad \sum_{h=0}^{\infty} \{\alpha_X(h)\}^{\nu/(2+\nu)} < \infty, \qquad \textit{for some } \nu > 0.$$

We have

$$E\overline{X}_n^2 = O\left(\frac{1}{n}\right).$$

Proof Let $K = K_0 \sup_t \|X_t\|_{2+\nu}^2$. From (A.8), we obtain

$$|\, EX_{t_1} X_{t_2} \,| = |\, \mathrm{Cov}(X_{t_1}, X_{t_2}) \,| \leq K\{\alpha_X(|\, t_2 - t_1 \,|)\}^{\nu/(2+\nu)}.$$

[6] This would not be true if we added 0 to the set of the possible values of u_t. For instance, $X_t = 0.4999\ldots = 0.5000\ldots$ would not tell us whether X_{t-1} is equal to $1 = 0.999\ldots$ or to 0.

We thus have

$$E\overline{X}_n^2 \leq \frac{2}{n^2} \sum_{1 \leq t_1 \leq t_2 \leq n} | EX_{t_1} X_{t_2} | \leq \frac{2}{n} K \sum_{h=0}^{\infty} \{\alpha_X(h)\}^{\nu/(2+\nu)} = O\left(\frac{1}{n}\right).$$

□

Corollary A.3 *Let $X = (X_t)$ be a centred process such that*

$$\sup_t \|X_t\|_{4+2\nu} < \infty, \quad \sum_{\ell=0}^{\infty} \{\alpha_X(\ell)\}^{\nu/(2+\nu)} < \infty, \quad \textit{for some } \nu > 0.$$

We have

$$\sup_t \sup_{h,k \geq 0} \sum_{\ell=-\infty}^{+\infty} |\text{Cov}(X_t X_{t+h}, X_{t+\ell} X_{t+k+\ell})| < \infty.$$

Proof Consider the case $0 \leq h \leq k$. We have

$$\sum_{\ell=-\infty}^{+\infty} |\text{Cov}(X_t X_{t+h}, X_{t+\ell} X_{t+k+\ell})| = d_1 + d_2 + d_3 + d_4,$$

where

$$d_1 = \sum_{\ell=h}^{+\infty} |\text{Cov}(X_t X_{t+h}, X_{t+\ell} X_{t+k+\ell})|,$$

$$d_2 = \sum_{\ell=-\infty}^{-k} |\text{Cov}(X_t X_{t+h}, X_{t+\ell} X_{t+k+\ell})|,$$

$$d_3 = \sum_{\ell=0}^{h-1} |\text{Cov}(X_t X_{t+h}, X_{t+\ell} X_{t+k+\ell})|,$$

$$d_4 = \sum_{\ell=-k+1}^{-1} |\text{Cov}(X_t X_{t+h}, X_{t+\ell} X_{t+k+\ell})|.$$

Inequality (A.8) shows that d_1 and d_2 are uniformly bounded in t, h and k by

$$K_0 \sup_t \|X_t\|_{4+2\nu}^4 \sum_{\ell=0}^{\infty} \{\alpha_X(\ell)\}^{\nu/(2+\nu)}.$$

To handle d_3, note that $d_3 \leq d_5 + d_6$, where

$$d_5 = \sum_{\ell=0}^{h-1} |\text{Cov}(X_t X_{t+\ell}, X_{t+h} X_{t+k+\ell})|,$$

$$d_6 = \sum_{\ell=0}^{h-1} |EX_t X_{t+h} EX_t X_{t+k} - EX_t X_{t+\ell} EX_{t+h} X_{t+k+\ell}|.$$

With the notation $K = K_0 \sup_t \|X_t\|_{4+2\nu}^4$, we have

$$d_5 \leq K_0 \sup_t \|X_t\|_{4+2\nu}^4 \sum_{\ell=0}^{h-1} \{\alpha_X(h-\ell)\}^{\nu/(2+\nu)} \leq K \sum_{\ell=0}^{\infty} \{\alpha_X(\ell)\}^{\nu/(2+\nu)},$$

$$d_6 \leq \sup_t \|X_t\|_2^2 \left(h \mid EX_tX_{t+h} \mid + \sum_{\ell=0}^{h-1} \mid EX_tX_{t+\ell} \mid \right)$$

$$\leq K \left\{ \sup_{\ell \geq 0} \ell \{\alpha_X(\ell)\}^{\nu/(2+\nu)} + \sum_{\ell=0}^{\infty} \{\alpha_X(\ell)\}^{\nu/(2+\nu)} \right\}.$$

For the last inequality, we used

$$\sup_t \|X_t\|_2^2 \mid EX_tX_{t+h} \mid \leq K_0 \sup_t \|X_t\|_2^2 \sup_t \|X_t\|_{2+\nu}^2 \{\alpha_X(h)\}^{\nu/(2+\nu)}$$

$$\leq K\{\alpha_X(h)\}^{\nu/(2+\nu)}.$$

Since the sequence of the mixing coefficients is decreasing, it is easy to see that $\sup_{\ell \geq 0} \ell \{\alpha_X(\ell)\}^{\nu/(2+\nu)} < \infty$ (cf. Exercise 3.9). The term d_3 is thus uniformly bounded in t, h and k. We treat d_4 as d_3 (see Exercise 3.10). □

Corollary A.4 *Under the conditions of Corollary A.3, we have*

$$E\overline{X}_n^4 = O\left(\frac{1}{n}\right).$$

Proof Let $t_1 \leq t_2 \leq t_3 \leq t_4$. From (A.8), we obtain

$$EX_{t_1}X_{t_2}X_{t_3}X_{t_4} = \text{Cov}(X_{t_1}, X_{t_2}X_{t_3}X_{t_4})$$

$$\leq K_0 \|X_{t_1}\|_{4+2\nu} \|X_{t_2}X_{t_3}X_{t_4}\|_{(4+2\nu)/3} \{\alpha_X(\mid t_2 - t_1 \mid)\}^{\nu/(2+\nu)}$$

$$\leq K\{\alpha_X(\mid t_2 - t_1 \mid)\}^{\nu/(2+\nu)}.$$

We thus have

$$E\overline{X}_n^4 \leq \frac{4!}{n^4} \sum_{1 \leq t_1 \leq t_2 \leq t_3 \leq t_4 \leq n} EX_{t_1}X_{t_2}X_{t_3}X_{t_4}$$

$$\leq \frac{4!}{n} K \sum_{h=0}^{\infty} \{\alpha_X(h)\}^{\nu/(2+\nu)} = O\left(\frac{1}{n}\right).$$

In the last inequality, we use the fact that the number of indices $1 \leq t_1 \leq t_2 \leq t_3 \leq t_4 \leq n$ such that $t_2 = t_1 + h$ is less than n^3. □

A.3.3 Central Limit Theorem

Herrndorf (1984) showed the following CLT.

Theorem A.4 (CLT for α-mixing processes) *Let $X = (X_t)$ be a centred process such that*

$$\sup_t \|X_t\|_{2+\nu} < \infty, \qquad \sum_{h=0}^{\infty} \{\alpha_X(h)\}^{\nu/(2+\nu)} < \infty, \qquad \textit{for some } \nu > 0.$$

If $\sigma^2 = \lim_{n\to\infty} \text{Var}\left(n^{-1/2} \sum_{t=1}^{n} X_t\right)$ exists and is not zero, then

$$n^{-1/2} \sum_{t=1}^{n} X_t \xrightarrow{\mathcal{L}} \mathcal{N}(0, \sigma^2).$$

Appendix B

Autocorrelation and Partial Autocorrelation

B.1 Partial Autocorrelation

Definition B.1 *The (theoretical) partial autocorrelation at lag $h > 0$, $r_X(h)$, of a second-order stationary process $X = (X_t)$ with non-degenerate linear innovations,*[1] *is the correlation between*

$$X_t - EL(X_t \mid X_{t-1}, X_{t-2}, \dots, X_{t-h+1})$$

and

$$X_{t-h} - EL(X_{t-h} \mid X_{t-1}, X_{t-2}, \dots, X_{t-h+1}),$$

where $EL(Y \mid Y_1, \dots, Y_k)$ denotes the linear regression of a square integrable variable Y on variables $Y_1, \dots, Y_k$. Let

$$r_X(h) = \text{Corr}(X_t, X_{t-h} \mid X_{t-1}, X_{t-2}, \dots, X_{t-h+1}). \tag{B.1}$$

The number $r_X(h)$ can be interpreted as the residual correlation between X_t and X_{t-h}, after the linear influence of the intermediate variables $X_{t-1}, X_{t-2}, \dots, X_{t-h+1}$ has been subtracted. Assume that (X_t) is centred, and consider the linear regression of X_t on $X_{t-1}, \dots, X_{t-h}$:

$$X_t = a_{h,1}X_{t-1} + \cdots + a_{h,h}X_{t-h} + u_{h,t}, \quad u_{h,t} \perp X_{t-1}, \dots, X_{t-h}. \tag{B.2}$$

We have

$$EL(X_t \mid X_{t-1}, \dots, X_{t-h}) = a_{h,1}X_{t-1} + \cdots + a_{h,h}X_{t-h}, \tag{B.3}$$

$$EL(X_{t-h-1} \mid X_{t-1}, \dots, X_{t-h}) = a_{h,1}X_{t-h} + \cdots + a_{h,h}X_{t-1}, \tag{B.4}$$

and

$$r_X(h) = a_{h,h} \tag{B.5}$$

[1] Thus the variance of $\epsilon_t := X_t - EL(X_t \mid X_{t-1}, \dots)$ is not equal to zero.

Proof of (B.3) and (B.4) We obtain (B.3) from (B.2), using the linearity of $EL(\cdot \mid X_{t-1}, \dots, X_{t-h})$ and $a_{h,1}X_{t-1} + \cdots + a_{h,h}X_{t-h} \perp u_{h,t}$. The vector of the coefficients of the theoretical linear regression of X_{t-h-1} on $X_{t-1}, \dots, X_{t-h}$ is given by

$$\left\{ E \begin{pmatrix} X_{t-1} \\ \vdots \\ X_{t-h} \end{pmatrix} (X_{t-1} \dots X_{t-h}) \right\}^{-1} EX_{t-h-1} \begin{pmatrix} X_{t-1} \\ \vdots \\ X_{t-h} \end{pmatrix}. \tag{B.6}$$

Since

$$E \begin{pmatrix} X_{t-1} \\ \vdots \\ X_{t-h} \end{pmatrix} (X_{t-1} \dots X_{t-h}) = E \begin{pmatrix} X_{t-h} \\ \vdots \\ X_{t-1} \end{pmatrix} (X_{t-h} \dots X_{t-1})$$

and

$$EX_{t-h-1} \begin{pmatrix} X_{t-1} \\ \vdots \\ X_{t-h} \end{pmatrix} = EX_t \begin{pmatrix} X_{t-h} \\ \vdots \\ X_{t-1} \end{pmatrix},$$

this is also the vector of the coefficients of the linear regression of X_t on $X_{t-h}, \dots, X_{t-1}$, which gives (B.4). □

Proof of (B.5) From (B.2) we obtain

$$\begin{aligned} EL(X_t \mid X_{t-1}, \dots, X_{t-h+1}) = a_{h,1}X_{t-1} + \cdots + a_{h,h-1}X_{t-h+1} \\ + a_{h,h}E(X_{t-h} \mid X_{t-1}, \dots, X_{t-h+1}). \end{aligned}$$

Thus

$$X_t - EL(X_t \mid X_{t-1}, \dots, X_{t-h+1}) = a_{h,h}\{X_{t-h} - EL(X_{t-h} \mid X_{t-1}, \dots, X_{t-h+1})\} + u_{h,t}.$$

This equality being of the form $Y = a_{h,h}X + u$ with $u \perp X$, we obtain $\mathrm{Cov}(Y, X) = a_{h,h}\mathrm{Var}(X)$, which gives

$$a_{h,h} = \frac{\mathrm{Cov}\{X_t - EL(X_t \mid X_{t-1}, \dots, X_{t-h+1}), X_{t-h} - EL\left(X_{t-h} \mid X_{t-1}, \dots, X_{t-h+1}\right)\}}{\mathrm{Var}\{X_{t-h} - EL(X_{t-h} \mid X_{t-1}, \dots, X_{t-h+1})\}}.$$

To conclude, it suffices to note that, using the parity of $\gamma_X(\cdot)$ and (B.4),

$$\begin{aligned} &\mathrm{Var}\{X_t - EL(X_t \mid X_{t-1}, \dots, X_{t-h+1})\} \\ &= \mathrm{Var}\{X_t - a_{h-1,1}X_{t-1} - \cdots - a_{h-1,h-1}X_{t-h+1}\} \\ &= \mathrm{Var}\{X_{t-h} - a_{h-1,1}X_{t-h+1} - \cdots - a_{h-1,h-1}X_{t-1}\} \\ &= \mathrm{Var}\{X_{t-h} - EL(X_{t-h} \mid X_{t-1}, \dots, X_{t-h+1})\}. \end{aligned}$$

□

B.1.1 Computation Algorithm

From the autocorrelations $\rho_X(1), \dots, \rho_X(h)$, the partial autocorrelation $r_X(h)$ can be computed rapidly with the aid of Durbin's algorithm:

$$a_{1,1} = \rho_X(1), \tag{B.7}$$

$$a_{k,k} = \frac{\rho_X(k) - \sum_{i=1}^{k-1} \rho_X(k-i)a_{k-1,i}}{1 - \sum_{i=1}^{k-1} \rho_X(i)a_{k-1,i}}, \tag{B.8}$$

$$a_{k,i} = a_{k-1,i} - a_{k,k}a_{k-1,k-i}, \quad i = 1, \dots, k-1. \tag{B.9}$$

Steps (B.8) and (B.9) are repeated for $k=2, \dots, h-1$, and then $r_X(h)=a_{h,h}$ is obtained by step (B.8); see Exercise 1.14.

Proof of (B.9) In view of (B.2),

$$EL(X_t \mid X_{t-1}, \dots, X_{t-k+1}) = \sum_{i=1}^{k-1} a_{k,i}X_{t-i} + a_{k,k}EL(X_{t-k} \mid X_{t-1}, \dots, X_{t-k+1}).$$

Using (B.4), we thus have

$$\sum_{i=1}^{k-1} a_{k-1,i}X_{t-i} = \sum_{i=1}^{k-1} a_{k,i}X_{t-i} + a_{k,k}\sum_{i=1}^{k-1} a_{k-1,k-i}X_{t-i},$$

which gives (B.9) (the variables $X_{t-1}, \dots, X_{t-k+1}$ are not almost surely linearly dependent because the innovations of (X_t) are non-degenerate). □

Proof of (B.8) The vector of coefficients of the linear regression of X_t on $X_{t-1}, \dots, X_{t-h}$ satisfies

$$E\begin{pmatrix} X_{t-1} \\ \vdots \\ X_{t-h} \end{pmatrix}(X_{t-1} \dots X_{t-h})\begin{pmatrix} a_{h,1} \\ \vdots \\ a_{h,h} \end{pmatrix} = EX_t\begin{pmatrix} X_{t-1} \\ \vdots \\ X_{t-h} \end{pmatrix}. \tag{B.10}$$

The last row of (B.10) yields

$$\sum_{i=1}^{h} a_{h,i}\gamma(h-i) = \gamma(h).$$

Using (B.9), we thus have

$$\begin{aligned} a_{h,h} &= \rho(h) - \sum_{i=1}^{h-1} \rho(h-i)a_{h,i} \\ &= \rho(h) - \sum_{i=1}^{h-1} \rho(h-i)(a_{h-1,i} - a_{h,h}a_{h-1,h-i}) \\ &= \frac{\rho(h) - \sum_{i=1}^{h-1} \rho(h-i)a_{h-1,i}}{1 - \sum_{i=1}^{h-1} \rho(h-i)a_{h-1,h-i}}, \end{aligned}$$

which gives (B.8). □

B.1.2 Behaviour of the Empirical Partial Autocorrelation

The empirical partial autocorrelation, $\hat{r}(h)$, is obtained from the algorithm (B.7)–(B.9), replacing the theoretical autocorrelations $\rho_X(k)$ by the empirical autocorrelations $\hat{\rho}_X(k)$, defined by

$$\hat{\rho}_X(h) = \frac{\hat{\gamma}_X(h)}{\hat{\gamma}_X(0)}, \quad \hat{\gamma}_X(h) = \hat{\gamma}_X(-h) = n^{-1}\sum_{t=1}^{n-h} X_tX_{t+h},$$

for $h=0, 1, \dots, n-1$. When (X_t) is not assumed to be centred, X_t is replaced by $X_t - \overline{X}_n$. In view of (B.5), we know that, for an AR(p) process, we have $r_X(h)=0$, for all $h>p$. When the noise is strong, the asymptotic distribution of the $\hat{r}(h)$, $h>p$, is quite simple.

Theorem B.1 (Asymptotic distribution of the $\hat{r}(h)$s for a strong AR(p) model) *If X is the non-anticipative stationary solution of the AR(p) model*

$$X_t - \sum_{i=1}^{p} a_i X_{t-i} = \eta_t, \quad \eta_t \text{ iid}(0, \sigma^2), \quad \sigma^2 \neq 0, \quad 1 - \sum_{i=1}^{p} a_i z^i \neq 0 \quad \forall \mid z \mid \leq 1,$$

then

$$\sqrt{n}\hat{r}(h) \xrightarrow{\mathcal{L}} \mathcal{N}(0, 1), \quad \forall h > p.$$

Proof Let $a_0 = (a_1, \dots, a_p, 0, \dots, 0)$ be the vector of coefficients of the AR(h) model, when $h > p$. Consider

$$\underline{X} = \begin{pmatrix} X_{n-1} & \dots & X_{n-h} \\ X_{n-2} & \dots & X_{n-h-1} \\ \vdots & & \\ X_0 & \dots & X_{1-h} \end{pmatrix}, \quad \underline{Y} = \begin{pmatrix} X_n \\ X_{n-1} \\ \vdots \\ X_1 \end{pmatrix} \quad \text{and} \quad \hat{a} = \{\underline{X}'\underline{X}\}^{-1}\underline{X}'\underline{Y},$$

the coefficient of the empirical regression of X_t on $X_{t-1}, \dots, X_{t-h}$ (taking $X_t = 0$ for $t \leq 0$). It can be shown that, as for a standard regression model, $\sqrt{n}(\hat{a} - a_0) \xrightarrow{\mathcal{L}} \mathcal{N}(0, \Sigma)$, where

$$\Sigma \overset{\text{a.s.}}{=} \sigma^2 \lim_{n \to \infty} n^{-1} \{\underline{X}'\underline{X}\}^{-1} = \sigma^2 \begin{pmatrix} \gamma_X(0) & \gamma_X(1) & \cdots & \gamma_X(h-1) \\ \gamma_X(1) & \gamma_X(0) & \cdots & \gamma_X(h-2) \\ \vdots & & & \vdots \\ \gamma_X(h-1) & \cdots & \gamma_X(1) & \gamma_X(0) \end{pmatrix}^{-1}.$$

Since $\hat{r}_X(h)$ is the last component of $\hat{a}$ (by (B.5)), we have

$$\sqrt{n}\hat{r}(h) \xrightarrow{\mathcal{L}} \mathcal{N}(0, \Sigma(h, h)),$$

with

$$\Sigma(h, h) = \sigma^2 \frac{\Delta(0, h-1)}{\Delta(0, h)}, \quad \Delta(0, j) = \begin{vmatrix} \gamma_X(0) & \gamma_X(1) & \cdots & \gamma_X(j-1) \\ \gamma_X(1) & \gamma_X(0) & \cdots & \gamma_X(j-2) \\ \vdots & & & \vdots \\ \gamma_X(j-1) & \cdots & \gamma_X(1) & \gamma_X(0) \end{vmatrix}.$$

Applying the relations

$$\gamma_X(0) - \sum_{i=1}^{h-1} a_i \gamma_X(i) = \sigma^2, \quad \gamma_X(k) - \sum_{i=1}^{h-1} a_i \gamma_X(k-i) = 0,$$

for $k = 1, \dots, h-1$, we obtain

$$\Delta(0, h) = \begin{vmatrix} \gamma_X(0) & \gamma_X(1) & \cdots & \gamma_X(h-2) & 0 \\ \gamma_X(1) & \gamma_X(0) & \cdots & \gamma_X(h-3) & 0 \\ \vdots & \vdots & & \vdots & \vdots \\ \gamma_X(h-2) & \gamma_X(h-1) & \cdots & \gamma_X(0) & 0 \\ \gamma_X(h-1) & \gamma_X(h-2) & \cdots & \gamma_X(1) & \gamma_X(0) - \sum_{i=1}^{h-1} a_i \gamma_X(i) \end{vmatrix}$$

$$= \sigma^2 \Delta(0, h-1).$$

Thus $\Sigma(h, h) = 1$, which completes the proof.

The result of Theorem B.1 is no longer valid without the assumption that the noise η_t is iid. We can, however, obtain the asymptotic behaviour of the $\hat{r}(h)$s from that of the $\hat{\rho}(h)$s. Let

$$\rho_m = (\rho_X(1), \dots, \rho_X(m)), \quad \hat{\rho}_m = (\hat{\rho}_X(1), \dots, \hat{\rho}_X(m)),$$
$$r_m = (r_X(1), \dots, r_X(m)), \quad \hat{r}_m = (\hat{r}_X(1), \dots, \hat{r}_X(m)).$$

Theorem B.2 (Distribution of the $\hat{r}(h)$s from that of the $\hat{\rho}(h)$s) *When $n \to \infty$, if*

$$\sqrt{n}(\hat{\rho}_m - \rho_m) \xrightarrow{\mathcal{L}} \mathcal{N}(0, \Sigma_{\hat{\rho}_m}),$$

then

$$\sqrt{n}(\hat{r}_m - r_m) \xrightarrow{\mathcal{L}} \mathcal{N}(0, \Sigma_{\hat{r}_m}), \quad \Sigma_{\hat{r}_m} = J_m \Sigma_{\hat{\rho}_m} J'_m,$$

where the elements of the Jacobian matrix J_m are defined by $J_m(i, j) = \partial r_X(i)/\partial \rho_X(j)$ and are recursively obtained for $k = 2, \dots, m$ by

$$\partial r_X(1)/\partial \rho_X(j) = a_{1,1}^{(j)} = \mathbb{1}_{\{1\}}(j),$$

$$\partial r_X(k)/\partial \rho_X(j) = a_{k,k}^{(j)} = \frac{d_k n_k^{(j)} - n_k d_k^{(j)}}{d_k^2},$$

$$n_k = \rho_X(k) - \sum_{i=1}^{k-1} \rho_X(k-i) a_{k-1,i},$$

$$d_k = 1 - \sum_{i=1}^{k-1} \rho_X(i) a_{k-1,i},$$

$$n_k^{(j)} = \mathbb{1}_{\{k\}}(j) - a_{k-1,k-j} - \sum_{i=1}^{k-1} \rho_X(k-i) a_{k-1,i}^{(j)},$$

$$d_k^{(j)} = -a_{k-1,j} - \sum_{i=1}^{k-1} \rho_X(i) a_{k-1,i}^{(j)},$$

$$a_{k,i}^{(j)} = a_{k-1,i}^{(j)} - a_{k,k}^{(j)} a_{k-1,k-i} - a_{k,k} a_{k-1,k-i}^{(j)}, \quad i = 1, \dots, k-1,$$

where $a_{i,j} = 0$ for $j \le 0$ or $j > i$.

Proof It suffices to apply the delta method,[2] considering $r_X(h)$ as a differentiable function of $\rho_X(1), \dots, \rho_X(h)$. □

It follows that for a GARCH, and more generally for a weak white noise, $\hat{\rho}(h)$ and $\hat{r}(h)$ have the same asymptotic distribution.

Theorem B.3 (Asymptotic distribution of $\hat{r}(h)$ and $\hat{\rho}(h)$ for weak noises) *If X is a weak white noise and*

$$\sqrt{n}\hat{\rho}_m \xrightarrow{\mathcal{L}} \mathcal{N}(0, \Sigma_{\hat{\rho}_m}),$$

then

$$\sqrt{n}\hat{r}_m \xrightarrow{\mathcal{L}} \mathcal{N}(0, \Sigma_{\hat{\rho}_m}).$$

[2] If $\sqrt{n}(X_n - \mu) \xrightarrow{\mathcal{L}} \mathcal{N}(0, \Sigma)$, for X_n in $\mathbb{R}^m$, and $g : \mathbb{R}^m \to \mathbb{R}^k$ is of class C^1 in a neighbourhood of μ, then $\sqrt{n}\{g(X_n) - g(\mu)\} \xrightarrow{\mathcal{L}} \mathcal{N}(0, J\Sigma J')$, where $J = \{\partial g(x)/\partial x'\}(\mu)$.

Proof Consider the calculation of the derivatives $a_{k,i}^{(j)}$ when $\rho_X(h)=0$ for all $h\neq 0$. It is clear that $a_{k,i}=0$ for all k and all i. We then have $d_k=1$, $n_k=0$ and $n_k^{(j)}=\mathbb{1}_{\{k\}}(j)$. We thus have $a_{k,k}^{(j)}=\mathbb{1}_{\{k\}}(j)$, and then $J_m=I_m$. □

The following result is stronger because it shows that, for a white noise, $\hat{\rho}(h)$ and $\hat{r}(h)$ are asymptotically equivalent.

Theorem B.4 (Equivalence between $\hat{r}(h)$ and $\hat{\rho}(h)$ for a weak noise) *If (X_t) is a weak white noise satisfying the condition of Theorem B.3 and, for all fixed h,*

$$\sqrt{n}(\hat{a}_{h-1,1},\dots,\hat{a}_{h-1,h-1})=O_P(1), \tag{B.11}$$

where $(\hat{a}_{h-1,1},\dots,\hat{a}_{h-1,h-1})'$ is the vector of estimated coefficients of the linear regression of X_t on $X_{t-1},\dots,X_{t-h+1}$ $(t=h,\dots,n)$, then

$$\hat{\rho}(h)-\hat{r}(h)=O_p(n^{-1}).$$

Proof The result is straightforward for $h=1$. For $h>1$, we have by (B.8),

$$\hat{r}(h)=\frac{\hat{\rho}(h)-\sum_{i=1}^{h-1}\hat{\rho}(h-i)\hat{a}_{h-1,i}}{1-\sum_{i=1}^{h-1}\hat{\rho}(i)\hat{a}_{h-1,i}}.$$

In view of the assumptions,

$$\hat{\rho}(k)=o_P(1),\quad (\hat{a}_{h-1,1},\dots,\hat{a}_{h-1,h-1})'=o_P(1)$$

and

$$\hat{\rho}(k)\hat{a}_{h-1,i}=O_P(n^{-1}),$$

for $i=1,\dots,h-1$ and $k=1,\dots,h$. Thus

$$n\{\hat{\rho}(h)-\hat{r}(h)\}=\frac{n\sum_{i=1}^{h-1}\hat{a}_{h-1,i}\{\hat{\rho}(h-i)-\hat{\rho}(i)\hat{\rho}(h)\}}{1-\sum_{i=1}^{h-1}\hat{\rho}(i)\hat{a}_{h-1,i}}=O_p(1).$$

Under mild assumptions, the left-hand side of (B.11) tends in law to a non-degenerate normal, which entails (B.11). □

B.2 Generalised Bartlett Formula for Non-linear Processes

Let $X_1,\dots,X_n$ be observations of a centred second-order stationary process $X=(X_t)$. The empirical autocovariances and autocorrelations are defined by

$$\hat{\gamma}_X(h)=\hat{\gamma}_X(-h)=\frac{1}{n}\sum_{t=1}^{n-h}X_tX_{t+h},\quad \hat{\rho}_X(h)=\hat{\rho}_X(-h)=\frac{\hat{\gamma}_X(h)}{\hat{\gamma}_X(0)}, \tag{B.12}$$

for $h=0,\dots,n-1$. The following theorem provides an expression for the asymptotic variance of these estimators. This expression, which will be called Bartlett's formula, is relatively easy to compute. For the empirical autocorrelations of (strongly) linear processes, we obtain the standard Bartlett formula, involving only the theoretical autocorrelation function of the observed process. For non-linear processes admitting a weakly linear representation, the generalised Bartlett formula involves the autocorrelation function of the observed process, the kurtosis of the linear innovation process, and the autocorrelation function of its square. This formula is obtained under a symmetry assumption on the linear innovation process.

Theorem B.5 (Bartlett's formula for weakly linear processes) *We assume that $(X_t)_{t\in\mathbb{Z}}$ admits the Wold representation*

$$X_t = \sum_{i=-\infty}^{\infty} \psi_i \epsilon_{t-i}, \quad \sum_{i=-\infty}^{\infty} |\psi_i| < \infty,$$

where $(\epsilon_t)_{t\in\mathbb{Z}}$ is a weak white noise such that $E\epsilon_t^4 := \kappa_\epsilon (E\epsilon_t^2)^2 < \infty$, and

$$E\epsilon_{t_1}\epsilon_{t_2}\epsilon_{t_3}\epsilon_{t_4} = 0 \quad \text{when } t_1 \neq t_2,\ t_1 \neq t_3 \text{ and } t_1 \neq t_4. \tag{B.13}$$

With the notation $\rho_{\epsilon^2} = \sum_{h=-\infty}^{+\infty} \rho_{\epsilon^2}(h)$, we have

$$\begin{aligned}\lim_{n\to\infty} n\mathrm{Cov}\{\hat{\gamma}_X(i), \hat{\gamma}_X(j)\} &= (\kappa_\epsilon - 3)\gamma_X(i)\gamma_X(j)\\ &+ \sum_{\ell=-\infty}^{\infty} \gamma_X(\ell)\{\gamma_X(\ell+j-i) + \gamma_X(\ell-j-i)\}\\ &+ (\rho_{\epsilon^2} - 3)(\kappa_\epsilon - 1)\gamma_X(i)\gamma_X(j)\\ &+ (\kappa_\epsilon - 1)\sum_{\ell=-\infty}^{\infty} \gamma_X(\ell-i)\{\gamma_X(\ell-j) + \gamma_X(\ell+j)\}\rho_{\epsilon^2}(\ell).\end{aligned} \tag{B.14}$$

Let $\gamma_{0:m} = (\gamma_X(0), \ldots, \gamma_X(m))'$ and $\hat{\gamma}_{0:m} = (\hat{\gamma}_X(0), \ldots, \hat{\gamma}_X(m))'$. If

$$\sqrt{n}(\hat{\gamma}_{0:m} - \gamma_{0:m}) \xrightarrow{\mathcal{L}} \mathcal{N}(0, \Sigma_{\hat{\gamma}_{0:m}}) \quad \text{as } n \to \infty,$$

where the elements of $\Sigma_{\hat{\gamma}_{0:m}}$ are given by (B.14), then

$$\sqrt{n}(\hat{\rho}_m - \rho_m) \xrightarrow{\mathcal{L}} \mathcal{N}(0, \Sigma_{\hat{\rho}_m}),$$

where the elements of $\Sigma_{\hat{\rho}_m}$ are given by the generalised Bartlett formula

$$\lim_{n\to\infty} n\mathrm{Cov}\{\hat{\rho}_X(i), \hat{\rho}_X(j)\} = v_{ij} + v_{ij}^*, \quad i > 0,\ j > 0, \tag{B.15}$$

$$\begin{aligned}v_{ij} = \sum_{\ell=-\infty}^{\infty} \rho_X(\ell)[&2\rho_X(i)\rho_X(j)\rho_X(\ell) - 2\rho_X(i)\rho_X(\ell+j)\\ &- 2\rho_X(j)\rho_X(\ell+i) + \rho_X(\ell+j-i) + \rho_X(\ell-j-i)],\end{aligned}$$

$$\begin{aligned}v_{ij}^* = (\kappa_\epsilon - 1)\sum_{\ell=-\infty}^{\infty} \rho_{\epsilon^2}(\ell)[&2\rho_X(i)\rho_X(j)\rho_X^2(\ell) - 2\rho_X(j)\rho_X(\ell)\rho_X(\ell+i)\\ &- 2\rho_X(i)\rho_X(\ell)\rho_X(\ell+j) + \rho_X(\ell+i)\{\rho_X(\ell+j) + \rho_X(\ell-j)\}].\end{aligned}$$

Remark B.1 (Symmetry condition (B.13) for a GARCH process) In view of (7.24), we know that if (ϵ_t) is a GARCH process with a symmetric distribution for η_t and if $E\epsilon_t^4 < \infty$, then (B.13) is satisfied (see Exercise 7.12).

Remark B.2 (The generalised formula contains the standard formula) The right-hand side of (B.14) is a sum of four terms. When the sequence (ϵ_t^2) is uncorrelated, the sum of the last terms is equal to

$$-2(\kappa_\epsilon - 1)\gamma_X(i)\gamma_X(j) + (\kappa_\epsilon - 1)\gamma_X(i)\{\gamma_X(j) + \gamma_X(-j)\} = 0.$$

In this case, we retrieve the standard Bartlett formula (1.1). We also have $v_{ij}^* = 0$, and we retrieve the standard Bartlett formula for the $\hat{\rho}(h)$s.

Example B.1 (Bartlett's formula for a weak white noise) If we have a white noise $X_t = \epsilon_t$ that satisfies the assumptions of the theorem, then we have the generalised Bartlett formula (B.15) with

$$\begin{cases} v_{i,j} = v^*_{i,j} = 0, & \text{for } i \neq j \\ v_{i,i} = 1 \quad \text{and} \quad v^*_{i,i} = \frac{\gamma_{\epsilon^2}(i)}{\gamma^2_\epsilon(0)}, & \text{for } i > 0. \end{cases}$$

Proof Using (B.13) and the notation $\psi_{i_1,i_2,i_3,i_4} = \psi_{i_1}\psi_{i_2}\psi_{i_3}\psi_{i_4}$, we obtain

$$\begin{aligned} EX_tX_{t+i}X_{t+h}X_{t+j+h} &= \sum_{i_1,i_2,i_3,i_4} \psi_{i_1,i_2,i_3,i_4} E\epsilon_{t-i_1}\epsilon_{t+i-i_2}\epsilon_{t+h-i_3}\epsilon_{t+j+h-i_4} \\ &= \sum_{i_1,i_3} \psi_{i_1,i_1+i,i_3,i_3+j} E\epsilon^2_{t-i_1}\epsilon^2_{t+h-i_3} + \sum_{i_1,i_2} \psi_{i_1,i_1+h,i_2,i_2+h+j-i} E\epsilon^2_{t-i_1}\epsilon^2_{t+i-i_2} \\ &\quad + \sum_{i_1,i_2} \psi_{i_1,i_1+h+j,i_2,i_2+h-i} E\epsilon^2_{t-i_1}\epsilon^2_{t+i-i_2} - 2E\epsilon^4_t \sum_{i_1} \psi_{i_1,i_1+i,i_1+h,i_1+h+j}. \end{aligned} \tag{B.16}$$

The last equality is obtained by summing over the i_1, i_2, i_3, i_4 such that the indices of $\{\epsilon_{t-i_1}, \epsilon_{t+i-i_2}, \epsilon_{t+h-i_3}, \epsilon_{t+j+h-i_4}\}$ are pairwise equal, which corresponds to three sums, and then by subtracting twice the sum in which the indices of $\{\epsilon_{t-i_1}, \epsilon_{t+i-i_2}, \epsilon_{t+h-i_3}, \epsilon_{t+j+h-i_4}\}$ are all equal (the latter sum being counted three times in the first three sums). We also have

$$\gamma_X(i) = \sum_{i_1,i_2} \psi_{i_1}\psi_{i_2} E\epsilon_{t-i_1}\epsilon_{t+i-i_2} = \gamma_\epsilon(0) \sum_{i_1} \psi_{i_1}\psi_{i_1+i}. \tag{B.17}$$

By stationarity and the dominated convergence theorem, it is easily shown that

$$\lim_{n\to\infty} n\text{Cov}\{\hat{\gamma}_X(i), \hat{\gamma}_X(j)\} = \sum_{h=-\infty}^{\infty} \text{Cov}\{X_tX_{t+i}, X_{t+h}X_{t+j+h}\}.$$

The existence of this sum is guaranteed by the conditions

$$\sum_i |\psi_i| < \infty \quad \text{and} \quad \sum_h |\rho_{\epsilon^2}(h)| < \infty.$$

In view of (B.16) and (B.17), this sum is equal to

$$\begin{aligned} &\sum_{i_1,i_3} \psi_{i_1,i_1+i,i_3,i_3+j} \sum_h (E\epsilon^2_{t-i_1}\epsilon^2_{t+h-i_3} - \gamma^2_\epsilon(0)) \\ &\quad + \sum_{h,i_1,i_2} \psi_{i_1,i_1+h,i_2,i_2+h+j-i} E\epsilon^2_{t-i_1}\epsilon^2_{t+i-i_2} \\ &\quad + \sum_{h,i_1,i_2} \psi_{i_1,i_1+h+j,i_2,i_2+h-i} E\epsilon^2_{t-i_1}\epsilon^2_{t+i-i_2} - 2E\epsilon^4_t \sum_{h,i_1} \psi_{i_1,i_1+i,i_1+h,i_1+h+j} \\ &= \rho_{\epsilon^2}\left(\sum_{i_1} \psi_{i_1}\psi_{i_1+i} \sum_{i_3} \psi_{i_3}\psi_{i_3+j}\right) \times \gamma_{\epsilon^2}(0) \\ &\quad + \sum_{i_1,i_2} \psi_{i_1}\psi_{i_2} E\epsilon^2_{t-i_1}\epsilon^2_{t+i-i_2} \sum_h \psi_{i_1+h}\psi_{i_2+h+j-i} \\ &\quad + \sum_{i_1,i_2} \psi_{i_1}\psi_{i_2} E\epsilon^2_{t-i_1}\epsilon^2_{t+i-i_2} \sum_h \psi_{i_1+h+j}\psi_{i_2+h-i} \\ &\quad - 2E\epsilon^4_t \sum_{i_1} \psi_{i_1}\psi_{i_1+i} \sum_h \psi_{i_1+h}\psi_{i_1+h+j}, \end{aligned}$$

using Fubini's theorem. With the notation $E\epsilon_t^4 = \kappa_\epsilon \gamma_\epsilon^2(0)$, and using once again (B.17) and the relation $\gamma_{\epsilon^2}(0) = (\kappa_\epsilon - 1)\gamma_\epsilon^2(0)$, we obtain

$$\begin{aligned}
\lim_{n\to\infty} n\mathrm{Cov}\{\hat{\gamma}_X(i), \hat{\gamma}_X(j)\} &= \gamma_{\epsilon^2}(0)\rho_{\epsilon^2}\gamma_\epsilon^{-2}(0)\gamma_X(i)\gamma_X(j) \\
&+ \sum_{i_1,i_2} \psi_{i_1}\psi_{i_2} E\epsilon_{t-i_1}^2 \epsilon_{t+i-i_2}^2 \gamma_\epsilon^{-1}(0)\gamma_X(i_2 + j - i - i_1) \\
&+ \sum_{i_1,i_2} \psi_{i_1}\psi_{i_2} E\epsilon_{t-i_1}^2 \epsilon_{t+i-i_2}^2 \gamma_\epsilon^{-1}(0)\gamma_X(i_2 - j - i - i_1) \\
&- 2E\epsilon_t^4 \gamma_\epsilon^{-2}(0)\gamma_X(i)\gamma_X(j) \\
&= \{(\kappa_\epsilon - 1)\rho_{\epsilon^2} - 2\kappa_\epsilon\}\gamma_X(i)\gamma_X(j) \\
&+ \gamma_\epsilon^{-1}(0) \sum_{i_1,i_2} \psi_{i_1}\psi_{i_2} \{\gamma_X(i_2 + j - i - i_1) + \gamma_X(i_2 - j - i - i_1)\} \\
&\times \{\gamma_{\epsilon^2}(i - i_2 + i_1) + \gamma_\epsilon^2(0)\}.
\end{aligned}$$

With the change of index $\ell = i_2 - i_1$, we finally obtain

$$\begin{aligned}
\lim_{n\to\infty} n\mathrm{Cov}\{\hat{\gamma}_X(i), \hat{\gamma}_X(j)\} &= \{(\kappa_\epsilon - 1)\rho_{\epsilon^2} - 2\kappa_\epsilon\}\gamma_X(i)\gamma_X(j) \\
+ \gamma_\epsilon^{-2}(0) \sum_{\ell=-\infty}^{\infty} \gamma_X(\ell)\{\gamma_X(\ell + j - i) &+ \gamma_X(\ell - j - i)\}\{\gamma_{\epsilon^2}(i - \ell) + \gamma_\epsilon^2(0)\},
\end{aligned} \tag{B.18}$$

which can also be written in the form (B.14) (see Exercise 1.11). The vector $(\hat{\rho}_X(i), \hat{\rho}_X(j))$ is a function of the triplet $(\hat{\gamma}_X(0), \hat{\gamma}_X(i), \hat{\gamma}_X(j))$. The Jacobian of this differentiable transformation is

$$J = \begin{pmatrix} -\frac{\gamma_X(i)}{\gamma_X^2(0)} & \frac{1}{\gamma_X(0)} & 0 \\ -\frac{\gamma_X(j)}{\gamma_X^2(0)} & 0 & \frac{1}{\gamma_X(0)} \end{pmatrix}.$$

Let Σ be the matrix of asymptotic variance of the triplet, whose elements are given by (B.18). By the delta method, we obtain

$$\begin{aligned}
\lim_{n\to\infty} n\mathrm{Cov}\{\hat{\rho}(i), \hat{\rho}(j)\} &= J\Sigma J'(1,2) \\
&= \frac{\gamma_X(i)\gamma_X(j)}{\gamma_X^4(0)}\Sigma(1,1) - \frac{\gamma_X(i)}{\gamma_X^3(0)}\Sigma(1,3) - \frac{\gamma_X(j)}{\gamma_X^3(0)}\Sigma(2,1) + \frac{1}{\gamma_X^2(0)}\Sigma(2,3) \\
&= \{(\kappa_\epsilon - 1)\rho_{\epsilon^2} - 2\kappa_\epsilon\}\left\{2\frac{\gamma_X(i)\gamma_X(j)}{\gamma_X^2(0)} - 2\frac{\gamma_X(i)\gamma_X(j)}{\gamma_X^2(0)}\right\} \\
&+ \gamma_\epsilon^{-2}(0) \sum_{\ell=-\infty}^{\infty} \left[\frac{\gamma_X(i)\gamma_X(j)}{\gamma_X^4(0)} 2\gamma_X^2(\ell)\{\gamma_{\epsilon^2}(-\ell) + \gamma_\epsilon^2(0)\}\right. \\
&- \frac{\gamma_X(i)}{\gamma_X^3(0)}\gamma_X(\ell)\{\gamma_X(\ell + j) + \gamma_X(\ell - j)\}\{\gamma_{\epsilon^2}(-\ell) + \gamma_\epsilon^2(0)\} \\
&- \frac{\gamma_X(j)}{\gamma_X^3(0)}\gamma_X(\ell)\{\gamma_X(\ell - i) + \gamma_X(\ell - i)\}\{\gamma_{\epsilon^2}(i - \ell) + \gamma_\epsilon^2(0)\} \\
&\left. + \frac{1}{\gamma_X^2(0)}\gamma_X(\ell)\{\gamma_X(\ell + j - i) + \gamma_X(\ell - j - i)\}\{\gamma_{\epsilon^2}(i - \ell) + \gamma_\epsilon^2(0)\}\right].
\end{aligned}$$

Simplifying and using the autocorrelations, the previous quantity is equal to

$$\sum_{\ell=-\infty}^{\infty} [2\rho_X(i)\rho_X(j)\rho_X^2(\ell) - \rho_X(i)\rho_X(\ell)\{\rho_X(\ell+j) + \rho_X(\ell-j)\}$$
$$- \rho_X(j)\rho_X(\ell)\{\rho_X(\ell-i) + \rho_X(\ell-i)\} + \rho_X(\ell)\{\rho_X(\ell+j-i) + \rho_X(\ell-j-i)\}]$$
$$+ (\kappa_\epsilon - 1) \sum_{\ell=-\infty}^{\infty} \rho_{\epsilon^2}(\ell)[2\rho_X(i)\rho_X(j)\rho_X^2(\ell) - \rho_X(i)\rho_X(\ell)\{\rho_X(\ell+j) + \rho_X(\ell-j)\}$$
$$- \rho_X(j)\rho_X(\ell-i)\{\rho_X(\ell) + \rho_X(\ell)\} + \rho_X(i-\ell)\{\rho_X(-\ell+j) + \rho_X(-\ell-j)\}].$$

Noting that

$$\sum_\ell \rho_X(\ell)\rho_X(\ell+j) = \sum_\ell \rho_X(\ell)\rho_X(\ell-j),$$

we obtain

$$\lim_{n\to\infty} n\mathrm{Cov}\{\hat{\rho}(i), \hat{\rho}(j)\}$$
$$= \sum_{\ell=-\infty}^{\infty} \rho_X(\ell)[2\rho_X(i)\rho_X(j)\rho_X(\ell) - 2\rho_X(i)\rho_X(\ell+j)$$
$$- 2\rho_X(j)\rho_X(\ell+i) + \rho_X(\ell+j-i) + \rho_X(\ell-j-i)]$$
$$+ (\kappa_\epsilon - 1) \sum_{\ell=-\infty}^{\infty} \rho_{\epsilon^2}(\ell)[2\rho_X(i)\rho_X(j)\rho_X^2(\ell) - 2\rho_X(i)\rho_X(\ell)\rho_X(\ell+j)$$
$$- 2\rho_X(j)\rho_X(\ell)\rho_X(\ell+i) + \rho_X(\ell+i)\{\rho_X(\ell+j) + \rho_X(\ell-j)\}].$$

□

Appendix C

Markov Chains on Countable State Spaces

We review the basic elements of the Markov Chain theory in the case where the state space is countable or finite.

We start by recalling the definition of a Markov chain process $(\Delta_t)_{t\geq 0}$ with discrete state space $\mathcal{E}$. The state space $\mathcal{E}$ may be infinite – such as the set of the non-negative integers $\mathbb{N} = \{0, 1, \dots\}$ – or finite as $\{1, \dots, d\}$. Otherwise stated, we assume that $\mathcal{E} = \mathbb{N}$. The following definitions can be easily adapted when the state space is finite.

C.1 Definition of a Markov Chain

To define a Markov Chain we need an initial distribution $\pi_0 = \{\pi_{0i}\}$ so that at time 0, the chain belongs to the ith state with probability π_{0i}:

$$P[\Delta_0 = i] = \pi_{0i}, \quad i \geq 0, \quad \pi_{0i} \geq 0, \quad \sum_{i=0}^{\infty} \pi_{0i} = 1. \tag{C.1}$$

It is not restrictive to assume $\pi_{0i} > 0$ (otherwise, state i can be withdrawn from the state space). We need a set of transition probabilities $\{p(i,j)\}$ such that, for any $t \geq 0$,

$$P[\Delta_{t+1} = j \mid \Delta_t = i] = p(i,j). \tag{C.2}$$

The transition probabilities obviously satisfy $p(i,j) \geq 0$ for all $i, j \in \mathcal{E}$, and $\sum_{j\in\mathcal{E}} p(i,j)$ for any $i \in \mathcal{E}$. The Markov property states that instead of conditioning by $[\Delta_t = i]$ in the latter probability, we may condition on the entire history without changing the conditional probability:

$$P[\Delta_{t+1} = j \mid \Delta_t = i, \Delta_{t-1} = i_{t-1}, \dots, \Delta_0 = i_0] = p(i,j), \tag{C.3}$$

for any integers $i_0, \dots, i_{t-1}, i, j$ in $\mathcal{E}$ (provided $P[\Delta_t = i, \Delta_{t-1} = i_{t-1}, \dots, \Delta_0 = i_0] > 0$).

GARCH Models: Structure, Statistical Inference and Financial Applications, Second Edition. Christian Francq and Jean-Michel Zakoian.

Definition C.1 *A process $(\Delta_t)_{t\geq 0}$ satisfying (C.1)–(C.3) is called a Markov chain with initial distribution $\{\pi_i\}$ and transition probabilities $\{p(i, j)\}$.*

C.2 Transition Probabilities

The matrix

$$\mathbb{P} = (p(i,j))_{i,j\in\mathcal{E}} = \begin{pmatrix} p(0,0) & p(0,1) & \dots \\ p(1,0) & p(1,1) & \dots \\ \vdots & \ddots & \end{pmatrix}$$

is called the transition matrix of the Markov chain. Matrix powers of $\mathbb{P}$ are defined by $\mathbb{P}^2 = \mathbb{P}\mathbb{P} = (p^{(2)}(i,j))_{i,j\in\mathcal{E}}$ where $p^{(2)}(i,j) = \sum_{k=0}^{\infty} p(i,k)p(k,j)$.[1] More generally, we define $\mathbb{P}^n = (p^{(n)}(i,j))_{i,j\in\mathcal{E}}$ recursively, for any non-negative integer n, with by convention $p^{(0)}(i,j) = \mathbb{1}_{i=j}$ and $p^{(1)}(i,j) = p(i,j)$. It can be shown that, for any states i, j, and any $n \geq 0$,

$$p^{(n)}(i,j) = P[\Delta_{t+n} = j|\Delta_t = i].$$

The Chapman–Kolmogorov equation, for $m, n \geq 0$,

$$p^{(n+m)}(i,j) = \sum_{k=0}^{\infty} p^{(n)}(i,k)p^{(m)}(k,j), \quad i,j \in \mathcal{E}$$

is a consequence of the matrix equality $\mathbb{P}^{n+m} = \mathbb{P}^n\mathbb{P}^m$.

C.3 Classification of States

For $i, j \in \mathcal{E}$, we say that j is *accessible* from i, written $i \to j$, if $p^{(n)}(i,j) > 0$ for some non-negative integer n. States i and j *communicate*, written $i \leftrightarrow j$, if $i \to j$ and $j \to i$. It can be checked that *communication is an equivalence relation.*[2]

A Markov chain is called *irreducible* if all states communicate.

The notion of recurrence is related to how often a chain returns to a given state. A state i is called *recurrent* if, starting from i, the chain returns to i with probability 1 in a finite number of states. More precisely, define the random variable

$$\tau_i = \inf\{m \geq 1, \Delta_m = i\},$$

with $\tau_i = \infty$ if the latter set is empty. State i is recurrent if $P[\tau_i < \infty | \Delta_0 = i] = 1$. On the contrary, state i is called *transient* if $P[\tau_i = \infty | \Delta_0 = i] > 0$ (there is a positive probability of never returning to state i). A state i is called *positive recurrent* if $E[\tau_i | \Delta_0 = i] < \infty$. The following result is a useful criterion for recurrence/transience (see Resnick 1992, Propositions 2.6.2 and 2.6.3).

Proposition C.1 *We have*

$$i \text{ is recurrent if and only if } \sum_{n=0}^{\infty} p^{(n)}(i,i) = \infty,$$

[1] Note that the infinite sum is well defined, since $\sum_{k=0}^{\infty} p(i,k)p(k,j) \leq \sum_{k=0}^{\infty} p(i,k) = 1$.

[2] That is, communication is (i) reflexive ($i \leftrightarrow i$ for any state i); (ii) symmetric ($i \leftrightarrow j$ if and only if $j \leftrightarrow i$); (iii) transitive ($i \leftrightarrow j$ and $j \leftrightarrow k$ entails $i \leftrightarrow k$).

and thus

$$i \text{ is transient if and only if } \sum_{n=0}^{\infty} p^{(n)}(i,i) < \infty.$$

Moreover, if i is transient, then for all states j, we have $\sum_{n=0}^{\infty} p^{(n)}(j,i) < \infty$.

It follows that if $\mathcal{E}$ is finite, not all states can be transient.[3]

States of a Markov chain can also be classified as either periodic or aperiodic. Define the period of state i as

$$d(i) = \gcd\{n \geq 1, p^{(n)}(i,i) > 0\},$$

where gcd stands for greatest common divisor. By convention, take $d(i) = 1$ if the latter set is empty. State i is called *aperiodic* if $d(i) = 1$. Otherwise, state i is called *periodic* with period $d(i) > 1$.

Recurrence, transience and the period of a state are called *solidarity properties* of the Markov chain: whenever state i has one of such properties, any state j such that $i \leftrightarrow j$ also has the property (see Resnick 1992, Proposition 2.8.1). The Markov chain is called recurrent (resp. aperiodic) if all the states are recurrent (resp. aperiodic).

C.4 Invariant Probability and Stationarity

Let $\pi = \{\pi_j, j \in \mathcal{E}\} = (\pi_0, \pi_1, \dots)'$ be a probability distribution on the states of the Markov chain. It is called an *invariant distribution* if

$$\boldsymbol{\pi} = \mathbb{P}'\boldsymbol{\pi} \tag{C.4}$$

where $\mathbb{P}'$ is the transpose of the transition matrix of the Markov chain. In other words,

$$\pi_i = \sum_{j \in \mathcal{E}} p(j,i)\pi_j, \quad i \in \mathcal{E}.$$

If $\nu = \{\nu_j, j \in \mathcal{E}\}$ is a sequence of non-negative constants satisfying $\nu = \mathbb{P}'\nu$, call ν an *invariant measure*. Note that if $\sum_{j \in \mathcal{E}} \nu_j = \infty$, it is not possible to scale such a measure ν to get an invariant distribution π.

If an invariant probability distribution π exists, and if the initial distribution is such that $\pi_0 = \pi$, then the Markov chain is a stationary process, that is

$$P[\Delta_0 = i_0, \dots, \Delta_k = i_k] = P[\Delta_n = i_0, \dots, \Delta_{n+k} = i_k], \quad \text{for any } n, k \geq 0$$

and any states $i_0, \dots, i_k$ (see Resnick 1992, Proposition 2.12.1).

When the state space $\mathcal{E}$ is finite, an invariant distribution π always exists[4] and can be computed by solving C.4.

For infinite state spaces, an invariant distribution π may or may not exist/be unique. We have the following result (see Resnick 1992, Proposition 2.12.3).

Proposition C.2 *If the Markov chain is irreducible and recurrent, an invariant measure* ν *exists and satisfies* $0 < \nu_j < \infty$ *for all* $j \in \mathcal{E}$. *This invariant measure is unique up to multiplication by constants. Moreover, if the Markov chain is positive recurrent and irreducible, there exists a unique stationary distribution* π.

[3] Indeed, if any state is transient, we have $p^{(n)}(j,i) \to 0$ as $n \to \infty$, for all $i, j \in \mathcal{E}$. But since $\sum_{i \in \mathcal{E}} p^{(n)}(j,i) = 1$ this is not possible when $\mathcal{E}$ is finite.

[4] To see this, first note that, as a stochastic matrix, $\mathbb{P}$ admits 1 as an eigenvalue which is also its largest real eigenvalue. By the Perron–Frobenius lemma, a corresponding eigenvector can be chosen to have strictly positive components. It is then possible to scale such a vector to obtain a probability distribution.

C.5 Ergodic Results

By the strong law of large numbers, the empirical mean of a sequence of iid integrable variables converges to the theoretical mean. For stationary and ergodic processes, such a convergence holds in virtue of the ergodic theorem (see Appendix A). For a Markov chain on a countable state space, one may ask whether or not, for some function $f : \mathcal{E} \to \mathbb{R}$, an empirical mean like

$$\frac{1}{n}\sum_{t=0}^{n} f(\Delta_t),$$

converges when $n \to \infty$. For instance, if $f(\Delta_t) = \mathbb{1}_{\{\Delta_t=i\}}$ for $i \in \mathcal{E}$, the empirical mean is the relative frequency that the chain visits state i. The following result is a sufficient condition for convergence of such empirical means (see Resnick 1992, Proposition 2.12.4).

Proposition C.3 *Suppose the Markov chain is irreducible and positive recurrent, and let π be the unique stationary distribution. Then, if f is positive or bounded,*

$$\frac{1}{n}\sum_{t=0}^{n} f(\Delta_t) \to \pi(f) := \sum_{j\in\mathcal{E}} f(j)\pi_j, \quad \text{a.s.} \quad \text{as } n \to \infty,$$

for any initial distribution.

By the dominated convergence theorem, for f bounded, it follows that $\frac{1}{n}\sum_{t=0}^{n} E[f(\Delta_t)|\Delta_0 = i] \to \pi(f)$ as $n \to \infty$. In particular, for $f(k) = \mathbb{1}_{\{j=k\}}$ we get

$$\frac{1}{n}\sum_{t=0}^{n} p^{(t)}(i,j) \to \pi_j, \quad \text{as } n \to \infty, \tag{C.5}$$

under the conditions of Proposition C.3.

C.6 Limit Distributions

A *limit distribution* is any probability distribution π satisfying

$$\lim_{t\to\infty} p^{(t)}(i,j) = \pi_j, \quad i,j \in \mathcal{E}.$$

For such a distribution, we have

$$\pi_j = \lim_{t\to\infty} p^{(t+1)}(i,j) = \lim_{t\to\infty} \sum_{k\in\mathcal{E}} p^{(t)}(i,k)p(k,j).$$

Thus, when $\mathcal{E}$ is finite, $\pi_j = \sum_{k=0}^{d} \pi_k p(k,j)$, which shows that π is a stationary distribution. In the general case where $\mathcal{E}$ is infinite, the limit and the sum cannot be inverted, but the result continues to hold (see Resnick 1992, Proposition 2.13.1).

Proposition C.4 *A limit distribution is a stationary distribution.*

The converse of this property requires assumptions on the Markov chain. Under irreducibility and positive recurrence, we only have a Cesro mean limit in (C.5). If, in addition, the chain is aperiodic, we have the following result (see Resnick 1992, Proposition 2.13.2).

Proposition C.5 *Suppose the Markov chain is irreducible and aperiodic and that a stationary distribution π exists, then (i) the Markov chain is positive recurrent; (ii) π is a limit distribution; (iii) for all $j \in \mathcal{E}$, $\pi_j > 0$; (iv) the stationary distribution is unique.*

A Markov chain which is positive recurrent, aperiodic and irreducible is called *ergodic*.

C.7 Examples

The example of the Ehrenfest chain (see Exercise 12.6) shows that the existence of a limit distribution may be lost when aperiodicity is relaxed.
The next example shows that, for random walks, recurrence depends on the dimension.

Example C.1 (Random walks in $\mathbb{Z}^m$) Let e_j the jth column of the $m \times m$ identity matrix I_m. A symmetric random walk on $\mathbb{Z}^m$ is defined by $\Delta_0 = 0$ and

$$\Delta_t = \Delta_{t-1} + \epsilon_t, \quad t > 0$$

with (ϵ_t) an iid sequence which is uniformly distributed over $\{e_1, -e_1, \dots, e_m, -e_m\}$. The process (Δ_t) is clearly a Markov chain with state space $\mathbb{Z}^m$. It is possible to reach any point in $\mathbb{Z}^m$ from any starting point, so the chain is irreducible. We now prove that (Δ_t) is recurrent for $m \leq 2$ but not for $m \geq 3$. We compute $p^{(n)}(i, i)$ for $m = 1$ and $m = 2$, and we will admit that the result extends for $m \geq 3$.

- When $m = 1$, we have $p^{(2n+1)}(i, i) = 0$ and

$$\begin{aligned} p^{(2n)}(i,i) &= P(\{n \text{ moves to the left and } n \text{ moves to the right}\}) \\ &= \binom{2n}{n} \frac{1}{2^{2n}} \sim \frac{1}{\sqrt{n\pi}} \end{aligned}$$

  as $n \to \infty$, using the Stirling formula $n! \sim \sqrt{2\pi} n^{n+1/2} e^{-n}$. It follows from Proposition C.1 that the Markov chain (Δ_t) is recurrent.

- When $m = 2$, we have $p^{(2n+1)}(i, i) = 0$ and

$$\begin{aligned} p^{(2n)}(i,i) &= \sum_{u=0}^{n} P(\{u \text{ steps to the west}, u \text{ to the east}, n-u \text{ to the north}, \\ &\qquad\qquad n-u \text{ to the south}\}) \\ &= \sum_{u=0}^{n} \binom{2n}{2u} \binom{2(n-u)}{n-u} \binom{2u}{u} \frac{1}{4^{2n}} \\ &= \sum_{u=0}^{n} \frac{(2n)!}{u!u!(n-u)!(n-u)!} \frac{1}{4^{2n}} \\ &= \frac{1}{4^{2n}} \binom{2n}{n} \sum_{u=0}^{n} \binom{n}{u} \binom{n}{n-u} = \frac{1}{4^{2n}} \binom{2n}{n} \binom{2n}{n} \sim \frac{1}{n\pi}. \end{aligned}$$

  It follows from Proposition C.1 that the Markov chain (Δ_t) is recurrent.

- In the general case, we still have $p^{(2n+1)}(i, i) = 0$, and it can be shown that $p^{(2n)}(i, i) = O(n^{-m/2})$, which entails $\sum_{n=1}^{\infty} p_{ii}^{(2n)} < \infty$. By Proposition C.1, the Markov chain (Δ_t) is transient.

In the following example, the convergence to the limit distribution can be fully characterised.

Example C.2 (two-state Markov chain) Any Markov chain on $\mathcal{E} = \{0, 1\}$ has transition matrix

$$\mathbb{P} = \begin{pmatrix} 1-p & p \\ q & 1-q \end{pmatrix}, \quad 0 \leq p, q \leq 1.$$

The Markov chain is irreducible and aperiodic if and only if $0 < p,\ q < 1$. It can be checked that, for $n \geq 0$,

$$\mathbb{P}^n = (p+q)^{-1} \left\{ \begin{pmatrix} q & p \\ q & p \end{pmatrix} + (1-p-q)^n \begin{pmatrix} p & -p \\ -q & q \end{pmatrix} \right\}.$$

When $0 < p,\ q < 1$ we have $|1 - p - q| < 1$, thus

$$\mathbb{P}^n \to (p+q)^{-1} \begin{pmatrix} q & p \\ q & p \end{pmatrix},$$

so $\pi = (p+q)^{-1}(q,\ p)'$ is a limit distribution of the Markov chain. By Propositions C.4 and C.5, the Markov chain is positive recurrent and π is the unique stationary distribution. We also have

$$\sup_{i,j} \mid p^{(n)}(i,j) - \pi_j \mid \leq K(1-p-q)^n$$

where K is a constant and $\pi = (\pi_1,\ \pi_2)$, showing that the speed of convergence of the n-step transition probabilities to the limit distribution is geometric.

Appendix D

The Kalman Filter

The state-space representation is a useful tool for analysing many dynamic models. When this representation exists, the Kalman filter can be applied to estimate the model parameters, to compute predictions or to smooth the series. This technique was introduced by Kalman (1960) in the field of engineering but has been used in various domains, in particular in economics.

We first introduce state-space representations and useful notations. Let (y_t) denote a $\mathbb{R}^N$-valued *observable* process, and let (α_t) a $\mathbb{R}^m$-valued *latent* (in general non-observable, or only partially observable) process. A state-space model is defined by

$$\begin{cases} y_t = M_t\alpha_t + d_t + u_t \\ \alpha_t = T_t\alpha_{t-1} + c_t + R_t v_t, \end{cases} \tag{D.1}$$

where M_t, d_t, T_t, c_t and R_t are deterministic matrices of appropriate dimensions, (u_t) and (v) are white noises, respectively, valued in $\mathbb{R}^N$ and $\mathbb{R}^m$. The vector α_t is called space vector. The first equation is called *measurement equation*, and the second equation *transition equation*.

The Kalman filter is an algorithm for

(i) *predicting* the period-t space vector from observations of y up to time $t-1$;

(ii) *filtering*, that is predicting the period-t space vector from observations of y up to time t;

(iii) *smoothing*, that is estimating the value of α_t from observations of y up to time T, with $T > t$.

In this appendix, we present basic properties of the Kalman filter. The reader is referred to Harvey (1989) for a comprehensive presentation of structural time series models and the Kalman filter.

In order to implement the algorithm, we make the following assumptions.

- The process (u_t, v) is an iid Gaussian white noise such that

$$\begin{pmatrix} u_t \\ v_t \end{pmatrix} \sim \mathcal{N}\left(\begin{pmatrix} 0 \\ 0 \end{pmatrix}, \begin{pmatrix} H_t & 0 \\ 0 & Q_t \end{pmatrix}\right). \tag{D.2}$$

GARCH Models: Structure, Statistical Inference and Financial Applications, Second Edition. Christian Francq and Jean-Michel Zakoian.

- The initial state vector is Gaussian and is independent from the noises (u_t) and (v_t):

$$\alpha_0 \sim \mathcal{N}(a_0, P_0), \quad \alpha_0 \perp (u_t), (v_t).$$

- For all t, the matrix H_t is positive-definite.

These assumptions imply, in particular, that the variables u_t and v are independent from $(y_1, y_2, \dots, y_{t-1})$. Assuming non-correlation of u_t and v is not essential, as we will see, but allows us to simplify presentation. The assumption that H_t is positive-definite can be very restrictive (in particular when the noise of the measurement equation is degenerate), but it is not necessary. It is introduced to ensure the positive-definiteness of the conditional variance of y_t given $y_1, \dots, y_{t-1}$. The algorithm allows to recursively compute the conditional distribution of α_t given $y_1, \dots, y_t$. This distribution is Gaussian and its mean provides "estimator" of de α_t which is optimal (in the L^2 sense). When the Gaussian assumption is in failure, the Kalman filter no longer provides the conditional expectation of α_t. The resulting estimator is no longer optimal, but only optimal among the linear estimators.

D.1 General Form of the Kalman Filter

The derivation of the algorithm requires the following notations:

$$\alpha_{t|t} = E(\alpha_t | y_1, \dots, y_t), \quad P_{t|t} = \text{Var}(\alpha_t | y_1, \dots, y_t),$$

$$\alpha_{t|t-1} = E(\alpha_t | y_1, \dots, y_{t-1}), \quad P_{t|t-1} = \text{Var}(\alpha_t | y_1, \dots, y_{t-1}).$$

The first two equalities hold $t \geq 1$, the other two ones for $t > 1$. Let $\alpha_{1|0} = E(\alpha_1)$ and $P_{1|0} = \text{Var}(\alpha_1)$.

First Step

By taking the conditional expectation with respect to $y_1, \dots, y_{t-1}$ in the transition equation, we get

$$\alpha_{t|t-1} = T_t \alpha_{t-1|t-1} + c_t \tag{D.3}$$

and then, by taking the conditional variance,

$$P_{t|t-1} = T_t P_{t-1|t-1} T_t' + R_t Q_t R_t'. \tag{D.4}$$

Such equations are called *prediction equations*.

The conditional moments of y_t follow:

$$y_{t|t-1} := E(y_t | y_1, \dots, y_{t-1}) = M_t \alpha_{t|t-1} + d_t \tag{D.5}$$

and

$$F_{t|t-1} := \text{Var}(y_t | y_1, \dots, y_{t-1}) = M_t P_{t|t-1} M_t' + H_t. \tag{D.6}$$

We will also use

$$\text{Cov}(\alpha_t, y_t | y_1, \dots, y_{t-1}) = P_{t|t-1} M_t'. \tag{D.7}$$

Second Step

Once the observation y_t becomes available, the preceding quantities are *updated*:

$$\alpha_{t|t} = \alpha_{t|t-1} + P_{t|t-1} M_t' F_{t|t-1}^{-1} (y_t - M_t \alpha_{t|t-1} - d_t) \tag{D.8}$$

and

$$P_{t|t} = P_{t|t-1} - P_{t|t-1}M_t'F_{t|t-1}^{-1}M_tP_{t|t-1}. \tag{D.9}$$

Such equations are called *updating equations.*

The normality assumption is only involved in the second step. Indeed, we use the fact that the distribution of (y_t, α_t) is Gaussian conditional on $y_1, \ldots, y_{t-1}$[1]

$$\begin{pmatrix} \alpha_t \\ y_t \end{pmatrix} \sim \mathcal{N}\left(\begin{pmatrix} \alpha_{t|t-1} \\ y_{t|t-1} \end{pmatrix}, \begin{pmatrix} P_{t|t-1} & P_{t|t-1}M_t' \\ M_tP_{t|t-1} & F_{t|t-1} \end{pmatrix}\right),$$

which allows to derive the law of α_t conditional on $y_1, \ldots, y_{t-1}, y_t$.[2]

Initial Values of the First Step

The starting values for the Kalman filter can be specified by noting that the conditional and unconditional moments coincide:

$$\alpha_{1|0} = T_1a_0 + c_1, \quad P_{1|0} = T_1P_0T_1' + R_1Q_1R_1'. \tag{D.10}$$

The quantities $\alpha_{t|t-1}, P_{t|t-1},$ and $\alpha_{t|t}, P_{t|t}$ can thus be recursively computed for $t = 1, \ldots, n$.

To summarise, the Kalman filter is an algorithm for computing the sequences $(\alpha_{t|t-1})_{t=1}^n$ and $(P_{t|t-1})_{t=1}^n$, where $\alpha_{t|t-1}$ denotes the optimal prediction of the state vector α_t given the observations $y_1, \ldots, y_{t-1}$. The mean squared error for this prediction is $(P_{t|t-1})_{t=1}^n$. These sequences can be directly obtained from the following formulas, which follow from substituting formulas (D.8) and (D.9), taken at $t-1$, in (D.3) and (D.4):

$$\alpha_{t|t-1} = T_t\alpha_{t-1|t-2} + K_t(y_{t-1} - M_{t-1}\alpha_{t-1|t-2} - d_{t-1}) + c_t \tag{D.11}$$

and

$$P_{t|t-1} = T_tP_{t-1|t-2}T_t' - K_tF_{t-1|t-2}K_t' + R_tQ_tR_t' \tag{D.12}$$

where

$$F_{t-1|t-2} = M_{t-1}P_{t-1|t-2}M_{t-1}' + H_{t-1}, \quad K_t = T_tP_{t-1|t-2}M_{t-1}'F_{t-1|t-2}^{-1}. \tag{D.13}$$

It can be noted that the sequence $(P_{t|t-1})_{t=1}^n$ does not depend on the observable variables, and can be computed independently from the sequence $(\alpha_{t|t-1})_{t=1}^n$. Matrix K_t is called the *gain matrix.*

[1] At period 1, we have

$$\begin{cases} y_1 = M_1\alpha_1 + d_1 + u_1, \\ \alpha_1 = T_1\alpha_0 + c_1 + R_1v_1. \end{cases}$$

Because (y_1, α_1) can be written as a linear combination of normal variables, its (unconditional) law is Gaussian. Thus, the law of α_1 conditional on y_1 is Gaussian. At time 2, we have

$$\begin{cases} y_2 = M_2\alpha_2 + d_2 + u_2, \\ \alpha_2 = T_2\alpha_1 + c_2 + R_2v_2. \end{cases}$$

The vector (y_2, α_2) can be written as a linear combination of α_1, u_2, and v_2. The law of (α_1, u_2, v_2) being Gaussian conditionally on y_1, the same property holds for (y_2, α_2). It follows that the law of α_2 conditional on (y_1, y_2) is normal. By iterating the argument, we get the announced result.

[2] If a vector $(x, y)'$ is Gaussian, with a positive-definite variance-covariance matrix, letting $\mu_x = E(x)$, $\mu_y = E(y)$, $\Sigma_{xx} = \mathrm{Var}(x)$, $\Sigma_{yy} = \mathrm{Var}(y)$, $\Sigma_{xy} = \Sigma_{yx}' = \mathrm{Cov}(x, y)$, the law of x conditional on y is the Gaussian law:

$$\mathcal{N}(\mu_x + \Sigma_{xy}\Sigma_{yy}^{-1}(y - \mu_y), \Sigma_{xx} - \Sigma_{xy}\Sigma_{yy}^{-1}\Sigma_{yx}).$$

Remark D.1 One can easily get rid of the assumption that the components of the noise, u_t and v, are uncorrelated. Replace assumption (D.2) by

$$\begin{pmatrix} u_t \\ v_t \end{pmatrix} \sim \mathcal{N}\left(\begin{pmatrix} 0 \\ 0 \end{pmatrix}, \begin{pmatrix} H_t & G_t' \\ G_t & Q_t \end{pmatrix}\right), \tag{D.14}$$

where G_t is a $m \times N$ matrix. Then the prediction equations – (D.3) and (D.4) – are unchanged, as well as the first conditional moment of y_t, (D.5). In contrast (D.6) and (D.7) are replaced by

$$F^*_{t|t-1} = \text{Var}(y_t|y_1, \dots, y_{t-1}) = M_t P_{t|t-1} M_t' + H_t + M_t R_t G_t + G_t' R_t' M_t',$$
$$\text{Cov}(\alpha_t, y_t|y_1, \dots, y_{t-1}) = P_{t|t-1} M_t' + R_t G_t.$$

It follows that the updating equations have to be replaced by

$$\alpha_{t|t} = \alpha_{t|t-1} + (P_{t|t-1} M_t' + R_t G_t) F^{*-1}_{t|t-1} (y_t - M_t \alpha_{t|t-1} - d_t)$$

and

$$P_{t|t} = P_{t|t-1} - (P_{t|t-1} M_t' + R_t G_t) F^{*-1}_{t|t-1} (M_t P_{t|t-1} + G_t' R_t').$$

Remark D.2 When the noise is not Gaussian, the formulas of the Kalman filter remain valid, in some sense, but they no longer provide the optimal updating.[3] Formulas (D.8) and (D.9) provide the optimal linear prediction of α_t, and the variance of this prediction. Formulas (D.5) and (D.6) are no longer valid, but the Kalman filter allows to compute the optimal linear prediction of y_t:

$$EL(y_t|y_1, \dots, y_{t-1}) = M_t \alpha_{t|t-1} + d_t$$

and the variance of the prediction error

$$\text{Var}\{y_t - EL(y_t|y_1, \dots, y_{t-1})\} = M_t P_{t|t-1} M_t' + H_t.$$

D.2 Prediction and Smoothing with the Kalman Filter

Prediction

The Kalman filter can be used for prediction at horizon larger than 1. To simplify presentation, let us assume that $c_t = d_t = 0$, $T_t = T$ and $M_t = M$ for all t. We thus have, for any integer h,

$$\alpha_{t+h} = T^{h+1} \alpha_{t-1} + \sum_{i=0}^{h} T^{h-i} R_{t+i} v_{t+i}$$

and, consequently,

$$\alpha_{t+h|t-1} = E(\alpha_{t+h}|y_1, \dots y_{t-1}) = T^{h+1} \alpha_{t-1|t-1}.$$

[3] If a vector (x, y) belongs to the space L^2 of the square integrable variables, endowed with the norm $\|\cdot\|_2$, the conditional expectation $E(x\,|\,y)$ minimises $\|x - z\|_2$ over the set of variables which are functions of y. If, instead, the minimisation is performed over the linear functions of the form $Ay + b$, the solution to the minimisation problem is called the linear conditional expectation, denoted $EL(x\,|\,y)$. We have $EL(x|y) = \mu_x + \Sigma_{xy} \Sigma_{yy}^{-1} (y - \mu_y)$ where $\mu_x = E(x)$, $\mu_y = E(y)$, $\Sigma_{xx} = \text{Var}(x)$, $\Sigma_{yy} = \text{Var}(y)$, $\Sigma_{xy} = \Sigma_{yx}' = \text{Cov}(x, y)$. The variance of the error is $\text{Var}(x - EL(x|y)) = \Sigma_{xx} - \Sigma_{xy} \Sigma_{yy}^{-1} \Sigma_{yx}$.

The variance of the prediction error at horizon $h+1$ is given by

$$P_{t+h|t-1} = \mathrm{Var}(\alpha_{t+h} - \alpha_{t+h|t-1}) = T^{h+1}P_{t-1|t-1}(T^{h+1})' + \sum_{i=0}^{h} T^{h-i}R_{t+i}Q_{t+i}(T^{h-i}R_{t+i})'.$$

From these equations, we deduce the predictions of the observed process. We have $y_{t+h} = M\alpha_{t+h} + u_{t+h}$ and thus

$$y_{t+h|t-1} = E(y_{t+h}|y_1, \dots, y_{t-1}) = M\alpha_{t+h|t-1} = MT^{h+1}\alpha_{t-1|t-1}.$$

The prediction error is $y_{t+h} - y_{t+h|t-1} = M(\alpha_{t+h} - \alpha_{t+h|t-1}) + u_{t+h}$ and the corresponding mean-square error is

$$\mathrm{Var}(y_{t+h} - y_{t+h|t-1}) = MP_{t+h|t-1}M' + H_{t+h}.$$

Smoothing

Formula (D.8) gives the smoothed value $\alpha_{t|t}$ of α_t, that is, its prediction given the observations up to period t. In some applications, the value of the state vector is of prime interest, and one may want to predict its values *a posteriori*. Smoothing techniques use observations posterior to period t to predict α_t. Let

$$\alpha_{t|n} = E(\alpha_t|y_1, \dots, y_n), \quad P_{t|n} = \mathrm{Var}(\alpha_t|y_1, \dots, y_n),$$

where n is the sample size.

The smoothed values can be computed as follows. First, apply the Kalman filter to the data, and compute the sequences $(\alpha_{t|t})$ and $(P_{t|t})$ from (D.8) and (D.9), as well as the sequences $(\alpha_{t|t-1})$ and $(P_{t|t-1})$, obtained from (D.3) and (D.4). Then, the following algorithm – initialised at $\alpha_{n|n}$ – is used to compute in a descending recursion the $\alpha_{t|n}$'s. The equations are

$$\alpha_{t|n} = \alpha_{t|t} + \tilde{F}_t(\alpha_{t+1|n} - \alpha_{t+1|t}), \quad t < n \tag{D.15}$$

and

$$P_{t|n} = P_{t|t} + \tilde{F}_t(P_{t+1|n} - P_{t+1|t})\tilde{F}'_t, \quad t < n \tag{D.16}$$

where

$$\tilde{F}_t = P_{t|t}T'_{t+1}P^{-1}_{t+1|t}, \quad t < n.$$

Note that the variance-covariance matrices $P_{t|n}$ can be computed independently from the data, and that the smoothed vectors $\alpha_{t|n}$ are linear combinations of the observations (with coefficients depending on t).

To establish such formulas, we use the fact that the law of $(y_t, \alpha_t, \alpha_{t+1})$ conditional on $y_1, \dots, y_{t-1}$ is Gaussian. It follows that, using again the property of multivariate normal vectors, the conditional law of α_t given $\alpha_{t+1}, y_1, \dots, y_t$ is Gaussian with mean

$$E(\alpha_t|\alpha_{t+1}, y_1, \dots, y_t) = \alpha_{t|t} + \tilde{F}_t(\alpha_{t+1} - \alpha_{t+1|t}),$$

because

$$\mathrm{Cov}(\alpha_t, \alpha_{t+1}|y_1, \dots, y_t) = \mathrm{Cov}(\alpha_t, T_{t+1}\alpha_t|y_1, \dots, y_t) = P_{t|t}T'_{t+1}.$$

Next, we note that, for predicting α_t, the knowledge of $y_{t+1}, \dots, y_n$ does not convey additional information with respect to $\alpha_{t+1}, y_1, \dots, y_t$. Indeed, the variables $y_{t+j}, j > 0$ can be written as linear combinations of $\alpha_{t+1}, u_{t+j}, v_{t+2}, \dots, v_{t+j}$. The prediction error $\alpha_t - E(\alpha_t \mid \alpha_{t+1}, y_1, \dots, y_t)$ is – by definition – orthogonal to α_{t+1}, and also to the future noise values. We thus have,

$$E(\alpha_t|\alpha_{t+1}, y_1, \dots, y_n) = \alpha_{t|t} + \tilde{F}_t(\alpha_{t+1} - \alpha_{t+1|t}). \tag{D.17}$$

By the iterated projection formula, it now suffices to take the expectation of both sides of this equality conditional on $y_1, \dots, y_n$ to get (D.15) (noting that the variables $\alpha_{t|t}$ and $\alpha_{t+1|t}$ are functions of the observables).

In order to compute the mean squared error, we note that the smoothing error is

$$\alpha_t - \alpha_{t|n} = \alpha_t - \alpha_{t|t} - \tilde{F}_t(\alpha_{t+1|n} - \alpha_{t+1|t}).$$

Therefore,

$$\alpha_t - \alpha_{t|n} + \tilde{F}_t\alpha_{t+1|n} = \alpha_t - \alpha_{t|t} + \tilde{F}_t\alpha_{t+1|t}$$

and hence,

$$\text{Var}(\alpha_t - \alpha_{t|n}) + \tilde{F}_t\text{Var}(\alpha_{t+1|n})\tilde{F}_t' = \text{Var}(\alpha_t - \alpha_{t|t}) + \tilde{F}_t\text{Var}(\alpha_{t+1|t})\tilde{F}_t'. \tag{D.18}$$

In this expression, the autocovariances are equal to zero because $\alpha_{t+1|n}$ (resp. $\alpha_{t+1|t}$) is a linear combination of the observations, and thus is orthogonal to the error $\alpha_t - \alpha_{t|n}$ (resp. $\alpha_t - \alpha_{t|t}$). Moreover, the smoothing error $\alpha_{t+1} - \alpha_{t+1|n}$ being orthogonal to $\alpha_{t+1|n}$, we have $\text{Cov}(\alpha_{t+1}, \alpha_{t+1|n}) = \text{Var}(\alpha_{t+1|n})$ thus

$$\begin{aligned}\text{Var}(\alpha_{t+1|n} - \alpha_{t+1}) &= \text{Var}(\alpha_{t+1|n}) + \text{Var}(\alpha_{t+1}) - \text{Cov}(\alpha_{t+1|n}, \alpha_{t+1}) - \text{Cov}(\alpha_{t+1}, \alpha_{t+1|n})\\ &= \text{Var}(\alpha_{t+1}) - \text{Var}(\alpha_{t+1|n}).\end{aligned}$$

Similarly, $\text{Var}(\alpha_{t+1|t} - \alpha_{t+1}) = \text{Var}(\alpha_{t+1}) - \text{Var}(\alpha_{t+1|t})$, therefore,

$$\text{Var}(\alpha_{t+1|n} - \alpha_{t+1}) - \text{Var}(\alpha_{t+1|t} - \alpha_{t+1}) = \text{Var}(\alpha_{t+1|t}) - \text{Var}(\alpha_{t+1|n}).$$

Combining the equation with (D.18) we get (D.16).

Note that, as for the Kalman filter formulas, the smoothing formulas remain valid without normality of the errors but only provide, in this case, linear expectations of the state vector.

D.3 Kalman Filter in the Stationary Case

It is worth considering the asymptotic behaviour, when t goes to infinity, of the Kalman filter formulas. Let us focus on the model with constant coefficients and constant variance-covariance matrices

$$\begin{cases} y_t = M\alpha_t + d + u_t, \\ \alpha_t = T\alpha_{t-1} + c + Rv_t, \end{cases} \qquad \begin{pmatrix} u_t \\ v_t \end{pmatrix} \sim \mathcal{N}\left(\begin{pmatrix} 0 \\ 0 \end{pmatrix}, \begin{pmatrix} H & 0 \\ 0 & Q \end{pmatrix}\right) \tag{D.19}$$

The state vector α_t is thus the solution of a first-order Vectoriel Autogregressive (VAR(1)) model. This model admits a second-order stationary solution if the spectral radius $\rho(T)$ of matrix T is strictly less than 1. Under this assumption, the first two moments of the stationary solution satisfy

$$E\alpha_t = (I - T)^{-1}c, \quad \text{Var}(\alpha_t) = T\text{Var}(\alpha_t)T' + RQR', \tag{D.20}$$

where

$$\text{vec}\{\text{Var}(\alpha_t)\} = (I - T \otimes T)^{-1}\text{vec}(RQR').$$

If the initial distribution of the state vector, $\alpha_0 \sim \mathcal{N}(a_0, P_0)$, is such that $a_0 = (I - T)^{-1}c$ and $\text{vec}(P_0) = (I - T \otimes T)^{-1}\text{vec}(RQR')$, then (D.20) holds for any $t \geq 0$.

Interestingly, under the stationarity assumption, the sequence $(P_{t|t-1})$ defined by (D.12) converges. Indeed, let us first note that this formula reduces to

$$P_{t|t-1} = T\{P_{t-1|t-2} - P_{t-1|t-2}M'F_{t-1|t-2}^{-1}MP_{t-1|t-2}\}T' + RQR' \tag{D.21}$$

where

$$F_{t-1|t-2} = MP_{t-1|t-2}M' + H.$$

If the latter sequence converges, the limit $P^* = \lim P_{t|t-1}$ necessarily satisfies

$$P^* = T\{P^* - P^*M'(MP^*M' + H)^{-1}MP^*\}T' + RQR'. \tag{D.22}$$

Proposition D.1 *If $\rho(T) < 1$, and if at least one of the two variance-covariance matrices H and Q is positive-definite, the sequence $(P_{t|t-1})_{t\geq 1}$ defined by (D.21) and initialised at any positive-semidefinite matrix $P_{1|0}$, converges to a unique matrix P^*, which is independent of $P_{1|0}$ and satisfies Eq. (D.22).*

See for instance dans Hamilton (1994, Section 13.5) for a proof. Note that (D.22) is called algebraic Ricatti equation. An explicit solution for this equation is seldom available and, moreover, the solution may not be unique. It can be shown that when P^* is the unique solution, the sequence $(P_{t|t-1})$ converges at exponential rate (see Harvey 1989 and references, Section 3.3.3).

From a numerical point of view, the convergence of the sequence $(P_{t|t-1})_{t\geq 1}$ may be worthwhile. Notice that the sequence $(F_{t|t-1})$ and (K_t) defined in (D.13) also converge in this case, with respective limits

$$F^* = MP^*M' + H, \quad K^* = TP^*M'F^{*-1}. \tag{D.23}$$

When $P_{t|t-1}$ is sufficiently close to the limit P^*, formula (D.11) updating the predictions of α_t can be approximated by

$$\alpha_{t|t-1} = T\alpha_{t-1|t-2} + K^*(y_{t-1} - M\alpha_{t-1|t-2} - d) + c. \tag{D.24}$$

The saving in the computation time may be very large when this approximation is used, because it allows to avoid the inversion of the – possibly high-dimensional – matrix $F_{t|t-1}$, at every step of the algorithm. Once the limit is close to be reached, the inverse of the matrix F^* can be used instead of $F^{-1}_{t|t-1}$. Equation (D.21) thus become useless. A criterion for stoping the computations of $P_{t|t-1}$ can be based on the determinant of this matrix: for instance, the approximation can be used if $|\det P_{t+1|t} - \det P_{t|t-1}| < \tau$ where τ is a very small positive number.

D.4 Statistical Inference with the Kalman Filter

In this section, we assume that the matrices M_t, d_t, T_t, c_t, and R_t of the state space model are constant and are parameterised by a vector θ belonging to a parameter set $\Theta \in \mathbb{R}^d$. The state-space representation thus has the form

$$\begin{cases} y_t = M(\theta)\alpha_t + d(\theta) + u_t, \\ \alpha_t = T(\theta)\alpha_{t-1} + c(\theta) + R(\theta)v_t. \end{cases} \tag{D.25}$$

We also assume that the joint Gaussian distribution of the noise is time-independent, but may depend on θ:

$$\begin{pmatrix} u_t \\ v_t \end{pmatrix} \sim \mathcal{N}\left(\begin{pmatrix} 0 \\ 0 \end{pmatrix}, \begin{pmatrix} H(\theta) & 0 \\ 0 & Q(\theta) \end{pmatrix}\right). \tag{D.26}$$

From observations $y_1, \ldots, y_n$, and for given functions M, d, T, c, H and Q, the problem is to estimate θ. Conditional on initial values $\epsilon_1(\theta)$ and $F_1(\theta)$, the Gaussian likelihood $L_n(\theta)$ writes

$$L_n(\theta) = L_n(\theta; y_1, \ldots, y_n) = \prod_{t=1}^{n} \frac{1}{\sqrt{(2\pi)^N |F_t(\theta)|}} \exp\left(-\frac{1}{2}\epsilon_t'(\theta)F_t(\theta)^{-1}\epsilon_t(\theta)\right),$$

where, for $t > 1$, $\epsilon_t(\theta) = y_t - E_\theta(y_t \mid y_1, \ldots, y_{t-1})$, $F_t(\theta) = \mathrm{Var}_\theta(y_t \mid y_1, \ldots, y_{t-1})$ and $|A|$ denotes the determinant of a square matrix A. Expectations and variances indexed by θ mean that they are computed as if θ was the true parameter value.

A maximum likelihood estimator (MLE) of θ is defined as any measurable solution $\hat{\theta}_n$ of

$$\hat{\theta}_n = \arg\max_{\theta\in\Theta} L_n(\theta).$$

By taking the logarithm, maximising the likelihood with respect to θ amounts to minimising

$$\tilde{\mathrm{l}}_n(\theta) = n^{-1}\sum_{t=1}^{n} \tilde{\ell}_t, \quad \text{where} \quad \tilde{\ell}_t = \tilde{\ell}_t(\theta) = \epsilon_t'(\theta)F_t(\theta)^{-1}\epsilon_t(\theta) + \log|F_t(\theta)|. \tag{D.27}$$

The Kalman filter allows us to compute $\epsilon_t(\theta)$ and $F_t(\theta)$, for any value of θ, when such quantities cannot be easily obtained. Numerical optimisation procedures can be used to obtain the optimum parameter value. The theoretical properties of the MLE (consistency, asymptotic normality) require additional assumptions on the observed process and the parameter space, which will not be detailed here.

Appendix E

Solutions to the Exercises

Chapter 1

1.1 1. (a) We have the stationary solution $X_t = \sum_{i\geq 0} 0.5^i(\eta_{t-i} + 1)$, with mean $EX_t = 2$ and autocorrelations $\rho_X(h) = 0.5^{|h|}$.

(b) We have an 'anticipative' stationary solution

$$X_t = -1 - \frac{1}{2}\sum_{i\geq 0} 0.5^i \eta_{t+i+1},$$

which is such that $EX_t = -1$ and $\rho_X(h) = 0.5^{|h|}$.

(c) The stationary solution

$$X_t = 2 + \sum_{i\geq 0} 0.5^i(\eta_{t-i} - 0.4\eta_{t-i-1})$$

is such that $EX_t = 2$ with $\rho_X(1) = 2/19$ and $\rho_X(h) = 0.5^{h-1}\rho_X(1)$ for $h > 1$.

2. The compatible models are, respectively, ARMA(1, 2), MA(3) and ARMA(1, 1).
3. The first noise is strong, and the second is weak because

$$\text{Cov}\{(\eta_t\eta_{t-1})^2, (\eta_{t-1}\eta_{t-2})^2\} = E\eta_t^2\eta_{t-1}^4\eta_{t-2}^2 - 1 \neq 0.$$

Note that, by Jensen's inequality, this correlation is positive.

1.2 Without loss of generality, assume $X_t = \overline{X}_n$ for $t < 1$ or $t > n$. We have

$$\sum_{h=-n+1}^{n-1} \hat{\gamma}(h) = \frac{1}{n}\sum_{h,t}(X_t - \overline{X}_n)(X_{t+h} - \overline{X}_n) = \frac{1}{n}\left\{\sum_{t=1}^{n}(X_t - \overline{X}_n)\right\}^2 = 0,$$

which gives $1 + 2\sum_{h=1}^{n-1}\hat{\rho}(h) = 0$, and the result follows.

1.3 Consider the degenerate sequence $(X_t)_{t=0,1,\ldots}$ defined, on a probability space $(\Omega, \mathcal{A}, \mathbb{P})$, by $X_t(\omega) = (-1)^t$ for all $\omega \in \Omega$ and all $t \geq 0$. With probability 1, the sequence $\{(-1)^t\}$ is the realisation of the process (X_t). This process is non-stationary because, for instance, $EX_0 \neq EX_1$.

Let U be a random variable, uniformly distributed on $\{0, 1\}$. We define the process $(Y_t)_{t=0,1,\ldots}$ by

$$Y_t(\omega) = (-1)^{t+U(\omega)}$$

for any $\omega \in \Omega$ and any $t \geq 0$. The process (Y_t) is stationary. We have in particular $EY_t = 0$ and $\mathrm{Cov}(Y_t, Y_{t+h}) = (-1)^h$. With probability 1/2, the realisation of the stationary process (Y_t) will be the sequence $\{(-1)^t\}$ (and with probability 1/2, it will be $\{(-1)^{t+1}\}$).

This example leads us to think that it is virtually impossible to determine whether a process is stationary or not, from the observation of only one trajectory, even of infinite length. However, practitioners do not consider $\{(-1)^t\}$ as a potential realisation of the stationary process (Y_t). It is more natural, and simpler, to suppose that $\{(-1)^t\}$ is generated by the non-stationary process (X_t).

1.4 The sequence $0, 1, 0, 1, \ldots$ is a realisation of the process $X_t = 0.5(1 + (-1)^t A)$, where A is a random variable such that $P[A = 1] = P[A = -1] = 0.5$. It can easily be seen that (X_t) is strictly stationary.

Let $\Omega^* = \{\omega \mid X_{2t} = 1, X_{2t+1} = 0, \forall t\}$. If (X_t) is ergodic and stationary, the empirical means $n^{-1}\sum_{t=1}^{n} \mathbb{1}_{X_{2t+1}=1}$ and $n^{-1}\sum_{t=1}^{n} \mathbb{1}_{X_{2t}=1}$ both converge to the same limit $P[X_t = 1]$ with probability 1, by the ergodic theorem. For all $\omega \in \Omega^*$ these means are, respectively, equal to 1 and 0. Thus $P(\Omega^*) = 0$. The probability of such a trajectory is thus equal to zero for any ergodic and stationary process.

1.5 We have $E\epsilon_t = 0$, $\mathrm{Var}\,\epsilon_t = 1$, and $\mathrm{Cov}(\epsilon_t, \epsilon_{t-h}) = 0$ when $h \neq 0$, thus (ϵ_t) is a weak white noise. We also have $\mathrm{Cov}(\epsilon_t^2, \epsilon_{t-1}^2) = E\eta_t^2\eta_{t-1}^4 \cdots \eta_{t-k}^4\eta_{t-k-1}^2 - 1 = 3^k - 1 \neq 0$, thus ϵ_t and ϵ_{t-1} are not independent, which shows that (ϵ_t) is not a strong white noise.

1.6 Assume $h > 0$. Define the random variable $\tilde{\rho}(h) = \tilde{\gamma}(h)/\tilde{\gamma}(0)$, where $\tilde{\gamma}(h) = n^{-1}\sum_{t=1}^{n} \epsilon_t\epsilon_{t-h}$. It is easy to see that $\sqrt{n}\hat{\rho}(h)$ has the same asymptotic variance (and also the same asymptotic distribution) as $\sqrt{n}\tilde{\rho}(h)$. Using $\tilde{\gamma}(0) \to 1$, stationarity, and Lebesgue's theorem, this asymptotic variance is equal to

$$\begin{aligned}
\mathrm{Var}\sqrt{n}\tilde{\gamma}(h) &= n^{-1}\sum_{t,s=1}^{n} \mathrm{Cov}(\epsilon_t\epsilon_{t-h}, \epsilon_s\epsilon_{s-h}) \\
&= n^{-1}\sum_{\ell=-n+1}^{n-1} (n - |\ell|)\mathrm{Cov}(\epsilon_1\epsilon_{1-h}, \epsilon_{1+\ell}\epsilon_{1+\ell-h}) \\
&\to \sum_{\ell=-\infty}^{\infty} \mathrm{Cov}(\epsilon_1\epsilon_{1-h}, \epsilon_{1+\ell}\epsilon_{1+\ell-h}) \\
&= E\epsilon_1^2\epsilon_{1-h}^2 = \begin{cases} 3^{k-h+1} & \text{if } 0 < h \leq k \\ 1 & \text{if } h > k. \end{cases}
\end{aligned}$$

This value can be arbitrarily larger than 1, which is the value of the asymptotic variance of the empirical autocorrelations of a strong white noise.

1.7 It is clear that (ϵ_t^2) is a second-order stationary process. By construction, ϵ_t and ϵ_{t-h} are independent when $h > k$, thus $\gamma_{\epsilon^2}(h) := \mathrm{Cov}(\epsilon_t^2, \epsilon_{t-h}^2) = 0$ for all $h > k$. Moreover, $\gamma_{\epsilon^2}(h) = 3^{k+1-h} - 1$, for $h = 0, \ldots, k$. In view of Theorem 1.2, $\epsilon_t^2 - 1$ thus follows an MA(k) process. In the case $k = 1$, we have

$$\epsilon_t^2 = 1 + u_t + bu_{t-1},$$

where $|b| < 1$ and (u_t) is a white noise of variance σ^2. The coefficients b and σ^2 are determined by

$$\gamma_{\epsilon^2}(0) = 8 = \sigma^2(1 + b^2), \quad \gamma_{\epsilon^2}(1) = 2 = b\sigma^2,$$

which gives $b = 2 - \sqrt{3}$ and $\sigma^2 = 2/b$.

1.8 Reasoning as in Exercise 1.6, the asymptotic variance is equal to

$$\frac{E\epsilon_1^2\epsilon_{1-h}^2}{(E\epsilon_1^2)^2} = E\frac{\eta_1^2}{\eta_{1-k}^2}\frac{\eta_{1-h}^2}{\eta_{1-h-k}^2}\left(E\frac{\eta_1^2}{\eta_{1-k}^2}\right)^{-2} = \begin{cases}(E\eta_1^2 E\eta_1^{-2})^{-1} & \text{if } 0 < h = k \\ 1 & \text{if } 0 < h \neq k.\end{cases}$$

Since $E(\eta_1^{-2}) \geq (E\eta_1^2)^{-1}$, for $k \neq h$ the asymptotic variance can be arbitrarily smaller than 1, which corresponds to the asymptotic variance of the empirical autocorrelations of a strong white noise.

1.9 1. We have

$$\left\|\sum_{k=m}^{n} a^k\eta_{t-k}\right\|_2 \leq \sum_{k=m}^{n} |a|^k\sigma \to 0$$

when $n > m$ and $m \to \infty$. The sequence $\{u_t(n)\}_n$ defined by $u_n = \sum_{k=0}^{n} a^k\eta_{t-k}$ is a Cauchy sequence in L^2, and thus converges in quadratic mean. A priori,

$$\sum_{k=0}^{\infty}|a^k\eta_{t-k}| := \lim_n \uparrow \sum_{k=0}^{n}|a^k\eta_{t-k}|$$

exists in $\mathbb{R} \cup +\{\infty\}$. Using Beppo Levi's theorem,

$$E\sum_{k=0}^{\infty}|a^k\eta_{t-k}| = (E|\eta_t|)\sum_{k=0}^{\infty}|a^k| < \infty,$$

which shows that the limit $\sum_{k=0}^{\infty}|a^k\eta_{t-k}|$ is finite almost surely. Thus, as $n \to \infty$, $u_t(n)$ converges, both almost surely and in quadratic mean, to $u_t = \sum_{k=0}^{\infty} a^k\eta_{t-k}$. Since

$$u_t(n) = au_{t-1}(n-1) + \eta_t, \quad \forall n,$$

we obtain, taking the limit as $n \to \infty$ of both sides of the equality, $u_t = au_{t-1} + \eta_t$. This shows that $(X_t) = (u_t)$ is a stationary solution of the AR(1) equation.

Finally, assume the existence of two stationary solutions to the equation $X_t = aX_{t-1} + \eta_t$ and $u_t = au_{t-1} + \eta_t$. If $u_{t_0} \neq X_{t_0}$, then

$$0 < |u_{t_0} - X_{t_0}| = |a^n||u_{t_0-n} - X_{t_0-n}|, \quad \forall n,$$

which entails

$$\limsup_{n\to\infty}|u_{t_0-n}| = +\infty \quad \text{or} \quad \limsup_{n\to\infty}|X_{t_0-n}| = +\infty.$$

This is in contradiction to the assumption that the two sequences are stationary, which shows the uniqueness of the stationary solution.

2. We have $X_t = \eta_t + a\eta_{t-1} + \cdots + a^k\eta_{t-k} + a^{k+1}X_{t-k-1}$. Since $|a| = 1$,

$$\text{Var}(X_t - a^{k+1}X_{t-k-1}) = (k+1)\sigma^2 \to \infty$$

as $k \to \infty$. If (X_t) were stationary,

$$\text{Var}(X_t - a^{k+1}X_{t-k-1}) = 2\{\text{Var}X_t \pm \text{Cov}(X_t, X_{t-k-1})\},$$

and we would have

$$\lim_{k\to\infty}|\text{Cov}(X_t, X_{t-k-1})| = \infty.$$

This is impossible, because by the Cauchy–Schwarz inequality,

$$|\text{Cov}(X_t, X_{t-k-1})| \leq \text{Var}X_t.$$

3. The argument used in Part 1 shows that

$$v_t(n) := -\sum_{k=1}^{n} \frac{1}{a^k}\eta_{t+k} \overset{n\to\infty}{\to} v_t = -\sum_{k=1}^{\infty} \frac{1}{a^k}\eta_{t+k}$$

almost surely and in quadratic mean. Since

$$v_t(n) = av_{t-1}(n+1) + \eta_t$$

for all n, (v_t) is a stationary solution (which is called anticipative, because it is a function of the future values of the noise) of the AR(1) equation. The uniqueness of the stationary solution is shown as in Part 1.

4. The autocovariance function of the stationary solution is

$$\gamma(0) = \sigma^2 \sum_{k=1}^{\infty} \frac{1}{a^{2k}} = \frac{\sigma^2}{a^2-1}, \quad \gamma(h) = \frac{1}{a}\gamma(h-1), \ \ h > 0.$$

We thus have $E\epsilon_t = 0$ and, for all $h > 0$,

$$\mathrm{Cov}(\epsilon_t, \epsilon_{t-h}) = \gamma(h) - \frac{1}{a}\gamma(h-1) - \frac{1}{a}\gamma(h+1) + \frac{1}{a^2}\gamma(h) = 0,$$

which confirms that ϵ_t is a white noise.

1.10 In Figure 1.6a, we note that several empirical autocorrelations are outside the 95% significance band, which leads us to think that the series may not be the realisation of a strong white noise. Inspection of Figure 1.6b confirms that the observed series $\epsilon_1, \ldots, \epsilon_n$ cannot be generated by a strong white noise; otherwise, the series $\epsilon_1^2, \ldots, \epsilon_n^2$ would also be uncorrelated. Clearly, this is not the case, because several empirical autocorrelations go far beyond the significance band. By contrast, it is plausible that the series is a weak noise. We know that Bartlett's formula giving the limits $\pm 1.96/\sqrt{n}$ is not valid for a weak noise (see Exercises 1.6 and 1.8). On the other hand, we know that the square of a weak noise can be correlated (see Exercise 1.7).

1.11 Using the relation $\gamma_{\epsilon^2}(0) = (\eta_\epsilon - 1)\gamma_\epsilon^2(0)$, formula (B.18) can be written as

$$\begin{aligned}\lim_{n\to\infty} n\mathrm{Cov}\{\hat{\gamma}_X(i), \hat{\gamma}_X(j)\} &= \{(\eta_\epsilon - 1)\rho_{\epsilon^2} - 2\eta_\epsilon\}\gamma_X(i)\gamma_X(j) \\ &\quad + \gamma_\epsilon^{-2}(0) \sum_{\ell=-\infty}^{\infty} \gamma_X(\ell)\{\gamma_X(\ell + j - i) + \gamma_X(\ell - j - i)\}\{\gamma_{\epsilon^2}(i-\ell) + \gamma_\epsilon^2(0)\} \\ &= (\rho_{\epsilon^2} - 3)(\eta_\epsilon - 1)\gamma_X(i)\gamma_X(j) + (\eta_\epsilon - 3)\gamma_X(i)\gamma_X(j) \\ &\quad + \sum_{\ell=-\infty}^{\infty} \gamma_X(\ell)\{\gamma_X(\ell + j - i) + \gamma_X(\ell - j - i)\} \\ &\quad + (\eta_\epsilon - 1) \sum_{\ell=-\infty}^{\infty} \gamma_X(\ell)\{\gamma_X(\ell + j - i) + \gamma_X(\ell - j - i)\}\rho_{\epsilon^2}(i-\ell).\end{aligned}$$

With the change of index $h = i - \ell$, we obtain

$$\begin{aligned}&\sum_{\ell=-\infty}^{\infty} \gamma_X(\ell)\{\gamma_X(\ell + j - i) + \gamma_X(\ell - j - i)\}\rho_{\epsilon^2}(i-\ell) \\ &\quad = \sum_{h=-\infty}^{\infty} \gamma_X(-h+i)\{\gamma_X(-h+j) + \gamma_X(-h-j)\}\rho_{\epsilon^2}(h),\end{aligned}$$

which gives (B.14), using the parity of the autocovariance functions.

1.12 We can assume $i \geq 0$ and $j \geq 0$. Since $\gamma_X(\ell) = \gamma_\epsilon(\ell) = 0$ for all $\ell \neq 0$, formula (B.18) yields

$$\lim_{n\to\infty} n\mathrm{Cov}\{\hat{\gamma}_X(i), \hat{\gamma}_X(j)\} = \gamma_\epsilon^{-1}(0)\{\gamma_X(j-i) + \gamma_X(-j-i)\}\{\gamma_{\epsilon^2}(i) + \gamma_\epsilon^2(0)\}$$

for $(i, j) \neq (0, 0)$ and

$$\lim_{n\to\infty} n\mathrm{Cov}\{\hat{\gamma}_X(0), \hat{\gamma}_X(0)\} = \{(\eta_\epsilon - 1)\rho_{\epsilon^2} - 2\eta_\epsilon\}\gamma_\epsilon^2(0) + 2\{\gamma_{\epsilon^2}(i) + \gamma_\epsilon^2(0)\}.$$

Thus

$$\lim_{n\to\infty} n\mathrm{Cov}\{\hat{\gamma}_X(i), \hat{\gamma}_X(j)\} = \begin{cases} 0 & \text{if } i \neq j \\ E\epsilon_t^2\epsilon_{t+i}^2 & \text{if } i = j \neq 0 \\ \gamma_{\epsilon^2}(0)\rho_{\epsilon^2} - 2E\epsilon_t^4 + 2E\epsilon_t^2\epsilon_{t+i}^2 & \text{if } i = j = 0. \end{cases}$$

In formula (B.15), we have $v_{ij} = 0$ when $i \neq j$ and $v_{ii} = 1$. We also have $v_{ij}^* = 0$ when $i \neq j$ and $v_{ii}^* = (\eta_\epsilon - 1)\rho_{\epsilon^2}(i)$ for all $i \neq 0$. Since $(\eta_\epsilon - 1) = \gamma_{\epsilon^2}(0)\gamma_\epsilon^{-2}(0)$, we obtain

$$\lim_{n\to\infty} n\mathrm{Cov}\{\hat{\rho}_X(i), \hat{\rho}_X(j)\} = \begin{cases} 0 & \text{if } i \neq j \\ \frac{E\epsilon_t^2\epsilon_{t+i}^2}{\gamma_\epsilon^2(0)} & \text{if } i = j \neq 0. \end{cases}$$

For significance intervals C_h of asymptotic level $1 - \alpha$, such that $\lim_{n\to\infty} P[\hat{\rho}(h) \in C_h] = 1 - \alpha$, we have

$$M = \sum_{h=1}^{m} \mathbb{1}_{\hat{\rho}(h)\in C_h}.$$

By definition of C_h,

$$E(M) = \sum_{h=1}^{m} P[\hat{\rho}(h) \in C_h] \to m(1 - \alpha), \quad \text{as } n \to \infty.$$

Moreover,

$$\begin{aligned} \mathrm{Var}(M) &= \sum_{h=1}^{m} \mathrm{Var}(\mathbb{1}_{\hat{\rho}(h)\in C_h}) + \sum_{h\neq h'} \mathrm{Cov}(\mathbb{1}_{\hat{\rho}(h)\in C_h}, \mathbb{1}_{\hat{\rho}(h')\in C_{h'}}) \\ &= \sum_{h=1}^{m} P(\hat{\rho}(h) \in C_h)\{1 - P(\hat{\rho}(h) \in C_h)\} \\ &\quad + \sum_{h\neq h'} \{P(\hat{\rho}(h) \in C_h, \hat{\rho}(h') \in C_{h'}) - P(\hat{\rho}(h) \in C_h)P(\hat{\rho}(h') \in C_{h'})\} \\ &\to m\alpha(1 - \alpha), \quad \text{as } n \to \infty. \end{aligned}$$

We have used the convergence in law of $(\hat{\rho}(h), \hat{\rho}(h'))$ to a vector of independent variables. When the observed process is not a noise, this asymptotic independence does not hold in general.

1.13 The probability that all the empirical autocorrelations stay within the asymptotic significance intervals (with the notation of the solution to Exercise 1.12) is, by the asymptotic independence,

$$P[\hat{\rho}(1) \in C_1, \ldots, \hat{\rho}(m) \in C_m] \to (1 - \alpha)^m, \quad \text{as } n \to \infty.$$

For $m = 20$ and $\alpha = 5\%$, this limit is equal to 0.36. The probability of not rejecting the right model is thus low.

1.14 In view of (B.7), we have $r_X(1) = \rho_X(1)$. Using step (B.8) with $k = 2$ and $a_{1,1} = \rho_X(1)$, we obtain

$$r_X(2) = a_{2,2} = \frac{\rho_X(2) - \rho_X^2(1)}{1 - \rho_X^2(1)}.$$

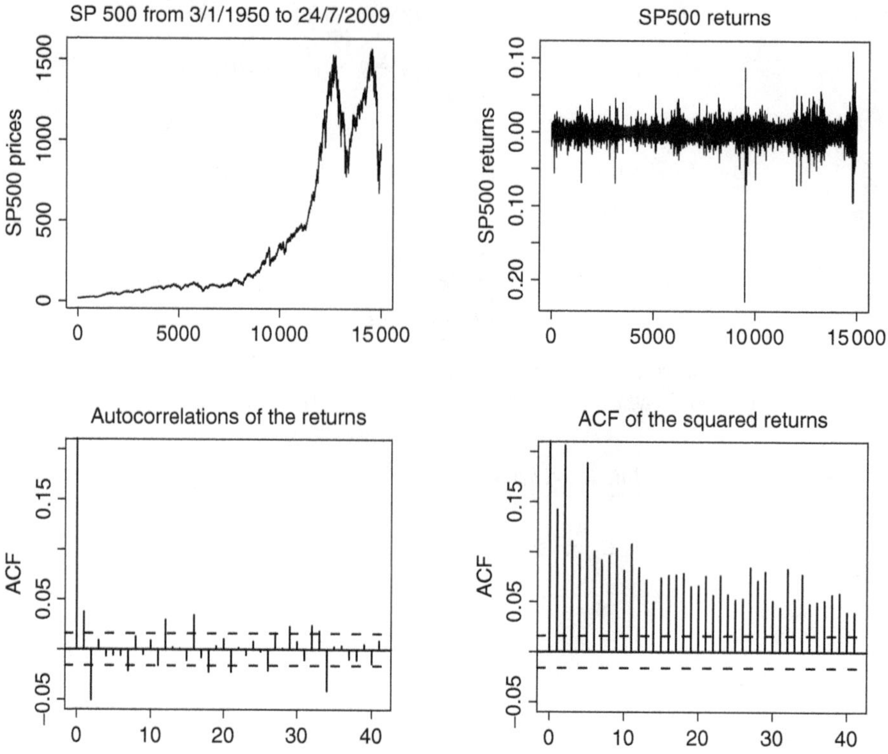

Figure E.1 Closing prices and returns of the S&P 500 index from 3 January 1950 to 24 July 2009.

Then, step (B.9) yields

$$\begin{aligned} a_{2,1} &= \rho_X(1) - a_{2,2}\rho_X(1) \\ &= \rho_X(1) - \frac{\rho_X(2) - \rho_X^2(1)}{1 - \rho_X^2(1)}\rho_X(1) = \rho_X(1)\frac{1 - \rho_X(2)}{1 - \rho_X^2(1)}. \end{aligned}$$

Finally, step (B.8) yields

$$\begin{aligned} r_X(3) &= \frac{\rho_X(3) - \rho_X(2)a_{2,1} - \rho_X(1)a_{2,2}}{1 - \rho_X(1)a_{2,1} - \rho_X(2)a_{2,2}} \\ &= \frac{\rho_X(3)\{1 - \rho_X^2(1)\} - \rho_X(2)\rho_X(1)\{1 - \rho_X(2)\} - \rho_X(1)\{\rho_X(2) - \rho_X^2(1)\}}{1 - \rho_X^2(1) - \rho_X^2(1)\{1 - \rho_X(2)\} - \rho_X(2)\{\rho_X(2) - \rho_X^2(1)\}}. \end{aligned}$$

1.15 The historical data from 3 January 1950 to 24 July 2009 can be downloaded via the URL: http://fr.finance.yahoo.com/q/hp?s = %5EGSPC. We obtain Figure E.1 with the following R code:

```
> # reading the SP500 data set
> sp500data <- read.table("sp500.csv",header=TRUE,sep=",")
> sp500<-rev(sp500data$Close) # closing price
> n<-length(sp500)
> rend<-log(sp500[2:n]/sp500[1:(n-1)]); rend2<-rend^2
```

```
> op <- par(mfrow = c(2, 2)) # 2 x 2 figures per page
> plot(ts(sp500),main="SP 500 from 1/3/50 to 7/24/09",
+                              ylab="SP500 Prices",xlab="")
> plot(ts(rend),main="SP500 Returns",ylab="SP500 Returns",
+                              xlab="")
> acf(rend, main="Autocorrelations of the returns",xlab="",
+                              ylim=c(-0.05,0.2))
> acf(rend2, main="ACF of the squared returns",xlab="",
+                              ylim=c(-0.05,0.2))
> par(op)
```

Chapter 2

2.1 This covariance is meaningful only if $E\epsilon_t^2 < \infty$ and $Ef^2(\epsilon_{t-h}) < \infty$. Under these assumptions, the equality is true and follows from $E(\epsilon_t \mid \epsilon_u, u < t) = 0$.

2.2 In case (i) the strict stationarity condition becomes $\alpha + \beta < 1$. In case (ii) elementary integral computations show that the condition is

$$2\sqrt{\frac{\beta}{3\alpha}} \arctan \sqrt{\frac{3\alpha}{\beta}} + \log(3\alpha + \beta) < 2.$$

2.3 Let $\lambda_1, \ldots, \lambda_m$ be the eigenvalues of A. If A is diagonalisable, there exists an invertible matrix P and a diagonal matrix D such that $A = P^{-1}DP$. It follows that, taking a multiplicative norm,

$$\begin{aligned}\log \|A^t\| &= \log \|P^{-1}D^tP\| \leq \log \|P^{-1}\|\|D^t\|\|P\| \\ &= \log \|P^{-1}\| + \log \|D^t\| + \log \|P\|.\end{aligned}$$

For the multiplicative norm $\|A\| = \sum \mid a_{ij}\mid$, we have $\log \|D^t\| = \log \sum_{i=1}^m \lambda_i^t$. The result follows immediately.

When A is any square matrix, the Jordan representation can be used. Let n_i be the multiplicity of the eigenvalue λ_i. We have the Jordan canonical form $A = P^{-1}JP$, where P is invertible, and J is the block-diagonal matrix with a diagonal of m matrices $J_i(\lambda_i)$, of size $n_i \times n_i$, with λ_i on the diagonal, 1 on the superdiagonal, and 0 elsewhere. It follows that $A^t = P^{-1}J^tP$, where J^t is the block-diagonal matrix whose blocks are the matrices $J_i^t(\lambda_i)$. We have $J_i(\lambda_i) = \lambda_i I_{n_i} + N_i$, where N_i is such that $N_i^{n_i} = 0_{n_i \times n_i}$. It can be assumed that $|\lambda_1 \mid > \mid \lambda_2 \mid > \cdots > \mid \lambda_m \mid$. It follows that

$$\begin{aligned}\frac{1}{t}\log \|J^t\| &= \frac{1}{t}\log \sum_{i=1}^m \|J_i^t(\lambda_i)\| = \frac{1}{t}\log \sum_{i=1}^m \|(\lambda_i I_{n_i} + N_i)^t\| \\ &= |\lambda_1| + \frac{1}{t}\log \sum_{i=1}^m \left\| \left(\frac{\lambda_i}{\lambda_1} I_{n_i} + \frac{1}{\lambda_1} N_i\right)^t \right\| \\ &= |\lambda_1| + \frac{1}{t}\log \sum_{i=1}^m \sum_{k=0}^{n_i-1} \binom{t}{k} \left|\frac{\lambda_i}{\lambda_1}\right|^{t-k} \frac{1}{|\lambda_1|^k}\|N_i^k\| \\ &\to |\lambda_1|\end{aligned}$$

as $t \to \infty$, and the proof easily follows.

2.4 We use the multiplicative norm $\|A\| = \sum |a_{ij}|$. Thus $\log\|Az_t\| \le \log\|A\| + \log\|z_t\|$; therefore, $\log^+\|Az_t\| \le \log^+\|A\| + \log^+|z_t|$, which admits a finite expectation by assumption. It follows that γ exists. We have

$$\log(\|A_tA_{t-1}\dots A_1\|) = \left(\log\|A^t\|\prod_{i=1}^t z_i\right) = \log\|A^t\| + \sum_{i=1}^t \log|z_i|$$

and thus

$$\gamma = \lim_{t\to\infty} \text{a.s.}\left(\frac{1}{t}\log\|A^t\| + \frac{1}{t}\sum_{i=1}^t \log|z_i|\right).$$

Using Eq. (2.21) and the ergodic theorem, we obtain

$$\gamma = \log\,\rho(A) + E\,\log|z_t|.$$

Consequently, $\gamma < 0$ if and only if $\rho(A) < \exp(-E\log|z_t|)$.

2.5 To show 1, first note that, by stationarity, we have $E\|A_tA_{t-1}\dots A_1\| = E\|A_{-1}A_{-2}\dots A_{-t}\|$. The replacement can thus be done in (2.22). To show that it can also be done in (2.23), let us apply Theorem 2.3 to the sequence (B_t) defined by $B_t = A'_{-t}$. Noting that $\|B'_1B'_2\dots B'_t\| = \|B_tB_{t-1}\dots B_1\|$, we have

$$\lim_{t\to\infty}\text{a.s.}\ \frac{1}{t}\log\|A_{-1}A_{-2}\dots A_{-t}\| = \lim_{t\to\infty}\text{a.s.}\ \frac{1}{t}\log\|B_tB_{t-1}\dots B_1\|$$

$$= \inf_{t\in\mathbb{N}^*}\frac{1}{t}E(\log\|B_tB_{t-1}\dots B_1\|) = \inf_{t\in\mathbb{N}^*}\frac{1}{t}E(\log\|A_{-1}A_{-2}\dots A_{-t}\|) = \gamma,$$

which completes the proof of 1.

We have shown that, for any t, the stationary sequences $(A_n)_{n\in\mathbb{Z}}$ and $(A_{n+t})_{n\in\mathbb{Z}}$ have the same top Lyapunov exponent, *i.e.*

$$\gamma = \lim_{n\to\infty}\text{a.s.}\ \frac{1}{n}\log\|A_{t-1}A_{-2}\dots A_{t-n}\|.$$

The convergence follows by showing that $\lim_{n\to\infty}\text{a.s.}\ n^{-1}\log\|\underline{b}_{t-n-1}\| = 0$.

2.6 For the Euclidean norm, multiplicativity follows from the Cauchy–Schwarz inequality. Since $N(A) = \sup_{x\neq 0}\|Ax\|/\|x\|$, we have

$$N(AB) = \sup_{x\neq 0, Bx\neq 0}\frac{\|ABx\|}{\|Bx\|}\frac{\|Bx\|}{\|x\|} \le \sup_{x\neq 0, Bx\neq 0}\frac{\|ABx\|}{\|Bx\|}\sup_{x\neq 0}\frac{\|Bx\|}{\|x\|} \le N(A)N(B).$$

To show that the norm N_1 is not multiplicative, consider the matrix A whose elements are all equal to 1: we then have $N_1(A) = 1$ but $N_1(A^2) > 1$.

2.7 We have

$$A_t^* = \begin{pmatrix}
\beta_1+\alpha_1\eta_{t-1}^2 & \beta_2 & \cdots & & \beta_p & \alpha_2 & & \cdots & & \alpha_q\\
1 & 0 & \cdots & & 0 & 0 & & \cdots & & 0\\
0 & 1 & \cdots & & 0 & 0 & & \cdots & & 0\\
\vdots & \ddots & \ddots & & \vdots & \vdots & \ddots & \ddots & & \vdots\\
0 & & \dots & 1 & 0 & 0 & & \dots & 0 & 0\\
\eta_{t-1}^2 & 0 & \cdots & & & 0 & & \cdots & & 0\\
0 & & \cdots & & 0 & 1 & 0 & \cdots & & 0\\
0 & & \cdots & & 0 & 0 & 1 & \cdots & & 0\\
\vdots & \ddots & \ddots & & \vdots & \vdots & \ddots & \ddots & & \vdots\\
0 & & \dots & 0 & 0 & 0 & & \dots & 1 & 0
\end{pmatrix}$$

and $\underline{b}_t^* = (\omega, 0, \dots, 0)'$.

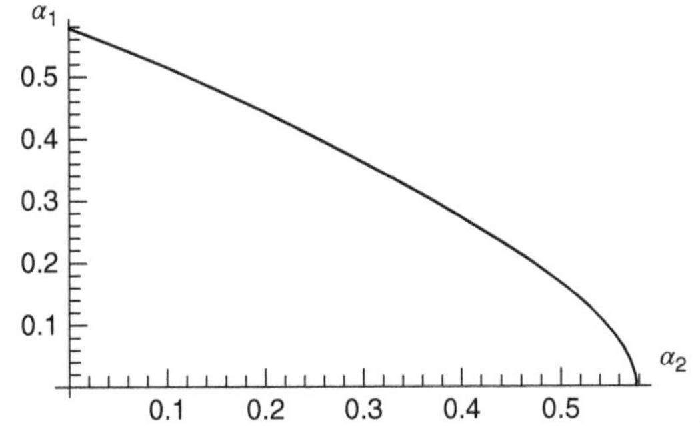

Figure E.2 Region of existence of the fourth-order moment for an ARCH(2) model (when $\mu_4 = 3$).

2.8 We have $\epsilon_t = \sqrt{\omega + \alpha_1 \epsilon_{t-1}^2 + \alpha_2 \epsilon_{t-2}^2}\eta_t$, therefore, under the condition $\alpha_1 + \alpha_2 < 1$, the moment of order 2 is given by

$$E\epsilon_t^2 = \frac{\omega}{1 - \alpha_1 - \alpha_2}$$

(see Theorem 2.5 and Remark 2.6(1)). The strictly stationary solution satisfies

$$\begin{aligned} E\epsilon_t^4 &= \mu_4 E(\omega + \alpha_1 \epsilon_{t-1}^2 + \alpha_2 \epsilon_{t-2}^2)^2 \\ &= \mu_4 \{\omega^2 + (\alpha_1^2 + \alpha_2^2) E\epsilon_t^4 + 2\omega(\alpha_1 + \alpha_2) E\epsilon_t^2 + 2\alpha_1\alpha_2 E\epsilon_t^2 \epsilon_{t-1}^2\} \end{aligned}$$

in $\mathbb{R} \cup \{+\infty\}$. Moreover,

$$E\epsilon_t^2 \epsilon_{t-1}^2 = E(\omega + \alpha_1 \epsilon_{t-1}^2 + \alpha_2 \epsilon_{t-2}^2)\epsilon_{t-1}^2 = \omega E\epsilon_t^2 + \alpha_1 E\epsilon_t^4 + \alpha_2 E\epsilon_{t-2}^2 \epsilon_{t-1}^2,$$

which gives

$$(1 - \alpha_2) E\epsilon_t^2 \epsilon_{t-1}^2 = \omega E\epsilon_t^2 + \alpha_1 E\epsilon_t^4.$$

Using this relation in the previous expression for $E\epsilon_t^4$, we obtain

$$\begin{aligned} E\epsilon_t^4 &\left[1 - \frac{\mu_4}{1 - \alpha_2}\{\alpha_1^2(1 + \alpha_2) + \alpha_2^2(1 - \alpha_2)\}\right] \\ &= \mu_4 \left[\omega^2 + \frac{2\omega^2}{(1 - \alpha_2)(1 - \alpha_1 - \alpha_2)}\{\alpha_1 + \alpha_2(1 - \alpha_2)\}\right]. \end{aligned}$$

If $E\epsilon_t^4 < \infty$, then the term in brackets on the left-hand side of the equality must be strictly positive, which gives the condition for the existence of the fourth-order moment. Note that the condition is not symmetric in α_1 and α_2. In Figure E.2, the points (α_1, α_2) under the curve correspond to ARCH(2) models with a fourth-order moment. For these models,

$$E\epsilon_t^4 = \frac{\mu_4 \omega^2 (1 + \alpha_1 + \alpha_1\alpha_2 - \alpha_2^2)}{(1 - \alpha_1 - \alpha_2)[1 - \alpha_2 - \mu_4\{\alpha_1^2(1 + \alpha_2) + \alpha_2^2(1 - \alpha_2)\}]}.$$

2.9 We have seen that (ϵ_t^2) admits the ARMA(1, 1) representation

$$\epsilon_t^2 - (\alpha + \beta)\epsilon_{t-1}^2 = \omega + \nu_t - \beta\nu_{t-1},$$

where $\nu_t = \epsilon_t^2 - E(\epsilon_t^2 \mid \epsilon_u, u < t)$ is a (weak) white noise. The author correlation of ϵ_t^2 thus satisfies

$$\rho_{\epsilon^2}(h) = (\alpha + \beta)\rho_{\epsilon^2}(h - 1), \quad \forall h > 1. \tag{E.1}$$

Using the MA(∞) representation

$$\epsilon_t^2 = \frac{\omega}{1-\alpha-\beta} + v_t + \alpha \sum_{i=1}^{\infty} (\alpha+\beta)^{i-1} v_{t-i},$$

we obtain

$$\gamma_{\epsilon^2}(0) = Ev_t^2 \left(1 + \alpha^2 \sum_{i=1}^{\infty} (\alpha+\beta)^{2(i-1)} \right) = Ev_t^2 \left(1 + \frac{\alpha^2}{1-(\alpha+\beta)^2} \right)$$

and

$$\gamma_{\epsilon^2}(1) = Ev_t^2 \left(\alpha + \alpha^2(\alpha+\beta) \sum_{i=1}^{\infty} (\alpha+\beta)^{2(i-1)} \right) = Ev_t^2 \left(\alpha + \frac{\alpha^2(\alpha+\beta)}{1-(\alpha+\beta)^2} \right).$$

It follows that the lag 1 autocorrelation is

$$\rho_{\epsilon^2}(1) = \frac{\alpha(1-\beta^2-\alpha\beta)}{1-\beta^2-2\alpha\beta}.$$

The other autocorrelations are obtained from (E.1) and $\rho_{\epsilon^2}(1)$. To determine the autocovariances, all that remains is to compute

$$Ev_t^2 = E(\epsilon_t^2 - \sigma_t^2)^2 = E(\eta_t^2 - 1)^2 E\sigma_t^4 = 2E\sigma_t^4,$$

which is given by

$$\begin{aligned} E\sigma_t^4 &= E(\omega + \alpha\epsilon_t^2 + \beta\sigma_t^2)^2 \\ &= \omega^2 + 3\alpha^2 E\sigma_t^4 + \beta^2 E\sigma_t^4 + 2\omega(\alpha+\beta)E\sigma_t^2 + 2\alpha\beta E\sigma_t^4 \\ &= \frac{\omega^2 + 2\omega(\alpha+\beta)\frac{\omega}{1-\alpha-\beta}}{1-3\alpha^2-\beta^2-2\alpha\beta} = \frac{\omega^2(1+\alpha+\beta)}{(1-\alpha-\beta)(1-3\alpha^2-\beta^2-2\alpha\beta)}. \end{aligned}$$

2.10 The vectorial representation $\underline{z}_t = \underline{b}_t + A_t \underline{z}_{t-1}$ is

$$\begin{pmatrix} \epsilon_t^2 \\ \sigma_t^2 \end{pmatrix} = \begin{pmatrix} \omega\eta_t^2 \\ \omega \end{pmatrix} + \begin{pmatrix} \alpha\eta_t^2 & \beta\eta_t^2 \\ \alpha & \beta \end{pmatrix} \begin{pmatrix} \epsilon_{t-1}^2 \\ \sigma_{t-1}^2 \end{pmatrix}.$$

We have

$$\underline{z}^{(1)} = \frac{\omega}{1-\alpha-\beta} \begin{pmatrix} 1 \\ 1 \end{pmatrix}, \quad \underline{b}^{(1)} = \omega \begin{pmatrix} 1 \\ 1 \end{pmatrix},$$

$$EA_t \otimes b_t = Eb_t \otimes A_t = \omega \begin{pmatrix} 3\alpha & 3\beta \\ \alpha & \beta \\ \alpha & \beta \\ \alpha & \beta \end{pmatrix}, \quad A^{(1)} = \begin{pmatrix} \alpha & \beta \\ \alpha & \beta \end{pmatrix}$$

$$\underline{b}^{(2)} = \omega^2 \begin{pmatrix} 3 \\ 1 \\ 1 \\ 1 \end{pmatrix}, \quad A^{(2)} = \begin{pmatrix} 3\alpha^2 & 3\alpha\beta & 3\alpha\beta & 3\beta^2 \\ \alpha^2 & \alpha\beta & \alpha\beta & \beta^2 \\ \alpha^2 & \alpha\beta & \alpha\beta & \beta^2 \\ \alpha^2 & \alpha\beta & \alpha\beta & \beta^2 \end{pmatrix}.$$

The eigenvalues of $A^{(2)}$ are 0, 0, 0 and $3\alpha^2+2\alpha\beta+\beta^2$, thus $I_4 - A^{(2)}$ is invertible (0 is an eigenvalue of $I_4 - A^{(2)}$ if and only if 1 is an eigenvalue of $A^{(2)}$), and the system (2.63) admits a unique solution. We have

$$\underline{b}^{(2)} + (EA_t \otimes \underline{b}_t + E\underline{b}_t \otimes A_t)\underline{z}^{(1)} = \omega^2 \frac{1+\alpha+\beta}{1-\alpha-\beta}\begin{pmatrix}3\\1\\1\\1\end{pmatrix}.$$

The solution to Eq. (2.63) is

$$\underline{z}^{(2)} = \frac{\omega^2(1+\alpha+\beta)}{(1-\alpha-\beta)(1-3\alpha^2-2\alpha\beta-\beta^2)}\begin{pmatrix}3\\1\\1\\1\end{pmatrix}.$$

As first component of this vector, we recognise $E\epsilon_t^4$, and the other three components are equal to $E\sigma_t^4$. Equation (2.64) yields

$$E\underline{z}_t \otimes \underline{z}_{t-h} = \frac{\omega^2}{1-\alpha-\beta}\begin{pmatrix}1\\1\\1\\1\end{pmatrix} + \begin{pmatrix}\alpha & 0 & \beta & 0\\ 0 & \alpha & 0 & \beta\\ \alpha & 0 & \beta & 0\\ 0 & \alpha & 0 & \beta\end{pmatrix} E\underline{z}_t \otimes \underline{z}_{t-h+1},$$

which gives $\gamma_{\epsilon^2}(\cdot)$, but with tedious computations, compared to the direct method utilised in Exercise 2.8.

2.11 It suffices to show that $\limsup_{n\to\infty} a_n/n \le a_k/k$ for all fixed $k \ge 1$. Let $M_1 = 0$ and $M_k = \max\{a_1, \dots, a_{k-1}\}$ for $k \ge 2$. For all n, write $n = p_n k + q_n$ with $0 \le q_n < k$. We have

$$\frac{a_n}{n} \le \frac{a_{p_n k}}{n} + \frac{a_{q_n}}{n} \le \frac{p_n a_k}{p_n k} + \frac{M_k}{n} \to \frac{a_k}{k} \text{as } n \to \infty,$$

and the result follows.

2.12 1. Subtracting the $(q+1)$th line of $(\lambda I_{p+q} - A)$ from the first, then expanding the determinant along the first row, and using Eq. (2.32), we obtain

$$\det(\lambda I_{p+q} - A) = \lambda^q \begin{vmatrix} \lambda-\beta_1 & -\beta_2 & \cdots & & -\beta_p \\ -1 & \lambda & \cdots & & 0 \\ 0 & -1 & \cdots & & 0 \\ \vdots & \ddots & \ddots & & \vdots \\ 0 & & \cdots & -1 & \lambda \end{vmatrix}$$

$$+ (-1)^q(-\lambda) \begin{vmatrix} -1 & \lambda & \cdots & & 0 \\ 0 & -1 & \cdots & & 0 \\ \vdots & \ddots & \ddots & & \vdots \\ 0 & & \cdots & -1 & \lambda \\ -\alpha_1 & -\alpha_2 & \cdots & & -\alpha_q \end{vmatrix} \lambda^{p-1}$$

$$= \lambda^{p+q}\left(1 - \sum_{j=1}^{p} \beta_j \lambda^{-j}\right)$$

$$+(-1)^{q+1}(-1)^{q-1}\lambda^p \begin{vmatrix} -\alpha_1 & -\alpha_2 & \cdots & & -\alpha_q \\ -1 & \lambda & \cdots & & 0 \\ 0 & -1 & \cdots & & 0 \\ \vdots & \ddots & \ddots & & \vdots \\ 0 & & \cdots & -1 & \lambda \end{vmatrix}$$

$$= \lambda^{p+q}\left(1 - \sum_{j=1}^{p} \beta_j \lambda^{-j}\right)$$

$$+ \lambda^{p+q}\left(1 - (\alpha_1 + \lambda)\lambda^{-1} - \sum_{i=2}^{q} \alpha_i \lambda^{-i}\right),$$

and the result follows.

2. When $\sum_{i=1}^{q} \alpha_i + \sum_{j=1}^{p} \beta_j = 1$ the previous determinant is equal to zero at $\lambda = 1$. Thus $\rho(A) \geq 1$. Now, let λ be a complex number of modulus strictly greater than 1. Using the inequality $|a-b| \geq |a| - |b|$, we then obtain

$$|\det(A - \lambda I_{p+q})| > 1 - \sum_{i=1}^{r} (\alpha_i + \beta_i) = 0.$$

It follows that $\rho(A) \leq 1$ and thus $\rho(A) = 1$.

2.13 For all $\epsilon > 0$, noting that the function $f(t) = P(t^{-1}|X_1| > \epsilon)$ is decreasing, we have

$$\sum_{n=1}^{\infty} P(n^{-1}|X_n| > \epsilon) = \sum_{n=1}^{\infty} P(n^{-1}|X_1| > \epsilon) \leq \int_0^{\infty} P(t^{-1}|X_1| > \epsilon)dt$$

$$= \int_0^{\infty} P(\epsilon^{-1}|X_1| > t)dt = \epsilon^{-1}E|X_1| < \infty.$$

The convergence follows from the Borel–Cantelli lemma.

Now, let (X_n) be an iid sequence of random variables with density $f(x) = x^{-2}\mathbb{1}_{x \geq 1}$. For all $K > 0$, we have

$$\sum_{n=1}^{\infty} P(n^{-1}X_n > K) = \sum_{n=1}^{\infty} \frac{1}{nK} = +\infty.$$

The events $\{n^{-1}X_n > K\}$ being independent, we can use the counterpart of the Borel–Cantelli lemma: the event $\{n^{-1}X_n > K$ for an infinite number of $n\}$ has probability 1. Thus, with probability 1, the sequence $(n^{-1}X_n)$ does not tend to 0.

2.14 First note that the last $r-1$ lines of B_tA are the first $r-1$ lines of A, for any matrix A of appropriate size. The same property holds true when B_t is replaced by $E(B_t)$. It follows that the last $r-1$ lines of $E(B_tA)$ are the last $r-1$ lines of $E(B_t)E(A)$. Moreover, it can be shown, by induction on t, that the ith line $\ell_{i,t-i}$ of $B_t \dots B_1$ is a measurable function of the η_{t-j}, for $j \geq i$. The first line of $B_{t+1}B_t \dots B_1$ is thus of the form $a_1(\eta_t)\ell_{1,t-1} + \cdots + a_r(\eta_{t-r})\ell_{r,t-r}$. Since

$$E\{a_1(\eta_t)\ell_{1,t-1} + \cdots + a_r(\eta_{t-r})\ell_{r,t-r}\} = Ea_1(\eta_t)E\ell_{1,t-1} + \cdots + Ea_r(\eta_{t-r})E\ell_{r,t-r},$$

the first line of $EB_{t+1}B_t \dots B_1$ is thus the product of the first line of EB_{t+1} and of $EB_t \dots B_1$. The conclusion follows.

2.15 1. For any fixed t, the sequence $(\underline{z}_t^{(K)})_K$ converges almost surely (to $\underline{z}_t$) as $K \to \infty$. Thus

$$\|\underline{z}_t^{(K)} - \underline{z}_t^{(K-1)}\| \to 0 \text{ almost surely}$$

and the first convergence follows. Now note that we have

$$\begin{aligned} E\|\underline{z}_t^{(K)} - \underline{z}_t^{(K-1)}\|^s &\leq E(\|\underline{z}_t^{(K)}\| + \|\underline{z}_t^{(K-1)}\|)^s \\ &\leq E\|\underline{z}_t^{(K)}\|^s + E\|\underline{z}_t^{(K-1)}\|^s < \infty. \end{aligned}$$

The first inequality uses $(a+b)^s \leq a^s + b^s$ for $a, b \geq 0$ and $s \in (0, 1]$. The second inequality is a consequence of $E\epsilon_t^{2s} < \infty$. The second convergence then follows from the dominated convergence theorem.

2. We have $\underline{z}_t^{(K)} - \underline{z}_t^{(K-1)} = A_t A_{t-1} \dots A_{t-K+1} \underline{b}_{t-K}$. The convergence follows from the previous question, and from the strict stationarity, for any fixed integer K, of the sequence $\{\underline{z}_t^{(K)} - \underline{z}_t^{(K-1)}, t \in \mathbb{Z}\}$.
3. We have

$$\|X_n Y\|^s = \left\{ \sum_{i,j} |X_{n,ij} Y_j| \right\}^s \geq |X_{n,i'j'} Y_{j'}|^s$$

for any $i' = 1, \dots, \ell, j' = 1, \dots, m$. In view of the independence between X_n and Y, it follows that $E|X_{n,i'j'} Y_{j'}|^s = E|X_{n,i'j'}|^s E|Y_{j'}|^s \to 0$ almost surely as $n \to \infty$. Since $E|Y_{j'}|^s$ is a strictly positive number, we obtain $E|X_{n,i'j'}|^s \to 0$ almost surely, for all i', j'. Using $(a+b)^s \leq a^s + b^s$ once again, it follows that

$$E\|X_n\|^s = E\left\{ \sum_{i,j} |X_{n,ij}| \right\}^s \leq \sum_{i,j} E|X_{n,ij}|^s \to 0.$$

4. Note that the previous question does not allow us to affirm that the convergence to 0 of $E(\|A_k A_{k-1} \dots A_1 \underline{b}_0\|^s)$ entails that of $E(\|A_k A_{k-1} \dots A_1\|^s)$, because $\underline{b}_0$ has zero components. For k large enough, however, we have

$$E(\|A_k A_{k-1} \dots A_1 \underline{b}_0\|^s) = E(\|A_k A_{k-1} \dots A_{N+1} Y\|^s)$$

where $Y = A_N \dots A_1 \underline{b}_0$ is independent of $A_k A_{k-1} \dots A_{N+1}$. The general term $a_{i,j}$ of $A_N \dots A_1$ is the (i, j)th term of the matrix A^N multiplied by a product of η_t^2 variables. The assumption $A^N > 0$ entails $a_{i,j} > 0$ almost surely for all i and j. It follows that the ith component of Y satisfies $Y_i > 0$ almost surely for all i. Thus $EY_i^s > 0$. Now the previous question allows to affirm that $E(\|A_k A_{k-1} \dots A_{N+1}\|^s) \to 0$ and, by strict stationarity, that $E(\|A_{k-N} A_{k-N-1} \dots A_1\|^s) \to 0$ as $k \to \infty$. It follows that there exists k_0 such that $E(\|A_{k_0} A_{k_0-1} \dots A_1\|^s) < 1$
5. If α_1 or β_1 is strictly positive, the elements of the first two lines of the vector $A^2 \underline{b}$ are also strictly positive, together with those of the $(q+1)$th and $(q+2)$th lines. By induction, it can be shown that $A^{\max(p,q)} \underline{b}_0 > 0$ under this assumption.
6. The condition $A_N \underline{b}_0 > 0$ can be satisfied when $\alpha_1 = \beta_1 = 0$. It suffices to consider an ARCH(3) process with $\alpha_1 = 0$, $\alpha_2 > 0$, $\alpha_3 > 0$, and to check that $A_4 \underline{b}_0 > 0$.

2.16 In the case $p = 1$, the condition on the roots of $1 - \beta_1 z$ implies $|\beta| < 1$. The positivity conditions on the ϕ_i yield

$$\begin{aligned} \phi_0 &= \omega/(1-\beta) > 0, \\ \phi_1 &= \alpha_1 \geq 0, \\ \phi_2 &= \beta_1 \alpha_1 + \alpha_2 \geq 0, \\ \phi_{q-1} &= \beta_1^{q-1} \alpha_1 + \beta_1^{q-2} \alpha_2 + \dots + \beta_1 \alpha_{q-1} + \alpha_q \geq 0, \\ \phi_k &= \beta_1^{k-q+1} \phi_{q-1} \geq 0, \quad k \geq q. \end{aligned}$$

The last inequalities imply $\beta_1 \geq 0$. Finally, the positivity constraints are

$$\omega > 0, \qquad 0 \leq \beta_1 < 1, \qquad \sum_{i=1}^{k+1} \alpha_i \beta_1^{k+1-i}, \quad k = 0, \dots, q-1.$$

If $q = 2$, these constraints reduce to

$$\omega > 0, \qquad 0 \leq \beta_1 < 1, \qquad \alpha_1 \geq 0, \qquad \beta_1\alpha_1 + \alpha_2 \geq 0.$$

Thus, we can have $\alpha_2 < 0$.

2.17 Using the ARCH(q) representation of the process (ϵ_t^2), together with Proposition 2.2, we obtain

$$\rho_{\epsilon^2}(i) = \alpha_1\rho_{\epsilon^2}(i-1) + \dots + \alpha_{i-1}\rho_{\epsilon^2}(1) + \alpha_i + \alpha_{i+1}\rho_{\epsilon^2}(1) + \dots + \alpha_q\rho_{\epsilon^2}(q-i) \geq \alpha_i.$$

2.18 Since $\rho_{\epsilon^2}(h) = \alpha_1\rho_{\epsilon^2}(h-1) + \alpha_2\rho_{\epsilon^2}(h-2)$, $h > 0$, we have $\rho_{\epsilon^2}(h) = \lambda r_1^h + \mu r_2^h$ where λ, μ are constants and r_1, r_2 satisfy $r_1 + r_2 = \alpha_1$, $r_1 r_2 = -\alpha_2$. It can be assumed that $r_2 < 0$ and $r_1 > 0$, for instance. A simple computation shows that, for all $h \geq 0$,

$$\rho_{\epsilon^2}(2h+1) < \rho_{\epsilon^2}(2h) \quad \Longleftrightarrow \quad \mu(r_2 - 1)r_2^{2h} < \lambda(1 - r_1)r_1^{2h}.$$

If the last equality is true, it remains true when h is replaced by $h+1$ because $r_2^2 < r_1^2$. Since $\rho_{\epsilon^2}(1) < \rho_{\epsilon^2}(0)$, it follows that $\rho_{\epsilon^2}(2h+1) < \rho_{\epsilon^2}(2h)$ for all $h \geq 0$. Moreover,

$$\rho_{\epsilon^2}(2h) > \rho_{\epsilon^2}(2h-1) \quad \Longleftrightarrow \quad \mu(r_2 - 1)r_2^{2h-1} > \lambda(1 - r_1)r_1^{2h-1}.$$

Since $r_2^2 < r_1^2$, if $\rho_{\epsilon^2}(2) < \rho_{\epsilon^2}(1)$ then we have, for all $h \geq 1$, $\rho_{\epsilon^2}(2h) < \rho_{\epsilon^2}(2h-1)$. We have thus shown that the sequence $\rho_{\epsilon^2}(h)$ is decreasing when $\rho_{\epsilon^2}(2) < \rho_{\epsilon^2}(1)$. If $\rho_{\epsilon^2}(2) \geq \rho_{\epsilon^2}(1) > 0$, it can be seen that for h large enough, say $h \geq h_0$, we have $\rho_{\epsilon^2}(2h) < \rho_{\epsilon^2}(2h-1)$, again because of $r_2^2 < r_1^2$. Thus, the sequence $\{\rho_{\epsilon^2}(h)\}_{h \geq h_0}$ is decreasing.

2.19 Since $X_n + Y_n \to -\infty$ in probability, for all K we have

$$\begin{aligned}
&\mathbb{P}(X_n + Y_n < K) \\
&\leq \mathbb{P}(X_n < K/2 \text{ or } Y_n < K/2) \\
&= \mathbb{P}(Y_n < K/2) + \mathbb{P}(X_n < K/2)\{1 - \mathbb{P}(Y_n < K/2)\} \to 1.
\end{aligned}$$

Since $X_n \nrightarrow -\infty$ in probability, there exist $K_0 \in \mathbb{R}$ and $n_0 \in \mathbb{N}$ such that $P(X_n < K_0/2) \leq \varsigma < 1$ for all $n \geq n_0$. Consequently,

$$\mathbb{P}(Y_n < K/2) + \varsigma\{1 - \mathbb{P}(Y_n < K/2)\} = 1 + (\varsigma - 1)\mathbb{P}(Y_n \geq K/2) \to 1$$

as $n \to \infty$, for all $K \leq K_0$, which entails the result.

2.20 We have

$$\begin{aligned}
&[a(\eta_{t-1}) \dots a(\eta_{t-n})\omega(\eta_{t-n-1})]^{1/n} \\
&= \exp\left[\frac{1}{n}\sum_{i=1}^{n} \log\{a(\eta_{t-i})\} + \omega(\eta_{t-n-1})\right] \to e^{\gamma} \quad \text{a.s.}
\end{aligned}$$

as $n \to \infty$. If $\gamma < 0$, the Cauchy rule entails that

$$h_t = \omega(\eta_{t-1}) + \sum_{i=1}^{\infty} a(\eta_{t-1}) \dots a(\eta_{t-i})\omega(\eta_{t-i-1}),$$

converges almost surely, and the process (ϵ_t), defined by $\epsilon_t = \sqrt{h_t}\eta_t$, is a strictly stationary solution of model (2.7). As in the proof of Theorem 2.1, it can be shown that this solution is unique, non-anticipative and ergodic. The converse is proved by contradiction, assuming that there exists a strictly stationary solution (ϵ_t, σ_t^2). For all $n > 0$, we have

$$\sigma_0^2 \geq \omega(\eta_{-1}) + \sum_{i=1}^{n} a(\eta_{-1}) \ldots a(\eta_{-i})\omega(\eta_{-i-1}).$$

It follows that $a(\eta_{-1}) \ldots a(\eta_{-n})\omega(\eta_{-n-1})$ converges to zero, almost surely, as $n \to \infty$, or, equivalently, that

$$\sum_{i=1}^{n} \log a(\eta_i) + \log \omega(\eta_{-n-1}) \to -\infty \quad \text{a.s. as } n \to \infty. \tag{E.2}$$

We first assume that $E\log\{a(\eta_t)\} > 0$. Then the strong law of large numbers entails $\sum_{i=1}^{n} \log a(\eta_i) \to +\infty$ almost surely. For (E.2) to hold true, it is then necessary that $\log \omega(\eta_{-n-1}) \to -\infty$ almost surely, which is precluded since (η_t) is iid and $\omega(\eta_0) > 0$ almost surely. Assume now that $E\log\{a(\eta_t)\} = 0$. By the Chung–Fuchs theorem, we have $\limsup \sum_{i=1}^{n} \log a(\eta_i) = +\infty$ with probability 1 and, using Exercise 2.17, the convergence (E.2) entails $\log \omega(\eta_{-n-1}) \to -\infty$ in probability, which, as in the previous case, entails a contradiction.

2.21 Letting $a(z) = \lambda + (1 - \lambda)z^2$, we have

$$\sigma_t^2 = a(\eta_{t-1})\sigma_{t-1}^2 = a(\eta_{t-1}) \cdots a(\eta_1)\{\lambda\sigma_0^2 + (1 - \lambda)\sigma_0^2\eta_0^2\}.$$

Regardless of the value of $\sigma_0^2 > 0$, fixed or even random, we have almost surely

$$\begin{aligned} \frac{1}{t} \log \sigma_t^2 &= \frac{1}{t} \log\{\lambda\sigma_0^2 + (1 - \lambda)\sigma_0^2\eta_0^2\} + \frac{1}{t} \sum_{k=1}^{t-1} \log a(\eta_k) \\ &\to E \log a(\eta_k) < \log Ea(\eta_k) = 0 \end{aligned}$$

using the law of large numbers and Jensen's inequality. It follows that $\sigma_t^2 \to 0$ almost surely as $t \to \infty$.

2.22 1. Since the ϕ_i are positive and $A_1 = 1$, we have $\phi_i \leq 1$, which shows the first inequality. The second inequality follows by convexity of $x \mapsto x \log x$ for $x > 0$.

2. Since $A_1 = 1$ and $A_p < \infty$, the function f is well defined for $q \in [p, 1]$. We have

$$f(q) = \log \sum_{i=1}^{\infty} \phi_i^q + \log E|\eta_0|^{2q}.$$

The function $q \mapsto \log E|\eta_0|^{2q}$ is convex on $[p,1]$ if, for all $\lambda \in [0, 1]$ and all $q, q^* \in [p, 1]$,

$$\log E|\eta_0|^{2\lambda q+2(1-\lambda)q^*} \leq \lambda \log E|\eta_0|^{2q} + (1 - \lambda) \log E|\eta_0|^{2q^*},$$

which is equivalent to showing that

$$E[X^\lambda Y^{1-\lambda}] \leq [EX]^\lambda [EY]^{1-\lambda},$$

with $X = |\eta_0|^{2q}$, $Y = |\eta_0|^{2q^*}$. This inequality holds true by Hölder's inequality. The same argument is used to show the convexity of $q \mapsto \log \sum_{i=1}^{\infty} \phi_i^q$. It follows that f is convex, as a sum of convex functions. We have $f(1) = 0$ and $f(p) < 0$, thus the left derivative of f at 1 is negative, which gives the result.

3. Conversely, we assume that there exists $p^* \in (0, 1]$ such that $A_{p^*} < \infty$ and that condition (2.52) is satisfied. The convexity of f on $[p^*, 1]$ and (2.52) implies that $f(q) < 0$ for q sufficiently close to 1. Thus condition (2.41) is satisfied. By convexity of f and since $f(1) = 0$, we have $f(q) < 0$ for all $q \in [p, 1]$. It follows that, by Theorem 2.6, $E|\epsilon_t|^q < \infty$ for all $q \in [0, 2]$.

2.23 Since $E(\epsilon_t^2 \mid \epsilon_u, u < t) = \sigma_t^2$, we have $a=0$ and $b=1$. Using condition (2.60), we can easily see that

$$\frac{\text{Var}(\sigma_t^2)}{\text{Var}(\epsilon_t^2)} = \frac{\alpha_1^2}{1 - 2\alpha_1\beta_1 - \beta_1^2} < \frac{1}{\kappa_\eta},$$

since the condition for the existence of $E\epsilon_t^4$ is $1 - 2\alpha_1\beta_1 - \beta_1^2 > \alpha_1^2\kappa_\eta$. Note that when the GARCH effect is weak (that is, α_1 is small), the part of the variance that is explained by this regression is small, which is not surprising. In all cases, the ratio of the variances is bounded by $1/\kappa_\eta$, which is largely less than 1 for most distributions (1/3 for the Gaussian distribution). Thus, it is not surprising to observe disappointing R^2 values when estimating such a regression on real series.

Chapter 3

3.1 Given any initial measure, the sequence $(X_t)_{t\in\mathbb{N}}$ clearly constitutes a Markov chain on $(\mathbb{R}, \mathcal{B}(\mathbb{R}))$, with transition probabilities defined by $P(x, B) = \mathbb{P}(X_1 \in B \mid X_0 = x) = P_\varepsilon(B - \theta x)$.

(a) Since P_ε admits a positive density on $\mathbb{R}$, the probability measure $P(x, .)$ is, for all $x \in E$, absolutely continuous with respect to λ and its density is positive on $\mathbb{R}$. Thus any measure φ which is absolutely continuous with respect to λ is a measure of irreducibility: $\forall x \in E$,

$$\varphi(B) > 0 \Rightarrow \lambda(B) > 0 \Rightarrow P(x, B) > 0.$$

Moreover, λ is a maximal measure of irreducibility.

(b) Assume, for example, that ε_t is uniformly distributed on $[-1, 1]$. If $\theta > 1$ and $X_0 = x_0 > 1/(\theta - 1)$, we have $x_0 < X_1 < X_2 < \dots$, regardless of the ε_t. Thus there exists no irreducibility measure: such a measure should satisfy $\varphi([-\infty, x]) = 0$, for all $x \in \mathbb{R}$, which would imply $\varphi = 0$.

3.2 If (X_n) is strictly stationary, X_1 and X_0 have the same distribution, μ, satisfying

$$\forall B \in \mathcal{B}, \quad \mu(B) = \mathbb{P}(X_0 \in B) = \mathbb{P}(X_1 \in B) = \int P(x, B) d\mu(x).$$

Thus μ is an invariant probability measure.

Conversely, suppose that μ is invariant. Using the Chapman–Kolmogorov relation, by which $\forall t \in \mathbb{N}, \ \forall s, \ 0 \le s \le t, \ \forall x \in E, \ \forall B \in \mathcal{E}$,

$$P^t(x, B) = \int_{y\in E} P^s(x, dy) P^{t-s}(y, B),$$

we obtain

$$\mu(B) = \mathbb{P}[X_1 \in B] = \int_x \left[\int_y P(y, dx)\mu(dy)\right] P(x, B)$$
$$= \int_y \mu(dy) \int_x P(y, dx) P(x, B) = \int \mu(dy) P^2(y, B) = \mathbb{P}[X_2 \in B].$$

Thus, by induction, for all t, $\mathbb{P}[X_t \in B] = \mu(B)$ $(\forall B \in \mathcal{B})$. Using the Markov property, this is equivalent to the strict stationarity of the chain: the distribution of the process $(X_t, X_{t+1}, \dots, X_{t+k})$ is independent of t, for any integer k.

3.3 We have

$$\begin{aligned}\pi(B) &= \lim_{t\to+\infty}\int \mu(dx)P^t(x,B)\\ &= \lim_{t\to+\infty}\int_x \mu(dx)\int_y P^{t-1}(x,dy)P(y,B)\\ &= \int_y P(y,B)\lim_{t\to+\infty}\int_x P^{t-1}(x,dy)\mu(dx)\\ &= \int P(y,B)\pi(dy).\end{aligned}$$

Thus π is invariant. The third equality is an immediate consequence of the Fubini and Lebesgue theorems.

3.4 Assume, for instance, $\theta > 0$. Let $C = [-c, c]$, $c > 0$, and let $\delta = \inf\{f(x); x \in [-(1+\theta)c,\ (1+\theta)c]\}$ We have, for all $A \subset C$ and all $x \in C$,

$$P(x,A) = \int_A f(y-\theta x)d\lambda(y) \geq \delta\lambda(A).$$

Now let $B \in \mathcal{E}$. Then for all $x \in C$,

$$\begin{aligned}P^2(x,B) &= \int_E P(x,dy)P(y,B)\\ &\geq \int_C P(x,dy)P(y,B)\\ &\geq \delta\int_C \lambda(dy)P(y,B) := \nu(B).\end{aligned}$$

The measure ν is non-trivial since $\nu(E) = \delta\lambda(C) = 2\delta c > 0$.

3.5 It is clear that (X_t) constitutes a Feller chain on $\mathbb{R}$. The λ-irreducibility follows from the assumption that the noise has a density which is everywhere positive, as in Exercise 3.1. In order to apply Theorem 3.1, a natural choice of the test function is $V(x) = 1 + |x|$. We have

$$\begin{aligned}E(V(X_t \mid X_{t-1} = x)) &\leq 1 + E(|\theta + b\epsilon_t|)|x| + E(|\epsilon_t|)\\ &:= 1 + K_1|x| + K_2 = K_1 g(x) + K_2 + 1 - K_1.\end{aligned}$$

Thus if $K_1 < 1$, we have, for $K_1 < K < 1$ and for $g(x) > (K_2 + 1 - K_1)/(K - K_1)$,

$$E(V(X_t \mid X_{t-1} = x)) \leq Kg(x).$$

If we put $A = \{x; g(x) = 1 + |x| \leq (K_2 + 1 - K_1)/(K - K_1)\}$, the set A is compact and the conditions of Theorem 3.1 are satisfied, with $1 - \delta = K$.

3.6 By summing the first n inequalities of (3.11) we obtain

$$\sum_{t=0}^{n-1} \mathbf{P}^{t+1}V(x_0) \leq (1-\delta)\sum_{t=0}^{n-1}\mathbf{P}^t V(x_0) + b\sum_{t=0}^{n-1} P^t(x_0, A).$$

It follows that

$$\begin{aligned}b\sum_{t=0}^{n-1} P^t(x_0,A) &\geq \delta\sum_{t=1}^{n-1}\mathbf{P}^t V(x_0) + \mathbf{P}^n V(x_0) - (1-\delta)V(x_0)\\ &\geq (n-1)\delta + 1 - (1-\delta)M,\end{aligned}$$

because $V \geq 1$. Thus, there exists $\kappa > 0$ such that

$$Q_n(x_0, A) = \frac{1}{n}\sum_{t=1}^{n} P^t(x_0, A) \geq \frac{n-1}{nb}\delta + \frac{1-(1-\delta)M}{nb} - \frac{1}{n} > \kappa.$$

Note that the positivity of δ is crucial for the conclusion.

3.7 We have, for any positive continuous function f with compact support,

$$\begin{aligned}\int_E f(y)\pi(dy) &= \lim_{k\to\infty}\int_E f(y)Q_{n_k}(x_0, dy)\\ &= \lim_{k\to\infty}\left\{\int_E \mathbf{P}f(y)Q_{n_k}(x_0, dy)\right.\\ &\quad\left.+\frac{1}{n_k}\int_E f(y)(P(x_0, dy) - P^{n_k+1}(x_0, dy))\right\}\\ &= \lim_{k\to\infty}\int_E \mathbf{P}f(y)Q_{n_k}(x_0, dy)\\ &\geq \int_E \mathbf{P}f(y)\pi(dy).\end{aligned}$$

The inequality is justified by (i) and the fact that $\mathbf{P}f$ is a continuous positive function. It follows that for $f = \mathbb{1}_C$, where C is a compact set, we obtain

$$\pi(C) \geq \int_E P(y, C)\pi(dy)$$

which shows that,

$$\forall B \in \mathcal{E}, \quad \pi(B) \geq \int_E P(y, B)\pi(dy)$$

(that is, π is subvarient) using (ii). If there existed B such that the previous inequality were strict, we should have

$$\begin{aligned}\pi(E) &= \pi(B) + \pi(B^c)\\ &> \int_E P(y, B)\pi(dy) + \int_E P(y, B^c)\pi(dy) = \int_E P(y, E)\pi(dy) = \pi(E),\end{aligned}$$

and since $\pi(E) < \infty$ we arrive at a contradiction. Thus

$$\forall B \in \mathcal{E}, \quad \pi(B) = \int_E P(y, B)\pi(dy),$$

which signifies that π is invariant.

3.8 See Francq and Zakoïan (2006b).

3.9 If $\sup_n nu_n$ were infinite then, for any $K > 0$, there would exist a subscript n_0 such that $n_0 u_{n_0} > K$. Then, using the decrease in the sequence, one would have $\sum_{k=1}^{n_0} u_k \geq n_0 u_{n_0} > K$. Since this should be true for all $K > 0$, the sequence would not converge. This applies directly to the proof of Corollary A.3 with $u_n = \{\alpha_X(n)\}^{\nu/(2+\nu)}$, which is indeed a decreasing sequence in view of point (v) on Section A.3.1.

3.10 We have

$$d_4 = \sum_{\ell=1}^{k-1} |\mathrm{Cov}(X_t X_{t+h}, X_{t-\ell} X_{t+k-\ell})| \leq d_7 + d_8,$$

where

$$d_7 = \sum_{\ell=1}^{k-1} |\mathrm{Cov}(X_{t-\ell}, X_t X_{t+h} X_{t+k-\ell})|,$$

$$d_8 = \sum_{\ell=1}^{k-1} |EX_t X_{t+h} EX_{t-\ell} X_{t+k-\ell}|.$$

Inequality (A.8) shows that d_7 is bounded by

$$K \sum_{\ell=0}^{\infty} \{\alpha_X(\ell)\}^{\nu/(2+\nu)}.$$

By an argument used to deal with d_6, we obtain

$$d_8 \leq K \sup_{k\geq 1} (k-1)\{\alpha_X(k)\}^{\nu/(2+\nu)} < \infty,$$

and the conclusion follows.

3.11 The chain satisfies the Feller condition (i) because

$$E\{g(X_t) \mid X_{t-1} = x\} = \sum_{k=1}^{9} g\left(\frac{x}{10} + \frac{k}{10}\right)\frac{1}{9}$$

is continuous at x when g is continuous.

To show that the irreducibility condition (ii) is not satisfied, consider the set of numbers in [0,1] such that the sequence of decimals is periodic after a certain lag:

$$\mathbb{B} = \{x = 0, u_1 u_2 \ldots \text{ such that } \exists n \geq 1, (u_t)_{t\geq n} \text{ periodic}\}.$$

For all $h \geq 0$, $X_t \in \mathbb{B}$ if and only if $X_{t+h} \in \mathbb{B}$. We thus have,

$$\forall t, \quad P^t(X_t \in \mathbb{B} \mid X_0 = x) = 0 \quad \text{for } x \notin \mathbb{B}$$

and,

$$\forall t, \quad P^t(X_t \in \overline{\mathbb{B}} \mid X_0 = x) = 0 \quad \text{for } x \in \mathbb{B}.$$

This shows that there is no non-trivial irreducibility measure.

The drift condition (iii) is satisfied with, for instance, a measure ϕ such that $\phi([-1, 1]) > 0$, the energy $V(x) = 1 + |x|$ and the compact set $A = [-1, 1]$. Indeed,

$$E(|X_t| + 1 \mid X_{t-1} = x) = E\left(\left|\frac{x}{10} + \frac{u_t}{10}\right| + 1\right) \leq \frac{3}{2} + \frac{|x|}{10} \leq (1-\delta)(1+|x|)$$

provided

$$0 < \delta \leq \frac{\frac{9}{10}|x| - \frac{1}{2}}{1+|x|}.$$

Chapter 4

4.1 Note that σ_t is a measurable function of $\eta^2_{t-1}, \dots, \eta^2_{t-h}$ and of $\epsilon^2_{t-h-1}, \epsilon^2_{t-h-2}, \dots$, that will be denoted by

$$\mathfrak{h}(\eta^2_{t-1}, \dots, \eta^2_{t-h}, \epsilon^2_{t-h-1}, \epsilon^2_{t-h-2}, \dots).$$

Using the independence between η_{t-h} and the other variables of $\mathfrak{h}$, we have, for all h,

$$\begin{aligned}\mathrm{Cov}(\sigma_t, \epsilon_{t-h}) &= E\{E(\sigma_t\epsilon_{t-h} \mid \eta_{t-1}, \dots, \eta_{t-h+1}, \epsilon_{t-h-1}, \epsilon_{t-h-2}, \dots)\}\\ &= E\left\{\int \mathfrak{h}(\eta^2_{t-1}, \dots, \eta^2_{t-h+1}, x^2, \epsilon^2_{t-h-1}, \epsilon^2_{t-h-2}, \dots)x\sigma_{t-h}d\mathbb{P}_\eta(x)\right\} = 0\end{aligned}$$

when the distribution $\mathbb{P}_\eta$ is symmetric.

4.2 A sequence X_i of independent real random variables such that $X_i = 0$ with probability $1/(i+1)$ and $X_i = (i+1)/i$ with probability $1 - 1/(i+1)$ is suitable, because $Y := \prod_{i=1}^{\infty} X_i = 0$ a.s., $EY = 0, EX_i = 1$ and $\prod_{i=1}^{\infty} EX_i = 1$. We have used $P(\lim_n \downarrow A_n) = \lim_n \downarrow P(A_n)$ for any decreasing sequence of events, in order to show that

$$P(Y \neq 0) = \prod_{i=1}^{\infty} P(X_i \neq 0) = \exp\left\{\sum_{i=1}^{\infty} \log(1 - 1/i)\right\} = 0.$$

4.3 By definition,

$$\prod_{i=1}^{\infty} E\exp\{|\lambda_i g(\eta_t)|\} = \lim_{n\to\infty} \prod_{i=1}^{n} E\exp\{|\lambda_i g(\eta_t)|\}$$

and, by continuity of the exponential,

$$\lim_{n\to\infty} \prod_{i=1}^{n} E\exp\{|\lambda_i g(\eta_t)|\} = \exp\left\{\lim_{n\to\infty} \sum_{i=1}^{n} \log E\exp\{|\lambda_i g(\eta_t)|\}\right\}$$

is finite if and only if the series of general term $\log E\exp\{|\lambda_i g(\eta_t)|\}$ converges. Using the inequalities $\exp(EX) \le E\exp(X) \le E\exp(|X|)$, we obtain

$$\lambda_i Eg(\eta_t) \le \log g_\eta(\lambda_i) \le \log E\exp\{|\lambda_i g(\eta_t)|\}.$$

Since the λ_i tend to 0 at an exponential rate and $E|g(\eta_t)| < \infty$, the series of general term $\lambda_i Eg(\eta_t)$ converges absolutely, and we finally obtain

$$\sum_{i=1}^{\infty} |\log g_\eta(\lambda_i)| \le \sum_{i=1}^{\infty} |\lambda_i Eg(\eta_t)| + \sum_{i=1}^{\infty} \log E\exp\{|\lambda_i g(\eta_t)|\},$$

which is finite under condition (4.12).

4.4 Note that (4.13) entails that

$$\epsilon_t^2 = e^{\omega^*}\eta_t^2 \lim_{n\to\infty} \prod_{i=1}^{n} \exp\{\lambda_i g(\eta_{t-i})\} = e^{\omega^*}\eta_t^2 \exp\left\{\lim_{n\to\infty} \sum_{i=1}^{n} \lambda_i g(\eta_{t-i})\right\}$$

with probability 1. The integral of a positive measurable function being always defined in $\mathbb{R}^+ \cup \{+\infty\}$, using Beppo Levi's theorem and then the independence of the η_t, we obtain

$$\begin{aligned}E\epsilon_t^2 &\le Ee^{\omega^*}\eta_t^2 \exp\left\{\lim_{n\to\infty} \uparrow \sum_{i=1}^{n} |\lambda_i g(\eta_{t-i})|\right\}\\ &= e^{\omega^*} \lim_{n\to\infty} \uparrow \prod_{i=1}^{n} E\exp\{|\lambda_i g(\eta_{t-i})|\},\end{aligned}$$

which is of course finite under condition (4.12). Applying the dominated convergence theorem, and bounding the variables $\exp\left\{\sum_{i=1}^{n}\lambda_i g(\eta_{t-i})\right\}$ by the integrable variable $\exp\left\{\sum_{i=1}^{\infty}|\lambda_i g(\eta_{t-i})|\right\}$, we then obtain the desired expression for $E\epsilon_t^2$.

4.5 Denoting by ϕ the density of $\eta \sim \mathcal{N}(0,1)$,

$$Ee^{\lambda|\eta|} = 2\int_0^{\infty} e^{\lambda x}\phi(x)dx = 2\int_0^{\infty} e^{\lambda^2/2}\phi(x-\lambda)dx = 2e^{\lambda^2/2}\Phi(\lambda)$$

and $E|\eta| = \sqrt{2/\pi}$. With the notation $\tau = |\theta| + |\varsigma|$, it follows that

$$Ee^{|\lambda g_\eta(\eta)|} \leq e^{|\lambda\varsigma|E|\eta|}Ee^{|\lambda|\tau|\eta|} = \exp\left(|\lambda\varsigma|\sqrt{\frac{2}{\pi}} + \frac{\lambda^2\tau^2}{2}\right)2\Phi(|\lambda|\tau).$$

It then suffices to use the fact that $\Phi(x)$ is equivalent to $1/2 + x\phi(0)$, and that $\log 2\Phi(x)$ is thus equivalent to $2x\phi(0)$, in a neighborhood of 0.

4.6 We can always assume $\varsigma = 1$. In view of the discussion on pages 79–80, the process $X_t = \log\epsilon_t^2$ satisfies an ARMA(1, 1) representation of the form

$$X_t - \beta_1 X_{t-1} = \omega + u_t + bu_{t-1},$$

where u_t is a white noise with variance σ_u^2. Using $\mathrm{Var}(\log\eta_t^2) = \pi^2/2$ and

$$\mathrm{Var}\{g(\eta_t)\} = 1 - \frac{2}{\pi} + \theta^2, \quad \mathrm{Cov}\{g(\eta_t), \log\eta_t^2\} = \frac{\log 16}{\sqrt{2\pi}},$$

the coefficients $|b| < 1$ and σ_u^2 are such that

$$\begin{aligned}\sigma_u^2(1+b^2) &= \mathrm{Var}\{\log\eta_t^2 + \alpha_1 g(\eta_{t-1}) - \beta_1\log\eta_{t-1}^2\}\\ &= \frac{\pi^2\,(1+\beta_1{}^2)}{2} + \alpha_1^2\left(\theta^2 + 1 - \frac{2}{\pi}\right) - 2\alpha_1\beta_1\frac{\log 16}{\sqrt{2\pi}}\end{aligned}$$

and

$$b\sigma_u^2 = \alpha_1\mathrm{Cov}\{g(\eta_t), \log\eta_t^2\} - \beta_1\mathrm{Var}\{\log\eta_t^2\} = \alpha_1\frac{\log 16}{\sqrt{2\pi}} - \frac{\pi^2\beta_1}{2}.$$

When, for instance, $\omega = 1$, $\beta_1 = 1/2$, $\alpha_1 = 1/2$ and $\theta = -1/2$, we obtain

$$X_t - 0.5X_{t-1} = 1 + u_t - 0.379685u_{t-1}, \quad \mathrm{Var}(u_t) = 0.726849.$$

4.7 In view of Exercise 3.5, an AR(1) process $X_t = aX_{t-1} + \eta_t$, $|a| < 1$, in which the noise η_t has a strictly positive density over $\mathbb{R}$, is geometrically β-mixing. Under the stationarity conditions given in Theorem 4.1, if η_t has a density $f > 0$ and if g (defined in (4.10)) is a continuously differentiable bijection (that is, if $\theta \neq 0$) then $(\log\sigma_t^2)$ is a geometrically β-mixing stationary process. Reasoning as in step (iv) of the proof of Theorem 3.4, it is shown that (η_t, σ_t^2), and then (ϵ_t), are also geometrically β-mixing stationary processes.

4.8 Since $\epsilon_t^+ = \sigma_t\eta_t^+$ and $\epsilon_t^- = \sigma_t\eta_t^-$, we have

$$\sigma_t = \omega + a(\eta_t)\sigma_{t-1}, \quad a(\eta_t) = \alpha_{1,+}\eta_t^+ - \alpha_{1,-}\eta_t^- + \beta_1.$$

If the volatility σ_t is a positive function of $\{\eta_u, u < t\}$ that possesses a moment of order 2, then

$$\{1 - Ea^2(\eta_t)\}E\sigma_t^2 = \omega^2 + 2\omega E\sigma_t Ea(\eta_t) > 0$$

under conditions (4.34). Thus, condition (4.38) is necessarily satisfied. Conversely, under (4.38) the strict stationarity condition is satisfied because

$$E\log a(\eta_t) = \frac{1}{2}E\log a^2(\eta_t) \leq \frac{1}{2}\log Ea^2(\eta_t) < 0,$$

and, as in the proof of Theorem 2.2, it is shown that the strictly stationary solution possesses a moment of order 2.

4.9 Assume the second-order stationarity condition (4.39). Let

$$\begin{aligned} a(\eta_t) &= \alpha_{1,+}\eta_t^+ - \alpha_{1,-}\eta_t^- + \beta_1, \\ \mu_{|\eta|} &= E|\eta_t| = \sqrt{2/\pi}, \\ \mu_{\eta^+} &= E\eta_t^+ = -E\eta_t^- = 1/\sqrt{2\pi}, \\ \mu_{\sigma,\,|\epsilon|}(h) &= E(\sigma_t|\epsilon_{t-h}|), \\ \mu_{|\epsilon|} &= E|\epsilon_t| = \mu_{|\eta|}\mu_\sigma \end{aligned}$$

and

$$\begin{aligned} \mu_\sigma &= E\sigma_t = \frac{\omega}{1 - \frac{1}{\sqrt{2\pi}}(\alpha_{1,+} + \alpha_{1,-}) - \beta_1}, \\ \mu_a &= Ea(\eta_t) = \frac{1}{\sqrt{2\pi}}(\alpha_{1,+} + \alpha_{1,-}) + \beta_1, \\ \mu_{a^2} &= \frac{1}{2}(\alpha_{1,+}^2 + \alpha_{1,-}^2) + \frac{2\beta_1}{\sqrt{2\pi}}(\alpha_{1,+} + \alpha_{1,-}) + \beta_1^2, \\ \mu_{\sigma^2} &= \frac{\omega^2 + 2\omega\mu_a\mu_\sigma}{1 - \mu_{a^2}}. \end{aligned}$$

Using $\sigma_t = \omega + a(\eta_{t-1})\sigma_{t-1}$, we obtain

$$\begin{aligned} \mu_{\sigma,\,|\epsilon|}(h) &= \omega\mu_{|\epsilon|} + \mu_a\mu_{\sigma,\,|\epsilon|}(h-1), \quad \forall h \geq 2, \\ \mu_{\sigma,\,|\epsilon|}(1) &= \omega\mu_{|\eta|}\mu_\sigma + \left\{\frac{1}{2}(\alpha_{1,+} + \alpha_{1,-}) + \sqrt{\frac{2}{\pi}}\beta_1\right\}\mu_{\sigma^2}, \\ \mu_{\sigma,\,|\epsilon|}(0) &= \sqrt{\frac{2}{\pi}}\mu_{\sigma^2}. \end{aligned}$$

We then obtain the autocovariances

$$\gamma_{|\epsilon|}(h) := \mathrm{Cov}(|\epsilon_t|, |\epsilon_{t-h}|) = \mu_{|\eta|}\mu_{\sigma,\,|\epsilon|}(h) - \mu_{|\epsilon|}^2$$

and the autocorrelations $\rho_{|\epsilon|}(h) = \gamma_{|\epsilon|}(h)/\gamma_{|\epsilon|}(0)$. Note that $\gamma_{|\epsilon|}(h) = \mu_a\gamma_{|\epsilon|}(h-1)$ for all $h > 1$, which shows that $(|\epsilon_t|)$ is a weak ARMA(1, 1) process. In the standard GARCH case, the calculation of these autocorrelations would be much more complicated because σ_t is not a linear function of σ_{t-1}.

4.10 This is obvious because an APARCH(1, 1) with $\delta = 1$, $\alpha_{1,+} = \alpha_1(1 - \varsigma_1)$ and $\alpha_{1,-} = \alpha_1(1 + \varsigma_1)$ corresponds to a TGARCH(1, 1).

4.11 This is an EGARCH(1, 0) with $\varsigma = 1$, $\alpha_1(\theta + 1) = \alpha_+$, $\alpha_1(\theta - 1) = -\alpha_-$ and $\omega - \alpha_1 E|\eta_t| = \alpha_0$. It is natural to impose $\alpha_+ \geq 0$ and $\alpha_- \geq 0$, so that the volatility increases with $|\eta_{t-1}|$. It is also natural to impose $\alpha_- > \alpha_+$ so that the effect of a negative shock is more important than the effect of a positive shock of the same magnitude. There always exists a strictly stationary solution

$$\epsilon_t = \eta_t \exp(\alpha_0)\{\exp(\alpha_+ \eta_{t-1})\mathbb{1}_{\{\eta_{t-1}\geq 0\}} + \exp(-\alpha_- \eta_{t-1})\mathbb{1}_{\{\eta_{t-1}<0\}}\},$$

and this solution possesses a moment of order 2 when

$$E\exp(\alpha_+\eta_t)\mathbb{1}_{\{\eta_t>0\}} < \infty \quad \text{and} \quad E\exp(-\alpha_-\eta_t)\mathbb{1}_{\{\eta_t<0\}} < \infty,$$

which is the case, in particular, for $\eta_t \sim \mathcal{N}(0, 1)$. In the Gaussian case, we have

$$E\sigma_t = e^{\alpha_0}\left\{e^{\alpha_+^2/2}\Phi(\alpha_+) + e^{\alpha_-^2/2}\Phi(\alpha_-)\right\},$$

$$E\epsilon_t^2 = E\sigma_t^2 = e^{2\alpha_0}\left\{e^{2\alpha_+^2}\Phi(2\alpha_+) + e^{2\alpha_-^2}\Phi(2\alpha_-)\right\}$$

and

$$\begin{aligned}\mathrm{Cov}(\sigma_t, \epsilon_{t-1}) &= E\sigma_t\eta_{t-1}E\sigma_{t-1}\\ &= e^{\alpha_0}\left[e^{\alpha_+^2/2}\{\phi(\alpha_+) + \alpha_+\Phi(\alpha_+)\} - e^{\alpha_-^2/2}\{\phi(\alpha_-) + \alpha_-\Phi(\alpha_-)\}\right]E\sigma_t,\end{aligned}$$

using the calculations of Exercise 4.5 and

$$\int_0^\infty xe^{\lambda x}\phi(x)dx = e^{\lambda^2/2}\int_{-\lambda}^\infty (y+\lambda)\phi(y)dy = e^{\lambda^2/2}\{\phi(\lambda) + \lambda\Phi(\lambda)\}.$$

Since $x \mapsto \phi(x) + x\Phi(x)$ is an increasing function, provided $\alpha_- > \alpha_+$, we observe the leverage effect $\mathrm{Cov}(\sigma_t, \epsilon_{t-1}) < 0$.

4.12 For all $\epsilon > 0$, we have

$$\sum_{n=1}^{\infty} P\left(\frac{X_n}{n} > \epsilon\right) = \sum_{n=1}^{\infty} P\left(\frac{X_1^+}{\epsilon} > n\right) \leq \int_0^\infty P\left(\frac{X_1^+}{\epsilon} > t\right)dt = \frac{EX_1^+}{\epsilon},$$

and the conclusion follows from the Borel–Cantelli lemma.

4.13 Because the increments of the Brownian motion are independent and Gaussian, we have

$$\begin{aligned}E\sum_{k=1}^{[1/h]}[R_t(kh) - R_t(kh-h)]^2 &= \sigma_t^2\sum_{k=1}^{[1/h]}E[W_t(kh) - W_t(kh-h)]^2\\ &= \sigma_t^2\sum_{k=1}^{[1/h]}h \to \sigma_t^2\end{aligned}$$

and

$$\mathrm{Var}\sum_{k=1}^{[1/h]}[R_t(kh) - R_t(kh-h)]^2 = 2\sigma_t^4\sum_{k=1}^{[1/h]}h^2 = o(1)$$

as $h \to 0$. The conclusion follows.

4.14 We have

$$\frac{\partial \ln f_\sigma(x)}{\partial \sigma^2} = -\frac{1}{2\sigma^2} + \frac{\nu+1}{2\sigma^2}\frac{1}{1+\frac{x^2}{\sigma^2(\nu-2)}}\frac{x^2}{\sigma^2(\nu-2)}.$$

Using the indication, one can check that $E\frac{\partial \ln f_{\sigma_t}(\epsilon_t)}{\partial \sigma_t^2} = 0$ and

$$I_{t-1} = \frac{\nu}{2\sigma_t^4(\nu+3)}.$$

We thus conclude by noting that

$$\frac{1}{I_{t-1}}\frac{\partial \ln f_{\sigma_t}(\epsilon_t)}{\partial \sigma_t^2} = \frac{\nu+3}{\nu}\left\{\frac{\nu+1}{\nu-2+\frac{\epsilon_t^2}{\sigma_t^2}}\epsilon_t^2 - \sigma_t^2\right\}.$$

Chapter 5

5.1 Let $(\mathcal{F}_t)$ be an increasing sequence of σ-fields such that $\epsilon_t \in \mathcal{F}_t$ and $E(\epsilon_t \mid \mathcal{F}_{t-1}) = 0$. For $h > 0$, we have $\epsilon_t\epsilon_{t+h} \in \mathcal{F}_{t+h}$ and

$$E(\epsilon_t\epsilon_{t+h} \mid \mathcal{F}_{t+h-1}) = \epsilon_t E(\epsilon_{t+h} \mid \mathcal{F}_{t+h-1}) = 0.$$

The sequence $(\epsilon_t\epsilon_{t+h}, \mathcal{F}_{t+h})_t$ is thus a stationary sequence of square integrable martingale increments. We thus have

$$n^{1/2}\tilde{\gamma}(h) \xrightarrow{\mathcal{L}} \mathcal{N}(0, E\epsilon_t^2\epsilon_{t+h}^2),$$

where $\tilde{\gamma}(h) = n^{-1}\sum_{t=1}^n \epsilon_t\epsilon_{t+h}$. To conclude,[1] it suffices to note that

$$n^{1/2}\tilde{\gamma}(h) - n^{1/2}\hat{\gamma}(h) = n^{-1/2}\sum_{t=n-h+1}^{n} \epsilon_t\epsilon_{t+h} \to 0$$

in probability (and even in L^2).

5.2 This process is a stationary martingale difference, whose variance is

$$\gamma(0) = E\epsilon_t^2 = \frac{\omega}{1-\alpha}.$$

Its fourth-order moment is

$$E\epsilon_t^4 = \mu_4(\omega^2 + \alpha^2 E\epsilon_t^4 + 2\alpha\omega E\epsilon_t^2).$$

Thus,

$$E\epsilon_t^4 = \frac{\mu_4(\omega^2 + 2\alpha\omega E\epsilon_t^2)}{1-\mu_4\alpha^2} = \frac{\mu_4\omega^2(1+\alpha)}{(1-\alpha)(1-\mu_4\alpha^2)}.$$

Moreover,

$$E\epsilon_t^2\epsilon_{t-1}^2 = E(\omega + \alpha\epsilon_{t-1}^2)\epsilon_{t-1}^2 = \frac{\omega^2}{1-\alpha} + \alpha E\epsilon_t^4.$$

Using Exercise 5.1, we thus obtain

$$n^{1/2}\hat{\gamma}(1) \xrightarrow{\mathcal{L}} \mathcal{N}\left\{0, \frac{\omega^2(1+\alpha\mu_4)}{(1-\alpha)(1-\mu_4\alpha^2)}\right\}.$$

[1] If $X_n \to x$, x constant, and $Y_n \xrightarrow{\mathcal{L}} Y$, then $X_n + Y_n \xrightarrow{\mathcal{L}} x + Y$.

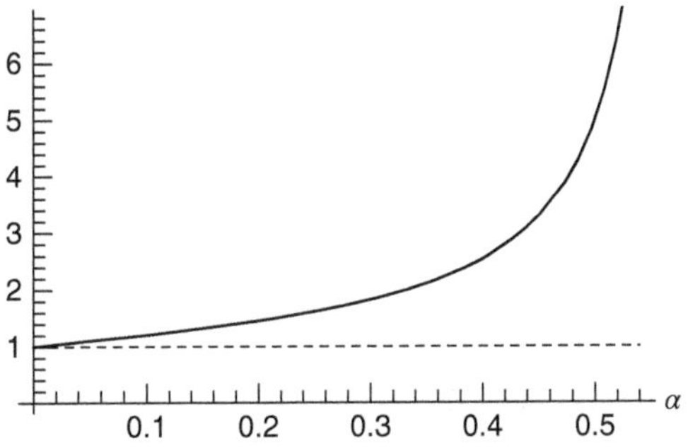

Figure E.3 Comparison between the asymptotic variance of $\sqrt{n}\hat{\rho}(1)$ for the ARCH(1) process (5.28) with (η_t) Gaussian (solid line) and the asymptotic variance of $\sqrt{n}\hat{\rho}(1)$ when ϵ_t is a strong white noise (dashed line).

5.3 We have

$$n^{1/2}\hat{\rho}(1) = \frac{n^{1/2}\hat{\gamma}(1)}{\hat{\gamma}(0)}.$$

By the ergodic theorem, the denominator converges in probability (and even a.s.) to $\gamma_\epsilon(0) = \omega/(1-\alpha) \neq 0$. In view of Exercise 5.2, the numerator converges in law to $\mathcal{N}\left\{0, \frac{\omega^2(1+\alpha\mu_4)}{(1-\alpha)(1-\mu_4\alpha^2)}\right\}$. Cramér's theorem[2] then entails

$$n^{1/2}\hat{\rho}(1) \xrightarrow{\mathcal{L}} \mathcal{N}\left\{0, \frac{(1-\alpha)(1+\alpha\mu_4)}{(1-\mu_4\alpha^2)}\right\}.$$

The asymptotic variance is equal to 1 when $\alpha = 0$ (that is, when ϵ_t is a strong white noise). Figure E.3 shows that the asymptotic distribution of the empirical autocorrelations of a GARCH can be very different from those of a strong white noise.

5.4 Using Exercise 2.8, we obtain

$$E\epsilon_t^2\epsilon_{t+h}^2 = \gamma_{\epsilon^2}(h) + \gamma^2(0).$$

In view of Exercises 5.1 and 5.3,

$$n^{1/2}\hat{\rho}(h) \xrightarrow{\mathcal{L}} \mathcal{N}\{0, E\epsilon_t^2\epsilon_{t+h}^2/\gamma^2(0)\}$$

for any $h \neq 0$.

5.5 Let $\mathcal{F}_t$ be the σ-field generated by $\{\eta_u, u \leq t\}$. If $s+2 > t+1$, then

$$E\epsilon_t\epsilon_{t+1}\epsilon_s\epsilon_{s+2} = E\{E(\epsilon_t\epsilon_{t+1}\epsilon_s\epsilon_{s+2} \mid \mathcal{F}_{s+1})\} = 0.$$

Similarly, $E\epsilon_t\epsilon_{t+1}\epsilon_s\epsilon_{s+2} = 0$ when $t+1 > s+2$. When $t+1 = s+2$, we have

$$\begin{aligned} E\epsilon_t\epsilon_{t+1}\epsilon_s\epsilon_{s+2} &= E\epsilon_{t-1}\epsilon_t\epsilon_{t+1}^2 = E\epsilon_{t-1}\epsilon_t(\omega + \alpha\epsilon_t^2 + \beta\sigma_t^2)\eta_{t+1}^2 \\ &= E\epsilon_{t-1}\sigma_t\eta_t(\omega + \alpha\sigma_t^2\eta_t^2 + \beta\sigma_t^2) = 0, \end{aligned}$$

because $\epsilon_{t-1}\sigma_t \in \mathcal{F}_{t-1}$, $\epsilon_{t-1}\sigma_t^3 \in \mathcal{F}_{t-1}$, $E(\eta_t \mid \mathcal{F}_{t-1}) = E\eta_t = 0$ and $E(\eta_t^3 \mid \mathcal{F}_{t-1}) = E\eta_t^3 = 0$. Using (7.24), the result can be extended to show that $E\epsilon_t\epsilon_{t+h}\epsilon_s\epsilon_{s+k} = 0$ when $k \neq h$ and (ϵ_t) follows a GARCH(p, q), with a symmetric distribution for η_t.

[2] If $Y_n \xrightarrow{\mathcal{L}} Y$ and $T_n \to t$ in probability, t constant, then $T_nY_n \xrightarrow{\mathcal{L}} Yt$.

5.6 Since $E\epsilon_t\epsilon_{t+1} = 0$, we have $\mathrm{Cov}\{\epsilon_t\epsilon_{t+1}, \epsilon_s\epsilon_{s+2}\} = E\epsilon_t\epsilon_{t+1}\epsilon_s\epsilon_{s+2} = 0$ in view of Exercise 5.5. Thus

$$\mathrm{Cov}\{n^{1/2}\hat{\rho}(1), n^{1/2}\hat{\rho}(2)\} = n^{-1}\sum_{t=1}^{n-1}\sum_{s=1}^{n-2}\mathrm{Cov}\{\epsilon_t\epsilon_{t+1}, \epsilon_s\epsilon_{s+2}\} = 0.$$

5.7 In view of Exercise 2.8, we have

$$\gamma_\epsilon(0) = \frac{\omega}{1+\alpha+\beta},$$

$$\gamma_{\epsilon^2}(0) = 2\frac{\omega^2(1+\alpha+\beta)}{(1-\alpha-\beta)(1-3\alpha^2-\beta^2-2\alpha\beta)}\left(1+\frac{\alpha^2}{1-(\alpha+\beta)^2}\right),$$

$$\rho_{\epsilon^2}(1) = \frac{\alpha(1-\beta^2-\alpha\beta)}{1-\beta^2-2\alpha\beta}, \quad \rho_{\epsilon^2}(h) = (\alpha+\beta)\rho_{\epsilon^2}(h-1), \quad \forall h > 1.$$

with $\omega = 1$, $\alpha = 0.3$ and $\beta = 0.55$. Thus $\gamma_\epsilon(0) = 6.667$, $\gamma_{\epsilon^2}(0) = 335.043$, $\rho_{\epsilon^2}(1) = 0.434694$. Thus

$$\frac{E\epsilon_t^2\epsilon_{t+i}^2}{\gamma_\epsilon^2(0)} = \frac{0.85^{i-1}\gamma_{\epsilon^2}(0)\rho_{\epsilon^2}(1) + \gamma_\epsilon^2(0)}{\gamma_\epsilon^2(0)}$$

for $i = 1, \ldots, 5$. Finally, using Theorem 5.1,

$$\lim_{n\to\infty}\mathrm{Var}\sqrt{n}\hat{\rho}_5 = \begin{pmatrix} 4.27692 & 0 & 0 & 0 & 0 \\ 0 & 3.78538 & 0 & 0 & 0 \\ 0 & 0 & 3.36758 & 0 & 0 \\ 0 & 0 & 0 & 3.01244 & 0 \\ 0 & 0 & 0 & 0 & 2.71057 \end{pmatrix}.$$

5.8 Since $\gamma_X(\ell) = 0$, for all $|\ell| > q$, we clearly have

$$v_{i,i} = \sum_{\ell=-q}^{+q}\rho_X^2(\ell).$$

Since $\rho_{\epsilon^2}(\ell) = \alpha^{|\ell|}$, $\gamma_\epsilon(0) = \omega/(1-\alpha)$ and

$$\gamma_{\epsilon^2}(0) = 2\frac{\omega^2}{(1-\alpha)^2(1-3\alpha^2)}$$

(see, for instance, Exercise 2.8), we have

$$v_{i,i}^* = \frac{2}{1-3\alpha^2}\sum_\ell \alpha^{|\ell|}\rho_X(\ell+i)\{\rho_X(\ell+i)+\rho_X(\ell-i)\}$$

$$= \frac{2}{1-3\alpha^2}\sum_{\ell=-q}^{q}\alpha^{i-\ell}\rho_X^2(\ell).$$

Note that $v_{i,i}^* \to 0$ as $i \to \infty$.

5.9 Conditionally on initial values, the score vector is given by

$$\frac{\partial \ell_n(\theta)}{\partial\theta} = -\frac{1}{2}\sum_{t=1}^{n}\frac{\partial}{\partial\theta}\left\{\frac{\epsilon_t^2(\theta)}{\sigma^2} + \log\sigma^2\right\}$$

$$= \begin{pmatrix} \frac{1}{\sigma^2}\sum_{t=1}^n \epsilon_t(\theta)\frac{\partial F_\theta(W_t)}{\partial\beta} + \frac{1}{2}\left(\frac{\epsilon_t^2(\theta)}{\sigma^2} - 1\right)\frac{\partial\sigma^2}{\partial\beta} \\ \frac{1}{\sigma^2}\sum_{t=1}^n \epsilon_t(\theta)\frac{\partial F_\theta(W_t)}{\partial\psi} \end{pmatrix}$$

where $\epsilon_t(\theta) = Y_t - F_\theta(W_t)$. We thus have

$$I_{22} = \frac{1}{\sigma^2} E \left\{ \frac{\partial F_{\theta_0}(W_t)}{\partial \psi} \right\} \left\{ \frac{\partial F_{\theta_0}(W_t)}{\partial \psi'} \right\},$$

and, when σ^2 does not depend on θ,

$$I_1 = \frac{1}{\sigma^2} \left\{ \frac{\partial F_{\theta_0}(W_t)}{\partial \beta} \right\} \left\{ \frac{\partial F_{\theta_0}(W_t)}{\partial \beta'} \right\}, \quad I_{12} = \frac{1}{\sigma^2} E \left\{ \frac{\partial F_{\theta_0}(W_t)}{\partial \beta} \right\} \left\{ \frac{\partial F_{\theta_0}(W_t)}{\partial \psi'} \right\}.$$

5.10 In the notation of Section 5.4.1 and denoting by $\beta = (\beta_1', \beta_2')'$ the parameter of interest, the log-likelihood is equal to

$$\ell_n(\beta) = -\frac{1}{2\sigma^2} \|Y - X_1\beta_1 - X_2\beta_2\|^2,$$

up to a constant. The constrained estimator is $\hat{\beta}^c = (\hat{\beta}_1^{c\prime}, 0')'$, with $\hat{\beta}_1^c = (X_1'X_1)^{-1}X_1'Y$. The constrained score and the Lagrange multiplier are related by

$$\frac{\partial}{\partial \beta} \ell_n(\hat{\beta}^c) = \begin{pmatrix} 0 \\ \hat{\lambda} \end{pmatrix}, \quad \hat{\lambda} = \frac{1}{\sigma^2} X_2'\hat{U}^c, \quad \hat{U}^c = Y - X_1\hat{\beta}_1^c.$$

On the other hand, the exact laws of the estimators under H_0 are given by

$$\sqrt{n}(\hat{\beta} - \beta) \sim \mathcal{N}(0, I^{-1}), \quad I = \frac{1}{n\sigma^2} X'X, \quad X = (X_1, X_2)$$

and

$$\frac{1}{\sqrt{n}} \hat{\lambda} \sim \mathcal{N}\{0, (I^{22})^{-1}\}, \quad I^{22} = (I_{22} - I_{21}I_{11}^{-1}I_{12})^{-1}$$

with

$$I_{ij} = \frac{1}{n\sigma^2} X_i'X_j.$$

For the case $X_1'X_2 = 0$, we can estimate I^{22} by

$$\hat{I}^{22} = \hat{I}_{22}^{-1}, \quad \hat{I}_{22} = \frac{1}{n\hat{\sigma}^{2c}} X_2'X_2, \quad n\hat{\sigma}^{2c} = \|\hat{U}^c\|^2.$$

The test statistic is then equal to

$$\mathrm{LM}_n = \hat{\lambda}'\hat{I}^{22}\hat{\lambda} = n\frac{\hat{U}^{c\prime}X_2(X_2X_2')^{-1}X_2'\hat{U}^c}{\hat{U}^{c\prime}\hat{U}^c} = n\frac{\hat{\sigma}^{2c} - \hat{\sigma}^2}{\hat{\sigma}^{2c}} = n(1 - R^2), \tag{E.3}$$

with

$$n\hat{\sigma}^{2c} = \|\hat{U}^c\|^2, \quad \hat{U}^c = Y - X\hat{\beta},$$

and where R^2 is the coefficient of determination (centred if X_1 admits a constant column) in the regression of $\hat{U}^c$ on the columns of X_2. For the first equality of (E.3), we use the fact that in a regression model of the form $Y = X\beta + U$, with obvious notation, Pythagoras's theorem yields

$$Y'X(X'X)^{-1}X'Y = \|\hat{Y}\|^2 = \|Y\|^2 - \|\hat{U}\|^2.$$

In the general case, we have

$$\mathrm{LM}_n = n\frac{\hat{U}^{c\prime}X_2\{X_2X_2' - X_2X_1'(X_1X_1')^{-1}X_1X_2'\}^{-1}X_2'\hat{U}^c}{\hat{U}^{c\prime}\hat{U}^c} = n\frac{\hat{\sigma}^{2c} - \hat{\sigma}^2}{\hat{\sigma}^{2c}}.$$

Since the residuals of the regression of Y on the columns of X_1 and X_2 are also the residuals of the regression of $\hat{U}^c$ on the columns of X_1 and X_2, we obtain LM_n by:

1. computing the residuals $\hat{U}^c$ of the regression of Y on the columns of X_1;
2. regressing $\hat{U}^c$ on the columns of X_2 and X_1, and setting $\mathrm{LM}_n = nR^2$, where R^2 is the coefficient of determination of this second regression.

5.11 Since $\overline{y} = 0$, it is clear that $R^2_{nc} = R^2$, with R^2_{nc} and R^2 defined in representations (5.29) and (5.30). Since T is invertible, we have $\mathrm{Col}(X) = \mathrm{Col}(\tilde{X})$, where $\mathrm{Col}(Z)$ denotes the vectorial subspace generated by the columns of the matrix Z, and

$$P_{\tilde{X}} = XT(T'X'XT)^{-1}T'X' = P_X.$$

If $e \in \mathrm{Col}(X)$ then $e \in \mathrm{Col}(\tilde{X})$ and

$$P_{\tilde{X}}\tilde{Y} = cP_{\tilde{X}}Y + dP_{\tilde{X}}e = c\hat{Y} + de.$$

Noting that $\overline{\tilde{y}} := n^{-1}\sum_{t=1}^n \tilde{Y}_t = c\overline{y} + d = d$, we conclude that

$$\frac{\|P_{\tilde{X}}\tilde{Y} - \overline{\tilde{y}}e\|^2}{\|\tilde{Y} - \overline{\tilde{y}}e\|^2} = \frac{c^2\|\hat{Y}\|^2}{c^2\|Y\|^2} = R^2_{nc}.$$

Chapter 6

6.1 From the observations $\epsilon_1, \ldots, \epsilon_n$, we can compute $\overline{\epsilon^2} = n^{-1}\sum_{t=1}^n \epsilon_t^2$ and

$$\hat{\rho}_{\epsilon^2}(h) = \frac{\hat{\gamma}_{\epsilon^2}(h)}{\hat{\gamma}_{\epsilon^2}(0)}, \quad \hat{\gamma}_{\epsilon^2}(h) = \hat{\gamma}_{\epsilon^2}(-h) = \frac{1}{n}\sum_{t=1+h}^{n}(\epsilon_t^2 - \overline{\epsilon^2})(\epsilon_{t-h}^2 - \overline{\epsilon^2}),$$

for $h = 0, \ldots, q$. We then put

$$\alpha_{1,1} = \hat{\rho}_{\epsilon^2}(1)$$

and then, for $k = 2, \ldots, q$ (when $q > 1$),

$$\alpha_{k,k} = \frac{\hat{\rho}_{\epsilon^2}(k) - \sum_{i=1}^{k-1}\hat{\rho}_{\epsilon^2}(k-i)\alpha_{k-1,i}}{1 - \sum_{i=1}^{k-1}\hat{\rho}_{\epsilon^2}(i)\alpha_{k-1,i}},$$

$$\alpha_{k,i} = \alpha_{k-1,i} - \alpha_{k,k}\alpha_{k-1,k-i}, \quad i = 1, \ldots, k-1.$$

With standard notation, the OLS estimators are then

$$\hat{\alpha}_i = \alpha_{i,q}, \quad \hat{\omega} = \left(1 - \sum \hat{\alpha}_i\right)\overline{\epsilon^2}.$$

6.2 The assumption that X has full column rank implies that $X'X$ is invertible. Denoting by $\langle \cdot, \cdot \rangle$ the scalar product associated with the Euclidean norm, we have

$$\langle Y - X\hat{\theta}_n, X(\hat{\theta}_n - \theta)\rangle = Y'\{X - X(X'X)^{-1}X'X\}(\hat{\theta}_n - \theta) = 0$$

and

$$\begin{aligned} Q_n(\theta) &= \|Y - X\theta\|^2 \\ &= \|Y - X\hat{\theta}_n\|^2 + \|X(\hat{\theta}_n - \theta)\|^2 + 2\langle Y - X\hat{\theta}_n, X(\hat{\theta}_n - \theta)\rangle \\ &\geq \|Y - X\hat{\theta}_n\|^2 = nQ_n(\hat{\theta}_n), \end{aligned}$$

with equality if and only if $\theta = \hat{\theta}_n$, and we are done.

6.3 We can take $n=2$, $q=1$, $\epsilon_0=0$, $\epsilon_1=1$, $\epsilon_2=0$. The calculation yields $(\hat{\omega},\hat{\alpha})'=(1,-1)$.

6.4 Case 3 is not possible, otherwise we would have

$$\epsilon_t^2<\epsilon_t^2-\hat{\omega}-\hat{\alpha}_1\epsilon_{t-1}^2-\hat{\alpha}_2\epsilon_{t-2}^2$$

for all t, and consequently $\|Y\|^2<\|Y-X\hat{\theta}_n\|^2$, which is not possible.

Using the data, we obtain $\hat{\theta}=(1,-1,-1/2)$, and thus $\hat{\theta}^c\neq\hat{\theta}$. Therefore, the constrained estimate must coincide with one of the following three constrained estimates: that constrained by $\alpha_2=0$, that constrained by $\alpha_1=0$, or that constrained by $\alpha_1=\alpha_2=0$. The estimate constrained by $\alpha_2=0$ is $\tilde{\theta}=(7/12,-1/2,0)$, and thus does not suit. The estimate constrained by $\alpha_1=0$ yields the desired estimate $\hat{\theta}^c=(1/4,0,1/4)$.

6.5 First note that $\epsilon_t^2>\omega_0\eta_t^2$. Thus $\epsilon_t^2=0$ if and only if $\eta_t=0$. The nullity of the ith column of X, for $i>1$, implies that $\eta_{n-i+1}=\cdots=\eta_2=\eta_1=0$. The probability of this event tends to 0 as $n\to\infty$ because, since $E\eta_t^2=1$, we have $P(\eta_t=0)<1$.

6.6 Introducing an initial value X_0, the OLS estimator of ϕ_0 is

$$\hat{\phi}_n=\left(\frac{1}{n}\sum_{t=1}^{n}X_{t-1}^2\right)^{-1}\frac{1}{n}\sum_{t=1}^{n}X_tX_{t-1},$$

and this estimator satisfies

$$\sqrt{n}(\hat{\phi}_n-\phi_0)=\left(\frac{1}{n}\sum_{t=1}^{n}X_{t-1}^2\right)^{-1}\frac{1}{\sqrt{n}}\sum_{t=1}^{n}\epsilon_tX_{t-1}.$$

Under the assumptions of the exercise, the ergodic theorem entails the almost sure convergence

$$\frac{1}{n}\sum_{t=1}^{n}X_{t-1}^2\to EX_t^2,\quad \frac{1}{n}\sum_{t=1}^{n}X_tX_{t-1}\to EX_tX_{t-1}=\phi_0EX_t^2,$$

and thus the almost sure convergence of $\hat{\phi}_n$ to ϕ_0. For the consistency, the assumption $E\epsilon_t^2<\infty$ suffices.

If $E\epsilon_t^4<\infty$, the sequence $(\epsilon_tX_{t-1},\mathcal{F}_t)$ is a stationary and ergodic square integrable martingale difference, with variance

$$\mathrm{Var}(\epsilon_tX_{t-1})=E(\sigma_t^2X_{t-1}^2).$$

We can see that this expectation exists by expanding the product

$$\sigma_t^2X_{t-1}^2=\left(\omega_0+\sum_{i=1}^{q}\epsilon_{t-i}^2\right)\left(\sum_{i=0}^{\infty}\phi_0^i\epsilon_{t-1-i}\right)^2.$$

The CLT of Corollary A.1 then implies that

$$\frac{1}{\sqrt{n}}\sum_{t=1}^{n}\epsilon_tX_{t-1}\xrightarrow{\mathcal{L}}\mathcal{N}(0,E(\sigma_t^2X_{t-1}^2)),$$

and thus

$$\sqrt{n}(\hat{\phi}_n-\phi_0)\xrightarrow{\mathcal{L}}\mathcal{N}\{0,(EX_t^2)^{-2}E(\sigma_t^2X_{t-1}^2)\}.$$

When $\sigma_t^2=\omega_0$, the condition $E\epsilon_t^2<\infty$ suffices for asymptotic normality.

6.7 By direct verification, $A^{-1}A=I$.

6.8 1. Let $\tilde{\epsilon}_t = \epsilon_t/\sqrt{\omega_0}$. Then $(\tilde{\epsilon}_t)$ solves the model

$$\tilde{\epsilon}_t = \left(1 + \sum_{i=1}^{q} \alpha_{0i}\tilde{\epsilon}_{t-i}^2\right)^{1/2} \eta_t.$$

The parameter ω_0 vanishing in this equation, the moments of $\tilde{\epsilon}_t$ do not depend on it. It follows that $E\epsilon_t^{2m} = E(\sqrt{\omega_0}\tilde{\epsilon}_t)^{2m} = K\omega_0^m$.

2. and 3. Write $M = M(\omega_0^k)$ to indicate that a matrix M is proportional to ω_0^k. Partition the vector $Z_{t-1} = (1, \epsilon_{t-1}^2, \dots, \epsilon_{t-q}^2)'$ into $Z_{t-1} = (1, W_{t-1})'$ and, accordingly, the matrices A and B of Theorem 6.2. Using the previous question and the notation of Exercise 6.7, we obtain

$$A_1 = A_1(1), \quad A_{12} = A_{21}' = A_{12}(\omega_0), \quad A_{22} = A_{22}(\omega_0^2).$$

We then have

$$A^{-1} = \begin{bmatrix} A_1^{-1} + A_1^{-1}A_{12}FA_{21}A_1^{-1}(1) & -A_1^{-1}A_{12}F(\omega_0^{-1}) \\ -FA_{21}A_1^{-1}(\omega_0^{-1}) & F(\omega_0^{-2}) \end{bmatrix}.$$

Similarly,

$$B_1 = B_1(\omega_0^2), \quad B_{12} = B_{21}' = B_{12}(\omega_0^3), \quad B_{22} = B_{22}(\omega_0^4).$$

It follows that $C = A^{-1}BA^{-1}$ is of the form

$$C = \begin{bmatrix} C_1(\omega_0^2) & C_{12}(\omega_0) \\ C_{21}(\omega_0) & C_{22}(1) \end{bmatrix}.$$

6.9 1. Let $\alpha = \inf_{y\in C} \|x - y\|$. Let us show the existence of x^*. Let (x_n) be a sequence of elements of C such that, for all $n > 0$, $\|x - x_n\|^2 < \alpha^2 + 1/n$. Using the parallelogram identity $\|a + b\|^2 + \|a - b\|^2 = 2\|a\|^2 + 2\|b\|^2$, we have

$$\begin{aligned} \|x_n - x_m\|^2 &= \|x_n - x - (x_m - x)\|^2 \\ &= 2\|x_n - x\|^2 + 2\|x_m - x\|^2 - \|x_n - x + (x_m - x)\|^2 \\ &= 2\|x_n - x\|^2 + 2\|x_m - x\|^2 - 4\|x - (x_m + x_n)/2\|^2 \\ &\le 2(\alpha^2 + 1/n) + 2(\alpha^2 + 1/m) - 4\alpha^2 = 2(1/n + 1/m), \end{aligned}$$

the last inequality being justified by the fact that $(x_m + x_n)/2 \in C$, the convexity of C and the definition of α. It follows that (x_n) is a Cauchy sequence and, E being a Hilbert space and therefore a complete metric space, x_n converges to some point x^*. Since C is closed, $x^* \in C$ and $\|x - x^*\| \ge \alpha$. We have also $\|x - x^*\| \le \alpha$, taking the limit on both sides of the inequality which defines the sequence (x_n). It follows that $\|x - x^*\| = \alpha$, which shows the existence.

Assume that there exist two solutions of the minimisation problem in C, x_1^* and x_2^*. Using the convexity of C, it is then easy to see that $(x_1^* + x_2^*)/2$ satisfies

$$\|x - (x_1^* + x_2^*)/2\| = \|x - x_1^*\| = \|x - x_2^*\|.$$

This is possible only if $x_1^* = x_2^*$ (once again using the parallelogram identity).

2. Let $\lambda \in (0, 1)$ and $y \in C$. Since C is convex, $(1 - \lambda)x^* + \lambda y \in C$. Thus

$$\|x - x^*\|^2 \le \|x - \{(1 - \lambda)x^* + \lambda y\}\|^2 = \|x - x^* + \lambda(x^* - y)\|^2$$

and, dividing by λ,

$$\lambda\|x^* - y\|^2 - 2\langle x^* - x, x^* - y\rangle \ge 0.$$

Taking the limit as λ tends to 0, we obtain inequality (6.17).

Let z such that, for all $y \in C$, $\langle z-x, z-y\rangle \le 0$. We have

$$\begin{aligned}\|x-z\|^2 &= \langle z-x, (z-y)+(y-x)\rangle \\ &\le \langle z-x, y-x\rangle \le \|x-z\|\,\|x-y\|,\end{aligned}$$

the last inequality being simply the Cauchy–Schwarz inequality. It follows that $\|x-z\| \le \|x-y\|, \forall\, y \in C$. This property characterising x^* in view of part 1, it follows that $z = x^*$.

6.10 1. It suffices to show that when $C = K$, (6.17) is equivalent to (6.18). Since $0 \in K$, taking $y = 0$ in (6.17) we obtain $\langle x - x^*, x^*\rangle \le 0$. Since $x^* \in K$ and K is a cone, $2x^* \in K$. For $y = 2x^*$ in (6.17) we obtain $\langle x - x^*, x^*\rangle \ge 0$, and it follows that $\langle x - x^*, x^*\rangle = 0$. The second equation of (6.18) then follows directly from (6.17). The converse, (6.18) $\Rightarrow$ (6.17), is trivial.

2. Since $x^* \in K$, then $z = \lambda x^* \in K$ for $\lambda \ge 0$. By (6.18), we have

$$\begin{cases}\langle \lambda x - z, z\rangle = \lambda^2\langle x - x^*, x^*\rangle = 0, \\ \langle \lambda x - z, y\rangle = \lambda\langle x - x^*, y\rangle \le 0, \quad \forall y \in K.\end{cases}$$

It follows that $(\lambda x)^* = z$ and (a) is shown. The properties (b) are obvious, expanding $\|x^* + (x - x^*)\|^2$ and using the first equation of (6.18).

6.11 The model is written as $Y = X^{(1)}\theta^{(1)} + X^{(2)}\theta^{(2)} + U$. Thus, since $M_2X^{(2)} = 0$, we have $M_2Y = M_2X^{(1)}\theta^{(1)} + M_2U$. Note that this is a linear model, of parameter $\theta^{(1)}$. Noting that $M_2'M_2 = M_2$, since M_2 is an orthogonal projection matrix, the form of the estimator follows.

6.12 Since J_n is symmetric, there exists a diagonal matrix D_n and an orthonormal matrix P_n such that $J_n = P_nD_nP_n'$. For n large enough, the eigenvalues of J_n are positive since $J = \lim_{n\to\infty} J_n$ is positive definite. Let λ_n be the smallest eigenvalue of J_n. Denoting by $\|\cdot\|$ the Euclidean norm, we have

$$X_n'J_nX_n = X_n'P_nD_nP_n'X_n \ge \lambda_n X_n'P_nP_n'X_n = \lambda_n\|X_n\|^2.$$

Since $\lim_{n\to\infty} X_n'J_nX_n = 0$ and $\lim_{n\to\infty}\lambda_n > 0$, it follows that $\lim_{n\to\infty}\|X_n\| = 0$, and thus that X_n converges to the zero vector of $\mathbb{R}^k$.

6.13 Applying the method of Section 6.3.2, we obtain $X^{(1)} = (1, 1)'$ and thus, by Theorem 6.8, $\hat{\theta}_n^c = (2, 0)'$.

Chapter 7

7.1 1. When $j < 0$, all the variables involved in the expectation, except ϵ_{t-j}, belong to the σ-field generated by $\{\epsilon_{t-j-1}, \epsilon_{t-j-2}, \dots\}$. We conclude by taking the expectation conditionally on the previous σ-field and using the martingale increment property.

2. For $j \ge 0$, we note that ϵ_t^2 is a measurable function of $\eta_t^2, \dots, \eta_{t-j+1}^2$ and of $\epsilon_{t-j}^2, \epsilon_{t-j-1}^2, \dots$ Thus $E\{g(\epsilon_t^2, \epsilon_{t-1}^2, \dots) \mid \epsilon_{t-j}, \epsilon_{t-j-1}, \dots\}$ is an even function of the conditioning variables, denoted by $h(\epsilon_{t-j}^2, \epsilon_{t-j-1}^2, \dots)$.

3. It follows that the expectation involved in the property can be written as

$$\begin{aligned}&E\{E(h(\eta_{t-j}^2\sigma_{t-j}^2, \epsilon_{t-j-1}^2, \dots)\eta_{t-j}\sigma_{t-j}f(\epsilon_{t-j-1}, \epsilon_{t-j-2}, \dots) \\ &\qquad\qquad \mid \epsilon_{t-j-1}, \epsilon_{t-j-2}, \dots)\} \\ &= E\left\{\int h(x^2\sigma_{t-j}^2, \epsilon_{t-j-1}^2, \dots)x\sigma_{t-j}f(\epsilon_{t-j-1}, \epsilon_{t-j-2}, \dots)\, d\mathbb{P}_\eta(x)\right\} \\ &= 0.\end{aligned}$$

The latter equality follows from of the nullity of the integral, because the distribution of η_t is symmetric.

7.2 By the Borel–Cantelli lemma, it suffices to show that for all real $\delta > 0$, the series of general terms $\mathbb{P}(\rho^t \epsilon_t^2 > \delta)$ converges. That is to say,

$$\sum_{t=0}^{\infty} \mathbb{P}(\rho^t \epsilon_t^2 > \delta) \leq \sum_{t=0}^{\infty} \frac{E(\rho^t \epsilon_t^2)^s}{\delta^s} = \frac{E(\epsilon_t^{2s})}{(1-\rho)\delta^s} < \infty,$$

using Markov's inequality, strict stationarity and the existence of a moment of order $s > 0$ for ϵ_t^2.

7.3 For all $\kappa > 0$, the process (X_t^κ) is ergodic and admits an expectation. This expectation is finite since $X_t^\kappa \leq \kappa$ and $(X_t^\kappa)^- = X_t^-$. We thus have, by the standard ergodic theorem,

$$\frac{1}{n}\sum_{t=1}^{n} X_t \geq \frac{1}{n}\sum_{t=1}^{n} X_t^\kappa \to E(X_1^\kappa), \quad \text{a.s.} \quad \text{as } n \to \infty.$$

When $\kappa \to \infty$, the variable X_1^κ increases to X_1. Thus by Beppo Levi's theorem $E(X_1^\kappa)$ converges to $E(X_1) = +\infty$. It follows that $n^{-1}\sum_{t=1}^{n} X_t$ tends almost surely to infinity.

7.4
1. The assumptions made on f and Θ guarantee that $Y_t = \{\inf_{\theta\in\Theta} X_t(\theta)\}$ is a measurable function of $\eta_t, \eta_{t-1}, \ldots$. By Theorem A.1, it follows that (Y_t) is stationary and ergodic.
2. If we remove condition (7.94), the property may not be satisfied. For example, let $\Theta = \{\theta_1, \theta_2\}$ and assume that the sequence $(X_t(\theta_1), X_t(\theta_2))$ is iid, with zero mean, each component being of variance 1 and the covariance between the two components being different when t is even and when t is odd. Each of the two processes $(X_t(\theta_1))$ and $(X_t(\theta_2))$ is stationary and ergodic (as iid processes). However, $Y_t = \inf_\theta (X_t(\theta)) = \min(X_t(\theta_1), X_t(\theta_2))$ is not stationary in general because its distribution depends on the parity of t.

7.5
1. In view of (7.30) and of the second part of assumption A1, we have

$$\begin{aligned}
&\sup_{\theta\in\Theta}|Q_n(\theta) - \tilde{Q}_n(\theta)| \\
&\quad = \sup_{\theta\in\Theta} n^{-1}\left|\sum_{t=1}^{n}\{(2\sigma_t^2 + \tilde{\sigma}_t^2 - \sigma_t^2)(\sigma_t^2 - \tilde{\sigma}_t^2) - 2\epsilon_t^2(\sigma_t^2 - \tilde{\sigma}_t^2)\}(\theta)\right| \\
&\quad \leq \sup_{\theta\in\Theta} Kn^{-1}\sum_{t=1}^{n}\{(2\sigma_t^2 + K\rho^t)\rho^t + 2\epsilon_t^2\rho^t\} \to 0 \qquad \text{(E.4)}
\end{aligned}$$

almost surely. Indeed, on a set of probability 1, we have for all $\iota > 0$,

$$\begin{aligned}
&\limsup_{n\to\infty} \sup_{\theta\in\Theta} Kn^{-1}\sum_{t=1}^{n}\{(2\sigma_t^2 + K\rho^t)\rho^t + 2\epsilon_t^2\rho^t\} \\
&\quad \leq \iota \limsup_{n\to\infty} n^{-1}\sum_{t=1}^{n}\{\sup_{\theta\in\Theta}\sigma_t^2 + \epsilon_t^2\} \\
&\quad = \iota\{E_{\theta_0}\sup_{\theta\in\Theta}\sigma_t^2 + E_{\theta_0}\sigma_t^2(\theta_0)\}. \qquad \text{(E.5)}
\end{aligned}$$

Note that $E\epsilon_t^2 < \infty$ and (7.29) entail that $E_{\theta_0}\sup_{\theta\in\Theta}\sigma_t^2(\theta) < \infty$. The limit superior (E.5) being less than any positive number, it is null.
2. Note that $\nu_t := \epsilon_t^2 - \sigma_t^2(\theta_0) = \epsilon_t^2 - E_{\theta_0}(\epsilon_t^2 \mid \epsilon_{t-1}, \ldots)$ is the strong innovation of ϵ_t^2. We thus have orthogonality between ν_t and any integrable variable which is measurable with respect to

the σ-field generated by θ_0:

$$\lim_{n\to\infty} Q_n(\theta) = E_{\theta_0}\{\epsilon_t^2 - \sigma_t^2(\theta_0) + \sigma_t^2(\theta_0) - \sigma_t^2(\theta)\}^2$$
$$= \lim_{n\to\infty} Q_n(\theta_0) + E_{\theta_0}\{\sigma_t^2(\theta_0) - \sigma_t^2(\theta)\}^2 + 2E_{\theta_0}[v_t\{\sigma_t^2(\theta_0) - \sigma_t^2(\theta)\}]$$
$$= \lim_{n\to\infty} Q_n(\theta_0) + E_{\theta_0}\{\sigma_t^2(\theta_0) - \sigma_t^2(\theta)\}^2 \geq \lim_{n\to\infty} Q_n(\theta_0),$$

with equality if and only if $\sigma_t^2(\theta) = \sigma_t^2(\theta_0)$ P_{θ_0}-almost surely, that is, $\theta = \theta_0$ (by assumptions A3 and A4; see the proof of Theorem 7.1).

3. We conclude that $\hat{\theta}_n$ is strongly consistent, as in (d) in the proof of Theorem 7.1, using a compactness argument and applying the ergodic theorem to show that, at any point θ_1, there exists a neighbourhood $V(\theta_1)$ of θ_1 such that

$$\text{if } \theta_1 \in \Theta, \quad \theta_1 \neq \theta_0, \quad \liminf_{n\to\infty} \inf_{\theta\in V(\theta_1)} Q(\theta) > \lim_{n\to\infty} Q(\theta_0) \quad \text{a.s.}$$

4. Since all we have done remains valid when Θ is replaced by any smaller compact set containing θ_0, for instance Θ^c, the estimator $\hat{\theta}_n^c$ is strongly consistent.

7.6 We know that $\hat{\theta}_n$ minimises, over Θ,

$$\tilde{\mathrm{l}}_n(\theta) = n^{-1}\sum_{t=1}^{n} \frac{\epsilon_t^2}{\tilde{\sigma}_t^2} + \log \tilde{\sigma}_t^2.$$

For all $c > 0$, there exists $\hat{\theta}_{n,c}$ such that $\tilde{\sigma}_t^2(\hat{\theta}_{n,c}) = c\tilde{\sigma}_t^2(\hat{\theta}_n)$ for all $t \geq 0$. Note that $\hat{\theta}_{n,c} \neq \hat{\theta}_n$ if and only if $c \neq 1$. For instance, for a GARCH(1, 1) model, if $\hat{\theta}_n = (\hat{\omega}, \hat{\alpha}_1, \hat{\beta}_1)$ we have $\hat{\theta}_{n,c} = (c\hat{\omega}, c\hat{\alpha}_1, \hat{\beta}_1)$. Let $f(c) = \tilde{\mathrm{l}}_n(\hat{\theta}_{n,c})$. The minimum of f is obtained at the unique point

$$c_0 = n^{-1}\sum_{t=1}^{n} \frac{\epsilon_t^2}{\tilde{\sigma}_t^2(\hat{\theta}_n)}.$$

If $\hat{\theta}_{n,c_0} \in \Theta$, we have $\hat{\theta}_{n,c_0} = \hat{\theta}_n$. It follows that $c_0 = 1$ with probability 1, which proves the result.

7.7 The expression for I_1 is a trivial consequence of (7.74) and $\mathrm{Cov}(1 - \eta_t^2, \eta_t) = 0$. Similarly, the form of I_2 directly follows from (7.38). Now consider the non-diagonal blocks. Using (7.38) and (7.74), we obtain

$$E_{\varphi_0}\left\{\frac{\partial \ell_t}{\partial \theta_i}\frac{\partial \ell_t}{\partial \vartheta_j}(\varphi_0)\right\} = E(1 - \eta_t^2)^2 E_{\varphi_0}\left\{\frac{1}{\sigma_t^4}\frac{\partial \sigma_t^2}{\partial \theta_i}\frac{\partial \sigma_t^2}{\partial \vartheta_j}(\varphi_0)\right\}.$$

In view of (7.41), (7.42), (7.79) and (7.24), we have

$$E_{\varphi_0}\left\{\frac{1}{\sigma_t^4}\frac{\partial \sigma_t^2}{\partial \omega}\frac{\partial \sigma_t^2}{\partial \vartheta_j}(\varphi_0)\right\}$$
$$= \sum_{k_1,k_2=0}^{\infty} B_0^{k_1}(1,1)B_0^{k_2}(1,1)\sum_{i=1}^{q} 2\alpha_{0i}E_{\varphi_0}\left\{\sigma_t^{-4}\epsilon_{t-k_2-i}\frac{\partial \epsilon_{t-k_2-i}}{\partial \vartheta_j}(\varphi_0)\right\} = 0,$$

$$E_{\varphi_0}\left\{\frac{1}{\sigma_t^4}\frac{\partial \sigma_t^2}{\partial \alpha_{i_0}}\frac{\partial \sigma_t^2}{\partial \vartheta_j}(\varphi_0)\right\}$$
$$= \sum_{k_1,k_2=0}^{\infty} B_0^{k_1}(1,1)B_0^{k_2}(1,1)\sum_{i=1}^{q} 2\alpha_{0i}E_{\varphi_0}\left\{\sigma_t^{-4}\epsilon_{t-k_1-i_0}^2\epsilon_{t-k_2-i}\frac{\partial \epsilon_{t-k_2-i}}{\partial \vartheta_j}(\varphi_0)\right\} = 0n$$

and

$$E_{\varphi_0}\left\{\frac{1}{\sigma_t^4}\frac{\partial\sigma_t^2}{\partial\beta_{j_0}}\frac{\partial\sigma_t^2}{\partial\vartheta_j}(\varphi_0)\right\}$$

$$=\sum_{k_1,k_2=0}^{\infty}\left\{\sum_{\ell=1}^{k_1}B_0^{\ell-1}B_0^{(j_0)}B_0^{k_1-\ell}\right\}(1,1)B_0^{k_2}(1,1)\sum_{i=1}^{q}2\alpha_{0i}$$

$$\times E_{\varphi_0}\left\{\sigma_t^{-4}\left(\omega_0+\sum_{i'=1}^{q}\alpha_{0i'}\epsilon_{t-k_1-i'}^2\right)\epsilon_{t-k_2-i}\frac{\partial\epsilon_{t-k_2-i}}{\partial\vartheta_j}(\varphi_0)\right\}=0.$$

It follows that

$$\forall i,j,\quad E_{\varphi_0}\left\{\frac{1}{\sigma_t^4}\frac{\partial\sigma_t^2}{\partial\theta_i}\frac{\partial\sigma_t^2}{\partial\vartheta_j}(\varphi_0)\right\}=0 \tag{E.6}$$

and $\mathcal{I}$ is block-diagonal. It is easy to see that $\mathcal{J}$ has the form given in the theorem. The expressions for J_1 and J_2 follow directly from (7.39) and (7.75). The block-diagonal form follows from (7.76) and (E.6).

7.8 1. We have $\epsilon_t = X_t - aX_{t-1}, \sigma_t^2 = 1+\alpha\epsilon_{t-1}^2$. The parameters to be estimated are a and α, ω being known. We have

$$\frac{\partial\epsilon_t}{\partial a}=-X_{t-1},\quad \frac{\partial\sigma_t^2}{\partial a}=-2\alpha\epsilon_{t-1}X_{t-2},\quad \frac{\partial\sigma_t^2}{\partial\alpha}=\epsilon_{t-1}^2,$$

$$\frac{\partial^2\sigma_t^2}{\partial a^2}=2\alpha X_{t-2}^2,\quad \frac{\partial^2\sigma_t^2}{\partial a\partial\alpha}=-2\alpha\epsilon_{t-1}X_{t-2},\quad \frac{\partial^2\sigma_t^2}{\partial\alpha^2}=0.$$

It follows that

$$\frac{\partial\ell_t}{\partial a}(\varphi_0)=-(1-\eta_t^2)\frac{2\alpha_0\epsilon_{t-1}X_{t-2}}{\sigma_t^2}-\eta_t\frac{2X_{t-1}}{\sigma_t},$$

$$\frac{\partial\ell_t}{\partial\alpha}(\varphi_0)=(1-\eta_t^2)\frac{\epsilon_{t-1}^2}{\sigma_t^2},$$

$$\frac{\partial^2\ell_t}{\partial\alpha^2}(\varphi_0)=\frac{\epsilon_{t-1}^4}{\sigma_t^4},$$

$$\frac{\partial^2\ell_t}{\partial a^2}(\varphi_0)=-\eta_t\frac{8\alpha_0\epsilon_{t-1}X_{t-1}X_{t-2}}{\sigma_t^3}+(2\eta_t^2-1)\frac{(2\alpha_0\epsilon_{t-1}X_{t-2})^2}{\sigma_t^4}$$
$$+(1-\eta_t^2)\frac{2\alpha_0X_{t-2}^2}{\sigma_t^2}+\frac{2X_{t-1}^2}{\sigma_t^4},$$

$$\frac{\partial^2\ell_t}{\partial a\partial\alpha}(\varphi_0)=\eta_t^2\frac{2\alpha_0\epsilon_{t-1}^3X_{t-2}}{\sigma_t^4}-(1-\eta_t^2)\frac{2\alpha_0\epsilon_{t-1}^3X_{t-2}}{\sigma_t^4}+\eta_t\frac{2\epsilon_{t-1}^2X_{t-1}}{\sigma_t^3}.$$

Letting $\mathcal{I}=(\mathcal{I}_{ij})$, $\mathcal{J}=(\mathcal{J}_{ij})$ and $\mu_3=E\eta_t^3$, we then obtain

$$\mathcal{I}_{11}=E_{\varphi_0}\left(\frac{\partial\ell_t}{\partial a}(\varphi_0)\right)^2$$
$$=4\alpha_0^2(\kappa_\eta-1)E_{\varphi_0}\left(\frac{\epsilon_{t-1}^2X_{t-2}^2}{\sigma_t^4}\right)+4E_{\varphi_0}\left(\frac{X_{t-1}^2}{\sigma_t^2}\right)$$

$$- 4\alpha_0\mu_3 E_{\varphi_0}\left(\frac{\epsilon_{t-1}X_{t-1}X_{t-2}}{\sigma_t^3}\right),$$

$$\mathcal{I}_{12} = \mathcal{I}_{12} = E_{\varphi_0}\left(\frac{\partial \ell_t}{\partial a}\frac{\partial \ell_t}{\partial \alpha}(\varphi_0)\right)$$

$$= 2\alpha_0(\kappa_\eta - 1)E_{\varphi_0}\left(\frac{\epsilon_{t-1}^3 X_{t-2}}{\sigma_t^4}\right) - 2\mu_3 E_{\varphi_0}\left(\frac{\epsilon_{t-1}^2 X_{t-1}}{\sigma_t^3}\right),$$

$$\mathcal{I}_{22} = E_{\varphi_0}\left(\frac{\partial \ell_t}{\partial \alpha}(\varphi_0)\right)^2 = (\kappa_\eta - 1)E_{\theta_0}\left(\frac{\epsilon_{t-1}^4}{\sigma_t^4}\right),$$

$$\mathcal{J}_1 = E_{\varphi_0}\left(\frac{\partial^2 \ell_t}{\partial a^2}\right) = 4\alpha_0^2 E_{\varphi_0}\left(\frac{\epsilon_{t-1}^2 X_{t-2}^2}{\sigma_t^4}\right) + 2E_{\varphi_0}\left(\frac{X_{t-1}^2}{\sigma_t^2}\right),$$

$$\mathcal{J}_{12} = \mathcal{J}_{21} = E_{\varphi_0}\left(\frac{\partial^2 \ell_t}{\partial a \partial \alpha}\right) = 2\alpha_0 E_{\varphi_0}\left(\frac{\epsilon_{t-1}^3 X_{t-2}}{\sigma_t^4},\right)$$

$$\mathcal{J}_{22} = E_{\varphi_0}\left(\frac{\partial^2 \ell_t}{\partial \alpha^2}\right) = E_{\theta_0}\left(\frac{\epsilon_{t-1}^4}{\sigma_t^4}\right).$$

2. In the case where the distribution of η_t is symmetric we have $\mu_3 = 0$ and, using (7.24), $\mathcal{I}_{12} = \mathcal{J}_{12} = 0$. It follows that

$$\Sigma = \begin{pmatrix} \mathcal{I}_1/\mathcal{J}_1^2 & 0 \\ 0 & \mathcal{I}_{22}/\mathcal{J}_{22}^2 \end{pmatrix}.$$

The asymptotic variance of the ARCH parameter estimator is thus equal to $(\kappa_\eta - 1)\{E_{\theta_0}(\epsilon_{t-1}^4/\sigma_t^4)\}^{-1}$: it does not depend on a_0 and is the same as that of the QMLE of a pure ARCH(1) (using computations similar to those used to obtain (7.1.2)).

3. When $\alpha_0 = 0$, we have $\sigma_t^2 = 1$, and thus $EX_t^2 = 1/(1 - a_0^2)$. It follows that

$$\mathcal{I} = \begin{pmatrix} 4/(1-a_0^2) & -2\mu_3^2 \\ -2\mu_3^2 & (\kappa_\eta - 1)\kappa_\eta \end{pmatrix}, \quad \mathcal{J} = \begin{pmatrix} 2/(1-a_0^2) & 0 \\ 0 & \kappa_\eta \end{pmatrix},$$

$$\Sigma = \begin{pmatrix} 1 - a_0^2 & -\mu_3^2(1-a_0^2)/\kappa_\eta \\ -\mu_3^2(1-a_0^2)/\kappa_\eta & (\kappa_\eta - 1)/\kappa_\eta \end{pmatrix}.$$

We note that the estimation of too complicated a model (since the true process is AR(1) without ARCH effect) does not entail any asymptotic loss of accuracy for the estimation of the parameter a_0: the asymptotic variance of the estimator is the same, $1 - a_0^2$, as if the AR(1) model were directly estimated. This calculation also allows us to verify the '$\alpha_0 = 0$' column in Table 7.3: for the $\mathcal{N}(0, 1)$ law we have $\mu_3 = 0$ and $\kappa_\eta = 3$; for the normalized $\chi^2(1)$ distribution we find $\mu_3 = 4/\sqrt{2}$ and $\kappa_\eta = 15$.

7.9 Let $\varepsilon > 0$ and $V(\theta_0)$ be such that (7.95) is satisfied. Since $\hat{\theta}_n \to \theta_0$ almost surely, for n large enough $\hat{\theta}_n \in V(\theta_0)$ almost surely. We thus have almost surely

$$\left\|\frac{1}{n}\sum_{t=1}^n J_t(\hat{\theta}_n) - J\right\| \le \left\|\frac{1}{n}\sum_{t=1}^n J_t(\theta_0) - J\right\| + \frac{1}{n}\sum_{t=1}^n \|J_t(\hat{\theta}_n) - J_t(\theta_0)\|$$

$$\le \left\|\frac{1}{n}\sum_{t=1}^n J_t(\theta_0) - J\right\| + \frac{1}{n}\sum_{t=1}^n \sup_{\theta \in V(\theta_0)} \|J_t(\theta) - J_t(\theta_0)\|.$$

It follows that

$$\lim_{n\to\infty}\left\|\frac{1}{n}\sum_{t=1}^{n}J_t(\hat{\theta}_n)-J\right\|\leq\varepsilon$$

and, since ε can be chosen arbitrarily small, we have the desired result.

In order to give an example where (7.95) is not satisfied, let us consider the autoregressive model $X_t=\theta_0X_{t-1}+\eta_t$ where $\theta_0=1$ and (η_t) is an iid sequence with mean 0 and variance 1. Let $J_t(\theta)=X_t-\theta X_{t-1}$. Then $J_t(\theta_0)=\eta_t$ and the first convergence of the exercise holds true, with $J=0$. Moreover, for all neighbourhoods of θ_0,

$$\frac{1}{n}\sum_{t=1}^{n}\sup_{\theta\in V(\theta_0)}|J_t(\theta)-J_t(\theta_0)|=\left(\frac{1}{n}\sum_{t=1}^{n}|X_{t-1}|\right)\sup_{\theta\in V(\theta_0)}|\theta-\theta_0|\to+\infty,$$

almost surely because the sum in brackets converges to $+\infty$, X_t being a random walk and the supremum being strictly positive. Thus (7.95) is not satisfied. Nevertheless, we have

$$\begin{aligned}\frac{1}{n}\sum_{t=1}^{n}J_t(\hat{\theta}_n)&=\frac{1}{n}\sum_{t=1}^{n}(X_t-\hat{\theta}_nX_{t-1})\\&=\frac{1}{n}\sum_{t=1}^{n}\eta_t+\sqrt{n}(\hat{\theta}_n-1)\frac{1}{n^{3/2}}\sum_{t=1}^{n}X_{t-1}\\&\to J=0,\ \text{ in probability.}\end{aligned}$$

Indeed, $n^{-3/2}\sum_{t=1}^{n}X_{t-1}$ converges in law to a non-degenerate random variable (see, for instance, Hamilton 1994, p. 406) whereas $\sqrt{n}(\hat{\theta}_n-1)\to0$ in probability since $n(\hat{\theta}_n-1)$ has a non-degenerate limit distribution.

7.10 It suffices to show that $J^{-1}-\theta_0\theta_0'$ is positive semi-definite. Note that $\theta_0'(\partial\sigma_t^2(\theta_0)/\partial\theta)=\sigma_t^2(\theta_0)$. It follows that

$$\theta_0'J=E(Z_t),\qquad\text{where }Z_t=\frac{1}{\sigma_t^2(\theta_0)}\frac{\partial\sigma_t^2(\theta_0)}{\partial\theta}.$$

Therefore $J-J\theta_0\theta_0'J=\mathrm{Var}(Z_t)$ is positive semi-definite. Thus

$$y'J(J^{-1}-\theta_0\theta_0')Jy=y'(J-J\theta_0\theta_0'J)y\geq0,\quad\forall y\in\mathbb{R}^{q+1}.$$

Setting $x=Jy$, we then have

$$x'(J^{-1}-\theta_0\theta_0')x\geq0,\quad\forall x\in\mathbb{R}^{q+1},$$

which proves the result.

7.11 1. In the ARCH case, we have $\theta_0'(\partial\sigma_t^2(\theta_0)/\partial\theta)=\sigma_t^2(\theta_0)$. It follows that

$$E\left\{\frac{1}{\sigma_t^2(\theta_0)}\frac{\partial\sigma_t^2(\theta_0)}{\partial\theta'}\left(1-\frac{\theta_0'}{\sigma_t^2(\theta_0)}\frac{\partial\sigma_t^2(\theta_0)}{\partial\theta}\right)\right\}=0,$$

or equivalently $\Omega'=\theta_0'J$, that is, $\Omega'J^{-1}=\theta_0'$. We also have $\theta_0'J\theta_0=1$, and thus

$$1=\theta_0'J\theta_0=\Omega'J^{-1}JJ^{-1}\Omega=\Omega'J^{-1}\Omega.$$

2. Introducing the polynomial $\mathcal{B}_\theta(z) = 1 - \sum_{j=1}^{p} \beta_j z^j$, the derivatives of $\sigma_t^2(\theta)$ satisfy

$$\mathcal{B}_\theta(L)\frac{\partial \sigma_t^2}{\partial \omega}(\theta) = 1,$$

$$\mathcal{B}_\theta(L)\frac{\partial \sigma_t^2}{\partial \alpha_i}(\theta) = \epsilon_{t-i}^2, \quad i = 1, \dots, q$$

$$\mathcal{B}_\theta(L)\frac{\partial \sigma_t^2}{\partial \beta_j}(\theta) = \sigma_{t-j}^2, \quad j = 1, \dots, p.$$

It follows that

$$\mathcal{B}_\theta(L)\frac{\partial \sigma_t^2(\theta)}{\partial \theta'}\overline{\theta} = \omega + \sum_{i=1}^{q} \alpha_i \epsilon_{t-i}^2 = \mathcal{B}_\theta(L)\sigma_t^2(\theta).$$

In view of assumption A2 and Corollary 2.2, the roots of $\mathcal{B}_\theta(L)$ are outside the unit disk, and the relation follows.

3. It suffices to replace θ_0 by $\overline{\theta_0}$ in 1.

7.12 Only three cases have to be considered, the other ones being obtained by symmetry. If $t_1 < \min\{t_2, t_3, t_4\}$, the result is obtained from (7.24) with $g = 1$ and $t-j = t_1$. If $t_2 = t_3 < t_1 < t_4$, the result is obtained from (7.24) with $g(\epsilon_t^2, \dots) = \epsilon_t^2$, $t = t_2$ and $t-j = t_1$. If $t_2 = t_3 = t_4 < t_1$, the result is obtained from (7.24) with $g(\epsilon_t^2, \dots) = \epsilon_t^2, j = 0, t_2 = t_3 = t_4 = t$ and $f(\epsilon_{t-j-1}, \dots) = \epsilon_{t_1}$.

7.13 1. It suffices to apply (7.38), and then to apply Corollary 2.1.

2. The result follows from the Lindeberg central limit theorem of Theorem A.3.

3. Using Eq. (7.39) and the convergence of ϵ_{t-1}^2 to $+\infty$,

$$\frac{1}{n}\sum_{t=1}^{n}\frac{\partial^2}{\partial \alpha^2}\ell_t(\alpha_0) = \frac{1}{n}\sum_{t=1}^{n}(2\eta_t^2 - 1)\left(\frac{\epsilon_{t-1}^2}{1+\alpha_0\epsilon_{t-1}^2}\right)^2 \to \frac{1}{\alpha_0^2} \quad \text{a.s.}$$

4. In view of Eq. (7.50) and the fact that $\partial^2\sigma_t^2(\alpha)/\partial\alpha^2 = \partial^3\sigma_t^2(\alpha)/\partial\alpha^3 = 0$, we have

$$\left|\frac{\partial^3}{\partial \alpha^3}\ell_t(\alpha)\right| = \left|\left\{2 - 6\frac{(1+\alpha_0\epsilon_{t-1}^2)\eta_t^2}{1+\alpha\epsilon_{t-1}^2}\right\}\left(\frac{\epsilon_{t-1}^2}{1+\alpha\epsilon_{t-1}^2}\right)^3\right|$$
$$\leq \left\{2 + 6\left(1 + \frac{\alpha_0}{\alpha}\right)\eta_t^2\right\}\frac{1}{\alpha^3}.$$

5. The derivative of the criterion is equal to zero at $\hat{\alpha}_n^c$. A Taylor expansion of this derivative around α_0 then yields

$$0 = \frac{1}{\sqrt{n}}\sum_{t=1}^{n}\frac{\partial}{\partial\alpha}\ell_t(\alpha_0) + \frac{1}{n}\sum_{t=1}^{n}\frac{\partial^2}{\partial\alpha^2}\ell_t(\alpha_0)\sqrt{n}(\hat{\alpha}_n^c - \alpha_0)$$
$$+ \frac{1}{n}\sum_{t=1}^{n}\frac{\partial^3}{\partial\alpha^3}\ell_t(\alpha^*)\frac{\sqrt{n}(\hat{\alpha}_n^c - \alpha_0)^2}{2},$$

where α^* is between $\hat{\alpha}_n^c$ and α_0. The result easily follows from the previous questions.

6. When $\omega_0 \neq 1$, we have

$$\frac{\partial}{\partial\alpha}\ell_t(\alpha_0) = \left(1 - \frac{\epsilon_t^2}{1+\alpha_0\epsilon_{t-1}^2}\right)\frac{\epsilon_{t-1}^2}{1+\alpha_0\epsilon_{t-1}^2}$$
$$= (1-\eta_t^2)\frac{\epsilon_{t-1}^2}{1+\alpha_0\epsilon_{t-1}^2} + d_t,$$

with

$$d_t = \frac{\epsilon_{t-1}^2(1-\omega_0)}{(1+\alpha_0\epsilon_{t-1}^2)^2}\eta_t^2.$$

Since $d_t \to 0$ almost surely as $t \to \infty$, the convergence in law of part 2 always holds true. Moreover,

$$\begin{aligned}\frac{\partial^2}{\partial\alpha^2}\ell_t(\alpha_0) &= \left(2\frac{\epsilon_t^2}{1+\alpha_0\epsilon_{t-1}^2} - 1\right)\left(\frac{\epsilon_{t-1}^2}{1+\alpha_0\epsilon_{t-1}^2}\right)^2 \\ &= (2\eta_t^2 - 1)\left(\frac{\epsilon_{t-1}^2}{1+\alpha_0\epsilon_{t-1}^2}\right)^2 + d_t^*,\end{aligned}$$

with

$$d_t^* = 2\frac{(\omega_0-1)\eta_t^2}{(1+\alpha_0\epsilon_{t-1}^2)}\left(\frac{\epsilon_{t-1}^2}{1+\alpha_0\epsilon_{t-1}^2}\right)^2 = o(1) \quad \text{a.s.},$$

which implies that the result obtained in Part 3 does not change. The same is true for Part 4 because

$$\begin{aligned}\left|\frac{\partial^3}{\partial\alpha^3}\ell_t(\alpha)\right| &= \left|\left\{2 - 6\frac{(\omega_0+\alpha_0\epsilon_{t-1}^2)\eta_t^2}{1+\alpha\epsilon_{t-1}^2}\right\}\left(\frac{\epsilon_{t-1}^2}{1+\alpha\epsilon_{t-1}^2}\right)^3\right| \\ &\le \left\{2 + 6\left(\omega_0 + \frac{\alpha_0}{\alpha}\right)\eta_t^2\right\}\frac{1}{\alpha^3}.\end{aligned}$$

Finally, it is easy to see that the asymptotic behaviour of $\hat{\alpha}_n^c(\omega_0)$ is the same as that of $\hat{\alpha}_n^c(\omega)$, regardless of the value that is fixed for ω.

7. In practice ω_0 is not known and must be estimated. However, it is impossible to estimate the whole parameter (ω_0, α_0) without the strict stationarity assumption. Moreover, under condition (7.14), the ARCH(1) model generates explosive trajectories which do not look like typical trajectories of financial returns.

7.14 1. Consider a constant $\underline{\alpha} \in (0, \alpha_0)$. We begin by showing that $\hat{\alpha}_n^c(1) > \underline{\alpha}$ for n large enough. Note that

$$\hat{\alpha}_n^c(1) = \arg\min_{\alpha\in[0,\infty)} Q_n(\alpha), \quad Q_n(\alpha) = \frac{1}{n}\sum_{t=1}^{n}\{\ell_t(\alpha) - \ell_t(\alpha_0)\}.$$

We have

$$\begin{aligned}Q_n(\alpha) &= \frac{1}{n}\sum_{t=1}^{n}\eta_t^2\left\{\frac{\sigma_t^2(\alpha_0)}{\sigma_t^2(\alpha)} - 1\right\} + \log\frac{\sigma_t^2(\alpha)}{\sigma_t^2(\alpha_0)} \\ &= \frac{1}{n}\sum_{t=1}^{n}\eta_t^2\frac{(\alpha_0-\alpha)\epsilon_{t-1}^2}{1+\alpha\epsilon_{t-1}^2} + \log\frac{1+\alpha\epsilon_{t-1}^2}{1+\alpha_0\epsilon_{t-1}^2}.\end{aligned}$$

In view of the inequality $x \ge 1 + \log x$ for all $x > 0$, it follows that

$$\inf_{\alpha<\underline{\alpha}} Q_n(\alpha) \ge \log\frac{1}{n}\sum_{t=1}^{n}\eta_t^2 + \log\frac{(\alpha_0-\underline{\alpha})\epsilon_{t-1}^2}{1+\alpha_0\epsilon_{t-1}^2} + 1.$$

For all $M > 0$, there exists an integer t_M such that $\epsilon_t^2 > M$ for all $t > t_M$. This entails that

$$\liminf_{n\to\infty}\inf_{\alpha<\underline{\alpha}} Q_n(\alpha) \ge \log\frac{(\alpha_0-\underline{\alpha})M}{1+\alpha_0 M} + 1.$$

Since M is arbitrarily large,

$$\liminf_{n\to\infty} \inf_{\alpha<\underline{\alpha}} Q_n(\alpha) \geq \log \frac{\alpha_0 - \underline{\alpha}}{\alpha_0} + 1 > 0 \tag{E.7}$$

provided that $\underline{\alpha} < (1 - e^{-1})\alpha_0$. If $\underline{\alpha}$ is chosen so that the constraint is satisfied, the inequalities

$$\limsup_{n\to\infty} Q_n(\hat{\alpha}_n) \leq \limsup_{n\to\infty} Q_n(\alpha_0) = 0$$

and (E.7) show that

$$\lim_{n\to\infty} \hat{\alpha}_n \geq \underline{\alpha} \quad \text{a.s.} \tag{E.8}$$

We will define a criterion O_n asymptotically equivalent to the criterion Q_n. Since $\epsilon_{t-1}^2 \to \infty$ a. s. as $t \to \infty$, we have for $\alpha \neq 0$,

$$\lim_{n\to\infty} Q_n(\alpha) = \lim_{n\to\infty} O_n(\alpha),$$

where

$$O_n(\alpha) = \frac{1}{n}\sum_{t=1}^{n} \eta_t^2 \frac{(\alpha_0 - \alpha)}{\alpha} + \log \frac{\alpha}{\alpha_0}.$$

On the other hand, we have

$$\lim_{n\to\infty} O_n(\alpha) = \frac{\alpha_0}{\alpha} - 1 + \log \frac{\alpha}{\alpha_0} > 0$$

when $\alpha_0/\alpha \neq 1$. We will now show that $Q_n(\alpha) - O_n(\alpha)$ converges to zero uniformly in $\alpha \in (\underline{\alpha}, \infty)$. We have

$$Q_n(\alpha) - O_n(\alpha) = \frac{1}{n}\sum_{t=1}^{n} \eta_t^2 \frac{\alpha - \alpha_0}{\alpha(1 + \alpha\epsilon_{t-1}^2)} + \log \frac{(1 + \alpha\epsilon_{t-1}^2)\alpha_0}{(1 + \alpha_0\epsilon_{t-1}^2)\alpha}.$$

Thus for all $M > 0$ and any $\varepsilon > 0$, almost surely

$$|Q_n(\alpha) - O_n(\alpha)| \leq (1 + \varepsilon)\frac{|\alpha - \alpha_0|}{\alpha^2 M} + \frac{|\alpha - \alpha_0|}{\alpha\alpha_0 M},$$

provided n is large enough. In addition to the previous constraints, assume that $\underline{\alpha} < 1$. We have $|\alpha - \alpha_0|/\alpha^2 M \leq \alpha_0/\underline{\alpha}^2 M$ for any $\underline{\alpha} < \alpha \leq \alpha_0$, and

$$\frac{|\alpha - \alpha_0|}{\alpha^2 M} \leq \frac{\alpha}{\alpha^2 M} \leq \frac{1}{\underline{\alpha} M} \leq \frac{1}{\underline{\alpha}^2 M}$$

for any $\alpha \geq \alpha_0$. We then have

$$\lim_{n\to\infty} \sup_{\alpha>\underline{\alpha}} |Q_n(\alpha) - O_n(\alpha)| \leq (1 + \varepsilon)\frac{1 + \alpha_0}{\underline{\alpha}^2 M} + \frac{1 + \alpha_0}{\underline{\alpha}\alpha_0 M}.$$

Since M can be chosen arbitrarily large and ε arbitrarily small, we have almost surely

$$\lim_{n\to\infty} \sup_{\alpha>\underline{\alpha}} |Q_n(\alpha) - O_n(\alpha)| = 0. \tag{E.9}$$

For the last step of the proof, let α_0^- and α_0^+ be two constants such that $\alpha_0^- < \alpha_0 < \alpha_0^+$. It can always be assumed that $\underline{\alpha} < \alpha_0^-$. With the notation $\hat{\sigma}_\eta^2 = n^{-1}\sum_{t=1}^n \eta_t^2$, the solution of

$$\alpha_n^* = \arg\min_{\alpha} O_n(\alpha).$$

is $\alpha_n^* = \alpha_0 \hat{\sigma}_\eta^2$. This solution belongs to the interval (α_0^-, α_0^+) when n is large enough. In this case

$$\alpha_n^{**} = \arg \min_{\alpha \notin (\alpha_0^-, \alpha_0^+)} O_n(\alpha)$$

is one of the two extremities of the interval (α_0^-, α_0^+), and thus

$$\lim_{n\to\infty} O_n(\alpha_n^{**}) = \min\{\lim_{n\to\infty} O_n(\alpha_0^-), \lim_{n\to\infty} O_n(\alpha_0^+)\} > 0.$$

This result, (E.9), the fact that $\min_\alpha Q_n(\alpha) \le Q_n(\alpha_0) = 0$ and (E.8) show that

$$\lim_{n\to\infty} \arg \min_{\alpha\ge 0} Q_n(\alpha) \in (\alpha_0^-, \alpha_0^+).$$

Since (α_0^-, α_0^+) is an arbitrarily small interval that contains α_0 and $\hat{\alpha}_n = \arg\min_\alpha Q_n(\alpha)$, the conclusion follows.

2. It can be seen that the constant 1 does not play any particular role and can be replaced by any other positive number ω. However, we cannot conclude that $\hat{\alpha}_n \to \alpha_0$ almost surely because $\hat{\alpha}_n = \hat{\alpha}_n^c(\hat{\omega}_n)$, but $\hat{\omega}_n$ is not a constant. In contrast, it can be shown that under the strict stationarity condition $\alpha_0 < \exp\{-E(\log \eta_t^2)\}$ the constrained estimator $\hat{\alpha}_n^c(\omega)$ does not converge to α_0 when $\omega \ne \omega_0$.

Chapter 8

8.1 Let the Lagrange multiplier $\lambda \in \mathbb{R}^p$. We have to maximise the Lagrangian

$$\mathcal{L}(x, \lambda) = (x - x_0)' J (x - x_0) + \lambda' K x.$$

Since at the optimum

$$0 = \frac{\partial \mathcal{L}(x, \lambda)}{\partial x} = 2Jx - 2Jx_0 + K'\lambda,$$

the solution is such that $x = x_0 - \frac{1}{2} J^{-1} K' \lambda$. Since $0 = Kx = Kx_0 - \frac{1}{2} K J^{-1} K' \lambda$, we obtain $\lambda = \left(\frac{1}{2} K J^{-1} K'\right)^{-1} K x_0$, and then the solution is

$$x = x_0 - J^{-1} K' (K J^{-1} K')^{-1} K x_0.$$

8.2 Let K be the $p \times n$ matrix such that $K(1, i_1) = \cdots = K(p, i_p) = 1$ and whose the other elements are 0. Using Exercise 8.1, the solution has the form

$$x = \{I_n - J^{-1} K' (K J^{-1} K')^{-1} K\} x_0. \tag{E.10}$$

Instead of the Lagrange multiplier method, a direct substitution method can also be used.

The constraints $x_{i_1} = \cdots = x_{i_p} = 0$ can be written as

$$x = Hx^*$$

where H is $n \times (n-p)$, of full column rank, and x^* is $(n-p) \times 1$ (the vector of the non-zero components of x). For instance: (i) if $n = 3$, $x_2 = x_3 = 0$ then $x^* = x_1$ and $H = \begin{pmatrix} 1 \\ 0 \\ 0 \end{pmatrix}$; (ii) if $n = 3$, $x_3 = 0$ then $x^* = (x_1, x_2)'$ and $H = \begin{pmatrix} 1 & 0 \\ 0 & 1 \\ 0 & 0 \end{pmatrix}$.

If we denote by Col(H) the space generated by the columns of H, we thus have to find

$$\min_{x \in \mathrm{Col}(H)} \|x - x_0\|_J$$

where $\|.\|_J$ is the norm $\|z\|_J = \sqrt{z'Jz}$.

This norm defines the scalar product $\langle z, y\rangle_J = z'Jy$. The solution is thus the orthogonal (with respect to this scalar product) projection of x_0 on Col(H). The matrix of such a projection is

$$P = H(H'JH)^{-1}H'J.$$

Indeed, we have $P^2 = P$, $PHz = Hz$, thus Col(H) is P-invariant, and $\langle Hy, (I-P)z\rangle_J = y'H'J(I-P)z = y'H'Jz - y'H'JH(H'JH)^{-1}H'Jz = 0$, thus $z - Pz$ is orthogonal to Col(H).

It follows that the solution is

$$x = Px_0 = H(H'JH)^{-1}H'Jx_0. \tag{E.11}$$

This last expression seems preferable to (E.10) because it only requires the inversion of the matrix $H'JH$ of size $n-p$, whereas in (E.11) the inverse of J, which is of size n, is required.

8.3 In case (a), we have

$$K = (0, 0, 1) \quad \text{and} \quad H = \begin{pmatrix} 1 & 0 \\ 0 & 1 \\ 0 & 0 \end{pmatrix},$$

and then

$$(H'JH)^{-1} = \begin{pmatrix} 2 & -1 \\ -1 & 2 \end{pmatrix}^{-1} = \begin{pmatrix} 2/3 & 1/3 \\ 1/3 & 2/3 \end{pmatrix},$$

$$H(H'JH)^{-1}H' = \begin{pmatrix} 2/3 & 1/3 & 0 \\ 1/3 & 2/3 & 0 \\ 0 & 0 & 0 \end{pmatrix}$$

and, using (E.11),

$$P = H(H'JH)^{-1}H'J = \begin{pmatrix} 1 & 0 & 1/3 \\ 0 & 1 & 2/3 \\ 0 & 0 & 0 \end{pmatrix}$$

which gives a constrained minimum at

$$x = \begin{pmatrix} x_{01} + x_{03}/3 \\ x_{02} + x_{03}/3 \\ 0 \end{pmatrix}.$$

In case (b), we have

$$K = \begin{pmatrix} 0 & 1 & 0 \\ 0 & 0 & 1 \end{pmatrix} \quad \text{and} \quad H = \begin{pmatrix} 1 \\ 0 \\ 0 \end{pmatrix}, \tag{E.12}$$

and, using (E.11), a calculation, which is simpler than the previous one (we do not have to invert any matrix since $H'JH$ is scalar), shows that the constrained minimum is at

$$x = \begin{pmatrix} x_{01} - x_{02}/2 \\ 0 \\ 0 \end{pmatrix}.$$

The same results can be obtained with formula (E.10), but the computations are longer, in particular because we have to compute

$$J^{-1} = \begin{pmatrix} 3/4 & 1/2 & -1/4 \\ 1/2 & 1 & -1/2 \\ -1/4 & -1/2 & 3/4 \end{pmatrix}. \tag{E.13}$$

8.4 Matrix J^{-1} is given by (E.13). With the matrix $K_1 = K$ defined by (E.12), and denoting by K_2 and K_3 the first and second rows of K, we then obtain

$$I_3 - J^{-1}K'(KJ^{-1}K')^{-1}K = \begin{pmatrix} 1 & -1/2 & 0 \\ 0 & 0 & 0 \\ 0 & 0 & 0 \end{pmatrix},$$

$$I_3 - J^{-1}K_2'(K_2J^{-1}K_2')^{-1}K_2 = \begin{pmatrix} 1 & -1/2 & 0 \\ 0 & 0 & 0 \\ 0 & 1/2 & 1 \end{pmatrix},$$

$$I_3 - J^{-1}K_3'(K_3J^{-1}K_3')^{-1}K_3 = \begin{pmatrix} 1 & 0 & 1/3 \\ 0 & 1 & 2/3 \\ 0 & 0 & 0 \end{pmatrix}.$$

It follows that the solution will be found among (a) $\lambda = Z$,

$$\text{(b) } \lambda = \begin{pmatrix} Z_1 - Z_2/2 \\ 0 \\ 0 \end{pmatrix}, \quad \text{(c) } \lambda = \begin{pmatrix} Z_1 - Z_2/2 \\ 0 \\ Z_3 + Z_2/2 \end{pmatrix}, \quad \text{(d) } \lambda = \begin{pmatrix} Z_1 + Z_3/3 \\ Z_2 + 2Z_3/3 \\ 0 \end{pmatrix}.$$

The value of $Q(\lambda)$ is 0 in case (a), $3Z_2^2/2 + 2Z_2Z_3 + 2Z_3^2$ in case (b), Z_2^2 in case (c) and $Z_3^2/3$ in case (d).

To find the solution of the constrained minimisation problem, it thus suffices to take the value λ which minimizes $Q(\lambda)$ among the subset of the four vectors defined in (a)–(d) which satisfy the positivity constraints of the two last components.

We thus find the minimum at $\lambda^{\Lambda} = Z = (-2, 1, 2)'$ in case (i), at

$$\lambda^{\Lambda} = \begin{pmatrix} -3/2 \\ 0 \\ 3/2 \end{pmatrix} \text{ in case (ii) where } Z = \begin{pmatrix} -2 \\ -1 \\ 2 \end{pmatrix},$$

$$\lambda^{\Lambda} = \begin{pmatrix} -5/2 \\ 0 \\ 0 \end{pmatrix} \text{ in case (iii) where } Z = \begin{pmatrix} -2 \\ 1 \\ -2 \end{pmatrix}$$

and

$$\lambda^{\Lambda} = \begin{pmatrix} -3/2 \\ 0 \\ 0 \end{pmatrix} \text{ in case (iv) where } Z = \begin{pmatrix} -2 \\ -1 \\ -2 \end{pmatrix}.$$

8.5 Recall that for a variable $Z \sim \mathcal{N}(0, 1)$, we have $EZ^+ = -EZ^- = (2\pi)^{-1/2}$ and $\text{Var}(Z^+) = \text{Var}(Z^-) = \frac{1}{2}(1 - 1/\pi)$. We have

$$Z = \begin{pmatrix} Z_1 \\ Z_2 \\ Z_3 \end{pmatrix} \sim \mathcal{N}\left\{0, \Sigma = (\kappa_\eta - 1)J^{-1} = \begin{pmatrix} (\kappa_\eta + 1)\omega_0^2 & -\omega_0 & -\omega_0 \\ -\omega_0 & 1 & 0 \\ -\omega_0 & 0 & 1 \end{pmatrix}\right\}.$$

It follows that

$$E\lambda^{\Lambda} = \begin{pmatrix} -\omega\sqrt{\frac{2}{\pi}} \\ \frac{1}{\sqrt{2\pi}} \\ \frac{1}{\sqrt{2\pi}} \end{pmatrix}.$$

The coefficient of the regression of Z_1 on Z_2 is $-\omega_0$. The components of the vector $(Z_1 + \omega_0 Z_2, Z_2)$ are thus uncorrelated and, this vector being Gaussian, they are independent. In particular $E(Z_1 + \omega_0 Z_2)Z_2^- = 0$, which gives $\mathrm{Cov}(Z_1, Z_2^-) = EZ_1Z_2^- = -\omega_0 E(Z_2 Z_2^-) = -\omega_0 E(Z_2^-)^2 = -\omega_0/2$. We thus have

$$\mathrm{Var}(Z_1 + \omega_0 Z_2^- + \omega_0 Z_3^-) = (\kappa_\eta + 1)\omega_0^2 + \omega_0^2\left(1 - \frac{1}{\pi}\right) - 2\omega_0^2 = \left(\kappa_\eta - \frac{1}{\pi}\right)\omega_0^2.$$

Finally,

$$\mathrm{Var}(\lambda^{\Lambda}) = \frac{1}{2}\left(1 - \frac{1}{\pi}\right)\begin{pmatrix} 2\frac{\kappa_\eta \pi - 1}{\pi - 1}\omega_0^2 & -\omega_0 & -\omega_0 \\ -\omega_0 & 1 & 0 \\ -\omega_0 & 0 & 1 \end{pmatrix}.$$

It can be seen that

$$\mathrm{Var}(Z) - \mathrm{Var}(\lambda^{\Lambda}) = \frac{1}{2}\left(1 + \frac{1}{\pi}\right)\begin{pmatrix} 2\omega_0^2 & -\omega_0 & -\omega_0 \\ -\omega_0 & 1 & 0 \\ -\omega_0 & 0 & 1 \end{pmatrix}$$

is a positive semi-definite matrix.

8.6 At the point $\theta_0 = (\omega_0, 0, \ldots, 0)$, we have

$$\frac{\partial \sigma_t^2(\theta_0)}{\partial \theta} = (1, \omega_0 \eta_{t-1}^2, \ldots, \omega_0 \eta_{t-q}^2)'$$

and the information matrix (written for simplicity in the ARCH(3) case) is equal to

$$J(\theta_0) := E_{\theta_0}\left(\frac{1}{\sigma_t^4(\theta_0)} \frac{\partial \sigma_t^2(\theta_0)}{\partial \theta} \frac{\partial \sigma_t^2(\theta_0)}{\partial \theta'}\right)$$

$$= \frac{1}{\omega_0^2}\begin{pmatrix} 1 & \omega_0 & \omega_0 & \omega_0 \\ \omega_0 & \omega_0^2\kappa_\eta & \omega_0^2 & \omega_0^2 \\ \omega_0 & \omega_0^2 & \omega_0^2\kappa_\eta & \omega_0^2 \\ \omega_0 & \omega_0^2 & \omega_0^2 & \omega_0^2\kappa_\eta \end{pmatrix}.$$

This matrix is invertible (which is not the case for a general GARCH(p, q)). We finally obtain

$$\Sigma(\theta_0) = (\kappa_\eta - 1)J(\theta_0)^{-1} = \begin{pmatrix} (\kappa_\eta + q - 1)\omega^2 & -\omega \ \cdots - \omega \\ -\omega & \\ \vdots & I_q \\ -\omega & \end{pmatrix}.$$

8.7 We have $\sigma_t^2 = \omega + \alpha\epsilon_{t-1}^2$,and

$$J := E_{\theta_0}\frac{1}{\sigma_t^4}\frac{\partial \sigma_t^2}{\partial \theta}\frac{\partial \sigma_t^2}{\partial \theta'}(\theta_0) = E_{\theta_0}\frac{1}{\sigma_t^4(\theta_0)}\begin{pmatrix} 1 & \epsilon_{t-1}^2 \\ \epsilon_{t-1}^2 & \epsilon_{t-1}^4 \end{pmatrix} = \frac{1}{\omega_0^2}\begin{pmatrix} 1 & \omega_0 \\ \omega_0 & \omega_0^2\kappa_\eta \end{pmatrix},$$

and thus

$$J^{-1} = \frac{1}{\kappa_\eta - 1}\begin{pmatrix} \omega_0^2\kappa_\eta & -\omega_0 \\ -\omega_0 & 1 \end{pmatrix}.$$

In view of Theorem 8.1 and (8.15), the asymptotic distribution of $\sqrt{n}(\hat{\theta} - \theta_0)$ is that of the vector λ^Λ defined by

$$\lambda^\Lambda = \begin{pmatrix} Z_1 \\ Z_2 \end{pmatrix} - Z_2^- \begin{pmatrix} -\omega_0 \\ 1 \end{pmatrix} = \begin{pmatrix} Z_1 + \omega_0 Z_2^- \\ Z_2^+ \end{pmatrix}.$$

We have $EZ_2^+ = -EZ_2^- = (2\pi)^{-1/2}$, thus

$$E\lambda^\Lambda = \frac{1}{\sqrt{2\pi}}\begin{pmatrix} -\omega_0 \\ 1 \end{pmatrix}.$$

Since the components of the Gaussian vector $(Z_1 + \omega_0 Z_2, Z_2)$ are uncorrelated, they are independent, and it follows that

$$\mathrm{Cov}(Z_1, Z_2^-) = -\omega_0 \mathrm{Cov}(Z_2, Z_2^-) = -\omega_0/2.$$

We then obtain

$$\mathrm{Var}\lambda^\Lambda = \begin{pmatrix} \omega_0^2\left\{\kappa_\eta - \frac{1}{2}\left(1 + \frac{1}{\pi}\right)\right\} & -\omega_0\frac{1}{2}\left(1 - \frac{1}{\pi}\right) \\ -\omega_0\frac{1}{2}\left(1 - \frac{1}{\pi}\right) & \frac{1}{2}\left(1 - \frac{1}{\pi}\right) \end{pmatrix}.$$

Let $f(z_1, z_2)$ be the density of Z, that is, the density of a centred normal with variance $(\kappa_\eta - 1)J^{-1}$. It is easy to show that the distribution of $Z_1 + \omega_0 Z_2^-$ admits the density $h(x) = \int_0^\infty f(x, z_2)dz_2 + \int_{-\infty}^0 f(x - \omega_0 z_2, z_2)dz_2$ and to check that this density is asymmetric.

A simple calculation yields $E(Z_2^-)^3 = -2/\sqrt{2\pi}$. From $E(Z_1 + \omega_0 Z_2)(Z_2^-)^2 = 0$, we then obtain $EZ_1(Z_2^-)^2 = 2\omega_0/\sqrt{2\pi}$. And from $E(Z_1 + \omega_0 Z_2)^2(Z_2^-) = E(Z_1 + \omega_0 Z_2)^2 E(Z_2^-)$ we obtain $EZ_1^2 Z_2^- = -\omega_0^2(\kappa_\eta + 1)/\sqrt{2\pi}$. Finally, we obtain

$$\begin{aligned} E(Z_1 + \omega_0 Z_2^-)^3 &= 3\omega_0 EZ_1^2 Z_2^- + 3\omega_0^2 EZ_1(Z_2^-)^2 + \omega_0^3 E(Z_2^-)^3 \\ &= \frac{\omega_0^3}{\sqrt{2\pi}}(1 - 3\kappa_\eta). \end{aligned}$$

8.8 The statistic of the C test is $\mathcal{N}(0, 1)$ distributed under H_0. The p-value of C is thus $1 - \Phi\left(n^{-1/2}\sum_{i=1}^n X_i\right)$. Under the alternative, we have almost surely $n^{-1/2}\sum_{i=1}^n X_i \sim \sqrt{n}\theta$ as $n \to +\infty$. It can be shown that $\log\{1 - \Phi(x)\} \sim -x^2/2$ in the neighbourhood of $+\infty$. In Bahadur's sense, the asymptotic slope of the C test is thus

$$c(\theta) = \lim_{n\to\infty} \frac{-2}{n} \times \frac{-(\sqrt{n}\theta)^2}{2} = \theta^2, \quad \theta > 0.$$

The p-value of C^* is $2(1 - \Phi\left(\left|n^{-1/2}\sum_{i=1}^n X_i\right|\right)$. Since $\log 2\{1 - \Phi(x)\} \sim -x^2/2$ in the neighbourhood of $+\infty$, the asymptotic slope of C^* is also $c^*(\theta) = \theta^2$ for $\theta > 0$. The C and C^* tests having the same asymptotic slope, they cannot be distinguished by the Bahadur approach.

We know that C is uniformly more powerful than C^*. The local power of C is thus also greater than that of C^* for all $\tau > 0$. It is also true asymptotically as $n \to \infty$, even if the sample is not Gaussian. Indeed, under the local alternatives $\tau/\sqrt{n}$, and for a regular statistical model, the statistic $n^{-1/2}\sum_{i=1}^n X_i$ is asymptotically $\mathcal{N}(\tau, 1)$ distributed. The local asymptotic power of C is thus $\gamma(\tau) = 1 - \Phi(c - \tau)$ with $c = \Phi^{-1}(1 - \alpha)$. The local asymptotic power of C^* is

$\gamma^*(\tau) = 1 - \Phi(c^* - \tau) + \Phi(-c^* - \tau)$, with $c^* = \Phi^{-1}(1 - \alpha/2)$. The difference between the two asymptotic powers is

$$D(\tau) = \gamma(\tau) - \gamma^*(\tau) = -\Phi(c - \tau) + \Phi(c^* - \tau) - \Phi(-c^* - \tau)$$

and, denoting the $\mathcal{N}(0, 1)$ density by $\phi(x)$, we have

$$D'(\tau) = \phi(c - \tau) - \phi(c^* - \tau) + \phi(-c^* - \tau) = \phi(\tau)e^{c\tau}g(\tau),$$

where

$$g(\tau) = e^{-c^2/2} + e^{c^{*2}/2}(-e^{(c^*-c)\tau} + e^{-(c^*+c)\tau}).$$

Since $0 < c < c^*$, we have

$$g'(\tau) = e^{c^{*2}/2}\{-(c^* - c)e^{(c^*-c)\tau} - (c^* + c)e^{-(c^*+c)\tau}\} < 0.$$

Thus, $g(\tau)$ is decreasing on $[0, \infty)$. Note that $g(0) > 0$ and $\lim_{\tau\to+\infty} g(\tau) = -\infty$. The sign of $g(\tau)$, which is also the sign of $D'(\tau)$, is positive when $\tau \in [0, a]$ and negative when $\tau \in [a, \infty)$, for some $a > 0$. The function D thus increases on $[0, a]$ and decreases on $[a, \infty)$. Since $D(0) = 0$ and $\lim_{\tau\to+\infty} D(\tau) = 0$, we have $D(\tau) > 0$ for all $\tau > 0$. This shows that, in Pitman's sense, the test C is, as expected, locally more powerful than C^* in the Gaussian case, and locally asymptotically more powerful than C^* in a much more general framework.

8.9 The Wald test uses the fact that

$$\sqrt{n}(\overline{X}_n - \theta) \sim \mathcal{N}(0, \sigma^2) \quad \text{and} \quad S_n^2 \to \sigma^2 \text{ a.s.}$$

To justify the score test, we remark that the log-likelihood constrained by H_0 is

$$-\frac{n}{2}\log\sigma^2 - \frac{1}{2\sigma^2}\sum_{i=1}^{n} X_i^2$$

which gives $\sum X_i^2/n$ as constrained estimator of σ^2. The derivative of the log-likelihood satisfies

$$\frac{1}{\sqrt{n}}\frac{\partial}{\partial(\theta, \sigma^2)}\log L_n(\theta, \sigma^2) = \left(\frac{n^{-1/2}\sum_i X_i}{n^{-1}\sum_i X_i^2}, 0\right)$$

at $(\theta, \sigma^2) = \left(0, \sum X_i^2/n\right)$. The first component of this score vector is asymptotically $\mathcal{N}(0, 1)$ distributed under H_0. The third test is of course the likelihood ratio test, because the unconstrained log-likelihood at the optimum is equal to $-(n/2)\log S_n^2 - (n/2)$ whereas the maximal value of the constrained log-likelihood is $-(n/2)\log\sum X_i^2/n - (n/2)$. Note also that $\mathbf{L}_n = n\log(1 + \overline{X}_n^2/S_n^2) \sim \mathbf{W}_n$ under H_0.

The asymptotic level of the three tests is of course α, but using the inequality $\frac{x}{1+x} < \log(1 + x) < x$ for $x > 0$, we have

$$\mathbf{R}_n \le \mathbf{L}_n \le \mathbf{W}_n,$$

with almost surely strict inequalities in finite samples, and also asymptotically under H_1. This leads us to think that the Wald test will reject more often under H_1.

Since S_n^2 is invariant by translation of the X_i, S_n^2 tends almost surely to σ^2 both under H_0 and under H_1, as well as under the local alternatives $H_n(\tau) : \theta = \tau/\sqrt{n}$. The behaviour of $\sum X_i^2/n$ under $H_n(\tau)$ is the same as that of $\sum (X_i + \tau/\sqrt{n})^2/n$ under H_0, and because

$$\frac{1}{n}\sum_{i=1}^{n}\left(X_i + \frac{\tau}{\sqrt{n}}\right)^2 = \frac{1}{n}\sum_{i=1}^{n} X_i^2 + \frac{\tau^2}{n} + 2\frac{\tau}{\sqrt{n}}\overline{X}_n \to \sigma^2 \text{ a.s.}$$

under H_0, we have $\sum X_i^2/n \to \sigma^2$ both under H_0 and under $H_n(\tau)$. Similarly, it can be shown that $\overline{X}_n/S_n \to 0$ under H_0 and under $H_n(\tau)$. Using these two results and $x/(1+x) \sim \log(1+x)$ in the neighbourhood of 0, it can be seen that the statistics $\mathbf{L}_n$, $\mathbf{R}_n$ and $\mathbf{W}_n$ are equivalent under $H_n(\tau)$. Therefore, the Pitman approach cannot distinguish the three tests.

Using $\log P(\chi_r^2 > x) \sim -x/2$ for x in the neighbourhood of $+\infty$, the asymptotic Bahadur slopes of the tests C_1, C_2 and C_3 are, respectively

$$c_1(\theta) = \lim_{n\to\infty} \frac{-2}{n} \log P(\chi_1^2 > \mathbf{W}_n) = \lim_{n\to\infty} \frac{\overline{X}_n^2}{S_n^2} = \frac{\theta^2}{\sigma^2}$$

$$c_2(\theta) = \frac{\theta^2}{\sigma^2+\theta^2}, \quad c_3(\theta) = \log\left(\frac{\sigma^2+\theta^2}{\sigma^2}\right) \quad \text{under } H_1$$

Clearly

$$c_2(\theta) = \frac{c_1(\theta)}{1+c_1(\theta)} < c_3(\theta) = \log\{1+c_1(\theta)\} < c_1(\theta).$$

Thus the ranking of the tests, in increasing order of relative efficiency in the Bahadur sense, is

score ≺ likelihood ratio ≺ Wald.

All the foregoing remains valid for a regular non-Gaussian model.

8.10 In Example 8.2, we saw that

$$\lambda^{\Lambda} = Z - Z_d^- \mathbf{c}, \quad \mathbf{c} = \begin{pmatrix} \gamma_1 \\ \vdots \\ \gamma_d \end{pmatrix}, \quad \gamma_i = \frac{E(Z_d Z_i)}{\mathrm{Var}(Z_d)}.$$

Note that $\mathrm{Var}(Z_d)\mathbf{c}$ corresponds to the last column of $\mathrm{Var}Z = (\kappa_\eta - 1)J^{-1}$. Thus $\mathbf{c}$ is the last column of J^{-1} divided by the (d, d)th element of this matrix. In view of Exercise 6.7, this element is $(J_{22} - J_{21}J_1^{-1}J_{12})^{-1}$. It follows that $J\mathbf{c} = (0, \dots, 0, J_{22} - J_{21}J_1^{-1}J_{12})'$ and $\mathbf{c}'J\mathbf{c} = J_{22} - J_{21}J_1^{-1}J_{12}$. By (8.24), we thus have

$$\begin{aligned} \mathbf{L} &= -\frac{1}{2}(Z_d^-)^2 \mathbf{c}'J\mathbf{c} + \frac{1}{2}Z_d^2(J_{22} - J_{21}J_1^{-1}J_{12}) \\ &= \frac{1}{2}(Z_d^+)^2(J_{22} - J_{21}J_1^{-1}J_{12}) = \frac{\kappa_\eta - 1}{2}\frac{(Z_d^+)^2}{\mathrm{Var}(Z_d)}. \end{aligned}$$

This shows that the statistic $2/(\kappa_\eta - 1)\mathbf{L}_n$ has the same asymptotic distribution as the Wald statistic $\mathbf{W}_n$, that is, the distribution $\delta_0/2 + \chi_1^2/2$ in the case $d_2 = 1$.

8.11 Using (8.29) and Exercise 8.6, we have

$$\begin{aligned} \mathbf{L} &= -\frac{1}{2}\left\{\left(\sum_{i=2}^d Z_i^-\right)^2 + \kappa_\eta \sum_{i=2}^d (Z_i^-)^2 - 2\left(\sum_{i=2}^d Z_i^-\right)^2 + \sum_{\substack{i,j=2\\ i\neq j}}^d Z_i^- Z_j^- - (\kappa_\eta - 1)\sum_{i=2}^d Z_i^2\right\} \\ &= \frac{\kappa_\eta - 1}{2}\sum_{i=2}^d (Z_i^+)^2. \end{aligned}$$

The result then follows from (8.30).

8.12 Since $XY = 0$ almost surely, we have $P(XY \neq 0) = 0$. By independence, we have $P(XY \neq 0) = P(X \neq 0$ and $Y \neq 0) = P(X \neq 0)P(Y \neq 0)$. It follows that $P(X \neq 0) = 0$ or $P(Y \neq 0) = 0$.

Chapter 9

9.1   Substituting $y = x/\sigma_t$, and then integrating by parts, we obtain

$$\int f(y)dy + \int yf'(y)dy = 1 + \lim_{a,b\to\infty} [yf(y)]^a_{-b} - \int f(y)dy = 0.$$

Since σ_t^2 and $\partial\sigma_t^2/\partial\theta$ belong to the σ-field $\mathcal{F}_{t-1}$ generated by $\{\epsilon_u : u < t\}$, and since the distribution of ϵ_t given $\mathcal{F}_{t-1}$ has the density $\sigma_t^{-1}f(\cdot/\sigma_t)$, we have

$$E\left\{ \frac{\partial}{\partial\theta} \log L_{n,f}(\theta_0)\bigg| \mathcal{F}_{t-1} \right\} = 0,$$

and the result follows. We can also appeal to the general result that a score vector is centred.

9.2   It suffices to use integration by parts.

9.3   We have

$$\Lambda(\theta, \theta_0, x) = -\frac{(x-\theta)^2}{2\sigma^2} + \frac{(x-\theta_0)^2}{2\sigma^2} = \frac{\theta_0^2 - \theta^2}{2\sigma^2} + x\frac{(\theta - \theta_0)}{\sigma^2}.$$

Thus

$$\begin{pmatrix} aX + b \\ \Lambda(\theta, \theta_0, X) \end{pmatrix} \sim \mathcal{N}\left\{ \begin{pmatrix} a\theta_0 + b \\ -\frac{(\theta-\theta_0)^2}{2\sigma^2} \end{pmatrix}, \begin{pmatrix} a^2\sigma^2 & a(\theta-\theta_0) \\ a(\theta-\theta_0) & \frac{(\theta-\theta_0)^2}{\sigma^2} \end{pmatrix} \right\}$$

when $X \sim \mathcal{N}(\theta_0, \sigma^2)$, and

$$\begin{pmatrix} aX + b \\ \Lambda(\theta, \theta_0, X) \end{pmatrix} \sim \mathcal{N}\left\{ \begin{pmatrix} a\theta + b \\ \frac{(\theta_0-\theta)(\theta_0-3\theta)}{2\sigma^2} \end{pmatrix}, \begin{pmatrix} a^2\sigma^2 & a(\theta-\theta_0) \\ a(\theta-\theta_0) & \frac{(\theta-\theta_0)^2}{\sigma^2} \end{pmatrix} \right\}$$

when $X \sim \mathcal{N}(\theta, \sigma^2)$. Note that

$$E_\theta(aX + b) = E_{\theta_0}(aX + b) + \mathrm{Cov}_{\theta_0}\{aX + b, \Lambda(\theta, \theta_0, X)\},$$

as in Le Cam's third lemma.

9.4   Recall that

$$\frac{\partial}{\partial\theta} \log L_{n,f_\lambda}(\vartheta) = -\sum_{t=1}^n \lambda\left(1 - \left|\frac{\epsilon_t}{\sigma_t}\right|^\lambda\right) \frac{1}{2\sigma_t^2} \frac{\partial\sigma_t^2}{\partial\theta}$$

and

$$\frac{\partial}{\partial\lambda} \log L_{n,f_\lambda}(\vartheta) = \sum_{t=1}^n \left\{ \frac{1}{\lambda} + \left(1 - \left|\frac{\epsilon_t}{\sigma_t}\right|^\lambda\right) \log\left|\frac{\epsilon_t}{\sigma_t}\right| \right\}.$$

Using the ergodic theorem, the fact that $(1 - |\eta_t|^\lambda)$ is centred and independent of the past, as well as elementary calculations of derivatives and integrals, we obtain

$$-n^{-1}\frac{\partial^2}{\partial\theta\partial\theta'} \log L_{n,f_\lambda}(\vartheta_0) = n^{-1}\sum_{t=1}^n \lambda_0^2|\eta_t|^\lambda \frac{1}{4\sigma_t^4} \frac{\partial^2\sigma_t^2}{\partial\theta\partial\theta'}(\theta_0) + o(1) \to \mathfrak{I}_1,$$

$$-n^{-1}\frac{\partial^2}{\partial\lambda\partial\theta'} \log L_{n,f_\lambda}(\vartheta_0) = -n^{-1}\sum_{t=1}^n \lambda_0|\eta_t|^\lambda \log|\eta_t| \frac{1}{2\sigma_t^2} \frac{\partial\sigma_t^2}{\partial\theta}(\theta_0) + o(1) \to \mathfrak{I}_{12},$$

and

$$-n^{-1}\frac{\partial^2}{\partial\lambda^2} \log L_{n,f_\lambda}(\vartheta_0) = n^{-1}\sum_{t=1}^n \left\{ \frac{1}{\lambda_0^2} + |\eta_t|^\lambda(\log|\eta_t|)^2 \right\} \to \mathfrak{I}_{22}$$

almost surely.

9.5 Jensen's inequality entails that

$$E \log \sigma f(\eta\sigma) - E \log f(\eta) = E \log \frac{\sigma f(\eta\sigma)}{f(\eta)}$$
$$\leq \log E \frac{\sigma f(\eta\sigma)}{f(\eta)} = \log \int \sigma f(y\sigma)dy = 0,$$

where the inequality is strict if $\sigma f(\eta\sigma)/f(\eta)$ is non-constant. If this ratio of densities were almost surely constant, it would be almost surely equal to 1, and we would have

$$E|\eta|^r = \int |x|^r f(x)dx = \int |x|^r \sigma f(x\sigma)dx = \sigma^{-r} \int |x|^r f(x)dx = E|\eta|^r/\sigma^r,$$

which is possible only when $\sigma = 1$.

9.6 It suffices to note that $\tau^2_{\ell_{f_b}} = \tau^2_{f_b f_b} (= E_{f_b}\eta_t^2 - 1)$.

9.7 The second-order moment of the double $\Gamma(b, p)$ distribution is $p(p+1)/b^2$. Therefore, to have $E\eta_t^2 = 1$, the density f of η_t must be the double $\Gamma(\sqrt{p(p+1)}, p)$. We then obtain

$$1 + \frac{f'(x)}{f(x)}x = p - \sqrt{p(p+1)}|x|.$$

Thus $\tilde{I}_f = p$ and $\tau^2_{ff} = 1/p$. We then show that $\kappa_\eta := \int x^4 f_p(x)dx = (3+p)(2+p)/p(p+1)$. It follows that $\tau^2_{\phi f}/\tau^2_{ff} = (3+2p)/(2+2p)$.

To compare the ML and Laplace QML, it is necessary to normalise in such a way that $E|\eta_t| = 1$, that is, to take the double $\Gamma(p, p)$ as density f. We then obtain $1 + xf'(x)/f(x) = p - p|x|$. We always have $\tilde{I}_f = \tau^{-2}_{ff} = p^2(E\eta_t^2 - 1) = p$, and we have $\tau^2_{\ell f} = (E\eta_t^2 - 1) = 1/p$. It follows that $\tau^2_{\ell f}/\tau^2_{ff} = 1$, which was already known from Exercise 9.6. This allows us to construct a table similar to Table 9.5.

9.8 Consider the first instrumental density of the table, namely

$$h(x) = c|x|^{\lambda-1} \exp(-\lambda|x|^r/r), \quad \lambda > 0.$$

Denoting by c any constant whose value can be ignored, we have

$$g(x, \sigma) = \log \sigma + (\lambda - 1)\log \sigma|x| - \lambda\sigma^r \frac{|x|^r}{r} + c,$$
$$g_1(x, \sigma) = \frac{\lambda}{\sigma} - \lambda\sigma^{r-1}|x|^r,$$
$$g_2(x, \sigma) = -\frac{\lambda}{\sigma^2} - \lambda(r-1)\sigma^{r-2}|x|^r,$$

and thus

$$\tau^2_{h,f} = \frac{E(1 - |\eta_t|^r)^2}{r^2} = \frac{E|\eta_t|^{2r} - 1}{r^2}.$$

Now consider the second density,

$$h(x) = c|x|^{-\lambda-1} \exp(-\lambda|x|^{-r}/r).$$

We have

$$g(x, \sigma) = \log \sigma + (-\lambda - 1)\log \sigma|x| - \lambda\sigma^{-r} \frac{|x|^{-r}}{r} + c,$$
$$g_1(x, \sigma) = \frac{-\lambda}{\sigma} + \lambda\sigma^{-r-1}|x|^{-r},$$
$$g_2(x, \sigma) = \frac{\lambda}{\sigma^2} - \lambda(r+1)\sigma^{-r-2}|x|^{-r},$$

which gives

$$\tau_{h,f}^2 = \frac{E(1-|\eta_t|^{-r})^2}{r^2}.$$

Consider the last instrumental density,

$$h(x) = c|x|^{-1}\exp\{-\lambda(\log|x|)^2\}.$$

We have

$$g(x,\sigma) = \log\sigma - \log\sigma|x| - \lambda\log^2(\sigma|x|) + c,$$
$$g_1(x,\sigma) = -2\frac{\lambda}{\sigma}\log(\sigma|x|),$$
$$g_2(x,\sigma) = 2\frac{\lambda}{\sigma^2}\log(\sigma|x|) - 2\frac{\lambda}{\sigma^2},$$

and thus

$$\tau_{h,f}^2 = E(\log|\eta_t|)^2.$$

In each case, $\tau_{h,f}^2$ does not depend on the parameter λ of h. We conclude that the estimators $\hat{\theta}_{n,h}$ exhibit the same asymptotic behaviour, regardless of the parameter λ. It can even be easily shown that the estimators themselves do not depend on λ.

9.9 1. The Laplace QML estimator applied to a GARCH in standard form (as defined in Example 9.4) is an example of such an estimator.

2. We have

$$\sigma_t^2(\theta^*) = \varrho^2\omega_0 + \sum_{i=1}^{q}\varrho^2\alpha_{0i}\epsilon_{t-i}^2 + \sum_{j=1}^{p}\beta_{0j}\sigma_{t-j}^2(\theta^*)$$
$$= \varrho^2\left(1-\sum_{j=1}^{p}\beta_{0j}B^j\right)^{-1}\left(\omega_0 + \sum_{i=1}^{q}\alpha_{0i}\epsilon_{t-i}^2\right) = \varrho^2\sigma_t^2(\theta_0).$$

3. Since

$$\frac{\partial\sigma_t^2(\theta)}{\partial\theta} = \left(1-\sum_{j=1}^{p}\beta_jB^j\right)^{-1}\begin{pmatrix}1\\ \epsilon_{t-1}^2\\ \vdots\\ \epsilon_{t-q}^2\\ \sigma_{t-1}^2(\theta)\\ \vdots\\ \sigma_{t-p}^2(\theta)\end{pmatrix},$$

we have

$$\frac{\partial\sigma_t^2(\theta^*)}{\partial\theta} = \varrho^2\Lambda_\varrho\frac{\partial\sigma_t^2(\theta_0)}{\partial\theta}, \quad \frac{1}{\sigma_t^2(\theta^*)}\frac{\partial\sigma_t^2(\theta^*)}{\partial\theta} = \Lambda_\varrho\frac{1}{\sigma_t^2(\theta_0)}\frac{\partial\sigma_t^2(\theta_0)}{\partial\theta}.$$

It follows that, using obvious notation,

$$J(\theta^*) = \Lambda_\varrho J(\theta_0)\Lambda_\varrho = \begin{pmatrix}\varrho^{-4}J_{11}(\theta_0) & \varrho^{-2}J_{12}(\theta_0)\\ \varrho^{-2}J_{21}(\theta_0) & J_{22}(\theta_0)\end{pmatrix}.$$

9.10 After reparameterisation, the result (9.27) applies with η_t replaced by $\eta_t^* = \eta_t/\varrho$, and θ_0 by

$$\theta^* = (\varrho^2\omega_0, \varrho^2\alpha_{01}, \dots, \varrho^2\alpha_{0q}, \beta_{01}, \dots, \beta_{0p})',$$

where $\varrho = \int |x| f(x)dx$. Thus, using Exercise 9.9, we obtain

$$\sqrt{n}(\hat{\theta}_{n,\ell} - \Lambda_{\varrho}^{-1}\theta_0) \xrightarrow{\mathcal{L}} \mathcal{N}(0, 4\tau_{\ell,f}^2 \Lambda_{\varrho}^{-1} J^{-1} \Lambda_{\varrho}^{-1})$$

with $\tau_{\ell,f}^2 = E\eta_t^{*2} - 1 = E\eta_t^2/\varrho^2 - 1$.

Chapter 10

10.1 The number of parameters of the diagonal GARCH(p, q) model is

$$\frac{m(m+1)}{2}(1+p+q),$$

that of the vectorial model is

$$\frac{m(m+1)}{2}\left\{1+(p+q)\frac{m(m+1)}{2}\right\},$$

that of the CCC model is

$$m\{1+(p+q)m^2\} + \frac{m(m-1)}{2},$$

that of the BEKK model is

$$\frac{m(m+1)}{2} + Km^2(p+q).$$

For $p = q = 1$ and $K = 1$ we obtain Table E.1.

10.2 Assume (10.100) and define $U(z) = \mathcal{B}_\theta(z)\mathcal{B}_{\theta_0}^{-1}(z)$. We have (10.101) because

$$U(z)\mathcal{A}_{\theta_0}(z) = \mathcal{B}_\theta(z)\mathcal{B}_{\theta_0}^{-1}(z)\mathcal{A}_{\theta_0}(z) = \mathcal{B}_\theta(z)\mathcal{B}_\theta(z)^{-1}\mathcal{A}_\theta(z) = \mathcal{A}_\theta(z)$$

and

$$U(z)\mathcal{B}_{\theta_0}(z) = \mathcal{B}_\theta(z)\mathcal{B}_{\theta_0}^{-1}(z)\mathcal{B}_{\theta_0}(z) = \mathcal{B}_\theta(z).$$

Conversely, it is easy to check that (10.101) implies (10.100).

10.3 1. Since X and Y are independent, we have

$$0 = \text{Cov}(X, Y) = \text{Cov}(X, X) = \text{Var}(X),$$

which shows that X is constant.

2. We have

$$\begin{aligned} P(X = x_1)P(X = x_2) = P(X = x_1)P(Y = x_2) &= P(X = x_1, Y = x_2) \\ &= P(X = x_1, X = x_2) \end{aligned}$$

which is nonzero only if $x_1 = x_2$, thus X and Y take only one value.

3. Assume that there exist two events A and B such that $P(X \in A)P(X \in B) > 0$ and $A \cap B = \emptyset$. The independence then entails that

$$P(X \in A)P(X \in B) = P(X \in A)P(Y \in B) = P(X \in A, X \in B),$$

and we obtain a contradiction.

Table E.1 Number of parameters as a function of m.

Model	$m = 2$	$m = 3$	$m = 5$	$m = 10$
Diagonal	9	18	45	165
Vectorial	21	78	465	6105
CCC	19	60	265	2055
BEKK	11	24	96	186

10.4 For all $x \in \mathbb{R}^{m(m+1)/2}$, there exists a symmetric matrix X such that $x = \text{vech}(X)$, and we have

$$\mathrm{D}_m^+\mathrm{D}_m x = \mathrm{D}_m^+\mathrm{D}_m \text{vech}(X) = \mathrm{D}_m^+ \text{vec}(X) = \text{vech}(X) = x.$$

10.5 The matrix $A'A$ being symmetric and real, there exist an orthogonal matrix C ($C'C = CC' = I$) and a diagonal matrix D such that $A'A = CDC'$. Thus, denoting by λ_j the (positive) eigenvalues of $A'A$, we have

$$x'A'Ax = x'CDC'x = y'Dy = \sum_{j=1}^{d} \lambda_j y_j^2,$$

where $y = C'x$ has the same norm as x. Assuming, for instance, that $\lambda_1 = \rho(A'A)$, we have

$$\sup_{\|x\|\leq 1} \|Ax\|^2 = \sup_{\|y\|\leq 1} \sum_{j=1}^{d} \lambda_j y_j^2 \leq \lambda_1 \sum_{j=1}^{d} y_j^2 \leq \lambda_1.$$

Moreover, this maximum is reached at $y = (1, 0, \dots, 0)'$.

An alternative proof is obtained by noting that $\|A\|^2$ solves the maximization problem of the function $f(x) = x'A'Ax$ under the constraint $x'x = 1$. Introduce the Lagrangian

$$\mathcal{L}(x, \lambda) = x'A'Ax - \lambda(x'x - 1).$$

The first-order conditions yield the constraint and

$$\frac{\mathcal{L}(x, \lambda)}{\partial x} = 2A'Ax - 2\lambda x = 0.$$

This shows that the constrained optimum is located at a normalized eigenvector x_i associated with an eigenvalue λ_i of $A'A$, $i = 1, \dots, d$. Since $f(x_i) = x_i'A'Ax_i = \lambda_i x_i'x_i = \lambda_i$, we of course have $\|A\|^2 = \max_{i=1,\dots,d} \lambda_i$.

10.6 Since all the eigenvalues of the matrix $A'A$ are real and positive, the largest eigenvalue of this matrix is less than the sum of all its eigenvalues, that is, of its trace. Using the second equality of (10.67), the first inequality of (10.68) follows. The second inequality follows from the same arguments, and noting that there are d_2 eigenvalues. The last inequality uses the fact that the determinant is the product of the eigenvalues and that each eigenvalue is less than $\|A\|^2$.

The first inequality of (10.69) is a simple application of the Cauchy–Schwarz inequality. The second inequality of (10.69) is obtained by twice applying the second inequality of (10.68).

10.7 For the positivity of H_t for all $t > 0$ it suffices to require Ω to be symmetric positive definite, and the initial values $H_0, \dots, H_{1-p}$ to be symmetric positive semi-definite. Indeed, if the H_{t-j} are symmetric and positive semi-definite then H_t is symmetric if and only Ω is symmetric, and we have, for all $\lambda \in \mathbb{R}^m$,

$$\lambda' H_t \lambda = \lambda'\Omega\lambda + \sum_{i=1}^{q} \alpha_i \{\lambda' \epsilon_{t-i}\}^2 + \sum_{j=1}^{p} \beta_j \lambda' H_{t-j} \lambda \geq \lambda'\Omega\lambda.$$

We now give a second-order stationarity condition. If $H := E(\epsilon_t\epsilon_t')$ exists, then this matrix is symmetric positive semi-definite and satisfies

$$H = \Omega + \sum_{i=1}^{q} \alpha_i H + \sum_{j=1}^{p} \beta_j H,$$

that is,

$$\left(1 - \sum_{i=1}^{q} \alpha_i - \sum_{j=1}^{p} \beta_j\right) H = \Omega.$$

If Ω is positive definite, it is then necessary to have

$$\sum_{i=1}^{q} \alpha_i + \sum_{j=1}^{p} \beta_j < 1. \tag{E.14}$$

For the reverse we use Theorem 10.5. Since the matrices $C^{(i)}$ are of the form $c_i I_s$ with $c_i = \alpha_i + \beta_i$, the condition $\rho(\sum_{i=1}^{r} C^{(i)}) < 1$ is equivalent to (E.14). This condition is thus sufficient to obtain the stationarity, under technical condition (ii) of Theorem 10.5 (which can perhaps be relaxed). Let us also mention that, by analogy with the univariate case, it is certainly possible to obtain the strict stationarity under a condition weaker than (E.14).

10.8 For the convergence in L^p, it suffices to show that (u_n) is a Cauchy sequence:

$$\begin{aligned}\|u_n - u_m\|_p &\le \|u_{n+1} - u_n\|_p + \|u_{n+2} - u_{n+1}\|_p + \cdots + \|u_m - u_{m-1}\|_p\\ &\le (m-n)C^{1/p}\rho^{n/p} \to 0\end{aligned}$$

when $m > n \to \infty$. To show the almost sure convergence, let us begin by noting that, using Hölder's inequality,

$$E|u_n - u_{n-1}| \le \{E|u_n - u_{n-1}|^p\}^{1/p} \le C^* \rho^{*n},$$

with $C^* = C^{1/p}$ and $\rho^* = \rho^{1/p}$. Let $v_1 = u_1$, $v_n = u_n - u_{n-1}$ for $n \ge 2$, and $v = \sum_n |v_n|$ which is defined in $\mathbb{R} \cup \{+\infty\}$, *a priori*. Since

$$Ev \le C^* \sum_{n=1}^{\infty} \rho^{*n} < \infty,$$

it follows that v is almost surely defined in $\mathbb{R}^+$ and $u = \sum_{n=1}^{\infty} v_n$ is almost surely defined in $\mathbb{R}$. Since $u_n = v_1 + \cdots + v_n$, we have $u = \lim_{n\to\infty} u_n$ almost surely.

10.9 It suffices to note that $pR + (1-p)Q$ is the correlation matrix of a vector of the form $\sqrt{p}X + \sqrt{1-p}Y$, where X and Y are independent vectors of the respective correlation matrices R and Q.

10.10 Since the $\boldsymbol{\beta}_j$ are linearly independent, there exist vectors $\boldsymbol{\alpha}_k$ such that $\{\boldsymbol{\alpha}_1, \dots, \boldsymbol{\alpha}_m\}$ forms a basis of $\mathbb{R}^m$ and such that $\boldsymbol{\alpha}_k'\boldsymbol{\beta}_j = \mathbb{1}_{\{k=j\}}$ for all $j = 1, \dots, r$ and all $k = 1, \dots, m$.[3] We then have

$$\lambda_{jt}^* := \mathrm{Var}(\boldsymbol{\alpha}_j'\epsilon_t \mid \epsilon_u, u < t) = \boldsymbol{\alpha}_j' H_t \boldsymbol{\alpha}_j = \boldsymbol{\alpha}_j'\Omega\boldsymbol{\alpha}_j + \lambda_{jt},$$

and it suffices to take

$$\Omega^* = \Omega - \sum_{j=1}^{r} (\boldsymbol{\alpha}_j'\Omega\boldsymbol{\alpha}_j)\boldsymbol{\beta}_j\boldsymbol{\beta}_j'.$$

[3] The $\boldsymbol{\beta}_1, \dots, \boldsymbol{\beta}_r$ can be extended to obtain a basis of $\mathbb{R}^m$. Let B be the $m \times m$ matrix of these vectors in the canonical basis. This matrix B is necessarily invertible. We can thus take the lines of B^{-1} as vectors $\boldsymbol{\alpha}_k$.

The conditional covariance between the factors $\alpha_j'\epsilon_t$ and $\alpha_k'\epsilon_t$, for $j \neq k$, is

$$\alpha_j' H_t \alpha_k = \alpha_j' \Omega \alpha_k,$$

which is a nonzero constant in general.

10.11 As in the proof of Exercise 10.10, define vectors α_j such that

$$\alpha_j' H_t \alpha_j = \alpha_j' \Omega \alpha_j + \lambda_{jt}.$$

Denoting by $\mathbf{e}_j$ the jth vector of the canonical basis of $\mathbb{R}^m$, we have

$$H_t = \Omega + \sum_{j=1}^{r} \beta_j \{\omega_j + a_j \mathbf{e}_j' \epsilon_{t-1}\epsilon_{t-1}' \mathbf{e}_j + b_j(\alpha_j' H_{t-1}\alpha_j - \alpha_j'\Omega\alpha_j)\}\beta_j',$$

and we obtain the BEKK representation with $K = r$,

$$\Omega^* = \Omega + \sum_{j=1}^{r} \{\omega_j - b_j \alpha_j' \Omega \alpha_j\}\beta_j \beta_j', \quad A_k = \sqrt{a_k}\beta_k \mathbf{e}_k', \quad B_k = \sqrt{b_k}\beta_k \alpha_k'.$$

10.12 Consider the Lagrange multiplier λ_1 and the Lagrangian $u_1'\Sigma u_1 - \lambda_1(u_1'u_1 - 1)$. The first-order conditions yield

$$2\Sigma u_1 - 2\lambda_1 u_1 = 0,$$

which shows that u_1 is an eigenvector associated with an eigenvalue λ_1 of Σ. Left-multiplying the previous equation by u_1', we obtain

$$\lambda_1 = \lambda_1 u_1' u_1 = u_1' \Sigma u_1 = \text{Var} C^1,$$

which shows that λ_1 must be the largest eigenvalue of Σ. The vector u_1 is unique, up to its sign, provided that the largest eigenvalue has multiplicity order 1.

An alternative way to obtain the result is based on the spectral decomposition of the symmetric definite positive matrices

$$\Sigma = P\Lambda P', \quad PP' = I_m, \quad \Lambda = \text{diag}(\lambda_1, \dots, \lambda_m), \quad 0 \leq \lambda_m \leq \cdots \leq \lambda_1.$$

Let $v_1 = Pu_1$, that is, $u_1 = P'v_1$. Maximizing $u_1'\Sigma u_1$ is equivalent to maximizing $v_1'\Lambda v_1$. The constraint $u_1'u_1 = 1$ is equivalent to the constraint $v_1'v_1 = 1$. Denoting by v_{i1} the components of v_1, the function $v_1'\Lambda v_1 = \sum_{i=1}^m v_{i1}^2 \lambda_i$ is maximized at $v_1 = (\pm 1, 0, \dots, 0)$ under the constraint, which shows that u_1 is the first column of P, up to the sign. We also see that other solutions exist when $\lambda_1 = \lambda_2$. It is now clear that the vector $P'X$ contains the m principal components of the variance matrix Λ.

10.13 All the elements of the matrices $\mathrm{D}_m^+ A_{ik}$ and D_m are positive. Consequently, when A_{ik} is diagonal, using Exercise 10.4, we obtain

$$0 \leq \mathrm{D}_m^+(A_{ik} \otimes A_{ik})\mathrm{D}_m \leq \sup_{j \in \{1,\dots,m\}} A_{ik}^2(j,j) I_{m(m+1)/2}$$

element by element. This shows that $\mathrm{D}_m^+(A_{ik} \otimes A_{ik})\mathrm{D}_m$ is diagonal, and the conclusion easily follows.

10.14 With the abuse of notation $B - \lambda I_{mp} = C(\mathbf{B}_1, \dots, \mathbf{B}_p)$, the property yields

$$\det(B - \lambda I_{mp}) = \det(-\lambda I_m) \det \left\{ C(\mathbf{B}_1, \dots, \mathbf{B}_{p-1}) + \frac{1}{\lambda} \begin{pmatrix} \mathbf{B}_p \\ 0 \end{pmatrix} (0 \ I_m) \right\}$$
$$= \det(-\lambda I_m) \det \left\{ C\left(\mathbf{B}_1, \dots, \mathbf{B}_{p-2}, \mathbf{B}_{p-1} + \frac{1}{\lambda}\mathbf{B}_p\right) \right\}.$$

The proof is completed by induction on p.

10.15 Let $A^{1/2}$ be the symmetric positive definite matrix defined by $A^{1/2}A^{1/2} = A$. If X is an eigenvector associated with the eigenvalue λ of the symmetric positive definite matrix $S = A^{1/2}BA^{1/2}$, then we have $ABA^{1/2}X = \lambda A^{1/2}X$, which shows that the eigenvalues of S and AB are the same. Write the spectral decomposition as $S = P\Lambda P'$ where Λ is diagonal and $P'P = I_m$. We have $AB = A^{1/2}SA^{-1/2} = A^{1/2}P\Lambda P'A^{-1/2} = Q\Lambda Q^{-1}$, with $Q = A^{1/2}P$.

10.16 Let $c = (c_1', c_2')'$ be a nonzero vector such that

$$c'(A + B)c = (c_1'A_{11}c_1 + 2c_1'A_{12}c_2 + c_2'A_{22}c_2) + c_1'B_{11}c_1.$$

On the right-hand side of the equality, the term in parentheses is nonnegative and the last term is positive, unless $c_1 = 0$. But in this case $c_2 \neq 0$ and the term in parentheses becomes $c_2'A_{22}c_2 > 0$.

10.17 Take the random matrix $A = XX'$, where $X \sim \mathcal{N}(0, I_p)$ with $p > 1$. Obviously A is never positive definite because this matrix always possesses the eigenvalue 0 but, for all $c \neq 0$, $c'Ac = (c'X)^2 > 0$ with probability 1.

Chapter 11

11.1 For $\omega = 0$, we obtain the geometric Brownian motion whose solution, in view of (11.13), is equal to

$$X_t^0 = x_0 \exp\{(\mu - \sigma^2/2)t + \sigma W_t\}.$$

By Itô's formula, the SDE satisfied by $1/X_t^0$ is

$$d\left(\frac{1}{X_t^0}\right) = \frac{1}{X_t^0}\{(-\mu + \sigma^2)dt - \sigma dW_t\}.$$

Using the hint, we then have

$$dY_t = X_t d\left(\frac{1}{X_t^0}\right) + \frac{1}{X_t^0}dX_t + \sigma^2\frac{X_t}{X_t^0}dt$$
$$= \frac{\omega}{X_t^0}dt.$$

It follows that

$$X_t = X_t^0\left(1 + \omega \int_0^t \frac{1}{X_s^0}ds\right).$$

The positivity follows.

11.2 It suffices to check the conditions of Theorem 11.1, with the Markov chain $(X_{k\tau})$. We have

$$\tau^{-1}E(X_{k\tau}^{(\tau)} - X_{(k-1)\tau}^{(\tau)} \mid X_{(k-1)\tau}^{(\tau)} = x) = \mu(x),$$

$$\tau^{-1}\mathrm{Var}(X_{k\tau}^{(\tau)} - X_{(k-1)\tau}^{(\tau)} \mid X_{(k-1)\tau}^{(\tau)} = x) = \sigma^2(x),$$

$$\tau^{-\frac{2+\delta}{2}}E\left(\left|X_{k\tau}^{(\tau)} - X_{(k-1)\tau}^{(\tau)}\right|^{2+\delta} \mid X_{k\tau}^{(\tau)} = x\right) = E\left(\left|\mu(x)\sqrt{\tau} + \sigma(x)\epsilon_{(k+1)\tau}\right|^{2+\delta}\right) < \infty,$$

this inequality being uniform on any ball of radius r. The assumptions of Theorem 11.1 are thus satisfied.

11.3 One may take, for instance,

$$\omega_\tau = \omega\tau, \quad \alpha_\tau = \tau, \quad \beta_\tau = 1 - (1+\delta)\tau.$$

It is then easy to check that the limits in (11.23) and (11.25) are null. The limiting diffusion is thus

$$\begin{cases} dX_t &= f(\sigma_t)dt + \sigma_t dW_t^1 \\ d\sigma_t^2 &= \quad (\omega - \delta\sigma_t^2)dt \end{cases}$$

The solution of the σ_t^2 equation is, using Exercise 11.1 with $\sigma = 0$,

$$\sigma_t^2 = e^{-\delta t}(\sigma_0^2 - \omega/\delta) + \omega/\delta, \quad t \geq 0,$$

where σ_0^2 is the initial value. It is assumed that $\sigma_0^2 > \omega/\delta$ and $\delta > 0$, in order to guarantee the positivity. We have

$$\lim_{t\to\infty} \sigma_t^2 = \omega/\delta.$$

11.4 In view of (11.34) the put price is $P(S,t) = e^{-r(T-t)}E^\pi[(K - S_T)^+ \mid S_t]$. We have seen that the discounted price is a martingale for the risk-neutral probability. Thus $e^{-r(T-t)}E^\pi[S_T \mid S_t] = S_t$. Moreover,

$$(S_T - K)^+ - (K - S_T)^+ = S_T - K.$$

The result is obtained by multiplying this equality by $e^{-r(T-t)}$ and taking the expectation with respect to the probability π.

11.5 A simple calculation shows that $\frac{\partial C(S,t)}{\partial S_t} = \Phi(x_t + \sigma\sqrt{\tau}) \in (0,1)$.

11.6 In view of (11.36), Itô's formula applied to $C_t = C(S,t)$ yields

$$dC_t = \left(\frac{\partial C_t}{\partial t} + \frac{\partial C_t}{\partial S_t}\mu S_t + \frac{1}{2}(\sigma S_t)^2\frac{\partial^2 C_t}{\partial S_t^2}\right)dt + \frac{\partial C_t}{\partial S_t}\sigma S_t dW_t := \mu_t C_t dt + \sigma_t C_t dW_t,$$

with, in particular, $\sigma_t = \frac{1}{C_t}\frac{\partial C_t}{\partial S_t}\sigma S_t$. In view of Exercise 11.5, we thus have

$$\frac{\sigma_t}{\sigma} = \frac{S_t\Phi(x_t + \sigma\sqrt{\tau})}{C_t} = \frac{S_t\Phi(x_t + \sigma\sqrt{\tau})}{S_t\Phi(x_t + \sigma\sqrt{\tau}) - e^{-r\tau}K\Phi(x_t)} > 1.$$

11.7 Given observations $S_1, \dots, S_n$ of model (11.31), and an initial value S_0, the maximum likelihood estimators of $m = \mu - \sigma^2/2$ and σ^2 are, in view of (11.33),

$$\hat{m} = \frac{1}{n}\sum_{i=1}^n \log(S_i/S_{i-1}), \quad \hat{\sigma}^2 = \frac{1}{n}\sum_{i=1}^n \{\log(S_i/S_{i-1}) - \hat{m}\}^2.$$

The maximum likelihood estimator of μ is then

$$\hat{\mu} = \hat{m} + \frac{\hat{\sigma}^2}{2}.$$

11.8 Denoting by ϕ the density of the standard normal distribution, we have

$$\frac{\partial C(S,t)}{\partial \sigma} = S_t\phi(x_t + \sigma\sqrt{\tau})\left(\frac{\partial x_t}{\partial \sigma} + \sqrt{\tau}\right) - e^{-r\tau}K\phi(x_t)\frac{\partial x_t}{\partial \sigma}.$$

It is easy to verify that $S_t\phi(x_t + \sigma\sqrt{\tau}) = e^{-r\tau}K\phi(x_t)$. It follows that

$$\frac{\partial C(S,t)}{\partial \sigma} = S_t\sqrt{\tau}\phi(x_t + \sigma\sqrt{\tau}) > 0.$$

The option buyer wishes to be covered against the risk: he thus agrees to pay more if the asset is more risky.

11.9 We have $S_t = S_{t-1}e^{r-\sigma_t^2/2+\sigma_t\eta_t^*}$ where $(\eta_t^*) \overset{iid}{\sim} \mathcal{N}(0,1)$. It follows that

$$E(S_t \mid I_{t-1}) = S_{t-1}\exp\left(r - \frac{\sigma_t^2}{2} + \frac{\sigma_t^2}{2}\right) = e^r S_{t-1}.$$

The property immediately follows.

11.10 The volatility is of the form $\sigma_t^2 = \omega + a(\eta_{t-1})\sigma_{t-1}^2$ with $a(x) = \omega + \{\alpha(x-\gamma)^2 + \beta\}\sigma_{t-1}^2$. Using the results of Chapter 2, the strict and second-order stationarity conditions are

$$E\log a(\eta_t) < 0 \quad \textit{and} \quad Ea(\eta_t) < 1.$$

We are in the framework of model (11.44), with $\mu_t = r + \lambda\sigma_t - \sigma_t^2/2$. Thus the risk-neutral model is given by (11.47), with $\eta_t^* = \lambda + \eta_t$ and

$$\sigma_t^2 = \omega + \{\alpha(\eta_{t-1}^* - \lambda - \gamma)^2 + \beta\}\sigma_{t-1}^2.$$

11.11 The constraints (11.41) can be written as

$$\begin{aligned} e^{-r} &= E\exp(a_t + b_t\eta_{t+1} + c_t\eta_{t+1}^2 \mid I_t), \\ 1 &= E\exp\{a_t + b_t\eta_{t+1} + c_t\eta_{t+1}^2 + Z_{t+1} \mid I_t\} \\ &= E\exp\{a_t + \mu_{t+1} + \sigma_{t+1}\eta_{t+1} + b_t\eta_{t+1} + c_t\eta_{t+1}^2 \mid I_t\}. \end{aligned}$$

It can easily be seen that if $U \sim \mathcal{N}(0,1)$ we have, for $a < 1/2$ and for all b,

$$E[\exp\{a(U+b)^2\}] = \frac{1}{\sqrt{1-2a}}\exp\left(\frac{ab^2}{1-2a}\right).$$

Writing

$$b_t\eta_{t+1} + c_t\eta_{t+1}^2 = c_t\left\{\eta_{t+1} + \frac{b_t}{2c_t}\right\}^2 - \frac{b_t^2}{4c_t},$$

we thus obtain

$$1 = \frac{1}{\sqrt{1-2c_t}}\exp\left(a_t + r + \frac{b_t^2}{2(1-2c_t)}\right), \quad c_t < 1/2;$$

and writing

$$\sigma_{t+1}\eta_{t+1} + b_t\eta_{t+1}^2 = b_t\left\{\eta_{t+1} + \frac{\sigma_{t+1}}{2b_t}\right\}^2 - \frac{\sigma_{t+1}^2}{4b_t},$$

we have

$$1 = \frac{1}{\sqrt{1-2c_t}} \exp\left(a_t + \mu_{t+1} + \frac{(b_t + \sigma_{t+1})^2}{2(1-2c_t)}\right).$$

It follows that

$$\frac{(2b_t + \sigma_{t+1})\sigma_{t+1}}{2(1-2c_t)} = r - \mu_{t+1} = \frac{\sigma_{t+1}^2}{2} - \lambda\sigma_{t+1}.$$

Thus

$$\frac{2b_t + \sigma_{t+1}}{2(1-2c_t)} = \frac{\sigma_{t+1}}{2} - \lambda.$$

There are an infinite number of possible choices for b_t and c_t. For instance, if $\lambda = 0$, one can take $b_t = \nu\sigma_{t+1}$ and $c_t = -\nu$ with $\nu > -1/2$. Then a_t follows. The risk-neutral probability $\pi_{t,t+1}$ is obtained by calculating

$$\begin{aligned} E^{\pi_{t,t+1}}(e^{uZ_{t+1}} \mid I_t) &= E\left(e^{a_t + r + u\mu_{t+1} + (b_t + u\sigma_{t+1})\eta_{t+1} + c_t\eta_{t+1}^2} \mid I_t\right) \\ &= \exp\left\{u\left(r - \frac{\sigma_{t+1}^2}{2(1-2c_t)}\right) + u^2 \frac{\sigma_{t+1}^2}{2(1-2c_t)}\right\}. \end{aligned}$$

Under the risk-neutral probability, we thus have the model

$$\begin{cases} \log(S_t/S_{t-1}) = r - \frac{\sigma_t^2}{2(1-2c_{t-1})} + \epsilon_t^*, \\ \epsilon_t^* = \frac{\sigma_t}{\sqrt{1-2c_{t-1}}}\eta_t^*, \qquad (\eta_t^*) \overset{iid}{\sim} \mathcal{N}(0,1). \end{cases} \tag{E.15}$$

Note that the volatilities of the two models (under historical and risk-neutral probability) do not coincide unless $c_t = 0$ for all t.

11.12 We have $\text{VaR}_{t,1}(\alpha) = -\log(2\alpha)/\lambda$. It can be shown that the distribution of $L_{t,t+2}$ has the density $g(x) = 0.25\lambda\exp\{-\lambda|x|\}(1 + \lambda|x|)$. At horizon 2, the VaR is thus the solution u of the equation $(2 + \lambda u)\exp\{-\lambda u\} = 4\alpha$. For instance, for $\lambda = 0.1$ we obtain $\text{VaR}_{t,2}(0.01) = 51.92$, whereas $\sqrt{2}\text{VaR}_{t,1}(0.01) = 55.32$. The VaR is thus underevaluated when the incorrect rule is applied, but for other values of α the VaR may be overevaluated: $\text{VaR}_{t,2}(0.05) = 32.72$, whereas $\sqrt{2}\text{VaR}_{t,1}(0.05) = 32.56$.

11.13 We have

$$\Delta P_{t+i} - m = A^i(\Delta P_t - m) + U_{t+i} + AU_{t+i-1} + \cdots + A^{i-1}U_{t+1}.$$

Thus, introducing the notation $A_i = (I - A^i)(I - A)^{-1}$,

$$\begin{aligned} L_{t,t+h} &= -a' \sum_{i=1}^{h} \left(m + A^i(\Delta P_t - m) + \sum_{j=1}^{i} A^{i-j}U_{t+j}\right) \\ &= -a'mh - a'AA_h(\Delta P_t - m) - a' \sum_{j=1}^{h} \left(\sum_{i=j}^{h} A^{i-j}\right) U_{t+j} \\ &= -a'mh - a'AA_h(\Delta P_t - m) - a' \sum_{j=1}^{h} A_{h-j+1}U_{t+j}. \end{aligned}$$

The conditional law of $L_{t,t+h}$ is thus the $\mathcal{N}(a'\mu_{t,h}, a'\Sigma_h a)$ distribution, and (11.58) follows.

11.14 We have

$$\Delta P_{t+2} = \sqrt{\omega + \alpha_1 \Delta P_{t+1}^2}\,U_{t+2} = \sqrt{\omega + \alpha_1(\omega + \alpha_1 \Delta P_t^2)U_{t+1}^2}\,U_{t+2}.$$

At horizon 2, the conditional distribution of ΔP_{t+2} is not Gaussian if $\alpha_1 > 0$, because its kurtosis coefficient is equal to

$$\frac{E_t \Delta P_{t+2}^4}{(E_t \Delta P_{t+2}^2)^2} = 3\left(1 + \frac{2\theta_t^2}{(\omega + \theta_t)^2}\right) > 3, \qquad \theta_t = \alpha_1(\omega + \alpha_1 \Delta P_t^2).$$

There is no explicit formula for F_h when $h > 1$.

11.15 It suffices to note that, conditionally on the available information I_t, we have

$$\begin{aligned}\alpha &= P\left(\frac{p_{t+h} - p_t}{p_t} < \frac{-\text{VaR}_t(h,\alpha)}{p_t} \mid I_t\right)\\ &= P\left\{r_{t+1} + \cdots + r_{t+h} < \log\left(1 - \frac{\text{VaR}_t(h,\alpha)}{p_t}\right) \mid I_t\right\}.\end{aligned}$$

11.16 For simplicity, in this proof we will omit the indices. Since $L_{t,t+h}$ has the same distribution as $F^{-1}(U)$, where U denotes a variable uniformly distributed on $[0,1]$, we have

$$\begin{aligned}E[L_{t,t+h}\mathbb{1}_{L_{t,t+h}>\text{VaR}(\alpha)}] &= E[F^{-1}(U)\mathbb{1}_{F^{-1}(U)>F^{-1}(1-\alpha)}]\\ &= E[F^{-1}(U)\mathbb{1}_{U>1-\alpha}]\\ &= \int_{1-\alpha}^{1} F^{-1}(u)du = \int_0^{\alpha} F^{-1}(1-u)du\\ &= \int_0^{\alpha} \text{VaR}_{t,h}(u)du.\end{aligned}$$

Using (11.63), the desired equality follows.

11.17 The monotonicity, homogeneity and invariance properties follow from (11.62) and from the VaR properties. For $L_3 = L_1 + L_2$ we have

$$\begin{aligned}&\alpha\{\text{ES}_1(\alpha) + \text{ES}_2(\alpha) - \text{ES}_3(\alpha)\}\\ &= E[L_1(\mathbb{1}_{L_1\geq \text{VaR}_1(\alpha)} - \mathbb{1}_{L_3\geq \text{VaR}_3(\alpha)})] + E[L_2(\mathbb{1}_{L_2\geq \text{VaR}_1(\alpha)} - \mathbb{1}_{L_3\geq \text{VaR}_3(\alpha)})].\end{aligned}$$

Note that

$$(L_1 - \text{VaR}_1(\alpha))(\mathbb{1}_{L_1\geq \text{VaR}_1(\alpha)} - \mathbb{1}_{L_3\geq \text{VaR}_3(\alpha)}) \geq 0$$

because the two bracketed terms have the same sign. It follows that

$$\begin{aligned}\alpha\{\text{ES}_1(\alpha) + \text{ES}_2(\alpha) - \text{ES}_3(\alpha)\} &\geq \text{VaR}_1(\alpha))E[\mathbb{1}_{L_1\geq \text{VaR}_1(\alpha)} - \mathbb{1}_{L_3\geq \text{VaR}_3(\alpha)}]\\ &\quad + \text{VaR}_2(\alpha))E[\mathbb{1}_{L_2\geq \text{VaR}_2(\alpha)} - \mathbb{1}_{L_3\geq \text{VaR}_3(\alpha)}]\\ &= 0.\end{aligned}$$

The property is thus shown.

11.18 The volatility equation is

$$\sigma_t^2 = a(\eta_{t-1})\sigma_{t-1}^2, \quad \text{where } a(x) = \lambda + (1-\lambda)x^2.$$

It follows that

$$\sigma_t^2 = a(\eta_{t-1}) \ldots a(\eta_0)\sigma_0^2.$$

We have $E \log a(\eta_t) < \log Ea(\eta_t) = 0$, the inequality being strict because the distribution of $a(\eta_t)$ is nondegenerate. In view of Theorem 2.1, this implies that $a(\eta_{t-1}) \dots a(\eta_0) \to 0$ a.s., and thus that $\sigma_t^2 \to 0$ a.s., when t tends to infinity.

11.19 Given, for instance, $\sigma_{t+1} = 1$, we have $r_{t+1} = \eta_{t+1} \sim \mathcal{N}(0, 1)$ and the distribution of

$$r_{t+2} = \sqrt{\lambda + (1 - \lambda)\eta_{t+1}^2}\eta_{t+2}$$

is not normal. Indeed, $Er_{t+2} = 0$ and $\mathrm{Var}(r_{t+2}) = 1$, but $E(r_{t+2}^4) = 3\{1 + 2(1 - \lambda)^2\} \neq 3$. Similarly, the variable

$$r_{t+1} + r_{t+2} = \eta_{t+1} + \sqrt{\lambda + (1 - \lambda)\eta_{t+1}^2}\eta_{t+2}$$

is centered with variance 2, but is not normally distributed because

$$E(r_{t+1} + r_{t+2})^4 = 6\{1 + (2 - \lambda)^2\} \neq 12.$$

Note that the distribution is much more leptokurtic when λ is close to 0.

Chapter 12

12.1 Recall that the expectation of an infinite product of independent variables is not necessarily equal to the product of the expectations (see Exercise 4.2). This explains why it seems necessary to impose the finiteness of the product of the $E\{\exp(\sigma \mid \beta^i v_t|)\}$ (instead of the α_is).
We have, using the independence assumptions on the sequences (η_t) and (v_t),

$$E(\epsilon_t) = E\left\{\exp\left(\frac{\omega}{2(1-\beta)} + \frac{\sigma}{2}\sum_{i=0}^{\infty}\beta^i v_{t-i}\right)\right\}E(\eta_t) = 0,$$

provided that the expectation of the term between accolades exists and is finite. To show this, write

$$\begin{aligned} Z_{t,n} &:= \exp\left(\frac{\omega}{2(1-\beta)} + \frac{\sigma}{2}\sum_{i=0}^{n}\beta^i v_{t-i}\right) \\ &\leq \exp\left(\frac{\omega}{2(1-\beta)} + \frac{\sigma}{2}\sum_{i=0}^{n}|\beta^i v_{t-i}|\right) := W_{t,n} \leq W_{t,\infty}. \end{aligned}$$

We have

$$\begin{aligned} E(\lim \uparrow W_{t,n}) &= \lim \uparrow E(W_{t,n}) \\ &= \lim \uparrow \exp\left(\frac{\omega}{2(1-\beta)}\right) E\left\{\exp\left(\frac{\sigma}{2}\sum_{i=0}^{n}|\beta^i v_{t-i}|\right)\right\} \\ &= \lim \uparrow \exp\left(\frac{\omega}{2(1-\beta)}\right)\prod_{i=0}^{n} E\exp\left(\frac{\sigma}{2}|\beta^i v_{t-i}|\right) \\ &= \exp\left(\frac{\omega}{2(1-\beta)}\right)\prod_{i=0}^{\infty} E\exp\left(\frac{\sigma}{2}|\beta^i v_{t-i}|\right) < \infty. \end{aligned}$$

Thus $EZ_{t,\infty} < \infty$. Similarly, the same arguments show that

$$\begin{aligned}\mathrm{Var}(\epsilon_t) &= E\left\{\exp\left(\frac{\omega}{2(1-\beta)}+\frac{\sigma}{2}\sum_{i=0}^{\infty}\beta^i v_{t-i}\right)\right\}^2 E(\eta_t^2)\\ &= \exp\left(\frac{\omega}{1-\beta}\right)\prod_{i=0}^{\infty}E\{\exp(\sigma\beta^i v_{t-i})\} = e^{\frac{\omega}{1-\beta}}\prod_{i=0}^{\infty}\alpha_i.\end{aligned}$$

Moreover, for all $k > 0$,

$$\mathrm{Cov}(\epsilon_t, \epsilon_{t-k}) = 0$$

using again the independence between (η_t) and (v_t).

12.2 We have, for any $x \geq 0$:

$$\begin{aligned}P(\epsilon_t \leq x) &= P(h_t\eta_t^2 \leq x^2, \eta_t > 0) + P(\eta_t < 0)\\ &= E\{P(h_t\eta_t^2 \leq x^2, \eta_t > 0 \mid h_t)\} + P(\eta_t < 0)\\ &= E\{P(h_t\eta_t^2 \leq x^2 \mid \eta_t > 0, h_t)P(\eta_t > 0 \mid h_t)\} + P(\eta_t < 0)\\ &= E\left\{P(h_t\eta_t^2 \leq x^2 \mid h_t)\frac{1}{2}\right\} + \frac{1}{2} = \frac{1}{2}P(\ln \epsilon_t^2 \leq \ln x^2) + \frac{1}{2},\end{aligned}$$

where the last equality holds because η_t and (h_t) are independent, and because the law of η_t is symmetric. The same arguments show that $P(\epsilon_t \leq -x) = \frac{1}{2}P(\ln \epsilon_t^2 \geq \ln x^2)$ for $x \geq 0$. Thus, there is a one-to-one relation between the law of ϵ_t and that of $\ln \epsilon_t^2$. In addition the law of ϵ_t is symmetric. Similarly, for any $n \geq 1$, one can show that there is a one-to-one relation between the law of $(\epsilon_t, \dots, \epsilon_{t+n})$ and that of $(\ln \epsilon_t^2, \dots, \ln \epsilon_{t+n}^2)$. When the distribution of η_t is not symmetric, the fourth equality of the previous computation fails.

12.3 Let $Y_t = \log h_t$. The process (Y_t) being the solution of the AR(1) model $Y_t = \omega + \beta Y_{t-1} + \sigma v_t$ its mean and autocovariance function are given by

$$\mu_Y = E(Y_t) = \frac{\omega}{1-\beta}, \quad \gamma_Y(0) = \mathrm{Var}(Y_t) = \frac{\sigma^2}{1-\beta^2},$$

$$\gamma_Y(k) = \mathrm{Cov}(Y_t, Y_{t-k}) = \beta\gamma_Y(k-1), \quad k > 0.$$

From the independence between (Y_t) and (Z_t), (X_t) is a second-order process whose mean and autocovariance function are obtained as follows:

$$\mu_X = E(X_t) = E(Y_t) + E(Z_t) = \frac{\omega}{1-\beta} + \mu_Z,$$

$$\gamma_X(0) = Var(X_t) = Var(Y_t) + Var(Z_t) = \frac{\sigma^2}{1-\beta^2} + \sigma_Z^2,$$

$$\gamma_X(k) = \gamma_Y(k) = \beta\gamma_Y(k-1) = \beta^k\frac{\sigma^2}{1-\beta^2}, \quad k > 0.$$

Since $\gamma_X(k) = \beta\gamma_X(k-1), \quad \forall k > 1$, the process (X_t) admits an ARMA(1,1) representation of the form (12.7). The constant α is deduced from the first two autocovariances of (X_t). By Eq. (12.7) we have, denoting by σ_u^2, the variance of the noise in this representation

$$\gamma_X(0) = \beta\gamma_X(1) + \sigma_u^2\{1 + \alpha(\alpha - \beta)\}, \quad \gamma_X(1) = -\alpha\sigma_u^2 + \beta\gamma_X(0).$$

Hence, if $\sigma_Z^2 \neq 0$, the coefficient α is a solution of

$$\frac{1+\alpha(\alpha-\beta)}{\alpha} = \frac{\gamma_X(0) - \beta\gamma_X(1)}{\beta\gamma_X(0) - \gamma_X(1)} = \frac{\sigma^2 + \sigma_Z^2}{\beta\sigma_Z^2} \tag{E.16}$$

and the solution of modulus less than 1 is given by

$$\alpha = \frac{(1+\beta^2)\sigma_Z^2 + \sigma^2 - \{(1+\beta)^2\sigma_Z^2+\sigma^2\}^{1/2}\{(1-\beta)^2\sigma_Z^2+\sigma^2\}^{1/2}}{2\beta\sigma_Z^2}.$$

Moreover, the variance of the noise in model (12.7) is $\sigma_u^2 = \frac{\beta\sigma_Z^2}{\alpha}$ if $\beta \neq 0$ (and $\sigma_u^2 = \sigma^2 + \sigma_Z^2$ if $\beta = 0$). Finally, if $\sigma_Z^2 = 0$ the relation $\gamma_X(k) = \beta\gamma_X(k-1)$ also holds for $k=1$ and (X_t) is an AR(1) (i.e. $\alpha = 0$ in model (12.7)).

Now, when $\beta \neq 0$ and $\sigma \neq 0$, we get $(\alpha - \beta)(\beta - 1/\alpha) = \sigma^2/\sigma_Z^2 > 0$, using (E.16). It follows that either $0 < \alpha < \beta < 1/\alpha$ or $0 > \alpha > \beta > 1/\alpha$. In particular $|\alpha| < |\beta|$, which shows that the orders of the ARMA(1,1) representation for X_t are exact.

12.4 By expansion (12.3) and arguments used to prove Proposition 12.7, we obtain

$$\begin{aligned}
E(\epsilon_t) &= E\left(\exp\left\{\frac{\omega}{2(1-\beta)} + \frac{\sigma}{2}\sum_{i=0}^{\infty}\beta^i v_{t-i}\right\}\eta_t\right)\\
&= e^{\frac{\omega}{2(1-\beta)}} E\left(e^{\frac{\sigma}{2}v_t}\eta_t\right)\prod_{i=1}^{\infty} E\left(e^{\frac{\sigma}{2}\beta^i v_t}\right)\\
&= e^{\frac{\omega}{2(1-\beta)}} \frac{\sigma\rho}{2} e^{\frac{\sigma^2}{8}+\frac{\sigma^2\beta^2}{8(1-\beta^2)}} = e^{\frac{\omega}{2(1-\beta)}} \frac{\sigma\rho}{2} e^{\frac{\sigma^2}{8(1-\beta^2)}}.
\end{aligned}$$

We also have,

$$\begin{aligned}
E(\epsilon_{t-1}\epsilon_t) &= E\left(\exp\left\{\frac{\omega}{(1-\beta)} + \frac{\sigma}{2}\sum_{i=0}^{\infty}\beta^i v_{t-i-1} + \frac{\sigma}{2}\sum_{i=0}^{\infty}\beta^i v_{t-i}\right\}\eta_{t-1}\eta_t\right)\\
&= e^{\frac{\omega}{1-\beta}} E\left(e^{\frac{\sigma}{2}v_t}\eta_t\right) E\left(e^{\frac{\sigma}{2}(1+\beta)v_{t-1}}\eta_{t-1}\right)\prod_{i=2}^{\infty} E\left(e^{\frac{\sigma}{2}\beta^{i-1}(1+\beta)v_t}\right)\\
&= e^{\frac{\omega}{1-\beta}}\left(\frac{\sigma\rho}{2}\right)^2(1+\beta)e^{\frac{\sigma^2(3-\beta)}{8(1-\beta)}}.
\end{aligned}$$

Thus

$$\text{Cov}(\epsilon_{t-1}, \epsilon_t) = e^{\frac{\omega}{1-\beta}}\left(\frac{\sigma\rho}{2}\right)^2\left\{(1+\beta)e^{\frac{\sigma^2(3-\beta)}{8(1-\beta)}} - e^{\frac{\sigma^2}{4(1-\beta^2)}}\right\} > 0,$$

for $\rho \neq 0$ and $|\beta| < 1$.

12.5 The estimated models on the return series $\{r_t, t = 2, \ldots, 2122\}$ and $\{r_t, t = 2123, \ldots, 4245\}$ have the volatilities

$$(M1): \quad \sigma_t^2 = \underset{(0.015)}{0.098} + \underset{(0.012)}{0.087}\, r_{t-1}^2 + \underset{(0.020)}{0.840}\, \sigma_{t-1}^2,$$

$$(M2): \quad \sigma_t^2 = \underset{(0.004}{0.012} + \underset{(0.009)}{0.075}\, r_{t-1}^2 + \underset{(0.009)}{0.919}\, \sigma_{t-1}^2.$$

Denote by $\theta^{(1)} = (0.098, 0.087, 0.84)'$ and $\theta^{(2)} = (0.012, 0.075, 0.919)'$ the parameters of the two models. The estimated values of ω and β seem quite different. Denote by $\sigma_\omega^{(i)}$ and $\sigma_\beta^{(i)}$ the estimated standard deviations of the estimators of ω and β of Model Mi. It turns out that the confidence intervals

$$[\omega^{(1)} - 2\sigma_\omega^{(1)}, \omega^{(1)} + 2\sigma_\omega^{(1)}] = [0.068, 0.129]$$

and

$$[\omega^{(2)} - 2\sigma_{\omega}^{(2)}, \omega^{(2)} + 2\sigma_{\omega}^{(2)}] = [0.004, 0.020]$$

have empty intersection. The same holds true for the confidence intervals

$$[\beta^{(1)} - 2\sigma_{\beta}^{(1)}, \beta^{(1)} + 2\sigma_{\beta}^{(1)}] = [0.80, 0.88]$$

and

$$[\beta^{(2)} - 2\sigma_{\beta}^{(2)}, \beta^{(2)} + 2\sigma_{\beta}^{(2)}] = [0.90, 0.94].$$

The third graph of Figure E.4 displays the boxplot of the distribution of $\hat{\theta}^{(1)} - \theta^{(1)}$ on 100 independent simulations of Model *M1*. The difference $\theta^{(2)} - \theta^{(1)}$ between the parameters of *M1* and *M2* is marked by a diamond shape. The difference $\theta^{(2)} - \theta^{(1)}$ is an outlier for the distribution of $\hat{\theta}^{(1)} - \theta^{(1)}$, meaning that estimated GARCH on the two periods are significantly distinct.

12.6 Δ_n is a Markov chain, whose initial probability distribution is given by: $P(\Delta_0 = i) = \binom{d}{i}/2^d$, and whose transition matrix is:

$$\mathbb{P} = \begin{pmatrix} 0 & 1 & 0 & \cdots & 0 \\ 1/d & 0 & (d-1)/d & \cdots & 0 \\ 0 & 2/d & 0 & (d-2)/d & \\ \vdots & & & & \\ & & & \ddots & \\ 0 & \cdots & 0 & 1 & 0 \end{pmatrix}. \tag{E.17}$$

The number of balls in urn *A* changes successively from odd to even, and conversely, along the steps. For instance $p^{(2k+1)}(2i, 2i) = 0$. Thus the chain is irreducible but periodic.

Using the formula $\binom{d}{i} = \binom{d}{i-1}(d-i+1)/d + \binom{d}{i+1}(i+1)/d$, it can be seen that π_0 is an invariant law. It follows that $\pi_n = \pi_0$ for all n.

When the initial distribution is the Dirac mass at 0 we have $\pi_n(2i) = 0$ when n is odd, and $\pi_n(2i+1) = 0$ when n is even. Thus $\lim_{n\to\infty} \pi_n$ does not exist.

12.7 Let i and j be two different states, and let $d(i)$ be the period of state i. If the chain is irreducible, there exists an integer m_1 such that $p^{(m_1)}(i,j) > 0$ and m_2 such that $p^{(m_2)}(j, i) > 0$. The integer $d(i)$ divides $m_1 + m_2$ since $p^{(m_1+m_2)}(i, i) \geq p^{(m_1)}(i,j)p^{(m_2)}(j, i) > 0$. Similarly $d(i)$ divides $m_1 + m + m_2$ for all $m \in \{m : p^{(m)}(j, j) > 0\}$. Using $m = m + m_1 + m_2 - (m_1 + m_2) = k_1 d(i) - k_2 d(i) = (k_1 - k_2) d(i)$, it follows that $d(i)$ divides m for all $m \in \{m : p^{(m)}(j, j) > 0\}$. Since $d(j)$ is the gcd of $\{m : p^{(m)}(j, j) > 0\}$, and we have just shown that $d(i)$ is a common divisor of all the elements of this set, it follows that $d(i) \leq d(j)$. By symmetry, we also have $d(j) \leq d(i)$.

12.8 The key part of the code is the following:

```
# one iteration of the EM algorithm
EM <- function(omega,pi0,p,y){
d<-length(omega)
n <- length(y) # y contient les n observations
vrais<-0
pit.t<-matrix(0,nrow=d,ncol=n)
pit.tm1<-matrix(0,nrow=d,ncol=n+1)
vecphi<-rep(0,d)
pit.tm1[,1]<-pi0
for (t in 1:n)  {
 for (j in 1:d) vecphi[j]<-{dnorm(y[t],
          mean=0,sd=sqrt(abs(omega[j])))}
```

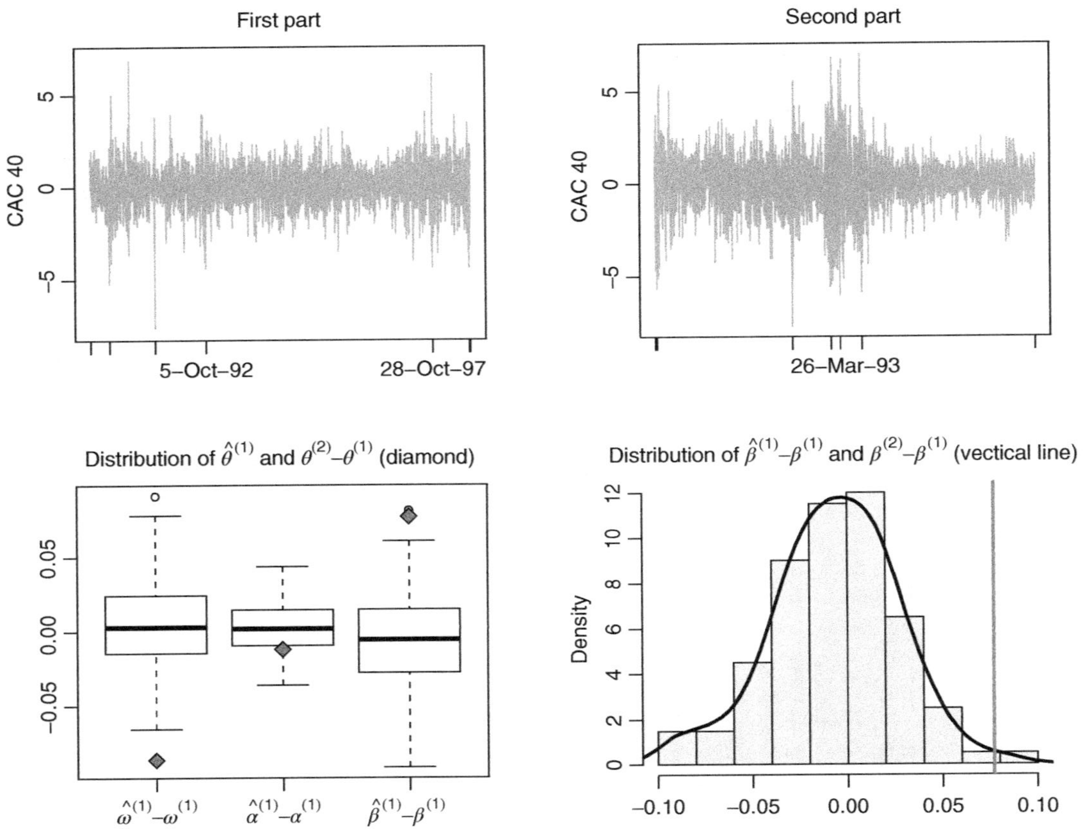

Figure E.4 The parameter $\theta^{(1)}$ (respectively, $\theta^{(2)}$) is that of a GARCH(1,1) fitted on the CAC 40 returns from March 1, 1990 to September 3, 1998 (respectively, from September 4, 1998 to December 29, 2006). The box plots display the empirical distributions of the estimated parameters $\hat{\theta}^{(1)}$ on 100 simulations of the model fitted on the first part of the CAC.

```
 den<-sum(pit.tm1[,t]*vecphi)
 if(den<=0)return(Inf)
 pit.t[,t]<-(pit.tm1[,t]*vecphi)/den
 pit.tm1[,t+1]<-t(p)%*%pit.t[,t]
 vrais<-vrais+log(den)
                   }
pit.n<-matrix(0,nrow=d,ncol=n)
pit.n[,n]=pit.t[,n]
for (t in n:2)  {
 for (i in 1:d)  {
 pit.n[i,t-1]<- {pit.t[i,t-1]*sum(p[i,1:d]*
                   pit.n[1:d,t]/pit.tm1[1:d,t])}
                   } }
pitm1et.n<-array(0,dim=c(d,d,n))
for (t in 2:n)  {
 for (i in 1:d)  {
  for (j in 1:d)  {
  pitm1et.n[i,j,t]<-p[i,j]*pit.t[i,t-1]*pit.n[j,t]/pit.tm1[j,t]
                   } } }
omega.final<-omega
pi0.final<-pi0
p.final<-p
for (i in 1:d)    {
omega.final[i]<-sum((y[1:n]^2)*pit.n[i,1:n])/sum(pit.n[i,1:n])
pi0.final[i]<-pit.n[i,1]
 for (j in 1:d) {
 p.final[i,j]<-sum(pitm1et.n[i,j,2:n])/sum(pit.n[i,1:(n-1)])
                  } }
liss<-{list(probaliss=pit.n,probatransliss=pitm1et.n,
vrais=vrais,omega.final=omega.final,pi0.final=pi0.final,
                 p.final=p.final)}
                              }
```

12.9 The last equality of Step 2 of the EM algoritm shows that $\pi_{t-1,t|n}(i_0, j_0) = 0$ for all t. Point 3, then shows that $p(i_0, j_0) \equiv 0$ in all the subsequent steps of the algorithm.

12.10 We have the Markovian representation $\underline{z}_t = \underline{b}_t + A_{0t}\underline{z}_{t-1}$, with

$$\underline{z}_t = (\epsilon_t^2, \dots, \epsilon_{t-q+1}^2, \sigma_t^2, \dots, \sigma_{t-p+1}^2)' \in \mathbb{R}^{p+q},$$

$$\underline{b}_t = (\omega(\Delta_t)\eta_t^2, 0, \dots, \omega, 0, \dots, 0)' \in \mathbb{R}^{p+q},$$

$$A_t = \begin{pmatrix} \alpha_1(\Delta_t)\eta_t^2 & \cdots & \alpha_q(\Delta_t)\eta_t^2 & \beta_1(\Delta_t)\eta_t^2 & \cdots & \beta_p(\Delta_t)\eta_t^2 \\ & I_{q-1} & 0 & & 0 & \\ \alpha_1(\Delta_t) & \cdots & \alpha_q(\Delta_t) & \beta_1(\Delta_t) & \cdots & \beta_p(\Delta_t) \\ & 0 & & & I_{p-1} & 0 \end{pmatrix}.$$

The proof of Theorem 2.4 in Chapter 2 applies directly with this sequence (A_t), showing that there exists a strictly stationary solution if and only if the top Lyapunov exponent of (A_t) is strictly negative. The solution is then unique, non-anticipative, and ergodic, and takes the form (2.18).

12.11 As in Example 2.1, the last exercise shows that the necessary and sufficient strict stationarity condition is

$$\gamma := E \log\{\alpha(\Delta_t)\eta_t^2 + \beta(\Delta_t)\} = \sum_{k=1}^{d} \pi(k)E_\eta \log\{\alpha(k)\eta_t^2 + \beta(k)\} < 0.$$

In the ARCH(1) case with d regimes, we obtain the necessary and sufficient condition

$$\sum_{d}^{k=1} \pi(k) \log \alpha(k) < -E \log \eta_t^2.$$

12.12 If (ϵ_t) is a strictly stationary and non-anticipative[4] solution and if the sequence (Δ_t) is iid, then $\alpha(\Delta_t)$ and $\beta(\Delta_t)$ are independent of ϵ_{t-1}^2 and σ_{t-1}^2. If in addition $E\epsilon_t^2 < \infty$ then, setting $a(\Delta_t, \eta_t) = \alpha(\Delta_t)\eta_t^2 + \beta(\Delta_t)$, we have

$$E\epsilon_t^2 = E\sigma_t^2 = E\omega(\Delta_t) + \{Ea(\Delta_t, \eta_t)\}E\sigma_t^2.$$

For the existence of a positive solution to this equation, it is necessary to have

$$Ea(\Delta_t, \eta_t) = E\alpha(\Delta_t) + E\beta(\Delta_t) = \sum_{k=1}^{d} \pi(k)\{\alpha(k) + \beta(k)\} < 1.$$

Conversely, under this condition, the process

$$\epsilon_t = \left\{ \omega(\Delta_t) + \sum_{i=1}^{\infty} a\left(\Delta_{t-1}, \eta_{t-1}\right) \dots a\left(\Delta_{t-i}, \eta_{t-i}\right) \omega(\Delta_{t-i-1}) \right\}^{1/2} \eta_t$$

is a strictly stationary and non-anticipative solution which satisfies

$$E\epsilon_t^2 = E\omega(\Delta_t) \left\{ 1 + \sum_{i=1}^{\infty} \{Ea(\Delta_t, \eta_t)\}^i \right\} < \infty.$$

12.13 Using the elementary inequality $\log x \leq x - 1$, we have

$$\begin{aligned}
0 &\leq Q(\theta^{(k+1)}, \pi_0^{(k+1)} \mid \theta^{(k)}, \pi_0^{(k)}) - Q(\theta^{(k)}, \pi_0^{(k)} \mid \theta^{(k)}, \pi_0^{(k)}) \\
&= \sum_{(e_1,\dots,e_n)\in\mathcal{E}^n} \log \frac{L_{\theta^{(k+1)},\pi_0^{(k+1)}}(\epsilon_1, \dots, \epsilon_n, e_1, \dots, e_n)}{L_{\theta^{(k)},\pi_0^{(k)}}(\epsilon_1, \dots, \epsilon_n, e_1, \dots, e_n)} \\
&\quad \times \mathbb{P}_{\theta^{(k)},\pi_0^{(k)}}(\Delta_1 = e_1, \dots, \Delta_n = e_n \mid \epsilon_1, \dots, \epsilon_n) \\
&\leq \sum_{(e_1,\dots,e_n)\in\mathcal{E}^n} \left\{ \frac{L_{\theta^{(k+1)},\pi_0^{(k+1)}}(\epsilon_1, \dots, \epsilon_n, e_1, \dots, e_n)}{L_{\theta^{(k)},\pi_0^{(k)}}(\epsilon_1, \dots, \epsilon_n, e_1, \dots, e_n)} - 1 \right\} \\
&\quad \times \frac{L_{\theta^{(k)},\pi_0^{(k)}}(\epsilon_1, \dots, \epsilon_n, e_1, \dots, e_n)}{L_{\theta^{(k)},\pi_0^{(k)}}(\epsilon_1, \dots, \epsilon_n)} \\
&= \frac{L_{\theta^{(k+1)},\pi_0^{(k+1)}}(\epsilon_1, \dots, \epsilon_n) - L_{\theta^{(k)},\pi_0^{(k)}}(\epsilon_1, \dots, \epsilon_n)}{L_{\theta^{(k)},\pi_0^{(k)}}(\epsilon_1, \dots, \epsilon_n)},
\end{aligned}$$

and the result follows.

[4] i.e. ϵ_t measurable with respect to $\{\Delta_u, \eta_u, u < t\}$.

12.14 Conditional to the initial variables $\epsilon_0^2, \dots, \epsilon_{1-q}^2$, Equations (12.25), (12.22)–(12.23), (12.26)–(12.28), and (12.29)–(12.31) remain valid, provided the density $\phi_k(\epsilon_t)$ is replaced by the density $\phi_k(\epsilon_t \mid \epsilon_{t-1}, \dots, \epsilon_{t-q})$ of the Gaussian $\mathcal{N}\left\{0, \omega(k) + \sum_{i=1}^q \alpha_i(k)\epsilon_{t-i}^2\right\}$ distribution (and by replacing the notation $M(\epsilon_t)$ by $M(\epsilon_t \mid \epsilon_{t-1}, \dots, \epsilon_{t-q})$).

The EM algorithm cannot be generalised trivially because the maximisation of Eq. (12.32) is replaced by that of

$$a_1 = a_1(\omega, \alpha, \beta) = \sum_{i=1}^{d} \sum_{t=1}^{n} \log \phi_i(\epsilon_t \mid \epsilon_{t-1}, \dots, \epsilon_{t-q}) \mathbb{1}_{\{\Delta_t = i\}},$$

which does not admit an explicit form like (12.35) but requires the use of an optimisation algorithm.

12.15 For the MS-GARCH(1,1) model $\epsilon_t = \sqrt{h_t}\eta_t$, with

$$h_t = \omega(\Delta_t) + \alpha(\Delta_t)\epsilon_{t-1}^2 + \beta(\Delta_t)h_{t-1},$$

we have

$$h_t = \omega(\Delta_t) + \alpha(\Delta_t)\epsilon_{t-1}^2 + \sum_{i=0}^{\infty} \beta(\Delta_t) \dots \beta(\Delta_{t-i})\{\omega(\Delta_{t-i-1}) + \alpha(\Delta_{t-i-1})\epsilon_{t-i-2}^2\}$$

under conditions entailing the existence of the series. For the alternative model, $\epsilon_t = \sigma_t(\Delta_t)\eta_t$ with

$$\sigma_t^2(\Delta_t) = \omega(\Delta_t) + \alpha(\Delta_t)\epsilon_{t-1}^2 + \beta(\Delta_t)\sigma_{t-1}^2(\Delta_t),$$

we have

$$\sigma_t^2(\Delta_t) = \{1 - \beta(\Delta_t)\}^{-1}\omega(\Delta_t) + \alpha(\Delta_t) \sum_{i=0}^{\infty} \beta^i(\Delta_t)\epsilon_{t-i-1}^2.$$

Let $\mathcal{F}_t$ be the sigma-field generated by the past observations $\{\epsilon_u, u < t\}$, and by the past and present value of the chain $\{\Delta_u, u \leq t\}$. We have

$$h_t = E(\epsilon_t^2 \mid \mathcal{F}_t) \quad \text{and} \quad \sigma_t^2(\Delta_t) = E(\epsilon_t^2 \mid \mathcal{F}_t),$$

but, given the past observations, $\sigma_t^2(\Delta_t)$ only depends on Δ_t, whereas h_t depends also on $\{\Delta_u, u < t\}$.

This entails differences between the two models, in terms of probabilistic properties (the stationarity conditions are easier to obtain for the standard MS-GARCH model, but they have also been obtained by Liu (2006) for the alternative model), of statistical inference (the fact that $E(\epsilon_t^2 \mid \mathcal{F}_t)$ only depends on Δ_t renders the alternative model much easier to estimate), and also on the dynamics behaviour and on the interpretation of the parameters.

For instance, for the MS-GARCH, $\beta(i)$ can be interpreted as a parameter of inertia of the volatility in regime i: if the volatility h_{t-1} is high and $\beta(i)$ is close to 1, the next volatility h_t will remain high in regime i. This interpretation is no more valid for the alternative model, since $\sigma_{t-1}^2(i)$ may not be equal to $\sigma_{t-1}^2(\Delta_{t-1})$.

References

Abramowitz, M. and Stegun, I.A. (1965). Handbook of mathematical functions: with formulas, graphs, and mathematical tables, vol. 55. Courier Corporation.

Abramson, A. and Cohen, I. (2008). Single-sensor audio source separation using classification and estimation approach and GARCH modelling. *IEEE Transactions on Audio Speech and Language Processing* 16: 1528–1540.

Aielli, G.P. (2013). Dynamic conditional correlation: on properties and estimation. *Journal of Business and Economic Statistics* 31: 282–299.

Aielli, G.P. and Caporin, M. (2014). Variance clustering improved dynamic conditional correlation MGARCH estimators. *Computational Statistics and Data Analysis* 76: 556–576.

Aït-Sahalia, Y., Amengual, D., and Manresa, E. (2015). Market-based estimation of stochastic volatility models. *Journal of Econometrics* 187: 418–435.

Aknouche, A. (2012). Multistage weighted least squares estimation of ARCH processes in the stable and unstable cases. *Statistical Inference for Stochastic Processes* 15: 241–256.

Alexander, C. (2002). Principal component models for generating large GARCH covariance matrices. *Economic Notes* 31: 337–359.

Almeida, D.D., Hotta, L., and Ruiz, E. (2018). MGARCH models: tradeoff between feasibility and flexibility. *International Journal of Forecasting* 34: 45–63.

Andersen, T.G. (1994). Stochastic autoregressive volatility: a framework for volatility modeling. *Mathematical Finance* 4: 103–119.

Andersen, T.G. and Bollerslev, T. (1998). Answering the skeptics: yes, standard volatility models do provide accurate forecasts. *International Economic Review* 39: 885–906.

Andersen, T.G., Bollerslev, T., Diebold, F.X., and Labys, P. (2003). Modeling and forecasting realized volatility. *Econometrica* 71: 579–625.

Andersen, T.G., Bollerslev, T., and Meddahi, N. (2011). Realized volatility forecasting and market microstructure noise. *Journal of Econometrics* 160: 220–234.

Andersen, T.G., Davis, R.A., Kreiss, J.-P., and Mikosch, T. (eds.) (2009). *Handbook of Financial Time Series*. Berlin: Springer.

Andersen, T.G. and Sørensen, B.E. (1996). GMM estimation of a stochastic volatility model: a Monte Carlo study. *Journal of Business and Economic Statistics* 14: 328–352.

Andrews, B. (2012). Rank based estimation for GARCH processes. *Econometric Theory* 28: 1037–1064.

Andrews, D.W.K. (1991). Heteroskedasticity and autocorrelation consistent covariance matrix estimation. *Econometrica* 59: 817–858.

Andrews, D.W.K. (1997). Estimation when a parameter is on a boundary of the parameter space: part II. Unpublished manuscript, Yale University.

Andrews, D.W.K. (1999). Estimation when a parameter is on a boundary. *Econometrica* 67: 1341–1384.

GARCH Models: Structure, Statistical Inference and Financial Applications, Second Edition. Christian Francq and Jean-Michel Zakoian.

Andrews, D.W.K. (2001). Testing when a parameter is on a boundary of the maintained hypothesis. *Econometrica* 69: 683–734.

Andrews, D.W.K. and Monahan, J.C. (1992). An improved heteroskedasticity and autocorrelation consistent covariance matrix estimator. *Econometrica* 60: 953–966.

Ango Nze, P. and Doukhan, P. (2004). Weak dependence models and applications to econometrics. *Econometric Theory* 20: 995–1045.

Ardia, D. (2008). *Financial Risk Management with Bayesian Estimation of GARCH Models: Theory and Applications*. Berlin: Springer.

Artzner, P., Delbaen, F., Eber, J.-M., and Heath, D. (1999). Coherent measures of risk. *Mathematical Finance* 9: 203–228.

Aue, A., Hörmann, S., Horváth, L., and Reimherr, M. (2009). Break detection in the covariance structure of multivariate time series models. *Annals of Statistics* 37: 4046–4087.

Aue, A., Horváth, L., and Pellatt, D. F. (2017). Functional generalized autoregressive conditional heteroskedasticity. *Journal of Time Series Analysis* 38: 3–21.

Avarucci, M., Beutner, E., and Zaffaroni, P. (2013). On moment conditions for quasi-maximum likelihood estimation of multivariate ARCH models. *Econometric Theory* 29: 545–566.

Avramov, D., Chordia, T., and Goyal, A. (2006). The impact of trades on daily volatility. *Review of Financial Studies* 19: 1241–1277.

Awartani, B.M.A. and Corradi, V. (2005). Predicting the volatility of the S&P-500 stock index via GARCH models: the role of asymmetries. *International Journal of Forecasting* 21: 167–183.

Azencott, R. and Dacunha-Castelle, D. (1984). *Séries d'observations irrégulières*. Paris: Masson.

Bachelier, L. (1900). Théorie de la spéculation. *Annales Scientifiques de l'École Normale Supérieure* 17: 21–86.

Bahadur, R.R. (1960). Stochastic comparison of tests. *Annals of Mathematical Statistics* 31: 276–295.

Bahadur, R.R. (1967). Rates of convergence of estimates and test statistics. *Annals of Mathematical Statistics* 38: 303–324.

Bai, X., Russell, J.R., and Tiao, G.C. (2004). Kurtosis of GARCH and stochastic volatility models with non-normal innovations. *Journal of Econometrics* 114: 349–360.

Baillie, R.T., Bollerslev, T., and Mikkelsen, H.O. (1996). Fractionally integrated generalized autoregressive conditional heteroscedasticity. *Journal of Econometrics* 74: 3–30.

Bardet, J.-M. and Wintenberger, O. (2009). Asymptotic normality of the quasi maximum likelihood estimator for multidimensional causal processes. *Annals of Statistics* 37: 2730–2759.

Barndorff-Nielsen, O.E. and Shephard, N. (2002). Econometric analysis of realized volatility and its use in stochastic volatility models. *Journal of the Royal Statistical Society B* 64: 253–280.

Bartholomew, D.J. (1959). A test of homogeneity of ordered alternatives. *Biometrika* 46: 36–48.

Basrak, B., Davis, R.A., and Mikosch, T. (2002). Regular variation of GARCH processes. *Stochastic Processes and their Applications* 99: 95–115.

Baum, L.E. (1972). An inequality and associated maximization technique in statistical estimation for probabilistic functions of Markov processes. *Inequalities* 3: 1–8.

Baum, L.E. and Petrie, T. (1966). Statistical inference for probabilistic functions of finite state Markov chains. *Annals of Mathematical Statistics* 30: 1554–1563.

Bauwens, L., Preminger, A., and Rombouts, J. (2010). Theory and inference for a Markov-switching GARCH model. *Econometrics Journal* 13: 218–244.

Bauwens, L., Braione, M., and Storti, G. (2016). Multiplicative conditional correlation models for realized covariance matrices. CORE Discussion Paper.

Bauwens, L. and Giot, P. (2003). Asymmetric ACD models: introducing price information in ACD models with a two state transition model. *Empirical Economics* 28: 123.

Bauwens, L., Hafner, C., and Laurent, S. (2012). *Handbook of Volatility Models and their Applications*. Wiley.

Bauwens, L., Laurent, S., and Rombouts, J.V.K. (2006). Multivariate GARCH models: a survey. *Journal of Applied Econometrics* 21: 79–109.

Bauwens, L., Preminger, A., and Rombouts, J. (2010). Theory and inference for a Markov-switching GARCH model. *Econometrics Journal* 13: 218–244.

Béguin, J.M., Gouriéroux, C., and Monfort, A. (1980). Identification of a mixed autoregressive-moving average process: the corner method. In: *Time Series* (ed. O.D. Anderson), 423–436. Amsterdam: North-Holland.

Bekaert, G. and Wu, G. (2000). Asymmetric volatility and risk in equity markets. *Review of Financial Studies* 13: 1–42.

Bera, A.K. and Higgins, M.L. (1997). ARCH and bilinearity as competing models for nonlinear dependence. *Journal of Business and Economic Statistics* 15: 43–50.

Beran, J. and Schützner, M. (2009). On approximate pseudo-maximum likelihood estimation for LARCH-processes. *Bernoulli* 15: 1057–1081.

Berk, K.N. (1974). Consistent autoregressive spectral estimates. *Annals of Statistics* 2: 489–502.

Berkes, I. and Horváth, L. (2003a). Asymptotic results for long memory LARCH sequences. *Annals of Applied Probability* 13: 641–668.

Berkes, I. and Horváth, L. (2003b). The rate of consistency of the quasi-maximum likelihood estimator. *Statistics and Probability Letters* 61: 133–143.

Berkes, I. and Horváth, L. (2004). The efficiency of the estimators of the parameters in GARCH processes. *Annals of Statistics* 32: 633–655.

Berkes, I., Horváth, L., and Kokoszka, P. (2003a). Asymptotics for GARCH squared residual correlations. *Econometric Theory* 19: 515–540.

Berkes, I., Horváth, L., and Kokoszka, P. (2003b). GARCH processes: structure and estimation. *Bernoulli* 9: 201–227.

Berlinet, A. (1984). Estimating the degrees of an ARMA model. *Compstat Lectures* 3: 61–94.

Berlinet, A. and Francq, C. (1997). On Bartlett's formula for nonlinear processes. *Journal of Time Series Analysis* 18: 535–555.

Bertholon, H., Monfort, A., and Pegoraro, F. (2008). Econometric asset pricing modelling. *Journal of Financial Econometrics* 6: 407–458.

Bickel, P.J. (1982). On adaptative estimation. *Annals of Statistics* 10: 647–671.

Bickel, P.J., Ritov, Y., and Ryden, T. (1998). Asymptotic normality of the maximum likelihood estimator for general hidden Markov models. *The Annals of Statistics* 26: 1614–1635.

Billingsley, P. (1961). The Lindeberg-Levy theorem for martingales. *Proceedings of the American Mathematical Society* 12: 788–792.

Billingsley, P. (1995). *Probability and Measure*. New York: Wiley.

Black, F. (1976). Studies of stock price volatility changes. Proceedings of the American Statistical Association, Business and Economics Statistics Section, pp. 177–181.

Black, F. and Scholes, M. (1973). The pricing of options and corporate liabilities. *Journal of Political Economy* 3: 637–654.

Blasques, F., Gorgi, P., Koopman, S.J., and Wintenberger, O. (2018). Feasible invertibility conditions and maximum likelihood estimation for observation-driven models. *Electronic Journal of Statistics* 12: 1019–1052.

Blasques, F., Koopman, S.J., and Lucas, A. (2015). Information-theoretic optimality of observation-driven time series models for continuous responses. *Biometrika* 102: 325–343.

Bollerslev, T. (1986). Generalized autoregressive conditional heteroskedasticity. *Journal of Econometrics* 31: 307–327.

Bollerslev, T. (1988). On the correlation structure for the generalized autoregressive conditional heteroskedastic process. *Journal of Time Series Analysis* 9: 121–131.

Bollerslev, T. (1990). Modelling the coherence in short-run nominal exchange rates: a multivariate generalized ARCH model. *Review of Economics and Statistics* 72: 498–505.

Bollerslev, T. (2008). Glossary to ARCH (GARCH). In: *Volatility and Time Series Econometrics: Essays in Honor of Robert F. Engle* (ed. T. Bollerslev, J.R. Russell and M. Watson). Oxford: Oxford University Press.

Bollerslev, T., Chou, R.Y., and Kroner, K.F. (1992). ARCH modeling in finance: a review of the theory and empirical evidence. *Journal of Econometrics* 52: 5–59.

Bollerslev, T., Engle, R.F., and Nelson, D.B. (1994). ARCH models. In: *Handbook of Econometrics*, vol. IV, Chapter 49 (ed. R.F. Engle and D.L. McFadden), 2959–3038. Amsterdam: North-Holland.

Bollerslev, T.P., Engle, R.F., and Woolridge, J. (1988). A capital asset pricing model with time varying covariances. *Journal of Political Economy* 96: 116–131.

Borkovec, M. and Klüppelberg, C. (2001). The tail of the stationary distribution of an autoregressive process with ARCH(1) errors. *Annals of Applied Probability* 11: 1220–1241.

Bose, A. and Mukherjee, K. (2003). Estimating the ARCH parameters by solving linear equations. *Journal of Time Series Analysis* 24: 127–136.

Boswijk, H.P. and van der Weide, R. (2006). Wake me up before you GO-GARCH. Discussion paper. University of Amsterdam.

Boubacar Maïnassara, Y. and Saussereau, B. (2018). Diagnostic checking in multivariate ARMA models with dependent errors using normalized residual autocorrelations. *Journal of the American Statistical Association* DOI: 10.1080/01621459.2017.1380030.

Bougerol, P. (1993). Kalman filtering with random coefficients and contractions. *SIAM Journal on Control and Optimization* 31: 942–959.

Bougerol, P. and Picard, N. (1992a). Strict stationarity of generalized autoregressive processes. *Annals of Probability* 20: 1714–1729.

Bougerol, P. and Picard, N. (1992b). Stationarity of GARCH processes and of some nonnegative time series. *Journal of Econometrics* 52: 115–127.

Boussama, F. (1998). Ergodicité, mélange et estimation dans les modèles GARCH. Doctoral thesis. Université Paris 7.

Boussama, F. (2000). Normalité asymptotique de l'estimateur du pseudo-maximum de vraisemblance d'un modèle GARCH. *Comptes Rendus de L'Académie de Sciences Paris* 331: 81–84.

Boussama, F. (2006). Ergodicité des chaînes de Markov à valeurs dans une variété algébrique: application aux modèles GARCH maultivariés. *Comptes Rendus del'Académie de Sciences Paris* 343: 275–278.

Boussama, F., Fuchs, F., and Stelzer, R. (2011). Stationarity and geometric ergodicity of BEKK multivariate GARCH models. *Stochastic Processes and their Applications* 121: 2331–2360.

Box, G.E.P. and Jenkins, G.M. (1970). *Time Series Analysis: Forecasting and Control*. San Francisco, CA: Holden-Day.

Box, G.E.P., Jenkins, G.M., and Reinsel, G.C. (1994). *Time Series Analysis: Forecasting and Control*. Englewood Cliffs, NJ: Prentice Hall.

Box, G.E.P. and Pierce, D.A. (1970). Distribution of residual autocorrelations in autoregressive-integrated moving average time series models. *Journal of the American Statistical Association* 65: 1509–1526.

Bradley, R.C. (1986). Basic properties of strong mixing conditions. In: *Dependence in Probability and Statistics: A Survey of Recent Results* (ed. E. Eberlein and M.S. Taqqu), 165–192. Boston, MA: Birkhäuser.

Bradley, R.C. (2005). Basic properties of strong mixing conditions. A survey and some open questions. *Probability Surveys* 2: 107–144.

Brandt, A. (1986). The stochastic equation $Y_{n+1} = A_n Y_n + B_n$ with stationary coefficients. *Advance in Applied Probability* 18: 221–254.

Breidt, F.J. and Carriquiry, A.L. (1996). Improved quasi-maximum likelihood estimation for stochastic volatility models. In: *Modelling and Prediction: Honoring Seymour Geisser* (ed. J.C. Lee, W.O. Johnson and A. Zellner), 228–247. New York: Springer.

Breidt, F.J., Davis, R.A., and Trindade, A. (2001). Least absolute deviation estimation for all-pass time series models. *Annals of Statistics* 29: 919–946.

Breiman, L. (1965). On some limit theorems similar to the arc-sin law. *Theory of probability and their Applications* 10: 323–331.

Brockwell, P.J. and Davis, R.A. (1991). *Time Series: Theory and Methods*, 2e. New York: Springer.

Cai, J. (1994). A Markov model of switching-regime ARCH. *Journal of Business and Economic Statistics* 12: 309–316.

Campbell, S.D. and Diebold, F.X. (2005). Weather forecasting for weather derivatives. *Journal of the American Statistical Association* 100: 6–16.

Campbell, J.Y. and Hentschel, L. (1992). No news is good news: an asymmetric model of changing volatility in stock returns. *Journal of Financial Economics* 31: 281–318.

Caporin, M. and McAleer, M. (2012). Do we really need both BEKK and DCC? A tale of two multivariate GARCH models. *Journal of Economic Surveys* 26: 736–751.

Cappé, O., Moulines, E., and Rydén, T. (2005). *Inference in Hidden Markov Models Series*. New York: Springer.

Carrasco, M. and Chen, X. (2002). Mixing and moment properties of various GARCH and stochastic volatility models. *Econometric Theory* 18: 17–39.

Cerovecki, C., Francq, C., Hörmann, S., and Zakoïan, J-M. (2019). Functional GARCH models: the quasi-likelihood approach and its applications. Forthcoming in *Journal of Econometrics*.

Chan, K.S. (1990). Deterministic stability, stochastic stability, and ergodicity. In: *Nonlinear Time Series: A Dynamical System Approach* (ed. H. Tong), 448–466. Oxford: Oxford University Press.

Chen, M. and An, H.Z. (1998). A note on the stationarity and the existence of moments of the GARCH model. *Statistica Sinica* 8: 505–510.

Chen, C.W.S., Gerlach, R., and So, M.K.P. (2006). Comparison of nonnested asymmetric heteroskedastic models. *Computational Statistics and Data Analysis* 51: 2164–2178.

Chernoff, H. (1954). On the distribution of the likelihood ratio. *Annals of Mathematical Statistics* 54: 573–578.

Chib, S., Nardari, F., and Shephard, N. (2002). Markov chain Monte Carlo methods for stochastic volatility models. *Journal of Econometrics* 108: 281–316.

Chou, R.Y. (2005). Forecasting financial volatilities with extreme values: the conditional autoregressive range (CARR) model. *Journal of Money, Credit and Banking* 37: 561–582.

Chow, Y.S. and Teicher, H. (1997). *Probability Theory*, 3e. New York: Springer.

Christie, A.A. (1982). The stochastic behavior of common stock variances: value, leverage and interest rate effects. *Journal of Financial Economics* 10: 407–432.

Christoffersen, P.F. (1998). Evaluating interval forecasts. *International Economic Review* 39: 841–862.

Christoffersen, P.F. and Jacobs, K. (2004). Which GARCH model for option valuation? *Management Science* 50: 1204–1221.

Christoffersen, P.F. and Pelletier, D. (2004). Backtesting value-at-risk: a duration-based approach. *Journal of Financial Econometrics* 2: 84–108.

Clark, P.K. (1973). A subordinated stochastic process with fixed variance for speculative prices. *Econometrica* 41: 135–155.

Cline, D.H.B. and Pu, H.H. (1998). Verifying irreducibility and continuity of a nonlinear time series. *Statistics and Probability Letters* 40: 139–148.

Cochrane, J. (2001). *Asset Pricing*. Princeton, NJ: Princeton University Press.

Cohen, I. (2004). Modelling speech signal in the time-frequency domain using GARCH. *Signal Processing* 84: 2453–2459.

Cohen, I. (2006). Speech spectral modeling and enhancement based on autoregressive conditional heteroscedasticity models. *Signal Processing* 86: 698–709.

Comte, F. and Lieberman, O. (2003). Asymptotic theory for multivariate GARCH processes. *Journal of Multivariate Analysis* 84: 61–84.

Conrad, C. and Karanasos, M. (2010). Negative volatility spillovers in the unrestricted ECCC-GARCH model. *Econometric Theory* 26: 838–862.

Conrad, C. and Mammen, E. (2016). Asymptotics for parametric GARCH-in-mean models. *Journal of Econometrics* 194: 319–329.

Corradi, V. (2000). Reconsidering the continuous time limit of the GARCH(1,1) process. *Journal of Econometrics* 96: 145–153.

Cox, D.R. (1981). Statistical analysis of time series: some recent developments. *Scandinavian Journal of Statistics* 8: 93–115.

Cox, J., Ross, S., and Rubinstein, M. (1979). Option pricing: a simplified approach. *Journal of Financial Economics* 7: 229–263.

Creal, D., Koopman, S.J., and Lucas, A. (2011). A dynamic multivariate heavy-tailed model for time-varying volatilities and correlations. *Journal of Business and Economic Statistics* 29: 552–563.

Creal, D., Koopman, S.J., and Lucas, A. (2013). Generalized autoregressive score models with applications. *Journal of Applied Econometrics* 28: 777–795.

Dana, R.A. and Jeanblanc-Picqué, M. (1994). *Marchés financiers en temps continu. Valoriation et équilibre*. Paris: Economica.

Danielsson, J. (1994). Stochastic volatility in asset prices, estimation with simulated maximum likelihood. *J. Econometrics* 64: 375–400.

Das, D. and Yoo, B. (2004). A Bayesian MCMC algorithm for Markov switching GARCH Models. City University of New York and Rutgers University Working Paper.

Davidian, M. and Carroll, R.J. (1987). Variance function estimation. *Journal of the American Statistical Association* 82: 1079–1091.

Davis, R.A., Holan, S.H., Lund, R., and Ravishanker, N. (2016). *Handbook of Discrete-Valued Time Series*. CRC Press.

Davis, R.A., Knight, K., and Liu, J. (1992). M-estimation for autoregressions with infinite variance. *Stochastic Processes and their Applications* 40: 145–180.

Davis, R.A. and Mikosch, T. (2009a). Extreme value theory for GARCH processes. In: *Handbook of Financial Time Series* (ed. T.G. Andersen, R.A. Davis, J.-P. Kreiss and T. Mikosch), 187–200. Berlin: Springer.

Davis, R.A. and Mikosch, T. (2009b). Probabilistic properties of stochastic volatility models. In: *Handbook of Financial Time Series* (ed. T.G. Andersen, R.A. Davis, J.-P. Kreiss and T. Mikosch), 255–267. New-York: Springer.

Davydov, Y.A. (1968). Convergence of distributions generated by stationary stochastic processes. *Theory of Probability and Applications* 13: 691–696.

Davydov, Y.A. (1973). Mixing conditions for Markov chains. *Theory of Probability and Applications* 18: 313–328.

Delbaen, F. (2002). Coherent measures of risk on general probability spaces. In: *Advances in Finance and Stochastics: Essays in Honor of Dieter Sondermann* (ed. K. Sandmann and P.J. Schönbucher), 1–37. Berlin: Springer.

Dellaportas, P. and Pourahmadi, M. (2012). Cholesky-GARCH models with applications to finance. *Statistics and Computing* 22: 849–855.

Demos, A. and Sentana, E. (1998). Testing for GARCH effects: a one-sided approach. *Journal of Econometrics* 86: 97–127.

den Hann, W.J. and Levin, A. (1997). A practitioner's guide to robust covariance matrix estimation. In: *Handbook of Statistics*, vol. 15 (ed. C.R. Rao and G.S. Maddala), 291–341. Amsterdam: North-Holland.

Devijver, P.A. (1985). Baum's forward-backward algorithm revisited. *Pattern Recognition Letters* 3: 369–373.

Diebold, F.X. (1986). Testing for serial correlation in the presence of ARCH. Proceedings of the American Statistical Association, Business and Economics Statistics Section, pp. 323–328.

Diebold, F.X. (2004). The nobel memorial prize for Robert F. Engle. *Scandinavian Journal of Economics* 106: 165–185.

Diebolt, J. and Guégan, D. (1991). Le modèle de série chronologique autorégressive β-ARCH. *Comptes Rendus de l'Académie des Sciences de Paris* 312: 625–630.

Ding, Z. and Engle, R.F. (2001). Large scale conditional covariance matrix modeling, estimation and testing. *Academia Economic Papers* 29: 157–184.

Ding, Z., Granger, C., and Engle, R.F. (1993). A long memory property of stock market returns and a new model. *Journal of Empirical Finance* 1: 83–106.

Douc, R., Moulines, E., and Stoffer, D. (2014). *Nonlinear Time Series: Theory, Methods and Applications with R Examples*. Chapman & Hall/CRC Press.

Douc, R., Roueff, F., and Soulier, P. (2008). On the existence of some ARCH(∞) processes. *Stochastic Processes and their Applications* 118: 755–761.

Doukhan, P. (1994). *Mixing: Properties and Examples*, Lecture Notes in Statistics, vol. 85. New York: Springer.

Doukhan, P., Teyssière, G., and Winant, P. (2006). Vector valued ARCH infinity processes. In: *Dependence in Probability and Statistics* (ed. P. Bertail, P. Doukhan and P. Soulier). New York: Springer.

Drost, F.C. and Nijman, T.E. (1993). Temporal aggregation of GARCH processes. *Econometrica* 61: 909–927.

Drost, F.C. and Klaassen, C.A.J. (1997). Efficient estimation in semiparametric GARCH models. *Journal of Econometrics* 81: 193–221.

Drost, F.C., Klaassen, C.A.J., and Werker, B.J.M. (1997). Adaptive estimation in time series models. *Annals of Statistics* 25: 786–818.

Drost, F.C., Nijman, T.E., and Werker, B.J.M. (1998). Estimation and testing in models containing both jumps and conditional heteroskedasticity. *Journal of Business and Economic Statistics* 16: 237–243.

Drost, F.C. and Werker, B.J.M. (1996). Closing the GARCH gap: continuous time GARCH modeling. *Journal of Econometrics* 74: 31–57.

Duan, J.-C. (1995). The GARCH option pricing model. *Mathematical Finance* 5: 13–32.

Duan, J.-C. and Simonato, J.-G. (2001). American option pricing under GARCH by a Markov chain approximation. *Journal of Economic Dynamics and Control* 25: 1689–1718.

Duchesne, P. and Lalancette, S. (2003). On testing for multivariate ARCH effects in vector time series models. *Revue Canadienne de Statistique* 31: 275–292.

Dueker, M.J. (1997). Markov switching in GARCH processes and mean-reverting stock market volatility. *Journal of Business and Economic Statistics* 15: 26–34.

Duffie, D. (1994). *Modèles dynamiques d'évaluation*. Paris: Presses Universitaires de France.

Dufour, J.-M., Khalaf, L., Bernard, J.-T., and Genest, I. (2004). Simulation-based finite-sample tests for heteroskedasticity and ARCH effects. *Journal of Econometrics* 122: 317–347.

Dupuis, D.J. (2017). Electricity price dependence in New-York state zones: a robust detrended correlation approach. *The Annals of Applied Statistics* 11: 248–273.

El Babsiri, M. and Zakoïan, J.-M. (1990). *Approximation en temps continu d'un modèle ARCH à seuil. Document de travail 9011*. INSEE.

El Babsiri, M. and Zakoïan, J.-M. (2001). Contemporaneous asymmetry in GARCH processes. *Journal of Econometrics* 101: 257–294.

Elie, L. (1994). Les processus ARCH comme approximations de processus en temps continu. In: *Modélisation ARCH: théorie statistique et applications dans le domaine de la finance* (ed. F. Droesbeke, F. Fichet and P. Tassi). Paris: Ellipses.

Embrechts, P., Klüppelberg, C., and Mikosch, T. (1997). *Modelling Extremal Events*. Springer.

Engle, R.F. (1982). Autoregressive conditional heteroskedasticity with estimates of the variance of U.K. inflation. *Econometrica* 50: 987–1008.

Engle, R.F. (1984). Wald, likelihood ratio, and Lagrange multiplier tests in econometrics. In: *Handbook of Econometrics*, vol. 2 (ed. Z. Griliches and M.D. Intriligator), 775–826. Amsterdam: North-Holland.

Engle, R.F. (2001). GARCH 101: the use of ARCH/GARCH models in applied econometrics. *Journal of Economic Perspectives* 15: 157–168.

Engle, R.F. (2002a). Dynamic conditional correlation—a simple class of multivariate GARCH models. *Journal of Business and Economic Statistics* 20: 339–350.

Engle, R.F. (2002b). New frontiers for ARCH models. *Journal of Applied Econometrics* 17: 425–446.

Engle, R.F. (2004). Nobel lecture. Risk and volatility: econometric models and financial practice. *American Economic Review* 94: 405–420.

Engle, R.F. (2009). *Anticipating Correlations. A New Paradigm for Risk Management*. Princeton, NJ: Princeton University Press.

Engle, R.F. (2016). Dynamic conditional beta. *Journal of Financial Econometrics* 14: 643–667.

Engle, R.F. and Bollerslev, T. (1986). Modelling the persistence of conditional variances (with comments and a reply by the authors). *Econometric Reviews* 5: 1–87.

Engle, R.F. and González-Rivera, G. (1991). Semiparametric ARCH models. *Journal of Business and Econometric Statistics* 9: 345–359.

Engle, R.F., Ghysels, E., and Sohn, B. (2013). Stock market volatility and macroeconomic fundamentals. *Review of Economics and Statistics* 95: 776–797.

Engle, R.F., Granger, C.W.J., and Kraft, D. (1984). Combining competing forecasts of inflation based on a bivariate ARCH model. *Journal of Economic Dynamics and Control* 8: 151–165.

Engle, R.F. and Kroner, K. (1995). Multivariate simultaneous GARCH. *Econometric Theory* 11: 122–150.

Engle, R.F., Lilien, D.M., and Robins, R.P. (1987). Estimating time varying risk premia in the term structure: the ARCH-M model. *Econometrica* 55: 391–407.

Engle, R.F. and Manganelli, S. (2004). CAViaR: conditional autoregressive value at risk by regression quantiles. *Journal of Business and Economic Statistics* 22: 367–381.

Engle, R.F. and Mustafa, C. (1992). Implied ARCH models from option prices. *Journal of Econometrics* 52: 289–311.

Engle, R.F. and Ng, V.K. (1993). Measuring and testing the impact of news on volatility. *Journal of Finance* 48: 1749–1778.

Engle, R.F., Ng, V.K., and Rothschild, M. (1990). Asset pricing with a factor ARCH covariance structure: empirical estimates for treasury bills. *Journal of Econometrics* 45: 213–238.

Engle, R.F. and Patton, A. (2001). What good is a volatility model? *Quantitative Finance* 1: 237–245.

Engle, R.F. and Rangel, A. (2008). The spline-GARCH model for low-frequency volatility and its global macroeconomic causes. *Review of Financial Studies* 21: 1187–1222.

Engle, R.F. and Russell, J.R. (1998). Autoregressive conditional duration: a new model for irregularly spaced transaction data. *Econometrica* 66: 1127–1162.

Engle, R.F. and Sheppard, K. (2001). Theoretical and empirical properties of dynamic conditional correlation multivariate GARCH. University of California San Diego, Discussion paper.

Escanciano, J.C. (2009). Quasi-maximum likelihood estimation of semi-strong GARCH models. *Econometric Theory* 25: 561–570.

Escanciano, J.C. and Olmo, J. (2010). Backtesting parametric value-at-risk with estimation risk. *Journal of Business and Economics Statistics* 28 (1): 36–51.

Ewing, B.T., Kruse, J.B., and Schroeder, J.L. (2006). Time series analysis of wind speed with time-varying turbulence. *Environmetrics* 17: 119–127.

Fama, E.F. (1965). The behavior of stock market prices. *Journal of Business* 38: 34–105.

Fan, J., Qi, L., and Xiu, D. (2014). Quasi-maximum likelihood estimation of GARCH models with heavy-tailed likelihoods. *Journal of Business & Economic Statistics* 32: 178–191.

Feigin, P.D. and Tweedie, R.L. (1985). Random coefficient autoregressive processes: a Markov chain analysis of stationarity and finiteness of moments. *Journal of Time Series Analysis* 6: 1–14.

Ferland, R., Latour, A., and Oraichi, D. (2006). Integer-valued GARCH process. *Journal of Time Series Analysis* 27: 923–942.

Fermanian, J.D. and Malongo, H. (2016). On the stationarity of dynamic conditional correlation models. *Econometric Theory* 33: 636–663.

Fokianos, K., Rahbek, A., and Tjøstheim, D. (2009). Poisson autoregression. *Journal of the American Statistical Association* 104: 1430–1439.

Francq, C., Horváth, L., and Zakoïan, J.-M. (2011). Merits and drawbacks of variance targeting in GARCH Models. *Journal of Financial Econometrics* 9: 619–656.

Francq, C., Horváth, L., and Zakoïan, J.-M. (2016). Variance targeting estimation of multivariate GARCH models. *Journal of Financial Econometrics* 14: 353–382.

Francq, C., Lepage, G., and Zakoïan, J-M. (2011). Two-stage Non Gaussian QML Estimation of GARCH Models and Testing the Efficiency of the Gaussian QMLE. *Journal of Econometrics* 165: 246–257.

Francq, C. and Roussignol, M. (1997). On white noises driven by hidden Markov chains. *Journal of Time Series Analysis* 18: 553–578.

Francq, C., Roussignol, M., and Zakoïan, J.-M. (2001). Conditional heteroskedasticity driven by hidden Markov chains. *Journal of Time Series Analysis* 22: 197–220.

Francq, C., Roy, R., and Zakoïan, J.-M. (2005). Diagnostic checking in ARMA models with uncorrelated errors. *Journal of the American Statistical Association* 100: 532–544.

Francq, C. and Sucarrat, G. (2017). An equation-by-equation estimator of a multivariate log-GARCH-X model of financial returns. *Journal of Multivariate Analysis* 153: 16–32.

Francq, C., Wintenberger, O., and Zakoïan, J.-M. (2013). GARCH models without positivity constraints: exponential or log GARCH? *Journal of Econometrics* 177: 34–46.

Francq, C., Wintenberger, O., and Zakoïan, J.-M. (2017). Goodness-of-fit tests for log-GARCH and EGARCH models. *TEST* 27: 27–51.

Francq, C. and Thieu, L.Q. (2015). QML inference for volatility models with covariates. MPRA Paper No. 63198.

Francq, C. and Zakoïan, J.-M. (2000). Estimating weak GARCH representations. *Econometric Theory* 16: 692–728.

Francq, C. and Zakoïan, J.-M. (2004). Maximum likelihood estimation of pure GARCH and ARMA-GARCH processes. *Bernoulli* 10: 605–637.

Francq, C. and Zakoïan, J.-M. (2005). The L^2-structures of standard and switching-regime GARCH models. *Stochastic Processes and their Applications* 115: 1557–1582.

Francq, C. and Zakoïan, J.-M. (2006a). Linear-representation based estimation of stochastic volatility models. *Scandinavian Journal of Statistics* 33: 785–806.

Francq, C. and Zakoïan, J.-M. (2006b). Mixing properties of a general class of GARCH(1,1) models without moment assumptions on the observed process. *Econometric Theory* 22: 815–834.

Francq, C. and Zakoïan, J.-M. (2006c). On efficient inference in GARCH processes. In: *Statistics for Dependent Data* (ed. P. Bertail, P. Doukhan and P. Soulier), 305–327. New York: Springer.

Francq, C. and Zakoïan, J.-M. (2007). Quasi-maximum likelihood estimation in GARCH processes when some coefficients are equal to zero. *Stochastic Processes and their Applications* 117: 1265–1284.

Francq, C. and Zakoïan, J.-M. (2008). Deriving the autocovariances of powers of Markov-switching GARCH models, with applications to statistical inference. *Computational Statistics and Data Analysis* 52: 3027–3046.

Francq, C. and Zakoïan, J.-M. (2009a). A tour in the asymptotic theory of GARCH estimation. In: *Handbook of Financial Time Series* (ed. T.G. Andersen, R.A. Davis, J.-P. Kreiss and T. Mikosch). Berlin: Springer.

Francq, C. and Zakoïan, J.-M. (2009b). Testing the nullity of GARCH coefficients: correction of the standard tests and relative efficiency comparisons. *Journal of the American Statistical Association* 104: 313–324.

Francq, C. and Zakoïan, J.-M. (2009c). Bartlett's formula for a general class of nonlinear processes. *Journal of Time Series Analysis* 30: 449–465.

Francq, C. and Zakoïan, J.-M. (2010) Inconsistency of the QMLE and inference based on weighted LS for LARCH models. *Journal of Econometrics* 159: 151–165.

Francq, C. and Zakoïan, J.-M. (2012a). QML estimation of a class of multivariate asymmetric GARCH models. *Econometric Theory* 28: 179–206.

Francq, C. and Zakoïan, J.M. (2012b). Strict stationarity testing and estimation of explosive and stationary GARCH models. *Econometrica* 80: 821–861.

Francq, C. and Zakoïan, J.M. (2013a). Inference in non stationary asymmetric GARCH models. *Annals of Statistics* 41: 1970–1998.

Francq, C. and Zakoïan, J-M. (2013b). Optimal predictions of powers of conditionally heteroscedastic processes. *Journal of the Royal Statistical Society: Series B (Statistical Methodology)* 75: 345–367.

Francq, C. and Zakoïan, J.-M. (2013c). Estimating the marginal law of a time series with applications to heavy tailed distributions. *Journal of Business and Economic Statistics* 31: 412–425.

Francq, C. and Zakoïan, J.M. (2016). Estimating multivariate GARCH models equation by equation. *Journal of the Royal Statistical Society: Series B (Statistical Methodology)* 78: 613–635.

Franke, J., Härdle, W., and Hafner, C. (2004). *Statistics of Financial Markets*. Berlin: Springer.

Franses, P.H. and van Dijk, D. (2000). *Non-Linear Time Series Models in Empirical Finance*. Cambridge: Cambridge University Press.

Fridman, M. and Harris, L. (1998). A maximum likelihood approach for non-Gaussian stochastic volatility models. *Journal of Business and Economic Statistics* 16: 284–291.

Frühwirth-Schnatter, S. (2006). *Finite Mixture and Markov Switching Models*. New York: Springer.

Fryzlewicz, P. and Rao, S.S. (2011). Mixing properties of ARCH and time-varying ARCH processes. *Bernoulli* 17: 320–346.

Garcia, R., Ghysels, E., and Renault, E. (1998). A note on hedging in ARCH and stochastic volatility option pricing models. *Mathematical Finance* 8: 153–161.

Geweke, J. (1986). Modeling the persistence of conditional variances: a comment. *Econometric Review* 5: 57–61.

Geweke, J. (1989). Exact predictive densities for linear models with ARCH disturbances. *Journal of Econometrics* 40: 63–86.

Ghysels, E. and Jasiak, J. (1998). Garch for irregularly spaced financial data: the ACD-GARCH model. *Studies in Nonlinear Dynamics and Econometrics* 2: 133–149.

Giraitis, L., Kokoszka, P., and Leipus, R. (2000). Stationary ARCH models: dependence structure and central limit theorem. *Econometric Theory* 16: 3–22.

Giraitis, L., Leipus, R., Robinson, P.M., and Surgailis, D. (2004). LARCH, leverage, and long memory. *Journal of Financial Econometrics* 2: 177–210.

Giraitis, L., Leipus, R., and Surgailis, D. (2006). Recent advances in ARCH modelling. In: *Long-Memory in Economics* (ed. A. Kirman and G. Teyssiere), 3–38. Berlin: Springer.

Giraitis, L., Leipus, R., and Surgailis, D. (2009). ARCH(∞) and long memory properties. In: *Handbook of Financial Time Series* (ed. T.G. Andersen, R.A. Davis, J.-P. Kreiss and T. Mikosch). Berlin: Springer.

Giraitis, L. and Robinson, P.M. (2001). Whittle estimation of ARCH models. *Econometric Theory* 17: 608–631.

Giraitis, L., Robinson, P.M., and Surgailis, D. (2000). A model for long memory conditional heteroscedasticity. *Annals of Applied Probability* 10: 1002–1024.

Giraitis, L. and Surgailis, D. (2002). ARCH-type bilinear models with double long memory. *Stochastic Processes and their Applications* 100: 275–300.

Glasbey, C.A. (1982). A generalization of partial autocorrelation function useful in identifying ARMA models. *Technometrics* 24: 223–228.

Glosten, L.R., Jagannathan, R., and Runkle, D. (1993). On the relation between the expected values and the volatility of the nominal excess return on stocks. *Journal of Finance* 48: 1779–1801.

Godfrey, L.G. (1988). *Misspecification Tests in Econometrics: The Lagrange Multiplier Principle and Other Approaches*. Cambridge: Cambridge University Press.

Goldsheid, I.Y. (1991). Lyapunov exponents and asymptotic behavior of the product of random matrices. In: *Lyapunov Exponents. Proceedings of the Second Conference held in Oberwolfach* (May 28–June 2 1990), , Lecture Notes in Mathematics, vol. 1486 (ed. L. Arnold, H. Crauel and J.-P. Eckmann), 23–37. Berlin: Springer.

Gonçalves, E. and Mendes Lopes, N. (1994). The generalized threshold ARCH model: wide sense stationarity and asymptotic normality of the temporal aggregate. *Publications de l'Institut de Statistique de Paris* 38: 19–35.

Gonçalves, E. and Mendes Lopes, N. (1996). Stationarity of GTARCH processes. *Statistics* 28: 171–178.

González-Rivera, G. (1998). Smooth transition GARCH models. *Studies in Nonlinear Dynamics and Econometrics* 3: 61–78.

González-Rivera, G. and Drost, F.C. (1999). Efficiency comparisons of maximum-likelihood-based estimators in GARCH models. *Journal of Econometrics* 93: 93–111.

Gouriéroux, C. (1997). *ARCH Models and Financial Applications*. New York: Springer.

Gouriéroux, C., Holly, A., and Monfort, A. (1982). Likelihood ratio test, Wald test and Kuhn-Tucker test in linear models with inequality constraints on the regression parameters. *Econometrica* 50: 63–79.

Gouriéroux, C. and Jasiak, J. (2001). *Financial Econometrics*. Princeton, NJ: Princeton University Press.

Gouriéroux, C. and Jasiak, J. (2008). Dynamic quantile models. *Journal of Econometrics* 147: 198–205.

Gouriéroux, C. and Monfort, A. (1992). Qualitative threshold ARCH models. *Journal of Econometrics* 52: 159–200.

Gouriéroux, C. and Monfort, A. (1995). *Statistics and Econometric Models*. Cambridge: Cambridge University Press.

Gouriéroux, C. and Monfort, A. (1996). *Time Series and Dynamic Models*. Cambridge: Cambridge University Press.

Gouriéroux, C., Monfort, A., and Renault, E. (1993). Indirect inference. *Journal of Applied Econometrics* 8: S85–S118.

Gouriéroux, C., Monfort, A., and Trognon, A. (1984). Pseudo maximum likelihood methods: theory. *Econometrica* 52: 681–700.

Gouriéroux, C. and Tiomo, A. (2007). *Risque de crédit*. Paris: Economica.

Grammig, J. and Maurer, K.-O. (2000). Non-monotonic hazard functions and the autoregressive conditional duration model. *Econometrics Journal* 3: 16–38.

Haas, M., Mittnik, S., and Paolella, M.S. (2004). A new approach to markov-switching GARCH models. *Journal of Financial Econometrics* 2: 493–530.

Hafner, C. (2003). Fourth moment structure of multivariate GARCH models. *Journal of Financial Econometrics* 1: 26–54.

Hafner, C. and Linton, O.B. (2010). Efficient estimation of a multivariate multiplicative volatility model. *Journal of Econometrics* 159: 55–73.

Hafner, C. and Preminger, A. (2009a). Asymptotic theory for a factor GARCH model. *Econometric Theory* 25: 336–363.

Hafner, C. and Preminger, A. (2009b). On asymptotic theory for multivariate GARCH models. *Journal of Multivariate Analysis* 100: 2044–2054.

Hafner, C.M. and Preminger, A. (2015). An ARCH model without intercept. *Economic Letters* 129: 13–17.

Hafner, C. and Reznikova, O. (2012). On the estimation of dynamic correlation models. *Computational Statistics and Data Analysis* 56: 3533–3545.

Hagerud, G.E. (1997). A new non-linear GARCH model. PhD thesis, IFE, Stockholm School of Economics.

Hall, P. and Yao, Q. (2003). Inference in ARCH and GARCH models with heavy-tailed errors. *Econometrica* 71: 285–317.

Hamilton, J.D. (1989). A new approach to the economic analysis of nonstationary time series and the business cycle. *Econometrica* 57: 357–384.

Hamilton, J.D. (1994). *Time Series Analysis*. Princeton, NJ: Princeton University Press.

Hamilton, J.D. and Susmel, R. (1994). Autoregressive conditional heteroskedasticity and changes in regime. *Journal of Econometrics* 64: 307–333.

Han, H. and Kristensen, D. (2014). Asymptotic theory for the QMLE in GARCH-X models with stationary and nonstationary covariates. *Journal of Business and Economic Statistics* 32: 416–429.

Hannan, E.J. (1969). The identification of vector mixed autoregressive-moving average systems. *Biometrika* 56: 223–225.

Hannan, E.J. (1970). *Multiple Time Series*. New York: Wiley.

Hannan, E.J. (1976). The identification and parametrization of ARMAX and state space forms. *Econometrica* 44: 713–723.

Hannan, E.J. and Deistler, M. (1988). *The Statistical Theory of Linear Systems*. New York: Wiley.

Hansen, B.E. (1994). Autoregressive conditional density estimation. *International Economic Review* 35: 705–730.

Hansen, L.P. and Richard, S.F. (1987). The role of conditioning information in deducing testable restrictions implied by dynamic asset pricing models. *Econometrica* 55: 587–613.

Hansen, P.R. and Huang, Z. (2016). Exponential GARCH modeling with realized measures of volatility. *Journal of Business & Economic Statistics* 34: 269–287.

Hansen, P.R., Huang, Z., and Shek, H.H. (2012). Realized GARCH: a joint model for returns and realized measures of volatility. *Journal of Applied Econometrics* 27: 877–906.

Hansen, P.R. and Lunde, A. (2005). A forecast comparison of volatility models: does anything beat a GARCH(1,1)? *Journal of Applied Econometrics* 20: 873–889.

Härdle, W. and Hafner, C. (2000). Discrete time option pricing with flexible volatility estimation. *Finance and Stochastics* 4: 189–207.

Harvey, A.C. (1989). *Forecasting, Structural Time Series Models and the Kalman Filter*. Cambridge University Press.

Harvey, A.C. (2013). *Dynamic Models for Volatility and Heavy Tails*, Econometric Society Monograph. Cambridge University Press.

Harvey, A.C. and Chakravarty, T. (2008). *Beta-t-(E)GARCH*. University of Cambridge, Faculty of Economics.

Harvey, A.C., Ruiz, E., and Sentana, E. (1992). Unobserved component time series models with ARCH disturbances. *Journal of Econometrics* 52: 129–158.

Harvey, A.C., Ruiz, E., and Shephard, N. (1994). Multivariate stochastic variance models. *Review of Economic Studies* 61: 247–264.

Harvey, A.C. and Shephard, N. (1996). Estimation of an asymmetric stochastic volatility model for asset returns. *Journal of Business and Economic Statistics* 14: 429–434.

Harville, D. (1997). *Matrix Algebra from a Statistician's Perspective*. New York: Springer.

Hautsch, N., Malec, P., and Schienle, M. (2014). Capturing the zero: a new class of zero-augmented distributions and multiplicative error processes. *Journal of Financial Econometrics* 12: 89–121.

He, C. and Teräsvirta, T. (1999). Fourth-moment structure of the GARCH(p, q) process. *Econometric Theory* 15: 824–846.

He, C. and Teräsvirta, T. (2004). An extended constant conditional correlation GARCH model and its fourth-moment structure. *Econometric Theory* 20: 904–926.

Heinen, A. (2003). Modelling time series count data: an autoregressive conditional poisson model. MPRA Paper No. 8113.

Henneke, S., Rachev, S., Fabozzi, F., and Nikolov, M. (2011). MCMC-based estimation of Markov switching ARMA-GARCH models. *Applied Economics* 43: 259–271.

Hentschel, L. (1995). All in the family: nesting symmetric and asymmetric GARCH models. *Journal of Financial Economics* 39: 71–104.

Herrndorf, N. (1984). A functional central limit theorem for weakly dependent sequences of random variables. *Annals of Probability* 12: 141–153.

Heston, S. and Nandi, S. (2000). A closed form GARCH option pricing model. *Review of Financial Studies* 13: 585–625.

Higgins, M.L. and Bera, A.K. (1992). A class of nonlinear ARCH models. *International Economic Review* 33: 137–158.

Hill, J.B. (2015). Robust estimation and inference for heavy tailed GARCH. *Bernoulli* 21: 1629–1669.

Hill, J.B. and Renault, E. (2012). Variance targeting for heavy tailed time series. University of North Carolina at Chapel Hill Discussion paper.

Hobson, D.G. and Rogers, L.C.G. (1998). Complete models with stochastic volatility. *Mathematical Finance* 8: 27–48.

Hong, Y. (1997). One sided testing for autoregressive conditional heteroskedasticity in time series models. *Journal of Time Series Analysis* 18: 253–277.

Hong, Y. and Lee, Y.J. (2001). One-sided testing for ARCH effects using wavelets. *Econometric Theory* 17: 1051–1081.

Hörmann, S. (2008). Augmented GARCH sequences: dependence structure and asymptotics. *Bernoulli* 14: 543–561.

Hörmann, S., Horváth, L., and Reeder, R. (2013). A functional version of the ARCH model. *Econometric Theory* 29: 267–288.

Horváth, L. and Liese, F. (2004). L_p-estimators in ARCH models. *Journal of Statistical Planning and Inference* 119: 277–309.

Horváth, L., Kokoszka, P., and Teyssière, G. (2004). Bootstrap misspecification tests for ARCH based on the empirical process of squared residuals. *Journal of Statistical Computation and Simulation* 74: 469–485.

Horváth, L. and Zitikis, R. (2006). Testing goodness of fit based on densities of GARCH innovations. *Econometric Theory* 22: 457–482.

Hoti, S., McAleer, M., and Chan, F. (2005). Modelling the spillover effects in the volatility of atmospheric carbon dioxide concentrations. *Mathematics and Computers in Simulation* 69: 46–56.

Hsieh, K.C. and Ritchken, P. (2005). An empirical comparison of GARCH option pricing models. *Review of Derivatives Research* 8: 129–150.

Huang, H.-H., Shiu, Y.-M., and Lin, P.-S. (2008). HDD and CDD option pricing with market price of weather risk for Taiwan. *Journal of Futures Markets* 28: 790–814.

Hull, J. and White, A. (1987). The pricing of options on assets with stochastic volatilities. *Journal of Finance* 42: 281–300.

Hurlin, C. and Tokpavi, S. (2006). Backtesting value-at-risk accuracy: a simple new test. *Journal of Risk* 9: 19–37.

Hwang, S.Y. and Kim, T.Y. (2004). Power transformation and threshold modeling for ARCH innovations with applications to tests for ARCH structure. *Stochastic Processes and their Applications* 110: 295–314.

Jacquier, E., Polson, N.G., and Rossi, P.E. (1994). Bayesian analysis of stochastic volatility models. *Journal of Business and Economic Statistics* 12: 371–389.

Jacquier, E., Polson, N.G., and Rossi, P.E. (2004). Bayesian analysis of stochastic volatility models with fat-tails and correlated errors. *Journal of Econometrics* 122: 185–212.

Jeantheau, T. (1998). Strong consistency of estimators for multivariate ARCH models. *Econometric Theory* 14: 70–86.

Jeantheau, T. (2004). A link between complete models with stochastic volatility and ARCH models. *Finance and Stochastics* 8: 111–131.

Jensen, S.T. and Rahbek, A. (2004a). Asymptotic normality of the QMLE estimator of ARCH in the nonstationary case. *Econometrica* 72: 641–646.

Jensen, S.T. and Rahbek, A. (2004b). Asymptotic inference for nonstationary GARCH. *Econometric Theory* 20: 1203–1226.

Jordan, H. (2003). Asymptotic properties of ARCH(p) quasi maximum likelihood estimators under weak conditions. PhD thesis. University of Vienna.

Jungbacker, B. and Koopman, S.J. (2009). Parameter estimation and practical aspects of modeling stochastic volatility. In: *Handbook of Financial Time Series* (ed. T.G. Andersen, R.A. Davis, J.-P. Kreiss and T. Mikosch), 313–344. New-York: Springer.

Kahane, J.-P. (1998). Le mouvement brownien, un essai sur les origines de la théorie mathématique. *Séminaires et Congrès SMF* 123–155.

Kalman, R.E. (1960). A new approach of linear filtering and prediction problem. *Journal of Basic Engineering* 82: 34–35.

Karanasos, M. (1999). The second moment and the autocovariance function of the squared errors of the GARCH model. *Journal of Econometrics* 90: 63–76.

Karatzas, I. and Shreve, S.E. (1988). *Brownian Motion and Stochastic Calculus*. New York: Springer.

Kaufmann, S. and Fruhwirth-Schnatter, S. (2002). Bayesian analysis of switching arch models. *Journal of Time Series Analysis* 23: 425–458.

Kazakevičius, V. and Leipus, R. (2002). On stationarity in the ARCH(∞) model. *Econometric Theory* 18: 1–16.

Kazakevičius, V. and Leipus, R. (2003). A new theorem on existence of invariant distributions with applications to ARCH processes. *Journal of Applied Probability* 40: 147–162.

Kazakevičius, V. and Leipus, R. (2007). On the uniqueness of ARCH processes. *Lithuanian Mathematical Journal* 47: 53–57.

Kesten, H. and Spitzer, F. (1984). Convergence in distribution for products of random matrices. *Zeitschrift für Wahrscheinlichkeitstheorie und Verwandte Gebiete* 67: 363–386.

Kim, S., Shephard, N., and Chib, S. (1998). Stochastic volatility: likelihood inference and comparison with ARCH models. *Review of Economic Studies* 65: 361–393.

King, M.L. and Wu, P.X. (1997). Locally optimal one-sided tests for multiparameter hypotheses. *Econometric Reviews* 16: 131–156.

King, M., Sentana, E., and Wadhwani, S. (1994). Volatility and links between national stock markets. *Econometrica* 62: 901–933.

Kingman, J.F.C. (1973). Subadditive ergodic theory. *Annals of Probability* (6): 883–909.

Klüppelberg, C., Lindner, A., and Maller, R. (2004). A continuous time GARCH process driven by a Lévy process: stationarity and second order behaviour. *Journal of Applied Probability* 41: 601–622.

Koenker, R. and Xiao, Z. (2006). Quantile autoregression. *Journal of the American Statistical Society* 101: 980–990.

Kokoszka, P.S. and Taqqu, M.S. (1996). Parameter estimation for infinite variance fractional ARIMA. *Annals of Statistics* 24: 1880–1913.

Koopman, S.J., Lucas, A., and Scharth, M. (2016). Predicting time-varying parameters with parameter-driven and observation-driven models. *The Review of Economics and Statistics* 98: 97–110.

Lanne, M. and Saikkonen, P. (2006). Why is it so difficult to uncover the risk-return tradeoff in stock returns? *Economic Letters* 92: 118–125.

Lanne, M. and Saikkonen, P. (2007). A multivariate generalized orthogonal factor GARCH model. *Journal of Business and Economic Statistics* 25: 61–75.

Laurent, S. (2009). *GARCH 6: Estimating and Forecasting ARCH Models*. Timberlake Consultants Press.

Laurent, S., Rombouts, J.V.K., and Violante, F. (2012). On the forecasting accuracy of multivariate GARCH models. *Journal of Applied Econometrics* 27: 934–955.

Le Cam, L. (1990). Maximum likelihood: an introduction. *International Statistical Review* 58: 153–171.

Lee, S.W. and Hansen, B.E. (1994). Asymptotic theory for the GARCH(1,1) quasi-maximum likelihood estimator. *Econometric Theory* 10: 29–52.

Lee, J.H.H. and King, M.L. (1993). A locally most mean powerful based score test for ARCH and GARCH regression disturbances. *Journal of Business and Economic Statistics* 11: 17–27.

Lee, S. and Taniguchi, M. (2005). Asymptotic theory for ARCH-SM models: LAN and residual empirical processes. *Statistica Sinica* 15: 215–234.

Leroux, B.G. (1992). Maximum likelihood estimation for hidden Markov models. *Stochastic Process and their Applications* 40: 127–143.

Li, W.K. (2004). *Diagnostic Checks in Time Series*. Boca Raton, FL: Chapman & Hall/CRC.

Li, C.W. and Li, W.K. (1996). On a double-threshold autoregressive heteroscedastic time series model. *Journal of Applied Econometrics* 11: 253–274.

Li, W.K. and Mak, T.K. (1994). On the squared residual autocorrelations in non- linear time series with conditional heteroscedasticity. *Journal of Time Series Analysis* 15: 627–636.

Li, W.K., Ling, S., and McAleer, M. (2002). Recent theoretical results for time series models with GARCH errors. *Journal of Economic Surveys* 16: 245–269.

Li, D., Zhang, X., Zhu, K., and Ling, S. (2017). The ZD-GARCH model: A new way to study heteroscedasticity. *Journal of Econometrics* 202: 1–17.

Ling, S. (2005). Self-weighted LSE and MLE for ARMA-GARCH models. Unpublished paper, Hong-Kong University of Science and Technology.

Ling, S. (2007). Self-weighted and local quasi-maximum likelihood estimators for ARMA-GARCH/IGARCH models. *Journal of Econometrics* 140: 849–873.

Ling, S. and Li, W.K. (1997). On fractionally integrated autoregressive moving-average time series models with conditional heteroscedasticity. *Journal of the American Statistical Association* 92: 1184–1194.

Ling, S. and Li, W.K. (1998). Limiting distributions of maximum likelihood estimators for unstable ARMA models with GARCH errors. *Annals of Statistics* 26: 84–125.

Ling, S. and McAleer, M. (2002a). Necessary and sufficient moment conditions for the GARCH(r, s) and asymmetric power GARCH(r, s) models. *Econometric Theory* 18: 722–729.

Ling, S. and McAleer, M. (2002b). Stationarity and the existence of moments of a family of GARCH processes. *Journal of Econometrics* 106: 109–117.

Ling, S. and McAleer, M. (2003a). Asymptotic theory for a vector ARMA-GARCH model. *Econometric Theory* 19: 280–310.

Ling, S. and McAleer, M. (2003b). Adaptative estimation in nonstationary ARMA models with GARCH errors. *Annals of Statistics* 31: 642–674.

Lintner, J. (1965). The valuation of risk assets on the selection of risky investments in stock portfolios and capital budgets. *Review of Economics and Statistics* 47: 13–37.

Linton, O. (1993). Adaptive estimation in ARCH models. *Econometric Theory* 9: 539–564.

Liu, J.-C. (2006). Stationarity of a Markov-switching GARCH model. *Journal of Financial Econometrics* 4: 573–593.

Liu, J., Li, W.K., and Li, C.W. (1997). On a threshold autoregression with conditional heteroskedastic variances. *Journal of Statistical Planning and Inference* 62: 279–300.

Ljung, G.M. and Box, G.E.P. (1978). On the measure of lack of fit in time series models. *Biometrika* 65: 297–303.

Longerstaey, J. (1996). RiskMetrics Technical Document. Technical Report fourth edition. JP Morgan/Reuters.

Lumsdaine, R.L. (1996). Consistency and asymptotic normality of the quasi-maximum likelihood estimator in IGARCH(1,1) and covariance stationary GARCH(1,1) models. *Econometrica* 64: 575–596.

Lütkepohl, H. (1991). *Introduction to Multiple Time Series Analysis*. Berlin: Springer.

Mahieu, R.J. and Schotman, P.C. (1998). An empirical application of stochastic volatility models. *Journal of Applied Econometrics* 13: 333–360.

Mandelbrot, B. (1963). The variation of certain speculative prices. *Journal of Business* 36: 394–419.

Markowitz, H. (1952). Portfolio selection. *Journal of Finance* 7: 77–91.

McAleer, M. and Chan, F. (2006). Modelling trends and volatility in atmospheric carbon dioxide concentrations. *Environmental Modelling and Software* 21: 1273–1279.

McAleer, M., Chan, F., Hoti, S., and Liebermann, O. (2009). Generalized autoregressive conditional correlation. *Econometric Theory* 28: 422–440.

McLeod, A.I. and Li, W.K. (1983). Diagnostic checking ARMA time series models using squared-residual autocorrelations. *Journal of Time Series Analysis* 4: 269–273.

McNeil, A.J., Frey, R., and Embrechts, P. (2005). *Quantitative Risk Management*. Princeton, NJ: Princeton University Press.

Meddahi, N., Renault, E., and Werker, B. (2006). GARCH and irregularly spaced data. *Economics Letters* 90: 200–204.

Meitz, M. and Saikkonen, P. (2008a). Stability of nonlinear AR-GARCH models. *Journal of Time Series Analysis* 29: 453–475.

Meitz, M. and Saikkonen, P. (2008b). Ergodicity, mixing, and existence of moments of a class of Markov models with applications to GARCH and ACD models. *Econometric Theory* 24: 1291–1320.

Melino, A. and Turnbull, S. (1990). Pricing foreign currency options with stochastic volatility. *Journal of Econometrics* 45: 7–39.

Merton, R.C. (1973). An intertemporal capital asset pricing model. *Econometrica* 41: 867–887.

Meyn, S.P. and Tweedie, R.L. (1996). *Markov Chains and Stochastic Stability*, 3e. London: Springer.

Mikosch, T., Gadrich, T., Klüppelberg, C., and Adler, R.J. (1995). Parameter estimation for ARMA models with infinite variance innovations. *Annals of Statistics* 23: 305–326.

Mikosch, T. (2001). *Modeling Financial Time Series*, Lecture presented at 'New Directions in Time Series Analysis'. Luminy: CIRM.

Mikosch, T. and Stărică, C. (2000). Limit theory for the sample autocorrelations and extremes of a GARCH(1,1) process. *Annals of Statistics* 28: 1427–1451.

Mikosch, T. and Stărică, C. (2004). Non-stationarities in financial time series, the long-range dependence, and the IGARCH effects. *Review of Economics and Statistics* 86: 378–390.

Mikosch, T. and Straumann, D. (2002). Whittle estimation in a heavy-tailed GARCH(1,1) model. *Stochastic Processes and their Applications* 100: 187–222.

Milhøj, A. (1984). The moment structure of ARCH processes. *Scandinanvian Journal of Statistics* 12: 281–292.

Milhøj, A. (1987). *A Multiplicative Parameterization of ARCH Models*. Mimeo: University of Copenhagen, Department of Statistics.

Mills, T.C. (1993). *The Econometric Modelling of Financial Time Series*. Cambridge: Cambridge University Press.

Mokkadem, A. (1990). Propriétés de mélange des processus autorégressifs polynomiaux. *Annales de l'Institut Henri Poincaré* 26: 219–260.

Moulin, H. and Fogelman-Soulié, F. (1979). *La convexité dans les mathématiques de la décision: optimisation et théorie micro-économique*. Paris: Hermann.

Muler, N. and Yohai, V.J. (2008). Robust estimates for GARCH models. *Journal of Statistical Planning and Inference* 138: 2918–2940.

Nelson, D.B. (1990a). Stationarity and persistence in the GARCH(1,1) model. *Econometric Theory* 6: 318–334.

Nelson, D.B. (1990b). ARCH models as diffusion approximations. *Journal of Econometrics* 45: 7–38.

Nelson, D.B. (1991). Conditional heteroskedasticity in asset returns: a new approach. *Econometrica* 59: 347–370.

Nelson, D.B. (1992). Filtering and forecasting with misspecified ARCH models I: getting the right variance with the wrong model. *Journal of Econometrics* 52: 61–90.

Nelson, D.B. and Cao, C.Q. (1992). Inequality constraints in the univariate GARCH model. *Journal of Business and Economic Statistics* 10: 229–235.

Nelson, D.B. and Foster, D.P. (1994). Asymptotic filtering theory for univariate ARCH models. *Econometrica* 62: 1–42.

Nelson, D.B. and Foster, D.P. (1995). Filtering and forecasting with misspecified ARCH models II: making the right forecast with the wrong model. *Journal of Econometrics* 67: 303–335.

Newey, W.K. and Steigerwald, D.G. (1997). Asymptotic bias for quasi-maximum-likelihood estimators in conditional heteroskedasticity models. *Econometrica* 65: 587–599.

Nijman, T. and Sentana, E. (1996). Marginalization and contemporaneous aggregation in multivariate GARCH processes. *Journal of Econometrics* 71: 71–87.

Noureldin, D., Shephard, N. and Sheppard, K. (2014). Multivariate rotated ARCH models. *Journal of Econometrics* 179: 16–30.

Nummelin, E. (1984). *General Irreducible Markov Chains and Non-Negative Operators*. Cambridge: Cambridge University Press.

Omori, Y., Chib, S., Shephard, N., and Nakajima, J. (2007). Stochastic volatility with leverage: fast and efficient likelihood inference. *Jorunal of Econometrics* 140: 425–449.

Pagan, A.R. (1996). The econometrics of financial markets. *Journal of Empirical Finance* 3: 15–102.

Pagan, A.R. and Schwert, G.W. (1990). Alternative models for conditional stock volatility. *Journal of Econometrics* 45: 267–290.

Palm, F.C. (1996). GARCH models of volatility. In: *Handbook of Statistics*, vol. 14 (ed. G.S. Maddala and C.R. Rao), 209–240. Amsterdam: North-Holland.

Pantula, S.G. (1986). Modeling the persistence of conditional variances: a comment. *Econometric Reviews* 5: 71–74.

Pantula, S.G. (1989). Estimation of autoregressive models with ARCH errors. *Sankhya B* 50: 119–138.

Parkinson, M. (1980). The extreme value method for estimating the variance of the rate of return. *Journal of Business* 53: 61–65.

Pedersen, R.S. (2016). Targeting estimation of CCC-GARCH models with infinite fourth moments. *Econometric Theory* 32: 498–531.

Pedersen, R.S. (2017). Inference and testing on the boundary in extended constant conditional correlation GARCH models. *Journal of Econometrics* 196: 23–36.

Pedersen, R.S. and Rahbek, A. (2014). Multivariate variance targeting in the BEKK-GARCH model. *The Econometrics Journal* 17: 24–55.

Pedersen, R.S. and Rahbek, A. (2018). Testing GARCH-X type models. *Econometric Theory*, 1–36.

Pelletier, D. (2006). Regime switching for dynamic correlations. *Journal of Econometrics* 131: 445–473.

Peng, L. and Yao, Q. (2003). Least absolute deviations estimation for ARCH and GARCH models. *Biometrika* 90: 967–975.

Perlman, M.D. (1969). One-sided testing problems in multivariate analysis. *Annals of Mathematical Statistics* 40: 549–567. Corrections in (1971). *Annals of Mathematical Statistics* 42: 1777.

Petrie, T. (1969). Probabilistic functions of finite state Markov chains. *The Annals of Mathematical Statistics* 40: 97–115.

Pourahmadi, M. (1999). Joint mean-covariance models with applications to longitudinal data: unconstrained parameterisation. *Biometrika* 86: 677–690.

Priestley, M.B. (1988). *Non-linear and Non-stationary Time Series Analysis*. New York: Academic Press.

Rabemananjara, R. and Zakoïan, J.-M. (1993). Threshold ARCH models and asymmetries in volatility. *Journal of Applied Econometrics* 8: 31–49.

Rabiner, L.R. and Juang, B.H. (1986). An introduction to hidden Markov models. *IEEE ASSP Magazine* 3: 4–16.

Reinsel, G.C. (1997). *Elements of Multivariate Time Series Analysis*. New York: Springer.

Renault, E. (2009). Probabilistic properties of stochastic volatility models. In: *Handbook of Financial Time Series* (ed. T.G. Andersen, R.A. Davis, J.-P. Kreiss and T. Mikosch), 269–311. New York: Springer.

Resnick, S. (1992). *Adventures in Stochastic Processes*. Boston, MA: Birkhäuser.

Rich, R.W., Raymond, J., and Butler, J.S. (1991). Generalized instrumental variables estimation of autoregressive conditional heteroskedastic models. *Economics Letters* 35: 179–185.

Rio, E. (1993). Covariance inequalities for strongly mixing processes. *Annales de l'Institut Henri Poincaré, Probabilités et Statistiques* 29: 587–597.

Robinson, P.M. (1991). Testing for strong correlation and dynamic conditional heteroskedasticity in multiple regression. *Journal of Econometrics* 47: 67–84.

Robinson, P.M. and Zaffaroni, P. (2006). Pseudo-maximum likelihood estimation of ARCH(∞) models. *Annals of Statistics* 34: 1049–1074.

Rogers, A.J. (1986). Modified Lagrange multiplier tests for problems with one-sided alternatives. *Journal of Econometrics* 31: 341–361.

Romano, J.P. and Thombs, L.A. (1996). Inference for autocorrelations under weak assumptions. *Journal of the American Statistical Association* 91: 590–600.

Romilly, P. (2006). Time series modelling of global mean temperature for managerial decision-making. *Journal of Environmental Management* 76: 61–70.

Rosenberg, B. (1972). The behavior of random variables with nonstationary variance and the distribution of security prices. In: *Working Paper 11, Graduate School of Business Administration*. Berkeley: University of California. Also reproduced in Shephard, N. (ed.) (2005). *Stochastic Volatility*, 83–108. Oxford: Oxford University Press.

Rosenblatt, M. (1956). A central limit theorem and a strong mixing condition. *Proceedings of the National Academy of Sciences of the United States of America* 42: 43–47.

Ruiz, E. (1994). Quasi-maximum likelihood estimation of stochastic volatility models. *Journal of Econometrics* 63: 289–306.

Russell, J. and Engle, R.F. (2005). A discrete-state continuous-time model of transaction prices and times: the ACM-ACD model. *Journal of Business & Economic Statistics* 23: 166–180.

Rydberg, T.H. and Shephard, N. (2003). Dynamics of trade-by-trade price movements: decomposition and models. *Journal of Financial Econometrics* 1: 2–25.

Sabbatini, M. and Linton, O. (1998). A GARCH model of the implied volatility of the Swiss market index from option prices. *International Journal of Forecasting* 14: 199–213.

Schwert, G.W. (1989). Why does stock market volatility change over time. *Journal of Finance* 14: 1115–1153.

Sentana, E. (1995). Quadratic ARCH models. *Review of Economic Studies* 62: 639–661.

Sharpe, W. (1964). Capital asset prices: a theory of market equilibrium under conditions of risk. *Journal of Finance* 19: 425–442.

Shephard, N. (1996). Statistical aspects of ARCH and stochastic volatility. In: *Time Series Models in Econometrics, Finance and Other Fields* (ed. D.R. Cox, D.V. Hinkley and O.E. Barndorff-Nielsen), 1–67. London: Chapman & Hall.

Shephard, N. (2005). Introduction. In: *Stochastic Volatility* (ed. N. Shephard), 1–33. Oxford: Oxford University Press.

Shephard, N. and Andersen, T.G. (2009). Stochastic volatility: origins and overview. In: *Handbook of Financial Time Series* (ed. T.G. Andersen, R.A. Davis, J.-P. Kreiss and T. Mikosch), 233–254. New York: Springer.

Shephard, N. and Sheppard, K. (2010). Realising the future: forecasting with high-frequency-based volatility (HEAVY) models. *Journal of Applied Econometrics* 25: 197–231.

Silvapulle, M.J. and Silvapulle, P. (1995). A score test against one-sided alternatives. *Journal of the American Statistical Association* 90: 342–349.

Silvennoinen, A. and Teräsvirta, T. (2009). Multivariate GARCH models. In: *Handbook of Financial Time Series* (ed. T.G. Andersen, R.A. Davis, J.-P. Kreiss and T. Mikosch). New York: Springer.

Stărică, C. (2006). Is GARCH(1,1) as good a model as the accolades of the Nobel Prize would imply? Working paper, Chalmers University of Technology.

Stelzer, R. (2008). On the relation between the Vec and BEKK multivariate GARCH models. *Econometric Theory* 24: 1131–1136.

Stentoft, L. (2005). Pricing American options when the underlying asset follows GARCH processes. *Journal of Empirical Finance* 12: 576–611.

Stentoft, L. (2008). Option pricing using realized volatility. CREATES Research Paper 2008-13, Aarhus Business School.

Straumann, D. (2005). *Estimation in Conditionally Heteroscedastic Time Series Models*, Lecture Notes in Statistics, vol. 181. Berlin: Springer.

Straumann, D. and Mikosch, T. (2006). Quasi-maximum-likelihood estimation in heteroscedastic time series: a stochastic recurrence equation approach. *Annals of Statistics* 34: 2449–2495.

Stroock, D. and Varadhan, S. (1979). *Multidimensional Diffusion Processes*. Berlin: Springer.

Sucarrat, G. and Grønneberg, S. (2016). Models of financial return with time-varying zero probability. MPRA Working Paper.

Sucarrat, G., Grønneberg, S., and Escribano, A. (2016). Estimation and inference in univariate and multivariate LOG-GARCH-X models when the conditional density is unknown. *Computational Statistics & Data Analysis* 100: 582–594.

Taylor, J.W. (2004). Volatility forecasting with smooth transition exponential smoothing. *International Journal of Forecasting* 20: 273–286.

Taylor, S.J. (1986). *Modelling Financial Time Series*. Chichester: Wiley.

Taylor, S.J. (1994). Modelling stochastic volatility. *Mathematical Finance* 4: 183–204.

Taylor, S.J. (2007). *Asset Price Dynamics, Volatility and Prediction*. Princeton, NJ: Princeton University Press.

Tjøstheim, D. (1990). Non-linear time series and Markov chains. *Advances in Applied Probability* 22: 587–611.

Tjøstheim, D. (2012). Some recent theory for autoregressive count time series. *TEST* 21: 413–438.

Tol, R.S.J. (1996). Autoregressive conditional heteroscedasticity in daily temperature measurements. *Environmetrics* 7: 67–75.

Tong, H. (1978). On a threshold model. In: *Pattern Recognition and Signal Processing* (ed. C.H. Chen), 575–586. Amsterdam: Sijthoff and Noordhoff.

Tong, H. and Lim, K.S. (1980). Threshold autoregression, limit cycles and cyclical data. *Journal of the Royal Statistical Society B* 42: 245–292.

Truquet, L. (2014). On a family of contrasts for parametric inference in degenerate ARCH models. *Econometric Theory* 30: 1165–1206.

Tsai, H. and Chan, K.-S. (2008). A note on inequality constraints in the GARCH model. *Econometric Theory* 24: 823–828.

Tsay, R.S. (2010). *Analysis of Financial Time Series*, 3e. New York: Wiley.

Tsay, R.S. (2005). *Analysis of Financial Data*, 2e. Hoboken, NJ: Wiley.

Tse, Y.K. (2002). Residual-based diagnostics for conditional heteroscedasticity models. *Econometrics Journal* 5: 358–373.

Tse, Y.K. and Tsui, A. (2002). A multivariate GARCH model with time-varying correlations. *Journal of Business and Economic Statistics* 20: 351–362.

Tweedie, R.L. (2001). Markov Chains: Structure and Applications. *Handbook of statistics* 19: 817–851.

van der Vaart, A.W. (1998). *Asymptotic Statistics*. Cambridge: Cambridge University Press.

van der Weide, R. (2002). GO-GARCH: a multivariate generalized orthogonal GARCH model. *Journal of Applied Econometrics* 17: 549–564.

Vaynman, I. and Beare, B.K. (2014). Stable limit theory for the variance targeting estimator. In: *Essays in Honor of Peter C.B. Phillips*, Advances in Econometrics, Chapter 24 , vol. 33 (ed. Y. Chang, T.B. Fomby and J.Y. Park), 639–672. Emerald Group Publishing Limited.

Visser, M.P. (2009). Volatility proxies and GARCH models. PhD thesis. Amsterdam: Korteweg-de Vries Institute for Mathematics.

Visser, M.P. (2011). GARCH parameter estimation using high-frequency data. *Journal of Financial Econometrics* 9: 162–197.

Vrontos, I.D., Dellaportas, P., and Politis, D.N. (2003). A full-factor multivariate GARCH model. *Econometrics Journal* 6: 312–334.

Wang, S. and Dhaene, J. (1998). Comonotonicity, correlation order and premium principles. *Insurance: Mathematics and Economics* 22: 235–242.

Wang, F. and Ghysels, E. (2015). Econometric analysis of volatility component models. *Econometric Theory* 31: 362–393.

Wang, S., Young, V., and Panjer, H. (1997). Axiomatic characterization of insurance prices. *Insurance: Mathematics and Economics* 21: 173–183.

Weiss, A.A. (1984). ARMA models with ARCH errors. *Journal of Time Series Analysis* 5: 129–143.

Weiss, A.A. (1986). Asymptotic theory for ARCH models: estimation and testing. *Econometric Theory* 2: 107–131.

Wintenberger, O. (2013). Continuous invertibility and stable QML estimation of the EGARCH(1,1) model. *Scandinavian Journal of Statistics* 40: 846–867.

Wolak, F.A. (1989). Local and global testing of linear and nonlinear inequality constraints in nonlinear econometric models. *Econometric Theory* 5: 1–35.

Wold, H. (1938). *A Study in the Analysis of Stationary Time Series*. Uppsala: Almqvist & Wiksell.

Wong, E. (1964). The construction of a class of stationary Markov processes. In: *Stochastic Processes in Mathematical Physics and Engineering*, Proceedings of Symposia in Applied Mathematics, vol. 16 (ed. R. Bellman), 264–276. Providence, RI: American Mathematical Society.

Wright, S.P. (1992). Adjusted P-values for simultaneous inference. *Biometrics* 48: 1005–1013.

Wu, G. (2001). The determinants of asymmetric volatility. *Review of Financial Studies* 837–859.

Xekalaki, E. and Degiannakis, S. (2009). *ARCH Models for Financial Applications*. Chichester: Wiley.

Yu, J. (2005). On leverage in a stochastic volatility model. *Journal of Econometrics* 127: 165–178.

Zakoïan, J.-M. (1994). Threshold heteroskedastic models. *Journal of Economic Dynamics and Control* 18: 931–955.

Zarantonello, E.H. (1971). Projections on convex sets in Hilbert spaces and spectral theory. In: *Contributions to Nonlinear Functional Analysis* (ed. E.H. Zarantonello), 237–424. New York: Academic Press.

Zhu, K. and Ling, S. (2011). Global self-weighted and local quasi-maximum exponential likelihood estimators for ARMA–GARCH/IGARCH models. *The Annals of Statistics* 39: 2131–2163.

Index

GARCH Models: Structure, Statistical Inference and Financial Applications, Second Edition. Christian Francq and Jean-Michel Zakoian.

www.ingramcontent.com/pod-product-compliance
Lightning Source LLC
Chambersburg PA
CBHW082045280125
20788CB00044B/45